List of Elements with Their Symbols and Atomic Masses

W9-CPE-195

Element	Symbol	Atomic Number	Atomic Mass[a] (amu)
Actinium	Ac	89	(227)
Aluminum	Al	13	26.9815
Americium	Am	95	(243)
Antimony	Sb	51	121.75
Argon	Ar	18	39.948
Arsenic	As	33	74.9216
Astatine	At	85	(210)
Barium	Ba	56	137.33
Berkelium	Bk	97	(247)
Beryllium	Be	4	9.01218
Bismuth	Bi	83	208.9806
Boron	B	5	10.811
Bromine	Br	35	79.904
Cadmium	Cd	48	112.41
Calcium	Ca	20	40.08
Californium	Cf	98	(251)
Carbon	C	6	12.01115
Cerium	Ce	58	140.12
Cesium	Cs	55	132.9055
Chlorine	Cl	17	35.453
Chromium	Cr	24	51.996
Cobalt	Co	27	58.9332
Copper	Cu	29	63.546
Curium	Cm	96	(247)
Dysprosium	Dy	66	162.50
Einsteinium	Es	99	(254)
Element 110	—	110	(269)
Element 111	—	111	(272)
Erbium	Er	68	167.26
Europium	Eu	63	151.96
Fermium	Fm	100	(257)
Fluorine	F	9	18.998403
Francium	Fr	87	(223)
Gadolinium	Gd	64	157.25
Gallium	Ga	31	69.72
Germanium	Ge	32	72.59
Gold	Au	79	196.9665
Hafnium	Hf	72	178.49
Hahnium[b]	Ha	105	(262)
Hassium[b]	Hs	108	(265)
Helium	He	2	4.00260
Holmium	Ho	67	164.9303
Hydrogen	H	1	1.0080
Indium	In	49	114.82
Iodine	I	53	126.9045
Iridium	Ir	77	192.22
Iron	Fe	26	55.847
Krypton	Kr	36	83.80
Lanthanum	La	57	138.9055
Lawrencium	Lr	103	(257)
Lead	Pb	82	207.2
Lithium	Li	3	6.941
Lutetium	Lu	71	174.967
Magnesium	Mg	12	24.305
Manganese	Mn	25	54.9380
Meitnerium[b]	Mt	109	(266)
Mendelevium	Md	101	(256)
Mercury	Hg	80	200.59
Molybdenum	Mo	42	95.94
Neodymium	Nd	60	144.24
Neon	Ne	10	20.179
Neptunium	Np	93	237.0482
Nickel	Ni	28	58.70
Nielsbohrium[b]	Ns	107	(262)
Niobium	Nb	41	92.9064
Nitrogen	N	7	14.0067
Nobelium	No	102	(255)
Osmium	Os	76	190.2
Oxygen	O	8	15.9994
Palladium	Pd	46	106.4
Phosphorus	P	15	30.9738
Platinum	Pt	78	195.09
Plutonium	Pu	94	(244)
Polonium	Po	84	(209)
Potassium	K	19	39.0983
Praseodymium	Pr	59	140.9077
Promethium	Pm	61	(145)
Protactinium	Pa	91	231.0359
Radium	Ra	88	226.0254
Radon	Rn	86	(222)
Rhenium	Re	75	186.207
Rhodium	Rh	45	102.9055
Rubidium	Rb	37	85.4678
Ruthenium	Ru	44	101.07
Rutherfordium[b]	Rf	104	(261)
Samarium	Sm	62	150.4
Scandium	Sc	21	44.9559
Seaborgium[b]	Sg	106	(263)
Selenium	Se	34	78.96
Silicon	Si	14	28.0855
Silver	Ag	47	107.868
Sodium	Na	11	22.9898
Strontium	Sr	38	87.62
Sulfur	S	16	32.06
Tantalum	Ta	73	180.9479
Technetium	Tc	43	98.9062
Tellurium	Te	52	127.60
Terbium	Tb	65	158.9254
Thallium	Tl	81	204.37
Thorium	Th	90	232.0381
Thulium	Tm	69	168.9342
Tin	Sn	50	118.69
Titanium	Ti	22	47.90
Tungsten	W	74	183.85
Uranium	U	92	238.029
Vanadium	V	23	50.9415
Xenon	Xe	54	131.30
Ytterbium	Yb	70	173.04
Yttrium	Y	39	88.9059
Zinc	Zn	30	65.37
Zirconium	Zr	40	91.22

[a] Based on the assigned relative atomic mass of ^{12}C = exactly 12 amu; parentheses indicate the mass number of the isotope with the longest half-life.

[b] American Chemical Society (ACS) recommended names and symbols. International Union of Pure & Applied Chemistry (IUPAC) recommended names and symbols differ as follows: 104 (Dubnium, Db); 105 (Joliotium, Jl); 106 (Rutherfordium, Rf); 107 (Bohrium, Bh); 108 (Hahnium, Hn).

BASIC CHEMISTRY
Seventh Edition

G. William Daub

Department of Chemistry
Harvey Mudd College

William S. Seese

Emeritus
Department of Chemistry
Casper College

PRENTICE HALL, Upper Saddle River, New Jersey 07458

Library of Congress Cataloging-in-Publication Data

DAUB, G. WILLIAM.
    Basic chemistry/G. William Daub, William S. Seese. — 7th ed.
      p. cm.
    Includes index.
    ISBN 0-13-373630-X (hardcover)
    1. Chemistry.   I. Seese, William S.   II. Title.
QD33.S39   1996
540—dc20

                                    95-32224
                                      CIP

Editorial Director: *Tim Bozik*
Editor-in-Chief: *Paul Corey*
Acquisition Editor: *Ben Roberts*
Production Editors: *Susan Fisher and Rose Kernan*
Copy Editor: *Carol J. Dean*
Director of Production and Manufacturing: *David Riccardi*
Executive Managing Editor: *Kathleen Schiaparelli*
Creative Director: *Paula Maylahn*
Art Director: *Heather Scott*
Cover/Interior Designer: *Amy Rosen*
Manufacturing Buyer: *Trudy Pisciotti*
Editorial Assistant: *Ashley Scattergood*
Cover Photo: *Superstock*

© 1996, 1992, 1988, 1985, 1981, 1977, 1972 by Prentice-Hall, Inc.
Simon & Schuster/A Viacom Company
Upper Saddle River, New Jersey 07458

Printed in the United States of America
10 9 8 7 6 5 4 3 2 1

ISBN 0-13-373630-X

Prentice-Hall International (UK) Limited, *London*
Prentice-Hall of Australia Pty. Limited, *Sydney*
Prentice-Hall Canada Inc., *Toronto*
Prentice-Hall Hispanoamericana, S.A., *Mexico*
Prentice-Hall of India Private Limited, *New Delhi*
Prentice-Hall of Japan, Inc., *Tokyo*
Simon & Schuster Asia Pte. Ltd., *Singapore*
Editora Prentice-Hall do Brasil, Ltda., *Rio de Janeiro*

BRIEF CONTENTS

CONTENTS

ONE

TWO

THREE

*Chapter or section that may be omitted without loss of continuity in a brief course.

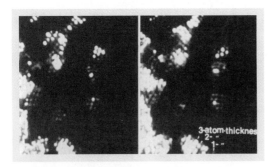

FOUR

The Structure of the Atom   79

FIVE

The Periodic Classification of the Elements   104

SIX

The Structure of Compounds   118

TEN

Calculations Involving Chemical Equations. Stoichiometry   248

ELEVEN

Gases   276

FIFTEEN

Acids, Bases, and Ionic Equations   405

SIXTEEN

Oxidation–Reduction Equations and Electrochemistry   440

Appendixes

PREFACE   To the Student

The goal of this book is simple: to help you learn chemistry as efficiently and painlessly as possible.

Like the previous six editions, we wrote this book for you, not for your instructor. We have assumed that you have little science background, let alone chemistry, and that your mathematics needs review. Therefore, we have used analogies and cartoons to help you understand some of the principles of chemistry. Throughout the text, we have placed new terms in **bold type** to make it easy for you to identify the important points you need to learn. These terms are defined in the text and also in a *running glossary* in the margin beside each term. Reviewing these marginal definitions is a good way to make sure you have grasped the main ideas of the chapter. To help you check definitions for terms in earlier chapters, all the terms appear alphabetically in the **Glossary** at the end of the text.

To help you learn chemistry, we have included a number of features in each chapter.

1. **Countdown** (not in Chapter 1).   This consists of five *review questions* from previous material which act as a *basis* for material to be introduced in the new chapter. These questions are keyed to the sections in the previous chapters. The answers are found in parentheses next to the questions.

2. **Goals.**   The goals specify *skills* that the chapter is designed to teach you. You must accomplish these goals in order to master the material in the chapter.

3. **Study Hints.**   These study hints are designed to *help* you understand the chemistry.

4. **Examples.**   These examples are *worked out* in the body of the chapter to help you solve a particular type of problem.

5. **Study Exercises.**   These study exercises are found in the body of the chapter and are there for *you to work*. They are similar to the examples and problems at the end of the chapter. The answers are found in parentheses next to the study exercise.

6. **Summary.**   A brief summary of the *important points* is given at the end of the chapter.

7. **Exercises and Problems.**   Exercises and problems are given at the end of the chapter and are *keyed* to the sections of the chapter. The answers are found in Appendix VII at the end of the book.

8. **Chapter Quiz.**   A chapter quiz is found near the end of the chapter and gives you an opportunity to *test yourself* on the chapter. Again, the answers are found in Appendix VII.

To relate what you are learning to the real world, we have included the following features:

1. **Elements** or **Compounds.**   This is a short, end-of-chapter discussion of an element or compound that may be of interest to you. It illustrates how *chemistry* is part of your *life*.

2. **Chemistry of the Atmosphere.**   This is an essay on *environmental* issues that affect your life.

3. **You and Chemistry.**   Throughout the book you will see an icon in the margin, , which brings to your attention how *chemistry* applies to your *real world*.

Before you start your study of chemistry, we would like to give you some friendly advice:

1. Don't read this book in bed. We don't think it will put you to sleep, but you will get a lot more out of it if you sit up and use paper and pencil to work the examples and study exercises as you go. Your book may also be in better shape at the end of the term.

2. Don't think you can learn everything the night before the quiz or exam. The easiest way to learn chemistry is to do a little bit every day. You should spend about two hours studying chemistry for every hour you are in chemistry class, that is, about six hours per week if your course has three hours of lecture per week.

We especially welcome your suggestions. You would be surprised how many students have written us over the past 24 years since the first edition appeared. Many of their suggestions have been incorporated in this edition. We, therefore, would greatly appreciate your writing to us about your reaction to this text.

G. William Daub
Department of Chemistry
Harvey Mudd College
Claremont, California 91711

William S. Seese
Emeritus
Casper College
Casper, Wyoming 82601

To the Instructor

In the seventh edition of *Basic Chemistry*, we have made the following changes:

1. **Countdown** (not in Chapter 1). This consists of five *review questions* from previous material which act as a *basis* for material to be introduced in the new chapter.

2. **Study Exercises.** Study exercises have been added *in the chapter* in an attempt to reinforce previously covered material.

3. **You and Chemistry.** These are applications of chemistry to the *real world*. They are found throughout the book and are indicated by an icon in the margin, 🖼️ .

4. **Text Changes from the Sixth Edition.** Five text changes have been made in the seventh edition from that of the sixth edition:

   ✔ 1. Calculating oxidation numbers has been introduced in Chapter 6 (Section 6.3) and reviewed again in Chapter 16 (Section 16.2).

   ✔ 2. Shapes of molecules and polyatomic ions has been added in Chapter 6 (Section 6.8).

   ✔ 3. Dilution of molar solutions has been added in Chapter 14 (Section 14.8).

   ✔ 4. Conversion of concentration of solutions has been deleted from Chapter 14.

   ✔ 5. Colloids and suspensions have been added in Chapter 14 (Section 14.12).

For the seventh edition we have written general **Goals** similar to those in the sixth edition. Specific **Objectives** including **Tasks** are found in the *Instructor's Resource Manual*.

We have retained the *General Problems* at the end of the chapter. These problems require the student to use previously covered material and are more difficult to solve. They stimulate the student to begin to think *critically*.

To provide more flexibility for you, an alternate edition is available, *Basic Chemistry*, 7th edition, Alternate Edition. This paperback version consists of Chapters 1 through 17. Another book based in part on *Basic Chemistry* is also available. In *Preparation for College Chemistry*, 5th edition, provides a quick primer for students approaching chemistry from the earliest starting point. It is also a paperback and consists of approximately 300 pages.

A number of supplements are available to accompany either version of the text. These include the highly successful *Laboratory Experiments* by Charles Corwin of American River College (Sacramento, California) and a *Student Workbook* and *Student Solutions Manual* by the authors of the text. Free upon adoption of the text are the following: *How to Study Chemistry* and *Text Bank for Basic Chemistry* by Vernon K. Burger of Cuyahoga Community College (Highland Hills, Ohio) and *Instructor's Resource Manual* by Gerald Ittenbach of Fayetteville Technical Community College (Fayetteville, North Carolina).

Many of you have made numerous suggestion for improving *Basic Chemistry*, and we have attempted to follow your suggestions in this seventh edition. These individuals are as follows:

David W. Ball, *Cleveland State University*
Hal Bender, *Clackamas Com. College*
Larry E. Bennett, *San Diego State University*
Gerald Berkowitz, *Erie Community College*
Donald J. Brown, *Western Michigan University*
Vernon K. Burger, *Coyahoga Community College*
B. Edward Cain, *Rochester Institute of Technology*
Rodney Cate, *Midwestern State University*
James Coke, *University of North Carolina*
Lorraine Deck, *University of New Mexico*
John M. DeKorte, *Glendale Community College*
Celia A. Dosmer, *Mohawk Valley Comm. College*
Paul D. Hooker, *Colby Community College*
Gerald Ittenbach, *Fayetteville Tech. Com. Col.*
T. G. Jackson, *University of Southern Alabama*
Floyd Kelly, *Casper College*
Ernest Kemnitz, *University of Nebraska*
Roy Kennedy, *Mass. Bay Community College*
Robley J. Light, *Florida State University*
Gene Lindsay, *St. Charles County Com. Col.*

Boon H. Loo, *University of Alabama*
Susanne M. Mathews, *Joilet Junior College*
Robert McDonald, *Valencia Community College*
Clark Most, *Delta College*
Paul O'Brien, *West Valley College*
Rosalie Rogiewicz Park, *Immaculata College*
Dennis R. Pettygrove, *College of Southern Idaho*
John Pressler, *student*
Fred Redmore, *Highland Community College*
Melissa S. Reeves, *Indiana University*
Samuel L. Rieger, *Naugatuch Val. Com. Tech. College*
Rene Rodriguez, *Idaho State University*
Allen Scism, *Central Missouri State University*
Dave Seapy, *Sultan Qaboos University*
Eugene H. Shannon, *Fayetteville Tech. Com. Col.*
Lee S. Sunderlin, *Northern Illinois University*
Richard Wheet, *Texas State Technical Institute*
David L. Winters, *Tidewater Community College*
John Youker, *Hudson Valley Com. Col.*

We would like to thank each of these individuals for their time and energy in helping to make the seventh edition the best ever.

We would also like to thank our wives Sandra Anne Hollenberg and Ann Reeves Seese for their advice, suggestions, and support. Finally, we would like to thank Mary Fuday Ginsburg and Carol J. Dean who contributed to the superb editing, Ben Roberts, chemistry editor, and Susan Fisher and Rose Kernan, our production editors.

We hope you and your students will enjoy this book and will continue to send us your suggestions on making this chemistry text "user-friendly."

G. William Daub
Department of Chemistry
Harvey Mudd College
Claremont, California 91711

William S. Seese
Emeritus
Casper College
Casper, Wyoming 82601

Themes of the Times

The New York Times and Prentice Hall are sponsoring *Themes of the Times*, a program designed to enhance student access to current information of relevance in the classroom.

Through this program, the core subject matter provided in the text is supplemented by a collection of time-sensitive articles from one of the world's most distinguished newpapers, *The New York Times*. These articles demonstrate the vital, ongoing connection between what is learned in the classroom and what is happening in the world around us.

To enjoy the wealth of information of *The New York Times* daily, a reduced subscription rate is available. For information, call toll-free: 1-800-631-1222.

Prentice Hall and *The New York Times* are proud to co-sponsor *The Themes of the Times*. We hope it will make the reading of both textbooks and newpapers a more dynamic, involving process.

CHAPTER 1

Introduction to Chemistry

Chemistry permeates our lives. How many examples of chemistry can you see in this street scene? Synthetic rubber (balloons), fibers (clothing), paints, dyes (clothing), helium gas (in balloons), mortar (buildings), photosynthesis (trees), human biological processes, soil (flower pots), wood (benches and tables), glass (windows), and the photographic process itself are all parts of the world of chemistry. Scene from Quincy Market, Boston.

GOALS FOR CHAPTER 1

1. To gain an understanding of science and the scientific method and of chemistry's place among the sciences (Section 1.1).

2. To classify a chemist's research interests or research papers as being in one of the five subfields of chemistry (Section 1.2).

3. To trace the development of chemistry as a science (Section 1.3).

What do you think of when you hear the word "chemical"? You probably think of some foul-smelling substance in a laboratory. Certainly such items are chemicals, but chemicals are everywhere. The soap and bleach you use to wash your clothes are chemicals. So are the metal and plastic that form the washer and dryer. Your body is a collection of chemicals, the most abundant of which is water (60% of your weight).

Studying chemistry, then, is studying *life*. As you study this book, you will learn about the most basic structures and interactions behind things that you take for granted: the water you drink, the air you breathe, and the food you eat. You will see how generations of chemists have given the world not only a better understanding of itself but also many products that improve our lives. Such products include plastics, polymers such as nylon and Orlon, synthetic rubber, fertilizers, and many pharmaceutical agents (see Figure 1.1).

Before you can appreciate the work that went into designing these products, however, you need to know more about chemistry: its methods, its history, and its study.

Science Organized or systematized knowledge gathered by the scientific method.

Scientific method Procedure for studying the world in three organized steps: experimentation, hypothesis proposal, and further experimentation.

Experimentation Collection of facts and data by observing natural events under carefully controlled conditions.

1.1 Science and the Scientific Method

You already know one of the most important things about chemistry: it is a science. **Science** is organized or systematized knowledge gathered by the scientific method. Chemists and other scientists approach their studies in an organized manner known as the scientific method. The **scientific method** involves three steps:

1. Collect facts and data by observing natural events under carefully controlled conditions—**experimentation.**

FIGURE 1.1
Insulin molecule built from crystal structure of porcine insulin. (Courtesy NIH/ Science Source Researchers)

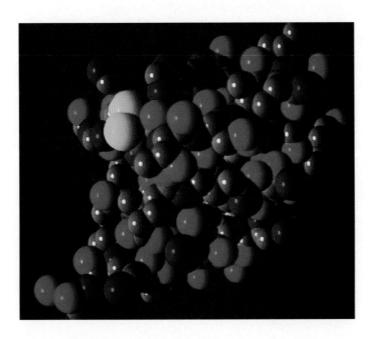

2. Examine and correlate these facts and propose a hypothesis. A **hypothesis** is a tentative theory to explain the data.

3. Plan and execute *further experimentation* to support or refute the hypothesis, and propose a scientific theory or law if possible.

Hypothesis Tentative explanation of the results of experimentation; it is subject to verification or rejection in further experiments.

You use a loose form of the scientific method whenever you play a video game. The first time you play the game you observe what happens as you move around the screen and push different combinations of buttons—you *experiment*. When the game ends, you consider the meaning of your observations in an attempt to figure out how to improve your score—you form a *hypothesis* to explain the events you noted. You then play again to see if you have made any progress in mastering the game. This *further experimentation* either supports your hypothesis or causes you to revise it.

Scientists follow and use these same three steps: experimentation, hypothesis formation, and further experimentation. They sometimes add one last step when repeated experiments confirm a hypothesis under certain conditions with no exceptions; they propose a *scientific law* (see Figure 1.2).

Some experiments result in hypotheses, a few in theories, and even fewer in scientific laws. Many times a hypothesis has to be modified or even discarded after further experimentation. Theories may stand unchallenged for years before new

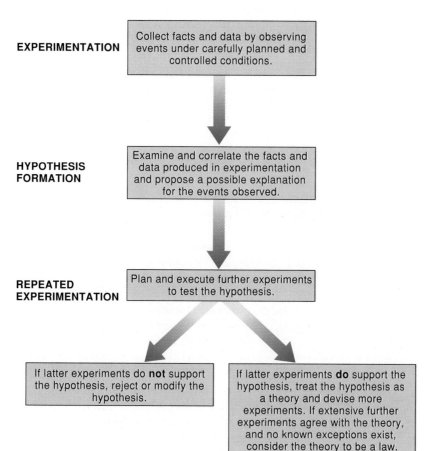

FIGURE 1.2
The scientific method and the development of a scientific law.

EXPERIMENTATION — Collect facts and data by observing events under carefully planned and controlled conditions.

HYPOTHESIS FORMATION — Examine and correlate the facts and data produced in experimentation and propose a possible explanation for the events observed.

REPEATED EXPERIMENTATION — Plan and execute further experiments to test the hypothesis.

If latter experiments do **not** support the hypothesis, reject or modify the hypothesis.

If latter experiments **do** support the hypothesis, treat the hypothesis as a theory and devise more experiments. If extensive further experiments agree with the theory, and no known exceptions exist, consider the theory to be a law.

experimental data reveal them to be unacceptable. New theories then arise, and the process continues.

Workers in all sciences formulate hypotheses and theories and rely on mathematics to express their findings as precisely as possible. Research workers in the *abstract sciences* (mathematics and logic), the *biological sciences* (botany, zoology, physiology, and microbiology), and the *physical sciences* (chemistry, geology, and physics) can often exert careful control of their experiments and come to precise conclusions. Those exploring the *social sciences* (archaeology, economics, history, political science, psychology, and sociology) cannot usually perform highly controlled experiments. The biological sciences, in their attempt to become more precise, have become more chemically oriented in the explanation of health and disease.

For example, a psychologist seeking to learn how people solve problems must allow for personal differences among the individuals tested. In contrast, a chemist studying the strength of a new material designed to be a glue can duplicate the glue and try it on different materials to determine how well it works.

Study Exercise 1.1
Suppose you decide to do some home cooking. You find a recipe for old-fashioned vegetable soup in your mom's cookbook. After reading the recipe, you think the soup sounds "flat" but that a little chili powder would spice it up. You make the soup and add a small amount of chili powder to it. You serve it to your family, and they consider it very good. You repeat the preparation of the soup but this time record the amount of chili powder you add to the recipe. Again, your family likes it. What is the experimentation? What is the hypothesis? What is the further experimentation? What would you need to do before making your conclusions a law?

Experimentation: Adding chili powder to the vegetable soup.
Hypothesis: Chili powder will improve the flavor of the vegetable soup.
Further experimentation: Repeat the preparation of the soup with a given amount of chili powder and see if your family continues to like it. Before making the recipe a law, vary the amount and type of chili powder and also serve the soup during various seasons of the year—summer, winter, and so on. You may also want to try it out on some of your friends.

Work Problem 3.

1.2   The Science of Chemistry

Chemistry   Study of the composition of substances and the changes they undergo.

Chemistry is the science concerned with the *composition* of substances and the *changes* they undergo. For example, chemistry is concerned with the components of water (composition) and the interactions between water and other substances (transformations).

Within the field of chemistry, chemists work on many types of problems. These problems can be categorized broadly as belonging to one of the five subfields of chemistry: (1) organic chemistry, (2) inorganic chemistry, (3) analytical chemistry, (4) physical chemistry, and (5) biochemistry. These subfields are presented in Table 1.1 along with some simple examples of problems within each subfield.

There is a certain amount of overlap among these subfields. In fact, some of the most exciting work in chemistry today involves problems related to more than one of the areas listed in Table 1.1. For example, a chemist studying the structure and nature of high-temperature superconductors draws on data and techniques from inorganic, analytical, and physical chemistry.

TABLE 1.1	The Five Subfields of Chemistry	
SUBFIELD	SUBJECT	EXAMPLE
Organic chemistry	Study of substances containing carbon, symbol C	Preparation of aspirin ($C_9H_8O_4$) or Tylenol ($C_8N_9NO_2$)
Inorganic chemistry	Study of all substances that do not contain carbon	Understanding how a car battery works
Analytical chemistry	Study of what is in a sample (*qualitative*) and how much of it is there (*quantitative*) (See Figure 1.3)	Measuring the amount of a particular pesticide in groundwater
Physical chemistry	Study of the structures of substances, how fast substances change (*kinetics*), and the role of heat in chemical change (*thermodynamics*)	Understanding the changes that occur when ice melts to give liquid water
Biochemistry	Study of the chemical reactions in living systems	Understanding how saliva breaks down some of the foods we eat as we chew

Organic chemistry Study of the substances containing carbon.

Inorganic chemistry Study of all substances other than those containing carbon.

Analytical chemistry Study of the quantitative and qualitative analysis (examination) of elements and compound substances.

Physical chemistry Study of the structures of substances, how fast they change. and the role of heat in chemical changes.

Biochemistry Study of the chemical reactions that occur in living organisms.

Study Exercise 1.2

Listed below are the research interests of some international chemists. Using the definitions of the branches of chemistry listed in Table 1.1, classify these research interests as organic, inorganic, analytical, physical, or biochemical.

a. Ring compounds of carbon that also contain sulfur
Gardner W. Stacy   (organic)

FIGURE 1.3
Chemists analyze new compounds under controlled conditions in order to better understand the properties of these materials.

b. Chemical composition of seeds of Ohio Buckeyes
Booker T. White   (biochemical)
c. Kinetics of transition metal reactions
Nancy Rowan   (physical or inorganic)

Work Problems 4 and 5.

1.3 Brief History of Chemistry

The many areas under investigation by chemists today reflect the explosion of interest in and understanding of chemical processes. Some of these processes have been known for a long time. For example, the ancient Egyptians and Chinese were well acquainted with the process of fermentation in the production of alcoholic beverages. Some of the foundations of modern chemistry trace back to the early Greeks and Arabs. Democritus (about 460–370 B.C.), a Greek, proposed a theory on the atomic structure of substances that preceded John Dalton's atomic theory by 2200 years. Al-Khowarizmi, an Arab mathematician, devised the number zero in about A.D. 825, providing the mathematical expression so essential to modern science. In the Medieval period, European alchemists experimented endlessly in a fruitless attempt to turn lead and other common metals into gold. Only recently, in the twentieth century, have scientists been able to transform platinum into gold (at great cost).

Two of the most important names in the founding of chemistry as a science are Robert Boyle (1627–1691) and Antoine Laurent Lavoisier (1743–1794). Boyle strongly advocated experimentation in the search for knowledge. Boyle's study of the effect of pressure on the volume of a gas is an example of the scientific method. Lavoisier is often considered the father of the science of chemistry. He demonstrated that combustion is the result of the combination of a fuel with oxygen from the air. Lavoisier used the scientific method to reveal that the popular theory of combustion was wrong and opened the door to a new way of looking at substances. We will study both Boyle's and Lavoisier's contributions to chemistry in Chapters 11 and 3, respectively.

Since the time of Lavoisier, individuals in virtually every nation in the world have made contributions to our understanding of chemistry. Table 1.2 lists just a few of the major discoveries since that time. Many of these individuals have received the Nobel prize in chemistry, the greatest single honor a scientist can receive.[*] These men and women came from many different nations and ethnic backgrounds and a variety of cultural and economic circumstances.

While Western Europeans dominated chemistry until World War I, American chemists have increasingly come to the forefront since 1920. In the aftermath of World War II, the United States became the world leader in all aspects of chemistry, from research and development to production. Efforts in Japan and the former Soviet Union have made both of these countries increasingly influential in the chemical community. The future influence of the countries of the former Soviet Union, particularly Russia, in the advancement of science remains to be seen. Possibly China will play a more influential role in the future.

FIGURE 1.4
Stocks of elemental sulfur. The chemical is a yellow solid that can be mined directly from the earth. The pure sulfur is then transported to industrial plants for use in various ways.

FIGURE 1.5
American postage stamp honoring Percy Julian.

*The Nobel prize was established in 1901 from the estate of Alfred Bernhard Nobel (1833–1896), the inventor of dynamite. Six Nobel prizes are customarily awarded annually for outstanding contributions to physics, chemistry, physiology or medicine, literature, economic sciences, and the promotion of world peace.

TABLE **1.2**		**Important Discoveries in Chemistry Since the Time of Lavoisier**	
NAME	DATES	COUNTRY OF ORIGIN	CONTRIBUTION
Charles Goodyear	1800–1860	United States	Discovered process for vulcanizing rubber using sulfur (see Figure 1.4).
Marie Curie[a]	1867–1934	Poland	Discovered radium
Gilbert Lewis	1875–1946	United States	Contributed to knowledge of chemical bonds
Otto Hahn[a]	1879–1968	Germany	Discovered nuclear fission, the process used in nuclear reactors
Niels Bohr[a]	1885–1962	Denmark	Developed model for the structure of the atom
Sir Robert Robinson[a]	1886–1975	United Kingdom	Prepared many useful medicinal agents
Albert von Szent-Gyorgi[a]	1893–1986	Hungary	Studied chemistry of muscle contractions
Wallace Carothers	1896–1937	United States	Discovered nylon
Percy Julian (see Figure 1.5)	1899–1975	United States	Developed process for mass-producing cortisone
Linus Pauling[a]	1901–1994	United States	Contributed to knowledge of chemical bonds and to world peace
Dorothy Crowfoot Hodgkin[a]	1910–1994	United Kingdom	Determined the structures of vitamin B_{12}, penicillin, and insulin
Melvin Calvin[a]	b. 1911	United States	Explained process of photosynthesis
Glenn Seaborg[a]	b. 1912	United States	Studied nuclear chemistry
Rosalind Franklin	1920–1958	United Kingdom	Studied structure of deoxyribonucleic acid (DNA) which controls the genetic code
Francis Crick[a]	b. 1916	United Kingdom	Proposed structure for DNA
Robert Woodward[a]	1917–1979	United States	Developed syntheses of various organic compounds such as quinine, lysergic acid, strychnine, reserpine, chlorophyll, vitamin B_{12}
Frederick Sanger[a]	b. 1918	United Kingdom	Identified chemical structure of insulin and investigated DNA
James Watson[a]	b. 1928	United States	Proposed structure of DNA

[a]Received the Nobel prize. Marie Curie, Linus Pauling, and Frederick Sanger all received the prize twice. Rosalind Franklin did not receive a Nobel prize because she died of cancer before her work could be recognized. Her research played an important part in the deoxyribonucleic acid structure proposed by Francis Crick and James Watson.

1.4   The Study of Chemistry

As you read the chapters to come, you will learn of the results of many chemists' work (see Figure 1.6). Whenever possible, we will attempt to tie their work to applications that affect your life (see Figure 1.7). The text also contains a variety of features to help you study chemistry efficiently and effectively.

In each chapter you should look for a number of features:

✔ **Countdown** (not in Chapter 1). This consists of five *review questions* on previous material which act as a *basis* for material to be introduced in the new chapter. These questions are keyed to sections in previous chapters, and the answers are given in parentheses next to the questions.

> **S**tudy Hint: Throughout the text you will find study hints in the margin like this. They will offer bits of help and wisdom to aid your understanding of the material.

Figure 1.6
Chemists are interested in many things. Here, someone is studying how light (in the form of a laser beam) interacts with a substance.

Figure 1.7
The synthesis of nylon. Materials that we take for granted today were invented by chemists.

✔ **Goals.** The goals specify skills that the chapter is designed to teach you, along with the number of the section related to that skill.

✔ **Study Hints.** These study hints are designed to help you understand the chemistry. Some of them are even humorous!

✔ **Examples.** These examples are worked out in the body of the chapter to help you solve a particular type of problem.

✔ **Study Exercises.** These study exercises are found in the body of the chapter and are there for *you* to work. The answers are found in parentheses next to the study exercise.

✔ **Summary.** A brief summary of the important points is given at the end of the chapter. The summary is usually not very long and is *not* a substitute for working through the chapter.

✔ **Exercises and Problems.** Exercises and problems are given at the end of the chapter and are keyed to sections of the chapter. *You* should work these to help you understand the material. The answers are found in Appendix VII at the end of the book. Under the **Problems** is a group of problems entitled *General Problems*. The information you need to solve these problems comes from previous chapters and not just the current chapter; you will need to recall material you have used previously.

✔ **Chapter Quiz.** The chapter quiz is found near the end of the chapter and gives you an opportunity to test yourself on the chapter. The answers are given in Appendix VII at the end of the book.

Three additional features are designed to show you applications of the principles you are learning. Each chapter is followed by a brief presentation on a featured chemical element or compound, such as gold, oxygen, carbon, salt, sugar, or soap.

Elements and **Compounds** give you an interesting look at how chemistry is part of your life. A second feature is a series of essays about chemistry in the atmosphere. **Chemistry of the Atmosphere** presents environmental issues that affect your life in a global way, from smog, pollution, and acid rain to the greenhouse effect and ozone chemistry. These essays show you how the chemistry you are learning is of real importance to the way you live. A third feature is **You and Chemistry**. Throughout the book you will see the icon shown here in the margin. This icon brings to your attention how chemistry applies to your real world.

You will get the most out of your efforts if you use a systematic approach to your studies. It is better to study six days a week for one hour than one day a week for six hours. As chemists, we have a prejudice: we want you not only to understand chemistry but to like it, too! Thus, we have written this text for you. We hope that the cartoons, analogies, and applications to everyday life will make chemistry interesting and fun to learn.

Finally, chemistry is not always easy. In fact, its challenges are what keep chemists excited about their field. By learning chemistry, you will be learning about a systematic approach to meeting challenges, which is a skill you can use whether your future in chemistry is as a teacher, a laboratory technician, a nurse, a technical sales representative, or a reader of the daily newspaper. Chemistry is an important part of your life and the future of our planet.

 ## Summary

This chapter considered the scientific method and its basis in experimentation; it described the various fields of science and classified chemistry as a physical science (Section 1.1). Chemistry is the study of the composition of substances and the changes they undergo. The chapter presented the various subfields of chemistry (Section 1.2) and a brief history of chemistry (Section 1.3). Finally, the chapter offered some hints on using this book and some insight into why you might want to learn chemistry (Section 1.4).

 ## Exercises

1. Define or explain the following terms (the number in parentheses refers to the section in the text where the term is mentioned):

   a. science (1.1)　　　　　　　　**b.** experimentation (1.1)

   c. hypothesis (1.1)　　　　　　　**d.** scientific law (1.1)

   e. scientific method (1.1)　　　　**f.** abstract sciences (1.1)

   g. physical sciences (1.1)　　　　**h.** biological sciences (1.1)

   i. chemistry (1.2)　　　　　　　**j.** organic chemistry (1.2)

   k. inorganic chemistry (1.2)　　　**l.** analytical chemistry (1.2)

   m. physical chemistry (1.2)　　　**n.** biochemistry (1.2)

2. Distinguish between

   a. biological and physical sciences　　**b.** hypothesis and scientific law

   c. inorganic and organic chemistry　　**d.** organic chemistry and biochemistry

✓ Problems

Science and the Scientific Method (See Section 1.1)

3. Imagine that you want to know the fastest route to the ice cream parlor. You begin by driving routes A, B, C, and D at the same speed on four successive evenings and seeing how long each takes. You compare the results and decide which is faster. You then repeat each of the measurements several times over the next few weeks to confirm your initial judgment. Which of the steps is experimentation? Hypothesis formation? Further experimentation? What would you need to do before making your conclusion a law?

The Science of Chemistry (See Section 1.2)

4. Listed below are the research interests of some international chemists. Using the definitions of the branches of chemistry listed in Table 1.1, classify each of these research interests as organic, inorganic, analytical, physical, or biochemical.

   a. Chemical kinetics and mechanisms of electrochemical reactions and dynamics of membrane function—William D. Weir

   b. Trace analysis of pollutants—Basil H. Vassos

   c. Bacterial metabolism, metabolism of radioactive purines in bacteria and animals; chemotherapy—Gertrude B. Elion

   d. Synthetic and theoretical organophosphorus chemistry—Sheldon Buckler

5. The following is a list of research papers published in various scientific journals. Using the definitions of the branches of chemistry, classify each of these research papers as organic, inorganic, analytical, physical, or biochemical.

   a. "Thermochromism in Copper(II) Halide Salts"—Darrell R. Bloomquist, Mark R. Pressprich, and Roger D. Willett

   b. "A New Procedure for Regiospecific Synthesis of Benzopyran-1-ones $[C_9H_6O_2]$"—Frank M. Hauser and Vaceli M. Baghdanov

   c. "Gas-Phase Chemistry of Pentacoordinate Silicon Hydride Ions"—David J. Hajdasz, Yeunghaiu Ho, and Robert R. Squires

   d. "Multiple Forms of the Nerve Growth Protein and Its Subunits"—Andrew P. Smith

   e. "The Nature of Soluble Copper(I) Hydride $[CuH]$"—J. A. Dilts

   f. "The Preparation and Properties of Peroxychromium(III) Species $[Cr_2O_2^{4+}$ and $Cr_3(O_2)_2^{5+}]$"—Edward L. King

   g. "Thermodynamics of Proton Ionization in Dilute Aqueous Solution"—James L. Christensen

(Answers to selected problems are in Appendix VII.)

✓ Chapter Quiz 1

1. What do we call the study of the composition of substances and the changes they undergo?

2. Listed below are the research interests of some international chemists. Using the definitions of the branches of chemistry, classify each of these research interests as organic, inorganic, analytical, physical, or biochemical.

   a. Propellant chemistry and thermodynamics—Frank I. Tanczos

   b. Structure of molecules by X-ray methods—Dorothy Hodgkin

   c. Chemistry of carbon-containing polymers—Carl S. Marvel

3. A child strikes out in a baseball game three times in a row. The child then changes to a lighter bat and gets a hit. The child subsequently uses the lighter bat and gets three more hits in five tries. Which of the steps is experimentation? Hypothis formation? Further experimentation? What would need to be done before making the conclusions a law?

(Answers for Chapter Quizzes are given in Appendix VII.)

Gold (Symbol: Au)

The Element GOLD: It Never Loses Its Shine

Gold is a bright and shiny metal that can be drawn into thin wires and cast into ingots.

Name: Symbol derives from the Latin *aurum* meaning "shining dawn." Gold was the first pure metal known to humans and has been valued ever since it was first discovered. Western settlers prospected for this valuable element, and fortunes were made and lost over it.

Appearance: Bright, lustrous, shiny yellow metal.

Occurrence: Gold is quite rare and makes up 0.0000005% (% = percent, parts per 100) of the earth's crust.

Source: Gold occurs as a metal in nature as a result of its remarkable stability. Gold is also found to a limited extent as the ores *calavarite* ($AuTe_2$) and *sylvanite* [$(AuAg)Te_2$].

Common Uses: The primary use of gold is as a currency standard for most nations of the world.

Gold and its alloys (mixtures, mostly with silver and copper) are used extensively in the making of fine jewelry. Gold alloys are also used in dentistry.

Gold has proved to be versatile in the electronics industry because of its excellent electrical conductivity, ductility, and resistance to corrosion.

Gold compounds are used in medicine in the treatment of arthritis.

Unusual Facts: The alchemists searched in vain for ways to change lead into gold. In doing so, however, they developed techniques and apparatus that led to the development of chemistry as a science. Other elements can now be changed into gold through nuclear chemistry, but not in an economical manner.

"There's gold in them there hills!"

CHAPTER 2

Measurements

The collection of samples is only the first step in the study of pollution of rivers and lakes. Once the samples are collected, they are taken to the laboratory where they are analyzed (examined) for pesticides and toxic wastes. Only by use of accurate measurements can water purity be compared over a given period of time.

GOALS FOR CHAPTER 2

1. To define matter and explain its chief characteristics (Section 2.1).

2. To be able to use the metric system of measurements (Section 2.2).

3. To understand the use and workings of measuring devices for various characteristics of matter, including temperature (Sections 2.3 and 2.4).

4. To determine the significant digits in a number and to perform mathematical operations involving them—including rounding off the answer (Sections 2.5 and 2.6).

5. To use exponential and scientific notations and to perform mathematical operations involving them (Section 2.7).

6. To use dimensional analysis to solve problems and to make metric system conversions, temperature conversions, and density conversions (Section 2.8).

7. To define specific gravity and use it in various calculations (Section 2.9).

Countdown

Perform the indicated mathematical operations. You may use a calculator if you wish. See Appendix II if you need help in using it.

5. Add:

$$\begin{array}{r} 32.15 \\ 7.42 \\ \underline{6.52} \end{array} \qquad (46.09)$$

4. Subtract:

$$\begin{array}{r} 17.64 \\ -\ 8.53 \end{array} \qquad (9.11)$$

3. Multiply: $18.0 \times 7.50 =$ (135)

2. Divide: $\dfrac{352}{11} =$ (32)

1. Divide: $\dfrac{2050}{25} =$ (82)

H ow many miles is it from your home to your college? How many ounces are in the cans of soda pop in the vending machine? How big is your waist?

Human beings have probably always been interested in somehow measuring the world around them. Records dating back thousands of years show the sizes of land holdings, the number of cattle owned, and the weight of crops harvested by members of earlier generations. Chemistry, with its focus on the composition of the substances that make up the world, relies heavily on measurement. This chapter thus considers the basic characteristics that chemists measure and the ways in which they make and use these measurements.

2.1   Matter and Its Characteristics

Matter   Any substance that has mass and occupies space.

Mass   Quantity of matter in a particular body.

Weight   Measure of the gravitational force of attraction between the body's mass and the mass of the planet or satellite on which it is weighed.

Chemists refer to substances that make up our universe as **matter**—anything that has mass and occupies space. In addition, matter has measurable characteristics such as weight, length, volume, temperature, and density.

We define **mass** as the quantity of matter in a particular body. *The mass of a body is constant and does not change,* whether it is measured in Colorado or New York or even on the moon. In contrast, the **weight** of a body is the *gravitational force of attraction* between the body's mass and the mass of the planet or satellite on which it is weighed. Thus, where we weigh matter affects its weight. For example, the earth is not spherical but is slightly pear-shaped, and because the gravitational attraction between the earth and a body varies with the distance between their centers, the weight of the body varies slightly depending on where it is measured. An object that weighs 10 pounds (lb) at the North Pole weighs only 9 lb 15 ounces (oz) at the Equator—a difference of 1 oz in weight. On the moon, low gravitational attraction causes objects to weigh less than on the earth. An object that weighs 10 lb at the North Pole would weigh only 1 lb 11 oz on the moon. Note that in each case *the mass of the object has not changed but the weight* (the attraction of the earth for the object) *has changed.*

Figure 2.1 summarizes the relations among matter, mass, and weight. Other characteristics of matter can also fluctuate from object to object, time to time, and

FIGURE 2.1
Mass and weight: (a) an astronaut is composed of matter and has a mass that is a measure of that matter; (b) the same astronaut also has a weight that is a measure of the earth's gravitational attraction for her; (c) the same astronaut weighs less on the moon because of the moon's weaker gravity. Note that her mass has not changed!

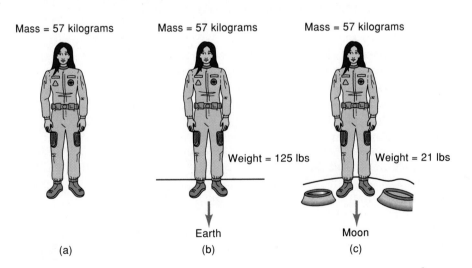

Mass = 57 kilograms   Mass = 57 kilograms   Mass = 57 kilograms

Weight = 125 lbs   Weight = 21 lbs

Earth   Moon

(a)   (b)   (c)

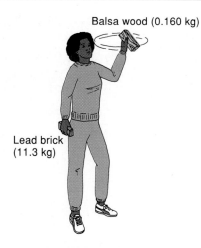

Balsa wood (0.160 kg)

Lead brick
(11.3 kg)

FIGURE 2.2
Balsa wood has a lower density than lead. Thus, a brick of balsa wood is much lighter than a lead brick of the same volume.

place to place. Two otherwise identical steel girders might differ in length, for example. Matter takes up space—it occupies *volume*. The **volume** of a block of wood—the cubic space it takes up—is measured as length times width times height. The volumes of other shapes can also be measured. The **temperature** of an object—its degree of hotness—can also vary considerably, as you will see if you are burned by steam from a hot iron or put your hand in the ice cube tray of a freezer.

Volume   Cubic space taken up by matter.

Temperature   Degree of hotness of matter.

Another measurable difference among objects is their **density** (*d*), which is defined as the mass of a substance occupying a unit volume, or

$$\textbf{density }(\textbf{\textit{d}}) \; = \; \frac{\text{mass }(m)}{\text{volume }(v)}$$

Density (*d*)   Mass of a substance occupying a unit volume, expressed as the mass divided by the volume.

As Figure 2.2 shows, because density takes into consideration both mass and volume, two objects with the same volume—a block of balsa wood and a lead brick—can differ greatly in mass and weight. Thus, they have different densities.

<hr>

2.2 Measuring Matter: A Focus on the Metric System

Mass, weight, length, volume, and density can be measured using two methods: (1) English-based units and (2) the metric system. In addition, scientists sometimes employ a system of units called the International System of Units, abbreviated SI from the French *Système International,* which is based on the metric system and described in Appendix I. English-based units (see Appendix I), such as the foot (ft) and the pound (lb), are used primarily in the nonscientific community in the United States. The **metric system**, developed in nineteenth-century France, is used throughout the rest of the world (even in the United Kingdom—England) and in the world scientific community. Civilian and business interests in the United States are gradually adopting the metric system, as shown in Figure 2.3. For example, the chemical industry is now in the process of converting to the metric system in the shipping and billing of industrial chemicals.

The metric system has as its basic units the gram (g), a measure of mass; the liter (or litre, L), a measure of volume; and the meter (or metre, m), a measure of

Metric system   System of weights and measures in which each unit is a tenth, hundredth, thousandth, and so on, of another unit: it is the standard system in use in every nation except the United States and is used extensively by scientists.

(a)

(b)

(c)

FIGURE 2.3
Even in the United States, the metric system is being adopted: (a) length. (William Felger/Grant Heilman Photography) (b) volume; (c) mass. (Dr. E. R. Degginger)

length. The units for mass, volume, and length in the metric system are expressed in multiples of 10, 100, 1000, 1,000,000, and so on, similar to some parts of our monetary system. For example, the prefix *centi-* represents $\frac{1}{100}$ of a basic metric unit, just as a cent represents $\frac{1}{100}$ of our basic monetary unit, the dollar.

Table 2.1 shows the prefixes used to define multiples or fractions of the basic units, as well as specific multiple metric units of mass (gram), volume (liter), and length (meter). You must learn these units and their equivalents in order to work problems. For example, you should know that $1,000,000$ (10^6) g $=$ 1 Mg, 1000 m $=$ 1 km, 10 dg $=$ 1 g, 100 cm $=$ 1 m, 1000 mL $=$ 1 L, $1,000,000$ (10^6) μg $=$ 1 g, $1,000,000,000$ (10^9) nm $=$ 1 m, and $1,000,000,000,000$ (10^{12}) pm $=$ 1 m. To give you a better sense of how much—or how little—these units represent, Table 2.2 shows the English-based equivalent of some common metric measurements.

Study Hint: The abbreviation for *mega-* is capital **M,** whereas the abbreviation for *milli-* is small **m.**

TABLE 2.1			Some Metric Units of Mass, Volume, and Length		
PREFIX	NUMBER OF BASIC UNITS[a]		MASS	VOLUME	LENGTH
mega-	1,000,000	(10^6)	megagram (Mg)	megaliter (ML)	megameter (Mm)
kilo-	1000	(10^3)	kilogram (kg)	kiloliter (kL)	kilometer (km)
basic unit	1		gram (g)	liter (L)	meter (m)
deci-	0.1	(10^{-1})	decigram (dg)	deciliter (dL)	decimeter (dm)
centi-	0.01	(10^{-2})	centigram (cg)	centiliter (cL)	centimeter (cm)
milli-	0.001	(10^{-3})	milligram (mg)	milliliter (mL)[b]	millimeter (mm)
micro-	0.000001	(10^{-6})	microgram (μg)	microliter (μL)	micrometer (μm)
nano-	0.000000001	(10^{-9})	nanogram (ng)	nanoliter (nL)	nanometer (nm)
pico-	0.000000000001	(10^{-12})	picogram (pg)	picoliter (pL)	picometer (pm)

[a] The notation, $10^{+ \text{ or } - \text{number}}$ is discussed in Section 2.7.

[b] The milliliter (mL) and the cubic centimeter (cm^3 or cc) are exactly equivalent because $1 \text{ L } = 1 \text{ dm}^3$ by definition:

$$1 \text{ mL } = 1 \text{ cm}^3 \text{ (cc)}$$
$$1 \text{ L } = 1 \text{ dm}^3$$

In the metric system, the units of density generally used for solids and liquids are g/mL (g/cm^3), and the units generally used for gases are g/L. Density has units of mass/volume, and whenever the density of a substance is expressed, the particular units of mass and volume must be given. For example, the density of water is 1.00 g/mL in the metric system and 1000 kg/m^3 in the SI. It is not enough to express the density of a substance as a pure number without units.

If two liquids that are not soluble in each other are placed in the same container, the liquid with the greater density will be on the bottom and the less dense liquid will be on top. An example of this is oil and water, as shown in Figure 2.4. When oil is spilled in the ocean, an oil slick is created.

You and Chemistry

Study Hint: From Table 2.1, a centi is equal to 0.01 of the basic unit; therefore, there are 100 centi in one basic unit. This is the same as saying that a cent is 0.01 of a dollar and that there are 100 cents in one dollar. A milli is equal to 0.001 of the basic unit; therefore, there are 1000 milli in one basic unit. A micro is equal to 0.000001 (10^{-6}) of the basic unit; therefore, there are 1,000,000 (10^6) micro in one basic unit.

TABLE 2.2		Some English-Metric Equivalents	
TYPE OF MEASUREMENT	ENGLISH SYSTEM		METRIC SYSTEM
Mass	{ 1.00 pound	⇔	454 grams
Length	⎧ 1.00 inch	⇔	2.54 centimeters
	⎨ 1.00 mile	⇔	1.61 kilometers
	⎩ 1.09 yard	⇔	1.00 meter
Volume	⎧ 1.06 quart	⇔	1.00 liter
	⎨ 1.00 pint	⇔	473 milliliters
	⎩ 1.00 gallon	⇔	3.78 liters

FIGURE 2.4
Oil and water. The density of water is greater than the density of oil, hence the water is on the bottom and the oil is on top. This occurs when oil is spilled in the ocean, creating oil slicks with the oil on top.

FIGURE 2.5
Comparison of the Celsius and Fahrenheit scales.

EXAMPLE 2.1 Calculate the density of a piece of metal that has a mass of 25 g and occupies a volume of 6.0 mL.

SOLUTION

$$\frac{25 \text{ g}}{6.0 \text{ mL}} = 4.2 \text{ g/mL} \qquad Answer$$

Study Exercise 2.1
Calculate the density of a piece of metal having a volume of 8.0 mL and a mass of 58.4 g.

(7.3 g/mL)

2.3 Temperature

Not all characteristics of matter are measured metrically. Temperature is measured using three common temperature scales:

1. Fahrenheit scale (°F)
2. Celsius (formerly called *centigrade*) scale (°C)
3. Kelvin scale (K) (used in the SI system)

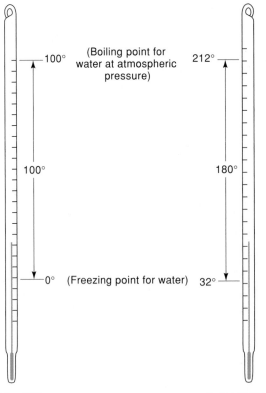

The Fahrenheit scale, named after German physicist Gabriel Daniel Fahrenheit (1686–1736), is probably the scale most familiar to you. On this scale, the freezing point of pure water is 32° and the normal boiling point of water is 212°. On the Celsius scale, named after Swedish astronomer Anders Celsius (1701–1744), these points correspond to 0° and 100°, respectively.

The Kelvin scale, named after British physicist and mathematician William Thomson (1824–1907), who was later titled Lord or Baron Kelvin, consists of a new scale with the zero point equal to −273°C (more accurately, −273.15°C). The lower limit of this scale is theoretically zero, with no upper limit. The temperature of some stars is estimated at many millions of kelvins.

Figure 2.5 compares the Fahrenheit and Celsius scales. On the Celsius scale there are 100° between the freezing point (fp) and the boiling point (bp) of water; however, on the Fahrenheit scale this difference is 180°. Thus, 180 Fahrenheit divisions equal 100 Celsius divisions, or there are 1.8°F to 1°C. In addition, the freezing point of water is 0° on the Celsius scale and 32° on the Fahrenheit scale.

(a)

(b)

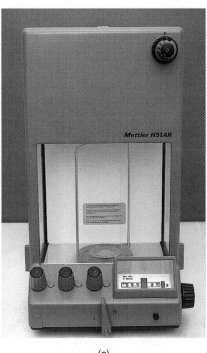

(c)

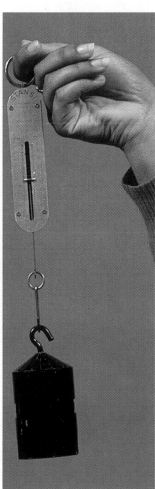

(d)

FIGURE 2.6
Measuring mass: (a) platform balance: note the unknown object (blue solid) and the known objects (brass weights); (b) triple-beam balance; (c) a single-pan balance. Measuring weight: (d) a spring scale.

2.4 Measuring Devices

A number of devices employing metric units and temperature scales assist chemists in measuring matter. Balances such as those shown in Figure 2.6a and b measure mass. In these balances, the body to be "massed" is placed on the left pan and objects of known mass ("weights") are placed on the right pan or on the sliding scale to counterbalance the unknown body. The mass of the unknown body and the masses of the weights are in balance when the pointer is at the center of the scale. A balance operates on the same principle as that involved in balancing a pencil on your finger. The force of gravity acts equally on the known masses (weights) and the unknown object when balance is achieved. As a result, the *mass* of the object is found to be the same regardless of where it is measured.

A modern balance is shown in Figure 2.6c. A sample is placed on the pan, and the mass generates a small electric current inside the balance. This current is measured and can be used to determine the mass of the object. Such balances are convenient and rapid and are used extensively.

Spring scales, like the one shown in Figure 2.6d, determine the *weight* of a body. The body to be weighed is attached to a hook, and the spring is stretched. How much the spring stretches depends on the gravitational attraction of the planet or satellite on which the body is being weighed. You have probably used a spring scale to weigh produce at the grocery store. People who fish also use spring scales to measure the weight of their catch.

FIGURE 2.7
Measuring volume: (a) measuring cup, (b) measuring teaspoon, (c) graduated cylinder, (d) buret, (e) pipet with a pipet filler used to fill the pipet.

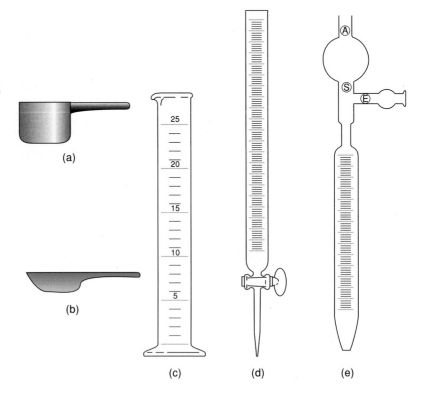

In chemistry, the balance is used exclusively, hence we measure the *mass* of a body. The two terms "mass" and "weight" are unfortunately used interchangeably, but if a balance is used, the correct term is "mass."

Other characteristics of matter call for other measuring devices. You may have used a measuring cup or measuring teaspoon (see Figure 2.7a and b) in cooking or taking certain liquid medications. Chemists measure volumes using *graduated cylinders, burets,* and *pipets* (see Figure 2.7c through e). You may measure length in inches or feet, but chemists measure length in centimeters or meters (see Figure 2.8). Temperature is commonly measured with a thermometer. Because density depends on the mass, volume, and temperature of an object, measurements of density require balances, volume measures such as cylinders, and thermometers.

FIGURE 2.8
Measuring length by a ruler in centimeters (cm) or inches (in.)

2.5 Significant Digits

In making measurements, chemists need to be accurate. But almost any measurement has limits to its precision (exactness). The digits in a measurement that are known to be precise (exact), along with a final digit about which there is some uncertainty, are known as **significant digits (figures).** How many significant digits a measurement contains depends on the nature of the measuring device used.

For example, suppose we measure a strip of metal using the two rulers shown in Figure 2.9. On rule A, we might estimate the length as $16\frac{1}{5}$ (16.2) cm. On rule B,

Significant digits (figures)
Digits in a measurement that are known to be precise, along with a final digit about which there is some uncertainty.

FIGURE 2.9
Measurement of a piece of metal. A portion of the centimeter scale of both rulers is magnified.

Ruler A	Metal	Ruler B
18		18
17		17
16		16
0		0

we might judge the length to be 16.26 cm. In both cases, the last digit, 2 in 16.2 and 6 in 16.26, is uncertain but significant.

To determine the number of significant digits in a measurement, we follow certain rules.

Rules for Significant Digits

1. *Nonzero digits:* 1, 2, 3, 4, 5, 6, 7, 8, and 9 are *always* significant.

6.2	two significant digits
16.2	three significant digits
16.26	four significant digits

2. *Leading zeros:* zeros that appear at the start of a number are *never* significant because they act only to fix the position of the decimal point in a number less than 1.

0.564	three significant digits
0.0564	three significant digits

3. *Confined zeros:* zeros that appear between nonzero numbers are *always* significant.

104	three significant digits
1004	four significant digits

4. *Trailing zeros:* zeros at the end of a number are significant *only* if the number (a) contains a decimal point *or* (b) contains an overbar.

154.00	five significant digits
154.0	four significant digits
15.40	four significant digits
1540.	four significant digits
56,$\overline{0}$00	five significant digits
56$\overline{0}$0	four significant digits
5$\overline{6}$00	three significant digits

EXAMPLE 2.2 Determine the number of significant digits in the following numbers:

Number	Answer [rule(s)]	Number	Answer [rule (s)]
a. 747	3 (1)	**g.** 7065	4 (1, 3)
b. 1011	4 (1, 3)	**h.** 0.604	3 (1, 2, 3)
c. 3.50	3 (1, 4a)	**i.** 10.04	4 (1, 3)
d. 0.056	2 (1, 2)	**j.** 122.0	4 (1, 4a)
e. 35$\overline{0}$	3 (1, 4b)	**k.** 7.0200	5 (1, 3, 4a)
f. 6.02	3 (1, 3)		

Study Exercise 2.2
Determine the number of significant digits in the following numbers:

a. 123   (3)   **b.** 0.074   (2)
c. 707   (3)   **d.** 7$\overline{0}$ (2)

How many significant digits are there in 85,000,000,000 hamburgers? (See Figure 2.10.)

Work Problems 5 and 6.

FIGURE 2.10
85,000,000,000 hamburgers! How many significant digits are there in this number?

2.6 Mathematical Operations Involving Significant Digits

The use of significant digits creates interesting problems in some cases. For example, suppose we want to express to three significant digits the percentage of water in a sample of matter and our calculator gives 64.06392? What if we are given measurements with different numbers of significant digits? Fortunately, scientists have certain simple rules that we can use in these situations.

Rounding Off

One method of dealing with significant digits is rounding off. You use rounding off constantly in your everyday life. When you say you have $100 in the bank, do you really have exactly $100.00? No, you have just estimated and rounded off your measurement to one or two significant digits. You may have learned to round off numbers in school. Chemists use much the same approach but must also allow for nonsignificant digits, which produces four rules:

Rules for Rounding Off

1. If the first nonsignificant digit is *less* than 5, drop it and the last significant digit remains the same. Thus, 47.21 is equal to 47.2 to three significant digits.
2. If the first nonsignificant digit is *more* than 5 or *is* 5 followed by *numbers other than zeros,* drop the nonsignificant digit(s) and increase the last significant digit by 1. Hence, 47.26 and 47.252 are both equal to 47.3 to three significant digits.
3. If the first nonsignificant digit is 5 and is followed by zeros, drop the 5 and
    a. increase the last significant digit *by one if it is odd,* or
    b. leave the last significant digit *the same if it is even.* Thus, 47.250 is equal to 47.2 to three significant digits, and 47.350 is equal to 47.4.
4. Nonsignificant digits to the *left of the decimal point are not discarded but are replaced by zeros.* Thus, 1781 becomes 1780 and not 178 when rounded off to three significant digits. Similarly, 25,369 is equal to 25,400 (not 254) to three significant digits.

| EXAMPLE 2.3 | Round off the following numbers to three significant digits. |

Number	Answer (rule)	Number	Answer (rule)
a. 462.2	462 (1)	**g.** 1248	1250 (2, 4)
b. 453.6	454 (2)	**h.** 12.750	12.8 (3)
c. 474.50	474 (3)	**i.** 0.027650	0.0276 (3)
d. 687.54	688 (2)	**j.** 0.027654	0.0277 (2)
e. 687.50	688 (3)	**k.** 0.027750	0.0278 (3)
f. 688.50	688 (3)	**l.** 93,483,291	93,500,000 (2, 4)

Study Exercise 2.3
Round off the following numbers to three significant digits.
a. 7.268   (7.27)   **b.** 4.365   (4.36)

Simple Math

Rounding can help you get the right number of digits. But what is "right" depends on rules about adding, subtracting, multiplying, and dividing numbers.

Rule for Addition and Subtraction

In addition and subtraction, *the answer must not contain a smaller place* (that is, decimal, units, tens, and so on) *than the number with the smallest place.*

The sum of

$$\begin{array}{r} 25.\mathbf{1} \\ + 22.11 \\ \hline 47.21 \end{array}$$

is

but the answer must be expressed to only the tenths decimal place because the tenths decimal place is the smallest place in the number 25.1; hence, the answer is 47.2. Likewise, the difference

$$\begin{array}{r} 4.732 \\ - 3.6\mathbf{2} \\ \hline 1.112 \end{array}$$

is

but the answer must be expressed to only the hundredths decimal place because the hundredths decimal place is the smallest place in the number 3.62; hence, the answer is 1.11.

The reason behind this rule is simple: the value of an unmeasured decimal place, such as the hundredth decimal place in 25.1 and the thousandth decimal place in 3.62, can be as little as 0 or as much as 9.

Study Hint: For addition and subtraction we consider the *places.*

Rule for Multiplication and Division

In multiplication and division, *the answer must not contain any more significant digits than the least number of significant digits in the numbers used in the multiplication or division.*

The product of

$$\begin{array}{r} 17.21 \\ \times 11.1 \\ \hline \text{is}\quad 191.031 \end{array}$$

but the answer must be expressed to only three significant digits because 11.1 has only three significant digits; hence, the answer is 191.

The quotient is

$$\frac{26.32}{2.23} = 11.80269$$

but the answer again must be expressed to only three significant digits because 2.23 has only three significant digits; therefore, the answer is 11.8.

> **Study Hint:** For multi-plication and division we consider the *significant digits*.

EXAMPLE 2.4   Perform the indicated mathematical operations and express the answer to the proper number of significant digits.

a. 17.8 + 14.73 + 16

SOLUTION   The smallest place is the units place in the number 16, and so we must express the answer to the units place.

$$\begin{array}{l} 17.8 \\ 14.73 \\ \underline{16} \quad \longleftarrow \text{number with the smallest place} \\ 48.53 \qquad 49 \qquad \textit{Answer} \end{array}$$

Rounded off to the units place, 48.53 is 49.

b. 0.647 + 0.03 + 0.31

> **Study Hint:** In addition and subtraction we consider *places, not significant digits.*

SOLUTION   The smallest place is the hundredths decimal place in the numbers 0.03 and 0.31; express the answer to the hundredths decimal place.

$$\begin{array}{l} 0.647 \\ 0.03 \quad \longleftarrow \\ \underline{0.31} \quad \longleftarrow \text{numbers with the smallest place} \\ 0.987 \qquad 0.99 \qquad \textit{Answer} \end{array}$$

Rounded off to the hundredths decimal place, 0.987 is 0.99.

c. 14.72 − 6.8

SOLUTION   The smallest place is the tenths decimal place in the number 6.8; thus, we must express the answer to the tenths decimal place.

$$\begin{array}{l} 14.72 \\ \underline{-6.8} \quad \longleftarrow \text{number with the smallest place} \\ 7.92 \qquad 7.9 \qquad \textit{Answer} \end{array}$$

Rounded off to the tenths decimal place, 7.92 is 7.9

d. $24.78 - 0.065$

SOLUTION The smallest place is the hundredths decimal place in the number 24.78; we must express the answer to the hundredths decimal place.

$$24.78 \longleftarrow \text{number with the smallest place}$$
$$\underline{-0.065}$$
$$24.715 \qquad 24.72 \qquad Answer$$

Rounded off to the hundredths decimal place, 24.715 is 24.72.

e. 752×13

SOLUTION The number with the least number of significant digits is 13, which has only two; thus the answer must have no more than two significant digits.

$$752 \times 13 = 9776 \qquad 9800 \qquad Answer$$

Rounded off to two significant digits, 9776 is 9800, not 98 (see rule 4).

f. 0.02×47

SOLUTION The number with the least number of significant digits is 0.02, which has only one; hence, we must express the answer to only one significant digit.

$$47 \times 0.02 = 0.94 \qquad 0.9 \qquad Answer$$

Rounded off to one significant digit, 0.94 is 0.9.

g. $\dfrac{181.8}{75}$

SOLUTION The number with the least number of significant digits is 75, which has two; therefore, we must express the answer to no more than two significant digits.

$$\frac{181.8}{75} = 2.424, \text{ from a calculator}$$

Rounded off to two significant digits, 2.424 is 2.4. *Answer*

h. $\dfrac{13.65}{2.26}$

SOLUTION The number with the least number of significant digits is 2.26, which has three; we must express the answer to no more than three significant digits.

$$\frac{13.65}{2.26} = 6.039823, \text{ from a calculator}$$

Rounded off to three significant digits, 6.039823 is 6.04. *Answer*

i. $\dfrac{9.74 \times 0.12}{1.28}$

SOLUTION The number with the least number of significant digits is 0.12, which has two, and so we must express the answer to two significant digits.

$$\frac{9.74 \times 0.12}{1.28} = 0.913125, \text{ from a calculator}$$

Rounded off to two significant digits, 0.913125 is 0.91. *Answer*

As you can see, you will be doing a lot of adding, subtracting, multiplying, and dividing as you study chemistry. A simple, inexpensive hand-held calculator will make these tasks much simpler. Your instructor may have a suggestion as to what model would be best; otherwise, buy one that has the following basic functions: $+$, $-$, $\times$, $\div$, and $\sqrt{}$. It would also be very useful if your calculator had log x and y^x functions as well. Appendix II will help you learn to use your calculator.

A Special Rule: Exact Numbers

Occasionally, a calculation involves an exact number. **Exact numbers** are precisely known and can have as many significant digits as a calculation requires, and so they are *not* used to determine the number of significant digits for an answer.

Exact numbers   Numbers for measurements that are precisely known and can have as many significant digits as a calculation requires, and so they are not used to determine the number of significant digits for an answer.

The most common of these cases is when a relationship has been precisely *defined* and *no measurement has been taken.* For example, there are exactly 12 in. in 1 ft. Thus, in calculations we can express a foot as 12 in., 12.0 in., 12.00 in., 12.000 in., and so on, using as many significant digits as we need to agree with the other numbers in our calculations.

Study Exercise 2.4
Perform the indicated mathematical operations and express your answer to the proper number of significant digits.
a. $23.452 + 2.73 + 0.7$ (26.9)
b. 62.1×0.3424 (21.3)   Work Problems 7 through 10.

2.7 Exponents, Exponential Notation, and Scientific Notation

In addition to a need for precision (exactness), scientists are often required to use extremely large or extremely small numbers and prefer easier ways to express them (see Figure 2.11). Most people find it faster to read about the world's 5.6 billion people than to count the zeros in 5,600,000,000. They understand that the United States makes up about 4.7% of that population more easily than that it is a 0.047 factor of everyone living. Scientists use an even more condensed way of expressing numbers by using exponents.

An **exponent** is a whole number or symbol written as a superscript to another number or symbol, the *base;* it denotes the number of times the base is to be multiplied by itself. For example,

Exponent   Whole number or symbol written as a superscript above a base and denoting the number of times the base is to be multiplied by itself.

$$10^6 = 10 \times 10 \times 10 \times 10 \times 10 \times 10 = 1{,}000{,}000$$

$$5^3 = 5 \times 5 \times 5 = 125 \quad \text{(exact)}$$

$$23^2 = 23 \times 23 = 529 \quad \text{(exact)}$$

 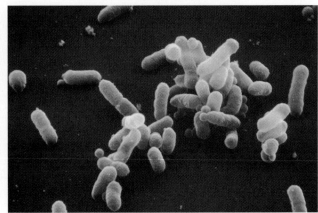

FIGURE 2.11
Scientific notation. (a) Mean distance of the moon from the earth—3.8×10^5 kilometers. (b) Bacteria—length, 10^{-6} meters.

Similarly, very small numbers (less than 1) are expressed with negative exponents. For example,

$$10^{-6} = \tfrac{1}{10} \times \tfrac{1}{10} \times \tfrac{1}{10} \times \tfrac{1}{10} \times \tfrac{1}{10} \times \tfrac{1}{10}$$

$$= \tfrac{1}{1,000,000} = 0.000001$$

$$5^{-3} = \tfrac{1}{5} \times \tfrac{1}{5} \times \tfrac{1}{5} = \tfrac{1}{125} = 0.008 \quad \text{(exact)}$$

$$23^{-2} = \tfrac{1}{23} \times \tfrac{1}{23} = \tfrac{1}{529} = 0.00189036$$

The number of times a base is repeated is called the *power of the base*. Thus, you can express 10^6 as "10 raised to the sixth power" and 10^{-6} as "10 to the negative sixth power."

Chemists and other scientists use positive and negative exponents, specifically powers of 10, to write large and small numbers in exponential notation. **Exponential notation** expresses any number as a product of two numbers, one a decimal and the other a power of 10. For example, the exponential notation for 241,000 is 2.41×10^5, with 2.41 the decimal and 10^5 the power of 10. It could also be 24.1×10^4, but as we will see later, 2.41×10^5 is the form used in science. Table 2.3 lists the values of some powers of 10.

In expressing a number in exponential notation, you may find the following guidelines helpful:

Exponential notation Form of mathematical expression in which a number is expressed as the product of two numbers, one a decimal and the other a power of 10.

TABLE 2.3
Powers of 10

1000	$=$	10^3
100	$=$	10^2
10	$=$	10^1
1	$=$	10^0
$\tfrac{1}{10}$	$=$	10^{-1}
$\tfrac{1}{100} = 0.01$	$=$	10^{-2}
$\tfrac{1}{1000} = 0.001$	$=$	10^{-3}

✔ A positive exponent means a number *larger* than 1. A negative exponent means a number *smaller* than 1.

✔ The exponent is equal numerically to the *number of places* the decimal point is moved.

✔ Changing a number by shifting the decimal point to the *left* of its original position results in a *positive* exponent. Changing a number by shifting the decimal point to the *right* of its original position results in a *negative* exponent.

EXAMPLE 2.5	Express the estimated world population of 5,600,000,000 people (or 5,600,000,000.) in exponential notation using 5.6 as the decimal.

SOLUTION   The decimal point in 5,600,000,000 must be moved *nine* places to the *left* to convert 5,600,000,000 to 5.6.

$$5,600,000,000$$
$$987\ 654\ 321$$

Because the decimal was moved *nine* places to the *left,* the exponent is $+9$. Hence, 5,600,000,000 is equivalent to 5.6×10^9.   *Answer*

EXAMPLE 2.6	Express the factor of the U.S. population to that of the world population, 0.047, in exponential notation using 4.7 as the decimal.

SOLUTION   The decimal point in 0.047 must be moved *two* places to the *right* to convert 0.047 to 4.7.

$$0.047$$
$$12$$

Because the decimal was moved *two* places to the *right,* the exponent is -2. Hence, 0.047 is equivalent to 4.7×10^{-2}.   *Answer*

Study Exercise 2.5
Express 0.000755 as 7.55×10^n. (7.55×10^{-4}) Work Problems 11 and 12.

Scientific Notation

Scientific notation is a more systematic form of exponential notation. In **scientific notation,** the decimal part must have *exactly one* nonzero digit to the left of the decimal point. Thus, instead of writing 31.0×10^6, we write 3.10×10^7 to express 31 million in scientific notation.

Scientific notation   Form of exponential notation in which the decimal part must have exactly one nonzero digit to the left of the decimal point; it is widely used by scientists.

EXAMPLE 2.7	Express the following in scientific notation to three significant digits:

Number	Answer	
a. 6,780,000	6.78×10^6	Move decimal point six places to the left.
b. 2170	2.17×10^3	Move decimal point three places to the left.
c. 0.0756	7.56×10^{-2}	Move decimal point two places to the right.
d. 10.7	1.07×10^1	Move decimal point one place to the left.

Study Exercise 2.6
Express the following in scientific notation to three significant digits:
a. 0.00000345 (3.45×10^{-6})
b. 7,297,000 (7.30×10^6)

Work Problems 13 through 15.

Mathematical Operations Involving Exponential Notation

Exponential and scientific notations not only make it easier to express very large or small numbers but also make it easier to manipulate these numbers mathematically. We need only follow a few general rules.

Rule for Addition and Subtraction of Exponential Numbers

To add and subtract exponential numbers, first express *each* quantity to the *same power of 10*. Then add or subtract decimals in the usual manner and record the powers of 10. That is, the power of 10 does *not* change at this point.

For example, to add 6.35×10^5 and 1.56×10^4, first convert the second number to 0.156×10^5. Then add and round off the decimals ($6.35 + 0.156 = 6.51$) and record the powers of 10 to get 6.51×10^5.

EXAMPLE 2.8 Carry out the operations indicated on the following exponential numbers:

a. $3.40 \times 10^3 + 2.10 \times 10^3$

SOLUTION   Both numbers have the same power of 10 (10^3); hence, we can add them:

$$
\begin{array}{r}
3.40 \times 10^3 \\
2.10 \times 10^3 \\
\hline
5.50 \times 10^3
\end{array}
\quad \textit{Answer}
$$

b. $4.20 \times 10^{-3} + 1.2 \times 10^{-4}$

SOLUTION   To add these numbers, first convert them to the *same power of 10*. The number 1.2×10^{-4} converts to 0.12×10^{-3}, following the guidelines given. We can now add these two numbers.

$$
\begin{array}{r}
4.20 \times 10^{-3} \\
0.12 \times 10^{-3} \\
\hline
4.32 \times 10^{-3}
\end{array}
\quad \textit{Answer}
$$

> **Study Hint:** Notice that the power of 10 *remains the same* in the answer.

Rule for Multiplication and Division of Exponential Numbers

For multiplying or dividing exponential numbers, the only requirement is that the numbers be expressed to the same *base*, which is 10 in exponential notation. In multiplication, multiply the decimals in the usual manner but *add* algebraically the *exponents* of the base 10. In division, divide the decimals in the usual manner and *subtract* algebraically the *exponents* of the base 10.

For example, to multiply 9.2×10^3 by 6.4×10^2, first multiply and round off the decimals ($9.2 \times 6.4 = 59$). Then add algebraically the exponents of the base $10^{(3 + 2)}$. The result is 59×10^5 or 5.9×10^6 in scientific notation.

To divide these numbers, first divide and round off the decimals (9.2 divided by 6.4 is 1.4). Then subtract algebraically the exponents of the base $10^{(3-2)}$. The result is 1.4×10^1.

EXAMPLE 2.9   Carry out the operations indicated on the following exponential numbers:

a. $1.70 \times 10^6 \times 2.40 \times 10^3$

SOLUTION   Multiply the decimals and then add the exponents algebraically.

$$(1.70 \times 2.40)(10^6 \times 10^3) = 4.08 \times 10^{6+3} = 4.08 \times 10^9 \quad \textit{Answer}$$

b. $1.70 \times 10^6 \times 2.40 \times 10^{-3}$

SOLUTION

$$(1.70 \times 2.40)(10^6 \times 10^{-3}) = 4.08 \times 10^{6-3} = 4.08 \times 10^3 \quad \textit{Answer}$$

c. $\dfrac{2.40 \times 10^5}{1.30 \times 10^3}$

SOLUTION   Divide the decimals and then subtract the exponents algebraically,

$$\frac{2.40 \times 10^5}{1.30 \times 10^3} = \frac{2.40}{1.30} \times \frac{10^5}{10^3} = 1.85 \times 10^{5-3} = 1.85 \times 10^2 \quad \textit{Answer}$$

d. $\dfrac{2.40 \times 10^5}{1.30 \times 10^{-3}}$

SOLUTION

$$\frac{2.40 \times 10^5}{1.30 \times 10^{-3}} = \frac{2.40}{1.30} \times \frac{10^5}{10^{-3}}$$

$$= 1.85 \times 10^{5-(-3)} = 1.85 \times 10^{5+3} = 1.85 \times 10^8 \quad \textit{Answer}$$

Rule for Square Roots of Exponential Numbers

To obtain the positive value of the square root of a number, first express the number in exponential notation in which the power of 10 has an *even* exponent. Next, obtain the square root of the decimal from a calculator. Finally, obtain the square root of the power of 10 by dividing the exponent by 2.

Study Hint: Some more expensive calculators move the decimal point automatically. Enter 8.12 and then press the EE↓ key followed by 5 as the exponent to display 8.2 05; next, press the √x key to display 9.0554   02. The rounded-off answer is 9.1×10^2.

For example, to find the square root of 8.2×10^5, first convert the expression to an even power: 0.82×10^6. Then take the square root of the decimal (0.91) and divide the exponent by 2 (6 divided by 2 is 3). The result is 0.91×10^3, or 9.1×10^2.

EXAMPLE 2.10 Determine the value of each of the following numbers:

a. $\sqrt{4.00 \times 10^{-4}}$

SOLUTION Take the positive square root of the decimal and then divide the exponent by 2.

$$\sqrt{4.00 \times 10^{-4}} = \sqrt{4.00} \times \sqrt{10^{-4}} = 2.00 \times 10^{-(4/2)} = 2.00 \times 10^{-2} \quad Answer$$

Study Hint: If your calculator does this automatically, then you must press the $+/-$ change sign key after the EE↓ key to display the negative exponent. The display should then read 5.6 -05. Continue by pressing the $\sqrt{x}$ key to display 7.4833 -03. The rounded-off answer is 7.48×10^{-3}.

b. $\sqrt{5.60 \times 10^{-5}}$

SOLUTION Change the number 5.60×10^{-5} to a number with an *even* exponent, following the guidelines previously mentioned, to obtain 56.0×10^{-6}. Then take the positive square root of 56.0×10^{-6}.

$$\sqrt{5.60 \times 10^{-5}} = \sqrt{56.0 \times 10^{-6}} = \sqrt{56.0} \times \sqrt{10^{-6}}$$
$$= 7.48 \times 10^{-(6/2)} = 7.48 \times 10^{-3} \quad Answer$$

(The 7.48 is obtained from a calculator.)

Rule for Positive Powers of Exponential Numbers

To raise an exponential number to a given positive power, first raise the decimal part to the power by using it the number of times indicated by the power and then multiply the exponent of 10 by the indicated power.

Study Hint: With a more expensive calculator, enter 4.3 and then press the EE↓ key followed by 3 as the exponent to display 4.3 03. Next, press the y^x key, followed by 4 (fourth power) and then the $=$ key to display 3.4188 14. The rounded-off answer is 3.4×10^{14}.

For example, to raise 4.3×10^3 to the fourth power, multiply and round off $4.3 \times 4.3 \times 4.3 \times 4.3$. Then multiply $10^{(3 \times 4)}$. The result is 340×10^{12}, or 3.4×10^{14}.

EXAMPLE 2.11 Perform the indicated operations on the following exponential numbers:

a. Raise 2.45×10^4 to the second power.

SOLUTION Multiply 2.45×2.45; then multiply the exponent (4) by 2 (the second power).

$$(2.45 \times 10^4)^2 = (2.45)^2 \times (10^4)^2 = 2.45 \times 2.45 \times 10^8 = 6.00 \times 10^8 \quad Answer$$

b. Raise 3.14×10^2 to the third power.

SOLUTION Multiply $3.14 \times 3.14 \times 3.14$; then multiply the exponent (2) by 3 (the third power).

$$(3.14 \times 10^2)^3 = (3.14)^3 \times (10^2)^3 = 3.14 \times 3.14 \times 3.14 \times 10^6$$
$$= 31.0 \times 10^6 \quad or \quad 3.10 \times 10^7 \quad Answer$$

Study Exercise 2.7
Carry out the operations indicated on the following exponential numbers:
a. $2.75 \times 10^3 + 3.2 \times 10^2$ $\qquad (3.07 \times 10^3)$
b. $4.72 \times 10^5 \times 1.83 \times 10^{-2}$ $\qquad (8.64 \times 10^3)$

Study Exercise 2.8
Perform the indicated operation on the following exponential numbers:

a. $\sqrt{7.43 \times 10^{-7}}$ $\hspace{4cm}$ (8.62×10^{-4})

b. Raise 3.12×10^2 to the third power $\hspace{2cm}$ (3.04×10^7) $\hspace{1cm}$ Work Problems 16 through 18.

2.8 Dimensional Analysis Method of Problem Solving for Conversions

You have probably heard the old expression, "You can't add apples and oranges." Thus, you need to learn ways to turn both apples and oranges into "fruit" to solve problems involving several different metric units of measurement. Chemists generally use **dimensional analysis** (also called the **factor-unit method**) to make such conversions.

$\hspace{2em}$ This simple method is based on developing a relationship between different units expressing the same physical dimension. For example, suppose you decide to make guacamole for dinner. You go the neighborhood supermarket and discover that avocados cost $0.80 each. If you have only $1.60, how many avocados can you buy to put in the guacamole?

$\hspace{2em}$ Simple algebra quickly shows that the answer is 2 ($1.60 divided by $0.80); but let's apply the factor-unit method of problem solving to this problem. We can express the relationship between dollars and the number of avocados as

$$\$0.80 \ = \ 1 \text{ avocado}$$

Dividing the equation by $0.80, we have

$$\frac{1 \text{ avocado}}{\$0.80}, \text{ which we will call factor } \mathbf{A}$$

Now, dividing the equation $0.80 $=$ 1 avocado by 1 avocado, we have

$$\frac{\$0.80}{1 \text{ avocado}}, \text{ which we will call factor } \mathbf{B}$$

$\hspace{2em}$ We can use one of these factors (**A** or **B**) to solve the problem. But which factor should we use? Since we are looking for an answer in terms of the number of avocados, we need to multiply the given quantity ($1.60) by a factor so that the dollars will cancel out. Thus, the correct factor is factor **A**, which has avocados in the numerator and dollars in the denominator.

$$\$1.60 \ \times \ \text{factor } \mathbf{A} \ = \ \text{number of avocados}$$

$$\$1.60 \ \times \ \frac{1 \text{ avocado}}{\$0.80} \ = \ 2 \text{ avocados}$$

If we had chosen factor **B** in error, the result would have been

$$\$1.60 \ \times \ \frac{\$0.80}{1 \text{ avocado}} \ = \ \frac{\$\$1.28}{1 \text{ avocado}}$$

which does not answer the original question and results in meaningless units.

Dimensional analysis (factor-unit method) Method of converting among measures expressed in different units by developing a relationship between these units and expressing this relationship as a factor of both units.

Study Hint: Pick the factor with the units you wish to remove in the *denominator*. Thus, factor **A** is correct because we want to remove the $ label.

Whenever possible in this text, we will use dimensional analysis in our problems, following a six-step approach.

1. Read the problem first very carefully to determine what is actually asked for.
2. Organize the data given, being sure to include *both* the *units* of the *given* quantity and the *units* of the *unknown* quantity.
3. Write down the *units* of the *given* quantity on the *left* side of a line. Write down the *units* of the *unknown* quantity on the *right* side of the line.
4. Apply the principles you have learned throughout the course to develop *factors* so that these factors, used properly, will give the correct units in the unknown.
5. Check your answer to see if it is reasonable by checking both the mathematics and the units.
6. Finally, check the number of significant digits.

Metric Conversions

Now that you understand the general principles involved, we are ready to make conversions within the metric system.

EXAMPLE 2.12   Convert 3.85 m to millimeters.

SOLUTION

$$3.85 \text{ m} \times \text{factor} = \text{millimeters} \qquad (\text{step 3 above})$$

We know that 1000 mm = 1 m (see Table 2.1); hence, our factors (step 4 above) are

$$\frac{1000 \text{ mm}}{1 \text{ m}} \qquad \text{and} \qquad \frac{1 \text{ m}}{1000 \text{ mm}} \qquad Answer$$
$$\qquad\quad \textbf{A} \qquad\qquad\qquad\qquad \textbf{B}$$

To obtain the correct units (millimeters), we must use factor **A**. Note how the units in the denominator of the factor must cancel out with the units of the given quantity (meters in this case).

$$3.85 \text{ m} \times \frac{1000 \text{ mm}}{1 \text{ m}} = 3850 \text{ mm} \qquad Answer$$

The answer is expressed to three significant digits since 3.85 has three significant digits.

EXAMPLE 2.13   Convert 75.2 mg to kilograms.

SOLUTION

$$75.2 \text{ mg} \times \text{factor} = \text{kilograms}$$

We do not know a factor converting milligrams directly to kilograms, but we do know factors that convert milligrams to grams and grams to kilograms.

$$1000 \text{ mg} = 1 \text{ g} \qquad \text{and} \qquad 1000 \text{ g} = 1 \text{ kg (see Table 2.1)}$$

Considering the first factor, 1000 mg = 1 g, we have

$$\frac{1000 \text{ mg}}{1 \text{ g}} \qquad \text{and} \qquad \frac{1 \text{ g}}{1000 \text{ mg}}$$
$$\textbf{A} \qquad\qquad\qquad \textbf{B}$$

If we use factor **B**, we will cancel the milligrams and our units will be in grams.

$$75.2 \text{ mg} \times \frac{1 \text{ g}}{1000 \text{ mg}} \times \text{factor} = \text{kg}$$

Now, considering the second factor, 1000 g = 1 kg, we have

$$\frac{1000 \text{ g}}{1 \text{ kg}} \qquad \text{and} \qquad \frac{1 \text{ kg}}{1000 \text{ g}}$$
$$\textbf{C} \qquad\qquad\qquad \textbf{D}$$

If we use factor **D**, we cancel the grams and our units will be in kilograms, which is the answer to the original question.

$$75.2 \text{ mg} \times \frac{1 \text{ g}}{1000 \text{ mg}} \times \frac{1 \text{ kg}}{1000 \text{ g}} = 0.0000752 \text{ kg} \quad \text{or} \quad 7.52 \times 10^{-5} \text{ kg} \quad \textit{Answer}$$

EXERCISE 2.14   The following masses were recorded in a laboratory experiment: 2.0000000 kg, 5.0000 g, 650.0 mg, 0.5 mg. What is the total mass in grams?

SOLUTION

$$2.0000000 \text{ kg} \times \frac{1000 \text{ g}}{1 \text{ kg}} = 2000.0000 \text{ g}$$

$$5.0000 \text{ g}$$

$$650.0 \text{ mg} \times \frac{1 \text{ g}}{1000 \text{ mg}} = 0.6500 \text{ g}$$

$$0.5 \text{ mg} \times \frac{1 \text{ g}}{1000 \text{ mg}} = 0.0005 \text{ g}$$

$$\overline{2005.6505 \text{ g}} \quad \textit{Answer}$$

In regard to significant digits note that all masses are recorded to a $\frac{1}{10,000}$ of a gram. For addition and subtraction we consider *places*.

EXAMPLE 2.15   Convert 0.0035 L to microliters.

SOLUTION   We know that 1,000,000 (10^6) μL = 1 L. Therefore, the solution is

$$0.0035 \text{ L} \times \frac{1,000,000 \text{ μL}}{1 \text{ L}} = 3500 \text{ μL or } 3.5 \times 10^3 \text{ μL} \quad \textit{Answer}$$

Study Hint: You may ask, Why can't I use exponents? Yes, you can, and it is simpler:

$$0.0035 \text{ L} = 3.5 \times 10^{-3} \text{ L}$$

$$3.5 \times 10^{-3} \text{ L} \times \frac{10^6 \text{ μL}}{1 \text{ L}}$$

$$= 3.5 \times 10^3 \text{ μL}$$

For factors larger than 1000 such as 1,000,000, it is simpler to use exponents, that is, 10^6.

Work Problems 19 through 23.

Study Hint: You may wish to simplify the problem if you recognize from Table 2.1 that 1000 (10^3) pg = 1 ng. Therefore, the solution is

$$7.6 \text{ ng} \times \frac{1000 \text{ pg}}{1 \text{ ng}} = 7600 \text{ pg}$$

$$= 7.6 \times 10^3 \text{ pg}$$

If this relationship is not that obvious to you, then you should go to the *basic unit* first as shown in the solution for Example 2.16.

EXAMPLE 2.16 Convert 7.6 ng to picograms.

SOLUTION From Table 2.1, we know the relationship between nanograms and grams (10^9 ng = 1 g) and between picograms and grams (10^{12} = 1 g). Therefore, the solution is

$$7.6 \text{ ng} \times \frac{1 \text{ g}}{10^9 \text{ ng}} \times \frac{10^{12} \text{ pg}}{1 \text{ g}} = 7.6 \times 10^3 \text{ pg} \qquad \textit{Answer}$$

It is better to go to the *basic unit* (gram, liter, or meter) *first* and then go to the desired unit.

Study Exercise 2.9

Carry out each of the following conversions:

a. 8.62 mg to kilograms   (8.62×10^{-6} kg)
b. 64.5 µg to grams   (6.45×10^{-5} g)
c. 4.5 µg to nanograms   (4500 ng or 4.5×10^3 ng)
d. 87.3 ng to picograms   (8.73×10^4 pg)

As you become more proficient with the metric system, you may wish to make these conversions by merely shifting the decimal point as you do in our monetary system. For example, in one of the preceding cases, 650.0 mg to grams involves shifting the decimal three places to the left, giving 0.6500 g, because 1 g = 1000 (or 10^3) mg.

Temperature Conversions

In Section 2.3 we noted that on the Celsius scale the difference between the freezing and boiling points of water is 100°, while on the Fahrenheit scale it is 180° (212 − 32). Using this relationship, we can derive a standard formula we can use to solve temperature-conversion problems.

To start, we divide both halves of this equation by the highest common denominator (20):

$$\frac{100 \text{ Celsius divisions}}{20} = \frac{180 \text{ Fahrenheit divisions}}{20}$$

$$5 \text{ Celsius divisions} = 9 \text{ Fahrenheit divisions}$$

The factor for converting from degrees Celsius to degrees Fahrenheit then becomes

$$°C \times \frac{9 \text{ Fahrenheit divisions}}{5 \text{ Celsius divisions}} = \left(\begin{array}{c} \text{Fahrenheit divisions above or below} \\ \text{the freezing point of water} \end{array} \right)$$

Because the freezing point of water is 32°F, we must add these divisions to 32 to get the temperature in degrees Fahrenheit:

$$\frac{9}{5} °C + 32 = °F \qquad (2.1)$$

$$\boxed{1.8°C + 32 = °F} \qquad (2.2)$$

Rearranging Equation 2.2 lets us convert degrees Fahrenheit to degrees Celsius:*

$$1.8°C = °F - 32$$

---

*Other formulas are also useful, such as °F = $\frac{9}{5}$(°C + 40) − 40 and °C = $\frac{5}{9}$(°F + 40) − 40. These formulas are based on the fact that the Celsius and Fahrenheit scales are equal at −40°.

$$°C = \frac{°F - 32}{1.8} \qquad (2.3)$$

In this case, we must remember to subtract 32 from the given temperature in degrees Fahrenheit to obtain the number of Fahrenheit divisions above or below the freezing point of water and then convert this to degrees Celsius by dividing by 1.8. Using the number 1.8 in Equations (2.2) and (2.3) simplifies the calculation if you are using a calculator.

To convert from degrees Celsius to kelvins, we need only add 273°. (In this text, we will use 273 instead of 273.15 to simplify calculations.)

$$K = °C + 273 \quad \text{or} \quad °C = K - 273 \qquad (2.4)$$

Study Hint: Both the Fahrenheit and Celsius temperature scales are arbitrary in that particular (but *arbitrary*) values were specified for freezing and boiling water.

EXAMPLE 2.17 Convert 25°C to degrees Fahrenheit.

SOLUTION   Substituting into our derived Equation 2.2, we get

$$25°C = [(1.8 \times 25) + 32]°F = 77°F \qquad Answer$$

EXAMPLE 2.18 Convert −25°F to degrees Celsius and to kelvins.

SOLUTION FOR DEGREES CELSIUS   Substituting into our derived Equation (2.3), we have

$$-25°F = \frac{-25 - 32}{1.8} °C = \frac{-57°C}{1.8} = -32°C \qquad Answer$$

SOLUTION FOR KELVINS   Substituting into Equation (2.4), we get

$$-32°C = (-32 + 273) \, K = 241 \, K \qquad Answer$$

EXAMPLE 2.19 Xenon has a freezing point of 133 K. What is its freezing point on the Fahrenheit scale?

SOLUTION   First, convert to degrees Celsius using Equation (2.4):

$$133 \, K = (133 - 273)°C = -14\overline{0}°C$$

Then convert from degrees Celsius to degrees Fahrenheit by using Equation (2.2):

$$-14\overline{0}°C = [1.8 \times (-14\overline{0}) + 32]°F = -22\overline{0}°F \qquad Answer$$

Study Hint: There are *exactly* 5°C for each 9°F by definition. Thus, 1.8 is an exact number and does *not* affect the number of significant digits in a calculation. Express your answer to the smallest place (units, tenths, and so on) as given to you in the number in the problem. If the number is given to you in the units place, then your answer should be in the units place. Regardless of the thermometer used, you should be able *to read each* with the *same degree of precision* (exactness).

Study Exercise 2.10
Convert −26°C to degrees Fahrenheit and to kelvins.   (−15°F, 247 K)

Study Exercise 2.11
Convert 235°F to degrees Celsius and to kelvins.   (−37°C, 236 K)

Work Problems 24 through 30.

Density Conversions

Density is often expressed as follows: $d^{20°} = 13.55$ g/mL for mercury. The 20° indicates the temperature in degrees Celsius at which the measurement was taken; hence, mercury at 20°C has a density of 13.55 g/mL. The reason for recording the temperature is that almost all substances expand when heated, and therefore the density decreases as the temperature is raised; for example, $d^{270°} = 12.95$ g/mL for mercury. Thus, density depends on temperature.

EXAMPLE 2.20 A cube of iron measures 2.00 cm on each edge and has a mass of 62.9 g. Calculate its density in kg/m³.

SOLUTION First calculate the volume of the cube of iron in cubic meters:

$$2.00 \text{ cm} \times \frac{1 \text{ m}}{100 \text{ cm}} \times 2.00 \text{ cm} \times \frac{1 \text{ m}}{100 \text{ cm}} \times 2.00 \text{ cm} \times \frac{1 \text{ m}}{100 \text{ cm}} = 8.00 \times 10^{-6} \text{ m}^3$$

Then find the density in kg/m³:

$$\frac{62.9 \text{ g}}{8.00 \times 10^{-6} \text{ m}^3} \times \frac{1 \text{ kg}}{1000 \text{ g}} = 7.86 \times 10^3 \text{ kg/m}^3 \qquad \textit{Answer}$$

EXAMPLE 2.21 Calculate the volume in liters at 20°C occupied by 880 g of benzene (a component of unleaded gasoline). For benzene, $d^{20°} = 0.88$ g/mL.

SOLUTION We have 880 g of benzene, which has a density of 0.88 g/mL at 20°C. We are asked to calculate the volume in liters. In other words, we wish to convert a given amount of benzene from mass units to volume units. We can do this readily by using the density of benzene as a conversion factor because 1 mL = 0.88 g.

$$880 \text{ g} \times \text{factor} = \text{volume units}$$

The choice of factors is

$$\frac{0.88 \text{ g}}{1 \text{ mL}} \qquad \text{and} \qquad \frac{1 \text{ mL}}{0.88 \text{ g}}$$
$$\textbf{A} \qquad\qquad\qquad \textbf{B}$$

If we use factor **A,** our units will be g²/mL, which have no meaning and do not answer our question. But let us consider factor **B:**

$$880 \text{ g} \times \frac{1 \text{ mL}}{0.88 \text{ g}}$$

Conversion from milliliters to liters yields the complete setup:

$$880 \text{ g} \times \frac{1 \text{ mL}}{0.88 \text{ g}} \times \frac{1 \text{ L}}{1000 \text{ mL}} = 1.0 \text{ L} \qquad \textit{Answer}$$

The answer is expressed to two significant digits since both 880 and 0.88 have two significant digits.

EXAMPLE 2.22 An experiment requires 0.156 kg of bromine. How many milliliters (20°C) should the chemist use? For bromine, $d^{20°} = 3.12$ g/mL.

SOLUTION

$$0.156 \text{ kg} \times \frac{1000 \text{ g}}{1 \text{ kg}} \times \frac{1 \text{ mL}}{3.12 \text{ g}} = 50.0 \text{ mL} \qquad \textit{Answer}$$

EXAMPLE 2.23 Calculate the volume in milliliters at 20°C occupied by 1.25 kg of chloroform. For chloroform, $d^{20°} = 1.49 \times 10^3$ kg/m^3.

SOLUTION   The density converts the mass to volume (meter cubed), but the unknown volume has units of milliliters. Therefore, to convert cubic meters to milliliters, we must know that 1 mL = 1 cm^3 (Table 2.1), and to convert from cubic meters to cubic centimeters, we must cube 100 cm = 1 m (Table 2.1). The complete solution is

$$1.25 \text{ kg} \times \frac{1 \text{ m}^3}{1.49 \times 10^3 \text{ kg}} \times \frac{(100)^3 (\text{cm})^3}{1 \text{ m}^3} \times \frac{1 \text{ mL}}{1 \text{ cm}^3} = 839 \text{ mL} \qquad Answer$$

> **S**tudy Hint: In cubing 100, remember to multiply it by itself *three times* (100 × 100 × 100). Alternatively, express 100 as 10^2 and cube it as $(10^2)^3 = 10^6$.

EXAMPLE 2.24 Calculate the mass in grams of 400 mL (20°C) of glycerine; $d^{20°} = 1.26$ g/mL for glycerine.

SOLUTION

$$400 \text{ mL} \times \frac{1.26 \text{ g}}{\text{mL}} = \begin{array}{l} 504 \text{ g, rounded off to one significant digit;} \\ \text{since 400 has only one significant digit,} \\ \text{the answer is 500 g} \qquad Answer \end{array}$$

Study Exercise 2.12
Calculate the volume in milliliters at 20°C occupied by 47.0 g of acetic acid; $d^{20°} = 1.05$ g/mL
(44.8 mL)

Study Exercise 2.13
Calculate the mass in grams of 275 mL (20°C) of acetic acid; $d^{20°} = 1.05$ g/mL
(289 g)   Work Problems 31 through 33.

2.9 Specific Gravity

One aspect of density, *specific gravity,* merits special consideration. The **specific gravity** of a substance is the density of the substance divided by the density of some substance taken as a standard. Thus, specific gravity is a useful way of looking at the density of a material because it allows us to compare the density of a substance relative to a standard value. For example, specific gravity is used in brewing (see Figure 2.12). In the wine industry it is used in two ways. The specific gravity of a batch of grape juice is a guide to the sugar content of the juice and helps the winemaker decide when to pick the grapes. The specific gravity of a batch of wine during the fermentation stage varies as the amount of alcohol increases, which helps the winemaker decide when to halt the fermentation.

Specific gravity is used in the medical field, too. The specific gravity of normal urine varies from 1.005 and 1.030, depending on the amounts of salts and waste products present. Urine that is largely water has a value of about 1.005, and urine with a lot of salts and waste products has values as high as 1.030. Along with other observations, the specific gravity of a patient's urine sample helps a physician to determine if something is wrong.

Specific gravity   Density of a substance divided by the density of some substance taken as a standard.

FIGURE 2.12
A technologist measures the specific gravity of beer being brewed.

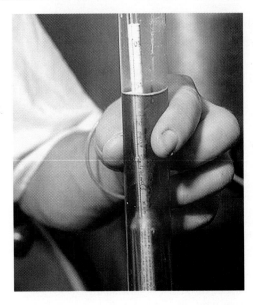

Chemists express the specific gravity of liquids and solids using water at 4°C as the standard:

$$\text{specific gravity} = \frac{\text{density of substance}}{\text{density of water at 4°C}}$$

$$\text{density of substance} = \text{specific gravity} \times \text{density of water at 4°C}$$

In calculating the specific gravity of a substance, we must express both densities in the *same units*. Specific gravity, therefore, has **no** *units*. To convert from specific gravity to density, we merely multiply specific gravity by the density of the reference substance (in most cases, water). We may thus find the density of any substance for which we have a reference density. Because the density of water at 4°C in the metric system is 1.00 g/mL, the density of solids or liquids expressed as g/mL is numerically equal to their specific gravities.

Specific gravity is often expressed as

$$\text{sp gr} = 0.708^{25°/4} \text{ of ether}$$

The 25° refers to the temperature in degrees Celsius at which the density of ether was measured, and the 4 refers to the temperature in degrees Celsius at which the density of water was measured. Table 2.4 lists the specific gravity of a few substances.

TABLE 2.4

Specific Gravity of Some Substances

SUBSTANCES	SPECIFIC GRAVITY
Water	$1.00^{4°/4}$
Ether	$0.708^{25°/4}$
Benzene	$0.880^{20°/4}$
Acetic acid	$1.05^{20°/4}$
Chloroform	$1.49^{20°/4}$
Carbon tetrachloride	$1.60^{20°/4}$
Sulfuric acid (concentrated)	$1.83^{18°/4}$

EXAMPLE 2.25 Calculate the mass in grams of $11\overline{0}$ mL of chloroform at 20°C.

SOLUTION From Table 2.4, the specific gravity of chloroform is 1.49 at 20°C. Therefore, in the metric system the density is 1.00 g/mL × 1.49 = 1.49 g/mL at 20°C.

$$11\overline{0} \text{ mL} \times \frac{1.49 \text{ g}}{1 \text{ mL}} = 163.9 \text{ g}$$

which must be rounded off to three significant digits since $11\overline{0}$ has three significant digits; the answer is 164 g.   *Answer*

EXAMPLE 2.26 The specific gravity of a certain organic liquid is 1.20. Calculate the number of liters in $84\overline{0}$ g of the liquid.

SOLUTION   The specific gravity is 1.20. Hence, in the metric system the density is 1.00 g/mL × 1.20 = 1.20 g/mL.

$$84\overline{0} \text{ g} \times \frac{1 \text{ mL}}{1.20 \text{ g}} \times \frac{1 \text{ L}}{1000 \text{ mL}} = 0.700 \text{ L} \qquad \textit{Answer}$$

EXAMPLE 2.27

The specific gravity of a certain organic liquid is 0.950. Calculate the number of kilograms in 3.75 L of the liquid.

SOLUTION   The specific gravity is 0.950. Therefore, in the metric system the density is 1.00 g/mL × 0.950 = 0.950 g/mL.

$$3.75 \text{ L} \times \frac{1000 \text{ mL}}{1 \text{ L}} \times \frac{0.950 \text{ g}}{1 \text{ mL}} \times \frac{1 \text{ kg}}{1000 \text{ g}} = 3.56 \text{ kg} \qquad \textit{Answer}$$

Study Exercise 2.14
The specific gravity of a certain organic liquid is 1.08. Calculate the number of liters in 925 g of the liquid.

(0.856 L)

Study Exercise 2.15
Calculate the number of grams in a 60.0-mL volume of sulfuric acid. (See Table 2.4.)

$(11\overline{0}$ g)   Work Problems 34 and 35.

 Summary

Matter is anything that has mass and occupies space. **Matter has a number of *measurable characteristics*** such as mass, weight, length, volume, and density (Section 2.1). **Chemists measure these properties using the metric system, which is based on units that are multiples of 10, 100, 1000, and so on (Section 2.2). The temperature of matter may be measured using three different temperature measurements, the Fahrenheit, Celsius, and Kelvin scales (Section 2.3). A variety of balances serve to measure mass, while scales measure weight, and cylinders, burets, and pipets are used to measure volume (Section 2.4).**

When a measurement is made, its precision (exactness) is related to the number of *significant digits* (figures) present in the measurement (Section 2.5). These significant digits affect the rounding off of answers after adding, subtracting, multiplying, and dividing numbers (Section 2.6). To express the very large and very small numbers often found in chemical measurements, chemists use a special form of exponential notation known as scientific notation, which expresses all numbers as a power of 10 times a number with exactly one nonzero digit to the left of the decimal point (Section 2.7).

The basic problem-solving technique used throughout this text, *dimensional analysis (factor-unit method),* is based on the fact that measured quantities have units and that we can use these units to solve quantitative problems in a logical way (Section 2.8). Dimensional analysis can be used to solve problems involving conversion within the metric system, calculations about density, and the expression of specific gravity (Section 2.9).

CHEMISTRY OF THE ATMOSPHERE

The Structure and Composition of the Atmosphere

The earth's atmosphere, the gaseous layer that surrounds the surface of the earth, is a boundary that separates the surface of the earth from the empty void of space. The atmosphere is generally divided into three parts, the *troposphere*, the *stratosphere*, and the *ionosphere* as the figure on the next page shows.

1. The troposphere is the first 9 or 10 mi (15 km) of the atmosphere above the earth's surface. It is characterized by a steady fall in temperature (from 25° to −55°C) as you go up. Most clouds and weather conditions develop in the troposphere.

2. The stratosphere is that part of the atmosphere that lies from about 10 to 25 mi (15 to 40 km) above the earth. The temperature of the stratosphere varies between −10° and −60°C. The boundary between the stratosphere and the troposphere is sometimes called the *tropopause*.

3. The ionosphere is the uppermost part of the atmos-

Weather conditions occur in the troposphere.

phere, from about 25 to 93 mi (40 to 150 km) above the earth. The temperature in the upper levels of the ionosphere reaches highs of 250° to 300°C. Once you leave the ionosphere you are in outer space. Most satellites and the space shuttle orbit the earth outside the ionosphere.

The density of the atmosphere drops quickly as you go to higher altitudes. For example, the

 Exercises

1. Define or explain the following terms (the number in parentheses refers to the section in the text where the term is mentioned):

    a. matter (2.1)   **b.** mass (2.1)

    c. weight (2.1)   **d.** volume (2.1)

atmosphere has a density of 1.22 g/L at sea level, while the average density of the stratosphere is about 0.1 g/L. This means that most gases are concentrated in the troposphere, and there is a gradual decrease in the mass of gases as you go up. In spite of this, the *overall* mass of all the gases in the atmosphere is enormous (5.2×10^{21} g) because the atmosphere is so large.

The decrease in the amount of air and the variation in temperature as you go up are the reasons why pilots and astronauts need space suits and breathing equipment to protect them. Even if you go up to only higher altitudes on the earth's surface, you may need protection. If you have climbed a mountain or gone camping at high altitude, you have certainly experienced the drop in temperature and the shortness of breath that results.

Air is composed of a number of gases mixed together as a solution. When you remove the water vapor from air (about 1% of air is water vapor), it contains mostly nitrogen (N_2, 78.09%), oxygen (O_2, 20.95%), and argon (Ar, 0.93%). A number of gases are also present in trace amounts (less than 0.1%). Although present in small quantities, these gases are crucial to the chemistry of the atmosphere. Among the many trace gases normally present are carbon dioxide (CO_2), neon (Ne), helium (He), krypton (Kr), methane (CH_4), hydrogen (H_2), xenon (Xe), ozone (O_3), and radon (Rn).

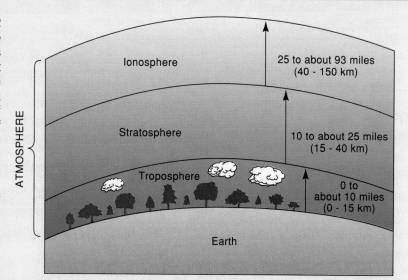

The atmosphere consists of the troposphere, the stratosphere, and the ionosphere.

We use the atmosphere to supply gases for various uses. Nitrogen, oxygen, and the noble gases helium, neon, argon, krypton, and xenon are all obtained from the atmosphere. Nitrogen is used in the production of ammonia (NH_3) see *The Element Nitrogen: Our Food Depends on It,* Chapter 17). Oxygen is used in steelmaking (see *The Element Oxygen: Chemistry and Life on Earth,* Chapter 11). Some atmospheric gases are used in the production of light sources. Argon and occasionally krypton gas are used to fill the space inside the glass bulb of incandescent light bulbs. Neon, helium, argon, and krypton are used to fill fluorescent, halogen, and "neon" lamp bulbs. The common flash attachments for 35-mm

cameras are based on a xenon lamp. A variety of lasers use neon, helium, argon, or xenon in the generation of laser beams.

Some trace gases are not normally present in the atmosphere and occur as a *result* of human activities. Such gases are called *pollutants*. They include most of the sulfur dioxide (SO_2), dinitrogen monoxide (N_2O), nitrogen monoxide (NO), nitrogen dioxide (NO_2), and **c**hlorofluorocarbons (CFCs) (for example, CF_2Cl_2, better known as Freon 12, which is used as a refrigerant). As you will see in later chapters, these pollutants are responsible for many of the environmental problems associated with the atmosphere.

e. temperature (2.1)

f. density (2.1)

g. metric system (2.1)

h. significant digits (figures) (2.5)

i. exact numbers (2.6)

j. exponents (2.7)

k. exponential notation (2.7)

l. scientific notation (2.7)

m. dimensional analysis (factor-unit method) (2.8)

n. specific gravity (2.9)

2. Distinguish between
    a. mass and weight
    b. a chemical balance and a spring scale
    c. the Celsius and Fahrenheit temperature scales
    d. the kelvin and Celsius temperature scales
    e. density and specific gravity

3. Explain how you would determine the mass of an object on
    a. a platform balance   b. a triple-beam balance

4. "The weight of a small medicine bottle on a triple-beam balance was found to be 8.53 g." Criticize this statement.

 # Problems

Significant Digits (See Section 2.5)

5. Determine the number of significant digits in the following numbers:
    a. 107   b. 265
    c. 828,060   d. 2809
    e. 37$\overline{0}$ f. 0.006

6. Determine the number of significant digits in the following numbers:
    a. 129   b. 0.0604
    c. 12.04   d. 0.0060180
    e. 125$\overline{0}$ f. 1250

Mathematical Operations Involving Significant Digits (See Section 2.6)

7. Round off the following numbers to three significant digits:
    a. 2.436   b. 2.4768
    c. 8.6850   d. 10.455
    e. 13.350   f. 96,750

8. Round off the following numbers to three significant digits:
    a. 10.62   b. 3.876
    c. 0.0045350   d. 0.78453
    e. 6.987   f. 3.462

9. The relation of the pound to the kilogram in the United States is 1 lb = 0.453592428 kg. Round this number off to three significant digits.

10. Perform the indicated mathematical operations and express your answer to the proper number of significant digits.
    a. $4.78 + 7.3654 + 0.5$ b. $0.423 + 76.720 + 4.6494$
    c. $14.745 - 2.60$ d. $0.5642 - 0.230$
    e. 6.02×3.0 f. 0.650×563
    g. 0.22×0.324 h. $\dfrac{194}{24}$
    i. $\dfrac{625}{17.5}$

Exponents (See Section 2.7)

11. Express 37,500 as 3.75×10^n.

12. Express 0.00325 as 3.25×10^n.

Scientific Notation (See Section 2.7)

13. Express the following in scientific notation to three significant digits:

   a. 975
   b. 9,840,000
   c. 0.000632
   d. 0.007275

14. Express the following in scientific notation to three significant digits:

   a. 0.000325
   b. 7,290,000
   c. 4778
   d. 0.0005265

15. On March 14, 1986, the European spacecraft *Giotta* intercepted Halley's comet some 93 million miles out in space. Express this distance to two significant digits in scientific notation.

Mathematical Operations Involving Exponential Notation (See Section 2.7)

16. Carry out the operations indicated on the following exponential numbers:

   a. $4.24 \times 10^3 + 1.50 \times 10^3$
   b. $4.73 \times 10^2 + 7.6 \times 10^1$
   c. $3.75 \times 10^3 - 2.63 \times 10^3$
   d. $6.54 \times 10^5 - 3 \times 10^3$
   e. $6.45 \times 10^3 \times 1.32 \times 10^2$
   f. $3.28 \times 10^6 \times 1.42 \times 10^{-2}$
   g. $\dfrac{6.62 \times 10^6}{2.82 \times 10^2}$
   h. $\dfrac{9.36 \times 10^{-5}}{2.32 \times 10^{-2}}$

17. Determine the value of each of the following numbers:

   a. $\sqrt{25.0 \times 10^8}$
   b. $\sqrt{1.25 \times 10^5}$
   c. $\sqrt{6.75 \times 10^9}$
   d. $\sqrt{3.74 \times 10^{-7}}$

18. Perform the indicated operation on the following exponential number:

   a. Raise 1.26×10^3 to the second power.
   b. Raise 1.83×10^8 to the second power.
   c. Raise 1.47×10^2 to the third power.
   d. Raise 3.65×10^3 to the third power.

Metric System Conversions (See Section 2.8)

19. Carry out each of the following conversions showing a solution setup:

   a. 3.1 kg to grams
   b. 13,000 m to kilometers
   c. 25 mg to kilograms
   d. 600 mL to liters
   e. 8750 mL to kiloliters
   f. 0.0042 m to nanometers

20. Carry out each of the following conversions showing a solution setup:

   a. 7.4 km to meters
   b. 18,000 g to kilograms
   c. 75 mg to kilograms
   d. 875 mL to liters
   e. 40.0 pm to nanometers
   f. 4.2 mL to cubic centimeters

21. Add the following masses: 375 mg, 0.500 g, 0.003000 kg, 200.0 cg, and 1.00 dg. What is the total mass in grams?

22. Add the following lengths: 6.0000000 km, 370.00 cm, 7.0000 m, and 0.4 mm. What is the total length in meters?

23. The relationship between a yard and a meter in the United States is 1 yd = 0.9144 m. What is the relationship between a yard and a centimeter to three significant digits?

Temperature Conversions (See Section 2.8)

24. Convert each of the following temperatures to degrees Fahrenheit and to kelvins:

    a. 38°C

    b. $12\bar{0}$°C

    c. −32°C

    d. −$11\bar{0}$°C

25. Convert each of the following temperatures to degrees Celsius and to kelvins:

    a. 68°F

    b. −12°F

    c. −45°F

    d. $32\bar{0}$°F

26. Liquid nitrogen has a boiling point of 77 K at 1 atmosphere (atm) pressure. What is its boiling point on the Fahrenheit scale?

27. At what temperature do the Celsius and Fahrenheit scales have the same numerical reading? Carry out your answer to three significant digits.

28. The official coldest temperature recorded in the United States was −79.8°F at Prospect Creek, Alaska, on January 23, 1971. What is this temperature on the Celsius scale?

29. The lowest recorded temperature in the world was recorded as −89.2°C at the Soviet Antarctic Station on July 21, 1983. What is the temperature in degrees Fahrenheit?

30. The highest recorded temperature in the world was recorded as 136.4°F at Aziza, Libya, in the Sahara desert on September 13, 1922. What is this temperature in degrees Celsius?

Density Conversions (See Sections 2.2 and 2.8)

31. Calculate the density in g/mL for each of the following:

    a. a piece of metal of volume $6\bar{0}$ mL and mass $3\bar{0}\bar{0}$ g

    b. a substance occupying a volume of $7\bar{0}$ mL and having a mass of 165 g

    c. a sample of metal occupying a volume of 5.0 mL and having a mass of 65 g

    d. a piece of metal measuring 2.0 cm × 0.10 dm × 25 mm and having a mass of 30.0 g

32. Calculate the volume in milliliters at 20°C occupied by each of the following:

    a. a sample of carbon tetrachloride having a mass of 65.0 g; $d^{20°}$ = 1.60 g/mL

    b. a sample of acetic acid having a mass of $32\bar{0}$ g; $d^{20°}$ = 1.05 g/mL

    c. a sample of chloroform having a mass of 37.5 g; $d^{20°}$ = 1.49 g/mL

    d. a sample of benzene having a mass of 2.5 kg; $d^{20°}$ = 8.8 × 10^2 kg/m^3

33. Calculate the mass in grams of each of the following:

    a. a 35.0-mL volume of ether; $d^{20°}$ = 0.708 g/mL

    b. a 185-mL volume of glycerine; $d^{20°}$ = 1.26 g/mL

   c. a 64.5-mL volume of carbon tetrachloride; $d^{20°} = 1.60$ g/mL

   d. a 0.150-L volume of bromine; $d^{20°} = 3.12$ g/mL

Specific Gravity (See Section 2.9)

34. Calculate the volume in liters occupied by each of the following (see Table 2.4):

   a. a sample of sulfuric acid (conc.) having a mass of 285 g

   b. a sample of acetic acid having a mass of 725 g

   c. a sample of chloroform having a mass of 0.560 kg

   d. a sample of carbon tetrachloride having a mass of 2.20 kg

35. Calculate the mass in grams of each of the following (see Table 2.4):

   a. a 20.0-mL volume of benzene

   b. a 15$\overline{0}$-mL volume of acetic acid

   c. a 1.65-L volume of chloroform

   d. a 3.35-L volume of carbon tetrachloride

General Problems

36. A football field from goal line to goal line is 10$\overline{0}$ yards (yd) long. What is the length of this part of the field expressed in meters? (3 ft = 1 yd, 1 ft = 0.305 m)

37. A piece of vanadium metal is machined into a cube that is 2.30 cm on an edge. The piece of metal has a mass of 72.5 g. Calculate the density of the vanadium metal in g/mL.

38. Legend tells us that Noah's ark was "three hundred cubits long, fifty cubits wide, and thirty cubits high." For the sake of the problem, assume these numbers to be 30$\overline{0}$ cubits long, 5$\overline{0}$ cubits wide, and 3$\overline{0}$ cubits high. A cubit is an ancient unit of measure based on the length of a person's forearm, and, in general, it is between 17 and 21 in. Calculate the maximum and minimum dimensions of the ark in feet. (*Hint:* How big was the ark if Noah had a 21-in. forearm? A 17-in. forearm?).

39. In November 1987, a massive iceberg broke loose from the Antarctic ice mass and floated free into the open ocean. The chunk of ice was estimated to be 98 mi long, 25 mi wide, and over 750 ft thick.

   a. A typical backyard swimming pool contains 24,000 gallons (gal) of water. How many of these pools could you fill from the iceberg? You may assume that the ice is roughly a rectangular block of these dimensions and that it consists of water only. Some useful factors are 5280 ft = 1 mi and 1 ft^3 = 7.48 gal.

   b. Estimate the mass of the iceberg in kilograms to two significant digits (1 ft^3 = 28.3 L, and the density of ice = 0.917 g/mL)

✓ Chapter Quiz 2A
(Sections 2.5 to 2.7)

1. Determine the number of significant digits in the following numbers:

   a. 1080   **b.** 0.06520

2. Round off the following numbers to three significant digits:

   a.   12.450   **b.**   3.749

3. Express the following numbers in scientific notation to three significant digits:

   a.   875,000   **b.**   0.00295

4. Perform the following operations and express your answer in scientific notation to three significant digits:

   a. $3.85 \times 10^3 + 2.11 \times 10^2$

   b. $3.42 \times 10^8 \times 2.15 \times 10^{-2}$

   c. $\sqrt{6.24 \times 10^9}$

   d.   Raise 1.23×10^2 to the third power

 ## Chapter Quiz 2B
(Sections 2.8 and 2.9)

1. Convert 925 mg to kilograms. Express your answer in scientific notation.

2. Convert 725 μg to grams. Express your answer in scientific notation.

3. Convert $-1\overline{0}0$°C to degrees Fahrenheit and to kelvins.

4. Calculate the mass in grams of 0.600 L (20°C) of acetic acid ($d^{20°}$ = 1.05 g/mL).

5. Calculate the volume in milliliters (20°C) of 0.400 kg of chloroform (sp gr = $1.49^{20°/4}$).

Sodium Chloride (Formula: NaCl)

Salt can be obtained by the evaporation of seawater, or it can be mined directly from the ground in salt deposits.

The Compound SODIUM CHLORIDE: An Old Friend—Salt

Name: You may know it as salt or table salt, but it is also called rock salt, sea salt, or sodium chloride. As a mineral, sodium chloride is called halite.

Appearance: Sodium chloride (NaCl) is a crystalline solid, melting at a high temperature. Finely divided salt or small crystals of salt appear white, but large crystals are nearly colorless.

Occurrence: Sodium chloride is an essential part of the diet of all mammals (including humans). Salt can be found in the oceans, where it comprises about 2.8 g out of every 100 g of seawater (2.8%). Salt can also be found in huge deposits on land, which presumably resulted from the drying up of inland seas that became isolated from the oceans. Thus, all salt comes from the oceans in one way or another.

Source: Virtually all salt is obtained by one of two methods. It is either mined directly from salt deposits, or it is obtained by evaporation of seawater by the sun in large ponds. The salt in some deposits contains other compounds, such as calcium chloride ($CaCl_2$) and magnesium chloride ($MgCl_2$), but some deposits yield salt that is 99.8% pure sodium chloride.

Its Role in Our World: Sodium chloride is one of the principal raw materials of the chemical industry. Virtually all compounds containing sodium or chlorine are derived from salt in some way or another. Important chemicals produced by using sodium chloride include sodium metal, chlorine gas, hydrochloric acid (swimming pool acid), sodium carbonate (soda ash), sodium bicarbonate (baking soda), sodium hypochlorite (chlorine bleach), and sodium hydroxide (caustic soda or lye). In addition, salt is an important part of the manufacture of soaps and dyes; it is used in tin metallurgy, glazing pottery, and tanning of leather.

Historically, salt has played an important role in the development of trade and the world economy. Because salt is an essential ingredient of human diets and proved to be a useful food preservative in ancient times, it was highly valued and was traded accordingly. The salt from central European mines was traded extensively throughout Europe, the Middle East, and the Far East. Early trade routes are sometimes referred to as "salt routes." In fact, the word "salary" derives from the Latin word *salarium,* which referred to the money given to Roman soldiers for the purpose of buying salt. Although salt is an important part of the chemical industry, it is far less valuable today than in ancient times since there are a variety of ways of obtaining salt now.

Unusual Facts: In more humid climates, salt absorbs moisture from the air and "cakes." Although these cakes are easy enough to break up, they can be a nuisance. This absorption of moisture is due to trace amounts of calcium chloride and magnesium chloride in the salt. Curiously, pure salt does not cake at all.

CHAPTER 3

Matter and Energy

The interaction of light with matter is seen in photosynthesis, a process which is vital to life.

GOALS FOR CHAPTER 3

1. To list and understand the characteristics of the three physical states of matter (Section 3.1).

2. To be able to differentiate between homogeneous and heterogeneous matter and among the different types of homogeneous matter. To name an element from its symbol and to identify the symbol by its name, and to classify common substances as compounds, elements, or mixtures (Section 3.2).

3. To determine the number of atoms and name of the element from one unit of a compound and to write the formula of a compound from the number of atoms and name of the element in one unit of the compound (Section 3.3).

4. To classify the physical and chemical properties and physical and chemical changes of substances (Section 3.4).

5. To calculate the specific heat of a substance and use it in various calculations (Section 3.5).

6. To define and understand the *laws* of *conservation* of *mass* and *energy* (Section 3.6).

7. To distinguish between metallic and nonmetallic elements (Section 3.7).

Countdown

5. Determine the number of significant digits in the following numbers (Section 2.5).
   a. 87.42   (4)   **b.** 0.0305   (3)
   c. 16.40   (4)   **d.** $36\overline{0}$ (3)

4. Perform the following operations and express your answer to the proper number of significant digits (Section 2.5 and 2.6).

   a.
   $$\begin{array}{r} 3.652 \\ 12.4 \\ \underline{1.25} \end{array}$$
   (17.3)

   b. $12.46 \times 0.0231 = $ (0.288)

   c. $\dfrac{14.85}{2.1} = $ (7.1)

   d. $\dfrac{3.42 \times 6.851}{13.2 \times 0.0124} = $ (143)

3. Calculate the density of silicon in g/mL if silicon has a volume of 11.5 cm^3 and a mass of 26.7 g (Sections 2.1, 2.2, and 2.5).
   (2.32 g/mL)

2. Carry out the operations indicated on the following numbers and express your answer in scientific notation to the proper number of significant digits (Section 2.5 and 2.7):

   a. $\dfrac{1.65 \times 10^4 \times 35\overline{0}}{100\overline{0}} = $ (5.78×10^3)

   b. $\dfrac{1.00 \times 68\overline{0}}{0.315 \times (65 - 45)} = $ (1.1×10^2)

1. The melting point of silicon is $257\overline{0}°F$. What is its melting point in degrees Celsius and in kelvins (Section 2.8)?
   $(141\overline{0}°C, 1683 \text{ K})$

W hat do ice cream and steel have in common? What do sand and iodine
have in common? What do krypton and car wax have in common?
Whether something is hot or cold, shiny or dull, hard or soft, wet or
dry—indeed, no matter what adjectives describe it—everything is matter. In Chapter 2, we defined matter as anything that has mass and occupies space. In this chapter, we will consider different types of matter and study the properties of matter and the changes it undergoes.

3.1 Physical States of Matter

Physical state   Any of three forms in which matter may exist, as a gas, as a liquid, or as a solid; the state depends on the surrounding temperature and atmospheric pressure, as well as on the specific characteristics of the particular type of matter.

Work Problem 3.

Matter exists in three **physical states**—solid, liquid, or gas—depending on the temperature, atmospheric pressure, and specific characteristics of the particular type of matter. Some matter can exist in all three physical states, as Figure 3.1 shows in the case of water. Other matter breaks down into new substances (decomposes) when an attempt is made to change its physical state. Common table sugar exists under normal conditions in only one physical state—solid. Attempts to change it to a liquid or a gas by heating at atmospheric pressure result in decomposition of the sugar; the sugar turns caramel brown to black as it breaks down into carbon and water vapor (see Figure 3.2).

3.2 Composition and Properties of Matter

Regardless of its physical state, all matter is either homogeneous or heterogeneous.

Heterogeneous and Homogeneous Matter

Homogeneous matter   Matter that is uniform in composition and properties throughout the sample.

Homogeneous and heterogeneous matter differ from one another in very clear ways. **Homogeneous matter** is *uniform* in composition and properties. It is the

FIGURE 3.1
The three physical states of water: solid (snow, ice), liquid, and gas (clouds—water in a gas).

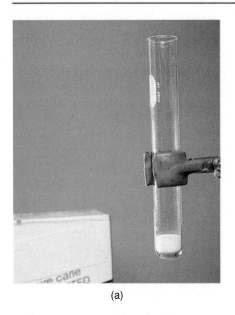 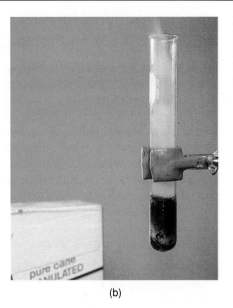

(a)   (b)

FIGURE 3.2
(a) Sugar before heating.
(b) Decomposition of sugar to produce carbon and water vapor.

same throughout. **Heterogeneous matter** is *not uniform* in composition and properties. It consists of two or more physically distinct *portions* or *phases* unevenly distributed. A class consisting of all women would be analogous to homogeneous matter, while a class of both men and women would be analogous to heterogeneous matter.

Heterogeneous matter is also commonly called a *mixture*. This type of **mixture** is composed of two or more pure substances, each of which retains its identity and specific properties. The properties of the mixture depend on what part of the mixture is being observed. In many mixtures, substances can be readily identified by visual observation. For example, in a mixture of salt and sand, the human eye or a hand lens can be used to distinguish between the white salt crystals and the tan sand crystals. Similarly, in a mixture of iron and sulfur, visual observation can identify the yellow sulfur and the black iron. Mixtures can usually be separated by a simple operation that does not change the composition of the several pure substances comprising the mixture. For example, a mixture of salt and sand can be separated by using water. The salt dissolves in water, but the sand does not. If, after removing the sand, we evaporate the water, we are then left with pure salt. A mixture of iron and sulfur can be separated by dissolving the sulfur in liquid carbon disulfide (the iron is insoluble) or by attracting the iron to a magnet (the sulfur is not attracted).

We can further divide homogeneous matter into three categories: homogeneous mixtures, solutions, and pure substances. A **homogeneous mixture** is homogeneous throughout and is composed of two or more pure substances whose proportions may be *varied* in some cases without limit. The properties of the substance do not depend on what part of the material is being observed; all samples of the substance look the same. One example of a homogeneous mixture is unpolluted air, which is a mixture of oxygen, nitrogen, and certain other gases.

Mixtures of gases are generally called homogeneous mixtures, but homogeneous mixtures composed of gases, liquids, or solids dissolved in *liquids* are called *solutions*. A **solution** is homogeneous throughout and is composed of two or more pure substances. However, its composition usually can be varied *within certain limits*. Some common examples of solutions are sugar solutions (sugar dissolved in

Heterogeneous matter Matter that is not uniform in composition and/or properties throughout the sample, but rather consists of two or more distinct substances unequally distributed.

Mixture Heterogeneous matter composed of two or more pure substances, each of which retains its identity and specific properties.

Homogeneous mixture Homogeneous throughout and composed of two or more pure substances whose proportions can be varied in some cases without limit.

Solution Homogeneous mixture involving two or more pure substances; its composition usually can be varied within certain limits.

water), salt solutions (salt dissolved in water), carbonated water (carbon dioxide dissolved in water), alcohol solutions (ethyl alcohol dissolved in water), and vinegar (acetic acid dissolved in water). In some cases, solids may dissolve in other solids to form homogeneous mixtures called solid solutions. Brass is a solid solution consisting of zinc dissolved in copper.

Pure substance Substance characterized by definite and constant composition and having definite and constant properties under a given set of conditions.

A **pure substance** is characterized by a *definite and constant composition*. It has *definite* and *constant properties* under a given set of conditions. A pure substance obeys our definition of homogeneous matter not only in that it is uniform throughout in both composition and properties but also in that it has a definite and constant composition and definite and constant properties. Examples of pure substances include water, salt (sodium chloride), sugar (sucrose), mercuric or mercury(II) oxide, gold, iron, and aluminum.

Homogeneous mixtures and solutions differ from pure substances in that these mixtures consist of two or more pure substances in *variable* proportions, whereas pure substances have *definite* and *constant compositions*.

Compounds and Elements

Compound Any pure substance that can be broken down by chemical means into two or more different, simpler substances.

Element Any pure substance that cannot be broken down by ordinary chemical means into two or more different, simpler substances.

Atom Smallest piece of an element that can exist and still exhibit the properties of that element including the ability to react with other atoms.

Pure substances are divided into two groups: compounds and elements. A **compound** is a pure substance that *can be broken down* by various chemical means into two or more different simpler substances. We noted above that water, salt, sugar, and mercuric or mercury(II) oxide are all pure substances. Yet they are also compounds that can be broken down. For example, the application of electric current can cause water to separate into hydrogen and oxygen and salt to separate into sodium and chlorine. Adding sulfuric acid to sugar breaks the sugar down into carbon and water (Figure 3.3), and heating mercuric or mercury(II) oxide breaks it down into mercury and oxygen.

An **element** is a pure substance that *cannot be decomposed* into simpler substances by ordinary chemical means. Gold, iron, aluminum, hydrogen, oxygen, sodium, chlorine, carbon, and mercury are elements. Elements are, in turn, composed of atoms. **Atoms** are the smallest "piece" of an element that can exist and still be that element. We might think of this as a sandy beach; the beach (element) is made up of individual grains (atoms) of sand. Figure 3.4 summarizes the classification of matter.

FIGURE 3.3
The addition of sulfuric acid to sugar to produce carbon and water.

(a)   (b)   (c)

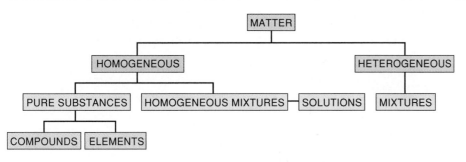

FIGURE 3.4
Classification of matter.

There are 111 elements at this writing, of which 90 have been found to occur naturally. The remaining 21 have been produced only synthetically by nuclear reactions (see Chapter 19). Minute amounts of some of these elements may also exist naturally. Table 3.1 gives the ranking by relative abundance (percent by mass) of the first 10 elements (see Figure 3.5 for photographs of some of these elements) in the earth's crust (the upper 10 mi, including the oceans and the atmosphere).

All the elements and their symbols are listed on the inside front cover of this book. Because many of the elements are rarely mentioned in a basic chemistry course, we have listed in Table 3.1 only the 47 most common. You *must* learn the *names and symbols* for the 47 elements in Table 3.1. To do this, we suggest that you make flash cards. For example, the symbol for copper is Cu, and so on a small card (cut a 3 × 5-in. file card in half) write "copper" on one side and "Cu" on the other side. Do this for all the elements listed in Table 3.1. Go over the cards until you know them all and then keep going over them so you will not forget them.

Work Problems 4 and 5.

FIGURE 3.5
Various elements in the earth's crust. Numbers in parentheses denote the ranking (percent by mass) in the earth's crust. (a) Aluminum (3); (b) iron (4); (c) calcium (5); (d) sodium (6); (e) potassium (7); (f) titanium (10).

| TABLE **3.1** | | | Some Common Elements and Their Symbols, and the Ranking of Relative Abundance (Percent by Mass) for the First 10 Elements in the Earth's Crust |

ELEMENT	SYMBOL[a]	RANKING (% BY MASS)[b]	ELEMENT	SYMBOL	RANKING (% BY MASS)
Aluminum	Al	3 (7.5)	Lithium	Li	
Antimony	Sb		Magnesium	Mg	8 (1.9)
Argon	Ar		Manganese	Mn	
Arsenic	As		Mercury	Hg	
Barium	Ba		Neon	Ne	
Beryllium	Be		Nickel	Ni	
Bismuth	Bi		Nitrogen	N	
Boron	B		Oxygen	O	1 (49.5)
Bromine	Br		Phosphorus	P	
Cadmium	Cd		Platinum	Pt	
Calcium	Ca	5 (3.4)	Potassium	K	7 (2.4)
Carbon	C		Radium	Ra	
Chlorine	Cl		Selenium	Se	
Chromium	Cr		Silicon	Si	2 (25.7)
Cobalt	Co		Silver	Ag	
Copper	Cu		Sodium	Na	6 (2.6)
Fluorine	F		Strontium	Sr	
Gold	Au		Sulfur	S	
Helium	He		Tin	Sn	
Hydrogen	H	9 (0.9)	Titanium	Ti	10 (0.6)
Iodine	I		Uranium	U	
Iron	Fe	4 (4.7)	Xenon	Xe	
Krypton	Kr		Zinc	Zn	
Lead	Pb				

[a] Some of these symbols do not appear to be related to the names of the elements. In these cases, the symbol has been obtained from the Latin name by which the element was known for centuries.

Name of Element	Latin Name (Symbol)
Antimony	*Stibium* (Sb)
Copper	*Cuprum* (Cu)
Gold	*Aurum* (Au)
Iron	*Ferrum* (Fe)
Lead	*Plumbum* (Pb)
Mercury	*Hydrargyrum* (Hg)
Potassium	*Kalium* (K)
Silver	*Argentum* (Ag)
Sodium	*Natrium* (Na)
Tin	*Stannum* (Sn)

[b] Percent (%) means parts per 100; that is, there are 7.5 parts by mass of aluminum (Al) in $\overline{100}$ parts by mass of total elements in the earth's crust.

3.3 Molecules. The Law of Definite Proportions

Combinations of atoms of the same or different elements form *molecules,* the structures that make up many compounds. **Molecules** are the smallest particle of a pure substance that can exist and undergo chemical changes. They consist of *atoms* of the same element or different elements held together by various forces. Examples of substances that exist as molecules include carbon monoxide, water, and oxygen gas.

Because the chemical symbol for an element can also stand for an atom of that element, chemists express the composition of compounds in terms of their atomic makeup by using *chemical formulas.* For compounds and elements that exist as molecules, we express their composition by a *molecular formula.* (A type of compound called an *ionic compound,* which is expressed in *formula units,* is discussed in Chapter 6.) A **molecular formula** is composed of an appropriate collection of elemental symbols representing *one* molecule of the compound or element. For example, consider the molecular formula of carbon monoxide, CO, which shows that one molecule of carbon monoxide is composed of one carbon atom (C) and one oxygen atom (O). When a molecule of a compound contains more than one atom of a particular element, the symbol for that element carries the appropriate subscript in the molecular formula. For example, water, H_2O (read "H two O"), has two hydrogen atoms (H) and one oxygen atom (O) in each molecule. Similarly, oxygen gas, O_2, has two oxygen atoms in each molecule. A subscript is not used when a single atom of an element is present.

If we are given the molecular formula for a compound, we can determine the atoms of each element present as well as the total number of atoms present in the molecule.

Molecule Smallest particle of a pure substance that can exist and undergo chemical changes.

Molecular formula Way of expressing the composition of one molecule of a compound or element by using elemental symbols for each element involved and subscripts reflecting the number of atoms (above 1) of that element in the molecule; it shows the actual number of atoms of each element present in one molecule of the compound.

EXAMPLE 3.1   Determine the number of atoms of each element, and write the name of the element and the total number of atoms in each of the following molecular formulas:

	Answer	
	Atoms of each element in one molecule	Total number of atoms
a. CO_2 (carbon dioxide)	1 carbon, 2 oxygen	3
b. NO (nitrogen monoxide)	1 nitrogen, 1 oxygen	2
c. H_2S (hydrogen sulfide)	2 hydrogen, 1 sulfur	3
d. H_2O_2 (hydrogen peroxide)	2 hydrogen, 2 oxygen	4
e. CH_4O (methyl alcohol)	1 carbon, 4 hydrogen, 1 oxygen	6

Likewise, if we know the number of atoms of each element in one molecule of the compound, we can write the molecular formula.

EXAMPLE 3.2   From the number of atoms of each element in one unit of the compound, write the molecular formula for each of the following compounds:

Answer

a. sulfur trioxide: SO_3
   1 sulfur, 3 oxygen

b. ethyl alcohol: C_2H_6O
   2 carbon, 6 hydrogen, 1 oxygen

c. ethyl ether: $C_4H_{10}O$
   4 carbon, 10 hydrogen, 1 oxygen

d. ethylene glycol (used in antifreeze): $C_2H_6O_2$
   2 carbon, 6 hydrogen, 2 oxygen

e. chlorophyll a: $C_{55}H_{72}MgN_4O_5$
   55 carbon, 72 hydrogen, 1 magnesium,
   4 nitrogen, 5 oxygen

Study Exercise 3.1

Determine the number of atoms of each element and write the name of the element and the total number of atoms in each of the following formulas:

a. SiO_2 (silicon dioxide, sand)   (1 silicon, 2 oxygen, 3)

b. $Mg(OH)_2$ (magnesium hydroxide, milk of magnesia)
   (*Hint:* Clear the parentheses. The subscript refers to everything in the
   parentheses.)   (1 magnesium, 2 oxygen, 2 hydrogen, 5)

Study Exercise 3.2

From the number of atoms of each element in one unit of the compound, write the formula for each of the following compounds:

a. benzene: 6 carbon, 6 hydrogen (C_6H_6)

b. methyl *tert*-(tertiary)-butyl ether: (an additive to gasoline):
   5 carbon, 12 hydrogen, 1 oxygen $(C_5H_{12}O)$

Work Problems 6 and 7.

Figure 3.6 summarizes the relations among atoms, molecules, formula units, elements, and compounds.

The molecular formula of water (H_2O) does not vary whether the water is Mississippi River water or pure distilled water. [Other impurities may be found in water such as contaminants in the form of sewage or industrial waste (regrettably found in large

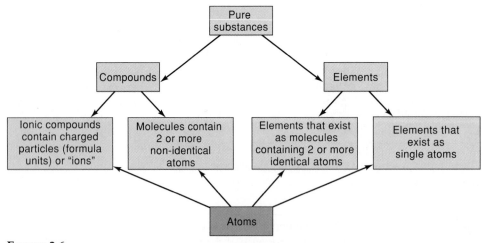

FIGURE 3.6
Summary of atoms, molecules, charged particles (formula units), elements, and compounds. Note that atoms are fundamental to compounds and elements.

rivers), but the formula of water is still H_2O.] This definite formula for water is expressed in the law of definite proportions or constant composition. The **law of definite proportions** or **constant composition** states that a given pure compound always contains the same elements in *exactly* the same proportions by *mass*. For example, exactly 1.0080 parts by mass of hydrogen combine with 7.9997 parts by mass of oxygen to form water. Also, 2.0160 g (2 × 1.0080) of hydrogen combines with 15.9994 g (2 × 7.9997) of oxygen to form water. Later (Section 8.5), we will consider the general approach to how the formula of water as H_2O was obtained, but for the present we can state that the composition by mass for a pure compound is invariable.

Law of definite proportions (constant composition) Principle that a given pure compound always contains the same elements in exactly the same proportions by mass.

The law of definite proportions or constant composition readily illustrates one of the chief differences between mixtures and compounds. A compound, which we previously classified as homogeneous matter, has a constant composition by mass of elements. But a mixture, classified as homogeneous and heterogeneous matter, has a *variable* composition by mass of elements or compounds. Two other important differences between mixtures and compounds are the following:

1. The individual substances (elements or compounds) in a mixture can often be identified, but the elements in a compound lose their elemental character when the compound is formed.

2. The components of a mixture can be separated by simple operations that do not change the identities of individual components, but a compound cannot be broken down into simpler substances without changing the identity of the material.

3.4 Properties and Changes of Pure Substances

Just as each person has his or her own appearance and personality, each pure substance has its own properties, distinguishing it from other substances. The properties of pure substances are divided into physical and chemical properties.

Physical properties are those properties that can be observed without changing the composition of the substance. They include color, odor, taste, solubility, density, melting point, and boiling point. The physical properties of a pure substance are similar to a person's physical appearance. We may note a person's hair or eye color and height or weight without changing the person in any way (see Figure 3.7).

Physical properties Properties of a substance that can be observed without the composition of the substance changing.

Chemical properties are those properties that can be observed only when a substance undergoes a change in composition. They include the fact that iron rusts, that coal or gasoline burns in air, that water undergoes electrolysis, and that chlorine reacts violently with sodium. The chemical properties of a pure substance are somewhat analogous to a person's personality. Is he or she kind or mean, friendly or grouchy, aggressive or shy? See Figure 3.7. Table 3.2 lists some physical and chemical properties of water and iron.

Chemical properties Properties of a substance that can be observed only when a substance undergoes a change in composition.

Study Exercise 3.3
The following are properties of the element silicon; classify each one as a physical or chemical property.
a. dark-gray, brittle element   (physical)
b. reacts with hydrofluoric acid (HF)   (chemical)   Work Problems 8 and 9.

FIGURE 3.7
Physical and chemical properties are analogous to a person's appearance and personality.

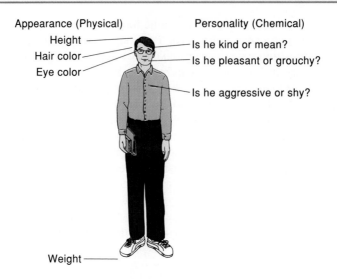

Appearance (Physical)
Height
Hair color
Eye color
Weight

Personality (Chemical)
Is he kind or mean?
Is he pleasant or grouchy?
Is he aggressive or shy?

Changes of Pure Substances

Just as pure substances have physical and chemical properties, so also can they go through physical and chemical changes.

Physical changes All changes in a substance other than changes in its chemical composition.

Physical changes are changes that occur *without* a change in the composition of the substance. The changes in the state of water between ice and liquid (melting and freezing) and liquid and water vapor (boiling and condensing) are examples of physical changes (see Figure 3.8).

$$\text{ice} \rightleftharpoons \text{liquid water} \rightleftharpoons \text{water vapor} \tag{3.1}$$

The difference between a property and a change should be noted here; a property distinguishes one *substance* from another substance, but a change is a *conversion* from one *form* to another. The *melting point* of a substance is a physical *prop-*

			PHYSICAL				
		DENSITY	**SPECIFIC HEAT**[a]		**MELTING**	**BOILING**	
SUBSTANCE	COLOR	(g/mL, 20°C)	cal/g·°C	J/kg·K	POINT (°C)	POINT (°C)[b]	CHEMICAL
Water (liquid)	Colorless	0.998	1.000	4.18×10^3	0	100	Undergoes electrolysis; yields hydrogen and oxygen
Iron (solid)	Gray-white	7.874	0.108	4.52×10^2	1535	3000	Rusts; reacts with oxygen in air to form an iron oxide [iron(III) oxide]

TABLE 3.2   **Some Physical and Chemical Properties of Water and Iron**

[a] See Section 3.5.

[b] At 1.00 atm pressure.

erty, while the process of *melting,* that is, going from a solid to a liquid, is a physical *change.* A physical change in a substance is similar to the change in a man's appearance when he puts on a sports coat. The change to different clothes has not changed the person he is.

Chemical changes are changes that can be observed only when a *change* in the composition of the substance occurs. *New substances are formed.* The properties of the new substances are different from those of the old substances. In a chemical change a gas may be produced, heat may be given off (the flask gets hot), a color change may occur, or an insoluble substance may appear. Our earlier examples (see Section 3.2) of the effect of electricity on water and salt, sulfuric acid on sugar, and heat on mercury(II) oxide are instances of chemical change. Elements can undergo chemical change to produce compounds, as when chlorine gas reacts violently when combined with sodium metal to yield sodium chloride, common table salt (see Figure 3.9). Another common example of a chemical change is rust on the fenders of your car caused by the salts (sodium chloride or calcium chloride) used to melt ice on the streets in winter. The metal (steel, iron) forms a red hydrated iron oxide and loses its metallic properties (see Figure 3.10). Chemical change in a substance is somewhat analogous to a personality change in a person. For example, suppose a grouchy person meets another person, possibly of the opposite sex, and becomes friendly, that is, a "new" person.

Table 3.3 lists various changes and classifies them as chemical or physical.

Study Exercise 3.4
Classify each of the following changes as physical or chemical:
a. cutting up bread to combine with ground beef to make a meat loaf
(physical)
b. cooking the ground beef with appropriate sauces at 350°F for $1\frac{1}{4}$ hours (h)
(chemical)

FIGURE 3.8
The change of state of water between ice and liquid is an example of a physical change.

Chemical changes   Changes that result in changes in the composition of the substance. New substances are formed.

Work Problems 10 and 11.

(a)

(b)

(c)

FIGURE 3.9
(a) Chlorine gas reacts with (b) sodium metal to form (c) sodium chloride. This is a chemical change.

FIGURE **3.10**
The rusting of the fenders on your car is an example of a chemical change.

TABLE **3.3**	**Classification of Changes as Physical or Chemical**
CHANGE	CLASSIFICATION
Boiling of water	Physical
Freezing of water	Physical
Electrolysis of water	Chemical
Reaction of chlorine with sodium	Chemical
Melting of iron	Physical
Rusting of iron	Chemical
Cutting of wood	Physical
Burning of wood	Chemical
Taking a bite of food	Physical
Digestion of food	Chemical

Energy Capacity to do work or transfer heat.

Potential energy Energy possessed by a substance by virtue of its position in space or its chemical composition.

Kinetic energy Energy possessed by a substance by virtue of its motion.

3.5 Energy

All changes and transformations in nature are accompanied by changes in energy. **Energy** is defined as the capacity to do work or to transfer heat. The primary types of energy are mechanical energy, heat energy, electrical energy, chemical energy, and light or radiant energy. Energy may also be potential or kinetic. **Potential energy** is the energy possessed by a substance by virtue of its *position* in space or its *chemical composition*. **Kinetic energy** is energy possessed by a substance by virtue of its *motion*. A rock high on a cliff has potential energy, but as it falls down the cliff its *potential energy decreases* and its *kinetic energy increases*.

Chemicals such as natural gas and gasoline have high potential energy (see Figure 3.11). As they are burned, this potential energy is transformed to heat energy to produce heat to warm our homes and mechanical energy to run our cars. Starting a car is an example of the transformation of forms of energy. When the ignition key is turned, energy from the lead storage battery (chemical energy) produces an elec-

FIGURE 3.11
Jet fuel has high potential energy. As it is burned to produce heat, the potential energy decreases and the kinetic energy increases.

> **S**tudy Hint: Our bodies also transform energy from one form to another. We convert food (chemical energy) to mechanical energy (movement), electrical energy (nerve impulses), and other forms of chemical energy (stored fat).

tric current (electrical energy) that is transmitted to the starting motor (mechanical energy) and to the spark plugs (electrical and heat energy), which ignite the compressed gas in the cylinders (chemical energy), which, in turn, is transferred to the crankshaft (mechanical energy). When the transmission is engaged, the car moves (kinetic energy).

The Measurement of Energy

Transformations of energy are at the heart of life. In photosynthesis, sunlight (radiant or light energy) initiates the chemical processes that create carbohydrates (chemical energy). The most common form of energy, **heat energy,** is energy transferred from one substance to another when there is a temperature difference between the substances. It is the energy associated with the random motion of molecules. The quantity of heat energy gained or lost by an object is measured in calories or joules. A **calorie (cal)** is equal to the quantity of heat required to raise the temperature of 1 g of water from 14.5° to 15.5°C. The SI uses the **joule (J,** pronounced "jool") as its standard unit of energy measurement. The relationship between the calorie and the joule is

$$1 \ cal \ = \ 4.184 \ J \tag{3.2}$$

Chemists currently use both calories and joules, and at different times we will use both sets of units.

In nutrition, the large calorie or Calorie (Cal), which is equal to 1000 small calories or 1 kilocalorie (kcal), is used. One tablespoon of sugar (12 g), when burned in the body, produces 45 Cal or kcal, which is equivalent to 45,000 cal of heat energy. A can of soft drink (355 mL) has a calorie value of 160 Cal. For an activity such as going to college, males require an average of about 3000 Cal/day and females require about 2200. Some activities require more Calories.

Heat energy   Energy transferred from one substance to another when there is a temperature difference between the substances: it is associated with the random motion of molecules.

Calorie (cal)   Standard unit for the measurement of heat energy; 1 cal is equal to the quantity of heat required to raise the temperature of 1 g of water from 14.5° to 15.5°C.

Joule (J)   Standard unit for the measurement of heat energy in the Système International (SI); 4.184 J = 1 cal.

Specific Heat

Specific heat Joules of heat required to raise the temperature of 1.00 kg of a substance by 1.00 K or calories to raise the temperature of 1.00 g of a substance by 1.00°C.

One physical property of matter is that it requires a certain quantity of heat to produce a given change in temperature per unit mass of a given substance. This is called the specific heat of the substance. **Specific heat** is defined as the number of joules required to raise the temperature of 1.00 kg of a substance 1.00 K *or* the number of calories required to raise the temperature of 1.00 g of a substance 1.00°C.

$$\text{specific heat} = \frac{\text{joules}}{\text{kg·K}} \quad \text{or} \quad \frac{\text{calories}}{\text{g·°C}} \tag{3.3}$$

Figure 3.12 illustrates the difference in the specific heats of water (liquid) and sodium chloride (salt). Table 3.4 lists the specific heats of a few substances in both units. Notice that water has a relatively high specific heat, while the other substances have relatively low values. That is, water can absorb more heat energy for each degree in temperature rise than can the other materials. This pattern is one reason why water is an effective cooling agent in automobile radiators. It can cool the engine without its temperature rising too much.

EXAMPLE 3.3 Exactly 75.0 cal of heat energy raises the temperature of 10.0 g of an unknown metal from 25.0° to 60.0°C. Calculate the specific heat of the metal in cal/g·°C.

SOLUTION The units of specific heat (cal/g·°C) help us to calculate the specific heat as follows:

$$\frac{75.0 \text{ cal}}{10.0 \text{ g} \times (60.0 - 25.0)\text{°C}} = 0.214 \text{ cal/g·°C} \qquad \textit{Answer}$$

Study Hint: The units guide us in the calculation. We need to remove the units of kilograms and kelvins from the denominator, and so they must appear in the numerator somewhere. When the units are "right," the answer usually is, too!

EXAMPLE 3.4 Calculate the number of joules required to raise the temperature of 0.120 kg of sodium chloride (solid) from 298 to 358 K.

SOLUTION From Table 3.4, the specific heat of sodium chloride (solid) is 8.54×10^2 J/kg·K. The number of joules is calculated as follows:

$$\frac{8.54 \times 10^2 \text{ J}}{1 \text{ kg·K}} \times 0.120 \text{ kg} \times (358 - 298) \text{ K} = 6150 \text{ J} \qquad \textit{Answer}$$

FIGURE 3.12
Comparison of specific heats of water (liquid) and sodium chloride (salt). Notice the difference in calories of heat energy.

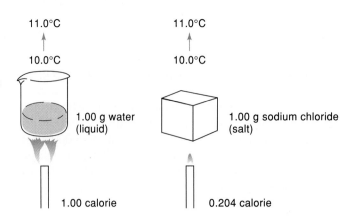

TABLE 3.4	Specific Heat of Some Substances	
	SPECIFIC HEAT	
SUBSTANCE	cal/g·°C	J/kg·K[a]
Water (liquid)	1.00	4.18×10^3
Aluminum (solid)	0.217	9.08×10^2
Lead (solid)	0.0310	1.30×10^2
Sugar (solid)	0.299	1.25×10^3
Silver (solid)	0.0570	2.38×10^2
Sodium chloride (salt, solid)	0.204	8.54×10^2

[a] See Section 2.7 for a discussion of exponents and scientific notation.

EXAMPLE 3.5 Calculate the number of joules required to raise the temperature of $25\overline{0}$ g of lead (solid) from 373 to 473 K.

SOLUTION From Table 3.4 the specific heat of lead (solid) is 1.30×10^2 J/kg·K. The number of joules is calculated as follows:

$$\frac{1.30 \times 10^2 \text{ J}}{1 \text{ kg·K}} \times \frac{1 \text{ kg}}{1000 \text{ g}} \times 25\overline{0} \text{ g} \times (473 - 373) \text{ K} = 3.25 \times 10^3 \text{ J} \qquad \textit{Answer}$$

EXAMPLE 3.6 Calculate the amount of aluminum (solid) in grams used if $8\overline{0}0$ cal of heat is required when the aluminum is heated from $4\overline{0}°$ to $8\overline{0}°$C.

SOLUTION From Table 3.4, the specific heat of aluminum (solid) is 0.217 cal/g·°C. We are asked for the amount of aluminum in grams, and so grams must appear in the answer. The factor 0.217 cal/g·°C must be inverted to solve for grams, and the solution is

$$\frac{1.00 \text{ g·°C}}{0.217 \text{ cal}} \times \frac{8\overline{0}0 \text{ cal}}{(8\overline{0} - 4\overline{0})°C} = 92 \text{ g} \qquad \textit{Answer}$$

Notice that the number (0.217) and the units (calories) are *both* inverted. The answer is given to two significant digits since there are two significant digits in the number $8\overline{0}0$.

EXAMPLE 3.7 Calculate the amount of water (liquid) in grams used if 3.25×10^3 J of heat is required when the water is heated from 293 to 313 K.

SOLUTION From Table 3.4, the specific heat of water (liquid) is 4.18×10^3 J/kg·K. We are asked for the amount of water in grams, and so the mass (in kilograms) must appear in the answer. The factor 4.18×10^3 J/kg·K must be inverted, and the solution is

$$\frac{1.00 \text{ kg·K}}{4.18 \times 10^3 \text{ J}} \times \frac{3.25 \times 10^3 \text{ J}}{(313 - 293) \text{ K}} \times \frac{1000 \text{ g}}{1 \text{ kg}} = 38.9 \text{ g} \qquad \textit{Answer}$$

Study Hint: If the heat is given or asked for in joules or kilojoules, use the specific heat of the substance in J/kg·K. If the heat is given or asked for in calories or kilocalories, use the specific heat of the substance in cal/g·°C.

Study Exercise 3.5
Calculate the number of kilocalories of water required to raise the temperature of 0.200 kg of water (liquid) from 15.0° to 90.0°C (see Table 3.4).

(15.0 kcal)

Study Exercise 3.6
Calculate the mass of lead (solid) in grams used if 90.0 cal of heat is absorbed when the lead is heated from 115° to 225°C (see Table 3.4).

Work Problems 12 through 17.

(26.4 g)

3.6 Laws of Conservation

Law of conservation of energy Principle that energy can neither be created nor destroyed, although it may be transformed from one form to another.

Law of conservation of mass Principle that mass is neither created nor destroyed and that the total mass of the substances involved in a physical or chemical change remains constant.

The relationships among the various forms of energy are described in the **law of conservation of energy**,* which states that *energy* can neither be created nor destroyed but may be transformed from one form to another. The **law of conservation of mass**,[†] which corresponds to the law of conservation of energy, states that *mass* is neither created nor destroyed and that the total mass of the substances involved in a physical or chemical change remains constant within our ability to detect changes in mass.

For example, consider the reaction of hydrogen and oxygen to produce water (a chemical change):

$$\text{hydrogen} + \text{oxygen} \rightarrow \text{water}$$
$$4.0320 \text{ g} + 31.9988 \text{ g} = 36.0308 \text{ g}$$

(3.4)

In this reaction, 4.0320 g of hydrogen combines with 31.9988 g of oxygen to yield 36.0308 g of water. The *sum* of the masses of the *reactants* (hydrogen and oxygen) is *equal* to the mass of the *product* (water), hence mass is neither created nor destroyed, as Figure 3.14 illustrates.

In 1905, the German-American physicist Albert Einstein (1879–1955) concluded that, under certain conditions, mass and energy may be interconverted. That is, energy can be changed into matter and matter can be changed into energy. Effectively combining the laws of conservation of mass and conservation of energy into the law of conservation of mass and energy, Einstein noted that, while energy and mass *may* be interconverted, the total of the mass and energy of a system remains constant.

When some mass is converted to energy, the energy liberated is equal to the mass consumed times the speed of light squared [$E = mc^2$, where c is the speed of light or 3.0×10^8 m/second (s)]. Thus, an enormous amount of heat would result from the conversion of a very small amount of mass to energy. For example, if 1 gram of mass is changed completely into heat (9×10^{13} J), it would be sufficient to heat a small mountain lake from 0° to 100°C! This process is the basis of the atomic bomb and nuclear power plants (see Chapter 19).

---

*James Prescott Joule (1818–1889), English physicist, is credited with formulating this law in the middle of the nineteenth century. The unit of energy, the joule, is named in his honor.
[†]Antoine Laurent Lavoisier [(1ǎ′vwǎ′zyā), 1743–1794; see Section 1.3], who published his results in 1789, is credited with formulating this law. His valued associate was his wife, Marie Paulze Lavoisier. She made drawings of the laboratory as her husband and his assistants worked (see Figure 3.13). Lavoisier was beheaded in 1794 during the French Revolution; Marie Lavoisier was not executed.

FIGURE 3.13
Antoine Laurent Lavoisier and his wife, Marie Paulze Lavoisier. She was his valued associate in his scientific achievements. (The Granger Collection, Metropolitan Museum of Art, New York.)

Fortunately, such large changes are very rare. In ordinary chemical reactions, the energy changes involved are relatively small (2×10^5 to 2×10^6 J), and the corresponding mass changes are extremely small (2×10^{-8} to 2×10^{-9} g). Such a change in mass is so small that it is undetectable, even by the finest balances available. Thus, *for all chemical changes, the laws of conservation of mass and conservation of energy are valid.*

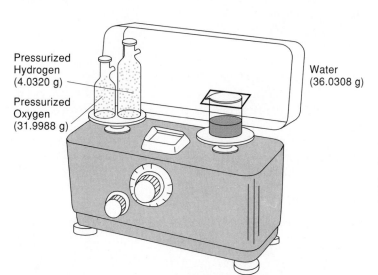

Pressurized Hydrogen (4.0320 g)

Pressurized Oxygen (31.9988 g)

Water (36.0308 g)

FIGURE 3.14
The law of conservation of mass illustrated through the reaction of hydrogen with oxygen to produce water. If the sum of the masses of the containers on the left is the same as the mass of the container on the right, then the mass of hydrogen and oxygen in the pressure vessels is equal to the mass of the water in the beaker.

3.7 Division of Elements. Metals and Nonmetals: Physical and Chemical Properties

Metals Elements that have a high luster, conduct electricity and heat well, are malleable and ductile, have high densities and melting points, are hard, and do not readily combine with one another.

Malleable Capable of being shaped by beating with a hammer.

Ductile Capable of being drawn out into a thin wire.

Chemists usually divide the elements into metals and nonmetals. The basis of this division is their physical and chemical properties.

In *general,* **metals** have the following physical properties.

✔ They have a high luster (shine), as in the case of silver.

✔ They conduct electricity and heat well, as in the case of copper.

✔ They are **malleable** (can be shaped by beating with a hammer), as in the case of tin.

✔ They are **ductile** (can be drawn out into a thin wire), as in the case of copper (see Figure 3.15).

✔ Most have high densities, as in the case of lead ($d^{20°} = 11.34$ g/mL).

FIGURE 3.15
Many metals are ductile—they can be drawn into a variety of wires. Here, rectangular copper wire is being prepared.

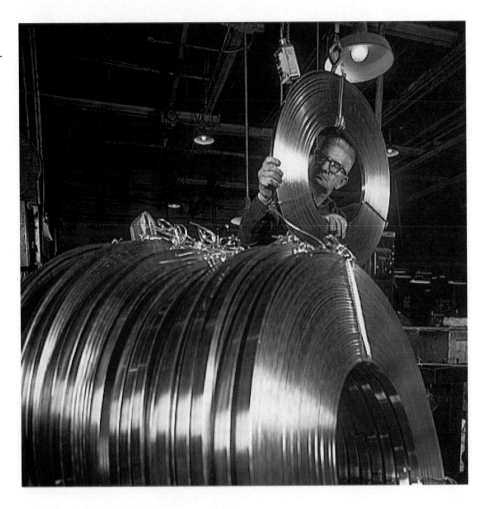

✔ Many have high melting points, as in the case of iron (mp = 1535°C). Hence, metals are generally solids at room temperature. The exceptions to this are mercury (Hg), gallium (Ga), and cesium (Cs), which are liquids at ordinary temperatures.

✔ Most are hard, as in the cases of iron, tungsten, and chromium. However, a few, such as sodium and lead, are soft.

In *general,* metals have the following chemical properties:

✔ They do not readily combine with each other.

✔ They do combine with nonmetals and hence are normally found in nature in the combined form. Iron is found combined with oxygen or sulfur, and aluminum is found combined with oxygen and silicon or just oxygen. A few relatively unreactive metals are found in nature in the **free state**, that is, not combined with any other element. Gold, silver, copper, and platinum are often found in the free state.

Free state Term describing an element existing on its own—not combined with any other element.

The physical and chemical properties of the metals just listed are *general* properties that vary from metal to metal. Metals usually exhibit many, but *not necessarily all,* of these properties.

In contrast, the physical properties of **nonmetals** generally differ from those of metals. The general physical properties of nonmetals are as follows:

Nonmetals Elements whose characteristics are the opposite of the metals and combine readily both with metals and with other nonmetals.

✔ They generally are not lustrous, but rather dull, as in the cases of sulfur and carbon (graphite).

✔ They are usually poor conductors of heat and electricity, as in the case of sulfur.

✔ They are not ductile or malleable, but rather brittle, as in the case of carbon (see Figure 3.16).

✔ They have low densities, as in the case of the gases nitrogen and oxygen.

FIGURE 3.16
Many nonmetals are quite brittle and fracture into chunks when struck. Chunks of coal (mostly carbon) may be burned to provide heat energy.

✔ They have low melting points, and so at least one nonmetal exists in each of the three physical states of matter. Sulfur, phosphorus, carbon, and iodine are solids, bromine is a liquid, and fluorine, chlorine, nitrogen, and oxygen are gases at room temperature and atmospheric pressure.

✔ They are soft, as in the cases of sulfur and phosphorus. One exception, however, is diamond, which is a form of carbon. Diamond is one of the hardest materials known.

Nonmetals also generally have the following chemical properties:

✔ They combine with metals. A few exist in nature in the free state (uncombined); these are oxygen and nitrogen (both in air), sulfur, and carbon (coal, graphite, and diamond).

✔ They may also combine with each other. Carbon dioxide, carbon monoxide, silicon dioxide (sand), sulfur dioxide, and carbon tetrachloride are examples of compounds formed from two nonmetals.

Periodic table Special table listing all the known elements arranged so that elements in a given column have similar chemical properties.

On the inside front cover of this text and in Figure 3.17 is a table of the elements, the **periodic table** or the *periodic chart.* Slightly to the right of the center of the table is a *colored* stair step line, which in general separates the metals from the nonmetals. To the *left* of this line lie the *metals,* and to the *right* are the *nonmetals.* Elements such as sodium, copper, iron, gold, lead, and platinum are metals. Nonmetals include such elements as carbon, nitrogen, oxygen, fluorine, sulfur, chlorine, and bromine.

Study Hint: Find sodium, copper, iron, gold, lead, and platinum in the periodic table. You need to know their symbols first!

The last column of elements is a special group of nonmetals called **noble gases**. They exist in nature in the free (uncombined) state. Helium is found in some gas wells, and a small amount is found in air along with argon. The noble gases are relatively *unreactive.* That is, they do not react much with other elements.

Noble gases Special type of nonmetals; these elements are relatively nonreactive and so were once called "inert."

Except for aluminum, elements that lie on either side of the colored stair step line are called **metalloids (semimetals)** and have both metallic and nonmetallic properties. Aluminum is not a metalloid but a metal because it has mostly metallic properties. Metalloids include boron, silicon, germanium (Ge), arsenic, antimony, tellurium (Te), polonium (Po), and astatine (At). Although hydrogen (H) lies far to the left of the stair step line, it sometimes exhibits both metallic and nonmetallic properties. Some metalloids are semiconductors (substances that conduct electricity to some degree but not as readily as metals do) and are used in the manufacture of transistors, computer chips, and solar cells. The electrical conduction of metalloids increases with temperature, in contrast to the electrical conduction of metals which decreases with temperature. Silicon is a metalloid that is extensively used as a semiconductor. More recently, gallium arsenide (GaAs) has become a very important material for specialized uses in the electronics industry. Gallium arsenide allows components such as computer processor chips to operate faster, which makes the computers themselves faster and more efficient.

Metalloids (semimetals) Elements (except aluminum) that lie on either side of the *colored* stair step line in the periodic table; they have both metallic and nonmetallic properties: boron, silicon, germanium, arsenic, antimony, tellurium, polonium, and astatine.

Work Problems 18 through 25.

GROUPS

PERIODS	1 IA	2 IIA	3 IIIB	4 IVB	5 VB	6 VIB	7 VIIB	8 VIII	9 VIII	10 VIII	11 IB	12 IIB	13 IIIA	14 IVA	15 VA	16 VIA	17 VIIA	18 VIIIA
1	1.008 **H** 1																	4.003 **He** 2
2	6.941 **Li** 3	9.012 **Be** 4											10.811 **B** 5	12.011 **C** 6	14.007 **N** 7	15.999 **O** 8	18.998 **F** 9	20.179 **Ne** 10
3	22.990 **Na** 11	24.305 **Mg** 12											26.982 **Al** 13	28.0855 **Si** 14	30.9738 **P** 15	32.06 **S** 16	35.453 **Cl** 17	39.948 **Ar** 18
4	39.0983 **K** 19	40.08 **Ca** 20	44.956 **Sc** 21	47.90 **Ti** 22	50.9415 **V** 23	51.996 **Cr** 24	54.938 **Mn** 25	55.847 **Fe** 26	58.933 **Co** 27	58.71 **Ni** 28	63.546 **Cu** 29	65.37 **Zn** 30	69.72 **Ga** 31	72.59 **Ge** 32	74.922 **As** 33	78.96 **Se** 34	79.904 **Br** 35	83.80 **Kr** 36
5	85.468 **Rb** 37	87.62 **Sr** 38	88.906 **Y** 39	91.22 **Zr** 40	92.9064 **Nb** 41	95.94 **Mo** 42	98.906 **Tc** 43	101.07 **Ru** 44	102.906 **Rh** 45	106.4 **Pd** 46	107.868 **Ag** 47	112.41 **Cd** 48	114.82 **In** 49	118.69 **Sn** 50	121.75 **Sb** 51	127.60 **Te** 52	126.904 **I** 53	131.30 **Xe** 54
6	132.906 **Cs** 55	137.33 **Ba** 56	138.906 ***La** 57	178.49 **Hf** 72	180.948 **Ta** 73	183.85 **W** 74	186.2 **Re** 75	190.2 **Os** 76	192.22 **Ir** 77	195.09 **Pt** 78	196.967 **Au** 79	200.59 **Hg** 80	204.37 **Tl** 81	207.2 **Pb** 82	208.981 **Bi** 83	(209) **Po** 84	(210) **At** 85	(222) **Rn** 86
7	(223) **Fr** 87	226.025 **Ra** 88	(227) ****Ac** 89	(261) **Rf** 104	(262) **Ha** 105	(263) **Sg** 106	(262) **Ns** 107	(265) **Hs** 108	(266) **Mt** 109	(269) 110	(272) 111							

TRANSITION ELEMENTS

*Lanthanide series

140.12 **Ce** 58	140.908 **Pr** 59	144.24 **Nd** 60	(145) **Pm** 61	150.4 **Sm** 62	151.96 **Eu** 63	157.25 **Gd** 64	158.925 **Tb** 65	162.50 **Dy** 66	164.930 **Ho** 67	167.26 **Er** 68	168.934 **Tm** 69	173.04 **Yb** 70	174.967 **Lu** 71

**Actinide series

232.038 **Th** 90	231.031 **Pa** 91	238.029 **U** 92	237.048 **Np** 93	(244) **Pu** 94	(243) **Am** 95	(247) **Cm** 96	(247) **Bk** 97	(251) **Cf** 98	(254) **Es** 99	(257) **Fm** 100	(256) **Md** 101	(255) **No** 102	(257) **Lr** 103

FIGURE 3.17
Periodic table of the elements. Note the *colored* stair step line, which, in general, separates the metals from the nonmetals.

Summary

The three physical states of matter are *solid, liquid,* and *gas* (Section 3.1). Matter is divided into *homogeneous* and *heterogeneous* matter, and homogeneous matter is further divided into *pure substances, homogeneous mixtures,* and *solutions.* Finally, pure substances are divided into *compounds* and *elements,* both of which are composed of atoms (Section 3.2).

Compounds may be composed of molecules and be represented by a *molecular formula.* The *law of definite proportions* states that a given pure compound always contains the same elements in exactly the same proportions by mass (Section 3.3). Compounds have physical properties such as color that can be observed without changing the composition of the substance. They also have chemical properties such as rusting that can be observed only when a substance undergoes change. Similarly, physical changes in a compound do not involve a change in the essential nature of the substance, while chemical changes do involve such changes (Section 3.4).

Changes or transformations in nature are often accompanied by changes in energy—the capacity to do work or transfer heat. Energy may be *potential* or *kinetic,* and energy may be transformed from one type (heat, electrical, mechanical, light, or chemical) into another. Heat energy is measured in calories or joules. The specific heat of a substance relates the amount of heat added to a substance, the mass of the substance, and the resulting change in temperature (Section 3.5).

The *law of conservation of energy* describes energy changes in nature, and the *law of conservation of mass* describes physical and chemical changes. These laws hold for all chemical changes, but in rare cases they must be combined because of the interconversion of mass and energy, as described by Einstein (Section 3.6).

Elements are divided into *metals* and *nonmetals* based on their physical and chemical properties. Metals are separated from nonmetals in the periodic table (see the inside front cover or Figure 3.17) by a *colored* stair step line. To the left of this line lie the metals, and to the right are the nonmetals. The last column of nonmetals is called the *noble gases.* Except for aluminum, elements that lie on either side of the colored stair step line are called *metalloids (semimetals)* (Section 3.7).

Exercises

1. Define or explain the following terms (the number in parentheses refers to the section in the text where the term is mentioned):

   a. physical state (3.1)   b. homogeneous matter (3.2)

   c. heterogeneous matter (3.2)   d. mixture (3.2)

   e. homogeneous mixture (3.2)   f. solution (3.2)

   g. pure substance (3.2)   h. compound (3.2)

   i. element (3.2)   j. atom (3.2)

   k. molecule (3.3)   l. molecular formula (3.3)

   m. law of definite proportions   n. physical properties (3.4)
      (constant composition) (3.3)

   o. chemical properties (3.4)   p. physical changes (3.4)

   q. chemical changes (3.4)   r. energy (3.5)

   s. potential energy (3.5)   t. kinetic energy (3.5)

u. heat energy (3.5)   v. calorie (cal) (3.5)

w. joule (J) (3.5)   x. specific heat (3.5)

y. Law of conservation of   z. Law of conservation of
    energy (3.6)   mass (3.6)

aa. metals (3.7)   bb. malleable (3.7)

cc. ductile (3.7)   dd. free state (3.7)

ee. nonmetals (3.7)   ff. periodic table (3.7)

gg. noble gases (3.7)   hh. metalloids (semimetals) (3.7)

2. Distinguish between

   a. physical and chemical properties b. physical and chemical changes

   c. calorie and temperature   d. calorie and joule

   e. metal and nonmetal   f. metal and metalloid (semimetal)

 Problems

States of Matter (See Section 3.1)

3. Classify the following as existing in one of the three physical states of matter at room temperature and atmospheric pressure:

   a. chalk   b. alcohol

   c. antifreeze   d. battery acid (dilute sulfuric acid)

   e. methane (natural gas)   f. oxygen

Symbols (See Section 3.2)

4. List in order the symbols for the 10 most abundant elements in the earth's crust.

Compounds, Elements, and Mixtures (See Section 3.2)

5. Classify each of the following as a compound, element, or mixture:

   a. calcium (Ca)   b. water (H_2O)

   c. silicon (Si)   d. salted popcorn

   e. salt (NaCl)   f. sugar ($C_{12}H_{22}O_{11}$)

   g. computer paper   h. gasoline

Formulas (See Section 3.3)

6. In each of the following formulas, determine the number of atoms of each element and write the name of the element. Indicate the total number of atoms.

   a. CH_4 (methane, natural gas)

   b. $C_6H_{12}O_6$ (glucose)

   c. CCl_2F_2 (Freon)

   d. $C_{16}H_{18}N_2O_5S$ (penicillin V)

   e. $C_{34}H_{32}FeN_4O_4$ (heme from hemoglobin)

   f. $C_{13}H_{18}O_2$ (ibuprofen, Motrin, Nuprin, or Medipren)

7. From the number of atoms of each element in one unit of the compound, write the formula for the following compounds:

   a. sulfur dioxide; 1 sulfur, 2 oxygen

   b. pyrite or fool's gold; 1 iron, 2 sulfur

   c. argentite; 2 silver, 1 sulfur

   d. caffeine; 8 carbon, 10 hydrogen, 4 nitrogen, 2 oxygen

   e. adenosine triphosphate (ATP); 10 carbon, 16 hydrogen, 5 nitrogen, 13 oxygen, 3 phosphorus

   f. 2,4,6-trinitrotoluene (TNT); 7 carbon, 5 hydrogen, 3 nitrogen, 6 oxygen

Physical and Chemical Properties (See Section 3.4)

8. Using a reference book such as the *Handbook of Chemistry and Physics* or any suitable reference book in your library, look up the melting and boiling points and the density of mercury.

9. The following are properties of the element cesium; classify them as physical or chemical properties.

   a. boiling point 678.4°C

   b. soft

   c. silvery white

   d. reacts with ice above −116°C

   e. prepared by heating cesium azide

   f. liquid at ordinary temperatures

   g. reacts explosively with cold water

   h. ductile

Physical and Chemical Changes (See Section 3.4)

10. Classify the following changes as physical or chemical:

    a. pumping oil out of a well

    b. separating components of oil by distillation

    c. burning gasoline

    d. flaring of gas from a well

    e. grinding beef in a meat grinder

    f. digestion of beef

11. Classify the following changes as physical or chemical:

    a. baking bread

    b. mixing flour with yeast

    c. fermentation to produce beer

    d. smashing a car against a tree

    e. burning your phone book

    f. burning toast

Energy and Specific Heat (See Section 3.5)

12. Heat energy (2930 J) raises the temperature of 0.100 kg of an unknown metal from 293.0 to 333.0 K. Calculate the specific heat of the metal.

13. Calculate the number of kilocalories required to raise the temperature of $15\overline{0}$ g of water (liquid) from 10.0° to 80.0°C (see Table 3.4).

14. Calculate the number of joules required to raise the temperature of $1\overline{00}$ g of aluminum (solid) from 293 to 363 K (see Table 3.4).

15. Calculate the mass of sodium chloride (solid) in grams used if 80.0 cal of heat is absorbed when the sodium chloride is heated from 35.0° to 85.0°C (see Table 3.4).

16. Calculate the mass of lead (solid) in grams used if 50.0 cal of heat is absorbed when lead is heated from 25.0° to 65.0°C (see Table 3.4).

17. Calculate the number of joules required to raise the temperature of 225 g of silver (solid) from 315 to 345 K (see Table 3.4).

Metals and Nonmetals (See Section 3.7)

18. Using the periodic table on the inside front cover of this text, write the names and symbols for 10 metals.

19. Using the periodic table on the inside front cover of this text, write the names and symbols for 10 nonmetals.

20. Write the names and symbols for three metals found in nature in the free state (uncombined).

21. Write the names and symbols for three nonmetals found in nature in the free state (uncombined).

22. For each of the three physical states of matter, list at least one nonmetal (name and symbol) that exists in that state at room temperature and atmospheric pressure.

23. Write the name and symbol of a metal that is a liquid at ordinary temperatures.

24. Write the names and symbols for two noble gases.

25. Write the names and symbols for two metalloids.

General Problems

26. Calculate the volume in liters at 20°C occupied by 3.40 lb of iron (see Table 3.2; 1 lb = 454 g).

27. Calculate the mass in pounds of 4.65 quarts (qt) (20°C) of water (see Table 3.2; 1 lb = 454 g, 1 L = 1.06 qt).

28. Calculate the number of (a) calories and (b) joules required to raise the temperature of $12\overline{0}$ g of sodium chloride (solid) from 35.0° to 65.0°F (see Table 3.4).

29. Calculate the specific heat in (a) cal/g·°C and (b) J/kg·K if 5.00 kcal of heat energy raises the temperature of 1.00 kg of a certain metal from 25.0° to 52.0°F.

30. Calculate the amount of heat in (a) joules and (b) kilocalories required to raise the temperature of $11\overline{0}$ g of iron from 535°C to its melting point (see Table 3.2).

31. Calculate the amount of iron (solid) in kilograms used if 15.0 kcal of heat energy is absorbed when the iron is heated from $137\overline{0}$°C to its melting point (see Table 3.2).

32. An empty cast-iron pot has a mass of 2.10 lb and is filled with 1.85 L of water. Calculate (a) the energy in joules required to heat the water from 23.0° to 90.0°C, and (b) the energy in joules required to heat the pot from 23.0° to 90.0°C (see Table 3.2; 1 lb = 454 g). (c) Which article, the pot or the water, absorbs more heat during the change? (d) How many total joules of energy are absorbed by the pot and the water during the process?

33. A piece of metal, which is a cube 1.25 cm on an edge, has a mass of 12.011 g. Calculate the density of the metal in g/mL.

34. A class of layered copper oxides that act as superconductors near 70 K has the general formula $Pb_2Sr_2LnCu_3O_{8+x}$. (a) How many atoms of lead are in this general formula? (b) Convert $7\overline{0}$ K to the nearest degree Fahrenheit.

✓ Chapter Quiz 3

1. Give the symbol for the following elements:

   a. cobalt   b. magnesium

2. Determine the number of atoms of each element and then write the name of the element and the total number of atoms in each of the following formulas:

   a. CuI_2 b. Al_2S_3

3. From the number of atoms of each element in one unit of the compound, write the formula for the following compounds:

   a. dinitrogen pentoxide: 2 nitrogen, 5 oxygen

   b. acetaminophen (Tylenol and Panadol): 8 carbon, 9 hydrogen, 1 nitrogen, 2 oxygen

4. The following are properties of the element zirconium; classify them as physical or chemical properties.

   a. specific gravity $= 6.506^{20°/4}$

   b. reacts with hot hydrogen chloride gas

   c. reacts with silicon at high temperature

   d. melting point $= 1852°C$

5. Classify the following changes as physical or chemical:

   a. tearing up on a piece of paper   b. burning a piece of paper

   c. dissolving sugar in water   d. burning sugar

6. Calculate the number of kilocalories required to raise the temperature of $18\overline{0}$ g of water (liquid) from $15.0°$ to $85.0°C$ [specific heat of water (liquid) $= 1.00$ cal/g·°C].

Silicon (Symbol: Si)

The Element SILICON: Chemistry and the Computer Revolution

Pieces of very pure silicon are very important in the electronic industry for the production of microprocessors, transistors, and memory chips.

Name: Derives from the Latin word *silex* for "flint."

Appearance: Dark-gray material with metallic luster. It is classified as a metalloid (semimetal), see periodic table, Figure 3.17.

Occurrence: Silicon is the second most abundant element of the earth's crust, making up 25.7% by mass (see Table 3.1). Most of it occurs as silicates and silica (quartz).

Source: Low-grade silicon is obtained by heating quartz $(SiO_2)_n$ and carbon to give silicon and carbon dioxide (CO_2).

Very pure silicon for use in the electronics industry is obtained by treating trichlorosilane $(SiHCl_3$, prepared from low-grade silicon) with hydrogen gas (H_2) to produce silicon and hydrogen chloride (HCl).

Common Uses: The largest use of silicon occurs in the steelmaking industry. Silicon is added to iron during the steelmaking process to give strength and other desirable properties to the end product. Iron–silicon–boron alloys are used to make bearings for heavy machinery.

Silicon dioxide $(SiO_2$, sand) is the principal ingredient in glass.

Silicate glasses, soluble silicate salts $(SiO_3^{2-}, Si_2O_7^{6-}$, and others), and gels of variable composition are components of many common products such as soaps, detergents, gel toothpastes, shampoos, antiperspirants, makeup, and paints. They are also used in water purification.

While the semiconductor industry does not consume large amounts of silicon, it is one of the most visible industrial users. Extremely pure crystals of silicon are required to produce modern transistors, integrated circuits, semiconductors, and other computer chips. The ability of computer manufacturers to make smaller and more powerful computers depends on the use of these silicon crystals. A group of cities in the Santa Clara Valley near San Jose, California, where this technology has been developed, is called Silicon Valley.

Silicones are organic silicon compounds. These compounds act as lubricants and sealants. Because of their lubricating properties, they have been made into body replacement parts such as hips, knees, and ankle joints. Silly Putty, a novel toy, is a silicone which on standing flows like a liquid but can be shaped into a ball and bounced.

Unusual Facts: Low-grade silicon is quite pure (95 to 99%), but not pure enough for the electronics industry. Computer chips and microprocessors require extremely pure silicon that contains less than a few atoms of impurity per billion atoms of silicon.

Silicosis is a disease of the lungs caused by inhalation of silica (SiO_2) dust over a long period of time. It occurs in mine workers. The silica accumulates in the lungs and produces shortness of breath. Other symptoms that can occur are dry cough, weakness, loss of appetite, hoarseness, and chest pains. Tuberculosis can result. There is no way of removing the silica from the lungs; the disease can be treated only by treating the symptoms.

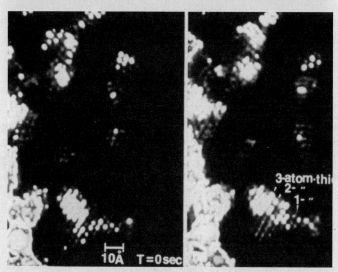

CHAPTER 4

The Structure of the Atom

Time-lapse micrographs of uranyl crystals in which uranium atoms can clearly be observed. (10 Å = 1000 pm = 1×10^{-9} m)

GOALS FOR CHAPTER 4

1. To define atomic mass scales and atomic mass units and explain their use in chemistry (Section 4.1).

2. To learn Dalton's theory of the atom and be able to describe how modern atomic theory differs from his (Section 4.2).

3. To identify the three major subatomic particles of the atom and describe a general arrangement of these particles in atoms (Sections 4.3 and 4.4).

4. To define the term "isotope" and to determine how the properties and structure of isotopes of a single element differ. To calculate the atomic mass of a given element from its isotopic mass and percent composition in nature (Section 4.5).

5. To describe the electronic principal energy levels of an atom, to calculate the maximum number of electrons in a principal energy level, and to arrange electrons in principal energy levels for an atom (Section 4.6).

6. To write electron-dot formulas of elements (Section 4.7).

7. To write the electronic configurations of an atom in sublevels and to describe the modern theory of orbital structure for sublevel electrons (Sections 4.8 and 4.9).

Countdown

5. Convert 0.256 nm to picometers (Sections 2.2 and 2.8).

(256 pm)

4. Express the number 0.005475 in scientific notation to three significant digits (Section 2.7).

(5.48×10^{-3})

3. Convert $2\overline{0}0$ pm to meters. Express your answer in scientific notation (Sections 2.2, 2.7, and 2.8).

(2.0×10^{-10} m)

2. The melting point of silver (Ag) is 962°C. Convert this temperature to degrees Fahrenheit and kelvins (Section 2.8).

(1764°F, 1235 K)

1. The density of silver ($d^{20°}$) is 10.5 g/cm^3. Calculate the volume in milliliters occupied by 0.460 kg of silver (Ag) (Sections 2.1 and 2.8).

(43.8 mL)

W e cannot see them with the naked eye or even with most microscopes, yet all matter on earth and in the universe is composed of atoms. Since the explosion of the first atomic bomb at the end of World War II, the world has become more aware of atoms, their nature, and their power. In this chapter, we will explore the structure of the atom in greater detail.

4.1 Atomic Mass

FIGURE 4.1
Atoms are very small. If atoms with a diameter of 100 pm were placed side by side, it would take 254,000,000 atoms to occupy a 1-in. length.

Atomic mass (atomic weight) scale Scale of the relative masses of atoms; it is based on an arbitrarily assigned value of exactly 12 atomic mass units for a single atom of carbon-12.

Atomic mass units (amu) Units used to express the relative masses of atoms on the atomic mass scale: 1 amu is equal to exactly one-twelfth the mass of a carbon-12 atom.

Atoms are very small. The diameter of an atom is in the range of 100 to 500 pm (see Table 2.1). If we were to place atoms with a diameter of 100 pm side by side, it would take 254,000,000 of them to occupy a 1-inch length, as illustrated in Figure 4.1. That's a lot of atoms!

The mass of an atom is also a very small quantity, too small to be determined with even the most sensitive balance. For example, by indirect methods, scientists have found the mass of a hydrogen atom to be 1.67×10^{-24} g, the mass of an oxygen atom to be 2.66×10^{-23} g, and the mass of a carbon atom to be 2.00×10^{-23} g. Because this mass is very small, chemists have devised a scale of relative masses of atoms called the **atomic mass (atomic weight) scale.** This scale is based on an arbitrarily assigned value of exactly 12 **atomic mass units (amu)** for an atom of carbon-12. (Section 4.5 will discuss the nature of carbon-12.) Hence, one atomic mass unit on the atomic mass scale is equal to one-twelfth the mass of a carbon-12 atom. An atom that is twice as heavy as a carbon-12 atom has a mass of 24 amu.

A simple analogy will help to illustrate the point. Suppose you wish to relate the "weight" of everyone in your class to one person—call him John Lode—as a standard. Today, John has a mass of 91 kg (weighs about 200 lb), and we assign him a relative mass of 12 units on our arbitrary scale. (Note that since a person's mass changes daily—John might be on a diet!—we must pick a particular day.) We could have assigned a value of 10, 15, 20, or some other value, but we arbitrarily chose a value of 12. A classmate, Jane Leit, has a mass of 53 kg (weighs about 117 lb) or 0.58 ($\frac{53}{91}$) times as much as John. Hence, Jane has a relative mass of 7 units (0.58 $\times$ 12) on our arbitrary scale. A similar relationship could be worked out for each member of the class, and a relative mass could be assigned to each based on the mass of John Lode on that particular day.

The inside front cover of this text lists all the elements and their relative atomic mass units based on carbon-12. As you can see, some of these numbers are very exact and are carried out even to the $\frac{1}{10,000}$ place, but others are expressed only to the units place. Therefore, for the calculations that you will be doing in this course we have developed a table of approximate atomic masses. You will find it on the inside back cover of this text, and *you should use it in all future calculations unless instructed otherwise.*

4.2 Dalton's Atomic Theory

In the early part of the nineteenth century, the English scientist John Dalton (1766–1844) (see Figure 4.2) proposed an atomic theory based on experimentation and the chemical laws known at that time. His five proposals, although somewhat modified, still form the framework of our knowledge of the atom.

FIGURE 4.2
John Dalton lived over 150 years ago, but his atomic structure formed the basis for how we think about atoms and molecules.

1. *Elements are composed of tiny, discrete particles called atoms.* This proposal has been verified experimentally. Single atoms of a variety of elements (see chapter opening photograph) have been photographed with an instrument called a *scanning transmission electron microscope.*

2. *Atoms are indivisible and indestructible and maintain their identity throughout physical and chemical changes.* Modern research has altered this proposal. Atoms are not indestructible and may lose their identity when split during nuclear reactions. Dalton's proposal remains true, however, for chemical reactions.

3. *Atoms of the same element are identical in mass and have the same chemical and physical properties. Atoms of different elements have different masses and different chemical and physical properties.* However, as you will learn in Section 4.5, atoms of the same element (called *isotopes*) can have different masses.

4. *When atoms of elements combine to form molecules of compounds, they do so in simple whole-number ratios.* For example, atoms might combine in ratios of 1:1, 1:2, or 2:3. This principle has been borne out experimentally. In Chapter 3 (Section 3.3), we mentioned that one molecule of water consists of two atoms of hydrogen and one atom of oxygen.

5. *Atoms of different elements can unite in different ratios to form more than one compound.* This is another experimentally proven principle. In the preceding example, two atoms of hydrogen united with one atom of oxygen to form a molecule of water, H_2O. Two atoms of hydrogen can also combine with two atoms of oxygen to form a molecule of hydrogen peroxide, H_2O_2. Carbon monoxide, CO, and carbon dioxide, CO_2, are other examples.

4.3 Subatomic Particles: Electrons, Protons, and Neutrons

Subatomic particles Particles that make up the atom of an element: the electron, proton, and neutron.

Electron Subatomic particle with a relative charge of -1 and a negligible mass (9.109×10^{-28} g); electrons exist outside the nucleus of the atom in one of many energy levels.

Study Hint: Just as atoms are the building blocks of compounds, protons, neutrons and electrons are the building blocks of atoms.

Proton Subatomic particle with a relative charge of $+1$ and a mass of approximately 1 amu (1.6726×10^{-24} g); it is located in the nucleus of the atom.

Neutron Subatomic particle with no charge but a mass of approximately 1 amu (1.6748×10^{-24} g); it is located in the nucleus of the atom.

As noted above, atoms are not indivisible as Dalton thought. Instead, every atom is composed of **subatomic particles:** *electrons, protons,* and *neutrons.*

These subatomic particles have two important properties that you must know, *mass* and *charge.* You are already familiar with the concept of mass. You may also be familiar with the concept of charge in automobile and household batteries. You know that a battery has a positive and a negative terminal. Charge also may be positive or negative. It can be measured in different ways, but we speak in terms of *relative charge.* Thus, subatomic particles can have a $+1$ relative charge, a -1 relative charge, or, of course, no charge. This charge is shown as a superscript.

The **electron,** abbreviated e$^-$, was discovered by the English chemist and physicist Sir William Crookes (1832–1919) in 1879. Over the next 30 years, the work of the English physicist Sir J. J. Thomson (1856–1940) and the American physicist Robert A. Millikan (1868–1953) established the mass and actual charge of the electron. The relative charge is -1, and the mass of a single electron is 9.109×10^{-28} g [5.486×10^{-4} (0.0005486) amu]. Thus, we can consider the mass of an electron to be negligible for all practical purposes.

Although you cannot see electrons, you are aware of their effects in your daily life. When you comb your hair with a hard rubber comb, electrons from your hair collect on the comb and can attract small pieces of paper. When you walk on a carpet and then approach certain objects, you get a shock. The electrons from the carpet accumulate in your body, and you may be shocked when you touch certain objects. Both phenomena occur most often when the humidity and temperature are low, and they are often described as the effects of "static electricity."

The **proton,** abbreviated p or p$^+$, was discovered by the German physicist Eugen Goldstein (1850–1930) in 1886. He showed that the proton has a positive charge, the opposite of the charge on an electron. Sir J. J. Thomson performed measurements that allowed the mass of the proton to be calculated. The relative charge on a proton is $+1$, and the mass of a single proton is 1.6726×10^{-24} g (1.0073 amu), a figure rounded off to 1 amu for most calculations.

The **neutron,** abbreviated, n or n^0, was first observed by the English physicist Sir James Chadwick (1891–1974) in 1932. The neutron has *no charge,* and the mass of a single neutron is 1.6748×10^{-24} g (1.0087 amu), again a figure rounded off to 1 amu for most calculations.

Table 4.1 summarizes the relative charges and masses of subatomic particles. You *must* be able to identify these particles and know their abbreviations, approximate masses in amu, and *relative charges.*

TABLE **4.1**	**Summary of Subatomic Particles**	
PARTICLE (ABBREV.)	APPROXIMATE MASS (amu)	RELATIVE CHARGE
Electron (e$^-$)	Negligible	-1
Proton (p or p$^+$)	1	$+1$
Neutron (n or n^0)	1	0

4.4 General Arrangement of Electrons, Protons, and Neutrons. Atomic Number

Now, how are these three subatomic particles arranged in an atom? Experiments conducted by the English physicist Ernest Rutherford (1871–1937) in 1911 showed that virtually all the mass of the atom is concentrated in a very small region at the center called the **nucleus.** Furthermore, he demonstrated that the nucleus is positively charged. Four observations form the framework for the *nuclear atom* as described by Rutherford:

1. *All the protons and neutrons in an atom are found at the center of the atom in the nucleus.* Because most of the mass of the atom is concentrated in this very small region, the nucleus of the atom has a very high density (1.0×10^{14} g/mL). One milliliter of nuclear matter would have a mass of 1.1×10^{8} tons! Also, because the protons are positively charged and the neutrons are neutral, the relative *charge* on the *nucleus* must be *positive* and *equal* to the *number of protons.*

2. *The number of protons* (mass of a proton, 1 amu) *plus the number of neutrons* (mass of a neutron, 1 amu) *equals the* **mass number** *of the atom* because the mass of the electron is negligible. Hence, the number of neutrons present is equal to the mass number minus the number of protons (neutrons = mass number − number of protons).

3. *An atom is electrically neutral.* The number of protons *equals* the number of electrons in a *neutral* atom. If an atom loses or gains electrons, it becomes an **ion.** In *ions* the number of electrons does not equal the number of protons, and the ion is positively or negatively charged. *Atoms* are *neutral; ions* possess *a charge.*

4. *Outside the nucleus is mostly empty space, but in this space are the electrons in certain energy levels.* **Energy levels** are a series of areas outside the nucleus of an atom in which the electrons are located. In these energy levels, the electrons are dispersed at a relatively great distance from the nucleus. The nucleus has a diameter of approximately 1×10^{-3} pm; the diameter of the entire atom is in the range of 100 to 500 pm (Section 4.1). Electrons are dispersed at distances that extend up to 100,000 times the diameter of the nucleus. Suppose that, as you sit at your chair reading this book, you represent the size of the nucleus of the atom. The electrons might be found up to 38 miles away, as Figure 4.3 illustrates.

Before we look at some examples of the general arrangement of the subatomic particles in the atoms of some elements, we must consider the symbols used to describe the atom. The following is a general symbol for an atom of an element, giving its mass number and atomic number:

$$_{Z}^{A}\text{E}$$

    A = mass number

    E = symbol of the element

    Z = atomic number

Nucleus Small, dense region at the center of an atom containing nearly all the mass of the atom—the protons and neutrons; it has a positive electric charge.

Mass number Sum of the number of protons and the number of neutrons in the nucleus of an atom.

Ion Charged entity resulting from loss or gain of electrons by an atom or group of bonded atoms. The number of electrons does not equal the number of protons, therefore it carries a positive or negative charge.

Energy levels Series of areas outside the nucleus of an atom in which the electrons are located.

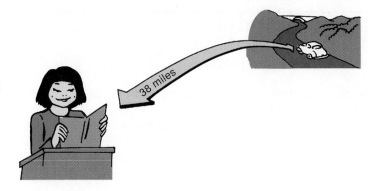

Figure 4.3
Distance between the nucleus and the electrons. If you are the size of the nucleus, the electrons would be spread out at distances as great as 38 miles away.

Atomic number Number of protons in the nucleus of an element's atom.

The **atomic number** is *equal* to the number of *protons* in the nucleus. The *mass number* is *equal* to the *sum of the number of protons and the number of neutrons* in the nucleus. For example, consider

$$^{1}_{1}\text{H}$$

The atomic number (lower left) is 1, and so there is 1 proton in the nucleus. The mass number (upper left) is also 1, and so

$$\text{mass number} = \text{protons} + \text{neutrons}$$
$$1 \quad = \quad 1 \quad + \quad 0$$

and there are 0 neutrons. Since the atom is neutral and has one positive charge (1 proton), it must also have one negative charge (1 electron). We can draw this relationship as

1p
0n
$1e^-$

Nucleus   **Outside nucleus**

EXAMPLE 4.1 For each of the following atoms, calculate the number of protons and neutrons in the nucleus and the number of electrons outside the nucleus.

a. $^{11}_{5}\text{B}$

SOLUTION

5p
6n $5e^-$

Nucleus   Outside nucleus

5 = atomic number = number of protons in nucleus
11 = mass number = sum of protons + neutrons
Neutrons = 11 − 5 = 6 neutrons in nucleus
Number of electrons = number of protons
　　　　　　　　　= 5 electrons outside the nucleus

b. $^{27}_{13}\text{Al}$

SOLUTION

13p
14n $13e^-$

Nucleus   Outside nucleus

13 = atomic number = number of protons in nucleus
27 = mass number = sum of protons + neutrons
Neutrons = 27 − 13 = 14 neutrons in nucleus
Number of electrons = number of protons
　　　　　　　　　= 13 electrons outside the nucleus

c. $^{197}_{79}\text{Au}$

SOLUTION

79p
118n   79e⁻

Nucleus   Outside nucleus

79 = atomic number = number of protons in nucleus
197 = mass number = sum of protons + neutrons
Neutrons = 197 − 79 = 118 neutrons in nucleus
Number of electrons = number of protons
                = 79 electrons outside the nucleus

Work Problems 5 and 6.

Study Exercise 4.1

For each of the following atoms, calculate the number of protons and neutrons in the nucleus and the number of electrons outside the nucleus.

a. $^{28}_{14}\text{Si}$ (14p / 14n)   14e⁻

b. $^{107}_{47}\text{Ag}$ (47p / 60n)   47e⁻

4.5 Isotopes

If you look closely at the atomic masses of the elements shown on the inside front cover of this book, you will note that the atomic masses of the elements are not whole numbers (carbon = 12.01115 amu and chlorine = 35.453 amu). Because the mass of each proton or neutron is equal to about 1, and because the mass of the electron is very slight, we would expect the atomic mass of an element to be very nearly a whole number—certainly not halfway between, as is the case with chlorine. The reason many atomic masses are not even close to whole numbers is that all atoms of the *same* element do not necessarily have the same mass, a contradiction of Dalton's third proposal. Atoms having different mass numbers but the same atomic numbers are called **isotopes.** Figure 4.4 illustrates the difference between the nuclei of the two naturally occurring isotopes of lithium.

    Carbon exists *in nature* as two isotopes: carbon-12, ^{12}C ($^{12}_{6}\text{C}$, exact atomic mass = 12.00000 amu, the atomic mass unit standard) and carbon-13, ^{13}C ($^{13}_{6}\text{C}$, exact atomic mass = 13.00335 amu). Structurally, the difference between these two isotopes is *one* neutron. ^{12}C has 6 neutrons, and ^{13}C has 7 neutrons, as follows:

Isotope Any atom having the same atomic number but a different mass number. They have the same number of protons and electrons but a *different* number of *neutrons.*

6p
6n   6e⁻   6p
7n   6e⁻

$^{12}_{6}\text{C}$ $^{13}_{6}\text{C}$

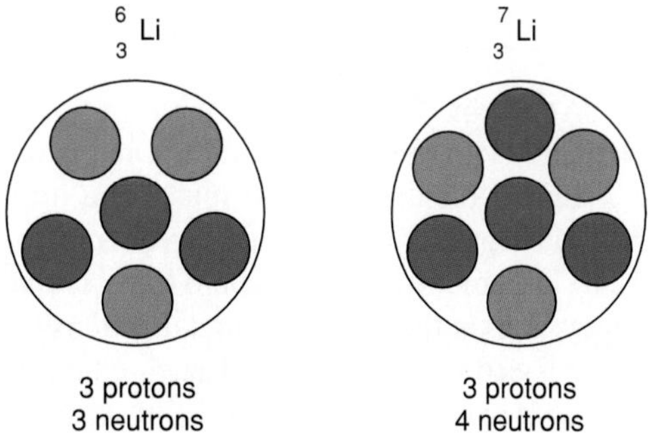

$^{6}_{3}\text{Li}$ $^{7}_{3}\text{Li}$

3 protons
3 neutrons   3 protons
4 neutrons

FIGURE 4.4
Nuclei of the two naturally occurring isotopes of lithium. The red circles denote protons, and the blue circles denote neutrons.

Study Hint: The slightly
different physical
properties and the same
chemical properties of isotopes of the same element
are analogous to a person
running 5K (5 km). If a
person whose mass is
58 kg (about 128 lb) can
run 5K in 17 minutes
(min), a gain in mass to
65 kg (about 143 lb) may
increase the running time
to 20 min. Running 5K is
like a physical property,
but it is still the same person. Therefore, in gaining
mass (neutrons), the time
needed to run 5K (physi-
cal property) changes, but
it is still the same person.

Isotopes of the same element have the same chemical properties but slightly different physical properties. For example, consider the monoxides of ^{12}C and ^{13}C, ^{12}CO and ^{13}CO. *Both* react with oxygen to form the dioxides, $^{12}CO_2$ and $^{13}CO_2$, respectively (chemical property). But ^{12}CO has a melting point of $-199°C$, and ^{13}CO had a melting point of $-207°C$ (physical property).

The atomic mass in amu for the elements C $=$ 12.01115 and Cl $=$ 35.453 is an *average mass* based on the *abundance of the isotopes in nature.* The atomic mass for the element may be obtained by multiplying the exact atomic mass of each isotope by the decimal of its percent abundance in nature and then taking the sum of the values obtained. This is similar to the calculation of your grade in a particular course. For example, if you get a score of 75 on an exam that counts 25% of your final grade and a score of 85 on an exam that counts 75%, your final average based on the weight of each exam is 82.5 (to three significant digits), *not* 80. The calculation is as follows:

$$75(0.25) + 85(0.75) = 18.75 + 63.75 = 82.5$$

(to three significant digits). Note that the percent, meaning "parts per 100," is converted to a decimal, meaning "parts per 1," by dividing by 100.

The following examples illustrate the calculation of atomic masses for elements.

EXAMPLE 4.2 Calculate the atomic mass to four significant digits for carbon, given the following data:

Isotope	Exact Atomic Mass (amu)	Abundance in Nature (%)
^{12}C	12.00000	98.89
^{13}C	13.00335	1.110

SOLUTION Convert the percents (98.89 and 1.110) to decimal form by dividing by 100 to obtain 0.9889 and 0.01110, respectively. Therefore,

12.00000 amu (0.9889) $+$ 13.00335 amu (0.01110) $=$ 12.01 amu   *Answer*

EXAMPLE 4.3 Calculate the atomic mass to four significant digits for chlorine, given the following data:

Isotope	Exact Atomic Mass (amu)	Abundance in Nature (%)
^{35}Cl	34.96885	75.53
^{37}Cl	36.96590	24.47

SOLUTION

34.96885 amu (0.7553) $+$ 36.96590 amu (0.2447) $=$ 35.46 amu   *Answer*

Appendix III contains a table of some naturally occurring stable isotopes of elements with their mass numbers and percent abundance in nature.

Based on the average mass, the atomic mass of carbon was found to be 12.01115 amu, but we would never find an atom of carbon with a relative mass of 12.01115 amu. Instead, it would have a relative mass of 12.00000 or 13.00335 amu, depending on the isotope found. Similarly, in the example of exam scores, your average based on the weight of each exam was 82.5. But you received scores of 75 and 85, never 82.5. In general, for an ordinary size sample of carbon atoms containing the isotopes in the proportions given, we find it convenient to use the average mass, 12.01115 amu. The same reasoning applies to all the other elements and their atomic mass units, which are given on the inside front cover of this text and are the average masses of the naturally occurring isotopes of the elements.

Study Exercise 4.2
The following are properties of silver-107. Which of these properties would be the same for silver-108?
a. reacts with sulfur in the air to form black silver sulfide (Ag_2S)   (same)
b. atomic mass $= 106.90509$ amu   (different)

Study Exercise 4.3
Calculate the atomic mass to four significant digits for silver, given the following data:

Isotope	Exact Atomic Mass (amu)	Abundance in Nature (%)
^{107}Ag	106.90509	51.82
^{109}Ag	108.9047	48.18

(107.9 amu)   Work Problems 7 through 10.

4.6 Arrangement of Electrons in Principal Energy Levels

Earlier in this chapter (Section 4.4) we stated that the electrons were located in mostly empty space outside the nucleus of the atom. You might have wondered, though, what keeps the electrons outside the nucleus. After all, opposite charges traditionally attract (a concept more scientifically known as the *law of electrostatics*), and a positively charged nucleus is the opposite of a negatively charged electron.

To explain this seeming contradiction, Danish physicist Niels Bohr (1885–1962) proposed a theory in 1914, based on a model of the hydrogen atom, that electrons in an atom have their energies restricted to certain specific *energy levels that increase in energy as they increase in distance from the nucleus.* Hence, the nearer the electron is to the nucleus, the less energy the electron has; the farther away from the nucleus it is, the more energy it has. These energy levels are a series of areas outside the nucleus in which the electrons move. They are called *principal energy levels* (first energy levels) and are designated by the whole numbers 1 through 7. There is a maximum number of electrons that can exist in a given principal energy level. This number depends on the value of the whole number (1 through 7) and is given by the following relation:

$$\text{maximum number of electrons in principal energy levels} = 2n^2$$

where $n =$ integers 1 to 7 of the principal energy levels.

EXAMPLE 4.4   Calculate the maximum number of electrons that may occupy the first ($n = 1$) and second ($n = 2$) principal energy levels.

SOLUTION
For 1: maximum number of electrons $= 2 \times 1^2 = 2 \times 1 = 2$ *Answer*
For 2: maximum number of electrons $= 2 \times 2^2 = 2 \times 4 = 8$ *Answer*

Table 4.2 lists the principal energy levels and the maximum number of electrons that may occupy these levels. These are the maximum numbers of electrons that may be accommodated in a given energy level, but an energy level *may have less than the maximum.*

Now let us consider the arrangement of the electrons within the principal energy levels. For the simplest elements—atomic numbers 1 through 18—electrons go into the *lowest principal energy level that has not been filled.* And so we begin by placing electrons in the lowest principal energy level and continue placing them in subsequent levels until the required number of electrons has been assigned. Remember not to exceed the maximum number of electrons for a given principal energy level.

$$^{4}_{2}\text{He} \ = \ \begin{array}{c} 2\text{p} \\ 2\text{n} \end{array} \quad 2\text{e}^{-}$$

$$\phantom{^{4}_{2}\text{He} \ = \ } 1 \qquad\qquad\quad \text{Principal energy level}$$

$$^{11}_{5}\text{B} \ = \ \begin{array}{c} 5\text{p} \\ 6\text{n} \end{array} \quad 2\text{e}^{-} \ \ 3\text{e}^{-}$$

$$\phantom{^{11}_{5}\text{B} \ = \ } 1 \quad\ 2 \qquad\qquad\quad \text{Principal energy level}$$

Note here that the maximum number of electrons in energy level 1 is 2, and so to place 5 electrons outside the nucleus, we must go to a higher energy level, level 2.

$$^{16}_{8}\text{O} \ = \ \begin{array}{c} 8\text{p} \\ 8\text{n} \end{array} \quad 2\text{e}^{-} \ \ 6\text{e}^{-}$$

$$\phantom{^{16}_{8}\text{O} \ = \ } 1 \quad\ 2 \qquad\qquad\quad \text{Principal energy level}$$

$$^{23}_{11}\text{Na} \ = \ \begin{array}{c} 11\text{p} \\ 12\text{n} \end{array} \quad 2\text{e}^{-} \ \ 8\text{e}^{-} \ \ 1\text{e}^{-}$$

$$\phantom{^{23}_{11}\text{Na} \ = \ } 1 \quad\ 2 \quad\ 3 \qquad\qquad\quad \text{Principal energy level}$$

Level 2 can accommodate a maximum of 8 electrons, and so to place 11 electrons outside the nucleus we must use not only levels 1 and 2 but also a higher energy level, level 3.

$$^{40}_{18}\text{Ar} \ = \ \begin{array}{c} 18\text{p} \\ 22\text{n} \end{array} \quad 2\text{e}^{-} \ \ 8\text{e}^{-} \ \ 8\text{e}^{-}$$

$$\phantom{^{40}_{18}\text{Ar} \ = \ } 1 \quad\ 2 \quad\ 3 \qquad\qquad\quad \text{Principal energy level}$$

TABLE 4.2	**Maximum Number of Electrons in Principal Energy Levels**
PRINCIPAL ENERGY LEVEL	MAXIMUM NUMBER OF ELECTRONS
1	2
2	8
3	18
4	32
5	50
6	72
7	98

Increasing energy

Thus, argon, a noble gas (see Section 3.7), has **eight** electrons in its highest principal energy level.

The need to fill energy levels sequentially explains why electrons at most levels stay put. For an electron to change its energy, it must shift from one energy level to another. For an electron to go to a *higher* energy level, the atom must *absorb* a small amount of energy. The amount of energy absorbed is equal to the difference in energy between the two levels. For an electron to go to a *lower* energy level, a *lower energy level must have a vacancy.* If a lower energy level has a vacancy, the electron may fall to the new energy level and the atom will *give off* energy. In this case, the amount of energy emitted by the atom is equal to the difference between the two energy levels. If the electrons in an atom are already arranged to fill their lowest energy levels and no lower levels have vacancies, then the electrons cannot drop in energy.

The movement of electrons among energy levels is similar to a person walking up a flight of stairs. To progress up the stairs, you can only go in whole numbers of steps at a time. You cannot raise yourself up between steps.

Study Exercise 4.4
Calculate the maximum number of electrons that can exist in the following principal energy levels:

a. 3 (18) **b.** 5 (50)

Study Exercise 4.5
Diagram the atomic structure for each of the following atoms. Indicate the number of protons and neutrons and arrange the electrons in principal energy levels.

Principal
energy level
1 2 3

a. $^{13}_{6}C$ $\left(\overset{6p}{\underset{7n}{\bigcirc}} \quad 2e^- \quad 4e^- \right)$

b. $^{37}_{17}Cl$ $\left(\overset{17p}{\underset{20n}{\bigcirc}} \quad 2e^- \quad 8e^- \quad 7e^- \right)$

Work Problems 11 through 13.

Study Hint: This filling of electrons is analogous to a basketball game in a gymnasium. On the court at any one time there can be only 10 players and 2 referees. If as the spectators arrive to watch the game, they are seated in *order* in the lower bleacher rows first and then as these seats fill, in the higher bleacher rows.

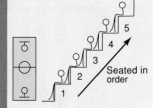

If the person in row 5 wishes to move down to row 4, he or she cannot because row 4 is filled. The court is analogous to the nucleus, and the people filling the bleacher rows are analogous to the electrons filling the principal energy levels with increasing energy the further the distance from the nucleus.

4.7 Electron-Dot Formulas of Elements

The electrons in the highest *principal* energy level in the preceding diagrams of the atom are usually called valence energy level electrons or **valence electrons**. The remainder of the atom (nucleus and other electrons) is called the *core* (kernel). The electrons in the valence energy level have higher energies than the inner electrons and are *gained, lost,* or *shared* when an atom unites or reacts with another atom to form a molecule or ion. These valence electrons are the most reactive electrons and are the electrons shown in electron-dot formulas. These electron-dot formulas do not represent a physical description of the position of the electrons. Rather, they are a "bookkeeping" device, a schematic representation of the electronic properties of the given atoms. This may be likened to keeping track of your calculator, pencils, and Magic Markers by placing them on your desk in a certain order.

Valence electrons Electrons occupying the highest principal energy level in an atom.

Rules for Writing Electron-Dot Formulas

To write electron-dot formulas of elements, you need only follow four simple rules:

1. Write the symbol for the element to represent the *core.*
2. Assign a maximum of *two* electrons to each of the four sides of the symbol to give a total of eight electrons around the symbol. A dot represents a single electron.
3. Arrange the valence electrons (highest principal energy level) around the four sides of the symbol, with *one* electron assigned to *each side up to a maximum of four electrons.*
4. If needed, pair up electrons on the four sides *up to a maximum of eight electrons.* Be sure not to exceed the actual number of valence electrons for the element. (Helium is an exception, with both valence electrons shown on the same side of the symbol, since it has completed principal energy level 1.)

Consider the electron-dot formulas for the following atoms (in each, be sure to determine the number of valence electrons):

1. $_1^1\text{H} = \text{H}\cdot$ or H, and so on (1 valence electron; the four sides are equivalent)
2. $_2^4\text{He} = \text{He}\colon$ or $\overset{\cdot\cdot}{\text{He}}$, and so on (exception—see rule 4, 2 valence electrons)
3. $_3^7\text{Li} = \text{Li}\cdot$ (1 valence electron)
4. $_5^{11}\text{B} = \cdot\text{B}\cdot$ or $\overset{\cdot}{\text{B}}\cdot$, and so on (3 valence electrons)
5. $_6^{12}\text{C} = \cdot\overset{\cdot}{\text{C}}\cdot$ (4 valence electrons)
6. $_7^{15}\text{N} = \cdot\overset{\cdot}{\text{N}}\colon$ (5 valence electrons, 2 paired up)
7. $_{10}^{20}\text{Ne} = \colon\overset{\cdot\cdot}{\text{Ne}}\colon$ (8 valence electrons, all sides filled)
8. $_{12}^{24}\text{Mg} = \text{Mg}\cdot$ (2 valence electrons)
9. $_{16}^{32}\text{S} = \cdot\overset{\cdot}{\text{S}}\colon$ (6 valence electrons)
10. $_{17}^{35}\text{Cl} = \colon\overset{\cdot\cdot}{\text{Cl}}\cdot$ (7 valence electrons)

Rule of eight (octet rule) Principle that, in forming molecules, most atoms attempt to obtain a stable configuration of eight valence electrons around the atom.

In the preceding examples, you may have noted that eight electrons would fill all four sides, as in the case of neon (Ne, example 7). There is a specific rule governing this, the **rule of eight** or the **octet rule.** In the formation of molecules from atoms, most atoms attempt to obtain this stable configuration of eight valence electrons around each atom. The elements helium (He), neon (Ne), argon (Ar), krypton (Kr), xenon (Xe), and radon (Rn) are called the *noble gases.* All except helium have eight valence electrons, and all (including helium, which has two valence electrons that complete its principal energy level 1) are relatively unreactive. In fact, they were once called inert gases because of their lack of reactivity, but scientists can now prepare compounds containing noble gases. We will refer to the rule of eight again in Section 6.1.

Study Exercise 4.6
Write the electron-dot formulas for the following atoms:

a. $^{28}_{14}\text{Si}$

$\left(\cdot\ddot{\underset{\cdot}{\text{Si}}}\cdot\right)$

b. $^{37}_{17}\text{Cl}$

$\left(\cdot\ddot{\underset{\cdot\cdot}{\text{Cl}}}{:}\right)$

Work Problems 14 and 15.

4.8 Arrangement of the Electrons in Sublevels

Experiments have shown that the arrangement of electrons is not quite as simple as we have described so far. In fact, the principal energy levels are divided further into sublevels. The sublevels, labeled *s, p, d,* and *f,* also have a limit to the number of electrons that they can contain. The *s, p, d,* and *f* sublevels may contain a maximum of 2, 6, 10, and 14 electrons, respectively, as shown below and in Table 4.3.

$$s = \mathbf{2}$$
$$p = \mathbf{6}$$
$$d = \mathbf{10}$$
$$f = \mathbf{14}$$

(Note that 4 electrons are added each time.) As you can see in Table 4.3, the number of sublevels equals the number of the principal energy level. For example, the first principal energy level has one sublevel (*s*), the second level has two sublevels (*s* and *p*), and the third level has three sublevels (*s, p,* and *d*).

   This arrangement of electrons in sublevels of principal energy levels is analogous to the arrangement of students in rooms on the various floors of the Electra Hostel. In the Electra Hostel, on the *first* (1) floor there is just one (1) room for guests, which can accommodate a maximum of two (2) students. On the *second* (2) floor there are two (2) rooms for guests. One room can accommodate a maximum of two (2) students, and the other room can accommodate a maximum of six (6) students, with a total maximum of eight (8) students on the second floor. On the *third* (3) floor there are three (3) rooms for guests. One room can accommodate a maximum of two (2) students, the second room a maximum of six (6) students, and the third room a maximum of ten (10) students, with a total maximum of eighteen (18) students on the third floor. On the *fourth* (4) floor there are four (4) rooms for guests. One room can accommodate a maximum of two (2) students, the second room a maximum of six (6) students, the third room a maximum of ten (10) students, and the fourth room a maximum of fourteen (14) students, with a total maximum of thirty-two (32) students on the fourth floor. From the analogy, the floors in the Electra Hostel represent the principal energy levels, the rooms on each floor represent the sublevels for each principal energy level, and the students the electrons.

   Each sublevel, with its respective principal energy level, has a *different energy.* The sublevels are ordered according to increasing energy in the following list (the < symbol is read "less than"):

$1s < 2s < 2p < 3s < 3p < 4s < 3d < 4p < 5s < 4d < 5p < 6s$

$< (4f < 5d) < 6p < 7s\,(5f < 6d)$

| TABLE 4.3 | | Maximum Number of Electrons in Principal Energy Levels 1 through 7 and Their Respective Sublevels | |

PRINCIPAL ENERGY LEVEL	SUBLEVEL[a]	MAXIMUM NUMBER OF ELECTRONS	
		SUBLEVEL[a]	PRINCIPAL ENERGY LEVEL
1	s	2	2
2	s	2	8
	p	6	
3	s	2	18
	p	6	
	d	10	
4	s	2	32
	p	6	
	d	10	
	f	14	
5	s	2	50 (actually 32^b)
	p	6	
	d	10	
	f	14	
	(g)	(18)	
6	s	2	72 (actually 16^b)
	p	6	
	d	10	
	f	14	
	(g)	(18)	
	(h)	(22)	
7	s	2	98 (actually 2^b)
	p	6	
	d	10	
	f	14	
	(g)	(18)	
	(h)	(22)	
	(i)	(26)	

[a] Sublevel letters in parentheses, along with the maximum number of electrons in that sublevel, are not used in the elements currently known.

[b] This is the actual maximum number of electrons found for the elements known at present; hence, these principal energy levels are incomplete.

As you can see, there are places where a 4 sublevel is lower in energy than a 3 sublevel (4s versus 3d) or a 5 or 6 sublevel is lower in energy than a 4 sublevel (5s versus 4d, 6s versus 4f).

In filling the sublevels, the *lowest-energy* sublevel is filled *first*. Figure 4.5 shows how, as sublevels become filled, the next lowest sublevel is then occupied. Figures 4.6 and 4.7 give two simplified ways of remembering the order of filling. You may use either one; they both give the same results. Follow the directions in

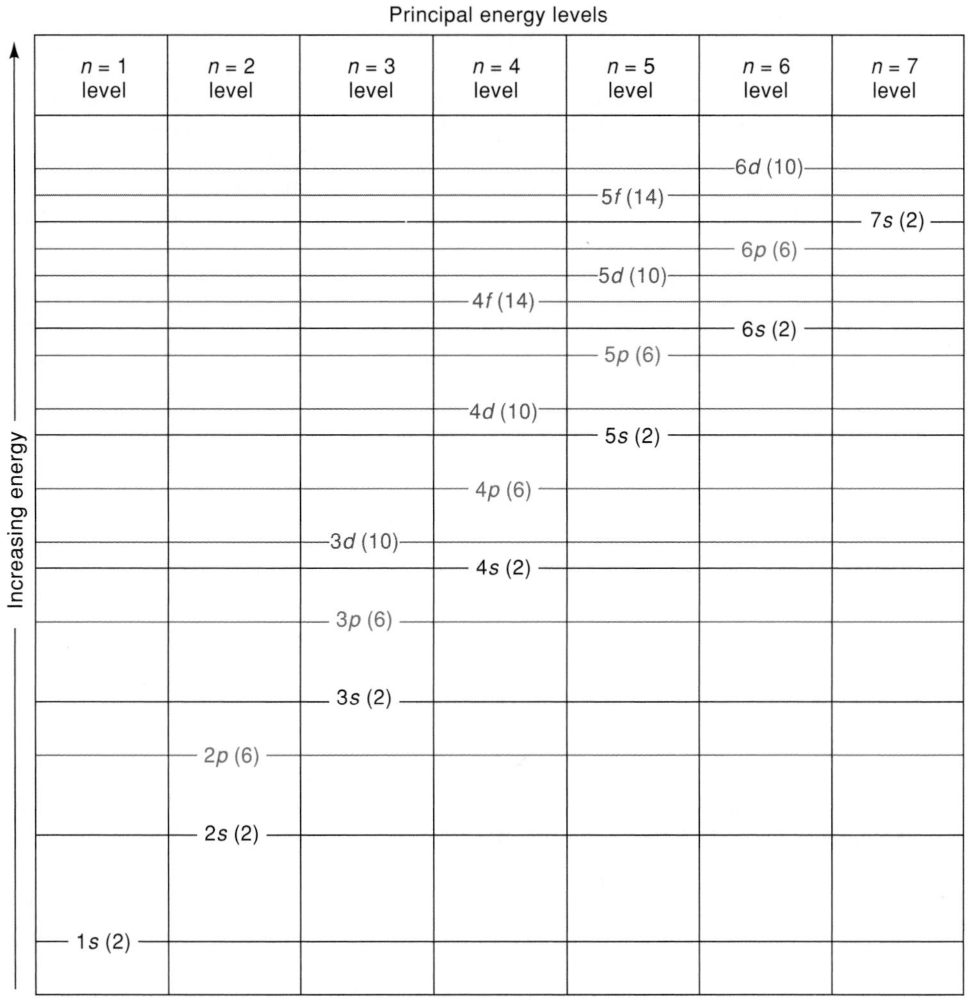

Principal energy levels

FIGURE 4.5
Diagram showing the relative energies of the different electronic sublevels. The numbers in parentheses are the maximum number of electrons in the sublevel. The s levels are shown in black, the p levels in red, the d levels in blue, and the f levels in green.

these figures to remember the order for filling the sublevels until you know them well. You should note that the $4s$ sublevel fills before the $3d$ sublevel. You should also note that the $4f$ and $5d$ and $5f$ and $6d$ sublevels appear in parentheses in the above order because the energies of these pairs of sublevels are quite close. From Figure 4.6 or 4.7, we see that after filling the $6s$, the $4f$ is skipped and then one electron is placed in the $5d$. We then go back and fill the $4f$. Once the $4f$ is filled we return to the $5d$ and completely fill it to its maximum of 10 electrons. The process repeats itself for the $6d$ and $5f$ sublevels. That is, after filling the $7s$, the $5f$ is skipped and then one electron is placed in the $6d$. We then go back and fill the $5f$. Once the $5f$ is filled we return to the $6d$. This return to fill the $6d$ represent the elements rutherfordium (Rf, 104), hahnium (Ha, 105), seaborgium (Sg, 106), nielsbohrium (Ns, 107), hassium (Hs, 108), meitnerium (Mt, 109), and the as yet unnamed elements 110 and 111. As new elements are prepared, their electronic

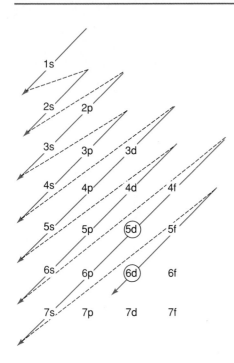

FIGURE 4.6
Order of filling the sublevels. (1) On separate lines, write the principal energy levels with their sublevels to the *f* sublevel. (2) Draw diagonal lines that follow the order of filling. (The diagonal lines need not be extended beyond the 6*d* sublevel because no elements at present have been prepared that have an electronic configuration beyond the 6*d* sublevel). (3) Circle the ⑤*d* and ⑥*d* sublevels because only one electron is placed in each of these sublevels initially. After filling the 6*s*, skip the 4*f* and place *one* electron in the 5*d*; then go back to the 4*f* and fill it. Once the 4*f* is filled, return to the 5*d* and completely fill it to its maximum of 10 electrons. Repeat the process for the 6*d* and 5*f* sublevels. After filling the 7*s*, skip the 5*f* and place *one* electron in the 6*d*; then go back to the 5*f* and fill it. Once the 5*f* is filled, return to the 6*d*.

configurations will fill the 6*d* to its maximum of 10 electrons. Exceptions to the order of filling of sublevels do occur, but we do not consider them in this text.

This order of filling of sublevels is not a random or contrived scheme. It is based upon *experimental observations* and *physical measurements* that have led to a standard procedure for generating the electronic configurations of atoms.

When writing the sublevel electron configuration of an atom, write the principal energy level *number* and the sublevel *letter*. Then write the number of electrons in the sublevel as a superscript. The sublevels of a given principal energy level may be *grouped together* or grouped as *they are filled*. In generating these configurations, draw one of the diagrams in either Figure 4.6 or 4.7 and then follow the procedure for filling the sublevels.

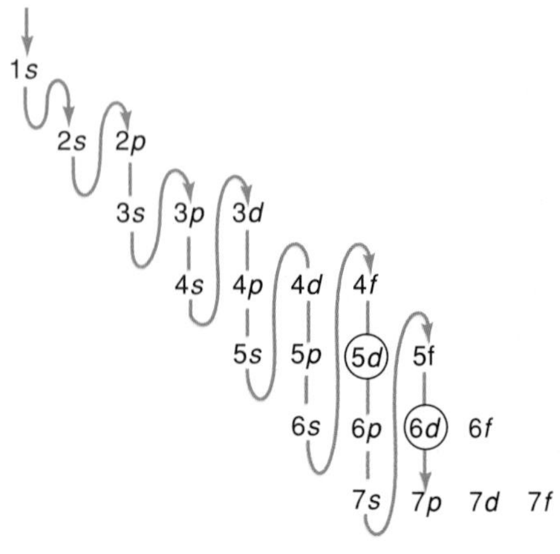

FIGURE 4.7
Order of filling the sublevels. (1) Write the principal energy levels with their sublevels to the *f* sublevel on separate lines. Place the next higher *s* sublevel *directly under* the lower *p* sublevel. (2) Draw curved lines that follow the order of filling. (The curved lines need not be extended beyond the 6*d* sublevel because no elements at present have been prepared that have an electronic configuration beyond the 6*d* sublevel). (3) Circle the ⑤*d* and ⑥*d* sublevels because only one electron is placed in each of these sublevels initially. After filling the 6*s*, skip the 4*f* and place *one* electron in the 5*d*; then go back to the 4*f* and fill it. Once the 4*f* is filled, return to the 5*d* and completely fill it to its maximum of 10 electrons. Repeat the process for the 6*d* and 5*f* sublevels. After filling the 7*s*, skip the 5*f* and place *one* electron in the 6*d*; then go back to the 5*f* and fill it. Once the 5*f* is filled, return to the 6*d*.

Consider the following atoms:

1. $^{1}_{1}\text{H} = 1s^{1}$

   — number of electrons in that sublevel
   — sublevel
   — principal energy level

   (**1** valence electron)

2. $^{4}_{2}\text{He} = 1s^{2}$: principal energy level 1 is now filled (**2** valence electrons)
3. $^{7}_{3}\text{Li} = 1s^{2}, 2s^{1}$: (**1** valence electron)
4. $^{11}_{5}\text{B} = 1s^{2}, 2s^{2}\, 2p^{1}$: the maximum in the $2s$ is 2, and so we next fill the $2p$ (**3** valence electrons; see Section 4.7)
5. $^{14}_{7}\text{N} = 1s^{2}, 2s^{2}\, 2p^{3}$: (**5** valence electrons)
6. $^{20}_{10}\text{Ne} = 1s^{2}, 2s^{2}\, 2p^{6}$: principal energy level 2 is now complete (**8** valence electrons)
7. $^{24}_{12}\text{Mg} = 1s^{2}, 2s^{2}\, 2p^{6}, 3s^{2}$: (**2** valence electrons)
8. $^{29}_{14}\text{Si} = 1s^{2}, 2s^{2}\, 2p^{6}, 3s^{2}\, 3p^{2}$: (**4** valence electrons)
9. $^{37}_{17}\text{Cl} = 1s^{2}, 2s^{2}\, 2p^{6}, 3s^{2}\, 3p^{5}$: (**7** valence electrons)
10. $^{39}_{19}\text{K} = 1s^{2}, 2s^{2}\, 2p^{6}, 3s^{2}\, 3p^{6}, 4s^{1}$: (**1** valence electron)

> **Study Hint:** To make it easier for you to keep track of the electrons we have used commas in the sublevel electronic configuration.

The next energy level after the $3p$ is the $4s,$ and so we go to that level before the $3d$.

11. $^{64}_{30}\text{Zn} = 1s^{2}, 2s^{2}\, 2p^{6}, 3s^{2}\, 3p^{6}\, 3d^{10}, 4s^{2}$ (**2** valence electrons); the $3d$ sublevel fills after the $4s$. We can group the $3d$ sublevel with the other sublevels of principal energy level 3, regardless of the order of filling, or *equally acceptable* is the electronic configuration $1s^{2}, 2s^{2}\, 2p^{6}, 3s^{2}\, 3p^{6},$ $4s^{2}, 3d^{10},$ following the order of filling. Using the first method, it is a little easier to identify the number of valence electrons.
12. $^{75}_{33}\text{As} = 1s^{2}, 2s^{2}\, 2p^{6}, 3s^{2}\, 3p^{6}\, 3d^{10}, 4s^{2}\, 4p^{3}$ (**5** valence electrons: the $4p$ sublevel fills after the $3d$. Or $1s^{2}, 2s^{2}\, 2p^{6}, 3s^{2}\, 3p^{6}, 4s^{2}, 3d^{10}, 4p^{3}$.
13. $^{138}_{56}\text{Ba} = 1s^{2}, 2s^{2}\, 2p^{6}, 3s^{2}\, 3p^{6}\, 3d^{10}, 4s^{2}\, 4p^{6}\, 4d^{10}, 5s^{2}\, 5p^{6}, 6s^{2}$ (**2** valence electrons); the $5s$ sublevel fills after the $4p$, then the $4d$, $5p$ fills, and the $6s$ fills last. Or $1s^{2}, 2s^{2}\, 2p^{6}, 3s^{2}\, 3p^{6}, 4s^{2}, 3d^{10}, 4p^{6}, 5s^{2}, 4d^{10}, 5p^{6},$ $6s^{2}$.
14. $^{158}_{64}\text{Gd} = 1s^{2}, 2s^{2}\, 2p^{6}, 3s^{2}\, 3p^{6}\, 3d^{10}, 4s^{2}\, 4p^{6}\, 4d^{10}\, 4f^{7}, 5s^{2}\, 5p^{6}\, 5d^{1}, 6s^{2}$ (*usually* **2** valence electrons); after filling the $6s,$ skip the $4f,$ place *one* electron in the $5d,$ and then place *seven* electrons in the $4f$. All sublevels of the same principal energy level can be grouped together, or $1s^{2}, 2s^{2}$ $2p^{6}, 3s^{2}\, 3p^{6}, 4s^{2}, 3d^{10}, 4p^{6}, 5s^{2}, 4d^{10}, 5p^{6}, 6s^{2}, 5d^{1}, 4f^{7}$.

Appendix IV gives the electronic configuration for all the elements, including the exceptions to the order of filling.

The filling of the sublevels correlates with the periodic table as Figure 4.8 shows. Note that there are blocks of elements that fill just the s sublevels, those that fill just the p sublevels, those that fill just the d sublevels, and finally those that fill just the f sublevels. When an element is in a particular block, this means that the last electron placed in that atom occupies the sublevel corresponding to that block. For example, titanium ($^{48}_{22}\text{Ti}$) has the electronic configuration $1s^{2}, 2s^{2}\, 2p^{6}, 3s^{2}\, 3p^{6}, 4s^{2},$ $3d^{2}$. Titanium is the second element in the d block, and so its last electron is the $3d^{2}$

> **Study Hint:** Study this material and Figure 4.8 carefully. They will help you understand a lot of the material discussed in this and subsequent chapters.

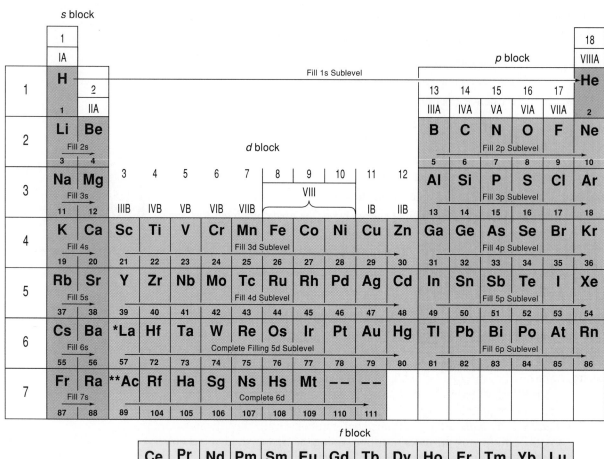

FIGURE 4.8
Correlation of the filling of the sublevels with the periodic table. Note the groups of elements that fill the *s, p, d,* and *f* sublevels. The periodic table indicates these groups of elements as follows: ■, *s* block; ■, *p* block; ■, *d* block; □, *f* block. The numbers at the bottom of each box refer to the atomic number of the element.

electron. Consider another example, that is, Gd (example 14 above). Gd is the seventh element in the *f* block, and so its last electron is the 4*f* electron. From example 14 above, Gd has a $4f^7$ electron. Hence, you can check that you have filled the sublevels correctly by comparing your answer to the periodic table.

Study Exercise 4.7

Write the electronic configuration for the sublevels of the following atoms and give the number of valence electrons in each.

		Valence electrons
a. $^{27}_{13}Al$	$[1s^2, 2s^2\,2p^6, 3s^2\,3p^1$	$(3)]$
b. $^{51}_{23}V$	$[1s^2, 2s^2\,2p^6, 3s^2\,3p^6\,3d^3, 4s^2$	(2)
	or $1s^2, 2s^2\,2p^6, 3s^2\,3p^6\,4s^2, 3d^3$	$(2)]$

Work Problems 16 and 17.

4.9 Orbitals

In the preceding discussion, we used a model of the atom that assumes that electrons exist in energy levels in discrete, definite orbital paths in accordance with the Bohr hydrogen atom model. However, chemists have modified this view of the arrangement of the electrons in atoms. We do not now visualize electrons traveling in fixed orbital paths. Instead, we consider them to occupy orbital *volumes* of space. An **orbital** is a region of space surrounding the nucleus of an atom in which there is a high probability of finding up to two electrons. Orbitals have *shape,* which is defined by a 90% probability of the two electrons being found in a specific region. The *s, p, d,* and *f* sublevel electrons are associated with the *s, p, d,* and *f* orbitals. Hence, when we define orbitals, we are simply assigning the *s, p, d,* and *f* sublevel electrons to a region in space around the nucleus. These *fast-moving* electrons have a 90% probability of being found *somewhere* within the shape of the orbitals. The orbitals are *not* hollow.

Orbital   Region of space surrounding the nucleus of an atom in which there is a high probability of finding up to two electrons.

Figure 4.9 shows the shapes of the *s* and *p* orbitals (the more complex *f* and *d* levels we leave to an advanced chemistry course). The shape of an *s* orbital is a sphere with the electrons—a maximum of two—somewhere within it. The *p* orbitals, of which there are three—p_x, p_y, p_z,—resemble dumbbells arranged on the *x, y,* and *z* axes. In each of these orbitals there are no more than two electrons; for example, $p_x = 2, p_y = 2, p_z = 2$, a total of six *p* electrons—the maximum for this sublevel.

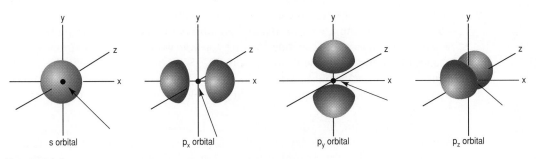

FIGURE 4.9

Shape of s, p_x, p_y, and p_z orbitals. The shape of the orbital is defined by a 90% probability of finding a maximum of two electrons somewhere within these orbitals. The arrows point to the nuclei. The p orbitals have two lobes, with the fast-moving electrons found somewhere within these lobes.

FIGURE 4.10
Relationships among the 1*s*, 2*s*, and 3*s* orbitals.

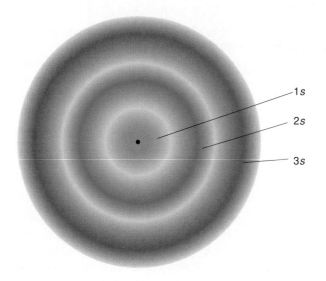

Study Hint: We use the idea of probability be-cause we can never really know *exactly* where an electron is. If you like, think of the electron as moving too fast for us to get to know.

To clarify the idea of a 90% probability of finding an electron somewhere in an orbital, suppose that every Monday afternoon you have chemistry labortory be-tween 1:00 and 4:00 in room 202. The probability of finding you in room 202 from 1:00 to 4:00 on Monday afternoon is very high, perhaps 90 times out of 100. The remaining 10 times that you are not there allows for the possibility that you might be ill (we assume you would not dream of cutting class). We say that the probabil-ity of finding you in room 202 is 90% (90 times in 100) and the probability of not finding you there is 10% (10 times in 100). Now you may be assigned a certain desk where you are to perform your experiments, but you may not always be at that desk for the entire time. You may leave your desk to get chemicals, to talk to a friend, or even to ask the laboratory instructor some questions, but you are still in room 202.

Now, suppose we checked on your exact position every 5 min for the 3-h pe-riod and placed pins in a drawing of the room wherever we found you. Assuming you completed your experiment, we would probably find most of the pins around your desk. Thus, the *region of highest probability* of finding you would be some-where near your desk. Someone looking for you would have an extremely high chance of finding you somewhere in room 202. Someone who looked in the general region of your desk would probably find you. The same reasoning applies to elec-trons: there is a 90% probability of finding the electrons somewhere within the des-ignated orbitals.

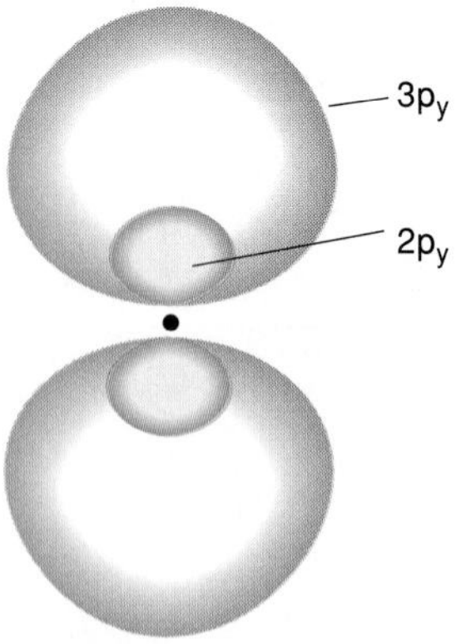

Within an atom, the various orbitals are laid out in repeating patterns. It may help you to understand the layout of the *s* orbitals shown in Figure 4.10 to think of it as a nest of balls. The 1*s* orbital is like a tennis ball suspended in the center of a volleyball (2*s* orbital), which, in turn, is suspended in the center of a basketball (3*s* orbital).

The *p* orbitals are similarly adjacent, as Figure 4.11 shows. The 2*p* orbital re-sembles a small dumbbell suspended inside a larger dumbbell, the 3*p* orbital.

FIGURE 4.11
Relationship of the 2*p*$_y$ and 3*p*$_y$ orbitals to each other.

Summary

Atoms are tiny, both in diameter and mass. Thus, it is easier to express the *relative atomic mass of an atom.* The *atomic mass scale* of the elements is based on the mass of a ^{12}C atom, which is assigned a value of exactly 12 atomic mass units (Section 4.1). Atoms are tiny, discrete particles that retain their identity throughout physical and chemical changes. They combine in simple whole-number ratios to form a molecule of a compound and in different ratios to form different compounds (Section 4.2).

Every atom is composed of three basic subatomic particles: *electrons, protons,* and *neutrons.* Electrons have a relative charge of -1, protons have a relative charge of $+1$, and neutrons have no charge (Section 4.3). The protons and neutrons are located in the nucleus of the atom, and the electrons are located outside the nucleus. The number of electrons is equal to the number of protons in neutral atoms. The number of protons in an atom is the *atomic number* of the element. The sum of the number of protons and the number of neutrons is the *mass number* (Section 4.4).

Isotopes of atoms have different mass numbers but the same atomic number. Isotopes of a given element have the same chemical properties but slightly different physical properties. Using the exact atomic mass and the percent abundance in nature of isotopes of an element, we can calculate the average atomic mass of the element (Section 4.5).

Electrons are arranged in principal energy levels, and there is a maximum number of electrons that may occupy each level. For an electron to rise one level, the atom must absorb energy. For an electron to drop one level, the atom must give off energy (Section 4.6).

We can draw an *electron-dot formula* for an element based on the number of *valence electrons* for that element. The *valence electrons* are the electrons that occupy the highest principal energy level (Section 4.7). Within any principal energy level, electrons occupy sublevels designated *s, p, d,* and *f* sublevels. It is important to know the sublevel filling order so that you can determine the *electronic configuration* of an element (Section 4.8). A more modern model of the atom identifies general regions in space called *orbitals* with each sublevel and places electrons in these orbitals (Section 4.9).

 Exercises

1. Define or explain the following terms (the number in parentheses refers to the section in the text where the term is mentioned):

   a. atomic mass (atomic weight) scale (4.1)

   b. atomic mass unit (4.1)

   c. subatomic particles (4.3)

   d. electron (4.3)

   e. proton (4.3)

   f. neutron (4.3)

   g. nucleus (4.4)

   h. ion (4.4)

   i. energy levels (4.4)

   j. atomic number (4.4)

   k. mass number (4.4)

   l. isotope (4.5)

   m. valence electrons (4.7)

   n. rule of eight (octet rule) (4.7)

   o. orbital (4.9)

2. Distinguish between

   a. subatomic particles in the nucleus and those outside the nucleus

   b. valence electrons and the core of an atom

   c. shape of s and p orbitals

   d. p_x and p_z orbitals

3. Explain the meaning of the following symbols or numbers:

   a. $^4_Z E$ b. $1s^2$

4. Draw the shape of

   a. an s orbital   b. a p_x orbital

   c. a p_y orbital   d. a p_z orbital

 # Problems

General Arrangements of Subatomic Particles (See Section 4.4)

5. For each of the following atoms, calculate the number of protons and neutrons in the nucleus and the number of electrons outside the nucleus.

   a. $^9_4 Be$ b. $^{40}_{18} Ar$ c. $^{46}_{22} Ti$

   d. $^{41}_{19} K$ e. $^{59}_{27} Co$ f. $^{96}_{44} Ru$

   (All these atoms exist, although some may not be the most abundant isotopes in nature.)

6. For each of the following atoms, calculate the number of protons and neutrons in the nucleus and the number of electrons outside the nucleus.

   a. $^{23}_{11} Na$ b. $^{63}_{29} Cu$ c. $^{105}_{46} Pd$

   d. $^{90}_{40} Zr$ e. $^{142}_{58} Ce$ f. $^{235}_{92} U$

   (All these atoms exist, although some may not be the most abundant isotopes in nature.)

Isotopes (See Section 4.5)

7. The following are properties of uranium-238. Which of these properties are the same for uranium-235?

   a. atomic mass = 238.0508 amu

   b. reacts rapidly with oxygen to form a uranium oxide (U_3O_8)

8. Boron (10.811 amu) consists of two isotopes: boron-10 (10.013 amu) and boron-11 (11.009 amu). Based on the average atomic mass of boron, which of the two isotopes is more abundant in nature?

9. Calculate the atomic mass to four significant digits for gallium, given the following data:

Isotope	Exact Atomic Mass (amu)	Abundance in Nature (%)
^{69}Ga	68.9257	60.40
^{71}Ga	70.9249	39.60

10. Calculate the atomic mass to four significant digits for antimony, given the following data:

Isotope	Exact Atomic Mass (amu)	Abundance in Nature (%)
^{121}Sb	120.9038	57.25
^{123}Sb	122.9041	42.75

Arrangement of Electrons in Principal Energy Levels (See Section 4.6)

11. Calculate the maximum number of electrons that can exist in the following principal energy levels:

    a. 1   b. 2   c. 3

    d. 4   e. 6   f. 7

12. Diagram the atomic structure for each of the following atoms. Indicate the number of protons and neutrons and arrange the electrons in principal energy levels.

    a. $^{7}_{3}$Li   b. $^{11}_{5}$B   c. $^{16}_{8}$O

    d. $^{23}_{11}$Na   e. $^{31}_{15}$P   f. $^{36}_{18}$Ar

    (All these atoms exist, although some may not be the most abundant isotopes in nature.)

13. Diagram the atomic structure for each of the following atoms. Indicate the number of protons and neutrons and arrange the electrons in principal energy levels.

    a. $^{11}_{4}$Be   b. $^{14}_{7}$N   c. $^{19}_{9}$F

    d. $^{20}_{10}$Ne   e. $^{24}_{12}$Mg   f. $^{34}_{16}$S

    (All these atoms exist, although some may not be the most abundant isotopes in nature.)

Electron-Dot Formulas of Elements (See Section 4.7)

14. Write the electron-dot formulas for the following atoms:

    a. $^{4}_{2}$He   b. $^{7}_{3}$Li   c. $^{9}_{4}$Be

    d. $^{16}_{8}$O   e. $^{19}_{9}$F   f. $^{40}_{18}$Ar

15. Write the electron-dot formulas for the following atoms:

    a. $^{12}_{6}$C   b. $^{20}_{10}$Ne   c. $^{23}_{11}$Na

    d. $^{24}_{12}$Mg   e. $^{31}_{15}$P   f. $^{32}_{16}$S

Arrangement of the Electrons in Sublevels (See Section 4.8)

16. (1) Write the electronic configuration in sublevels for the following atoms. (2) Give the number of valence electrons for each.

    a. $^{7}_{3}$Li   b. $^{9}_{4}$Be   c. $^{12}_{6}$C

    d. $^{16}_{8}$O   e. $^{31}_{15}$P   f. $^{78}_{34}$Se

17. (1) Write the electronic configuration in sublevels for the following atoms. (2) Give the number of valence electrons for each.

    a. $^{11}_{5}$B   b. $^{19}_{9}$F   c. $^{32}_{16}$S

    d. $^{51}_{23}$V   e. $^{75}_{33}$As   f. $^{112}_{48}$Cd

General Problems

18. The element osmium (Os) has the following physical properties.

$$mp = 3045°C, bp = 5027°C, density = 22.57 \text{ g/cm}^3$$

   a. Calculate its melting point in degrees Fahrenheit.
   b. Calculate its density in kg/m^3.
   c. Calculate its density in lb/ft^3 (1 lb = 454 g; 1 in. = 2.54 cm).

19. Write the electronic configuration in sublevels for an isotope of osmium (Os, atomic number 76) having a mass number of 192.

20. Antimony (atomic number 51) exists in nature as a mixture of two isotopes with mass numbers 121 and 123. What differences would you expect in their nuclear and electronic structures?

21. A single proton has a mass of 1.67×10^{-24} g. Calculate the number of protons in 1.00 lb of protons (1 lb = 454 g).

 Chapter Quiz 4

You may use the periodic table.

1. Diagram the atomic structure for the following atoms, indicating the numbers of protons and neutrons, and arrange the electrons in *principal energy* levels
   a. $^{13}_{6}C$ b. $^{28}_{14}Si$

2. Calculate the number of electrons that can exist in the following principal energy levels:
   a. 4   b. 6

3. Write the electronic configuration in sublevels for the following atoms:
   a. $^{35}_{17}Cl$ b. $^{55}_{25}Mn$

4. Write the electron-dot formulas for the following atoms:
   a. $^{31}_{15}P$ b. $^{40}_{18}Ar$

5. Calculate the atomic mass to four significant digits for element X, given the following data:

Isotope	Exact Atomic Mass (amu)	Abundance in Nature (%)
^{13}X	13.00	60.00
^{15}X	15.10	40.00

Silver (Symbol: Ag)

Silver compounds play a major role in our everyday life because they are key ingredients in photographic film.

The Element SILVER: Pretty as a Picture

Name: Silver is one of the oldest known metals (the others are gold and copper). The origins of the name for silver are not clear, but they probably derive from the German *silber* and the Old English *siolfor*. The symbol (Ag) comes from the Latin *argentum*.

Appearance: A lustrous, bright silver metal that is an excellent conductor of heat and electricity.

Occurrence: Sometimes silver occurs in the pure (uncombined) state in nature, although most of these deposits have been exhausted. Argentite (Ag_2S) is a valuable silver-rich ore.

Source: Currently, most silver is obtained as a by-product during the processing of copper, lead, and zinc containing ores [*argentites* $(Cu, Fe, Zn, Ag)_{12}Sb_4S_{13}$].

Its Role in Our World: The most important use of silver containing salts is in the photographic industry. Silver bromide (AgBr) and silver iodide (AgI) are used extensively in preparing photographic films.

The second biggest silver consumer is the electronics industry where silver and silver alloys (silver mixed with cadmium, copper, palladium, or gold) are used as electric contacts, solders, and brazing alloys.

Silver salts are also used in the manufacture of silver-based batteries. Such diverse devices as torpedoes, watches, aircraft, and rockets use batteries containing silver oxide (Ag_2O) and zinc.

Silver is used in jewelry, sterling silverware, mirrors, and dental fillings. Although silver was once used in coins and as an exchange medium, coins now account for less than 5% of world silver consumption.

Unusual Facts: Very dilute solutions containing silver salts are excellent disinfectants. Historically, humans have taken advantage of this property by storing food in silver containers to retard the rate of spoilage.

<table>
<tr><td>

CHAPTER 5

Melodies are often repeated in musical compositions. You hear a particular musical pattern repeated at different times during the song. Similarly, the chemical properties of the elements repeat in a periodic fashion when they are ordered by atomic number, as in the periodic table.

GOALS FOR CHAPTER 5

1. To define the term *periodic law* and to explain what this law means in terms of properties of elements (Section 5.1).

2. To describe the structure of the periodic table in terms of its periods and groups and distinguish between the representative and transition elements (Section 5.2).

3. To use the periodic table to identify common patterns within groups of elements (Section 5.3).

</td></tr>
</table>

The Periodic Classification of the Elements

Countdown

5. Chlorine has an atomic radius of 9.9×10^{-11} m. Convert this atomic radius to picometers (Section 2.2 and 2.8).
   (99 pm)

4. Chlorine has a melting point of $-101°C$. Convert this temperature to degrees Fahrenheit and to kelvins (Section 2.8).
   ($-15\overline{0}°F$, 172 K)

3. The following are properties of the element chlorine (Cl). Classify them as physical or chemical properties (Section 3.4).
   a. greenish-yellow gas   (physical)
   b. at 10°C, 1 volume of water dissolves 3.1 volumes of chlorine   (physical)
   c. reacts with hydrogen to form hydrogen chloride gas (HCl)   (chemical)
   d. density of 1.56 g/mL as a liquid   (physical)

2. For each of the following atoms, calculate the number of protons and neutrons in the nucleus and the number of electrons outside the nucleus (Section 4.4).

   a. $^{37}_{17}Cl$ $\left(\left(\begin{smallmatrix}17p\\20n\end{smallmatrix}\right)17e^-\right)$

   b. $^{133}_{55}Cs$ $\left(\left(\begin{smallmatrix}55p\\78n\end{smallmatrix}\right)55e^-\right)$

1. Write the electronic configuration in sublevels for the following atoms and give the number of valence electrons in each (Section 4.8).
   a. $^{35}_{17}Cl$ $[1s^2, 2s^2\, 2p^6, 3s^2\, 3p^5\ (7)]$
   b. $^{51}_{23}V$ $[1s^2, 2s^2\, 2p^6, 3s^2\, 3p^6\, 3d^3, 4s^2$ or $1s^2, 2s^2\, 2p^6, 3s^2\, 3p^6, 4s^2, 3d^3\ (2)]$

The 1800s was a time of enormous scientific discovery. By 1830, chemists had identified 55 elements and were struggling with different ways of categorizing them. The ultimate result of their work—the periodic table—is already somewhat familiar to you. In Section 3.7, we mentioned the periodic table in connection with the differences among the metals, the nonmetals, and the metalloids (semimetals). In Section 4.8, we showed how the periodic table correlates with the filling of the electronic sublevels of the different elements. In this chapter, we will consider the classification of the elements in the periodic table and some general characteristics of groups of elements.

5.1 The Periodic Law

The urge to group things—to find common connections—is a human drive. Since the beginning of recorded history, biologists and philosophers, mathematicians, historians, and even chemists have attempted to create sense out of the universe by such groupings. For example, one of the classifications of mammals is the cat family, whose characteristics are a round head, 28 to 30 teeth, eyes with vertically slit pupils, and retractable claws. The cat family includes not only domestic cats but also lions, tigers, leopards, jaguars, and bobcats, to name just a few. All have the *same* general characteristics mentioned. Similarly, many of the elements have general characteristics that classify them as belonging to a particular group or family.

The periodic table we know today originated in the work of two chemists who independently classified the elements known at that time. Lothar Meyer (1830–1895), a German, published an incomplete periodic table in 1864 and an extended version with 56 elements in 1869. That same year, Dmitri Mendeleev (1834–1907), a Russian, presented a paper describing a periodic table (see Figure 5.1). Mendeleev went further than Meyer in that he left gaps in his table and predicted that new elements would be discovered to fill them. He also predicted the properties of these yet undiscovered new elements—truly a bold undertaking in science. Mendeleev lived to see the discovery of some of the elements he predicted, with properties similar to those he had forecast.

Meyer's and Mendeleev's periodic tables differed from today's periodic table in some ways because they ordered the elements by *increasing atomic mass*. After the discovery of the proton, Henry G. J. Moseley (1888–1915), a British physicist, determined the nuclear charge on the atoms of the elements and concluded that the elements should be arranged by *increasing atomic number*. With this new arrangement of the periodic table, the discrepancies disappeared.

When the elements are arranged in order of increasing atomic number, those with similar chemical properties occur at regular, periodic intervals. This relationship is known as the **periodic law**. For example, in Figure 5.2, note that all the elements with the same number and kind of valence electrons are located in the same vertical column. That is, both Be and Mg have two valence electrons in an *s* sublevel (Section 4.8). The noble gases (He, Ne, Ar; see Section 4.7) all appear in the same vertical column, and all have eight electrons in their highest energy level (following the rule of eight), except helium, which has two (a completed first energy level).

Study Hint: Elements in the same vertical column have similar chemical properties. This is somewhat analogous to people in the same family looking alike! Do you somewhat resemble your sister or brother?

Periodic law Appearance of elements with similar chemical properties at regular, periodic intervals when they are listed in order of increasing atomic number.

FIGURE 5.1
Section of a manuscript of "Essay on the System of Elements," by Dmitri I. Mendeleev, dated February 17, 1869. In this preliminary version the periods were vertical and the groups were horizontal. Note the absence of the noble gases, which were unknown at the time. The elements were placed in order of increasing atomic masses, but in the case of discrepancies similarities in the chemical properties were used to place the elements. A discrepancy occurs with Te and I (see lines that begin with O = 16 and F = 19, respectively). Based on increasing atomic masses I should be where Te is and Te should be where I is. Mendeleev chose to reverse this because of the chemical properties of Te and I, and hence Te is with O, S, and Se, and I is with F, Cl, and Br. Note also the various question marks.

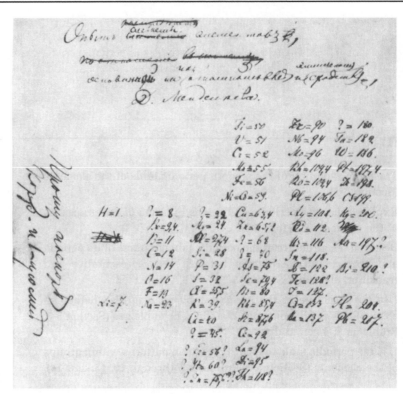

FIGURE 5.2
Abbreviated periodic classification of the elements, based on atomic number. Similar chemical properties recur at definite intervals. (The numbers in blue represent the atomic numbers of the elements.)

H							He
1							2
Li	Be	B	C	N	O	F	Ne
3	4	5	6	7	8	9	10
Na	Mg	Al	Si	P	S	Cl	Ar
11	12	13	14	15	16	17	18

Period   One of the seven horizontal rows in the periodic table.

Group (family)   One of the 18 vertical columns in the periodic table.

5.2 The Periodic Table. Periods and Groups

Following the periodic law and completing our abbreviated classification of the elements begun in Figure 5.2, we obtain a complete modern periodic table, as shown in Figure 5.3 and on the inside front cover of this text. This periodic table, first proposed in 1895 by Julius Thomsen (1826–1909), a Danish chemist, is arranged in 7 horizontal rows called **periods** and 18 vertical columns called **groups (families)**. The number of the period gives the number of the outermost principal energy level that the electrons *begin* to fill, while the elements in a given group resemble each other in that they have similar chemical properties.

GROUPS

PERIODS	1 IA	2 IIA	3 IIIB	4 IVB	5 VB	6 VIB	7 VIIB	8	9 VIII	10	11 IB	12 IIB	13 IIIA	14 IVA	15 VA	16 VIA	17 VIIA	18 VIIIA
1	H 1																	He 2
2	Li 3	Be 4											B 5	C 6	N 7	O 8	F 9	Ne 10
3	Na 11	Mg 12				TRANSITION ELEMENTS							Al 13	Si 14	P 15	S 16	Cl 17	Ar 18
4	K 19	Ca 20	Sc 21	Ti 22	V 23	Cr 24	Mn 25	Fe 26	Co 27	Ni 28	Cu 29	Zn 30	Ga 31	Ge 32	As 33	Se 34	Br 35	Kr 36
5	Rb 37	Sr 38	Y 39	Zr 40	Nb 41	Mo 42	Tc 43	Ru 44	Rh 45	Pd 46	Ag 47	Cd 48	In 49	Sn 50	Sb 51	Te 52	I 53	Xe 54
6	Cs 55	Ba 56	*La 57	Hf 72	Ta 73	W 74	Re 75	Os 76	Ir 77	Pt 78	Au 79	Hg 80	Tl 81	Pb 82	Bi 83	Po 84	At 85	Rn 86
7	Fr 87	Ra 88	**Ac 89	Rf 104	Ha 105	Sg 106	Ns 107	Hs 108	Mt 109	-- 110	-- 111							

*Lanthanide series	Ce 58	Pr 59	Nd 60	Pm 61	Sm 62	Eu 63	Gd 64	Tb 65	Dy 66	Ho 67	Er 68	Tm 69	Yb 70	Lu 71
**Actinide series	Th 90	Pa 91	U 92	Np 93	Pu 94	Am 95	Cm 96	Bk 97	Cf 98	Es 99	Fm 100	Md 101	No 102	Lr 103

FIGURE 5.3

Periodic table of the elements. The numbers below the symbols for the elements represent their atomic numbers. Black numbers are the representative elements, purple blocked numbers are the transition elements, red blocked numbers are the lanthanide series, and green blocked numbers are the actinide series.

Because the groups have similar properties, they also have special names (like the cat family in our example of the classification of the mammals). Group IA (1) elements (except hydrogen) are called the **alkali metals**. Hydrogen, although present in group IA (1), is not considered with the alkali metals because not all its properties resemble those of the alkali metals. The elements in group IIA (2) are called the **alkaline earth metals**; those in group VIA (16) are called the **chalcogens** (kal´-co-jens); those in group VIIA (17) are called the **halogens**; and those in group VIIIA (18) are called the **noble gases**. Recently, this table has been modified by using numbers for the 18 vertical columns in addition to Roman numerals. These numbers are shown above the Roman numerals in the period table (see Figure 5.3 and on the inside front cover of this text). When we refer to a group (vertical column), we use the Roman numeral and in parentheses the column number. The elements from left to right in a given period vary gradually from having very metallic properties, such as sodium (Na), to nonmetallic properties, such as chlorine (Cl). At the end of each period is group VIIIA (18), the noble gases.

Now, let us consider in detail each of the seven periods (horizontal rows). Follow this discussion by referring to the periodic table (Figure 5.3). Follow the addition of electrons by referring to Figure 4.8.

Alkali metals All the elements in group IA (1) of the periodic table (*except* for hydrogen): lithium, sodium, potassium, rubidium, cesium, and francium.

Alkaline earth metals All the elements in group IIA (2) of the periodic table: beryllium, magnesium, calcium, strontium, barium, and radium.

Chalcogens All the elements in group VIA (16) of the periodic table: oxygen, sulfur, selenium, tellurium, and polonium.

Halogens All the elements in group VIIA (17) of the periodic table: fluorine, chlorine, bromine, iodine, and astatine.

Noble gases All the elements in group VIIIA (18) of the periodic table; these relatively unreactive (inert) elements include helium, neon, argon, krypton, xenon, and radon.

Lanthanide series Elements 58 to 71 in the periodic table, so named because they follow the element lanthanum. They are also called the *rare-earth* elements.

Actinide series Elements 90 to 103 in the periodic table, so named because they follow the element actinium.

Representative elements All the A group elements in the periodic table.

Transition elements All the B group elements plus the group VIII (8, 9, 10) elements in the periodic table.

Period 1 contains only two elements, hydrogen (H) and helium (He). In this period, the first principal energy level (1*s* sublevel) fills with two electrons. Helium is placed in group VIIIA (18), the **noble gases**. The number of the period gives the principal energy level number that the electrons *begin* to fill.

Period 2 contains eight elements from lithium (Li) through neon (Ne). In this period, the second principal energy level (2*s* and 2*p* sublevels) fills, resulting in a completely filled second principal energy level in neon.

Period 3 also contains eight elements, from sodium (Na) through argon (Ar), with the third principal energy level (3*s* and 3*p* sublevels *only*) filling. Argon, the last element in the period, has eight electrons in its third principal energy level. Since they contain only eight elements each, periods 2 and 3 are called the *short periods.*

Period 4 contains 18 elements, from potassium (K) through krypton (Kr). In this period, the 4*s*, 4*p*, and 3*d* sublevels fill. The 3*d* fills with scandium (Sc) through zinc (Zn).

Period 5 contains 18 elements from rubidium (Rb) through xenon (Xe). In this period, the 5*s*, 5*p*, and 4*d* sublevels fill. The 4*d* fills with yttrium (Y) through cadmium (Cd).

Period 6 contains 32 elements, from cesium (Cs) through radon (Rn). In this period, the 6*s*, 6*p*, 5*d*, and 4*f* sublevels fill. The 5*d* fills with lanthanum (La) and hafnium (Hf) through mercury (Hg). Elements 58 to 71, cerium (Ce) to lutetium (Lu), are called the **lanthanide series** (after lanthanum, which they follow in atomic number) and are also referred to as the *rare-earth* elements. They correspond to the filling of the 4*f* sublevel. These elements are placed at the bottom of the table for convenience, because if they were placed in the main body, the table would be extremely wide and cumbersome.

Period 7 contains at present 25 elements from francium (Fr) to the newly discovered, but yet unnamed, element 111. In this period, the 7*s*, 6*d*, and 5*f* sublevels fill. The 6*d* partially fills with actinium (Ac) and rutherfordium (Rf) through element 111; it is incomplete. Elements 90 to 103, thorium (Th) to lawrencium (Lr), are called the **actinide series** (after actinium, which they follow in atomic number) and correspond to the filling of the 5*f* sublevel. Again, for convenience these elements are placed at the bottom of the table. This period is incomplete. It will end with element 118 (atomic number), which will be one of the noble gases and will probably have properties like those of radon (Rn). Periods 4 through 7 are called the *long periods* because they contain more elements than the other periods.

When the Roman numeral designation is used, most of the 18 groups or families (vertical columns) are classed as A or B groups. The A group elements are called the **representative elements.** The B group elements and the group **VIII** elements (groups 8, 9, and 10) are called the **transition elements.** Lanthanum (La) plus the lanthanide series and actinium (Ac) plus the actinide series are classed as transition elements in group **IIIB** (3).

The gradual change from metallic to nonmetallic properties from left to right within a given period is more evident in the representative elements than in the transition elements. The transition elements are all metals and have one or two electrons in their outermost level (an *s* sublevel). In addition, they also have electrons in the next lower *d* sublevel or the *f* sublevel that lies below that. These *d* and *f* electrons are very close in energy to the valence electrons. In this respect they differ markedly from the representative elements, whose *d* or *f* electrons are *not* similar in energy to the valence electrons. There are six transition series, each linked to periods in the periodic table and to the electron sublevels in the elements. The first transition series, which comprises elements 21 (Sc) through 30 (Zn), corresponds to the filling of the 3d sublevel. The second transition series, which includes elements 39 (Y) through 48 (Cd), corresponds to the filling of the 4d sublevel. The third transition series, elements 57 (La) and 72 (Hf) through 80 (Hg), corresponds to the filling of the 5d sublevel. The fourth transition series, 89 (Ac) and 104 (Rf) through the yet unnamed element 111, corresponds to the partial filling of the 6d sublevel. The fifth transition series [elements 58 (Ce) through 71 (Lu)] and the sixth transition series [elements 90 (Th) through 103 (Lr)] fill the 4f and 5f sublevels, respectively.

5.3 General Characteristics of the Groups

The use of the periodic table to correlate the general characteristics of the elements is one of the fundamental principles of chemistry. We will consider here five general characteristics of groups.

First, the periodic table separates the *metals* from the *nonmetals* with a solid *colored* stair step line. To the right of this line are the nonmetals, and to the left are the metals, with the most metallic elements on the *extreme lower left*. As you can see, most of the elements are considered metals. Elements that lie on either side of the *colored* stair step line are called **metalloids (semimetals)**, except for aluminum (Al), which is not a metalloid but a metal because it has mostly metallic properties. Metalloids have both metallic and nonmetallic properties. They are boron, silicon, germanium (Ge), arsenic, antimony, tellurium (Te), polonium (Po), and astatine (AL).

Metalloids (semimetals)   Elements (except aluminum) that lie on either side of the *colored* stair step line in the periodic table; they have both metallic and nonmetallic properties: boron, silicon, germanium, arsenic, antimony, tellurium, polonium, and astatine.

Work Problems 3 and 4.

Study Exercise 5.1
Using the periodic table, classify each of the following elements as a metal, nonmetal, or metalloid (semimetal):

a. silicon   (metalloid, semimetal)
b. chromium   (metal)

Second, in the *A group* elements (representative elements), the *group Roman numeral* or the *unit digit* in the numbered vertical column gives the number of valence electrons (see Section 4.7). For example, sodium (Na) is in group IA (*1*) because it has *1* valence electron ($1s^2, 2s^2 2p^6, 3s^1$). Aluminum is in group IIIA (*13*) because it has *3* valence electrons ($1s^2, 2s^2 2p^6, 3s^2 3p^1$). Sulfur (S) is in group VIA (*16*) because it has *6* valence electrons ($1s^2, 2s^2 2p^6, 3s^2 3p^4$). Neon (Ne) is in group VIIIA (*18*) because it has *8* valence electrons ($1s^2, 2s^2 2p^6, 3s^2 3p^6$). Helium (He), which is in group VIIIA (*18*) but has only 2 valence electrons ($1s^2$), is an exception to this rule. This general characteristic does not hold for the transition elements either [B group elements and group VIII elements (8, 9, and 10)].

Study Exercise 5.2
Using the periodic table, indicate the number of valence electrons for each of the following elements:

a. calcium   (2)   **b.** bromine   (7)

Work Problems 5 and 6.

Third, elements in the same group have *similar chemical properties* and *similar electronic configurations*. For example, all the alkali metals [group IA (1)] react rapidly with chlorine to form the metal chloride, MCl (see Section 3.4 and Figure 5.4). All of them have a similar electronic configuration in the valence energy level (s^1). The members of this group differ only in the number of core electrons, as shown below:

Li $\quad 1s^2, 2s^1$

Na $\quad 1s^2, 2s^2\, 2p^6, 3s^1$

K $\quad 1s^2, 2s^2\, 2p^6, 3s^2\, 3p^6, 4s^1$

Rb $\quad 1s^2, 2s^2\, 2p^6, 3s^2\, 3p^6\, 3d^{10}, 4s^2\, 4p^6, 5s^1$

Cs $\quad 1s^2, 2s^2\, 2p^6, 3s^2\, 3p^6\, 3d^{10}, 4s^2\, 4p^6\, 4d^{10}, 5s^2\, 5p^6, 6s^1$

Fr $\quad 1s^2, 2s^2\, 2p^6, 3s^2\, 3p^6\, 3d^{10}\, 4f^{14}, 5s^2\, 5p^6\, 5d^{10}, 6s^2\, 6p^6, 7s^1$

Because the electronic configurations of the elements in a group are similar, the formulas of compounds of elements in that group are also similar. Sodium hydroxide has the formula NaOH; hence, the formula for cesium (Cs) hydroxide is CsOH because cesium is in the same group as sodium. If there is an exception to this similarity of chemical properties in a given group, it is usually in the first element of the group. For example, lithium (Li) is not as similar to sodium in chemical properties as sodium is to potassium (K). Also, boron is not as similar to aluminum as aluminum is to gallium (Ga). In other words, if one of the elements in a group is "out of step," it is usually the first one in the group.

Study Exercise 5.3
Given are the electronic configurations for four elements. Pair the elements you would expect to show similar chemical properties.

a. $1s^2, 2s^2\, 2p^6, 3s^2\, 3p^6\, 3d^{10}, 4s^2\, 4p^4$

b. $1s^2, 2s^2\, 2p^6, 3s^2\, 3p^6\, 3d^{10}, 4s^2\, 4p^3$

c. $1s^2, 2s^2\, 2p^6, 3s^2\, 3p^3$

d. $1s^2, 2s^2\, 2p^6, 3s^2\, 3p^4$

(**a** and **d**; **b** and **c**)

FIGURE 5.4
FIGURE 5.4
Elements in the same group have similar chemical properties. Lithium, sodium, and potassium [group IA (1), clockwise from the wire] are all soft metals and react rapidly with chlorine to form the metal chloride (a chemical property).

Study Exercise 5.4

Using Appendix IV, determine the chemical symbols for the electronic configurations in Study Exercise 5.3 and then check your answer by referring to the periodic table to see if you placed the elements with similar chemical properties in the same group.

(**a.** Se; **b.** As; **c.** P; **d.** S)

Work Problems 7 through 10.

Fourth, within a given *A group*, the *metallic properties* of the elements *increase* and the *nonmetallic properties decrease* with *increasing atomic number*. In group VA (15) (see Figure 5.5), the first member of the group is nitrogen, a nonmetal, the second member of the group is phosphorus, also a nonmetal, the third member of the group is antimony, a metalloid, and the last member of the group is bismuth, a metal. Since the most metallic elements are on the extreme lower left of the table and the metallic properties increase with increasing atomic number in a given A group, the most metallic stable (nonradioactive) element is found in the lower-left corner and is cesium (Cs).* The most nonmetallic element [excluding the relatively unreactive group VIIIA (18), the noble gases] is found in the upper-right corner and is fluorine (F).

Study Exercise 5.5

Using the periodic table, indicate which element in each of the following pairs of elements is more metallic:

a. arsenic or bismuth   (bismuth)
b. tin or iodine   (tin)

Work Problems 11 and 12.

Fifth, there is a gradual change *of many physical and chemical properties* within a given group with increasing atomic number. As Table 5.1 shows, in group VIIA (17), the halogens, the melting and boiling points, the densities, and the atomic radii of the elements increase as the atomic number increases. Figure 5.6

FIGURE 5.5
The metallic properties of the elements increase and the nonmetallic properties decrease as you move down a given A group with increasing atomic number. The elements in group VA (15): nitrogen, a nonmetal, is a colorless gas (right); phosphorus, a nonmetal, is a red solid (top); and antimony, a metalloid, is a gray solid (left).

*Francium (Fr) is radioactive and is unstable, decomposing to other elements. It is not considered here because of its instability.

	MELTING POINT	BOILING POINT	DENSITY	ATOMIC
ELEMENT	(°C)	(°C)[b]	(g/mL)[c]	RADIUS (pm)[d]
F	−219.6	−188.1	1.11 at bp	72
Cl	−101.0	−34.6	1.56 at bp	99
Br	−7.2	58.8	2.93 at bp	114
I	113.5	184.4	4.93 at 20°C	133

TABLE 5.1 **Some Physical Properties of the Halogens**[a]

[a] Although astatine (At) is a halogen, it is not considered in this table because it is radioactive and so unstable that it is not found in nature. Hence, an insufficient amount of it is present at any one time to allow study of its properties in detail.

[b] At 1.00 atm pressure.

[c] All densities are for the liquid state, except for iodine, whose solid-state density is given.

[d] The atomic radius of a halogen (X) is one-half the distance between the two atoms in an X—X bond.

Study Hint: This increase in atomic radius is analogous to adding coverings to a ball. Suppose we wrap a ball in tissue paper, then in newspaper, and finally in a bedsheet; we find that the size (diameter) increases each time. This is analogous to adding principal energy levels.

shows the physical states of chlorine, bromine, and iodine at room temperature and ordinary pressure. Chlorine is a gas under these conditions, while bromine is a liquid and iodine is a solid. Compare these physical states with the boiling points given in Table 5.1. There is an increase in atomic radius with an increase in atomic number within a given group because a new principal energy level is being filled as we go down the group to the next period. Thus, the radius of the atom increases as shown in Figure 5.7. The chemical reactivity of the halogens also changes gradually as we go down the group. Fluorine is the most reactive, followed by chlorine, bromine, and iodine.

(a)

(b)

(c)

FIGURE 5.6
Physical states of chlorine, bromine, and iodine at room temperature and ordinary pressure. (a) Chlorine, a gas; (b) bromine, a liquid; (c) iodine, a solid.

Study Exercise 5.6
Using the periodic table, indicate which element in each of the following pairs of elements has the greater atomic radius:

a. antimony and phosphorus (antimony)
b. nickel and platinum (platinum)

Work Problems 13 and 14.

 Summary

The *periodic table* is a listing of the elements by increasing *atomic number.* Arranging the elements in this way causes elements with similar chemical properties to recur at regular intervals, the *periodic law* (Section 5.1).

The periodic table is arranged in **7** horizontal rows called *periods* and **18** vertical columns called *groups (families).* The number of a period gives the principal energy level number that the electrons *begin* to fill. Elements within a group have similar chemical properties. Elements in a period vary from very metallic [group IA (1)] to nonmetallic [group VIIIA (18)]. The *representative elements* include the A group elements. The *transition elements* consist of the B group elements and group VIII elements (groups 8, 9, and 10). Special names are given to some of the groups of elements: group IA (1), the *alkali metals*; group IIA (2), the *alkaline earth metals*; group VIA (16), the *chalcogens*; group VIIA (17), the *halogens*; and group VIIIA (18), the *noble gases* (Section 5.2).

Because each group of elements exhibits similar chemical properties, we can use the periodic table to predict the general characteristics and estimate the properties of the elements. The colored stair step line separates the metals from the nonmetals, with the metals falling to the left of the line and the nonmetals falling to the right of the line. Except for aluminum, all elements that lie on either side of this stair step line are called *metalloids (semimetals).*

For the A group elements in the periodic table the group Roman numeral is equal to the number of *valence electrons.* For group VIIIA (18) elements, the number of valence electrons is eight, except for helium, which has only two. Elements in the same group have similar chemical properties and similar electronic configurations. The metallic properties increase within a given A group with increasing atomic number, and nonmetallic properties correspondingly decrease. There is a uniform change of many physical and chemical properties within a given group with increasing atomic number. Examples of these properties are boiling point, melting point, density, and atomic radius (Section 5.3).

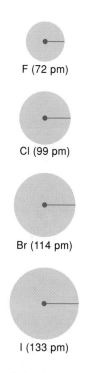

F (72 pm)

Cl (99 pm)

Br (114 pm)

I (133 pm)

FIGURE 5.7
Radii of group VIIA (17) elements (except astatine). As the atomic number increases in a given group, the radii of the atoms increase.

 Exercises

1. Define or explain the following terms (the number in parentheses refers to the section in the text where the term is mentioned):

   a. periodic law (5.1)
   b. period (5.2)
   c. group (family) (5.2)
   d. alkali metals (5.2)
   e. alkaline earth metals (5.2)
   f. chalcogens (5.2)
   g. halogens (5.2)
   h. noble gases (5.2)
   i. lanthanide series (5.2)
   j. actinide series (5.2)
   k. representative elements (5.3)
   l. transition elements (5.3)
   m. metalloids (semimetals) (5.3)

2. Distinguish between
   a. a period and a group
   b. lanthanides and actinides
   c. short and long periods
   d. representative and transition elements
   e. alkali and alkaline earth metals

✓ Problems

If in some of the following problems you are not familiar with the symbols for the elements, look them up on the inside front cover of this text.

Metals, Nonmetals, or Metalloids (See Section 5.3)

3. Using the periodic table, classify each of the following elements as a metal, nonmetal, or metalloid (semimetal):
   a. cesium   b. iridium   c. tellurium   d. selenium

4. Using the periodic table, classify each of the following elements as a metal, nonmetal, or metalloid (semimetal):
   a. chlorine   b. thallium   c. germanium   d. tin

Valence Electrons (See Section 5.3)

5. Using the periodic table, indicate the number of valence electrons for each of the following elements:
   a. potassium   b. germanium   c. tellurium   d. argon

6. Using the periodic table, indicate the number of valence electrons for each of the following elements:
   a. krypton   b. polonium   c. gallium   d. arsenic

Electronic Configuration (See Section 5.3)

7. Given are the electronic configurations for four elements. Pair the elements you would expect to show similar chemical properties.
   a. $1s^2, 2s^2\, 2p^6, 3s^2\, 3p^6\, 3d^{10}, 4s^2\, 4p^6\, 4d^{10}, 5s^2\, 5p^5$
   b. $1s^2, 2s^2\, 2p^6, 3s^2\, 3p^6, 4s^2$
   c. $1s^2, 2s^2\, 2p^6, 3s^2$
   d. $1s^2, 2s^2\, 2p^6, 3s^2\, 3p^5$

8. Using Appendix IV, determine the chemical symbols for the electronic configurations in Problem 7 and then check your answer by referring to the periodic table to see if you placed the elements with similar chemical properties in the same group.

9. Given are the electronic configurations for four elements. Pair the elements you would expect to show similar chemical properties.
   a. $1s^2, 2s^2\, 2p^6, 3s^2\, 3p^2$
   b. $1s^2, 2s^2\, 2p^6, 3s^2\, 3p^6\, 3d^{10}, 4s^2\, 4p^6\, 4d^{10}, 5s^2\, 5p^6, 6s^1$
   c. $1s^2, 2s^2\, 2p^6, 3s^2\, 3p^6, 4s^1$
   d. $1s^2, 2s^2\, 2p^6, 3s^2\, 3p^6\, 3d^{10}, 4s^2\, 4p^6\, 4d^{10}\, 4f^{14}, 5s^2\, 5p^6\, 5d^{10}, 6s^2\, 6p^2$

10. Using Appendix IV, determine the chemical symbols for the electronic configurations in Problem 9 and then check your answer by referring to the periodic table to see if you placed the elements with similar chemical properties in the same group.

Metallic Properties (See Section 5.3)

11. Using the periodic table, indicate which element in each of the following pairs of elements is more metallic:

    a. sulfur or selenium   **b.** cesium or sodium

    c. silicon or aluminum   **d.** lead or germanium

12. Using the periodic table, indicate which element in each of the following pairs of elements is more metallic:

    a. barium or calcium   **b.** magnesium or silicon

    c. silicon or lead   **d.** oxygen or polonium

Physical and Chemical Properties (See Section 5.3)

13. Using the periodic table, indicate which element in each of the following pairs of elements has the greater atomic radius:

    a. fluorine or bromine   **b.** sulfur or oxygen

    c. barium or magnesium   **d.** copper or gold

14. Using the periodic table, indicate which element in each of the following pairs of elements has the greater atomic radius:

    a. carbon or silicon   **b.** lead or tin

    c. barium or strontium   **d.** zinc or mercury

 # General Problems

15. Prior to the discovery of germanium (Ge) in 1886, Mendeleev predicted in 1869 the properties of this element, which he called "eka-silicon." Using the periodic table, determine the following for germanium (atomic number 32):

    a. Would this element be classified as a metal, a nonmetal, or a metalloid (semimetal)?

    b. How many valence electrons would it have?

    c. Write the electronic configuration in sublevels for both germanium and its group precursor, silicon.

    d. Would it be more metallic or more nonmetallic than its precursor, silicon?

    e. Mendeleev predicted a density of 5.5 g/mL for what we call germanium. The actual density for germanium is 5.3 g/ml. Convert the predicted and actual values to SI units, kg/m^3. (*Hint:* See Section 2.8, Density Conversions.)

16. The element iodine (I) has an atomic radius of 133 pm. Calculate its atomic radius in (a) centimeters and (b) inches (1 in. = 2.54 cm).

✓ Chapter Quiz 5

You may use the periodic table and the list of symbols of the elements.

1. Classify each of the following elements as a metal, nonmetal, or metalloid (semimetal):

   a. potassium　　**b.** arsenic　　**c.** nitrogen　　**d.** gold

2. Indicate the number of valence electrons for each of the following elements:

   a. gallium　　**b.** selenium　　**c.** barium　　**d.** astatine

3. Given are the electronic configurations for four elements. Pair the elements you would expect to show similar chemical properties.

   a. $1s^2, 2s^2 2p^6, 3s^1$　　　　　　**b.** $1s^2, 2s^2 2p^6, 3s^2 3p^3$

   c. $1s^2, 2s^2 2p^6, 3s^2 3p^6, 4s^1$　　**d.** $1s^2, 2s^2 2p^6, 3s^2 3p^6 3d^{10}, 4s^2 4p^3$

4. Indicate which element in each of the following pairs is more metallic:

   a. potassium or cesium　　　**b.** bismuth or phosphorus

5. Indicate which element in each of the following pairs has the greater atomic radius:

   a. polonium or selenium　　　**b.** zinc or cadmium

Chlorine (Symbol: Cl)

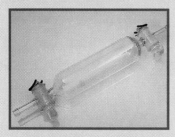

Chlorine is a greenish-yellow gas with a molecular formula of Cl_2.

The Element CHLORINE: Whiter Than White

Name: Chlorine was the first halogen element [group VIIA (17)] discovered. Its name derives from the Greek *chlor-*, meaning "green," and *-ine*, which means "made of" or "like."

Appearance: A greenish-yellow gas.

Occurrence: Chlorine is quite common as the chloride ion. While it comprises less than 1% of the earth's crust, it accounts for almost 2% (by mass) of the oceans as sodium chloride.

Source: Chlorine gas is prepared by passing an electric current through aqueous sodium chloride solutions (NaCl in water) or a molten metal chloride salt like sodium or magnesium chloride ($MgCl_2$).

Common Uses: Chlorine is used in the production of important bleaching compounds such as chlorinated lime [a mixture of $Ca(ClO)_2$, $CaCl_2$, $Ca(OH)_2$, and water] and sodium hypochlorite solutions (NaClO). Bleach is a 5.25% solution of sodium hypochlorite. These solutions are used to bleach fibers, wood pulp, and *clothing*.

Chlorinated lime is also used to disinfect sewage, drinking water, and the water in swimming pools. If the chlorine content in the pool is too high, your eyes may "burn."

Chlorine gas is used extensively in the production of important organic compounds, including a variety of pharmaceutical agents, pesticides, and synthetic rubbers and plastics [polyvinyl chloride (PVC) piping].

Chlorine is also used in the production of chloro-fluorocarbons, which are widely used as refrigerants and solvents and are believed to have a harmful effect on the ozone in the atmosphere (see the essays in Chapters 13 and 17, *Chemistry of the Atmosphere*).

Unusual Facts: Chlorine gas is extremely toxic to living systems. It was one of the poisonous gases used against soldiers during World War I. Its use is considered a violation of international law today.

CHAPTER 6

The Structure of Compounds

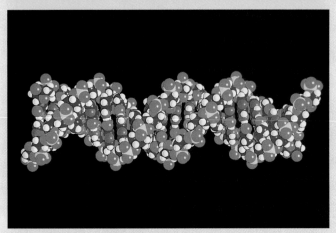

Deoxyribonucleic acid (DNA) space-filling computer graphic molecular model showing the double helix.

GOALS FOR CHAPTER 6

1. To explain the importance of chemical bonds and to comprehend how ionization energy and electron affinity influence chemical bonding (Sections 6.1 and 6.2).

2. To calculate the oxidation number of an element in a compound or ion (Section 6.3).

3. To describe how atoms form an ionic bond (Section 6.4).

4. To describe how atoms form a covalent bond (Section 6.5).

5. To describe how atoms form a coordinate covalent bond and to explain the similarity and difference between coordinate covalent bonds and covalent bonds (Section 6.7).

6. To write Lewis structures and structural formulas for molecules and polyatomic ions (Section 6.7).

7. To determine the shape and bond angles in molecules and polyatomic ions (Section 6.8).

8. To write chemical formulas for compounds (Section 6.9).

9. To predict oxidation numbers, properties, formulas, and types of bonding in compounds using the periodic table (Section 6.10).

Countdown

You may use the periodic table.

5. Give the symbol for each of the following elements (Section 3.1).
   a. phosphorus   (P)
   b. potassium   (K)
   c. antimony   (Sb)
   d. tin   (Sn)

4. Diagram the atomic structure for each of the following atoms. Indicate the number of protons and neutrons and arrange the electrons in principal energy levels (Section 4.6).

   Principal
   1   2   3   Energy Level

   a. $^{14}_{7}N$ $\left(\boxed{\begin{smallmatrix}7p\\7n\end{smallmatrix}}\, 2e^- \quad 5e^-\right)$

   b. $^{35}_{17}Cl$ $\left(\boxed{\begin{smallmatrix}17p\\18n\end{smallmatrix}}\, 2e^- \quad 8e^- \quad 7e^-\right)$

3. Write the electron-dot formula for each of the following atoms (Section 4.7).
   a. $^{14}_{7}N$ $(\cdot\ddot{N}\cdot)$ b. $^{35}_{17}Cl$ $(\cdot\ddot{\underset{\cdot\cdot}{Cl}}\!:)$

2. Write the electronic configuration in sublevels for the following atoms (Section 4.8).
   a. $^{32}_{16}S$ $(1s^2, 2s^2\, 2p^6, 3s^2\, 3p^4)$
   b. $^{51}_{22}Ti$ $(1s^2, 2s^2\, 2p^6, 3s^2\, 3p^6\, 3d^2, 4s^2$ or $1s^2, 2s^2\, 2p^6, 3s^2\, 3p^6, 4s^2, 3d^2)$

1. Using the periodic table, indicate the number of valence electrons for the following elements (Section 5.3).
   a. carbon   (4)   b. sulfur   (6)
   c. arsenic   (5)   d. strontium   (2)

The elements that make up the periodic table are the foundation of our universe. Yet seldom do we encounter these elements in their pure state. Indeed, the most basic substance in human life, water, is a compound (H_2O).

In this chapter, we will consider the forces that hold atoms of various elements together to form the compounds of our world. As we will see, compounds result from different forces. These differences, which influence the way chemists write down compounds, can be predicted from the periodic table.

6.1   Chemical Bonds

The attractive forces that hold atoms together are called **chemical bonds**. There are two general types of bonds between atoms in a compound: (1) ionic bonds and (2) covalent bonds. *These bonds are formed through interactions among the valence electrons of the atoms in the compound.* To understand these interactions, recall the **rule of eight (octet rule)** (Section 4.7). According to this rule, a stable configuration is achieved in many cases if 8 electrons are present in the valence energy level surrounding each atom. The atoms achieve these complete energy levels by *gaining, losing, or sharing valence electrons.* One exception to the rule of eight is helium, whose first principal energy level is completed with just 2 electrons. This exception gives us the **rule of two**: *a completed first principal energy level is also a stable configuration.*

In general, atoms having 1, 2, or 3 valence electrons tend to *lose* these electrons to become positively charged ions. Metals show this type of behavior. Alternatively, atoms with 5, 6, or 7 valence electrons tend to *gain* electrons and become negatively charged ions. Many nonmetals fall into this category. Many nonmetals may also *share* electrons to obtain 8 electrons in their valence energy level. Those elements with 4 valence electrons, for example, carbon, are the most apt to share their valence electrons.

Chemical bonds   The attractive forces that hold atoms together as compounds.

Rule of eight (octet rule) Principle that in forming molecules, atoms attempt to obtain a stable configuration of eight valence electrons around the atom.

Rule of two   An exception to the rule of eight (octet rule). A completed first principal energy level is also a stable configuration. Helium atoms and hydrogen atoms in the combined state obey this rule.

6.2   Ionization Energy and Electron Affinity

The formation of compounds and molecules depends not only on the existence of unfilled energy levels in many elements but also on the ionization energy of the atoms involved. The **ionization energy** of an atom is the amount of energy required to remove only the most loosely bound electron from the atom. The remaining portion of the atom is then a **cation**, a positively charged ion, because the atom now has *more protons than electrons.* That is,

$$\text{atom} + \text{energy} \longrightarrow \text{cation} + \text{electron} \qquad (6.1)$$

Chemists then express the ion by writing the element symbol with a plus sign and a number representing the number of electrons the atom lost. For example, a sodium atom can be converted to a sodium ion by the addition of 8.24×10^{-19} J of energy:

$$\text{Na} + 8.24 \times 10^{-19}\,\text{J} \longrightarrow \text{Na}^+ + 1\ \text{electron} \qquad (6.2)$$

Ionization energy   Amount of energy required to remove only the most loosely bound electron from an atom.

Cation   Any ion carrying a positive charge.

Note that the positive charge on the ion matches the number of electrons that are removed from the atom. Similarly, 2 electrons can be removed from a magnesium atom by the addition of 3.64×10^{-18} J of energy:

$$Mg + 3.64 \times 10^{-18}\,J \longrightarrow Mg^{2+} + 2\text{ electrons} \qquad (6.3)$$

As you will see, many cations of the elements exist, and the most common ones are summarized in Table 6.1. You must memorize the formulas of these ions because they are components of many compounds.

Study Hint: For the cations marked with an asterisk (*), the Roman group numeral in the periodic table represents the positive ionic charge on the element.

TABLE 6.1	Some Common Metals with the Formulas of the Cations and Their Names	
METAL (SYMBOL)	CATION[a]	NAME OF CATION[b]
+1 Ionic Charge		
Hydrogen[c] (H)	*H$^+$	Hydrogen
Lithium (Li)	*Li$^+$	Lithium
Potassium (K)	*K$^+$	Potassium
Silver (Ag)	*Ag$^+$	Silver
Sodium (Na)	*Na$^+$	Sodium
+2 Ionic Charge		
Barium (Ba)	*Ba^{2+}	Barium
Cadmium (Cd)	*Cd^{2+}	Cadmium
Calcium (Ca)	*Ca^{2+}	Calcium
Magnesium (Mg)	*Mg^{2+}	Magnesium
Strontium (Sr)	*Sr^{2+}	Strontium
Zinc (Zn)	*Zn^{2+}	Zinc
+3 Ionic Charge		
Aluminum (Al)	*Al^{3+}	Aluminum
+1 and +2 Ionic Charges		
Copper (Cu)	Cu$^+$	Copper(I) or cuprous
	Cu^{2+}	Copper(II) or cupric
Mercury (Hg)	Hg$_2^{2+}$	Mercury(I) or mercurous[d]
	Hg^{2+}	Mercury(II) or mercuric
+2 and +3 Ionic Charges		
Iron (Fe)	Fe^{2+}	Iron(II) or ferrous
	Fe^{3+}	Iron(III) or ferric
+2 and +4 Ionic Charges		
Lead (Pb)	Pb^{2+}	Lead(II) or plumbous
	Pb^{4+}	Lead(IV) or plumbic
Tin (Sn)	Sn^{2+}	Tin(II) or stannous
	Sn^{4+}	Tin(IV) or stannic

[a] For the cations marked with an asterisk (*), you can determine the ionic charge using the periodic table. You must memorize the ionic charge on all other cations.

[b] The Roman numeral in parentheses of the name indicates the ionic charge on each atom in the ion.

[c] Not a metal but often reacts as a metal.

[d] Experimental evidence indicates that mercury(I) or mercurous ion exists as a dimer (two units) with an ionic charge of 1^+ on *each* atom $[Hg^+]_2 = Hg_2^{2+}$. Hg_2^{2+} is a dimer ion.

TABLE 6.2	Some Common Nonmetals with the Formulas of the Anions and Their Names		
NONMETALS (SYMBOL)	ANION	NAME OF ANION	
−1 Ionic Charge			
Bromine (Br)	Br^-	Brom*ide* ion	
Chlorine (Cl)	Cl^-	Chlor*ide* ion	
Fluorine (F)	F^-	Fluor*ide* ion	
Hydrogen (H)	H^-	Hydr*ide* ion	
Iodine (I)	I^-	Iod*ide* ion	
−2 Ionic Charge			
Oxygen (O)	O^{2-}	Ox*ide* ion	
Sulfur (S)	S^{2-}	Sulf*ide* ion	
−3 Ionic Charge			
Nitrogen (N)	N^{3-}	Nitr*ide* ion	
Phosphorus (P)	P^{3-}	Phosph*ide* ion	

Study Hint: The ionic charge on the anions (except the hydride ion, H^-) can be determined by subtracting 8 (rule of eight) from the Roman group number in the periodic table. The hydride ion can be determined by subtracting 2 (rule of two) from the Roman group number [IA (1): $1 - 2 = -1$.]

The counterpart of ionization energy is an atom's **electron affinity**, which is the amount of energy *given off* when an atom gains an *extra* electron. By picking up an extra electron, the atom now has *more electrons than protons,* and it has a net negative charge. Such a species is called an **anion**.

$$\text{atom} + \text{electron} \longrightarrow \text{anion} + \text{energy} \qquad (6.4)$$

For example, a chlorine atom plus 1 electron can be converted to a chloride ion with 5.80×10^{-19} J of energy given off:

$$\text{Cl} + 1 \text{ electron} \longrightarrow Cl^- + 5.80 \times 10^{-19} \text{ J} \qquad (6.5)$$

Note that the charge on the ion matches the number of electrons added to the atom and that the charge on the atom is expressed as a superscript with a *minus* sign since there is a net negative charge. Anions with −2 or −3 charges may also exist in compounds. Table 6.2 contains the most common anions. Learn these ions now; you will need this information in the chapters to come.

Ionization energy and electron affinity are important concepts because they help us to understand bonding between atoms. Atoms gain, lose, or share electrons in the act of bonding. The ionization energy helps a chemist understand what happens when an atom loses an electron, while electron affinity helps describe what happens when an atom gains an electron.

Electron affinity   Amount of energy given off when an atom or ion gains an extra electron.

Anion   Any ion carrying a negative charge.

6.3 Oxidation Numbers. Calculating Oxidation Numbers

Before we consider how atoms bond together to form compounds and the structures of these compounds, we must understand the meaning of a new term "oxidation number."

Oxidation number (ox. no. or oxidation state) A positive or negative whole number assigned to an element in a compound or ion. It is based on certain rules.

Ionic charge Charge on an ion. The ion may consist of a single atom or a group of atoms bonded together.

Study Hint: Ionic charges or oxidation numbers are used later (Section 6.8) in writing formulas of compounds and then in nomenclature (Chapter 7). This is material you *must* know because you will use it again.

Oxidation number (ox. no. or **oxidation state)** is a positive or negative whole number assigned to an element in a compound or ion. A compound contains elements with both positive and negative oxidation numbers, and the *sum of the oxidation numbers of all of the atoms in a compound is zero.* This principle applies to *all* compounds.

The oxidation number is an assignment based on certain rules (see below) that provide chemists with a method of electronic "bookkeeping." It helps keep track of the electrons which are transferred or shared when an atom combines with another atom or atoms in forming compounds. For example, in HCl, we assign H an oxidation number of $+1$, which means that Cl has an oxidation number of -1 since the sum of the oxidation numbers totals zero. Now, we will look at compounds of several metals with chlorine: $NaCl$, $MgCl_2$, and $AlCl_3$. If Cl has an oxidation number of -1 (from the HCl example), we can immediately assign oxidation numbers of $+1$, $+2$, and $+3$ to Na, Mg, and Al in the series of compounds above.

Some elements have only one oxidation number or oxidation state. Sodium (Na), magnesium (Mg), and aluminum (Al) are examples of such elements. Other elements may have more than one oxidation state. An example is oxygen: in water (H_2O), oxygen has an oxidation number of -2 (remember, hydrogen is $+1$), while in hydrogen peroxide (H_2O_2), oxygen has an oxidation number of -1.

For ions containing a single atom, such as Na^+ and Cl^-, the oxidation number of the element is the same as the charge on the ion, or the **ionic charge**. Cations have positive oxidation numbers, and anions have negative oxidation numbers. *In general,* when combined, metals have positive oxidation numbers and nonmetals have negative oxidation numbers.

The following rules are used for assigning or determining oxidation numbers:

1. The algebraic sum of the oxidation numbers of all the atoms in the formula for a *compound* is zero.

2. The oxidation number of an element in the free or uncombined state is always zero, such as Cu and Ag.

3. The oxidation number of a monatomic ion (an ion containing a single atom) is the same as its ionic charge, such as Na^+ and Cl^-.

4. The algebraic sum of the oxidation numbers of all the atoms in a polyatomic ion (an ion containing two or more atoms) is equal to the ionic charge on the polyatomic ion. In ClO^- the sum of the oxidation numbers of Cl and O is equal to -1.

5. In general, metals have positive oxidation numbers when combined with nonmetals, and nonmetals have negative oxidation numbers when combined with metals. For example, in NaCl, Na is $+1$ and Cl is -1.

6. In compounds containing two nonmetals, a negative oxidation number is assigned to the more electronegative atom (see Section 6.5 and Figure 6.11). A positive oxidation number is assigned to the less electronegative atom. In the compound NO, the N nonmetal has the positive oxidation number and the O nonmetal has the negative oxidation number. The reason for this is that the electronegativity of N is less (3.0; see Figue 6.11) than the electronegativity of O (3.5).

7. In most compounds containing hydrogen, the oxidation number of hydrogen is $+1$. For example, in HCl, H has an oxidation number of $+1$; therefore, Cl has an oxidation number of -1. The exceptions to this rule are the hydrides of metals (NaH, LiH, CaH_2, AlH_3, and others) in which

hydrogen has an oxidation number of -1. Note that here the hydrogen atom is written second. In forming hydrides, hydrogen acts as a nonmetal.

8. In most oxygen compounds, the oxidation number of oxygen is -2. For example, in MgO, O has an oxidation number of -2; therefore, Mg has an oxidation number of $+2$. The exceptions to this rule include the peroxides (H_2O_2, Na_2O_2, BaO_2, and others) in which oxygen has an oxidation number of -1.

Figure 6.22 at the end of this chapter summarizes the oxidation numbers of elements in a given period. Now let's apply these rules.

EXAMPLE 6.1   Calculate the oxidation number for the element indicated in each of the following compounds or ions.

a. N in HNO_3

SOLUTION: The oxidation numbers of H and O in the compound are $+1$ and -2 (see rules 7 and 8, respectively). The sum of the oxidation numbers of all the elements in the compound must equal zero (rule 1). Therefore,

$$+1 + \text{ox. no. N} + 3(-2) = 0$$
$$+1 + \text{ox. no. N} - 6 = 0$$
$$\text{ox. no. N} - 5 = 0$$
$$\text{ox. no. N} = +5 \quad \textit{Answer}$$

b. N in NO_2^-

SOLUTION: The oxidation number of oxygen is -2 (rule 8), and the sum of the oxidation numbers of all the elements in the polyatomic ion *must equal the charge on the ion or* -1 (rule 4). Therefore,

$$\text{ox. no. N} + 2(-2) = -1$$
$$\text{ox. no. N} - 4 = -1$$
$$\text{ox. no. N} = +4 - 1$$
$$\text{ox. no. N} = +3 \quad \textit{Answer}$$

c. Cr in Cr_2O_3

SOLUTION: The oxidation number of oxygen is -2 (rule 8). The sum of the oxidation numbers of all the elements in the compound must equal zero (rule 1). There are two atoms of chromium, but we must solve for *one* atom of chromium; therefore,

$$2(\text{ox. no. Cr}) + 3(-2) = 0$$
$$2(\text{ox. no. Cr}) - 6 = 0$$
$$2(\text{ox. no. Cr}) = +6$$
$$\text{ox. no. Cr} = +6/2$$
$$\text{ox. no. Cr} = +3 \quad \textit{Answer}$$

d. Cr in $Cr_2O_7^{2-}$

SOLUTION: The oxidation number of oxygen is -2 (rule 8), and the sum of the oxidation numbers of all the elements in the polyatomic ion *must equal the charge on the ion or* -2 (rule 4). There are two atoms of chromium, but we must solve for *one* atom of chromium; therefore,

$$2(\text{ox. no. Cr}) + 7(-2) = -2$$
$$2(\text{ox. no. Cr}) - 14 = -2$$
$$2(\text{ox. no. Cr}) = +14 - 2 = +12$$
$$\text{ox. no. Cr} = +12/2$$
$$\text{ox. no. Cr} = +6 \quad \textit{Answer}$$

Study Exercise 6.1

Calculate the oxidation number for the element indicated in each of the following compounds or ions:

a. Cl in ClO_2 $(+4)$ **b.** Cl in Cl_2O_7 $(+7)$
c. Cl in ClO_2^- $(+3)$ **d.** Cl in ClO_4^- $(+7)$

Work Problems 4 and 5.

6.4 The Ionic Bond

Ionic bond The force of attraction between ions of opposite charge which holds them together in an ionic compound. These ions of opposite charge are formed by the transfer of electrons from one atom to another.

Ionization energy and electron affinities are important in understanding ionic bonds. An **ionic bond** is the force of attraction between ions of opposite charge which holds them together in an ionic compound. These ions of opposite charge are formed by the transfer of one or more electrons from one atom to another. As a result of this transfer, one of the atoms is a positively charged cation, while the other is a negatively charged anion. These ions form a bond because, as the law of electrostatics states, *particles with unlike charges attract each other and particles with like charges repel each other.* In ionic compounds there are *not* just two ions but many ions, and so the total force of attraction between the ions in an ionic bond is very strong.

Study Hint: The electrons are all the same. They do not have colored signs saying "I came from Na" or "I came from Cl." We just use colored dots to help you understand how an ionic bond is formed.

Examples of ionic compounds abound. Table salt or sodium chloride (NaCl) is formed when a sodium atom combines with a chlorine atom, as Figure 6.1 shows. The sodium atom has 1 valence electron, and the chlorine atom has 7 valence electrons. The 1 valence electron from sodium is lost to the chlorine atom, giving 8 electrons in the highest energy level of the sodium ion and 8 electrons in the highest energy level of the chloride ion. The rule of eight is satisfied for both the positive sodium and the negative chloride ions in this ionic bond.

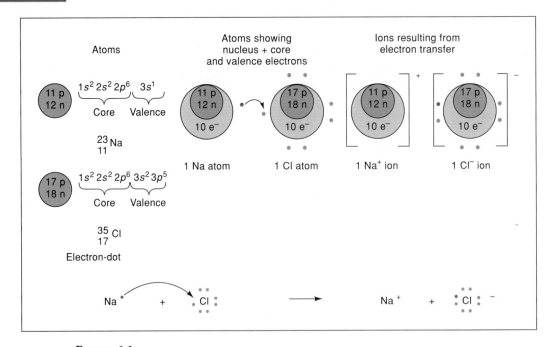

FIGURE 6.1

Formation of sodium chloride (NaCl) from a sodium atom and a chlorine atom, is an example of a compound formed by ionic bonding. The negative chloride ion attracts the positive sodium ion.

General Points about Compounds with Ionic Bonds

There are five important points to consider regarding compounds with ionic bonds. *First,* the transfer of electrons can result in great changes in properties. For example, sodium and chlorine atoms differ considerably from the sodium chloride they form (see Figure 6.2). Sodium is a soft, active metallic solid and can be cut with a knife. Chlorine is a greenish gas with a strong, irritating odor. Sodium chloride is a white, crystalline solid (see Figure 6.3). Sodium chloride is edible, but both sodium metal and chlorine gas are poisonous. Sodium reacts with water to give an explosive reaction, but sodium chloride dissolves in water. (Table 6.3 lists some physical properties of sodium, chlorine, and sodium chloride.)

Second, the charge of the ion is related to the number of *protons* and *electrons* in the ion. In the sodium atom, there are 11 protons in the nucleus and 11 electrons about the nucleus; the atom is neutral (see Figure 6.1). There are still 11 protons in the nucleus in the ion but only 10 electrons since 1 electron was lost to the chlorine atom. The result is a *net of one proton* or *one* positive charge in excess, giving it an ionic charge or oxidation number on the sodium ion of +1. In the chlorine atom there are 17 protons in the nucleus and 17 electrons about the nucleus; the atom is neutral (see Figure 6.1). After an electron is received from the sodium atom there are 18 electrons about the nucleus but still 17 protons in the nucleus, resulting in a *net of one electron* or *one* negative charge in excess, giving it an ionic charge or oxidation number on the chloride ion of −1. Therefore, the charges on the ions are directly related to their atomic structures.

Third, the radii of the ions differ from those of the atoms, as Figure 6.4 shows. The radius of the sodium atom is 186 pm, and the radius of the sodium ion is 95 pm. This decrease in radius results from (1) the loss of electrons in an energy level because the third principal energy level in the sodium atom has been emptied in the transfer of an electron to the chlorine atom, (2) a further decrease in size because of a somewhat greater nuclear attraction of the 11 positively charged protons for the remaining 10 electrons, and (3) decreased repulsion among the electrons now that only 10 are left surrounding the nucleus. This last effect allows the remaining electrons to pack in toward the nucleus a bit more than they could in the neutral atom. In contrast, the radius of the chlorine atom is 99 pm, but the radius of the chloride ion has increased to 181 pm. This increase is due to (1) a smaller nuclear attraction (17

FIGURE 6.2
When two substances react (sodium metal + chlorine gas) to form an ionic bond, the product (white sodium chloride) usually looks quite different.

FIGURE 6.3
Crystals of sodium chloride.

TABLE 6.3	Properties of Sodium, Chlorine, and Sodium Chloride		
ELEMENT OR COMPOUND	APPEARANCE AT ROOM TEMPERATURE	MELTING POINT (°C)	BOILING POINT (°C)[a]
Sodium	Soft, silvery, solid; cut with a knife	98	892
Chlorine	Greenish gas; strong irritating odor	−101	−35
Sodium chloride	White crystalline solid	801	1413

[a] At 1.00 atm pressure.

FIGURE 6.4
Radii of ions differ from those of the atoms as shown by a sodium atom and ion and by a chlorine atom and ion.

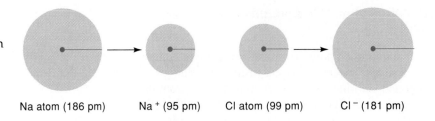

Na atom (186 pm)   Na$^+$ (95 pm)   Cl atom (99 pm)   Cl$^-$ (181 pm)

protons) on the 18 orbital electrons, which results in a slight expansion of the radius of the orbital, and (2) the increased amount of repulsion now that an extra orbital electron (18 electrons) has been added, forcing the electrons to occupy a larger volume.

Fourth, energy is *given off in bond formation.* This is true for *all* bonds. *Bond formation leads to a more stable substance.* In the formation of 1.00 g of sodium chloride from the constituent atoms (Na and Cl), 7.06 kJ or 1.69 kcal is given off. Therefore, to "break" these ionic bonds in 1.00 g of solid sodium chloride and to form the sodium and chlorine atoms requires 7.06 kJ or 1.69 kcal of energy.

Fifth, the smallest unit of an *ionic compound* is a **formula unit** since it is a combination of *ions* and *not* discrete molecules (see Section 3.3). Hence, one formula unit of NaCl consists of one sodium ion and one chloride ion.

Formula unit   Smallest unit of an ionically bonded substance that can exist and undergo chemical charges.

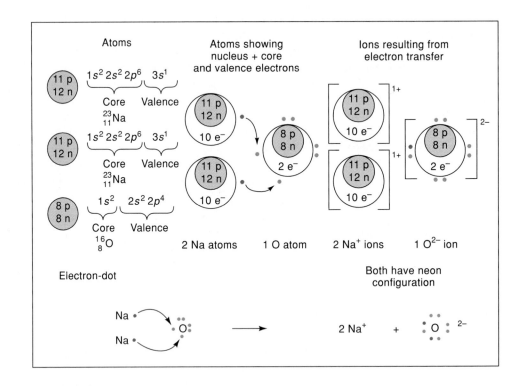

FIGURE 6.5
Formation of sodium oxide (Na$_2$O), from two sodium atoms and an oxygen atom is an example of a compound formed by ionic bonding. The negative oxide ion attracts *two* positive sodium ions.

Not all ionic bonds are one-to-one exchanges. The *formula unit* of sodium oxide (Na_2O), is formed when 2 atoms of sodium each transfer an electron to 1 atom of oxygen, as Figure 6.5 shows. For the oxygen atom to acquire 8 electrons in its highest energy level, it must gain 2 electrons because it already has 6 valence electrons. This requires 2 sodium atoms because each sodium atom has only 1 valence electron to donate. Therefore, each of the 2 sodium atoms *transfers* its valence electron to 1 oxygen atom to form the formula unit, Na_2O, consisting of 2 positive sodium ions and 1 negative oxide ion in a crystal of many such ions. The loss of an electron from the sodium atom results in a $+1$ ionic charge for each sodium ion, and the gain of 2 electrons by the oxygen atom results in a -2 ionic charge for the oxide ion.

Some properties of ionic compounds are as follows:

✔ They have relatively high melting points (above 300°C).

✔ They conduct an electric current in the liquid state or in aqueous solution.

Study Exercise 6.2

Diagram the ionic structure for each of the following ions, indicate the number of protons and neutrons in the nucleus, and arrange the electrons in principal energy levels.

a. $_3^7Li^+$

$$\left(\overset{3p}{\underset{4n}{\bigcirc}} \; 2e^- \quad +1 \right)$$

b. $_{17}^{35}Cl^-$

$$\left(\overset{17p}{\underset{18n}{\bigcirc}} \; 2e^- \quad 8e^- \quad 8e^- \quad -1 \right)$$

Study Exercise 6.3

Write the electronic configuration in *sublevels* for the following ions:

a. $_{13}^{27}Al^{3+}$ $\hspace{4cm}$ ($1s^2, 2s^2\, 2p^6$)

b. $_9^{19}F^-$ $\hspace{4.5cm}$ ($1s^2, 2s^2\, 2p^6$) $\hspace{1cm}$ Work Problems 6 through 9.

Ions are important to you. Table 6.4 lists some common ions in your body and their function.

TABLE 6.4	Some Common Ions, Their Main Site in the Body, and Their Function	
ION	MAIN SITE IN THE BODY	FUNCTION
Na^+	Outside cells	Control of body water and nerve and muscle function
K^+	Within cells	Nerve and muscle function
Ca^{2+}	Bones, muscle, teeth	Regulation of cell metabolism, nerve and muscle activity, and blood clotting; maintenance of bones
Mg^{2+}	Mainly within cells, bone	Regulation of neuromuscular function, metabolic activity
Cl^-	Outside cells	Nerve membrane function

6.5 The Covalent Bond

Covalent bond   Type of chemical bond formed by the sharing of electrons between two atoms.

Unlike ionic bonds, **covalent bonds** are formed by the *sharing* of electrons between atoms. The smallest unit of a *covalent compound* formed by such a bond is a molecule (see Section 3.3). Note that a formula unit—the smallest unit in ionic compounds—is *not* a molecule because a formula unit does not really exist as a discrete entity but as ions. Covalently bonded compounds have different properties than ionically bonded compounds. Covalent compounds have relatively lower melting points (less than 300°C) and do not conduct an electric current as a liquid or in aqueous solution as ionic compounds do.

The hydrogen molecule (H_2) is a simple example of a *covalent* compound. As Figure 6.6 shows, an isolated hydrogen atom is relatively unstable since it has only one valence electron. By sharing its valence electron with another hydrogen atom, it completes the first principal energy levels in both hydrogen atoms and gives a stable configuration to both atoms in the molecule. One way of representing the H_2 molecule (Figure 6.7) is to allow the $1s$ orbitals of both atoms to overlap and form a new peanut-shaped region in space (orbital) that contains both bonding electrons. These two negatively charged electrons now attract both positively charged nuclei and hold the molecule together.

General Points about Compounds with Covalent Bonds

In the hydrogen molecule, as in all compounds with covalent bonds, there are four important points. *First,* as in ionic compounds, the properties of the individual uncombined atoms differ markedly from the properties of the molecules. In fact, indi-

FIGURE 6.6
Formation of hydrogen (H_2) from two hydrogen atoms is an example of covalent bonding. The blue circles on the molecule show how each atom now has a completed principal energy level (two electrons) as a result of electron sharing.

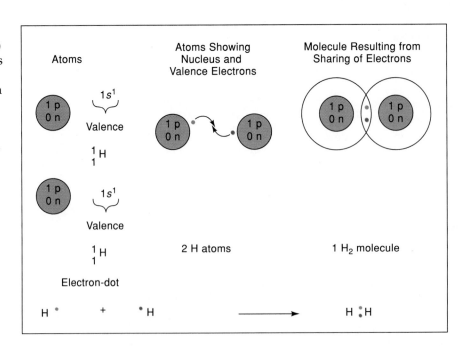

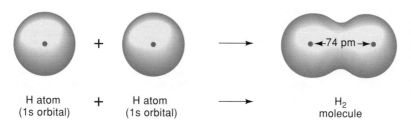

FIGURE 6.7
Orbital representation of a hydrogen molecule (H_2). The black dots represent the nuclei of the hydrogen atoms.

vidual hydrogen atoms are so unstable that they exist for only a very short time. Thus, when we write the formula for a *hydrogen molecule*, we must write it as H_2 (two atoms of hydrogen, a **diatomic molecule**) and not as H.

Second, the two positive nuclei attract each of the two electrons to produce a molecule more stable than the separate atoms. A bond has formed, and a more stable molecule results. This attraction by the nuclei for the two electrons counterbalances the repulsion of the two positive nuclei for each other; the greatest probability of finding the electrons is somewhere *between* the two nuclei. A simple analogy suggested by the late Henry Eyring of the University of Utah, and slightly modified here, may help to illustrate this point. Suppose we think of the nuclei of the two hydrogen atoms as "old potbellied stoves" and the two electrons as "children" running around each of these stoves trying to keep warm. When two atoms come together, the children (electrons) now have two sources of heat (nuclei), and these children can now run between the stoves and keep all parts, front and back, warm. The children (electrons) are now warmer and happier than they were when they had just one stove (nucleus), and a stable molecule results.

Third, the distance between the nuclei is such that the $1s$ orbitals of the hydrogen atoms have the maximum overlap without having the nuclei so close to each other that they repel each other (causing the molecule to fly apart). In the hydrogen molecule, the distance between the nuclei is 74 pm, as Figure 6.7 shows. The distance between the nuclei of covalently bonded atoms is called the **bond length**.

Fourth, during the process of covalent bond formation, energy is given off. In this case, 218 kJ or 52.0 kcal of energy is given off in the formation of 1.0 g of gaseous hydrogen, H_2. Therefore, to "break" these covalent bonds in 1.0 g of gaseous hydrogen and to form the hydrogen atoms requires 218 kJ or 52.0 kcal of energy.

Diatomic molecule Molecule composed of two atoms.

Bond length Distance between the nuclei of two atoms that are bonded together.

Study Exercise 6.4

When F atoms unite to form a fluorine molecule (F_2), 4.13×10^5 J of energy is given off in the formation of 1.00 g of gaseous fluorine. How many joules of energy are required to break the F—F bond in 1.00 g of gaseous fluorine and to form the fluorine atoms?

(4.13×10^5 J)   Work Problem 10.

More about Covalent Bonds

Most covalent bonds are more complex than the hydrogen molecule. Consider Figures 6.8 through 6.10 which show how two atoms of chlorine share their valence electrons to form a Cl_2 molecule. For each chlorine atom to raise its third principal energy level to a total of eight electrons (in keeping with the rule of eight), it must share one electron with the other atom. The result is one *shared pair* of electrons as well as three *unshared pairs* of electrons on each atom.

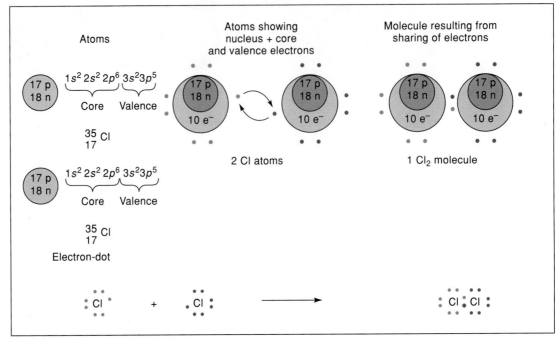

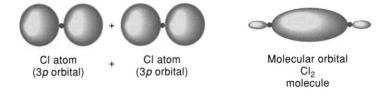

FIGURE 6.8
Formation of chlorine (Cl_2) from two chlorine atoms is an example of covalent bonding. The dots on the molecule show how each atom now has a completed principal energy level (eight electrons) as a result of electron sharing.

H_2
F_2
Cl_2
Br_2
I_2
O_2
N_2

Study Hint: A simple way to remember the *seven* diatomic elements is to look at the periodic table. Notice that the six *elements,* N, O, F, Cl, Br, and I trace the numeral 7, with the top of the 7 pointing toward H!

Electronegativity Degree to which an atom attracts a pair of covalently bonded electrons to itself.

In addition to H_2 and Cl_2, five other elements are written *only as diatomic molecules*: F_2, Br_2, I_2, O_2, and N_2. Because of the number of electrons in the valence energy levels of these atoms, none is stable unless coupled with another. Such molecules with equal sharing of electrons are often referred to as *nonpolar molecules*.

Electronegativity in Covalent Bonds

In the preceding examples, it was assumed that the electrons were shared *equally* by both atoms. This principle of equal sharing is not generally found in molecules that contain different atoms because some atoms have a greater attraction for electrons than others. The tendency for an atom to attract a pair of electrons in a covalent bond is called **electronegativity**. Figure 6.11 is a result of the work of the late

FIGURE 6.9
Orbital representation of a chlorine molecule (Cl_2). The blue dots represent the nuclei of the chlorine atoms. Note the overlap of the two *p* orbitals to form a new orbital that contains the two shared electrons.

Cl atom
(3*p* orbital)   +   Cl atom
(3*p* orbital)

Molecular orbital
Cl_2
molecule

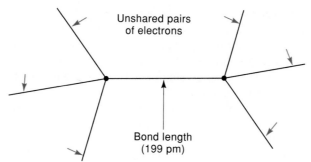

FIGURE 6.10
Model of the Cl_2 molecule. The dots represent the nuclei of the chlorine atoms, and the blue arrows indicate the unshared pairs of electrons on each chlorine atom. Note that the bond length (199 pm) is the distance between the two nuclei.

Linus C. Pauling, an American scientist, and gives the electronegativity values of most of the elements in the periodic table. The values are based on an arbitrary scale from 0.7 to 4.0, where a *low* value indicates an element that does *not* have a strong attraction for bonded electrons and a high value means an element does have a strong attraction for bonded electrons. Note that *metals* have *low* electronegativity values and *nonmetals* have *high* ones. A modified scale incorporating the most common elements is given below and is useful to learn.

$$F > O > Cl = N > Br > I = C = S > P = H$$

Why are some elements more electronegative than others? *First,* the smaller the radius of the atom, the greater the attraction between the nucleus and the outermost electrons. **Coulomb's law** states that the attractive force between a proton and an electron *increases* as the distance between these two particles *decreases*. The smaller atom often has fewer occupied energy levels and consequently has a greater attraction for the bonding electrons. Conversely, a larger atom with more energy levels occupied has less attraction for the bonding electrons. The nitrogen atom has

F
O
Cl = N
Br
I = C = S
P = H

Coulomb's law   Principle that the attractive force between a proton and an electron increases as the distance between these two particles decreases.

Electronegativity increases →

Electronegativity decreases ↓

	1 IA	2 IIA	3 IIIB	4 IVB	5 VB	6 VIB	7 VIIB	8	9 VIII	10	11 IB	12 IIB	13 IIIA	14 IVA	15 VA	16 VIA	17 VIIA	18 VIIA
1	**H** 2.1																	**He** —
2	**Li** 1.0	**Be** 1.5											**B** 2.0	**C** 2.5	**N** 3.0	**O** 3.5	**F** 4.0	**Ne** —
3	**Na** 0.9	**Mg** 1.2											**Al** 1.5	**Si** 1.8	**P** 2.1	**S** 2.5	**Cl** 3.0	**Ar** —
4	**K** 0.8	**Ca** 1.0	**Sc** 1.3	**Ti** 1.3	**V** 1.6	**Cr** 1.6	**Mn** 1.5	**Fe** 1.8	**Co** 1.8	**Ni** 1.8	**Cu** 1.9	**Zn** 1.6	**Ga** 1.6	**Ge** 1.8	**As** 2.0	**Se** 2.4	**Br** 2.8	**Kr** —
5	**Rb** 0.8	**Sr** 1.0	**Y** 1.2	**Zr** 1.2	**Nb** 1.6	**Mo** 1.8	**Tc** 1.9	**Ru** 2.2	**Rh** 2.2	**Pd** 2.2	**Ag** 1.9	**Cd** 1.7	**In** 1.7	**Sn** 1.8	**Sb** 1.9	**Te** 2.1	**I** 2.5	**Xe** —
6	**Cs** 0.7	**Ba** 0.9	***La** 1.1	**Hf** 1.3	**Ta** 1.5	**W** 1.7	**Re** 1.9	**Os** 2.2	**Ir** 2.2	**Pt** 2.2	**Au** 2.4	**Hg** 1.9	**Tl** 1.8	**Pb** 1.8	**Bi** 1.9	**Po** 2.0	**At** 2.2	**Rn** —

FIGURE 6.11
Electronegativities of a number of the elements.

a smaller radius than the carbon atom. Thus, nitrogen has a greater attraction for its outermost electrons than carbon does, and hence a greater electronegativity.

Second, atoms having fewer energy levels of electrons between the nucleus and the outermost energy level are more electronegative than those with intervening energy levels. The intervening levels of electrons *shield* the outer electrons from the full effect of the positively charged nucleus. Because of this *shielding* effect, fluorine is more electronegative than chlorine, and chlorine is more electronegative than bromine. Compare the electronic structures of these atoms.

Third, when filling the same energy level in a period, electronegativity increases as the nuclear charge increases. Therefore, fluorine (atomic number 9, nine protons) is more electronegative than oxygen (atomic number 8, eight protons).

Polar Bonds

Polar bond   Type of chemical bond, also known as a polar covalent bond, formed by the *unequal* sharing of electrons between two atoms whose electronegativities differ.

Differences in electronegativity cause *unequal* sharing of electrons in a covalent bond. Indeed, the greater the differences in electronegativities, the more unequal the sharing of electrons in the covalent bond. A typical example of such a **polar bond**, or *polar covalent bond,* is that in hydrogen chloride gas, shown in Figure 6.12. The electronegativity of hydrogen is 2.1; that of chlorine is 3.0 (see Figure 6.11). Hence, in a molecule of hydrogen chloride gas, the more electronegative

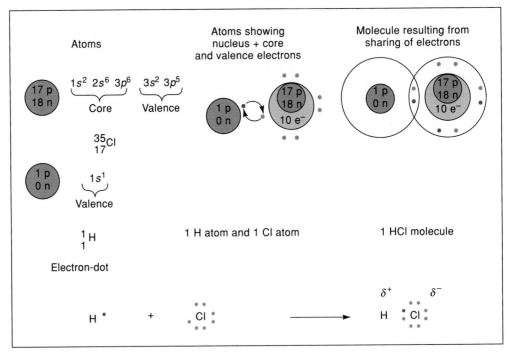

FIGURE 6.12
Formation of hydrogen chloride (HCl) from one hydrogen atom and one chlorine atom is an example of an unequal sharing of electrons in a covalent bond. The large, outer black circles on the molecule show how each atom now has a completed principal energy level (two electrons for hydrogen and eight electrons for chlorine) as a result of electron sharing.

chlorine has a greater attraction for the pair of electrons in the covalent bond than the hydrogen atom does. As Figure 6.13 illustrates, this unequal sharing of electrons in a covalent bond is often shown by placing a δ^- (lowercase Greek letter delta, δ, meaning "partially charged") above the relatively negative atom and a δ^+ above the partially positive atom. Thus, hydrogen chloride gas is depicted as

$$\overset{\delta^+}{H}\,\overset{\delta^-}{:\!\ddot{Cl}\!:}$$

As Figure 6.14 shows, unequal sharing of electrons in a polar bond is a situation intermediate between the equal sharing of electrons in covalent bonding and purely ionic bonding where the difference in the electronegativities of the atoms is great.

Study Exercise 6.5

Place a δ^+ above the atom or atoms that are relatively positive, and a δ^- above the atom or atoms that are relatively negative in the following covalently bonded molecules (see Figure 6.11):

a. ClF_3

$$\overset{\delta^+}{(Cl}\ \overset{\delta^-}{F_3)}$$

b. ClO_2

$$\overset{\delta^+}{(Cl}\ \overset{\delta^-}{O_2)}$$

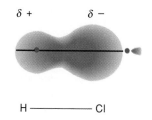

FIGURE 6.13
Hydrogen chloride molecule, showing the greater attraction for the electron pair in the covalent bond by the electronegative chlorine atom. Compare this unequal sharing of electrons with the equal sharing of electrons shown in Figure 6.7 with hydrogen. The dots represent the nuclei of the atoms.

Work Problem 11.

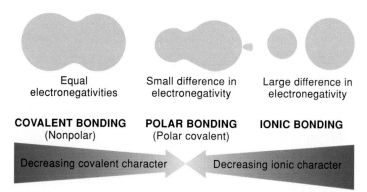

Equal electronegativities — Small difference in electronegativity — Large difference in electronegativity

COVALENT BONDING (Nonpolar) — **POLAR BONDING** (Polar covalent) — **IONIC BONDING**

Decreasing covalent character — Decreasing ionic character

FIGURE 6.14
Intermediate between the equal sharing of electrons and ionic bonding is polar bonding. When the difference in the electronegativities of the atoms is sufficiently large, the more electronegative atom gains essentially full possession of the shared electron pair and ions result.

6.6 The Coordinate Covalent Bond

In a covalent bond, *each* atom contributes one electron to form an electron pair between the *two atoms*. In **coordinate covalent bonding**, also called *coordinate* bonding, one atom supplies *both* the electrons of the electron-pair bond. The other atom offers only an empty orbital.

For example, the ammonium ion (NH_4^+) is formed from a hydrogen ion (a H atom without an electron, H^+) and an ammonia molecule (NH_3). The ammonia molecule is formed from 3 hydrogen atoms and 1 nitrogen atom, as Figure 6.15 shows. The nitrogen atom with 5 electrons in its second principal energy level shares 1 electron from each of the 3 hydrogen atoms, for a total of 8 electrons around the nitrogen atom. By sharing its 1 electron with the nitrogen atom, the hydrogen atom ends up with 2 electrons, completing its first principal energy level.

Coordinate covalent bond
Type of chemical bond, also known as a coordinate bond, formed when one atom supplies both electrons of the electron-pair bond, while the other atom offers only an empty orbital.

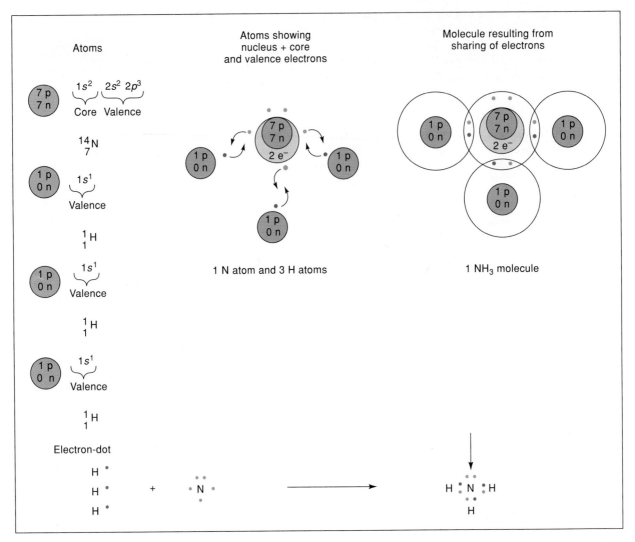

FIGURE 6.15
Formation of ammonia (NH_3) from three hydrogen atoms and one nitrogen atom.
(Note that in the electron-dot formula the unshared pair of electrons is shown by the arrow, ↓.) The black circles on the molecule show how each atom now has a completed principal energy level (two electrons for hydrogen and eight electrons for nitrogen) as a result of electron sharing.

Unshared pair of electrons
(*nonbonding pair, lone pair*) A *pair* of electrons on one atom.

The nitrogen of ammonia thus has an unshared pair of electrons, as shown in the electron-dot formula in Figure 6.15. An **unshared pair of electrons,** also called a *nonbonding pair* or a *lone pair,* is a *pair* of electrons on one atom. Figure 6.16 shows a model of the ammonia molecule.

When a hydrogen ion (H^+) is added to ammonia, the ion attaches to this unshared pair of electrons to form a *coordinate covalent bond,* as Figure 6.17 shows. The hydrogen now shares two electrons and is stabilized. The unshared pair of electrons acts like glue to form the coordinate covalent bond with the proton. The new

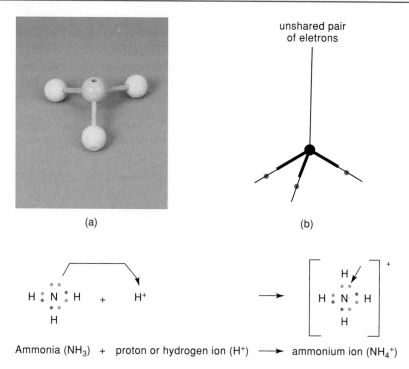

unshared pair
of eletrons

(a)

(b)

FIGURE 6.16
Models of ammonia (NH_3).
(a) Ball-and-stick model of
ammonia. The yellow balls
represent hydrogen atoms,
and the blue ball represents
the nitrogen atom; (b) the
blue dots represent the nuclei of the hydrogen atoms,
and the black dot, the nucleus of the nitrogen atom.
Notice the unshared pair of
electrons.

Ammonia (NH_3) + proton or hydrogen ion (H^+) ⟶ ammonium ion (NH_4^+)

FIGURE 6.17
Formation of an ammonium
ion (NH_4^+) is an example of
coordinate covalent bond-
ing. The arrow (↙) shows
the coordinate covalent
bond, and the positive ionic
charge is dispersed over the
entire ion. All four N—H
bonds are equivalent once
NH_4^+ is formed.

ion that is formed, the ammonium ion (NH_4^+) has a positive charge because it has
more protons than electrons. The ammonia molecule did not have a charge, but the
hydrogen ion had a +1 charge; therefore, the ammonion ion has a +1 charge. This
+1 charge is dispersed over the entire ion.

The four N—H bonds on the NH_4^+ are chemically the same. The descriptions
of ionic, covalent, and coordinate covalent bonds are only models that we use to visualize the actual phenomenon of chemical bonding. ***There is no real difference
between covalent and coordinate covalent bonds, but rather a difference in the
way we visualize their formation.*** A coordinate covalent bond is identical to a covalent bond; only its history is different.

> **S**tudy Hint: The un-
> shared pair of elec-
> trons on the nitrogen acts
> like an electrical plug or
> glue and forms a coordi-
> nate covalent bond by
> "plugging" or "gluing" to
> hydrogen. The +1 charge
> is over the entire ion.

6.7 Lewis Structures and Structural Formulas of More Complex Molecules and Polyatomic Ions

So far, we have considered only very simple molecules. Even in the most complicated case, NH_4^+, placing the dots representing valence electrons in the electron-
dot formulas and drawing the bonds between atoms and ions in the structural formulas were fairly easy. Real life is seldom so simple, though. Glucose, the basic
sugar your body uses to meet many of your energy needs, has the chemical formula
$C_6H_{12}O_6$, with many atoms, and consequently needs many more bonds to hold the
atoms together.

Lewis Structures

Electron-dot formulas are vitally important in depicting the structures of molecules. Indeed, Gilbert N. Lewis, an American chemist, first devised these formulas to help his general chemistry students understand compound formation. Thus, you need to understand how to write these formulas, called **Lewis structures**. The process is relatively simple if you follow three basic guidelines:

Lewis structures   Method of expressing electrons among atoms in a molecule using the rule of eight (octet rule) and dots (:) to represent electrons.

1. Write the electron-dot formulas for the elements that occur in the molecule. (Review Sections 4.7 and 5.3.)

2. Arrange the atoms so that each atom obeys the rule of eight (octet rule)* and hydrogen obeys the rule of two.

3. In molecules containing three or more atoms, the "central atom" acts as the starting point, with the other atoms arranged around it. The central atom is generally the *least* electronegative atom (excluding hydrogen). The least electronegative atom is most willing to share electrons with several other atoms.

Consider the following example for writing a Lewis structure using the periodic table.

Structural formula   Method of expressing chemical bonds among atoms in a molecule using lines to represent the bonds.

Bond angle   Angle defined by three atoms and two covalent bonds that connect them.

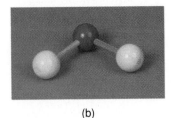

(a)

(b)

FIGURE 6.18
Structure of water (H_2O). (a) Structural formula for water showing the bond angle; (b) ball-and-stick model of water with the yellow balls representing hydrogen atoms and the red ball representing an oxygen atom.

| **EXAMPLE 6.2** | Write a Lewis structure for H_2O. |

SOLUTION:

Step 1:   Write the electron-dot formulas for the elements present (see the periodic table, guideline 1). There are 1 and 6 valence electrons for H and O, respectively. Each hydrogen must gain 1 electron, and oxygen must gain 2 electrons to complete their valence energy levels.

H·

·Ö:

H·

Step 2:   Arrange the atoms so that each obeys the rule of eight or two (guideline 2) and hydrogen does not occupy the central position (guideline 3). Each hydrogen atom now shares its valence electron with the oxygen atom to give a total of 8 electrons about the oxygen and 2 about each of the hydrogens.

H:Ö:
H   *Answer*

Structural Formulas and Bond Angles

Once a Lewis structure is drawn, it is convenient to convert it to a structural formula. A **structural formula** is a formula showing the arrangement of atoms within a molecule where a dash (—) is used to denote a pair of electrons shared between two atoms. Unshared pairs of electrons are not shown. Structural formulas show us the arrangement of atoms and the bonding between them without having so many dots. Figure 6.18a shows the structural formula for water.

The angle defined by the two O—H bonds is the bond angle. In water, this angle has been found to be 105°. A **bond angle** is the angle defined by three atoms and the two covalent bonds that connect them. Note that there is no such thing as a bond angle defined by two atoms alone; there must be two bonds and three atoms to define an angle. The ball-and-stick model in Figure 6.18b shows the bond angle for the water molecule.

---

*There are exceptions to this rule. The Lewis structures to these exceptions will be given to you and are covered in Section 6.8. You will not be asked to write Lewis structures for these exceptions.

Consider the following example for writing a Lewis structure and a structural formula for a compound.

EXAMPLE 6.3 | Write the structural formula for CH_4, methane (natural gas).

SOLUTION

Step 1: Write the electron-dot formulas for the elements present (see the periodic table, guideline 1). There are 4 and 1 valence electrons for C and H, respectively. Carbon must gain 4 electrons, and each hydrogen must gain 1 electron to complete their valence energy levels.

H·
H· ·Ċ·
H·
H·

Step 2: Arrange the atoms so that each obeys the rule of eight or two (guideline 2) and hydrogen does not occupy the central position (guideline 3). Each hydrogen atom now shares its valence electron with the carbon atom to give a total of 8 electrons about carbon and 2 electrons about each hydrogen.

H
H:C̈:H
H

> **Study Hint:** When you make electron-dot formulas, you can put the electrons anywhere you want to as long as you obey the rule of eight!

Step 3: Replace each shared pair of electrons with a dash.

$$
\begin{array}{c}
\text{H} \\
| \\
\text{H}-\text{C}-\text{H} \quad \textit{Answer} \\
| \\
\text{H}
\end{array}
$$

Figure 6.19a shows the ball-and-stick model for methane. This model does not show methane as a planar molecule as we have drawn it in the structural formula. Methane forms a tetrahedron. A tetrahedron is a three-sided-base pyramid as shown in Figure 6.19b.

Multiple Bonds

In some cases we may find that there are not enough electrons to supply every atom with eight electrons. In these cases we can often share more than two electrons between two atoms and solve the problem. Such bonds are called *multiple* bonds. If four electrons are shared, we have a **double bond**. If six electrons are shared, we have a **triple bond**. Examples 6.4 and 6.5 provide illustrations of these types of bonds.

(a)

(b)

FIGURE 6.19
(a) Ball-and-stick model of methane; the black ball represents a carbon atom; (b) a tetrahedron with the three-sided base shaded.

Double bond Chemical bond in which two atoms share two pairs of electrons (four electrons).

Triple bond Chemical bond in which two atoms share three pairs of electrons (six electrons).

EXAMPLE 6.4 | Write the structural formula for CO_2, carbon dioxide.

SOLUTION

Step 1: Write the electron-dot formulas for the elements present (see the periodic table, guideline 1). There are 4 and 6 valence electrons for C and O, respectively.

:Ö·
·Ċ·
:Ö·

Step 2: Arrange the atoms so that each obeys the rule of eight (guideline 2). Carbon is the central atom (guideline 3) because its electronegativity value, 2.5, is lower than the 3.5 value of oxygen (see Figure 6.11). By sharing *four* electrons between the carbon atom and each oxygen atom, we can form two *double bonds*. This sharing completes the valence energy level of eight for all the atoms as shown by the circles drawn in the second diagram.

Ö::C::Ö

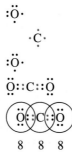
8 8 8

Step 3: Replace each shared pair of electrons by a dash. The *double bond* is represented by two dashes.

O=C=O   *Answer*

EXAMPLE 6.5   Write the structural formula for HCN, hydrogen cyanide.

SOLUTION

Step 1:   Write the electron-dot formulas for the elements present (see the periodic table, guideline 1). There are 1, 4, and 5 valence electrons for H, C, and N, respectively.

H·   ·Ċ·   ·N̈·

Step 2:   Arrange the atoms so that each obeys the rule of eight or two (guideline 2). Place the carbon atom in the center (guideline 3) because it has the lowest electronegativity value (2.5 versus 3.0 for nitrogen; see Figure 6.11). Bonding the hydrogen to the carbon by a single bond gives 2 electrons about the hydrogen. Sharing 3 electrons each from the carbon and nitrogen atoms creates a *triple bond* between the carbon and nitrogen.

H:Ċ·   ·N̈·

H:C:::N:

Step 3:   Replace each shared pair of electrons by a dash. The *triple bond* is represented by three dashes.

H—C≡N   *Answer*

Study Hint: The coordinate covalent bond is a situation where one atom provides both electrons for sharing—the other "mooches."

Coordinate Covalent Bonds

In some cases when drawing Lewis structures, we have more atoms than places to put them. In these cases one atom supplies *both* electrons, and the other atom, by shifting electrons, provides an empty orbital. This allows the formation of a *coordinate covalent bond.*

EXAMPLE 6.6   Write (a) the Lewis structure and (b) the structural formula for H_3PO_4, phosphoric acid.

SOLUTION

Step 1:   Write the electron-dot formulas for the elements present (see the periodic table, guideline 1). There are 1, 5, and 6 valence electrons for H, P, and O, respectively.

H°   ·P̈·   ·Ö:

H°   ·Ö:

H°   ·Ö:

            ·Ö:

Step 2:   Arrange the atoms so that each obeys the rule of eight (octet rule) or two (guideline 2). Place the phosphorus atom in the center (guideline 3) because it has the lowest electronegative value (2.1 versus 3.5 for oxygen; see Figure 6.11). Bond the 3 oxygen atoms to the phosphorus atom with covalent bonds and then bond the 3 hydrogen atoms to these 3 oxygen atoms using covalent bonds. This gives 8 electrons around the phosphorus and oxygen atoms and 2 electrons around each hydrogen atom. We still must account for *one more* oxygen atom. We can place this oxygen atom on the phosphorus atom after moving one of the *single* electrons on oxygen to form an empty orbital. This forms a coordinate covalent bond and obeys the rule of eight (octet rule) for both oxygen and phosphorus and gives (a) the Lewis structure.

H:Ö:P:Ö:H
    :Ö:
    H

:Ö:
H:Ö:P:Ö:H
    :Ö:
    H

Answer (a)

Step 3:   Replace each shared pair of electrons by a dash (—) to give (b) the structural formula.

$$
\begin{array}{c}
\text{O} \\
| \\
\text{H—O—P—O—H} \\
| \\
\text{O} \\
| \\
\text{H}
\end{array}
$$

Answer (b)

Lewis Structures and Structural Formulas for Polyatomic Ions

Up to now, we have looked only at Lewis structures and structural formulas of molecules. Many common compounds are composed of *polyatomic ions* such as the ammonium (NH_4^+) ion. A **polyatomic ion** is any ion made up of more than one atom. Table 6.5 lists various polyatomic ions. The second polyatomic ion in the

Polyatomic ion   Any ion made up of more than one atom.

TABLE 6.5	Some Common Polyatomic Ions and Their Formulas
FORMULA OF POLYATOMIC ION	**NAME OF POLYATOMIC ION**
+1 Ionic Charge	
NH_4^+	Ammonium
−1 Ionic Charge	
$C_2H_3O_2^{\,-}$	Acetate
ClO^-	Hypochlorite
$ClO_2^{\,-}$	Chlorite
$ClO_3^{\,-}$	Chlorate
$ClO_4^{\,-}$	Perchlorate
CN^-	Cyanide
$HCO_3^{\,-}$	Hydrogen carbonate or bicarbonate
$HSO_3^{\,-}$	Hydrogen sulfite or bisulfite
$HSO_4^{\,-}$	Hydrogen sulfate or bisulfate
OH^-	Hydroxide
$NO_2^{\,-}$	Nitrite
$NO_3^{\,-}$	Nitrate
$MnO_4^{\,-}$	Permanganate (see Figure 6.20)
−2 Ionic Charge	
CO_3^{2-}	Carbonate
$C_2O_4^{2-}$	Oxalate
CrO_4^{2-}	Chromate (see Figure 6.20)
$Cr_2O_7^{2-}$	Dichromate
SO_3^{2-}	Sulfite
SO_4^{2-}	Sulfate
−3 Ionic Charge	
PO_4^{3-}	Phosphate

FIGURE 6.20
Solutions of colored poly-atomic ions. Solutions of potassium permanganate ($KMnO_4$), left, and sodium chromate (Na_2CrO_4), right. (Courtesy Richard Megna. All rights reserved. Fundamental Photographs)

table is acetate ion. The formula for the acetate ion is $C_2H_3O_2^-$ (read "C-two H-three O-two minus). You *must* learn the names and formulas of these polyatomic ions so that you can use them to write formulas of compounds.

Once we know the formula for a polyatomic ion, we can easily construct a Lewis structure (and then a structural formula) for this substance. We follow the same three guidelines we used for Lewis structures of molecules and one additional guideline dealing with the charge on the ion:

Guideline 4:   When writing structures for *ions* (not molecules), be sure either to add an extra electron for each negative charge present or to take away an electron for each positive charge on the ion.

The reason for this guideline is that negative polyatomic ions have *extra* electrons present *beyond* the total number of valence electrons contributed by the neutral atoms that make up the polyatomic ion. Positively charged polyatomic ions have *fewer* electrons present than the total number of valence electrons contributed by the atoms in the ion. Thus, the ammonium ion (NH_4^+) has 8 electrons for bonding: 5 from nitrogen, 1 each from 4 hydrogens, less 1 electron for the $+1$ ionic charge. In contrast, the cyanide ion (CN^-) has 10 electrons: 4 from carbon, 5 from nitrogen, and 1 more for the -1 ionic charge.

The following examples illustrate the assignment of Lewis structures and structural formulas for polyatomic ions.

EXAMPLE 6.7   Write (a) the Lewis structure and (b) the structural formula for OH^-, the hydroxide ion.

SOLUTION

Step 1:   Write the electron-dot formulas for the elements present (see the periodic table, guideline 1). There are 1 and 6 valence electrons for H and O, respectively. Note that the addition of an *extra* electron (x) to account for the ionic charge of -1 gives a total of 8 bonding electrons (guideline 4).

$$
\begin{aligned}
1\,H &= 1\,e^- \\
1\,O &= 6\,e^- \\
-1 \text{ ionic charge} &= \underline{1\,e^-} \\
\text{total} &= 8\,e^-
\end{aligned}
$$

Step 2:   Arrange the hydrogen atom with the oxygen atom and distribute the 8 bonding electrons to follow the rule of eight for oxygen and the rule of two for hydrogen.

$H:\overset{..}{\underset{..}{O}}:^{x-}$ *Answer (a)*

Step 3:   Replace each shared pair of electrons by a dash. The ionic charge is generally placed as a superscript as shown. The ionic charge is dispersed over the entire ion.

$\left[O - H \right]^-$ *Answer (b)*

EXAMPLE 6.8   Write (a) the Lewis structure and (b) the structural formula for SO_4^{2-}, the sulfate ion.

SOLUTION

Step 1:   Write the electron-dot formulas for the elements present (see the periodic table, guideline 1). There are 6 valence electrons for both O and S. Note that the addition of 2 extra electrons ($\times$) to account for the ionic charge of -2 gives a total of 32 bonding electrons (guideline 4).

$$
\begin{aligned}
1\,S &= & 6\ e^- \\
4\,O = 6 \times 4 &= & 24\ e^- \\
-2\ \text{ionic charge} &= & \underline{2\ e^-} \\
\text{total} &= & 32\ e^-
\end{aligned}
$$

Step 2:   Arrange the atoms so that each obeys the rule of eight (guideline 2). Place the sulfur atom in the center (guideline 3) because it has the lower electronegativity value. The 32 bonding electrons are distributed to give 8 electrons about each oxygen and the sulfur. Note the coordinate covalent bonds to the oxygen above and below the sulfur.

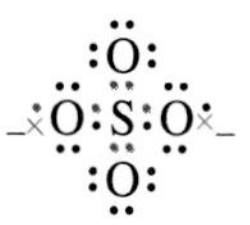

Answer (a)

Answer (b)

Step 3:   Replace each shared pair of electrons by a dash. The ionic charge again appears as a superscript. The -2 ionic charge is dispersed over the entire ion.

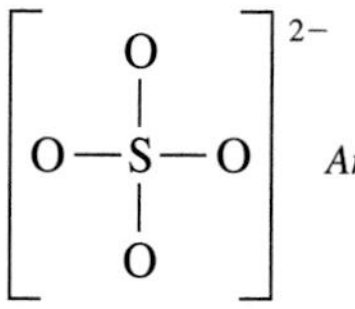

EXAMPLE 6.9   Write (a) the Lewis structure and (b) the structural formula for $NO_3{}^-$, the nitrate ion.

Step 1:   Write the electron-dot formulas for the elements present (see the periodic table, guideline 1). Note that the addition of one electron ($\times$) to account for the ionic charge of -1 gives a total of 24 bonding electrons (guideline 4).

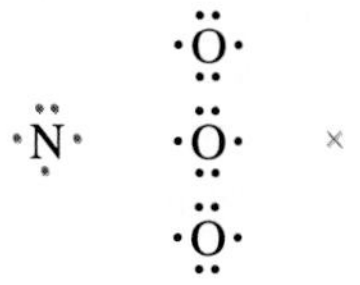

$$
\begin{aligned}
1\,N &= & 5\ e^- \\
3\,O = 6 \times 3 &= & 18\ e^- \\
-1\ \text{ionic charge} &= & \underline{1\ e^-} \\
\text{total} &= & 24\ e^-
\end{aligned}
$$

Step 2:   Arrange the atoms so that each obeys the rule of eight (guideline 2). Place the nitrogen atom in the center (guideline 3) because it has the lower electronegativity value and arrange the oxygens around the nitrogen. The formation of a double bond, a covalent bond, and a coordinate covalent bond completes the structure.

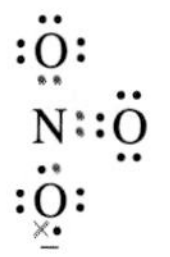

Step 3:   Replace each shared pair of electrons by a dash. The ionic charge again appears as a superscript. The -1 ionic charge is dispersed over the entire ion.

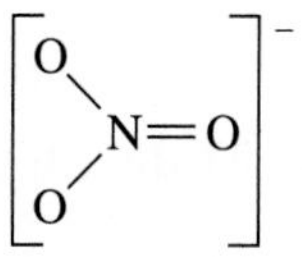

Study Exercise 6.6

Write the Lewis structures and structural formulas for the following molecules or polyatomic ions. (You may use the periodic table.)

a. CH_3Cl

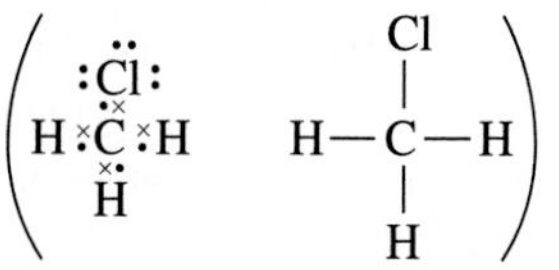

b. C_2H_6

$$\left(\begin{array}{cc} \text{H H} \\ \text{H}\overset{x}{\underset{o}{\text{C}}}\overset{\bullet}{\underset{\bullet}{\text{C}}}\text{:H} \\ \text{H H} \end{array} \qquad \begin{array}{c} \text{H H} \\ | \quad | \\ \text{H}-\text{C}-\text{C}-\text{H} \\ | \quad | \\ \text{H H} \end{array}\right)$$

c. C_2H_4

$$\left(\begin{array}{cc} \text{H H} \\ \text{H}\overset{x}{\underset{o}{\text{C}}}\overset{\bullet}{\underset{\bullet}{\text{:}}}\overset{\bullet}{\underset{o}{\text{C}}}\text{:H} \end{array} \qquad \begin{array}{c} \text{H}\diagdown \qquad \diagup\text{H} \\ \text{C}=\text{C} \\ \diagup \qquad \diagdown \\ \text{H} \qquad \text{H} \end{array}\right)$$

d. $PO_4{}^{3-}$

$$\left(\begin{array}{c} :\overset{\circ\circ}{\underset{\circ\circ}{\text{O}}}: \\ {}^{-}\overset{x}{\underset{\circ\circ}{\text{O}}}:\overset{\bullet\bullet}{\underset{\bullet\bullet}{\text{P}}}:\overset{\circ\circ}{\underset{\circ\circ}{\text{O}}}{}^{-} \\ :\overset{\circ\circ}{\underset{\circ x}{\text{O}}}:{}^{-} \end{array} \qquad \left[\begin{array}{c} \text{O} \\ | \\ \text{O}-\text{P}-\text{O} \\ | \\ \text{O} \end{array}\right]^{3-}\right)$$

Work Problems 12 and 13.

<table><tr><td>6.8</td><td># The Shapes of Molecules and Polyatomic Ions</td></tr></table>

Lewis structures and structural formulas are useful in helping us understand the bonds that hold atoms together to form molecules and polyatomic ions. They do not give us a feeling for the three-dimensional *shapes* these species take on. This is important to chemists, for it is the three-dimensional shapes of molecules and polyatomic ions that usually determine how these species interact with each other.

The model we use to determine the shape of molecules and polyatomic ions is sometimes called the ***valence shell electron pair repulsion*** (***VSEPR***) model. This model for understanding the factors that determine the shapes of molecules and ions relies on Coulomb's law (see Section 6.5). Most molecules or polyatomic ions have a central atom to which other atoms are attached. The shape of the molecule depends on the number of pairs of electrons (see Section 6.6) surrounding the central atom. Coulomb's law suggests that these pairs of electrons (both bonding pairs and unshared pairs) arrange themselves about the central atom in such a way as to maximize the distances among all of these electron pairs. In doing this, the electron pairs minimize the repulsive energy generated among them.

Linear Molecules

For any given number of electron pairs about a central atom, there is one arrangement that maximizes these distances and minimizes the repulsions among all of the electron pairs. An analogy may help to illustrate this concept. When two cartoon characters, say a cat and a mouse, have a chase scene, the pair invariable ends up on opposite sides of a table. If the cat moves in one direction (clockwise), the mouse responds by also moving in the same direction (clockwise) in order to keep the distance between the two at a maximum. The cat cannot get any closer.

This analogy applies to a molecule where there are two atoms attached to a central atom by means of two single bonds. For example, beryllium hydride (BeH_2) is represented as

$$H \overset{..}{.} Be \overset{..}{.} H \qquad H - Be - H$$

Lewis structure*   structural formula

The shape of BeH_2 can be modeled by considering that the *two* bonding pairs of electrons maintain a maximum distance between them, like the cat and the mouse. Thus, the $H - Be - H$ bond angle is 180°, and such a geometry is termed ***linear***.

$$H \overset{..}{.} Be \overset{..}{.} H$$
$$180°$$

Carbon dioxide (CO_2) is another example. Here, we have two atoms attached to a central carbon atom by means of two double bonds (see Section 6.7, Example 6.4).

$$\overset{..}{O} :: C :: \overset{..}{O}$$

Since the four electrons in the double bond act as a *unit* in bonding the C and O atoms together, we consider them to be *one group* of electrons for the purpose of shape. Thus, there are *two groups* of electrons ($4e^-$ in each group) about the central atom (C). Carbon dioxide is a ***linear*** molecule, and the $O - C - O$ bond angle is 180°.

$$\overset{..}{O} :: C :: \overset{..}{O}$$
$$180°$$

Trigonal Planar Molecules and Ions

Let's examine the optimal shape associated with *three* groups of electrons around a central atom. An example of such a species is boron trifluoride (BF_3). The central boron atom is surrounded by three pairs of bonding electrons which serve to attach the three fluorine atoms to boron.

Lewis structure*   structural formula

The best way to arrange three pairs of electrons is with 120° bond angles about the boron, with all four atoms lying in a plane with the solid lines representing bonds. The broken lines are intended to show that the atoms are all in a single plane. This shape is termed *trigonal planar.*

*This Lewis structure represents an exception to the rule of eight.

The carbonate ion is another example of a trigonal planar species. The Lewis structure is given below, and you can see that there are *three groups* of electrons about the central atom, two pairs of electrons associated with C — O single bonds, and a group of four electrons associated with the C = O double bond. This ion is **trigonal planar** and has a 120° O — C — O bond angle.

Tetrahedral Molecules

When *four* groups of electrons surround a central atom, the best arrangement for the groups is tetrahedral (see Figure 6.19b), where each pair of electrons points toward the corner of a tetrahedron. This means that the bond angles are all equal and are 109.5°. An example of such a shape is methane (CH_4) (see Section 6.7, Example 6.3, Figure 6.19). For methane, there are four C — H single bonds, with each pair of bonding electrons acting as a group of electrons. The four groups try to stay as far apart from each other as possible. Hence, the **tetrahedral** shape. The hydrogen atoms all occupy the corners of a tetrahedron.

Here --- indicates behind the plane of the page, ► indicates out from the plane of the page— toward you, and — indicates in the plane of the page.

tetrahedron

Special Cases of Tetrahedral: Bent and Pyramidal Molecules

Tetrahedral has two special cases: (1) bent and (2) pyramidal.

You might ask, "Why is water shown as a bent molecule in Figure 6.18, Example 6.2?" That is a good question. The answer lies in the fact that there are *four* groups of electrons around the central O atom, *two pairs* of bonding electrons (O — H, single bonds) and *two pairs* of unshared electrons. Thus, the four groups of electrons possess roughly the same shape as the four groups of electrons in methane. However, only two of the groups of electrons serve to attach hydrogen atoms. Hence, the only bond angle we see is the H — O — H bond, which is predicted by our model to be about 109.5°. The experimental value of 105° is quite close to this number. We will consider this again in Chapter 13 (Section 13.2). We refer to such a shape as *bent* since the molecule looks "bent."

unshared pairs of electrons

bent

A similar argument can be made for the pyramidal shape that ammonia exhibits (see Figure 6.16). There are also *four* groups of electrons about the central nitrogen atom, three pairs of electrons serving as N—H single bonds, and *one unshared pair* of electrons. The overall effect is that while the four pairs of electrons have a tetrahedral arrangement with bond angles of approximately 109.5°, the ammonia molecule looks like a pyramid since we can "see" only the nitrogen and hydrogen atoms. The experimental bond angle of 107° is quite close to our predicted value of 109.5°, and we classify the shape as *pyramidal.*

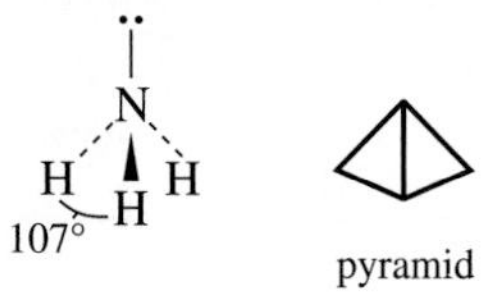

pyramid

In determining the shape and bond angle of molecules or polyatomic ions we follow these guidelines:

1. Write the Lewis structure for the molecule or polyatomic ion. (We will give you the Lewis structures for molecules or polyatomic ions that are exceptions to the rule of eight.)

2. Determine the total number of electron pairs or groups of electrons around the central atom.

3. Arrange these electrons, including the unshared pairs, around the central atom obeying Coulomb's law with maximum distance between all the electron pairs.

4. Use the total number of electron pairs or groups of electrons and unshared pairs to determine the shape and bond angle of the molecule or polyatomic ion as shown in Table 6.6.

Now let's consider some examples.

EXAMPLE 6.10

Determine the shape (linear, trigonal planar, tetrahedral, bent, or pyramidal) and the bond angle of the following molecules. (You may use the periodic table.)

a. PH_3 **b.** CS_2

SOLUTION

a. Using the periodic table, P has 5 valence electrons and H has 1, and so the Lewis structure (guideline 1) for PH_3 is

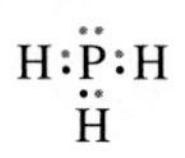

$$H{:}\overset{\cdot\cdot}{\underset{\overset{\displaystyle|}{H}}{P}}{:}H$$

The total number of electron pairs or groups (guideline 2) is 4. Based on Coulomb's law (guideline 3), the structures are

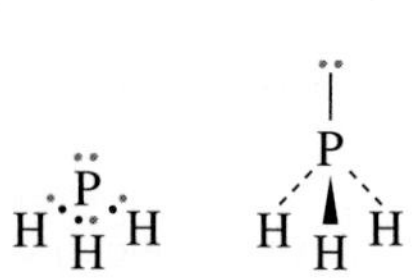

With 3 bonding pairs and 1 unshared pair of electrons (Table 6.6), the shape is pyramidal with a bond angle of about 109.5°.   *Answer*

| TABLE 6.6 | | | Shape and Bond Angle of Molecules and Polyatomic Ions[a] | | |

TOTAL ELECTRON PAIRS OR GROUPS	BONDING PAIRS OR GROUPS	UNSHARED PAIRS OF ELECTRONS	SHAPE	BOND ANGLE	EXAMPLE(S)
2	2	0	X—A—X **linear**	180°	H—Be—H O=C=O
3	3	0	trigonal planar	120°	F, F, B, F; $\left[\begin{array}{c} O, O \\ C \\ O \end{array}\right]^{2-}$
4	4	0	tetrahedral	109.5°	H, C, H, H, H
4	2	2	bent	109.5°	O, H, H
4	3	1	pyramidal	109.5°	N, H, H, H

[a] - - -, behind the plane of the page; ►, out from the plane of the page—toward you; —, in the plane of the page.

b. Using the periodic table, C has 4 valence electrons and S has 6, and so the Lewis structure (guideline 1) for CS_2 is

$$\ddot{S}::C::\ddot{S}$$

The total number of electron groups (guideline 2) is 2. There are 4 electrons in a group, and one group acts as a unit. Based on Coulomb's law (guideline 3), the structure is

$$S=C=S$$

There are 2 electron bonding pair groups and 0 unshared pairs of electrons (Table 6.6). The unshared pairs of electrons on the oxygen atoms do not count, as they are not attached to the central carbon atom. The shape is linear with a bond angle of 180°. *Answer*

Study Exercise 6.7
Determine the shape (linear, trigonal planar, tetrahedral, bent, or pyramidal) and the approximate bond angle of the following molecules. (You may use the periodic table).

a. CCl_4 (tetrahedral, 109.5°)   Work Problem 14.
b. H_2S (bent, 109.5°)

6.9 Writing Formulas

We will now use the names and formulas of the cations (Table 6.1), anions (Table 6.2), and polyatomic ions (Table 6.5 and Figure 6.22) to write formulas of compounds. To write the correct formula for a compound, we *must know* or have given to us the ionic charges of the cations and anions. In these formulas, *the sum of the total positive charges must be equal to the sum of the total negative charges*; that is, *the compound must **not** possess a net charge.* When the charge on the positive ion is *not* equal to the charge on the negative ion, we use subscripts to balance the positive charges with the negative charges. In most cases, we write the positive ion first and then the negative ion. For example, in iron(II) bromide, in order to have positive charges equal to the total negative charges, we need to have *one* Fe^{2+} for every *two* Br^-. The +2 charge of the iron is just balanced by the −2 charge on the two bromides. Iron(II) bromide can be written as $Fe^{2+}Br^-Br^-$ or as $Fe^{2+}(Br^-)_2$, using a subscript. To simplify the formula, delete the charges and write the formula $FeBr_2$. This result highlights two rules for writing formulas:

1. By convention, *the simplest whole-number ratio of the elements is usually used for writing formulas.* Thus, iron(II) bromide is written as $FeBr_2$ and not Fe_2Br_4 or Fe_3Br_6.

2. As a result of the use of the simplest whole-number ratio in writing formulas, *only a single formula can be written for any compound.*

We will now consider more examples to illustrate writing formulas. The names and formulas of the ions in Tables 6.1, 6.2, and 6.5 will be given here, but by the time we cover nomenclature (Chapter 7), you should have learned these names and formulas.

1. sodium (Na^+) and $(Na^+)(Cl^-)$, NaCl
    chloride (Cl^-) $+1 + (-1) = 0$

2. barium (Ba^{2+}) and $(Ba^{2+})(F^-)_2$, BaF_2
    fluoride (F^-) $+2 + 2(-1) = 0$

3. aluminum (Al^{3+}) and $(Al^{3+})(Br^-)_3$, $AlBr_3$
    bromide (Br^-) $+3 + 3(-1) = 0$

4. iron(III) (Fe^{3+}) and $(Fe^{3+})_2(S^{2-})_3$, Fe_2S_3
    sulfide (S^{2-}) $2(+3) + 3(-2) = 0$
    Note: The least common multiple is 6, hence 2(3) and 3(2).

Study Hint: After you have mastered nomenclature, you will be asked to write chemical equations (Chapter 9) using these formulas. Using these equations, you will then be asked to determine quantities used or obtained in a given chemical equation (Chapter 10). Therefore, to be successful in chemistry *you must memorize* the formulas of the ions in Table 6.1, 6.2, and 6.5.

5. copper(II) (Cu^{2+}) and nitrate (NO_3^-)

$(Cu^{2+})(NO_3^-)_2$, $Cu(NO_3)_2$

$+2 + 2(-1) = 0$

Note: There are two nitrate ions; thus, we must use parentheses. These parentheses mean that there are 2 atoms of nitrogen, 6 atoms of oxygen, and 1 atom of copper in *one* formula unit of copper(II) nitrate.

6. lithium (Li^+) and sulfate (SO_4^{2-})

$(Li^+)_2(SO_4^{2-})$, Li_2SO_4

$2(+1) + (-2) = 0$

7. mercury(I) (Hg_2^{2+}) and acetate ($C_2H_3O_2^-$)

$(Hg_2^{2+})(C_2H_3O_2^-)_2$, $Hg_2(C_2H_3O_2)_2$

$+2 + 2(-1) = 0$

Note: From Table 6.1, mercury(I) is Hg_2^{2+}, a dimer ion; therefore, the correct formula for mercury(I) acetate is $Hg_2(C_2H_3O_3)_2$.

8. ammonium (NH_4^+) and sulfite (SO_3^{2-})

$(NH_4^+)_2(SO_3^{2-})$, $(NH_4)_2SO_3$

$2(+1) + (-2) = 0$

9. strontium (Sr^{2+}) and phosphate (PO_4^{3-})

$(Sr^{2+})_3(PO_4^{3-})_2$, $Sr_3(PO_4)_2$

$3(+2) + 2(-3) = 0$

10. magnesium (Mg^{2+}) and hydrogen carbonate (HCO_3^-)

$(Mg^{2+})(HCO_3^-)_2$, $Mg(HCO_3)_2$

$+2 + 2(-1) = 0$

11. calcium (Ca^{2+}) and permanganate (MnO_4^-)

$(Ca^{2+})(MnO_4^-)_2$, $Ca(MnO_4)_2$

$+2 + 2(-1) = 0$

12. potassium (K^+) and dichromate ($Cr_2O_7^{2-}$)

$(K^+)_2(Cr_2O_7^{2-})$, $K_2Cr_2O_7$

$2(+1) + (-2) = 0$

13. aluminum (Al^{3+}) and oxalate ($C_2O_4^{2-}$)

$(Al^{3+})_2(C_2O_4^{2-})_3$, $Al_2(C_2O_4)_3$

$2(+3) + 3(-2) = 0$

You should begin memorizing the names and formulas in Tables 6.1, 6.2, and 6.5 *now* because you will need them in the next chapter when we discuss the nomenclature of compounds.

Study Exercise 6.8

Write the correct formula for the compound formed by the combination of the following ions:

a. potassium (K^+) and sulfide (S^{2-})   (K_2S)
b. barium (Ba^{2+}) and iodide (I^-)   (BaI_2)
c. aluminum (Al^{3+}) and sulfate (SO_4^{2-})   [$Al_2(SO_4)_3$]
d. lead(II) (Pb^{2+}) and phosphate (PO_4^{3-})   [$Pb_3(PO_4)_2$]

Work Problems 15 and 16.

6.10 Using the Periodic Table for Predicting Oxidation Numbers, Properties, Formulas, and Types of Bonding in Compounds

The periodic table can be most helpful to you in learning the ionic charges on the cations (Table 6.1) and anions (Table 6.2). It can also help you to predict properties, formulas, and types of bonding in compounds.

Oxidation Numbers

In general, the *Roman group numeral* represents the *maximum positive oxidation number* for the elements in that group.* For example, aluminum is in group IIIA (13) and hence has a +3 oxidation number or ionic charge (see Table 6.1). For the nonmetals, the Roman numeral represents the maximum positive oxidation number. Also, for the nonmetals we can calculate the *negative oxidation number* by *subtracting 8* from the Roman group number. For example, chlorine, in group VIIA (17), has a maximum positive oxidation number of +7 (group VII) in $KClO_4$ and a negative oxidation number of -1 (VII $-$ 8 $=$ -1) in KCl. See Table 6.2 for the ionic charge on the chloride ion. Sulfur, in group VIA (16), has a maximum positive oxidation number of +6 (group VI) in H_2SO_4, and a negative oxidation number of -2 (VI $-$ 8 $=$ -2) in H_2S. Review Section 6.3 on calculating oxidation numbers of elements.

Using the maximum positive and negative oxidation numbers, we can also predict the formulas of some compounds containing two different elements (*binary compounds*). When barium and iodine form a binary compound, the formula is BaI_2. Barium is in group IIA (2) and has a +2 oxidation number or ionic charge, while iodine is in group VIIA (17) and has a -1 negative oxidation number or ionic charge (VII $-$ 8 $=$ -1). The correct formula (see Section 6.9) is $(Ba^{2+})(I^-)_2$, BaI_2. This prediction of formulas applies primarily to the A group elements. We can determine the positive oxidation numbers or ionic charges on some of the cations in Table 6.1 by using the periodic table. And by using the periodic table, we can also determine the negative oxidation numbers or ionic charges on all the anions given in Table 6.2.

Study Exercise 6.9
Using the periodic table, indicate a maximum positive oxidation number for each of the following elements. For those that are nonmetals, give *both* the maximum positive oxidation number and the negative oxidation number. (If you do not know the symbol for the element, look it up on the inside front cover of this text.)

a. calcium   (+2)   **b.** phosphorus   (+5, -3)
c. rubidium   (+1)   **d.** tellurium   (+6, -2)

Study Exercise 6.10
Using the periodic table to determine the oxidation numbers, predict the formulas of the binary compounds formed from the following combinations of elements. (If you do not know the symbol for the element, look it up on the inside front cover of this text.)

a. calcium and bromine $(CaBr_2)$
b. calcium and nitrogen (Ca_3N_2)
c. indium and phosphorus (InP)
d. indium and selenium (In_2Se_3) Work Problems 17 and 18.

We can now use the general characteristics outlined in Section 5.3 for predicting properties of elements, formulas of compounds, and types of bonding in compounds.

*The maximum positive oxidation number is not always the most common oxidation number. In all cases, the oxidation number is also the ionic charge on the monatomic ion.

Properties

For example, consider the following list of atomic radii of elements in group VIA (16). Can we estimate the radius of the fourth element in the group, tellurium (Te)?

ELEMENT	RADIUS (pm)[a]
O	74
S	104
Se	117
Te	?

[a] For the picometer unit see Section 2.2.

The radii increase because an additional principal energy level is added each time we move down a member in the group. Thus, we expect the radius of tellurium to be larger than that of selenium (Se). We can estimate the value of this radius by taking the difference between the radii of sulfur and selenium and adding it to the radius of selenium. Hence, we estimate that the radius of tellurium will be $13\overline{0}$ pm [117 + (117 − 104)]. It has been found to be 137 pm.

We can apply this same general procedure to many of the properties of the elements with reasonable reliability. Consider the ionization energies (see Section 6.2) of the first three elements in group VIIA (17) and then see if we can predict the ionization energy of the fourth element in the group, iodine (I).

ELEMENT	FIRST IONIZATION ENERGY (kJ/mol)
F	1681
Cl	1251
Br	1140
I	?

The ionization energy decreases as we move down the group, and so we expect the ionization energy of I to be less than that of Br. This generalization applies to all the A group elements. It makes sense because each time we move down a group in the periodic table we add a new principal energy level. The electrons occupying this new level are farther from the nucleus and hence less tightly held. We can estimate the ionization energy for I by taking the difference between the ionization energy of Cl and Br and subtracting it from the value for Br. Hence, we would expect the ionization energy for iodine (I) to be about 1029 kJ/mol [$114\overline{0}$ − (1251 − $114\overline{0}$)]. This energy has been determined experimentally to be 1013 kJ/mol.

Study Exercise 6.11

Estimate the missing value for the following:

ELEMENT	BP (°C, 1 atm)
He	−269
Ne	−246
Ar	−186
Kr	−152
Xe	−107
Rn	?

(−62°C, actual −62°C)

Work Problem 19.

Formulas

In Section 5.3 we mentioned that because the electronic configurations of all elements in a group are similar, the formulas of compounds of elements in that group are similar. Consider the following examples.

✔ The formula for calcium bromide is CaBr$_2$. Radium is in the same group (IIA, 2) as calcium. The formula for radium (Ra) bromide is RaBr$_2$.

✔ The formula for water is H$_2$O. Tellurium is in the same group (VIA, 16) as oxygen. The formula for hydrogen telluride (Te) is H$_2$$Te$.

✔ The formula for magnesium sulfate is MgSO$_4$. Strontium is in the same group (IIA, 2) as magnesium, and selenium is in the same group (VIA, 16) as sulfur. The formula for strontium selenate (Sr is strontium and **Se** is selenium) is Sr**Se**O$_4$.

Study Exercise 6.12
The following are examples of compounds and their formulas:

calcium chloride, CaCl$_2$
calcium chlorate, Ca(ClO$_3$)$_2$

Using the periodic table, write the formulas for the following compounds. If you do not know the symbol for the element, look it up on the inside front cover of this text. (*Hint*: Note the ending of each name for the compound.)

a. magnesium chlorate   [Mg(ClO$_3$)$_2$]
b. barium chloride   (BaCl$_2$)
c. radium iodide   (RaI$_2$)
d. radium iodate   [Ra(IO$_3$)$_2$]

Work Problem 20.

Type of Bonding—Ionic or Covalent

Binary Compounds   Earlier in this chapter (Section 6.4) we stated that the term "molecule" is reserved for compounds bonded primarily by covalent bonds, and the term "formula unit" for compounds bonded with ionic bonds. In compounds consisting of only two *different* elements (**binary compounds**), the greater the *difference* in the electronegativity of the elements, the greater the ionic bonding of the compound. The halogens (group VIIA, 17) have high electronegativities. Therefore, if they combine with elements having relatively low electronegativities, an ionic compound is formed. In contrast, the alkali metals [group IA (1) *except* hydrogen] and alkaline earth metals [group IIA (2)] have low electronegativities. Therefore, we can make a general statement that if binary compounds are formed between elements in *group IA* (1, except hydrogen) or *group IIA* (2) with elements in *group VIIA* (17) or *group VIA* (16, *oxygen* and *sulfur* only), ionic compounds result. Because both fluorine and oxygen have high electronegativities, any compound formed with *fluorine* or *oxygen* and a *metal* is also classified as an ionic compound. Hence, the smallest unit in these ionic compounds is the *formula unit*. Figure 6.21 summarizes these statements. Consider some examples:

Binary compounds   Compounds containing two different elements.

✔ Strontium chloride (SrCl$_2$) is an ionic compound because strontium is in group IIA (2) and chloride is in group VIIA (17).

✔ Potassium oxide (K$_2$O) is an ionic compound since potassium is in group IA (1) and oxygen is in group VIA (16) and because a compound of any *metal* with *oxygen* is considered ionic.

GROUPS

PERIODS	1 IA	2 IIA	3 IIIB	4 IVB	5 VB	6 VIB	7 VIIB	8	9 VIII	10	11 IB	12 IIB	13 IIIA	14 IVA	15 VA	16 VIA	17 VIIA	18 VIIIA
1	H 1																	He 2
2	Li 3	Be 4											B 5	C 6	N 7	O 8	F 9	Ne 10
3	Na 11	Mg 12			TRANSITION ELEMENTS								Al 13	Si 14	P 15	S 16	Cl 17	Ar 18
4	K 19	Ca 20	Sc 21	Ti 22	V 23	Cr 24	Mn 25	Fe 26	Co 27	Ni 28	Cu 29	Zn 30	Ga 31	Ge 32	As 33	Se 34	Br 35	Kr 36
5	Rb 37	Sr 38	Y 39	Zr 40	Nb 41	Mo 42	Tc 43	Ru 44	Rh 45	Pd 46	Ag 47	Cd 48	In 49	Sn 50	Sb 51	Te 52	I 53	Xe 54
6	Cs 55	Ba 56	*La 57	Hf 72	Ta 73	W 74	Re 75	Os 76	Ir 77	Pt 78	Au 79	Hg 80	Tl 81	Pb 82	Bi 83	Po 84	At 85	Rn 86
7	Fr 87	Ra 88	**Ac 89	Rf 104	Ha 105	Sg 106	Ns 107	Hs 108	Mt 109	-- 110	-- 111							

*Lanthanide series	Ce 58	Pr 59	Nd 60	Pm 61	Sm 62	Eu 63	Gd 64	Tb 65	Dy 66	Ho 67	Er 68	Tm 69	Yb 70	Lu 71
**Actinide series	Th 90	Pa 91	U 92	Np 93	Pu 94	Am 95	Cm 96	Bk 97	Cf 98	Es 99	Fm 100	Md 101	No 102	Lr 103

FIGURE 6.21(a)

Binary *ionic* compounds are formed in two ways: **(a)** from elements in group VIIA (17, ▪) or group VIA (16, ▪ oxygen and sulfur only) with elements in group IA (1, ▪ except hydrogen) or group IIA (2, ▪); or **(b)** from any metal (▪) with fluorine (▪) or oxygen (▪).

✔ Iron(III) fluoride (FeF_3) is an ionic compound because any compound formed with *fluorine* and a *metal* is considered ionic.

Other binary compounds are considered to be covalent and hence are referred to as *molecules.* Examples are carbon dioxide (CO_2; carbon is a nonmetal), sulfur dioxide (SO_2; sulfur is a nonmetal), water (H_2O), and methane (CH_4). All general statements have exceptions, but knowing that a binary compound is considered to be ionic if it is formed from elements in certain groups in the periodic table will help you in further studying the properties of compounds.

Ternary compounds Compounds containing three different elements.

Ternary Compounds The preceding general statements about predicting the type of bonding in compounds applies only to *binary* compounds. Now let us consider **ternary compounds**, which contain three different elements, and higher compounds. These compounds involve polyatomic ions. In general, the combination of *any element* (*hydrogen* being the sole exception) with any *polyatomic ion* to form a

GROUPS

PERIODS	1 IA	2 IIA	3 IIIB	4 IVB	5 VB	6 VIB	7 VIIB	8	9 VIII	10	11 IB	12 IIB	13 IIIA	14 IVA	15 VA	16 VIA	17 VIIA	18 VIIIA
1	H 1																	He 2
2	Li 3	Be 4											B 5	C 6	N 7	O 8	F 9	Ne 10
3	Na 11	Mg 12				TRANSITION ELEMENTS							Al 13	Si 14	P 15	S 16	Cl 17	Ar 18
4	K 19	Ca 20	Sc 21	Ti 22	V 23	Cr 24	Mn 25	Fe 26	Co 27	Ni 28	Cu 29	Zn 30	Ga 31	Ge 32	As 33	Se 34	Br 35	Kr 36
5	Rb 37	Sr 38	Y 39	Zr 40	Nb 41	Mo 42	Tc 43	Ru 44	Rh 45	Pd 46	Ag 47	Cd 48	In 49	Sn 50	Sb 51	Te 52	I 53	Xe 54
6	Cs 55	Ba 56	*La 57	Hf 72	Ta 73	W 74	Re 75	Os 76	Ir 77	Pt 78	Au 79	Hg 80	Tl 81	Pb 82	Bi 83	Po 84	At 85	Rn 86
7	Fr 87	Ra 88	**Ac 89	Rf 104	Ha 105	Sg 106	Ns 107	Hs 108	Mt 109	-- 110	-- 111							

*Lanthanide series	Ce 58	Pr 59	Nd 60	Pm 61	Sm 62	Eu 63	Gd 64	Tb 65	Dy 66	Ho 67	Er 68	Tm 69	Yb 70	Lu 71
**Actinide series	Th 90	Pa 91	U 92	Np 93	Pu 94	Am 95	Cm 96	Bk 97	Cf 98	Es 99	Fm 100	Md 101	No 102	Lr 103

FIGURE 6.21(b)

ternary or higher compound results in an *ionic compound* because the polyatomic ion can readily accommodate the positive or negative ionic charge over its many atoms. Consider some examples:

✔ Sodium sulfate (Na_2SO_4) is an ionic compound because sulfate (SO_4^{2-}) is a polyatomic ion.

✔ Silver nitrate ($AgNO_3$) is an ionic compound because nitrate (NO_3^-) is a polyatomic ion.

✔ Ammonium chlorate (NH_4ClO_3) is an ionic compound because both ammonium (NH_4^+) and chlorate (ClO_3^-) are polyatomic ions.

Study Exercise 6.13
Using the periodic table, classify the following compounds as essentially ionic or covalent:

a. CsCl	(ionic)	**b.** SO_2	(covalent)	
c. SrI_2	(ionic)	**d.** CrF_3	(ionic)	
e. Ag_2O	(ionic)	**e.** $Ca(MnO_4)_2$	(ionic)	

Work Problem 21.

Take some time to study the periodic table shown in Figure 6.22. This figure summarizes *much* of the material in this chapter.

PERIODIC TABLE OF THE ELEMENTS

	alkali metals	alka-line earth metals																halo-gens	noble gases
maximum positive ox. no.	1^+	2^+	3^+	4^+	5^+	6^+	7^+		8^+		1^{+a}	2^+	3^+	4^+	5^+	6^+	7^+	
maxiumum negative ox. no.	$-$	$-$	$-$	$-$	$-$	$-$	$-$		$-$		$-$	$-$	$-$	$-$	3^-	2^-	1^-	$-$
	1	2	3	4	5	6	7	8	9	10	11	12	13	14	15	16	17	18
PERIODS	IA	IIA	IIIB	IVB	VB	VIB	VIIB		VIII		IB	IIB	IIIA	IVA	VA	VIA	VIIA	VIIIA

GROUPS

TRANSITION ELEMENTS

Period																		
1	1.008 H 1 hydrogen																	4.003 He 2 helium
2	6.941 Li 3 lithium	9.012 Be 4 beryllium											10.811 B 5 boron	12.011 C 6 carbon	14.007 N 7 nitrogen	15.999 O 8 oxygen	18.998 F 9 fluorine	20.179 Ne 10 neon
3	22.990 Na 11 sodium	24.305 Mg 12 magnesium											26.982 Al 13 aluminum	28.0855 Si 14 silicon	30.9738 P 15 phosphorous	32.06 S 16 sulfur	35.453 Cl 17 chlorine	39.948 Ar 18 argon
4	39.0983 K 19 potassium	40.08 Ca 20 calcium	44.956 Sc 21 scandium	47.90 Ti 22 titanium	50.9415 V 23 vandium	51.996 Cr 24 chromium	54.938 Mn 25 manganese	55.847 Fe 26 iron	58.933 Co 27 cobalt	58.71 Ni 28 nickel	63.546 Cu 29 copper	65.37 Zn 30 zinc	69.72 Ga 31 gallium	72.59 Ge 32 germanium	74.922 As 33 arsenic	78.96 Se 34 selenium	79.904 Br 35 bromine	83.80 Kr 36 kryton
5	85.468 Rb 37 rubidium	87.62 Sr 38 strontium	88.906 Y 39 yttrium	91.22 Zr 40 zirconium	92.9064 Nb 41 niobium	95.94 Mo 42 molybdenum	98.906 Tc 43 technetium	101.07 Ru 44 ruthenium	102.906 Rh 45 rhodium	106.4 Pd 46 palladium	107.868 Ag 47 silver	112.41 Cd 48 cadmium	114.82 In 49 indium	118.69 Sn 50 tin	121.75 Sb 51 antimony	127.60 Te 52 tellurium	126.90 I 53 iodine	131.30 Xe 54 xenon
6	132.906 Cs 55 cesium	137.33 Ba 56 barium	138.906 *La 57 lanthanum	178.49 Hf 72 hafnium	180.948 Ta 73 tantalum	183.85 W 74 tungsten	186.2 Re 75 rhenium	190.2 Os 76 osmium	192.22 Ir 77 iridium	195.09 Pt 78 platinum	196.967 Au 79 gold	200.59 Hg 80 mercury	204.37 Tl 81 thallium	207.2 Pb 82 lead	208.981 Bi 83 bismuth	(209) Po 84 polonium	(210) At 85 astatine	(222) Rn 86 radon
7	(223) Fr 87 francium	226.025 Ra 88 radium	(227) **Ac 89 actinium	(261) Rf 104 rutherfordium	(262) Ha 105 hahnium	(263) Sg 106 seaborgium	(262) Ns 107 neilsbohrium	(261) Hs 108 hassium	(266) Mt 109 meltnerium	(269) –– 110 ––	(272) –– 111 ––							

*Lanthanide series

140.12 Ce 58 cerium	140.908 Pr 59 praseo-dymium	144.24 Nd 60 neodymium	(145) Pm 61 promethium	150.4 Sm 62 samarium	151.96 Eu 63 europium	157.25 Gd 64 gadolinium	158.925 Tb 65 terbium	162.50 Dy 66 dysprosium	164.930 Ho 67 holmium	167.26 Er 68 erbium	168.934 Tm 69 thulium	173.04 Yb 70 ytterbium	174.967 Lu 71 lutetium

**Actinide series

232.038 Th 90 thorium	231.031 Pa 91 protactinium	238.029 U 92 uranium	237.048 Np 93 neptunium	(244) Pu 94 plutonium	(243) Am 95 americium	(247) Cm 96 curium	(247) Bk 97 berkelium	(251) Cf 98 californium	(254) Es 99 einsteinium	(257) Fm 100 fermium	(256) Md 101 mende-levium	(255) No 102 nobelium	(257) Lr 103 lawrencium

▭ metals;   ▭ metalloids;   ▭ nonmetals;   ▭ noble gases

Numbers below the symbol of the element indicate the atomic numbers. Atomic masses, above the symbol of the element, are based on the assigned relative atomic mass of ^{12}C = exactly 12; ( ) indicates the mass number of the isotope with the longest half-life.

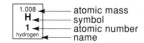

^{a}Certain elements in Group IB also form 2^+ and 3^+ oxidation numbers.

FIGURE 6.22

Periodic table summarizing most of the generalizations of the elements discussed in this chapter and previous chapters.

CHEMISTRY OF THE ATMOSPHERE

Sunlight, Energy for the Earth

Light (*electromagnetic radiation*) from the sun is the ultimate source of all energy on earth. Plant life absorbs sunlight and converts it to high-energy chemicals such as sugars, starches, and oils. Animals eat these plants and break down these chemicals to release the energy they need to live. Similarly, humans use fossil fuels (natural gas, oil, and gasoline), which result from the decomposition of ancient plant and animal life, to provide energy in the form of heat, electricity, and locomotion.

Understanding the chemistry of the atmosphere requires that you know something about the light energy that comes from the sun.

There are different kinds of light, depending on the *wavelength* of the light, as the figure shows. Short-wavelength light includes X-rays and ultraviolet light (the light that tans your skin and causes sunburn!). Long-wavelength light includes radio and television waves and infrared light, such as you get from a heat lamp. Both short- and long-wavelength light waves are invisible to the human eye. In between these types of light are middle-wavelength light waves, which include the visible light region (red, orange, yellow, green, blue, indigo, and violet). Short-wavelength light possesses a great deal more energy than long-wavelength light. As a result, a high dose of X-rays (short wavelength, high energy) can be dangerous to humans and other animals, while radio and television waves (long wavelength, low energy) are relatively harmless (except, of course, for the mental listlessness and poor physical shape typical of the true "couch potato").

Molecules and ions can absorb light energy. However, a particular type of matter does not absorb just any kind of light energy but rather *only selected wavelengths of light*. The nature of the atoms and how they combine with each other in the matter determines which wavelengths are absorbed. So, for example, carbon dioxide gas (CO_2) absorbs certain wavelengths of infrared light very well, but it does not absorb visible light. Oxygen and nitrogen, on the other hand, absorb only certain wavelengths of ultraviolet light.

Matter can emit light energy too. The wavelength of the light energy emitted depends on the

Energy in the form of sunlight makes life on earth possible.

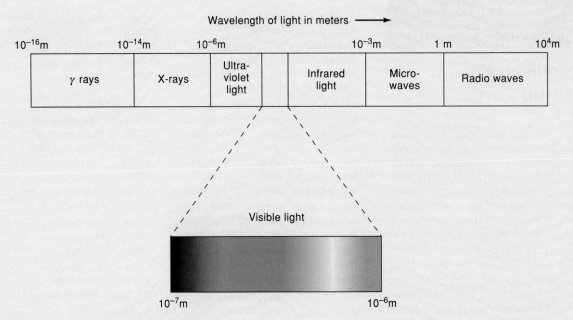

The spectrum of electromagnetic radiation (light).

temperature of the material. Molecules or ions at normal temperatures—and the earth itself—emit much of this light energy in the infrared wavelength region. Hotter bodies like the sun, however, emit light that contains a greater proportion of short-wavelength light like ultraviolet and visible light.

As you will see in the essay following Chapter 11, the earth's reemission of light from the sun determines our planet's overall temperature range. The following table allows you to compare the approximate temperatures of the planets in our solar system. It should be quite apparent that the earth is the only planet with a "reasonable" average temperature from a human perspective.

Approximate Average Temperatures of the Planets

Sun or Planet	Temperature or Temperature Range (K)	Comment
Sun	2×10^7	Not much chance of living here!
Mercury	700 (side facing sun)	The side away from the sun is no better at 110 K.
Venus	730	Many molecules fall apart at these temperatures.
Earth	250–300	Our planet looks comfortable by comparison.
Mars	150–250	Dry ice (solid CO_2) does well here.
Jupiter	110–150	Natural gas could liquefy on parts of Jupiter.
Saturn	92	Oxygen liquefies at about this temperature.
Uranus	58	This is lower than the temperature of *solid* nitrogen.
Neptune	56	Oxygen gas solidifies at about this temperature.
Pluto	40	Dress warmly!

✓ Summary

Chemical bonds are the forces that hold atoms together to make compounds (Section 6.1). Formation of such bonds depends in part on the *ionization energy* of an element—the amount of energy it takes to remove the most loosely held electron from an atom of the element to give a *cation*—and the *electron affinity* of an element—the amount of energy released when an atom of the element combines with an electron to give an *anion* (Section 6.2).

An *oxidation number* is an arbitrary assignment based on certain rules that provide chemists with a method of electronic "bookkeeping." The rules are used to calculate the oxidation numbers of elements in compounds and polyatomic ions (Section 6.3).

There are two general types of chemical bonds, *ionic bonds* and *covalent bonds*. Ionic bonds are formed by the *transfer* of electrons from one atom to another to produce charged ions that attract each other in keeping with the law of electrostatics (Section 6.4). Covalent bonds result from the sharing of electrons between atoms and the resulting attraction of the shared electrons for the two nuclei. Unequal sharing of electrons between atoms due to differences in electronegativities produces *polar bonds* (Sections 6.5). Coordinate covalent bonds are covalent bonds in which one atom supplies both electrons in the electron pair that is shared between the two atoms (Section 6.6).

Lewis structures and structural formulas representing molecules or polyatomic ions are obtained by applying the rule of eight and the rule of two (Section 6.7). Shapes of molecules and polyatomic ions can be determined from the valence electrons around the central atom (Section 6.8). Formulas for compounds can be written if we know the charges on the various metal cations (Table 6.1), nonmetal anions (Table 6.2), and polyatomic ions (Table 6.5). Only a single formula can be written for a compound (Section 6.9).

We can use the periodic table to predict oxidation numbers, properties of compounds, formulas of compounds, and the types of bonding in compounds (Section 6.10).

✓ Exercises

1. Define or explain the following terms (the number in parentheses refers to the section in the text where the term is mentioned):

a.	chemical bond (6.1)	**b.**	rule of eight (octet rule) (6.1)
c.	rule of two (6.1)	**d.**	ionization energy (6.2)
e.	cation (6.2)	**f.**	electron affinity (6.2)
g.	anion (6.2)	**h.**	oxidation number (6.3)
i.	ionic charge (6.3)	**j.**	ionic bond (6.4)
k.	formula unit (6.4)	**l.**	covalent bond (6.5)
m.	diatomic molecule (6.5)	**n.**	bond length (6.5)
o.	electronegativity (6.5)	**p.**	Coulomb's law (6.5)
q.	polar bond (6.5)	**r.**	coordinate covalent bond (6.6)
s.	Lewis structures (6.7)	**t.**	structural formula (6.7)
u.	bond angle (6.7)	**v.**	double bond (6.7)
w.	triple bond (6.7)	**x.**	polyatomic ion (6.7)
y.	binary compounds (6.9)	**z.**	ternary compounds (6.9)

2. Distinguish between:

   a. cations and anions

   b. an ionic bond and a covalent bond

   c. a formula unit and a molecule

   d. an unshared pair of electrons and an unpaired electron

   e. equal and unequal sharing of electrons in a covalent bond

3. Explain the meaning of the following symbols:

   a. — in regard to bonding

   b. 2+ in E^{2+}

   c. $\cdot\cdot$ in $-\overset{\cdot\cdot}{\underset{|}{E}}-$

 # Problems

Oxidation Numbers (See Section 6.3)

4. Calculate the oxidation number for the element indicated in each of the following compounds or ions:

   a. Br in HBrO   b. I in HIO_3

   c. N in HNO_2 d. S in H_2S

   e. S in HSO_3^{-} f. Bi in BiO_3^{-}

   g. S in SO_4^{2-} h. As in AsO_4^{3-}

   i. I in IO_2^{-} j. P in $P_2O_7^{4-}$

5. Calculate the oxidation number for the element indicated in each of the following compounds or ions:

   a. Cl in $HClO_4$ b. Cl in $HClO$

   c. P in H_3PO_4 d. B in $H_2B_4O_7$

   e. Mn in MnO_4^{-} f. Sb in SbO_3^{3-}

   g. Mn in MnO_4^{2-} h. Cr in CrO_4^{2-}

   i. Cl in ClO^{-} j. Ti in TiO_5^{4-}

Ionic Bond (See Section 6.4)

6. Diagram the ionic structure for each of the following ions, indicating the number of protons and neutrons in the nucleus, and arrange the electrons in principal energy levels.

   a. $^{1}_{1}H^{+}$ b. $^{9}_{4}Be^{2+}$ c. $^{24}_{12}Mg^{2+}$ d. $^{23}_{11}Na^{+}$ e. $^{27}_{13}Al^{3+}$

   f. $^{19}_{9}F^{-}$ g. $^{16}_{8}O^{2-}$ h. $^{32}_{16}S^{2-}$ i. $^{14}_{7}N^{3-}$ j. $^{31}_{15}P^{3-}$

7. Write the electronic configuration in *sublevels* for the following ions:

   a. $^{7}_{3}Li^{+}$ b. $^{9}_{4}Be^{2+}$ c. $^{23}_{11}Na^{+}$ d. $^{24}_{12}Mg^{2+}$ e. $^{40}_{20}Ca^{2+}$

   f. $^{14}_{7}N^{3-}$ g. $^{35}_{17}Cl^{-}$ h. $^{16}_{8}O^{2-}$ i. $^{32}_{16}S^{2-}$ j. $^{81}_{35}Br^{-}$

8. The radius of the Mg atom is $16\bar{0}$ pm, and that of Mg^{2+} is 65 pm. Explain this change in size.

9. The radius of the O atom is 66 pm, and that of O^{2-} is $14\overline{0}$ pm. Explain this change in size.

Covalent Bond (See Section 6.5)

10. When Cl atoms unite to form a chlorine molecule (Cl_2), 3.42×10^3 J of energy is given off in the formation of 1.00 g of gaseous chlorine. How many joules of energy are required to break the Cl—Cl bond in 1.00 g of gaseous chlorine and to form the chlorine atoms?

11. Place a δ^+ above the atom or atoms that are relatively positive, and a δ^- above the atom or atoms that are relatively negative in the following covalently bonded molecules (see Figure 6.11):

    a. HF   b. HCl   c. H_2O d. BrCl   e. BCl_3

    f. $SiCl_4$ g. PCl_5 h. NH_3 i. OF_2 j. Cl_2O

Lewis Structures and Structural Formulas of More Complex Molecules and Polyatomic Ions (See Section 6.7)

12. Write the Lewis structures and structural formulas for the following molecules or polyatomic ions. (You may use the periodic table.)

    a. HCl   b. H_2S c. CCl_4 d. CS_2 e. N_2

    f. C_2H_4 g. C_2H_2 h. SH^- i. CN^- j. SO_3^{2-}

13. Write the Lewis structures and structural formulas for the following molecules and polyatomic ions. (You may use the periodic table.)

    a. F_2 b. PCl_3 c. $CHCl_3$ d. Cl_2O

    e. Cl_2 f. H_2SO_3 g. H_2CO_3 h. HNO_3

    i. PO_4^{3-} j. $P_2O_7^{4-}$ (*Hint:* $[O_3-P-O-P-O_3]^{4-}$)

The Shapes of Molecules and Polyatomic Ions (See Section 6.8)

14. Determine the shape (linear, trigonal planar, tetrahedral, bent, or pyramidal) and the approximate bond angle of the following molecules or polyatomic ions. (You may use the periodic table).

    a. HCN   (*Hint:* See Example 6.5)

    b. SO_4^{2-} (*Hint:* See Example 6.8)

    c. NO_3^- (*Hint:* See Example 6.9)

    d. PCl_3

Writing Formulas (See Section 6.9)

15. Write the correct formula for the compound formed by the combination of the following ions:

    a. sodium (Na^+) and chloride (Cl^-)

    b. mercury (II) (Hg^{2+}) and iodide (I^-)

    c. magnesium (Mg^{2+}) and nitride (N^{3-})

    d. iron (III) (Fe^{3+}) and chloride (Cl^-)

    e. cadmium (Cd^{2+}) and oxide (O^{2-})

    f. calcium (Ca^{2+}) and phosphide (P^{3-})

    g. lithium (Li^+) and hydride (H^-)

    h. barium (Ba^{2+}) and nitrate (NO_3^-)

    i. aluminum (Al^{3+}) and perchlorate (ClO_4^-)

    j. barium (Ba^{2+}) and phosphate (PO_4^{3-})

16. Write the correct formula for the compound formed by the combination of the following ions:

   a.   silver (Ag^+) and chloride (Cl^-)

   b.   strontium (Sr^{2+}) and oxide (O^{2-})

   c.   copper (II) (Cu^{2+}) and bromide (Br^-)

   d.   tin (II) (Sn^{2+}) and hydrogen sulfite (HSO_3^-)

   e.   zinc (Zn^{2+}) and bicarbonate (HCO_3^-)

   f.   iron (III) (Fe^{3+}) and carbonate (CO_3^{2-})

   g.   iron (II) (Fe^{2+}) and phosphate (PO_4^{3-})

   h.   aluminum (Al^{3+}) and phosphate (PO_4^{3-})

   i.   mercury (I) (Hg_2^{2+}) and cyanide (CN^-)

   j.   ammonium (NH_4^+) and dichromate ($Cr_2O_7^{2-}$)

Using the Periodic Table, Predicting Oxidation Numbers (See Section 6.10)

17. Using the periodic table, indicate a maximum positive oxidation number for each of the following elements. For those that are nonmetals, give *both* the maximum positive oxidation number and the negative oxidation number. (If you do not know the symbol for the element, look it up on the inside front cover of this text.)

   a.   barium   b.   cesium

   c.   sulfur   d.   iodine

   e.   aluminum   f.   selenium

   g.   astatine   h.   nitrogen

   i.   gallium   j.   osmium

18. Using the periodic table to determine the oxidation numbers, predict the formulas of the binary compounds formed from the following combinations of elements. (If you do not know the symbol for the element, look it up on the inside front cover of this text.)

   a.   barium and oxygen

   b.   cesium and phosphorus

   c.   sodium and nitrogen

   d.   strontium and selenium

   e.   indium and oxygen

   f.   magnesium and arsenic

   g.   aluminum and sulfur

   h.   gallium and selenium

   i.   thallium and sulfur

   j.   sodium and tellurium

Using the Periodic Table, Predicting Properties (See Section 6.10)

19. Estimate the missing value for the following:

   a.
Element	Radius (pm)
K	202
Rb	216
Cs	?

b.

Element	Density (g/mL)
Ca	1.54
Sr	2.60
Ba	?

c.

Element	mp (°C)
Ca	845
Sr	?
Ba	$71\overline{0}$

d.

Compound	Density (g/mL)
B_2S_3	1.55
Al_2S_3	2.37
Ga_2S_3	3.50
In_2S_3	?

e.

Element	First Ionization Energy (kcal/mol)
O	314
S	$24\overline{0}$
Se	226
Te	?

Using the Periodic Table, Predicting Formulas (See Section 6.10)

20. The following are some examples of compounds and their formulas:

> sodium sulfate, Na_2SO_4
> magnesium phosphate, $Mg_3(PO_4)_2$
> aluminum oxide, Al_2O_3

Using the periodic table, write the formulas for the following compounds. (*Hint*: Note the ending of each name for the compound.)

a. potassium sulfate
b. gallium oxide
c. magnesium arsenate
d. aluminum sulfide
e. cesium selenate
f. barium arsenate
g. thallium(III) sulfide
h. indium selenide
i. rubidium selenate
j. indium sulfide

Using the Periodic Table, Predicting Type of Bonding—Ionic or Covalent (See Section 6.10)

21. Using the periodic table, classify the following compounds as essentially ionic or covalent:

a. NaI   **b.** Fe_2O_3 **c.** N_2O_3 **d.** BiF_3 **e.** C_2H_2
f. Na_2SO_4 **g.** BaS   **h.** P_4O_{10} **i.** CS_2 **j.** $Fe(NO_3)_3$

General Problems

22. Consider the undiscovered element X with atomic number 114.

a. In what group would it be placed?
b. How many valence electrons would it have?
c. Would it be more metallic or more nonmetallic than its predecessor in the same group?

   d. What element would it most likely resemble in properties?
   e. Suppose that element X, atomic number 114, forms the XO_3^{2-} ion. Write the Lewis structure and structural formula for the XO_3^{2-} ion.

23. Consider the undiscovered element Y with atomic number 117.
   a. In what group would it be placed?
   b. How many valence electrons would it have?
   c. Give the name and symbol of the element it would most likely resemble in properties.
   d. Would it be classified as a metal, nonmetal, or metalloid?
   e. What charge would you expect on the anion of Y?
   f. Suppose that element Y, atomic number 117, reacts with sodium. What would the formula of this compound be?
   g. Would you expect this element to have a larger or smaller atomic radius than iodine? Why?

24. Consider the undiscovered element Z with atomic number 119.
   a. In what group would it be placed?
   b. How many valence electrons would it have?
   c. What charge would you expect on the cation of Z?
   d. What element would it most likely resemble in properties?
   e. Suppose that element Z, atomic number 119, reacts with bromine. What would the formula of this compound be? Would the bonding in this compound be primarily ionic or covalent?

✓ Chapter Quiz 6

You may use the periodic table.

1. Calculate the oxidation number for the element indicated in each of the following compounds or ions:
   a. Se in SeO_2 **b.** Se in SeO_4^{2-}

2. Write the electronic configuration in *sublevels* for the following ions:
   a. $^{23}_{11}Na^+$ **b.** $^{16}_{8}O^{2-}$

3. Write the correct formula for the compound formed by the combination of the following ions:
   a. calcium (Ca^{2+}) and chloride (Cl^-) ions
   b. potassium (K^+) and sulfate (SO_4^{2-}) ions
   c. iron(III) (Fe^{3+}) and carbonate (CO_3^{2-}) ions
   d. tin(IV) (Sn^{4+}) and sulfite (SO_3^{2-}) ions

4. Write the Lewis structures and structural formulas for the following:
   a. CH_4 **b.** C_2H_4

5. Determine the shape (linear, trigonal planar, tetrahedral, pyramidal, or bent) and the approximate bond angle of the following molecules or polyatomic ions:
   a. AsH_3 **b.** PO_4^{3-}

6. Classify the following compounds as essentially ionic or covalent:
   a. SrO   **b.** CS_2 **c.** CdF_2 **d.** $AgNO_3$

Phosphoric Acid (Formula: H_3PO_4)

The Compound PHOSPHORIC ACID: From Rust Remover to Soft Drinks

FROM Tastes Good!

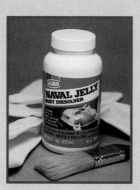

TO Gets the rust out!

Name: Phosphoric acid (H_3PO_4) is named for the parent element, phosphorus, with the *-ic* ending indicating the +5 oxidation number of phosphorus in phosphoric acid.

Appearance: Phosphoric acid is normally used as a colorless aqueous solution. However, pure phosphoric acid is a low-melting (mp = 41° to 44°C) white solid.

Occurrence: Phosphoric acid is not found as such in nature. It must be prepared industrially in order to be used for products and processes in our society. The parent polyatomic ion (phosphate, PO_4^{3-}) from which phosphoric acid is prepared is found in a vast array of minerals called phosphate rock. The most common component of phosphate rock is *fluorapatite* [$Ca_5(PO_4)_3F$].

Source: Almost all phosphoric acid is prepared by one of two methods: (1) the wet-acid process, or (2) the furnace process. In the wet-acid process the fluorapatite is reacted with aqueous sulfuric acid (H_2SO_4) to produce phosphoric acid. In the furnace process the fluorapatite is reacted with sand (SiO_2) and carbon to produce phosphorus. The phosphorus then reacts with oxygen to produce a phosphorus oxide (P_4O_{10}) which in turn reacts with water to produce phosphoric acid. The phosphoric acid produced by the furnace process is purer but more expensive to produce.

Uses: The most important use of phosphoric acid is in the manufacture of phosphate fertilizers. At one time, "triple superphosphate" [$Ca(H_2PO_4)_2 \cdot H_2O$] was a major fertilizer formulation. It was made from fluorapatite and phosphoric acid. The use of triple super phosphate has now declined, and ammonium phosphate [$(NH_4)_3PO_4$] is the major phosphate fertilizer. Ammonium phosphate is prepared from phosphoric acid and ammonia (NH_3). Phosphoric acid is used to make sodium phosphate salts, NaH_2PO_4 and Na_2HPO_4. Mixtures of these two salts are used in textile processing and in the manufacture of foods, where they serve to control the acidity in food products such as processed cheese. Another phosphate product prepared from these salts is sodium tripolyphosphate ($Na_5P_3O_{10}$). It is a common additive in synthetic detergents and cured meat products.

Unusual Facts: Phosphoric acid is a common ingredient in many soft drink formulations, especially colas and root beers. The acid serves three purposes. First, it imparts a tartness that is common to most soft drinks. All soft drinks have acid, called an *acidulant,* in them for this purpose. Second, the acid acts as preservative. Third, the acid has an effect on how our sense of taste perceives the sugar in soft drinks. Different acids have slightly different effects, and the beverage company chooses an acid that gives the flavor it thinks is best.

Phosphoric acid is also used as a rust remover in the form of a jelly applied to rusted car bodies. The phosphoric acid reacts with the rust after standing for several minutes, and the rusted spot is then washed with water to produce unrusted metal.

CHAPTER 7

Chemical Nomenclature of Inorganic Compounds

The White Cliffs of Dover, which overlook the English Channel, are composed of fine-grained limestone deposits that were produced by the decomposition of ancient fossil shells. The common name for this mineral is chalk. *However, a minerologist might call it a mixture of* calcite *and* dolomite, *while a chemist would refer to it as a mixture of calcium carbonate (CaCO$_3$) and calcium magnesium carbonate [CaMg(CO$_3$)$_2$].*

GOALS FOR CHAPTER 7

1. To understand the importance of systematic nomenclature in chemistry (Section 7.1).

2. To determine the name from the formula and the formula from the name for
   a. binary compounds containing two nonmetals (Section 7.2)
   b. binary compounds containing a metal and a nonmetal (Section 7.3)
   c. ternary compounds (Section 7.4)
   d. halogen-containing ternary compounds (Section 7.5)
   e. acids and bases (Section 7.6).

3. To determine whether a compound is (1) an acid, (2) a base, or (3) a salt from its formula (Section 7.6).

4. To determine the formula of a compound from its common name (Section 7.7).

Countdown

You may use the periodic table.

5. Give the name for each of the following symbols of elements (Section 3.1).
   a. Au   (gold)   **b.** As   (arsenic)
   c. Sr   (strontium)   **d.** Si   (silicon)

4. Give the symbol for each of the following elements (Section 3.1).
   a. fluorine   (F)   **b.** chlorine   (Cl)
   c. magnesium   (Mg)   **d.** manganese   (Mn)

3. Write the correct formula for the compound formed by the combination of the following ions (Section 6.9).
   a. calcium (Ca^{2+}) and fluoride (F^-)   (CaF$_2$)
   b. silver (Ag^+) and selenide (Se^{2-})   (Ag$_2$Se)
   c. aluminum (Al^{3+}) and sulfide (S^{2-})   (Al$_2$S$_3$)
   d. tin (IV) (Sn^{4+}) and sulfate (SO_4^{2-})   [Sn(SO$_4$)$_2$]

2. Using the periodic table to determine the oxidation numbers, predict the formulas of the binary compounds formed from the combination of the following elements (Section 6.10, Oxidation Numbers). (*Hint:* If you do not know the symbol of the element, look it up on the inside front cover of this text.)
   a. gallium and arsenic   (GaAs)
   b. gallium and selenium   (Ga$_2$Se$_3$)
   c. germanium and selenium   (GeSe$_2$)
   d. germanium and nitrogen   (Ge$_3$N$_4$)

1. The following are examples of compounds and their formulas:

   potassium chloride, KCl

   sodium bromate, NaBrO$_3$

   Using the periodic table, write the formulas for the following compounds. (*Hint:* If you do not know the symbol for the element, look it up on the inside front cover of this book. Also, note the endings of each name of the compounds.) (Section 6.10, Formulas)
   a. rubidium bromate   (RbBrO$_3$)
   b. cesium chlorate   (CsClO$_3$)
   c. potassium iodate   (KIO$_3$)
   d. cesium iodide   (CsI)

"word to the wise is enough," wrote Benjamin Franklin. If you are a wise chemistry student, we have one word for you—nomenclature. Nomenclature, of course, is actually many words. But if you wish to both learn about chemistry and be able to communicate what you have learned, you must understand the words that make up the language of chemistry.

One goal of a beginning chemistry course is to teach students chemical nomenclature—to name compounds and, given a name, to write the formula of the compound. Once you have this knowledge, you can detect patterns and predict the properties of many compounds from their names alone. For example, you should already know that table salt is sodium chloride (NaCl), but did you know that a Tums tablet works as an antacid (decreases acid in the stomach) because it is made of calcium carbonate?

In this chapter, you will learn to use the names and formulas of the cations (Table 6.1), anions (Table 6.2), and polyatomic ions (Table 6.5) from Chapter 6 to name compounds of these ions and write their formulas. As you will see, there are two kinds of names in chemical nomenclature: *systematic chemical names* and *common names*. Systematic chemical names are used most often, but there are still a few compounds, such as water (H_2O) and ammonia (NH_3), whose common names persist. In this chapter, we first consider systematic chemical names and then nonsystematic common names.

7.1 Systematic Chemical Names

Study Hint: The name of a compound is like the name of a person. It includes a first name (positive portion) and a last name (negative portion).

Today it seems obvious that we need a systematic way to name chemical compounds. Yet for decades chemists simply named compounds at their individual whim. The result was sometimes colorful (quicksilver, Hg and cinnabar, HgS) and sometimes amusing (laughing gas, N_2O). But it also created confusion and difficulties. Two chemists finding the same compound would give it two different names, slowing the spread of scientific knowledge. And imagine having to memorize hundreds or thousands of names of compounds with no rhyme or reason!

Fortunately, in 1921 the International Union of Pure and Applied Chemistry (IUPAC) stepped in and laid down the rules that, with a few revisions, govern modern chemical nomenclature. Following these rules, the names of **inorganic compounds**, those not generally containing carbon, are constructed so that every compound can be named from its formula and each formula has a name specific to that formula. The more *positive portion* (the metal, the positive polyatomic ion, the hydrogen ion, or the less electronegative nonmetal) is named and written *first*. The more *negative portion* (the more electronegative nonmetal or the negative polyatomic ion) is named and written *last*. Additional rules depend on whether the compounds are **binary** (contain two different elements), **ternary** (contain three different elements), or higher, or take the form of acids, bases, or salts.

Inorganic compounds
Those compounds not containing carbon.

Binary compounds   Compounds containing two different elements.

Ternary compounds   Compounds containing three different elements.

7.2 Binary Compounds Containing Two Nonmetals

For all binary compounds, the ending of the second element is *-ide*. When both elements are *nonmetals*, the number of atoms of *each* element is indicated in the name with Greek prefixes, as shown in Table 7.1, except in the case of mono (one), which is used only for the *second* nonmetal. When no prefix appears, one atom is assumed. In addition, when two vowels appear next to each other, like "oo" in monooxide, or "ao" in tetraoxide, pentaoxide, and heptaoxide, the vowel from the Greek prefix is dropped for better pronunciation.

These rules are used to name the following binary compounds:

FORMULA	NAME	ITS PLACE IN OUR WORLD
BCl_3	Boron *tri*chloride	Production of boron compounds and refining of alloys
PCl_5	Phosphorus *penta*chloride	Production of chlorine-containing materials
SO_2	Sulfur *di*oxide	A dangerous and destructive air pollutant
CO	Carbon *mon*oxide	A major air pollutant produced in automobile exhaust and faulty furnaces
N_2O_4	*Di*nitrogen *tetr*oxide	A component of the fuel in two small rocket engines on a space shuttle that put the shuttle into orbit and later cause it to leave orbit and return to earth (see Figure 7.1)
NO_2	Nitrogen *di*oxide	A serious air pollutant (see Figure 7.2)

TABLE 7.1

Greek Prefixes

GREEK PREFIX	NUMBER
mono-	1
di-	2
tri-	3
tetra-	4
penta-	5
hexa-	6
hepta-	7
octa-	8
nona- (or *ennea-*)[a]	9
deca-	10

[a] IUPAC prefers *ennea-* to the Latin *nona-*, but *nona-* is still used.

FIGURE 7.1
Dinitrogen tetroxide fuels the rocket engines that allow the space shuttle to maneuver.

FIGURE 7.2
Production of nitrogen dioxide (NO_2) from the reaction of copper with concentrated nitric acid. Nitrogen dioxide, a component of automobile and truck exhaust, is found in smog. The brown tinge that sometimes appears in polluted air on hot days is probably caused by nitrogen dioxide. (Courtesy Dr. E. R. Degginger)

Work Problems 6 and 7.

Working backward from the name, we can also write the formulas for binary compounds of nonmetals as follows:

NAME	FORMULA
Nitrogen trichloride	NCl_3
Carbon tetrachloride	CCl_4
Dichlorine monoxide	Cl_2O
Chlorine dioxide	ClO_2
Dichlorine heptoxide	Cl_2O_7
Dinitrogen monoxide	N_2O

Study Exercise 7.1
Write the correct name for each of the following compounds:
a. SF_6 (sulfur hexafluoride)
b. N_2O_3 (dinitrogen trioxide)

Study Exercise 7.2
Write the correct formula for each of the following compounds:
a. carbon disulfide   (CS_2)
b. tetraphosphorus decoxide   ($P_4O_{10})$

7.3 Binary Compounds Containing a Metal and a Nonmetal

Not all binary compounds are composed of two nonmetals, however. Some have both metal and nonmetal components. (Binary compounds are *never* made up of two metals.) The nomenclature for such compounds depends on whether the metal is said to have a *fixed ionic charge* or a *variable ionic charge*. Of the metals in Table 6.1, eleven have a fixed ionic charge. Hydrogen, although not a metal, also has a fixed ionic charge. The fixed ionic charge of these 11 metals and hydrogen can easily be predicted according to their location in the periodic table (see Section 6.10, oxidation numbers). The Roman numeral in the periodic table at the head of the column gives the fixed ionic charge (see Figure 7.3). For example, strontium is in group IIA (2) and hence has a fixed ionic charge of +2. *All other metals* in Table 6.1 *have a variable ionic charge.* You must memorize the ionic charges of the metal ions with *variable ionic charges* given in Table 6.1.

The rules for the nomenclature of binary compounds containing both metal and nonmetal components call for the metal to be named first, followed by the nonmetal with the ending *-ide,* as in all binary compounds. *No Greek prefixes are used* because only a single compound is possible for metals that carry a fixed ionic charge.

Metals with Fixed Ionic Charges

Following these rules produces names for binary compounds of metals with fixed ionic charges and nonmetals such as the following:

PERIOD	1 IA	2 IIA									11 IB	12 IIB	13 IIIA					
1	H 1+																	
2	Li 1+																	
3	Na 1+	Mg 2+											Al 3+					
4	K 1+	Ca 2+										Zn 2+						
5		Sr 2+									Ag 1+	Cd 2+						
6		Ba 2+																	
7																		

FIGURE 7.3
Periodic table showing only those elements from Table 6.1 that exist as cations with a fixed ionic charge.

FORMULA	NAME
KCl	Potassium chlor*ide*
Na_2S	Sodium sulf*ide*
LiBr	Lithium brom*ide*
MgO	Magnesium ox*ide*
CaH_2	Calcium hydr*ide*

In writing the formulas of compounds, we must know the ionic charges of the metal cations and the nonmetal anions. For example, consider the following:

NAME	FORMULA
Sodium fluoride	NaF (Na is +1; F is −1; see the periodic table or Tables 6.1 and 6.2. This compound is used in several toothpastes as a tooth decay preventative.
Strontium iodide	SrI_2 (Sr is +2; I is −1)
Cadmium phosphide	Cd_3P_2 (Cd is +2; P is −3)
Magnesium nitride	Mg_3N_2 (Mg is +2; N is −3)
Aluminum sulfide	Al_2S_3 (Al is +3; S is −2)

Study Exercise 7.3
Write the correct name for each of the following compounds:
a. $BaBr_2$ (barium bromide)
b. $AlCl_3$ (aluminum chloride)

Study Exercise 7.4

Write the correct formula for each of the following compounds:

a. zinc iodide (ZnI_2)

b. calcium phosphide (Ca_3P_2)

Metals with Variable Ionic Charges

To name a binary compound that contains a metal with a variable ionic charge, we must take into consideration the ionic charge of that metal ion. As their names imply, metals with variable ionic charges have the same element name but different ionic charges. For example, both Cu^+ and Cu^{2+} are ions of copper. How do we know which ion to use?

Fortunately, the *Stock system** of nomenclature in modern use inserts a Roman numeral in parentheses immediately following the name of the metal. This number represents the charge. For example, copper(II) has an ionic charge of $+2$. This system is a great improvement on the older use of Latin stems for the metal plus *-ous* and *-ic* suffixes. In that system, *-ous* was the ion with lower charge and *-ic* was the ion with the higher charge, but you had to know whether a particular ion's charges were $+1$ and $+2$ or $+2$ and $+3$. Table 7.2 compares these two systems. Although the Stock system is widely used today, some remnants of the older system remain in use, and so you should become familiar with both.

Using these systems and the preceding rules produces compound names such as the following:

FORMULA	NAME
$CuCl_2$	Copper(II) chlor*ide* or cupr*ic* chlor*ide*; because chloride is -1, copper must be $+2$
FeO	Iron(II) ox*ide* or ferr*ous* ox*ide*; because oxide is -2, iron must be $+2$
SnF_4	Tin(IV) fluor*ide* or stann*ic* fluor*ide*; because fluoride is -1, tin must be $+4$
PbS	Lead(II) sulf*ide* or plumb*ous* sulf*ide*; because sulfide is -2, lead must be $+2$
HgO	Mercury(II) ox*ide* or mercur*ic* ox*ide*; because oxide is -2; mercury must be $+2$

Working backward, we can also derive the formulas of binary compounds of nonmetals and metals with variable charges as in the following examples:

NAME	FORMULA
Cupric phosphide	Cu_3P_2 (Cu is $+2$; P is -3; see periodic table or Tables 6.1 and 6.2)
Iron(III) oxide	Fe_2O_3 (Fe is $+3$; O is -2)
Lead (IV) oxide	PbO_2 (Pb is $+4$; O is -2)
Copper(I) chloride	CuCl (Cu is $+1$; Cl is -1; see Figure 7.4)
Stannous fluoride	SnF_2 (Sn is $+2$; F is -1) (once used as a tooth decay preventative in toothpaste)

---

*The Stock system is named after German chemist Alfred Stock (1876–1946). The IUPAC prefers the Stock system.

TABLE 7.2		Naming the Ions of Some Common Metals	

| METAL (SYMBOL) | CATION | Name of Cation | |
		STOCK SYSTEM	-ous OR -ic SYSTEM
Copper (Cu)	Cu^+	Copper(I) ion	Cuprous ion
	Cu^{2+}	Copper(II) ion	Cupric ion
Iron (Fe)	Fe^{2+}	Iron(II) ion	Ferrous ion
	Fe^{3+}	Iron(III) ion	Ferric ion
Lead (Pb)	Pb^{2+}	Lead(II) ion	Plumbous ion
	Pb^{4+}	Lead(IV) ion	Plumbic ion
Mercury (Hg)	Hg_2^{2+}	Mercury(I) ion	Mercurous ion
	Hg^{2+}	Mercury(II) ion	Mercuric ion
Tin (Sn)	Sn^{2+}	Tin(II) ion	Stannous ion
	Sn^{4+}	Tin(IV) ion	Stannic ion

FIGURE 7.4
Copper(II) chloride (left) and copper(I) chloride (right) have different charges on the copper and different physical and chemical properties.

Study Exercise 7.5
Write the correct name for each of the following compounds:
a. CuS   [copper(II) sulfide or cupric sulfide]
b. $PbCl_2$ [lead(II) chloride or plumbous chloride]

Study Exercise 7.6
Write the correct formula for each of the following compounds:
a. iron(III) sulfide   (Fe_2S_3)
b. copper(II) phosphide   (Cu_3P_2)   Work Problems 8 and 9.

7.4 Ternary and Higher Compounds

In naming and writing the formulas of ternary and higher compounds, we follow the same procedure as for binary compounds, except that we use the name or formula of the polyatomic ion. This is why you need to memorize the names and formulas of all the polyatomic ions in Table 6.5. Some of the negative polyatomic ions have suffixes of **-ate** and **-ite**. The most observable difference in the formulas of the -ate and -ite negative polyatomic ions is that the -ate has *one more oxygen atom* than the -ite. For example, the formula for sulf*ite* is SO_3^{2-}; that of sulf*ate* is SO_4^{2-}. This rule holds for all the negative polyatomic ions in Table 6.5. Note that this table also contains three polyatomic ions that do not have an -ate or -ite ending: NH_4^+, ammonium ion, the only positive polyatomic ion in this table; OH^-, hydroxide, with an -ide ending, the same as that found in binary compounds (see Section 7.6); and CN^-, cyanide, which also has an -ide ending.

For metals that have a variable charge, either the Stock system or the -ous or -ic suffix system may be used, but the Stock system is preferred.

> **Study Hint:** "-*ite* is slight and -*ate* is great." Sulfite (SO_3^{2-}) has less oxygen and sulfate (SO_4^{2-}) has more.

Consider the following examples of naming ternary and higher compounds, specifically noting the number of oxygen atoms and the name endings:

FORMULA	NAME
$NaNO_3$	Sodium nit*rate*
$NaNO_2$	Sodium nit*rite*
$NaHSO_3$	Sodium hydrogen sulf*ite*
	Sodium bisulf*ite*
NH_4NO_3	Ammonium nit*rate*
$Cu_3(PO_4)_2$	Copper(II) phosph*ate* (because phosphate is −3, Cu is +2)
	Cupr*ic* phosph*ate*
CuCN	Copper(I) cyan*ide* (because cyanide is −1, Cu is +1)
	Cupr*ous* cyan*ide*
K_2CO_3	Potassium carbon*ate*
$Ba(C_2H_3O_2)_2$	Barium acet*ate*
$Fe_2(CrO_4)_3$	Iron(III) chrom*ate* (because chromate is −2, Fe is +3)
	Ferr*ic* chrom*ate*
$AgClO_3$	Silver chlor*ate*

The first four of these compounds are important in our daily life. Sodium nitrite and sodium nitrate are both used as color fixatives and food preservatives in various meat products such as frankfurters, bologna, and poultry. The nitrite (NO_2^-) ion is very effective in preventing the growth of bacteria that cause deadly botulism. Sodium nitrate is used in poultry primarily as a color fixative to produce a pink color to the meat rather than as a preservative. Recently, however, the value of both sodium nitrite and sodium nitrate has been questioned. The nitrite ion (NO_2^-) is believed to react with organic compounds in the body to produce new compounds that can cause cancer, although studies indicate that this evidence is inconclusive. The nitrate ion (NO_3^-) is not that harmful, but it may be converted to the nitrite ion in the body. The presence of sodium salts also poses a problem for people who have high blood pressure and are on restricted-sodium diets. A committee of the National Research Council has thus recommended that sodium nitrate be eliminated from all poultry and most meat products, with the possible exception of fermented sausages and dry-cured meats.

Similarly, sulfite salts such as sodium hydrogen sulfite or bisulfite were once used to preserve the fresh appearance of fruits and vegetables in salad bars in restaurants. The Food and Drug Administration (FDA) has banned their use since they can cause nausea, diarrhea, hives, shortness of breath, and even death in some individuals.

Even compounds not ingested can threaten human life. Ammonium nitate is used in the production of fireworks and fertilizers. It can produce violent explosions when heated. Seven firemen were killed in such an explosion in Kansas City, Missouri, in November 1988. It was also believed to be one of the components of the explosive used in the bomb that destroyed the Federal Building in Oklahoma City in April 1995.

Again working backward, we can also derive the formulas for ternary and higher compounds by specifically noting the name endings and the number of oxygen atoms if present.

NAME	FORMULA
Barium cyanide	$Ba(CN)_2$ (Ba is $+2$; CN is -1; see the periodic table and Table 6.5 or Tables 6.1 and 6.5)
Iron(II) phosphate	$Fe_3(PO_4)_2$ (Fe is $+2$; PO_4 is -3)
Iron(III) sulfate	$Fe_2(SO_4)_3$ (Fe is $+3$; SO_4 is -2)
Copper(II) sulfite	$CuSO_3$ (Cu is $+2$; SO_3 is -2)
Ammonium hydrogen carbonate	NH_4HCO_3 (NH_4 is $+1$; HCO_3 is -1)
Strontium chlorite	$Sr(ClO_2)_2$ (Sr is $+2$; ClO_2 is -1)
Tin(II) sulfate	$SnSO_4$ (Sn is $+2$; SO_4 is -2)
Calcium permanganate	$Ca(MnO_4)_2$ (Ca is $+2$; MnO_4 is -1)
Cadmium nitrate	$Cd(NO_3)_2$ (Cd is $+2$; NO_3 is -1)
Iron(II) hydrogen sulfite	$Fe(HSO_3)_2$ (Fe is $+2$; HSO_3 is -1)

Study Exercise 7.7
Write the correct name for each of the following compounds:
a. $CaCO_3$ (calcium carbonate)
b. $FePO_4$ [iron(III) phosphate or ferric phosphate]

Study Exercise 7.8
Write the correct formula for each of the following compounds:
a. magnesium oxalate (MgC_2O_4)
b. tin(IV) chromate $[Sn(CrO_4)_2]$ Work Problems 10 and 11.

7.5 Halogen-Containing Ternary Compounds

Table 6.5 lists four different polyatomic ions containing chlorine: perchlorate (ClO_4^-), chlorate (ClO_3^-), chlorite (ClO_2^-), and hypochlorite (ClO^-). We have previously mentioned the relationship of chlorite (ClO_2^-) to chlorate (ClO_3^-). Hypochlorite (ClO^-) is related to chlorite (ClO_2^-) by one less oxygen atom. The prefix **hypo-** is a Greek word meaning "under." Hence, *hypo*chlorite has one atom "under" the number of oxygen atoms of chlorite. Perchlorate (ClO_4^-) is related to chlorate (ClO_3^-) by one more oxygen atom. The prefix **per-** can be used to mean "over." Therefore, perchlorate has one atom "over" the number of oxygen atoms of chlorate.

These prefixes can also be applied to other compounds containing oxygen and halogen ions, such as those of bromine and iodine, since they are in the same group as chlorine. Consider the following oxybromine ions:

ClO_4^- is perchlorate, and so BrO_4^- is *perbromate*.

ClO_3^- is chlorate, and so BrO_3^- is *bromate*.

ClO_2^- is chlorite, and so BrO_2^- is *bromite*.

ClO^- is hypochlorite, and so BrO^- is *hypobromite*.

> **Study Hint:** You may remember the term *hypo-* by remembering that a hypodermic needle goes under (hypo) the skin (dermis).

The same reasoning can be applied to I. Consider some examples:

ClO_3^- is chlorate, and so IO_3^- is iodate.

ClO^- is hypochlorite, and so IO^- is hypoiodite.

See if you can write the formulas of the polyatomic ions periodate and iodite by relating them to the corresponding polyatomic ions containing chlorine. [Fluorine does not form polyatomic ions with oxygen because both elements have high electronegativities (see Figure 6.11)].

Consider the following examples of naming these special ternary compounds:

FORMULA	NAME
NH_4ClO_4	Ammonium perchlorate
$NaClO$	Sodium hypochlorite
$KBrO_2$	Potassium bromite
$Ca(IO_3)_2$	Calcium iodate

Ammonium perchlorate is used in the second stage of the space shuttle. This stage uses a solid-fuel propellant that is 70% ammonium perchlorate. A chemical plant in Henderson, Nevada, was destroyed by an ammonium perchlorate explosion in the spring of 1988.

Ordinary household chlorine bleach contains sodium hypochlorite mixed with water to create a solution of about 5% sodium hypochlorite and 95% water.

Consider the following examples of writing the formulas of these special ternary compounds:

NAME	FORMULA
Barium hypoiodite	$Ba(IO)_2$ (Ba is $+2$; IO is -1, as is ClO; see the periodic table and Table 6.5 or Tables 6.1 and 6.5)
Calcium perbromate	$Ca(BrO_4)_2$ (Ca is $+2$; BrO_4 is -1, as is ClO_4)
Potassium chlorate	$KClO_3$ (K is $+1$; ClO_3 is -1)
Iron(III) iodate	$Fe(IO_3)_3$ (Fe is $+3$; IO_3 is -1, as is ClO_3)

Study Exercise 7.9
Write the correct name for each of the following compounds:
a. $Sr(BrO_3)_2$ (strontium bromate)
b. $Hg(IO_3)_2$ [mercury(II) iodate or mercuric iodate]

Study Exercise 7.10
Write the correct formula for each of the following compounds:
a. barium hypoiodite $[Ba(IO)_2]$
b. iron(II) perchlorate $[Fe(ClO_4)_2]$

Work Problems 12 and 13.

7.6 Acids, Bases, and Salts

Previously, we did not classify compounds but only named them and wrote their formulas. Again, we are going to name them and write their formulas and also classify them as *acids, bases,* or *salts*. Many of these acids, bases, and salts are common household substances with which you are probably familiar. We will point out these common substances in our discussion of acids, bases, and salts.

Special rules come into play when a compound includes a hydrogen ion (H^+) instead of a metal ion or positive polyatomic ion. These hydrogen compounds have completely different properties in the gaseous or liquid state (pure compounds) than they do in **aqueous** (from the Latin *aqua* for "water") **solution** of a compound—a solution in which the compound is dissolved in water.

In the gaseous or liquid state, hydrogen compounds are named as hydrogen derivatives. For example, HCl is hydrogen chloride, HCN is hydrogen cyanide, and HBr is hydrogen bromide.

Aqueous solutions   Solutions in which a gas, solid, or liquid is dissolved in water.

Acid (simplified definition) Hydrogen compounds that yield hydrogen ions (H^+) in aqueous solution.

Acids

In aqueous solution, hydrogen compounds are called *acids*. You are probably already familiar with the sour taste of lemons which contain an acid known as citric acid. While we will develop a more precise definition of an acid in Chapter 15, for now you need only think of an **acid** as a hydrogen compound that yields hydrogen ion (H^+) in aqueous solution. For a *binary* compound, the prefix **hydro-,** meaning "hydrogen" or "in water," is added and the -**ide** of the anion name is replaced by -**ic acid**. Therefore, hydrogen chloride in aqueous solution is *hydro*chlor*ic acid*. The same procedure applies to other binary compounds and also to hydrogen cyanide (HCN), which is called *hydro*cyan*ic acid* in aqueous solution.

For *ternary* and higher compounds, the word "hydrogen" is dropped, and the name of the polyatomic ion—and any prefix (*per-, hypo-*)—is used. The **-ate** or **-ite** is also dropped, and **-ic** or **-ous acid**, respectively, is added. In addition, in each of the ternary acids involving phosphorus, "or" from phosph*or*us is reinserted in the acid name (the same applies to the *-ur* in sulfur). Therefore, hydrogen phosph*ate* (H_3PO_4) in aqueous solution is phosphor*ic acid,* and hydrogen phosph*ite* (H_3PO_3) is phosphor*ous acid*. Table 7.3 summarizes these changes.

Study Hint:

-*ate* = -*ic* acid
-*ite* = -*ous* acid

If you "*ate*" too much, you may **h-*ic*-cup!**

Study Hint: Acids have the general formula **HX**, where **X** is an anion (nonmetal ion or negative polyatomic ion) in aqueous solution.

TABLE 7.3		Summary of the Naming of Binary and Ternary Compounds of Hydrogen in the Gas or Liquid State and in Aqueous (Water) Solution		
GENERAL			EXAMPLE	
GAS OR LIQUID	AQUEOUS SOLUTION	FORMULA	NAME OF GAS OR LIQUID	NAME OF AQUEOUS SOLUTION
Binary and -ide endings				
Hydrogen _____-ide	hydro_____-ic acid	HCl	Hydrogen chloride	Hydrochloric acid
Ternary and higher				
Hydrogen _____-ate	_____-ic acid	H_3PO_4	Hydrogen phosphate	Phosphoric acid
Hydrogen _____-ite	_____-ous acid	H_3PO_3	Hydrogen phosphite	Phosphorous acid

Consider the naming of the following hydrogen compounds in aqueous solution:

FORMULA	NAME OF AQUEOUS SOLUTION
Binary	
HBr	*Hydro*brom*ic acid*
HI	*Hydr*iod*ic acid*[a]
H_2S	*Hydro*sulfur*ic acid*
Ternary and higher	
HNO_3	Nitr*ic acid*
$HC_2H_3O_2$	Acet*ic acid*
H_2SO_4	Sulfur*ic acid*
$HClO_2$	Chlor*ous acid*
$HBrO_4$	*Per*brom*ic acid*
HClO	*Hypo*chlor*ous acid*

[a] For ease of pronunciation, the "o" in *hydro-* is dropped when followed by a vowel.

Working backward again, we can get the formulas from the names of acids:

NAME	FORMULA
Hydrofluoric acid	HF (H is +1; F is −1; see the periodic table or Tables 6.1 and 6.2)
Sulfurous acid	H_2SO_3 (H is +1; SO_3 is −2; see Table 6.5)
Chloric acid	$HClO_3$ (H is +1; ClO_3 is −1)
Chromic acid	H_2CrO_4 (H is +1; CrO_4 is −2)

Study Exercise 7.11
Write the correct name for each of the following compounds:
a. $HBrO_2$ in aqueous solution   (bromous acid)
b. $HClO_4$ in aqueous solution   (perchloric acid)

Study Exercise 7.12
Write the correct formula for each of the following compounds:
a. bromic acid   ($HBrO_3$)
b. dichromic acid   ($H_2Cr_2O_7$)

Study Hint: Bases have the general formula **MOH,** where **M** is a metal cation.

Base (simplified definition)
A compound that contains a metal ion and one or more hydroxide (OH^-) ions.

Bases

Although you may not know them as *bases*, you have probably used such compounds at some time in your life. Bases have a bitter taste [for example, milk of magnesia, $Mg(OH)_2$] and have a soapy or slick feel (for example, lye solution, NaOH). We will develop a more precise definition of a base in Chapter 15, but for now you need only think of a **base** as a compound that contains a metal ion and one or more hydroxide (OH^-) ions. Even though these bases are not binary compounds, they have the ending *-ide* because of the name of the OH^- polyatomic ion.

Consider the naming of the following bases:

FORMULA	NAME
LiOH	Lithium hydrox*ide*
KOH	Potassium hydrox*ide*
Ca(OH)$_2$	Calcium hydrox*ide*
Al(OH)$_3$	Aluminum hydrox*ide*

And, again we can derive the formulas from the name:

NAME	FORMULA
Iron(III) hydroxide	Fe(OH)$_3$ (Fe is +3; OH is −1; see Table 6.1 and 6.5)
Barium hydroxide	Ba(OH)$_2$ (Ba is +2; OH is −1)
Magnesium hydroxide	Mg(OH)$_2$ (Mg is +2; OH is −1)
Sodium hydroxide	NaOH (Na is +1; OH is −1)

Study Exercise 7.13

Write the correct name for each of the following compounds:

a. LiOH (lithium hydroxide)
b. Pb(OH)$_2$ [lead(II) hydroxide or plumbous hydroxide]   Work Problems 14 and 15.

Study Exercise 7.14

Write the correct formula for each of the following compounds:

a. cadmium hydroxide [Cd(OH)$_2$]
b. iron(II) hydroxide [Fe(OH)$_2$]

Salts

When *one* or *more* of the hydrogen ions of an acid are replaced by a cation (a metal or a positive polyatomic ion), or when one or more of the hydroxide ions of a base are replaced by an anion (nonmetal or negative polyatomic ion), the result is a *salt*. Therefore, a **salt** is an ionic compound made up of a positively charged ion (cation) and a negatively charged ion (anion). Examples include the binary compounds of metal cations with nonmetal anions and the ternary compounds of metal cations or ammonium ions with negative polyatomic ions. Thus, potassium bromide (KBr), sodium nitrate (NaNO$_3$), and ammonium sulfate [(NH$_4$)$_2$SO$_4$] are all salts.

Figure 7.5 shows some common items that are acids, bases, or salts.

In the study hints for acids, bases, and salts we gave you general formulas for these compounds. Use these general formulas to classify each of the following compounds as (1) acid, (2) base, or (3) salt.

> **Study Hint:** Salts have the general formula **MX,** where **M** is a cation (metal ion or positively charged polyatomic ion); **X** is an anion (nonmetal ion or negative polyatomic ion).

Salt Ionic compound made up of a positively charged ion (cation) and a negatively charged ion (anion).

FORMULA	CLASSIFICATION
HC$_2$H$_3$O$_2$ in aqueous solution	1, Acid, where X = C$_2$H$_3$O$_2$ anion
Ca(C$_2$H$_3$O$_2$)$_2$	3, Salt, where M = Ca cation and X = C$_2$H$_3$O$_2$ anion
Mg(OH)$_2$	2, Base, where M = Mg cation
MgS	3, Salt, where M = Mg cation and X = S anion

> **Study Hint:** Briefly these general formulas are:
> **HX** = acid
> **MOH** = base
> **MX** = salt

FIGURE 7.5
Acid, base, or salt? Common items that are acidic include carbonated water (carbonic acid), fruits (citric acid), vinegar (acetic acid), aspirin (acetylsalicylic acid), and vitamin C (ascorbic acid). Common items that are basic include drain cleaner (sodium hydroxide), ammonia, bleach (sodium hypochlorite), and soap. Alka-Seltzer contains both an acid and a base, while plaster of paris is a salt $(CaSO_4 \cdot \frac{1}{2}H_2O)$.

Study Exercise 7.15
Classify each of the following compounds as (1) an acid, (2) a base, or (3) a salt.

a. $Ca(OH)_2$ (base)
b. $CaCrO_4$ (salt)
c. H_2CrO_4 in aqueous solution (acid)
d. K_2CrO_4 (salt)

Work Problems 16 and 17.

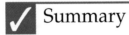 ## 7.7 Common Names

Despite the virtues of systematic names, common names for some compounds still persist because the systematic names are too long and complicated to use. Not even the most ardent chemist asks for sodium chloride instead of table salt at the dinner table. The same is true for hydrogen oxide—water. Table 7.4 lists the common names of some substances with which you may be familiar and some of their uses. Figure 7.6 attempts to illustrate why some of these compounds may be important to you some day.

✓ Summary

Nomenclature is part of the language of chemistry. The systematic naming of compounds derives from the names and formulas of cations, anions, and polyatomic ions (Section 7.1)

All binary compounds end in *-ide*. In naming binary compounds composed of two *nonmetals*, we use Greek prefixes to indicate the number of atoms of each element in the compound (Section 7.2). Other binary compounds contain a metal ion (with a fixed

TABLE 7.4	Some Common Names, Systematic Names, Their Formulas, and Some of Their Uses		

COMMON NAME	SYSTEMATIC NAME	FORMULA	USE
Ammonia	Hydrogen nitride	NH_3	Cleaner, commercial refrigerator, fertilizer
Baking soda, bicarbonate	Sodium hydrogen carbonate	$NaHCO_3$	In baking powder, some fire extinguishers, antacid, deodorizer
Dry ice (solid), carbonic gas (gas)	Carbon dioxide	CO_2	Fire extinguishers, freezing substances, greenhouse gas
Epsom salts	Magnesium sulfate heptahydrate	$MgSO_4 \cdot 7H_2O$	Strong laxative, bathing infected tissue
Laughing gas	Dinitrogen monoxide	N_2O	Anesthetic
Marble, chalk, limestone	Calcium carbonate	$CaCO_3$	To make cement and antacid and prevent diarrhea
Milk of magnesia	Magnesium hydroxide	$Mg(OH)_2$	Antacid and laxative
Muriatic acid	Hydrochloric acid	HCl	Cleaning metals, such as iron before galvanizing; stomach (digestive) acid; "swimming pool" acid
Natural gas	methane (major component)	CH_4	Heating for homes and buildings
Oil of vitriol	Sulfuric acid	H_2SO_4	Battery acid (dilute), cleaning metals
Quicklime, lime	Calcium oxide	CaO	To make slaked lime
Slaked lime	Calcium hydroxide	$Ca(OH)_2$	In mortar and plaster
Soda lye, caustic soda	Sodium hydroxide	$NaOH$	To make soap, drain cleaner
Sugar	Sucrose	$C_{12}H_{22}O_{11}$	Sweetener
Table salt	Sodium chloride	$NaCl$	Seasoning
Vinegar (dilute solution, about 5%)	Acetic acid	$HC_2H_3O_2$	Salad dressing, pickling of some foods
Water	Hydrogen oxide (dihydrogen monoxide)	H_2O	Drinking, washing

Study Hint: A common name is like a nickname. It is a shorter way of naming something.

Work Problem 18.

or variable ionic charge) and a nonmetal ion. These compounds are named by using the name of the metal ion followed by the name of the nonmetal ion. For compounds containing metal ions with variable charge, the Stock system or the *-ous* or *-ic* system is used to specify the charge on the metal ion. The Stock system is the preferred system (Section 7.3).

FIGURE 7.6
Baking soda (sodium hydrogen carbonate) puts out fires.

QUICK, Hand me the baking soda!

Ternary and higher compounds are composed of three or more types of atoms. They include a metal ion (with fixed or variable ionic charge) and a polyatomic ion (Section 7.4). Halogen-containing ternary compounds contain a polyatomic ion (an oxyhalogen ion) and a metal ion (Section 7.5).

Acids and bases are special types of compounds. An *acid* is a hydrogen compound that yields hydrogen (H^+) ions in aqueous solution. A *base* is a compound that contains a metal ion and one or more hydroxide ions. Salts are compounds that result when one or more of the hydrogen ions in an acid are replaced by another cation or when one or more of the hydroxide ions in a base are replaced by another anion. Therefore, a *salt* is an ionic compound made up of a positively charged ion (cation) and a negatively charged ion (anion) (Section 7.6).

The use of common names for some substances in chemistry persists. Examples include water, salt, and ammonia. A few of these are summarized in Table 7.4 (Section 7.7).

 Exercises

1. Define or explain the following terms (the number in parentheses refers to the section in the text where the term is mentioned):

   a. inorganic compounds (7.1)   **b.** binary compounds (7.1)

   c. ternary compounds (7.1)   **d.** aqueous solutions (7.6)

   e. acid (7.6)   **f.** base (7.6)

   g. salt (7.6)

2. Give the meaning of the following prefixes or suffixes:

   a. *-ide*   **b.** *-ate*   **c.** *-ic* acid   **d.** *-ite*

   e. *-ous* acid   **f.** *-ous* (cation) **g.** *-ic* (cation)   **h.** *hypo-*

   i. *hydro-*   **j.** *per-*

3. Distinguish between

   a. a binary and a ternary compound

   b. a polyatomic positive ion and a polyatomic negative ion

c. ammonia and ammonium ion

d. baking soda and soda lye

e. lime and slaked lime

f. *-ate* and *-ite*

Formulas of Ions

4. Write the formula for each of the following ions:

a. sulfate

b. nitrate

c. carbonate

d. chloride

e. chlorite

f. sulfite

g. bromate

h. bromite

i. perbromate

j. hydrogen carbonate

k. chromate

l. dichromate

m. iron(III)

n. iron(II)

o. barium

p. copper(II)

q. aluminum

r. potassium

s. tin(IV)

t. stannous

Names of Ions

5. Name each of the following ions:

a. ClO^- b. ClO_2^- c. S^{2-} d. OH^-

e. HSO_3^- f. CN^- g. $Cr_2O_7^{2-}$ h. Br^-

i. HCO_3^- j. HSO_4^- k. Mg^{2+} l. NH_4^+

m. Cd^{2+} n. Sn^{4+} o. Sn^{2+} p. Zn^{2+}

q. Pb^{2+} r. H^+ s. Hg_2^{2+} t. Hg^{2+}

 # Problems

Binary Compounds Containing Two Nonmetals (See Section 7.2)

6. Write the correct name for each of the following compounds:

a. ClO_2 b. P_2S_5 c. N_2O_5 d. N_2O e. SO_2

f. CO g. P_4O_6 h. CO_2 i. N_2O_4 j. SO_3

7. Write the correct formula for each of the following compounds:

a. sulfur trioxide

b. dinitrogen tetroxide

c. carbon dioxide

d. tetraphosphorus hexoxide

e. carbon monoxide

f. sulfur dioxide

g. dinitrogen monoxide

h. dinitrogen pentoxide

i. diphosphorus pentasulfide

j. chlorine dioxide

Binary Compounds Containing a Metal and a Nonmetal (See Section 7.3)

8. Write the correct name for each of the following compounds:

a. SnS_2 b. CuO c. CaO d. $HgBr_2$ e. SnF_2

f. LiI g. Hg_2Cl_2 h. PbS i. SnI_4 j. SrO

9. Write the correct formula for each of the following compounds:
   a. strontium oxide
   b. tin(IV) iodide
   c. lead(II) sulfide
   d. mercury(I) chloride
   e. lithium iodide
   f. tin(II) fluoride
   g. mercury(II) bromide
   h. calcium oxide
   i. copper(II) oxide
   j. tin(IV) sulfide

Ternary and Higher Compounds (See Section 7.4)

10. Write the correct name for each of the following compounds:
    a. $Fe(NO_2)_2$
    b. $Mg(CN)_2$
    c. $(NH_4)_2SO_4$
    d. $CdCO_3$
    e. $Al_2(Cr_2O_7)_3$
    f. $Al_2(SO_4)_3$
    g. Li_2SO_3
    h. Cu_2CO_3
    i. $Ba(NO_3)_2$
    j. Ag_3PO_4

11. Write the correct formula for each of the following compounds:
    a. silver phosphate
    b. barium nitrate
    c. copper(I) carbonate
    d. lithium sulfite
    e. aluminum sulfate
    f. aluminum dichromate
    g. cadmium carbonate
    h. ammonium sulfate
    i. magnesium cyanide
    j. iron(II) nitrite

Halogen-Containing Ternary Compounds (See Section 7.5)

12. Write the correct name for each of the following compounds:
    a. $LiBrO_4$
    b. $Cu(ClO_3)_2$
    c. $Mg(ClO_4)_2$
    d. $Ca(ClO)_2$
    e. $Mg(BrO_3)_2$
    f. $LiClO_4$
    g. $KClO$
    h. $Fe(ClO_4)_3$
    i. $Cd(IO_3)_2$
    j. $NaClO_2$

13. Write the correct formulas for each of the following compounds:
    a. sodium chlorite
    b. cadmium iodate
    c. iron (III) perchlorate
    d. potassium hypochlorite
    e. lithium perchlorate
    f. magnesium bromate
    g. calcium hypochlorite
    h. magnesium perchlorate
    i. copper(II) chlorate
    j. lithium perbromate

Acids, Bases, and Salts (See Section 7.6)

14. Write the correct name for each of the following compounds:
    a. H_2SO_4 in aqueous solution
    b. $Ca(OH)_2$
    c. $H_2C_2O_4$ in aqueous solution
    d. HNO_3 in aqueous solution
    e. $Ba(OH)_2$
    f. $HBrO_3$ in aqueous solution
    g. $LiOH$
    h. $HClO$ in aqueous solution

15. Write the correct formula for each of the following compounds:
    a. hypochlorous acid
    b. lithium hydroxide
    c. bromic acid
    d. barium hydroxide

e. nitric acid   f. oxalic acid

g. calcium hydroxide   h. sulfuric acid

16. Classify each of the following compounds as (1) an acid, (2) a base, or (3) a salt. Assume that all soluble compounds are in aqueous solution.

a. H_3PO_4 b. $SrCO_3$ c. $(NH_4)_2CO_3$ d. $AlCl_3$

e. $HMnO_4$ f. $Ca(OH)_2$ g. $HC_2H_3O_2$ h. $CaCl_2$

17. Classify each of the following compounds as (1) an acid, (2) a base, or (3) a salt. Assume that all soluble compounds are in aqueous solution.

a. $MgCl_2$ b. $CaSO_4$ c. $PbCO_3$ d. $Sr(OH)_2$

e. $Hg(C_2H_3O_2)_2$ f. $H_2Cr_2O_7$ g. NH_4Br h. $LiOH$

Common Names (See Section 7.7)

18. Write the correct formula for the "principal" chemical ingredient in each of the following:

a. vinegar   b. marble

c. table salt   d. baking soda

e. milk of magnesia   f. ammonia

g. laughing gas   h. muriatic acid

19. Complete the following table by writing the correct formulas for the compounds formed by the following cations and anions:

CATIONS	Anions			
	CHLORIDE	CARBONATE	SULFATE	PHOSPHATE
Potassium				
Barium				
Iron(III)				
Aluminum				

20. Write the correct formula for each of the following compounds:

a. tin(II) phosphate   b. silver permanganate

c. calcium hypoiodite   d. magnesium sulfate

e. sodium oxalate   f. perchloric acid

g. phosphorus trifluoride   h. cadmium nitrate

i. lead(II) phosphate   j. strontium hydrogen carbonate

k. calcium nitride   l. mercury(II) chloride

m. phosphorus pentachloride n. plumbic oxide

21. Write the correct name for each of the following compounds:

a. $Ca_3(PO_4)_2$ b. $Mg(ClO_3)_2$

c. $PbSO_4$ d. $CaCr_2O_7$

    e. $Ba(OH)_2$ f. $Sn(HCO_3)_2$

    g. $K_2C_2O_4$ h. $Li_2Cr_2O_7$

    i. $HC_2H_3O_2$ in aqueous solution   j. $ZnCl_2$

    k. Cd_3P_2 l. LiOH

    m. KH   n. SnS_2

22. The active ingredient in some commercial antacids is listed as "calcium carbonate." What is the formula of this substance?

23. Potassium iodide is used in cough syrups as an expectorant to "loosen the cough." What are the electronic configurations in sublevels of the potassium and iodide ions that make up potassium iodide?

✓ Chapter Quiz 7

You may use the periodic table.

1. Write the correct name for each of the following compounds:

    a. KI   b. CuI   c. $FeSO_3$ d. P_2S_5

    e. $KClO_3$ f. SnO_2 g. $(NH_4)_2SO_4$ h. Na_3PO_4

2. Write the correct formula for each of the following compounds:

    a. calcium nitride   b. copper(I) oxide

    c. ammonia   d. lithium chromate

    e. nitric acid   f. ammonium perchlorate

    g. tin(IV) sulfate   h. magnesium dichromate

3. Classify each of the following compounds as (1) an acid, (2) a base, or (3) a salt. Assume that all soluble compounds are in aqueous solution.

    a. $AlPO_4$ b. $HC_2H_3O_2$ c. $Mg(OH)_2$ d. $MgCl_2$

Sucrose (Formula: $C_{12}H_{22}O_{11}$)

For better or worse, sucrose (sugar) is part of our lives.

The Compound SUCROSE: How Sweet It Is!

Name: Sucrose is commonly called *sugar*. Some more common names for sucrose are *saccharose, cane sugar,* and *beet sugar*. Its chemical formula is $C_{12}H_{22}O_{11}$. Sucrose is a *carbohydrate* compound made up of carbon (*carbo-*) and water (*-hydrate*). A carbohydrate has a chemical formula of $(CH_2O)_n$. Sucrose is made up of two smaller carbohydrates, fructose $[(CH_2O)_6]$ and glucose [dextrose, also $(CH_2O)_6$]. In combining the two smaller units into sucrose, a molecule of water is lost; thus, sucrose has a formula of $(C)_{12}(H_2O)_{11}$ rather than $(CH_2O)_{12}$.

Appearance: Sugar is a white, crystalline solid that turns brown (caramelizes) and then black on heating. This is the basis for candy making. The slow cooking of sucrose turns the candy a light brown (soft caramels) and then a darker brown (toffee or peanut brittle) as the heating is prolonged. If you don't stir the mixture well enough, it gets too hot and scorches (turns black).

Occurrence: Sucrose is found universally throughout the plant kingdom in fruits, seeds, flowers, and roots. The primary sources of sucrose are sugarcane (15 to 20% of stalks) and sugar beet (10 to 17% of beet). Lesser amounts of sucrose (as syrups) are obtained from sugar maple and sorghum.

Source: Sucrose is the highest volume organic compound produced worldwide. The cane is crushed and washed with water to remove most of the sucrose. The resulting solution is evaporated to give a *syrup*. Further concentration of the brownish-black syrup and crystallization provide raw sucrose. More purification by crystallization yields the white solid with which we are familiar.

Its Role in Our World: Sucrose has been known as a sweetener throughout history. Sugarcane is believed to have been native to either India or New Guinea, and by 1000 B.C. it had spread throughout the South Pacific. By 400 B.C. sucrose could be found in the Middle East, and by the twelfth century it was used throughout most of

Europe. Venice became a center for the sugar trade, and during his travels Marco Polo recorded the advanced sugar-refining techniques used in China. Columbus brought sugar to the New World, and by 1750 sugar was being used worldwide.

The consumption of sucrose in the United States is over 45 kg (100 lb) per person per year. Sucrose is used not only as a sweetener but also as a preservative, a bulking agent, a flavor enhancer, and a texturizer. Another important characteristic of sucrose is that it can serve as a food for yeast. Yeast converts sucrose and water into ethyl alcohol (C_2H_6O) and carbon dioxide (see box on carbon dioxide in Chapter 9).

Many forms of sucrose are available. Granulated sugar is the typical form that is produced. Further grinding produces powdered sugar, to which a little cornstarch must be added to keep it from "caking." Sugar cubes are prepared by mixing sucrose and a sucrose syrup, pouring the mixture into molds, and allowing the cubes to harden by draining and evaporation. Sugar cubes are more expensive than granulated sugar because of the extra steps involved. Brown sugar can be made in two ways: (1) by mixing sucrose and an appropriate syrup to impart the desired taste and color, or (2) by concentrating a syrup and allowing it to crystallize rapidly to trap compounds that impart the color and flavor of brown sugar. True *molasses* is really a thick, black syrup with a strong taste that is used as cattle feed or for fermentation. What you buy in the store as "molasses" is a concentrated syrup with a more pleasing taste than true molasses.

Unusual Facts: The most important variety of sugarcane for the commercial production of sucrose prior to 1920 was the *noble cane*. In the 1920s, the mosaic virus decimated the commercial cane fields, and new strains had to be developed that were resistant to mosaic disease. Many such strains are now available and are used commercially. Another interesting aspect of the growing of sugarcane is that sucrose does not spoil in the canes as fruit on the tree or vine does. Thus, sugarcane does not "ripen" and can be harvested at almost any time, weather and season permitting. The canes are cut off at ground level, and another crop usually grows without replanting. Eventually, the yields decrease, however, and the old plants are replaced by new ones.

CHAPTER 8

Calculations Involving Elements and Compounds

Copper (Cu, 63.5 g), mercury (Hg, 200.0 g), and lead (Pb, 207.2 g). Although each substance has a different mass, they do have some things in common. They are all equal to one mole of their respective substance and have the same number of particles. In this chapter, we will discuss these points.

GOALS FOR CHAPTER 8

1. To calculate the formula or molecular mass of a compound (Section 8.1).

2. To calculate the molar mass of an element or compound and, using it and/or Avogadro's number, to interconvert among mass, quantity of particles, and moles. To be able to use molar relationships of elements in a mole of a compound (Section 8.2).

3. To be able to use molar volume of a gas at STP to interconvert among mass, moles, molecular mass and molar mass, and density of a gas at STP of an element or compound (Section 8.3).

4. To calculate the percent composition of a compound (Section 8.4).

5. To calculate the empirical and molecular formulas of a compound (Section 8.5).

Countdown

5. Determine the number of atoms of each element and write the name of the element in each of the following formulas (Section 3.3).
   a. $SnCl_2$ (1 atom tin, 2 atoms chlorine)
   b. $Sn(CrO_4)_2$ (1 atom tin, 2 atoms chromium, 8 atoms oxygen)

4. Determine the number of significant digits in the following numbers (Section 2.5).
   a. 12.60   (4)   **b.** 0.0750   (3)
   c. 4007   (4)   **d.** $89\overline{0}$ (3)

3. Perform the following operations and express your answer to the proper number of significant digits (Sections 2.5 and 2.6).
   a. $\dfrac{72.0}{12}$ **b.** $\dfrac{5.0 \times 1\overline{00}}{64.5} =$
               (84)   (7.8)
   c. $0.321 \times 142.1 =$ **d.** $\dfrac{8.752}{32.0} =$
            (45.6)   (0.274)

2. Carry out the operations indicated on the following exponential numbers. Express your answer to three significant digits in scientific notation (Sections 2.6 and 2.7).
   a. $\dfrac{7.43 \times 10^{26}}{6.02 \times 10^{23}} =$ (1.23×10^3)
   b. $1.23 \times 10^{-1} \times 6.02 \times 10^{23} = (7.40 \times 10^{22})$
   c. $\dfrac{35.453}{6.02 \times 10^{23}} =$ (5.89×10^{-23})
   d. $\dfrac{197.5 \times 10^{26}}{6.02 \times 10^{23}} =$ (3.28×10^4)

1. Calculate the density of metallic copper in g/cm^3 if a piece of copper has a volume of 25.0 mL and a mass of 224 g (Sections 2.1, 2.2, 2.5, and 2.6).
                             $(8.96 \ g/cm^3)$

S uppose for a moment that you are a research chemist trying to identify a poison used in a murder investigation. Or perhaps you are merely trying to develop a laundry additive that gets clothes "whiter than white." Like other practicing chemists in the world, you would perform experiments and do calculations in the process of solving the problem. These calculations might range from very simple ones to very complex ones that require computers. In this chapter we will consider some of the basic calculations made by chemists.

In the previous chapters, we considered a general description of elements and compounds with a few simple calculations. We will now consider more quantitative (how much?) concepts involving elements and compounds. In our calculations we will use the *dimensional analysis* method of problem solving introduced in Section 2.8. Before you read further we recommend that you review this section.

8.1 Calculation of Formula or Molecular Mass

One of the simplest calculations is that of the molecular mass or formula mass of a compound. We already know (from Section 3.3) that the subscripts in a formula of a compound represent the number of atoms of the respective elements in a molecule or formula unit of a compound. For example, in a molecule of sugar (sucrose, $C_{12}H_{22}O_{11}$) there are 12 atoms of carbon, 22 atoms of hydrogen, and 11 atoms of oxygen. From Section 4.1 we know that the relative atomic masses of the elements can be expressed in *amus*. To determine the molecular or formula mass of a compound, we first multiply the number of atoms of a particular element in a compound by the atomic mass of one such atom. We will use the table of approximate atomic masses on the inside back cover of this text. Then we add together the results for all types of atoms in the compound. That is,

$$\boxed{\begin{array}{c}\text{molecular mass}\\\text{or}\\\text{formula mass}\end{array}} = \boxed{\begin{array}{c}\text{number of atoms}\\\text{of element times}\\\text{atomic mass of}\\\text{1 atom of element}\end{array}} + \boxed{\begin{array}{c}\text{same for other}\\\text{elements in}\\\text{compound}\end{array}}$$

Note that, although the calculations for formula mass and molecular mass are identical, the term "molecular mass" applies only to compounds that exist as molecules and are held together by covalent bonds. The term *formula mass* is used to describe compounds that exist as ions and bond ionically because these compounds are expressed in *formula units* (see Section 6.3).

EXAMPLE 8.1   Calculate the formula mass of sodium sulfate (Na_2SO_4).

SOLUTION   In one formula unit there are 2 atoms of sodium, 1 atom of sulfur, and 4 atoms of oxygen. Hence, the formula mass is calculated as

$$2 \text{ atoms Na} \times \frac{23.0 \text{ amu}}{1 \text{ atom Na}} = 46.0 \text{ amu}$$

$$1 \text{ atom S} \times \frac{32.1 \text{ amu}}{1 \text{ atom S}} = 32.1 \text{ amu}$$

$$4 \text{ atoms O} \times \frac{16.0 \text{ amu}}{1 \text{ atom O}} = 64.0 \text{ amu}$$

$$\text{formula mass of } Na_2SO_4 = \overline{142.1 \text{ amu}} \qquad Answer$$

The answer is expressed to the smallest place present in all the numbers that are added (Section 2.6), which in this example is the tenths decimal place. The calculation can be simplified as

$$
\begin{aligned}
2 \times 23.0 \text{ amu} &= 46.0 \text{ amu} \\
1 \times 32.1 \text{ amu} &= 32.1 \text{ amu} \\
4 \times 16.0 \text{ amu} &= 64.0 \text{ amu} \\
\hline
\text{formula mass of } Na_2SO_4 &= 142.1 \text{ amu} \qquad \textit{Answer}
\end{aligned}
$$

EXAMPLE 8.2   Calculate the molecular mass of sugar (sucrose $C_{12}H_{22}O_{11}$).

SOLUTION

$$
\begin{aligned}
12 \times 12.0 \text{ amu} &= 144 \text{ amu} \\
22 \times 1.0 \text{ amu} &= 22 \text{ amu} \\
11 \times 60.0 \text{ amu} &= 176 \text{ amu} \\
\hline
\text{molecular mass of } C_{12}H_{22}O_{11} &= 342 \text{ amu} \qquad \textit{Answer}
\end{aligned}
$$

> **Study Hint:** Notice that we use the same method in solving for molecular mass and formula mass.

Study Exercise 8.1
Calculate the molecular mass of water (H_2O).   (18.0 amu)   Work Problems 3 and 4.

8.2   Calculation of Moles of Units. Avogadro's Number (*N*)

In our discussion of atomic masses, we compared all elements to a standard of ^{12}C. We will now use ^{12}C to define a new quantity, the mole. The **mole** (abbreviated **mol**) is the amount of a substance containing the *same number* of atoms, formula units, molecules, or ions as there are atoms in *exactly 12 g of carbon-12*. But just how many ^{12}C atoms are there in exactly 12 g of carbon-12? Chemical experiments have shown that there are 6.02×10^{23} atoms in exactly 12 g of carbon-12. That is, in 1 mol *of* ^{12}C *atoms there are* 6.02×10^{23} *atoms, and this number of atoms has a mass of exactly* 12 g, *the atomic mass for* ^{12}C *expressed in grams.* The number 6.02×10^{23}, or 602,000,000,000,000,000,000,000, is called **Avogadro's** (ä′vŏ·gä′dro) **number** (*N*) after the Italian chemist and physicist Amedeo Avogadro (1776–1856), seen in Figure 8.1. Figures 8.2, 8.3, and 8.4 may be of some help in understanding the meaning of this very large number.

Mole (mol)   Amount of a substance containing the same number of atoms, formula units, molecules, or ions as there are atoms in exactly 12 g of carbon-12 (approximately 6.02×10^{23} atoms).

Avogadro's number   Number of atoms in exactly 12 g of carbon-12 (approximately 6.02×10^{23}); it is equivalent to 1 mol of a substance.

FIGURE 8.1
Italian postage stamp honoring Amedeo Avogadro. (Courtesy C. Marvin Lang; photo by Gary Shuifer)

FIGURE 8.2
Counting Avogadro's number, (N) of peas. If all the people now alive on the earth (5.6 billion) started counting Avogadro's number of peas at a rate of two peas per second, it would take approximately 1.7 million (1,700,000) years. That's a lot of peas!

| 0.1 million (100,000) years | 1.0 million (1,000,000) years | 1.5 million (1,500,000) years | 1.7 million (1,700,000) years |

What does Avogadro's number mean for other elements? One mole (6.02×10^{23}) of oxygen atoms has a mass of 16.0 g (to three significant digits), the same as the atomic mass of oxygen expressed in grams. For the same number of atoms (as called for by the definition of a mole), oxygen has a greater mass than carbon. That is, an oxygen atom is heavier than a ^{12}C atom. The same kind of comparison can be made for any element. *One mole of atoms of any element contains* 6.02×10^{23}

FIGURE 8.3
Avogadro's number is larger than the number of grains of sand in sand dunes. (Sunrise Monument Valley, Arizona)

atoms of the element and is equal to the atomic mass of the element expressed in grams:

1 mol of atoms of an element	=	6.02×10^{23} atoms of the element	=	atomic mass of the element in grams

The mole is the "chemist's dozen." It is a quantity of matter based on a certain number (6.02×10^{23}) of elementary units per mole, just as a dozen is defined as 12 units per dozen or a gross as 144 units per gross. Can you imagine how inconvenient it would be if chemists had to constantly refer to 2.80×10^{25} atoms of carbon or 1.74×10^{22} atoms of gold? These amounts are much more conveniently expressed as 46.5 mol of carbon and 0.0289 mol of gold!

The reasoning we just applied to atoms of an element we can also apply to formula units and to molecules of a compound. Therefore, in *one mole of a compound there are* 6.02×10^{23} *formula units or molecules, and this number of formula units or molecules has a mass equal to the formula or molecular mass expressed in grams:*

1 mol of a compound	=	6.02×10^{23} formula units or molecules of the compound	=	formula or molecular mass in grams

Thus, 1 mol (6.02×10^{23} molecules) of water (H_2O) has a mass of 18.0 g ($2 \times 1.0 + 1 \times 16.0 = 18.0$ amu, to the tenths place), which is the molecular mass of water expressed in grams. Similarly, 1 mol (6.02×10^{23} formula units) of sodium sulfate (Na_2SO_4) has a mass of 142.1 g ($2 \times 23.0 + 1 \times 32.1 + 4 \times 16.0 = 142.1$ amu, to the tenths place), which is the formula mass of sodium sulfate expressed in grams.

The atomic mass, formula mass, or molecular mass expressed in *grams* has a special name, the *molar mass*. The **molar mass** is the mass in *grams* of *one* mole of *any* substance, element, or compound. This quantity is a most convenient one, and we will use it regularly in this text. Consider the following example:

FIGURE 8.4
A troy ounce of gold is worth about $380. A troy ounce contains 31.10 g, while the ounce we use contains 28.35 g. This comes out to about 4×10^{-19} cents per atom! See if you can do the calculation; calculate moles of gold in 31.10 g of gold and then atoms of gold.

Molar mass Mass in *grams* of 1 mol of any substance, element, or compound.

EXAMPLE 8.3	Calculate the molar mass of ethane (C_2H_6).

SOLUTION

$$2 \times 12.0 \text{ amu} = 24.0 \text{ amu}$$
$$6 \times 1.0 \text{ amu} = \underline{6.0 \text{ amu}}$$
$$\text{molecular mass of } C_2H_6 = \overline{30.0 \text{ amu}}$$

$$\text{molar mass of } C_2H_6 = 30.0 \text{ g} \qquad Answer$$

This amount of ethane (30.0 g) is equal to the mass of 6.02×10^{23} molecules of ethane.

Study Hint: The molecular mass is expressed in *amu*, while the molar mass is expressed in *grams*.

Study Exercise 8.2
Calculate the molar mass of methane (CH_4). (16.0 g)

Work Problem 5.

We can apply the same reasoning to *ions* or to *any unit*. Therefore, in *1 mol of ions there are* 6.02×10^{23} *ions, and this number of ions has a mass equal to the atomic or formula mass of the ion expressed in grams.*

1 mol of ions	=	6.02×10^{23} ions	=	atomic or formula mass of the ions in grams

Study Hint: Consider the following analogy. Each person normally has two arms and one head. If we have a dozen people, we would have two dozen arms and one dozen heads. The ratio of arms to heads (2:1) is the same for one person as it is for a group of people—a dozen. Therefore, *one person* is analogous to *one molecule* or *formula unit*, and *one dozen people* is analogous to *one mole* of molecules or formula units.

That is, 1 mol or 6.02×10^{23} sodium ions has a mass of 23.0 g (to the tenths place), which is the atomic mass of sodium expressed in grams. And 1 mol or 6.02×10^{23} sulfate ions has a mass of 96.1 g ($1 \times 32.1 + 4 \times 16.0 = 96.1$ amu, to the tenths place), which is the formula mass of sulfate ion expressed in grams.

All these relationships bring us back to the start of the chapter and the idea that the subscripts in the formulas of compounds represent the number of atoms of each element in a formula unit or molecule of the compound. These subscripts also represent the *number of moles of ions or atoms* of the elements in 1 mol of molecules or formula units of the compound. Consider the case of water (H_2O), which consists of 2 atoms of hydrogen and 1 atom of oxygen. Water has a molecular mass of 18.0 amu (2.0 amu of hydrogen atoms plus 16.0 amu of oxygen atoms). One mole of water molecules has a molar mass of 18.0 g, consisting of 2.0 g of hydrogen atoms and 16.0 g of oxygen atoms. Hence, the moles of atoms of each element in *one* mole of water molecules are

$$2.0 \text{ g H atoms} \times \frac{1 \text{ mol H atoms}}{1.0 \text{ g H atoms}} = 2 \text{ mol H atoms}$$

$$16.0 \text{ g O atoms} \times \frac{1 \text{ mol O atoms}}{16.0 \text{ g O atoms}} = 1 \text{ mol O atoms}$$

EXAMPLE 8.4   Calculate the number of moles of oxygen *atoms* in 24.0 g of oxygen atoms.

SOLUTION   The atomic mass of oxygen is 16.0 amu; therefore, 1 mol of oxygen *atoms* has a molar mass of 16.0 g and we calculate the number of moles of oxygen atoms as

$$24.0 \text{ g O} \times \frac{1 \text{ mol O atoms}}{16.0 \text{ g O}} = 1.50 \text{ mol oxygen atoms} \qquad \textit{Answer}$$

The answer is expressed to three significant digits since the number 24.0 in the given example has three significant digits. (For a review of significant digits, see Section 2.5.)

EXAMPLE 8.5   Calculate the number of moles of oxygen *molecules* in 24.0 g of oxygen gas (O_2).

SOLUTION   An O_2 molecule has a molecular mass of 32.0 amu (2×16.0). Therefore, 1 mol of oxygen *molecules* has a molar mass of 32.0 g. We calculate the moles of oxygen molecules in 24.0 g of oxygen as

$$24.0 \text{ g } O_2 \times \frac{1 \text{ mol } O_2 \text{ molecules}}{32.0 \text{ g } O_2} = 0.750 \text{ mol oxygen molecules} \qquad \textit{Answer}$$

EXAMPLE 8.6   Calculate the number of moles of sulfur dioxide (SO_2) in 24.5 g of sulfur dioxide.

SOLUTION   The molecular mass of sulfur dioxide is

$$
\begin{aligned}
1 \times 32.1 \text{ amu} &= 32.1 \text{ amu} \\
2 \times 16.0 \text{ amu} &= 32.0 \text{ amu} \\
\hline
\text{molecular mass of } SO_2 &= 64.1 \text{ amu}
\end{aligned}
$$

Therefore, 1 mol of SO_2 molecules has a molar mass of 64.1 g. The number of moles of SO_2 in 24.5 g of SO_2 is

$$24.5 \text{ g } SO_2 \times \frac{1 \text{ mol } SO_2}{64.1 \text{ g } SO_2} = 0.382 \text{ mol } SO_2 \qquad Answer$$

EXAMPLE 8.7 Calculate the number of moles of water in 9.65×10^{23} molecules of water.

SOLUTION   One mole of water molecules contains 6.02×10^{23} molecules. There-fore, the number of moles in 9.65×10^{23} molecules of water is

$$9.65 \times 10^{23} \text{ molecules } H_2O \times \frac{1 \text{ mol } H_2O}{6.02 \times 10^{23} \text{ molecules } H_2O} = 1.60 \text{ mol } H_2O \quad Answer$$

EXAMPLE 8.8 Calculate the number of moles of sodium ions in 1.3 mol of sodium sulfate.

SOLUTION   The formula of sodium sulfate is Na_2SO_4. Because there are 2 ions of Na in one formula unit of Na_2SO_4, there will be 2 mol of Na ion in 1 mol of Na_2SO_4. Calculate the number of moles of Na ions in 1.3 mol of Na_2SO_4 as

$$1.3 \text{ mol } Na_2SO_4 \times \frac{2 \text{ mol Na ions}}{1 \text{ mol } Na_2SO_4} = 2.6 \text{ mol } Na^+ \qquad Answer$$

(The moles of atoms or ions in 1 mol of formula units or molecules are regarded as exact values and are not considered in computing significant digits. The answer is expressed to two significant digits since the number 1.3 in the given example has two significant digits.)

Study Exercise 8.3
Calculate the number of moles of methane (CH_4) in the following:
a. 37.0 g CH_4 (2.31 mol)
b. 3.56×10^{23} molecules CH_4 (0.591 mol)

Study Exercise 8.4
Calculate the number of moles of hydrogen atoms in 1.8 mol of methane (CH_4).
                                                       (7.2 mol)   Work Problems 6 through 9.

EXAMPLE 8.9 Calculate the mass in grams of sodium sulfate in 1.30 mol of sodium sulfate.

SOLUTION   The formula mass of Na_2SO_4 is 142.1 amu (see Example 8.1). Therefore, 1 mol of Na_2SO_4 has a molar mass of 142.1 g, and we calculate the mass of 1.30 mol as

$$1.30 \text{ mol } Na_2SO_4 \times \frac{142.1 \text{ g } Na_2SO_4}{1 \text{ mol } Na_2SO_4} = 185 \text{ g } Na_2SO_4 \qquad Answer$$

EXAMPLE 8.10 Calculate the mass in grams of sodium sulfate in 4.54×10^{23} formula units of sodium sulfate.

SOLUTION   One mole of Na_2SO_4 has 6.02×10^{23} formula units with a molar mass of 142.1 g (see Examples 8.1 and 8.9). Therefore, in 4.54×10^{23} formula units of Na_2SO_4, there are

$$4.54 \times 10^{23} \text{ formula units } Na_2SO_4 \times \frac{1 \text{ mol } Na_2SO_4}{6.02 \times 10^{23} \text{ formula units } Na_2SO_4}$$

$$\times \frac{142.1 \text{ g } Na_2SO_4}{1 \text{ mol } Na_2SO_4} = 107 \text{ g } Na_2SO_4 \qquad \textit{Answer}$$

EXAMPLE 8.11   Calculate the mass in grams of oxygen present in 1.30 mol of sodium sulfate.

SOLUTION   Because there are 4 atoms of oxygen in one formula unit of Na_2SO_4, there will be 4 mol of oxygen atoms in 1 mol of Na_2SO_4. The atomic mass of oxygen is 16.0 amu, and 1 mol of oxygen atoms has a molar mass of 16.0 g. Therefore, calculate the number of grams of oxygen present in 1.30 mol of Na_2SO_4 as

$$1.30 \text{ mol } Na_2SO_4 \times \frac{4 \text{ mol O atoms}}{1 \text{ mol } Na_2SO_4} \times \frac{16.0 \text{ g O}}{1 \text{ mol O atoms}} = 83.2 \text{ g oxygen} \qquad \textit{Answer}$$

Study Exercise 8.5

Calculate the mass in grams of methane (CH_4) in the following:
a. 0.252 mol of CH_4 (4.03 g)
b. 3.65×10^{23} molecules CH_4 (9.70 g)

Study Exercise 8.6

Calculate the mass in grams of hydrogen present in 3.60 mol of methane (CH_4).
                                                                                    (14.4 g)

Work Problems 10 through 12.

EXAMPLE 8.12   Calculate the number of formula units of sodium sulfate in 1.30 mol of sodium sulfate.

SOLUTION   There are 6.02×10^{23} formula units in 1 mol of formula units of a compound, and so in 1.30 mol of sodium sulfate there are

$$1.30 \text{ mol } Na_2SO_4 \times \frac{6.02 \times 10^{23} \text{ formula units } Na_2SO_4}{1 \text{ mol } Na_2SO_4}$$

$$= 7.83 \times 10^{23} \text{ formula units } Na_2SO_4 \qquad \textit{Answer}$$

EXAMPLE 8.13   Calculate the number of water molecules in 4.50 g of water.

SOLUTION   The molecular mass of water (H_2O) is 18.0 amu. Therefore, 1 mol of water molecules has a mass of 18.0 g and 1 mol of water molecules contains 6.02×10^{23} molecules. We calculate the number of water molecules in 4.50 g of water as

$$4.50 \text{ g } H_2O \times \frac{1 \text{ mol } H_2O \text{ molecules}}{18.0 \text{ g } H_2O} \times \frac{6.02 \times 10^{23} \text{ molecules } H_2O}{1 \text{ mol } H_2O \text{ molecules}}$$

$$= 1.50 \times 10^{23} \text{ molecules } H_2O \qquad \textit{Answer}$$

Study Exercise 8.7
Calculate the number of molecules of methane (CH_4) in 1.80 mol of methane.

$(1.08 \times 10^{24}$ molecules$)$

Study Exercise 8.8
Calculate the number of molecules of methane (CH_4) in 30.2 g of methane.

$(1.14 \times 10^{24}$ molecules$)$

| **EXAMPLE 8.14** | Use the atomic mass of hydrogen to five significant digits (see the inside front cover of this text) to calculate the mass in grams of 1 atom of hydrogen. |

SOLUTION   One mole of hydrogen atoms contains 6.02×10^{23} atoms of hydrogen and has a molar mass of 1.0080 g. Hence, we calculate the mass of 1 atom as

$$1 \text{ atom H} \times \frac{1 \text{ mol H atoms}}{6.02 \times 10^{23} \text{ atoms H}} \times \frac{1.0080 \text{ g H}}{1 \text{ mol H atoms}} = 0.167 \times 10^{-23} \quad \text{g H}$$

$$= 1.67 \times 10^{-24} \quad \text{g H} \quad \text{(in scientific notation; see Section 2.7)} \quad \textit{Answer}$$

Study Exercise 8.9
Calculate the mass in grams of one atom of an isotope of lithium, atomic mass = 7.02 amu.

$(1.17 \times 10^{-23}$ g$)$

Work Problems 13 and 14.

> **S**tudy Hint: Section 4.1 also gives the mass of an oxygen atom and a carbon atom. See if you can calculate these values from the precise atomic masses of oxygen and carbon (inside front cover of this book).

Work Problem 15.

(a)

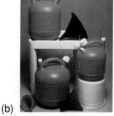

(b)

FIGURE 8.5
Molar volume of a gas. 6.02×10^{23} molecules of any gas (1 mol of gas molecules) occupy a volume of 22.4 L at 0°C and 76$\overline{0}$mm Hg (torr) (STP conditions). This volume is approximately the volume occupied by (a) three standard basketballs or (b) six gallons (two 2½ gallon and 1 gallon cans).

Standard temperature and pressure (STP)   Temperature of 0°C and an atmospheric pressure of 76$\overline{0}$ mm Hg (torr).

Molar volume of a gas   The 22.4 L occupied by 1 mol of any gas molecules at standard temperature (0°C) and pressure (76$\overline{0}$ mm Hg, torr).

8.3 Molar Volume of a Gas and Related Calculations

Avogadro's number is also useful in calculating the volume of a gas. Experiments have shown that 6.02×10^{23} molecules of any gas or 1 mol of molecules of that gas occupies a volume of 22.4 L *at* 0°C (273 K) *and a pressure of* 76$\overline{0}$ mm Hg (torr):* The conditions of 0°C and 76$\overline{0}$ mm Hg (torr) are defined as **standard temperature and pressure (STP)**. This volume of 22.4 L occupied by 1 mol of any gas molecules at STP is called the **molar volume of a gas** and is approximately the volume occupied by *three* standard basketballs or six gallons (see Figure 8.5). This molar volume of a gas relates the mass of a gas to its volume at STP and can be used to calculate various characteristics of compounds.

Moles or Mass

The molar volume of a gas allows us to determine the mass of gas or the number of moles of gas in a given volume of gas at STP. Since many common substances are gases at normal temperatures, this can be a very useful calculation.

---

* In Chapter 11 we consider the units of pressure and evaluate the effect of temperature and pressure on the volume of a gas.

| EXAMPLE 8.15 | Calculate the number of moles of oxygen *molecules* in 5.60 L of oxygen gas (O_2) at STP. |

SOLUTION   One mole of O_2 molecules occupies a volume of 22.4 L at STP. Hence, we calculate the number of moles of O_2 molecules in 5.60 L as

$$5.60 \, \cancel{L \, O_2} \times \frac{1 \, mol \, O_2}{22.4 \, \cancel{L \, O_2}} = 0.250 \, mol \, oxygen \, gas \qquad Answer$$

| EXAMPLE 8.16 | Calculate the number of grams of oxygen in 5.60 L of oxygen gas (O_2) at STP. |

SOLUTION   Calculate the number of moles of O_2 *molecules* as in Example 8.15 and then calculate the mass in grams of 5.60 L of oxygen molecules at STP from the molar mass of O_2 ($2 \times 16.0 \, amu = 32.0 \, amu = 32.0 \, g$):

$$5.60 \, \cancel{L \, O_2} \times \frac{1 \, \cancel{mol \, O_2}}{22.4 \, \cancel{L \, O_2}} \times \frac{32.0 \, g \, O_2}{1 \, \cancel{mol \, O_2}} = 8.00 \, g \, O_2 \qquad Answer$$

Molar volume and the calculation of moles in the previous section (Section 8.2) can be summarized in the following diagram:

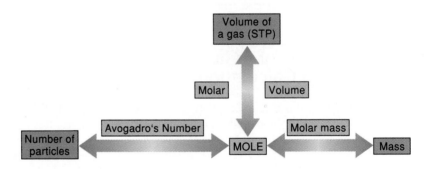

Study Exercise 8.10
Calculate the number of
a. moles of methane gas (CH_4) in 4.50 L of methane at STP   (0.201 mol)
b. grams of methane gas (CH_4) in 5.00 L of methane at STP   (3.57 g)

Work Problem 16.

Molecular Mass
and Molar Mass

This knowledge of the molar volume of a gas can help us determine the molecular mass and molar mass of any gas when they are unknown. This is a step in the identification of an unknown gas. We need only solve for the number of grams per mole (g/mol) of the gas from measurements of volume and mass. This is numerically equal to the molecular mass in amu and the molar mass in grams.

| EXAMPLE 8.17 | Calculate the molecular mass and molar mass of a gas if 5.00 L measured at STP has a mass of 9.85 g. |

SOLUTION   Solving for g/mol, calculate the molecular mass and molar mass as

$$\frac{9.85 \text{ g}}{5.00 \text{ L STP}} \times \frac{22.4 \text{ L STP}}{1 \text{ mol}} = 44.1 \text{ g/mol} \quad \text{molecular mass} = 44.1 \text{ amu} \qquad Answer$$

$$\text{molar mass} = 44.1 \text{ g} \qquad Answer$$

| EXAMPLE 8.18 | The density of a certain gas is 1.30 g/L at STP. Calculate the molecular mass and molar mass of this gas. |

SOLUTION   Solving for g/mol, calculate the molecular mass and molar mass as

$$\frac{1.30 \text{ g}}{1 \text{ L STP}} \times \frac{22.4 \text{ L STP}}{1 \text{ mol}} = 29.1 \text{ g/mol}$$

$$\text{molecular mass} = 29.1 \text{ amu} \qquad Answer$$

$$\text{molar mass} = 29.1 \text{ g} \qquad Answer$$

Study Exercise 8.11
Calculate the molecular mass and molar mass of the following gases:
a. 2.50 L at STP has a mass of 4.92 g   (44.1 amu, 44.1 g)
b. the density of a gas at STP is 1.23 g/L   (27.6 amu, 27.6 g)   Work Problem 17.

Density

Just as we can determine the molar mass of a gas from its density, so can we determine the density from the molar mass *and* molar volume. The density of any gas at STP is calculated by solving for the number of *grams per liter* of the gas (see Section 2.2). The volume of a gas at STP can also be calculated from the mass of the gas using a similar calculation.

| EXAMPLE 8.19 | Calculate the density of oxygen gas (O_2) at STP. |

SOLUTION   The units of density for a gas are g/L. Hence, from the molar mass of O_2 (32.0 g), calculate the density as

$$\frac{32.0 \text{ g } O_2}{1 \text{ mol } O_2} \times \frac{1 \text{ mol } O_2}{22.4 \text{ L STP}} = 1.43 \text{ g/L at STP} \qquad Answer$$

| EXAMPLE 8.20 | Calculate the volume in liters occupied by 5.00 g of oxygen gas (O_2) at STP. |

SOLUTION   By calculating the number of moles of oxygen *molecules* and then using the molar volume, we can calculate the volume occupied by 5.00 g of oxygen molecules at STP as

$$5.00 \text{ g } O_2 \times \frac{1 \text{ mol } O_2}{32.0 \text{ g } O_2} \times \frac{22.4 \text{ L } O_2 \text{ STP}}{1 \text{ mol } O_2} = 3.50 \text{ L oxygen at STP} \qquad Answer$$

Study Exercise 8.12
Calculate the density of methane gas (CH_4) at STP. (0.714 g/L)

Study Exercise 8.13
Calculate the volume in liters occupied by 3.25 g of methane gas (CH_4) at STP.

Work Problems 18 and 19.
(4.55 L)

8.4 Calculation of Percent Composition of Compounds

Scientists are not always interested in examining the total makeup of a compound. In fact, they are often more concerned with the percent composition of a compound, that is, the percent by mass of each element in it. For example, suppose that a medicinal tea can be brewed from a plant discovered deep in the jungles of Brazil. To identify the compound responsible for the healing effects, a chemist first needs to separate the various compounds in the tea and determine which one(s) is active. Then he or she must find out *what* elements are present in that compound and *how much* of each (the percent of each) is present.

Calculating the percent composition of elements in a compound is relatively straightforward. *Percent means parts per hundred.* For example, if your school has a student enrollment of 1000 and there are 400 men students, the percent of men students is 40 (400/1000 × 100 = 40%), or 40 men students *per hundred* (students). Because chemists generally express the composition of a compound in terms of *percent by mass*, they first determine the amu or mass of various elements in the compound. Then they determine the total amu or mass of the compound. Finally, they divide the amu or mass of each element by the total mass and multiply by 100.

Study Hint: The molar mass (64.5 g) and 24.0 g C, 5.0 g H, and 35.5 g Cl could have also been used to give the *same* results. Therefore, both amu and grams give percent by *mass*.

EXAMPLE 8.21 Calculate the percent composition of ethyl chloride (C_2H_5Cl).

SOLUTION Calculate the molecular mass of C_2H_5Cl and its components.

$$2 \times 12.0 \text{ amu} = 24.0 \text{ amu}$$
$$5 \times 1.0 \text{ amu} = 5.0 \text{ amu}$$
$$1 \times 35.5 \text{ amu} = 35.5 \text{ amu}$$
$$\text{molecular mass of } C_2H_5Cl = \overline{64.5 \text{ amu}}$$

Calculate the percent of each element in the compound by dividing the contribution of each element (amu) by the molecular mass (amu) and multiplying by 100.

% carbon: $\dfrac{24.0 \text{ amu}}{64.5 \text{ amu}} \times 100 = 37.2\% \text{ C}$

% hydrogen: $\dfrac{5.0 \text{ amu}}{64.5 \text{ amu}} \times 100 = 7.8\% \text{ H}$

% chlorine: $\dfrac{35.5 \text{ amu}}{64.5 \text{ amu}} \times 100 = 55.0\% \text{ Cl}$ *Answer*

EXAMPLE 8.22 A student found that 1.00 g of a metal combined with 0.65 g of oxygen to form an oxide of the metal. Calculate the percent of metal in the oxide.

SOLUTION   The total mass of the oxide is 1.65 g (1.00 g + 0.65 g = 1.65 g). Calculate the percent of metal in the oxide by dividing the mass of the metal in grams by the total mass of the oxide in grams and multiplying by 100.

$$\% \ metal: \quad \frac{1.00 \ g \ metal}{1.65 \ g \ oxide} \times 100 = 60.6\% \ metal \qquad Answer$$

EXAMPLE 8.23 A crude sample of zinc sulfide has a mass of 8.00 g. It contains 5.00 g of zinc sulfide and 3.00 g of impurities that contain no zinc. What is the percent of zinc in the crude sample?

SOLUTION   The chemical formula of zinc sulfide is ZnS (see Section 7.3), and the formula mass is 97.5 amu:

$$1 \times 65.4 \ amu = 65.4 \ amu$$
$$1 \times 32.1 \ amu = 32.1 \ amu$$
$$formula \ mass \ of \ ZnS = \overline{97.5 \ amu}$$

The molar mass of ZnS is 97.5 g. Therefore, in 97.5 g of ZnS there are 65.4 g of Zn, and so we calculate the percent of zinc in 8.00 g of crude zinc sulfide as

$$\frac{5.00 \ g \ pure \ ZnS}{8.00 \ g \ crude \ ZnS} \times \frac{65.4 \ g \ Zn}{97.5 \ g \ pure \ ZnS} \times 100 = 41.9\% \ Zn \qquad Answer$$

> **S**tudy Hint: A crude sample is a mixture of a pure compound and other materials that might happen to come with it. It is analogous to octane (C_8H_{18}) found in gasoline: the octane is a pure substance, and the other carbon-containing substances along with the additives in the gasoline are the "impurities."

EXAMPLE 8.24 Calculate the number of grams of carbon in 17.6 g of carbon dioxide.

SOLUTION   Calculate the molecular mass of CO_2 and its components.

$$1 \times 12.0 \ amu = 12.0 \ amu$$
$$2 \times 16.0 \ amu = 32.0 \ amu$$
$$molecular \ mass \ of \ CO_2 = \overline{44.0 \ amu}$$

The molar mass of CO_2 is 44.0 g. Therefore, in 44.0 g of CO_2 there are 12.0 g of C, and so we calculate the number of grams of carbon in 17.6 g of CO_2 as

$$17.6 \ g \ CO_2 \times \frac{12.0 \ g \ C}{44.0 \ g \ CO_2} = 4.80 \ g \ C \qquad Answer$$

Study Exercise 8.14
Calculate the percent composition of glucose ($C_6H_{12}O_6$).

(40.0% C, 6.7% H, 53.3% O)

Study Exercise 8.15
Calculate the percent of metal in an oxide if 0.450 g of metal combines with 0.375 g of oxygen to form the oxide.

(54.5%)

Study Exercise 8.16
Calculate the number of grams of sodium in 6.55 g of sodium chloride.

Work Problems 20 through 23.

(2.58 g)

8.5 Calculation of Empirical (Simplest) and Molecular Formulas

Empirical formula Simplest formula for a compound; it is the smallest whole-number ratio of the atoms present.

Molecular formula Way of expressing the composition of one molecule of a compound or element by using elemental symbols for each element involved and subscripts reflecting the number of atoms (above 1) of that element in the molecule; it shows the actual number of atoms of each element present in one molecule of the compound.

Just as in Chapter 7 where we worked from nomenclature to formulas and back again, so chemists sometimes need to translate knowledge about the percent composition of a compound into a formula for that compound.

The **empirical formula** (*simplest formula*) of a compound is the formula containing the *smallest whole-number ratio of the atoms* present in a molecule or formula unit of the compound. This empirical formula is found from the percent composition of the compound, which is determined *experimentally* from analysis of the compound in the laboratory. The empirical formula gives only the ratio of the atoms present expressed as the *smallest whole numbers*.

In contrast, the **molecular formula** of the compound is the formula containing the *actual* number of atoms of each element present in one molecule of the compound. The molecular formula is a *whole-number multiple* of the empirical formula. A simple analogy may help to illustrate these two types of formulas. In your school, the ratio of women to men may be 2:1 (empirical formula), but the actual number of women to men may be 800:400 (molecular formula). On a chemical note, the empirical formula of glucose (an important sugar in your body) is CH_2O, but the molecular formula is $6 \times (CH_2O)$ or $C_6H_{12}O_6$. The molecular formula is determined from the empirical formula *and* the experimentally determined molecular mass of the compound.

In some cases, both the empirical and molecular formulas are the same, as in the case of H_2O. The true formulas of compounds existing as *molecules* (covalent compounds) are always referred to as *molecular formulas*. Only for molecules for which the *molecular and empirical formulas are the same* does the empirical formula give the actual numbers of atoms present in the molecule. For compounds written as *formula units* (ionic compounds), there are no molecular formulas because these compounds do not exist as molecules. Hence, their formulas are always empirical formulas.

In real life, chemists use this relationship between the empirical and molecular formulas of compounds when they make new compounds. Generally, they begin by performing a chemical analysis of the new compound to determine its empirical formula. Then, using this formula and the molecular mass of the compound, they identify its molecular formula. Once chemists know the molecular formula, they can perform experiments that will help determine the structure of the compound. The process of determining the empirical and molecular formulas may sound time-consuming, and at one time it was. Happily, with modern instruments, it can usually be done in about 30 min. The following summarizes these steps:

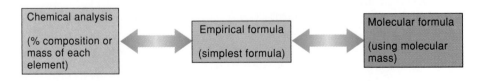

EXAMPLE 8.25   Determine the empirical formula for a compound composed of 32.4% sodium, 22.6% sulfur, and 45.1% oxygen.*

SOLUTION   In exactly $10\overline{0}$ g of the compound there are 32.4 g of Na, 22.6 g of S, and 45.1 g O. The first step is to calculate the moles of *atoms* of each element present.

Step I: Moles of atoms.

$$32.4 \text{ g Na} \times \frac{1 \text{ mol Na atoms}}{23.0 \text{ g Na}} = 1.41 \text{ mol Na atoms}$$

$$22.6 \text{ g S} \times \frac{1 \text{ mol S atoms}}{32.1 \text{ g S}} = 0.704 \text{ mol S atoms}$$

$$45.1 \text{ g O} \times \frac{1 \text{ mol O } atoms}{16.0 \text{ g O}} = 2.82 \text{ mol O } atoms$$

Combine the elements in a ratio of 1.41 mol of Na atoms to 0.704 mol of S atoms to 2.82 mol of 0 atoms as

$$\text{Na}_{1.41 \text{ mol of atoms}} : \text{S}_{0.704 \text{ mol of atoms}} : \text{O}_{2.82 \text{ mol of atoms}}$$

The empirical formula must express these relationships in terms of the *smallest whole numbers*. The second step is to express these *whole-number* relationships by *dividing* each value *by the smallest* of the three.

Step II: Divide by the smallest.

$$For\ Na: \quad \frac{1.41}{0.704} = 2.00 = 2$$

$$For\ S: \quad \frac{0.704}{0.704} = 1.00 = 1$$

$$For\ O: \quad \frac{2.82}{0.704} = 4.01 \simeq 4$$

where $\simeq$ stands for "is about equal to."

Finally, combine the elements in a ratio of 2 mol of Na atoms to 1 mol of S atoms to 4 mol of O atoms, and the empirical formula is Na_2SO_4.   *Answer*

> **Study Hint:** We could also assume that we have 10.00 g of compound. This sample would contain 3.24 g Na, 2.26 g of S, and 4.51 g of O. Calculate the empirical formula of Na_2SO_4 from this data (it should be the same).

EXAMPLE 8.26   Calculate the empirical formula for a compound composed of 26.6% potassium, 35.4% chromium, and 38.1% oxygen.

SOLUTION   First, calculate the moles of *atoms* of each element in exactly $10\overline{0}$ g of the compound.

Step I: Moles of atoms.

$$26.6 \text{ g K} \times \frac{1 \text{ mol K atoms}}{39.1 \text{ g K}} = 0.680 \text{ mol K atoms}$$

$$35.4 \text{ g Cr} \times \frac{1 \text{ mol Cr atoms}}{52.0 \text{ g Cr}} = 0.681 \text{ mol Cr atoms}$$

$$38.1 \text{ g O} \times \frac{1 \text{ mol O } atoms}{16.0 \text{ g O}} = 2.38 \text{ mol O } atoms$$

---

* The difference here of 0.1% between 100.1% (32.4 + 22.6 + 45.1) and exactly 100% emphasizes the experimental nature of these values and occurs because of experimental error.

Second, reduce these values to simpler numbers by dividing each one by the smallest value.

Step II: Divide by the smallest.

$$\text{For } K: \quad \frac{0.680}{0.680} = 1$$

$$\text{For } Cr: \quad \frac{0.681}{0.680} \simeq 1$$

$$\text{For } O: \quad \frac{2.38}{0.680} = 3.5$$

Third, convert these relative ratios to small whole numbers by *multiplying by 2.*

Step III: Ratio of whole numbers.

The empirical formula is $(K_1Cr_1O_{3.5})_2 = K_2Cr_2O_7$.   *Answer*

Study Hint: The ratio must be ± 0.05 for us to round up or down to get whole-number ratios. Examples: 1.95 and 2.05 would both be 2.

In cases like Example 8.26, where the ratio does not come out as a simple whole-number ratio, *we must multiply all the numbers by an integer to obtain small whole numbers.* If the ratio of the elements ends in 0.5, multiply *all* the numbers by 2 to obtain small whole numbers. If the ratio of these numbers ends in 0.33 (0.33 . . . , the fraction $\frac{1}{3}$), then we must multiply *all* the numbers by 3 to obtain small whole numbers.

EXAMPLE 8.27 An oxide of nitrogen gave the following analysis: 3.04 g of nitrogen combined with 6.95 g of oxygen. Experiments show that the molecular mass of this compound is 91.0 amu. Determine its molecular formula.

SOLUTION   We calculate the empirical formula from the analysis information in the same way we would from the percent composition. First, calculate the number of moles of atoms in 3.04 g and 6.95 g of nitrogen and oxygen, respectively:

Step I: Moles of atoms.

$$3.04 \text{ g N} \times \frac{1 \text{ mol N } atoms}{14.0 \text{ g N}} = 0.217 \text{ mol N } atoms$$

$$6.95 \text{ g O} \times \frac{1 \text{ mol O } atoms}{16.0 \text{ g O}} = 0.434 \text{ mol O } atoms$$

Second, reduce these values to small whole numbers by dividing by the smallest value:

Step II: Divide by the smallest.

$$\text{For } N: \quad \frac{0.217}{0.217} = 1$$

$$\text{For } O: \quad \frac{0.434}{0.217} = 2$$

Step III: Ratio of whole numbers.

The empirical formula is NO_2. The *molecular formula* is equal either to the empirical formula or to some multiple (2, 3, 4, and so on) of it.

Step IV: Molecular formula.

(1) Calculate the *empirical formula mass* of NO_2 as

$$
\begin{aligned}
1 \times 14.0 \text{ amu} &= 14.0 \text{ amu} \\
2 \times 16.0 \text{ amu} &= 32.0 \text{ amu} \\
\text{empirical formula mass} &= \overline{46.0 \text{ amu}}
\end{aligned}
$$

(2) The molecular mass as given in the problem is 91.0 amu.* The multiple of the empirical formula is thus approximately 2:

$$
\boxed{\frac{\text{molecular mass}}{\text{empirical formula mass}}} = \frac{91.0 \text{ amu}}{46.0 \text{ amu}} = 1.98 \text{ or approximately 2}
$$

Therefore, the molecular formula is

$$
(NO_2)_2 = N_2O_4 \qquad Answer
$$

EXAMPLE 8.28 A hydrocarbon has the following composition: carbon = 92.3% and hydrogen = 7.7%. Experiments show that the molecular mass of this compound is 78.0 amu. Determine its molecular formula.

SOLUTION   First, calculate the *empirical formula* from the percent composition by calculating the moles of atoms of carbon and hydrogen in exactly $10\overline{0}$ g of the compound.

Step I: Moles of atoms.

$$
92.3 \text{ g C} \times \frac{1 \text{ mol C atoms}}{12.0 \text{ g C}} = 7.69 \text{ mol C atoms}
$$

$$
7.7 \text{ g H} \times \frac{1 \text{ mol H atoms}}{1.0 \text{ g H}} = 7.7 \text{ mol H atoms}
$$

Second, reduce these values by dividing by the smallest value.

Step II: Divide by the smallest.

$$
For\ C: \quad \frac{7.69}{7.69} = 1.00 = 1
$$

$$
For\ H: \quad \frac{7.7}{7.69} = 1.00 = 1
$$

Step III: Ratio of whole numbers.

The empirical formula is CH. The *molecular formula* is equal either to the empirical formula or to some multiple (2, 3, 4, and so on) of it.

Step IV: Molecular formula.

(1) Calculate the *empirical formula mass* of CH as

$$
\begin{aligned}
1 \times 12.0 \text{ amu} &= 12.0 \text{ amu} \\
1 \times 1.0 \text{ amu} &= 1.0 \text{ amu} \\
\text{empirical formula mass} &= \overline{13.0 \text{ amu}}
\end{aligned}
$$

---

* The difference between an exact value of 92.0 amu and this value of 91.0 amu results from *experimental error* in the determination of the molecular mass.

(2) The molecular mass as given in the problem is 78.0 amu. The multiple of the empirical formula is thus 6:

$$\boxed{\dfrac{\text{molecular mass}}{\text{empirical formula mass}}} = \dfrac{78.0 \text{ amu}}{13.0 \text{ amu}} = 6$$

Therefore, the molecular formula is

$$(\text{CH})_6 = \text{C}_6\text{H}_6 \quad Answer$$

EXAMPLE 8.29   A hydrocarbon has the following composition: carbon = 82.7% and hydrogen = 17.4%. The density of its vapor at STP is 2.60 g/L. Calculate the molecular formula of the hydrocarbon.

SOLUTION   First, calculate the *empirical formula* from the percent composition by calculating the moles of atoms of carbon and hydrogen in exactly $\overline{100}$ g of the compound.

Step I: Moles of atoms.

$$82.7 \text{ g C} \times \dfrac{1 \text{ mol C atoms}}{12.0 \text{ g C}} = 6.89 \text{ mol C atoms}$$

$$17.4 \text{ g H} \times \dfrac{1 \text{ mol H atoms}}{1.0 \text{ g H}} = 17.4 \text{ mol H atoms}$$

Second, reduce these values by dividing by the smallest value.

Step II: Divide by the smallest.

$$For\ C: \quad \dfrac{6.89}{6.89} = 1.00 = 1$$

$$For\ H: \quad \dfrac{17.4}{6.89} = 2.53 \simeq 2.5$$

Convert these relative ratios to small whole numbers *by multiplying by 2.*

Step III: Ratio of whole numbers.

The empirical formula is $(\text{C}_1\text{H}_{2.5})_2 = \text{C}_2\text{H}_5$. The *molecular formula* is equal either to the empirical formula or to some multiple (2, 3, 4, and so on) of it.

Step IV: Molecular formula.

(1) Calculate the *empirical formula mass* of C_2H_5 as

$$\begin{aligned} 2 \times 12.0 \text{ amu} &= 24.0 \text{ amu} \\ 5 \times 1.0 \text{ amu} &= \underline{5.0 \text{ amu}} \\ \text{empirical formula mass} &= 29.0 \text{ amu} \end{aligned}$$

(2) The molecular mass is not given in the problem; therefore, we must calculate it as we did in Example 8.18:

$$\dfrac{2.60 \text{ g}}{1 \text{ L STP}} \times \dfrac{22.4 \text{ L STP}}{1 \text{ mol}} = 58.2 \text{ g/mol}$$

Hence, the molecular mass is 58.2 amu. The multiple of the empirical formula is approximately 2.

$$\frac{\text{molecular mass}}{\text{empirical formula mass}} \simeq \frac{58.2 \text{ amu}}{29.0 \text{ amu}} \simeq 2$$

Therefore, the molecular formula is

$$(C_2H_5)_2 = C_4H_{10} \qquad Answer$$

Study Exercise 8.17
Calculate the empirical formula for a compound with a composition of 52.9% aluminum and 47.1% oxygen.

$$(Al_2O_3)$$

Study Exercise 8.18
Calculate the molecular formula from the following experimental data: 56.4% phosphorus, 43.6% oxygen, and a molecular mass of $2\overline{2}0$ amu.

$$(P_4O_6) \qquad \text{Work Problems 24 through 30.}$$

 # Summary

The *molecular mass* of a compound is the sum of the atomic masses of all of the atoms present in one *molecule* of the compound. It represents the mass of one molecule of that compound in atomic mass units (amu). The *formula mass* of an ionic compound is calculated by adding up the atomic masses of all the atoms present in one *formula unit* of the compound (Section 8.1).

A *mole* is defined as the amount of a substance containing the same number of atoms, formula units, molecules, or ions as there are atoms in exactly 12 g of ^{12}C atoms. This number, 6.02×10^{23}, is *Avogadro's number*. One mole of an *element* contains 6.02×10^{23} atoms and has a mass equal to the *atomic mass* in grams of that element. One mole of a *compound* contains 6.02×10^{23} molecules or formula units and has a mass equal to the *molecular mass* or *formula mass* in grams of that compound. The molecular or formula mass expressed in grams is the *molar mass* and is the mass in grams of 1 mol of an element or compound (Section 8.2).

One mole of a gas occupies 22.4 L at standard temperature and pressure [STP, 0°C and $76\overline{0}$ mm Hg (torr)]. This number, the *molar volume of a gas*, allows us to calculate the number of moles of molecules in a particular volume of a gaseous compound. We can also use the molar volume to calculate the molecular mass and molar mass of a compound or its density at STP (Section 8.3).

We can determine the percent composition of a compound from the formula or from experimental analysis. This percent composition lets us figure out the amount of any element in a sample of the compound (Section 8.4).

The *empirical formula* of a compound gives the smallest whole-number ratio of atoms in a molecule or formula unit of the compound. It can be determined from chemical analysis (percent composition or mass of each element) of the compound. The *molecular formula* of a compound is the formula containing the actual number of atoms of each element in one molecule of the compound. It is a whole-number multiple of the empirical formula. We can determine the molecular formula from the empirical formula and the experimental molecular mass of the compound (Section 8.5).

Exercises

1. Define or explain the following terms (the number in parentheses refers to the section in the text where the term is mentioned):

   a. mole (8.2)

   b. Avogadro's number (8.2)

   c. molar mass (8.2)

   d. standard temperature and pressure (STP) (8.3)

   e. molar volume of a gas (8.3)

   f. empirical formula (8.5)

   g. molecular formula (8.5)

2. Distinguish between:

   a. moles of atoms and moles of molecules

   b. empirical formula and molecular formula

 Problems

Use the atomic masses given in the table of approximate atomic masses on the inside back cover of this text.

Formula and Molecular Mass (See Section 8.1)

3. Calculate the molecular mass of each of the following compounds:

   a. CO_2

   b. $C_6H_{12}O_6$

   c. NH_3

   d. CH_4

   e. SO_3

   f. N_2O_3

4. Calculate the formula mass of each of the following compounds:

   a. Al_2O_3

   b. ZnF_2

   c. $Ca(OH)_2$

   d. $Hg(NO_3)_2$

   e. $Ca_3(PO_4)_2$

   f. $Al_2(SO_4)_3$

Moles of Units and Avogadro's Number (See Section 8.2 and Chapter 7 to write the formulas from the names of the compounds)

5. Calculate the molar mass of each of the following compounds:

   a. CO

   b. C_2H_4

   c. $CaCl_2$

   d. $Ca(NO_3)_2$

6. Give the number of moles of atoms of each element in 1 mol of the following formula units or molecules of compounds:

   a. C_6H_6

   b. N_2O_4

   c. $Al_2(SO_4)_3$

   d. $Ca(OH)_2$

   e. K_2CO_3

   f. $Ba(C_2H_3O_2)_2$

7. Calculate the number of

   a. moles of aluminum in 3.60 g of aluminum

   b. moles of oxygen *atoms* in 50.0 g of oxygen atoms

   c. moles of oxygen *molecules* in 50.0 g of oxygen gas (O_2)

   d. moles of silver chloride in 55.0 g of silver chloride

   e. moles of calcium carbonate in 3.50 g of calcium carbonate

   f. moles of sulfuric acid in 0.125 kg of sulfuric acid

    g. moles of iron and chlorine atoms in 1.50 mol of iron(III) chloride

    h. moles of magnesium ions, phosphorus atoms, and oxygen atoms in 2.70 mol of magnesium phosphate

    i. moles of sodium in 1.60×10^{23} sodium atoms

    j. moles of water in 7.50×10^{24} molecules of water

8. Calculate the number of

    a. moles of sodium in 22.0 g of sodium

    b. moles of sulfur in 85.0 g of sulfur

    c. moles of methane (CH_4) in 108 g of methane

    d. moles of sodium chloride in 4.25 g of sodium chloride

    e. moles of calcium carbonate in 4.20 kg of calcium carbonate

    f. moles of aluminum ions in 12.6 g of aluminum ions

    g. moles of sulfur in 0.350 mol of aluminum sulfate

    h. moles of potassium bromide in 5.65×10^{24} formula units of potassium bromide

    i. moles of carbon dioxide in 1.50×10^{24} molecules of carbon dioxide

    j. moles of nitrogen dioxide in 6.85×10^{25} molecules of nitrogen dioxide

9. Calculate the number of

    a. moles of arsenic in 17.6 g of arsenic

    b. moles of potassium bromide in 18.9 g of potassium bromide

    c. moles of chlorine atoms in 3.20 mol of barium chloride

    d. moles of arsenic in 5.25×10^{24} atoms of arsenic

    e. moles of potassium bromide in 4.21×10^{25} formula units of potassium bromide

10. Calculate the mass in

    a. grams of carbon dioxide in 1.25 mol of carbon dioxide

    b. grams of sodium phosphate in 1.50 mol of sodium phosphate

    c. grams of sodium in 1.30 mol of sodium atoms

    d. grams of sodium chloride in 0.150 mol of sodium chloride

    e. milligrams of potassium sulfate in 0.00250 mol of potassium sulfate

    f. grams of sugar ($C_{12}H_{22}O_{11}$) in 1.30 mol of sugar

    g. milligrams of ammonia (NH_3) in 0.0200 mol of ammonia

    h. grams of phosphorus in 1.40 mol of sodium phosphate

    i. grams of oxygen in 1.25 mol of sodium phosphate

    j. grams of magnesium in 3.45×10^{23} atoms of magnesium

11. Calculate the mass in

    a. grams of nitrogen in 4.00 mol of nitrogen molecules (N_2)

    b. grams of copper in 2.50 mol of copper atoms

    c. grams of barium carbonate in 0.400 mol of barium carbonate

    d. milligrams of oxygen in 0.00300 mol of oxygen molecules (O_2)

  e. grams of phosphorus in 0.305 mol of phosphorus

  f. milligrams of carbon in 0.00240 mol of dextrose (glucose, $C_6H_{12}O_6$)

  g. grams of sulfuric acid in 2.00 mol of sulfuric acid

  h. grams of potassium in 6.70 mol of potassium chloride

  i. grams of methane (CH_4) in 1.27×10^{21} molecules of methane

  j. grams of carbon dioxide in 2.30×10^{21} molecules of carbon dioxide

12. Calculate the mass in

  a. grams of titanium in 0.132 mol of titanium

  b. grams of magnesium oxide in 0.362 mol of magnesium oxide

  c. grams of chlorine in 1.12 mol of magnesium chloride

  d. grams of titanium in 3.25×10^{24} atoms of titanium

  e. grams of magnesium oxide in 1.36×10^{25} formula units of magnesium oxide

13. Calculate the number of

  a. atoms in 0.600 mol of carbon atoms

  b. atoms in 0.0400 mol of phosphorus atoms

  c. molecules in 7.80 mol of methane (CH_4)

  d. molecules in 15.0 g of carbon dioxide

14. Calculate the number of

  a. molecules in 3.10 mol of hydrogen gas (H_2)

  b. molecules in 20.0 g of hydrogen gas (H_2)

  c. atoms of hydrogen in 5.00 g of hydrogen gas (H_2)

  d. atoms of oxygen in 7.80 g of carbon dioxide

15. Calculate the mass in grams of *one* atom of

  a. an isotope of helium, atomic mass = 4.00 amu

  b. an isotope of nickel, atomic mass = 61.9 amu

  c. an isotope of rubidium, atomic mass = 84.9 amu

  d. an isotope of mercury, atomic mass = 204 amu

Molar Volume and Related Problems (See Section 8.3)

16. Calculate the number of

  a. moles of helium gas in 15.0 L of helium at STP

  b. moles of oxygen gas molecules in 875 mL of oxygen (O_2) at STP

  c. moles of nitrogen gas molecules in 48.0 L of nitrogen (N_2) at STP

  d. grams of carbon dioxide gas in 14.0 L of carbon dioxide at STP

  e. grams of methane (CH_4) gas in 6.50 L of methane at STP

  f. grams of carbon monoxide gas in 5.65 L of carbon monoxide at STP

17. Calculate the molecular mass and molar mass of the following gases:

  a. 3.30 L at STP has a mass of 0.580 g

  b. 4.00 L at STP has a mass of 4.98 g

  c. 2.45 L at STP has a mass of 7.40 g

d. the density of a gas at STP is 0.725 g/L

e. the density of a gas at STP is 1.65 g/L

f. the density of a gas at STP is 1.80 g/L

18. Calculate the density in g/L of the following gases at STP:

    a. ammonia (NH_3)

    b. ethane (C_2H_6)

    c. acetylene (C_2H_2)

    d. propane (C_3H_8)

    e. hydrogen iodide (HI)

    f. X_2Y (atomic masses: X, 4.0 amu, Y, 3.2 amu)

19. Calculate the volume in liters at STP that the following gases occupy:

    a. 9.00 g of nitrogen (N_2)

    b. 7.50 g of oxygen (O_2)

    c. 2.50 g of carbon monoxide

    d. 4.60 g of chlorine (Cl_2)

    e. 8.40 g of hydrogen chloride

    f. 6.30 g of dinitrogen monoxide

Percent Composition (See Section 8.4)

20. Calculate the percent composition of the following compounds:

    a. NaCl   **b.** H_2S **c.** $BaCO_3$

    d. $Ca_3(PO_4)_2$ **e.** C_2H_6O **f.** $Fe(C_2H_3O_2)_3$

21. Calculate the percent of metal in the following compounds:

    a. 0.550 g of a metal combines with 0.400 g of oxygen

    b. 0.400 g of a metal combines with 0.380 g of oxygen

    c. 1.85 g of a metal combines with 1.30 g of sulfur

    d. 275 mg of a metal combines with 135 mg of sulfur

22. A crude sample of lye has a mass of 13.0 g. It contains 6.85 g of sodium hydroxide. What is the percent sodium in the crude sample? Assume the impurities contain no sodium.

23. Calculate the number of

    a. grams of cadmium in 28.0 g of cadmium sulfide

    b. grams of magnesium in 68.0 g of magnesium nitride

    c. grams of calcium sulfide containing 5.37 g of sulfur

    d. grams of mercury(II) oxide containing 6.40 g of mercury

Empirical and Molecular Formulas (See Section 8.5)

24. Determine the empirical formula for each of the following compounds:

    a. 48.0% zinc and 52.0% chlorine

    b. 19.0% tin and 81.0% iodine

    c. 25.9% iron and 74.1% bromine

    d. 62.6% lead, 8.5% nitrogen, and 29.0% oxygen

    e. 28.8% magnesium, 14.2% carbon, and 57.0% oxygen

    f. 38.8% calcium, 20.0% phosphorus, and 41.3% oxygen

    g. 36.5% sodium, 25.4% sulfur, and 38.1% oxygen

    h. 44.9% potassium, 18.4% sulfur, and 36.7% oxygen

    i. 7.2% phosphorus and 92.8% bromine

    j. 74.4% gallium and 25.6% oxygen

25. Determine the empirical formula for each of the following compounds from the experimental data:

    a. 1.99 g of aluminum combines with 1.76 g of oxygen

    b. 1.07 g of carbon combines with 1.43 g of oxygen

    c. 2.95 g of sodium combines with 2.05 g of sulfur

    d. 0.500 g of sulfur combines with 0.500 g of oxygen

26. Determine the molecular formula for each of the following compounds from the experimental data:

    a. 80.0% carbon, 20.0% hydrogen, and a molecular mass of 30.0 amu

    b. 83.7% carbon, 16.3% hydrogen, and a molecular mass of 86.0 amu

    c. 92.3% carbon, 7.7% hydrogen, and a molecular mass of 26.0 amu

    d. 41.4% carbon, 3.5% hydrogen, 55.1% oxygen, and a molecular mass of 116.0 amu

    e. 37.8% carbon, 6.3% hydrogen, 55.8% chlorine, and a molecular mass of 127.0 amu

27. Sulfadiazine, a sulfa drug used in the treatment of bacterial infections, on analysis gives 48.0% carbon, 4.0% hydrogen, 22.4% nitrogen, 12.8% sulfur, and 12.8% oxygen. The molecular mass is found to be $25\overline{0}$ amu. Calculate the molecular formula of sulfadiazine.

28. Estrone, a female sex hormone, on analysis gives 80.0% carbon, 8.2% hydrogen, and 11.8% oxygen. The molecular mass is found to be $27\overline{0}$ amu. Calculate the molecular formula for estrone.

29. Nicotine, a compound found from 2 to 8% in tobacco leaves, on analysis gives 74.0% carbon, 8.7% hydrogen, and 17.3% nitrogen. The molecular mass is found to be 162 amu. Calculate the molecular formula for nicotine.

30. Cyanogen, a highly poisonous gas with an almondlike odor, on analysis gives 46.2% carbon and 53.8% nitrogen. At STP, cyanogen has a density of 2.32 g/L. Calculate its molecular formula.

General Problems

31. Calculate the number of moles of H_2SO_4 in 695 g of 48.0% sulfuric acid solution (by mass). (*Hint:* A 48.0% by mass sulfuric acid solution means 48.0 g of pure H_2SO_4 in $10\overline{0}$ g of solution.)

32. How many milliliters of concentrated nitric acid will be needed to supply 4.00 mol of HNO_3? Concentrated nitric acid is 72.0% HNO_3 and has a specific gravity of 1.42. (*Hint:* A 72.0% concentrated HNO_3 solution means 72.0 g of pure HNO_3 in $10\overline{0}$ g of concentrated solution.)

33. When a person driving a motor vehicle has a blood alcohol level of $10\overline{0}$ mg of alcohol (C_2H_6O) per $10\overline{0}$ mL of blood, this person, in most states, is considered to be "driving while intoxicated" (DWI). How many alcohol molecules per milliliter of blood are required for the person to be considered DWI?

34. Liquid ammonia (100% ammonia) and pure ammonium nitrate are both used as fertilizers for their nitrogen content. Both sell for approximately $300 per ton. Based on the nitrogen content, which would be the better buy?

35. A millimole (mmol) is 0.001 mol. Calculate the number of

   a. millimoles of sugar ($C_{12}H_{22}O_{11}$) in 1.40 g of sugar

   b. grams of sugar in 11.2 mmol of sugar

   c. molecules of sugar in 8.25 mmol of sugar

36. The blood glucose level of a normal person is about $9\overline{0}$ mg of glucose ($C_6H_{12}O_6$) per $10\overline{0}$ mL of blood. On oral ingestion of 1.00 g of glucose per kilogram of body weight, the blood glucose level rises to about $14\overline{0}$ mg of glucose per $10\overline{0}$ mL of blood.

   a. Calculate the number of millimoles of glucose per milliliter of blood and the number of glucose molecules per milliliter of blood before and after consumption of the glucose.

   b. The average total blood volume in a person is 5.50 L. Calculate the total number of millimoles and grams of glucose in the blood before and after consumption of the glucose. (See Problem 35 for a definition of millimole.)

37. In a diabetic person, the blood glucose level is about 135 mg of glucose ($C_6H_{12}O_6$) per $10\overline{0}$ mL of blood. On oral ingestion of 1.00 g of glucose per kilogram of body weight, the blood glucose level rises to about $23\overline{0}$ mg per $10\overline{0}$ mL of blood.

   a. Calculate the number of millimoles of glucose per milliliter of blood and the number of glucose molecules per milliliter of blood before and after consumption of the glucose.

   b. The average total blood volume in a person is 5.50 L. Calculate the total number of millimoles and grams of glucose in the blood before and after consumption of the glucose. (See Problem 35 for a definition of millimole.)

38. A gaseous hydrocarbon has a density of 1.25 g/L at 0°C and $76\overline{0}$ torr. Its composition is 85.6% carbon and 14.4% hydrogen. Calculate its molecular formula.

39. Cyclopropane, a gaseous hydrocarbon used as an anesthetic, on analysis gives 85.6% carbon and 14.4% hydrogen. At STP, 7.52 L of cyclopropane has a mass of 14.1 g. Calculate its molecular formula.

40. The normal dose of the antiinflammatory ibuprofen is $20\overline{0}$ mg. Calculate the number of molecules in this dose (molar mass = 206 g).

41. All four members of a family became ill. The air in the house was thought to be the causative agent, and a sample of it was taken. After removing the oxygen and nitrogen from the sample, it was found that $10\overline{0}$ mL of the air, corrected to STP conditions, had a mass of 0.124 g. The two possible causes could be (a) formaldehyde (CH_2O) from improperly installed wall insulation or (b) carbon monoxide from an improperly installed gas heating furnace. Based on this information, what is the probable cause? Explain your answer.

42. The gas from a gas well known to contain both helium (He) and methane (CH₄) was analyzed. A 1$\overline{00}$-mL sample of the gas, corrected to STP conditions, had a mass of 0.067 g. Based on this information, was the gas from the well mostly helium, mostly methane, or about equal amounts of each? Explain your answer.

Chapter Quiz 8A
(Sections 8.1 to 8.3)

ELEMENT	ATOMIC MASS UNITS (AMU)
Cl	35.5
O	16.0
C	12.0

1. Calculate the mass in grams of dichlorine monoxide in 1.30 mol of dichlorine monoxide.

2. Calculate the number of dichlorine monoxide molecules in 13.5 g of dichlorine monoxide.

3. Calculate the number of liters of dichlorine monoxide gas at STP in 28.5 g of dichlorine monoxide.

4. Calculate the molecular mass and molar mass of a gas that has a density of 1.45 g/L at STP.

5. Calculate the density in g/L of carbon monoxide gas at STP.

Chapter Quiz 8B
(Sections 8.4 to 8.5)

ELEMENT	ATOMIC MASS UNITS (AMU)
C	12.0
H	1.0
O	16.0
P	31.0
Br	79.9

1. Calculate the percent composition of $C_6H_{12}O_6$ (dextrose or glucose).

2. Calculate the number of grams of oxygen in 30.0 g of $C_6H_{12}O_6$.

3. Determine the empirical formula for a compound that on analysis gives 7.20% phosphorus and 92.8% bromine.

4. Determine the molecular formula of a compound that on analysis gives 40.0% carbon, 6.7% hydrogen, and 53.3% oxygen and has a molecular mass of 90.0 amu.

5. Determine the molecular formula of a gas that on analysis gives 80.0% carbon and 20.0% hydrogen. At STP, 2.99 L of this gas has a mass of 4.00 g.

Copper (Symbol: Cu)

The Element COPPER: Electrical Conductivity and High-Speed Trains!

Copper is one of the few metals that can be found in the uncombined state. Its attractive color and luster have made it an important ornamental metal throughout history.

Name: The name derives from the Latin *aes cyprium*, meaning "Cyprian metal," because large amounts of copper were produced in ancient times on the island of Cyprus in the Mediterranean Sea. The symbol and the modern name evolved from the popular Latin word *cuprum*.

Appearance: A lustrous, beautiful, red-brown metal that is an excellent conductor of electricity and heat.

Occurrence: Copper is one of the few metals that can be found in nature uncombined with other elements. This is one reason why it was one of the first metals used by humans. It is also found combined in the minerals *chalcopyrite* ($CuFeS_2$) and *chalcocite* (Cu_2S). Along with iron and nickel, copper is one of the most important industrial metals.

Source: Metallic copper is produced from (1) copper(I) sulfide or (2) copper(I) oxide. The copper(I) sulfide is allowed to react with oxygen gas to produce the metallic copper and sulfur dioxide. The copper(I) oxide is allowed to react with carbon, carbon monoxide gas, or hydrogen gas to produce metallic copper and carbon dioxide gas or water.

Common Uses: Copper is an excellent metal for preparing alloys, and over 1000 different copper alloys have been produced. Metal alloys can have more strength and improved properties over pure metals. Common copper alloys include brasses (mixtures of copper and zinc with added traces of other metals), bronzes (mixtures of copper and tin with added traces of other metals), copper–nickel alloys, and copper–silver alloys.

The largest use of copper (50%) occurs in the electronics industry, where the electrical conductivity of copper is important. Common uses include electrical wires and contacts.

Other uses of copper and its alloys include silver-plating (Cu–Ag–Ni alloy), jewelry, and silverware.

The new superconducting materials (materials that conduct electricity with very little resistance) discovered in the middle 1980s contain copper as a component. While these materials have yet to make a big impact on industry, they may lead to very important products in the future. The material $YBa_2Cu_3O_x$ (Y = yttrium, atomic number = 39), where $x \approx 7$, loses virtually all resistance to the passage of electric current when cooled to 90 K ($-183°C$). There is speculation that these new materials may lead to high-speed trains in the future.

Unusual Facts: The large number of useful copper alloys derives in some part from the alchemists (see Section 1.3). While the alchemists never succeeded in changing other metals into gold, their experiments on mixing different metals led to the preparation of alloys and the first principles of metallurgy.

| CHAPTER 9 | Chemical Equations |

The precipitation of lead(II) iodide. Colorless potassium io-dide solution (left beaker) is poured into a colorless solution of lead(II) nitrate to produce a yellow precipitate (insoluble substance, right beaker) of lead(II) iodide.

GOALS FOR CHAPTER 9

1. To define the meaning of the term *chemical equation* and to explain why chemical equations are useful expressions of chemical reactions (Section 9.1).

2. To identify terms and symbols used in writing chemical equations (Section 9.2).

3. To use the guidelines for balancing chemical equations and to balance various chemical reaction equations (Sections 9.3 and 9.4).

4. To balance various combination reaction equations (Section 9.6).

5. To balance various decomposition reaction equations (Section 9.7).

6. To *complete* and balance various single-replacement reaction equations (Section 9.8).

7. To *complete* and balance various double-replacement reaction equations (Section 9.9).

8. To *complete* and balance various neutralization reaction equations (Section 9.10).

9. To balance various chemical reaction equations and *identify* them as (i) a combination reaction, (ii) a decomposition reaction, (iii) a single-replacement reaction, (iv) a double-replacement reaction, or (v) a neutralization reaction (Sections 9.5 through 9.10).

Countdown

You may use the periodic table.

5. Classify the following changes as physical or chemical (Section 3.4).
   a. building a box for your books   (physical)
   b. burning the box   (chemical)
   c. smashing your car's right front fender into a stop sign while trying to slow down on an icy road   (physical)
   d. rusting of your car's left rear fender caused by the "salting" of roads   (chemical)

4. Write the correct formula for the compound formed by the combination of the following ions (Section 6.9).
   a. copper(II) (Cu^{2+}) and chloride (Cl^-) ($CuCl_2$)
   b. copper(I) (Cu^+) and chloride (Cl^-)   ($CuCl$)
   c. copper(II) (Cu^{2+}) and selenide (Se^{2-}) ($CuSe$)
   d. copper(I) (Cu^+) and selenide (Se^{2-})   (Cu_2Se)

3. Write the correct formula for each of the following binary compounds (Sections 7.2 and 7.3).
   a. diphosphorus tetroxide   (P_2O_4)
   b. lead(IV) chloride   ($PbCl_4$)
   c. barium nitride   (Ba_3N_2)
   d. copper(I) phosphide   (Cu_3P)

2. Write the correct formula for each of the following ternary and higher compounds (Sections 7.4, 7.5, and 7.6).
   a. calcium nitrite   [$Ca(NO_2)_2$]
   b. chromic acid   (H_2CrO_4)
   c. tin(IV) sulfate   [$Sn(SO_4)_2$]
   d. tin(II) phosphate   [$Sn_3(PO_4)_2$]

1. Classify each of the following compounds as an acid, a base, or a salt. Assume that all soluble compounds are in aqueous solution (Section 7.6).
   a. $H_2Cr_2O_7$ (acid)   b. $CaCr_2O_7$ (salt)
   c. $Ca(OH)_2$ (base)   d. $Ca(C_2H_3O_2)_2$ (salt)

Did you ever wonder why, if you put both lemon and milk in tea, the milk curdles? Do you know how baking soda works? How about toilet bowl cleaner? And did you ever stop to wonder why a match lights when you strike it? In each case, the answer is that a chemical reaction is taking place.

In the last chapter, we considered how elements and compounds interact with one another to form new compounds. But we have not really considered how this interaction affects us and our world. In this chapter, we will describe some of the major types of chemical interactions and what they mean to our everyday life. For example, by the end of this chapter, you will understand why a Tums antacid tablet works on an upset stomach and why air pollutants such as sulfur dioxide are dangerous to your health and well-being. To understand these effects, however, you must first become proficient with an important chemical tool—balancing chemical equations.

9.1 Definition of a Chemical Equation

Chemical reaction Any interaction among chemical substances that brings about some change.

Scientists refer to the interactions among chemicals that bring about some change as **chemical reactions**. But how do we know a chemical reaction occurs? In Section 3.4 we considered *chemical changes* and stated that a chemical change could be observed only by a change in the composition of the substances. New substances are formed, and the properties of the new substances differ from the properties of the old substances. Therefore, a *chemical change* often can be recognized by such events as

1. the production of a gas (fizzes)
2. the production of heat (the flask gets hot) or the absorption of heat (the flask gets cold)
3. a permanent change in color, or
4. the appearance of an insoluble substance (see the photograph at the beginning of this chapter).

Chemical equation A shorthand method for expressing a chemical reaction in symbols and formulas.

A chemical change means that a chemical reaction occurs. For chemical reactions, we write chemical equations. A **chemical equation** is just a shorthand way of expressing these reactions in symbols and formulas. Two basic rules apply in writing chemical equations:

1. We cannot write an equation for a reaction unless we know how the substances react and what new substances they form.
2. Every chemical equation must be *balanced*. That is, the number of atoms of each element on the left side of the equation must be the same as the number of atoms of that element on the right side of the equation.

Actually, these rules are not much different from those governing the very first arithmetic problem you ever solved. When you learned to write $1 + 1 = 2$, you learned to use symbols (+ instead of "plus," for example). You learned that numbers "react" with one another (in this case, they combine). And you learned that failure to balance an equation (writing $1 + 1 = 3$) was not acceptable; 1 plus 1 does *not* equal 3.

In chemistry, you need to balance equations not only to make your instructor happy but also because of the *law of conservation of mass,* which was first shown experimentally by Antoine Laurent Lavoisier, a French chemist, in 1789. As noted in Section 3.6, this law states that mass is neither created nor destroyed in ordinary chemical changes. Hence, the *total* mass involved in a physical or chemical change remains the same. *The law of conservation of mass further requires that the number of atoms or moles of atoms of each element be the same on both sides of the equation.* This is why we balance chemical equations. A balanced chemical equation is one of the most fundamental and important things you can know about a chemical process.

Chemical equations may be written in two general ways: as *complete* (or *molecular*) *equations* (which apply to both elements and compounds that exist as molecules and also to those written as formula units) and as *ionic equations.* In this chapter, we consider only complete (or molecular) equations. The more complex balancing of ionic equations is covered in Chapter 15.

9.2 Terms, Symbols, and Their Meanings

Balancing chemical equations first requires that you learn the symbols and conventions chemists use in writing these equations. In a chemical equation, the **reactants**, the substances that interact with one another, are written on the left. The **products**, the substances that are formed, are written on the right. A single arrow ($\rightarrow$) or a double arrow ($\rightleftharpoons$) depending on the reaction conditions, separates the reactants from the products. A plus sign ($+$) separates the reactants from one another and the products from each other. The three physical states of substances involved in the reaction are sometimes indicated by single-letter abbreviations in parentheses following the formula of the substance.

Reactants   Substances that interact with one another in a chemical reaction; they are written on the left side of a chemical equation.

Products   Substances formed by a chemical reaction; they are written on the right side of a chemical equation.

1. *gas* by (*g*): $H_2(g)$
2. *liquid* by (*l*): $H_2O(l)$
3. *solid* by (*s*): $AgCl(s)$

Water is often used to dissolve solids and a substance dissolved in water is indicated by (*aq*), meaning *aq*ueous solution, such as $NaCl(aq)$. A delta sign (Δ) is chemical shorthand for "heat." It may appear above or below the arrow that separates the reactants and products, such as $\xrightarrow{\Delta}$, meaning that heat is necessary to make the reaction start or go. Other conditions or substances needed for the reaction may also be indicated above or below the arrow. For example, in $\xrightarrow{Pt}$, Pt is a catalyst. A **catalyst** is a substance that increases the rate of (speeds up) a chemical reaction but is recovered unchanged at the end of the reaction. One of nature's most important catalysts is chlorophyll, which is found only in green plants (see Figure 9.1). Through a process known as *photosynthesis*, chlorophyll turns carbon dioxide, water, and sunlight into food (glucose, a sugar) for these plants. The various enzymes used in digesting food are also catalysts. One example is ptyalin in saliva, which catalyzes the breakdown of large molecules such as starch to smaller molecules such as maltose.

Catalyst   Any substance that increases the rate of (speeds up) a chemical reaction but is recovered unchanged at the end of the reaction.

FIGURE 9.1
Chlorophyll is responsible for the green color of this vegetation. It catalyzes the conversion of carbon dioxide, water, and the energy from sunlight into glucose (a sugar) and oxygen:

$$6\ CO_2(g) + 6\ H_2O(l)$$

$$\xrightarrow[\text{chlorophyll}]{\text{sunlight}}\ C_6H_{12}O_6(s)$$
$$\text{glucose}$$
$$+\ 6\ O_2(g)$$

Catalysts and other chemical symbols are not found in every equation. In some equations you may see many of the symbols summarized in Table 9.1. In others you may find only a few. It all depends on the nature of the reaction involved.

TABLE 9.1 — **Terms, Symbols, and Their Meanings Used in Chemical Equations**

TERM OR SYMBOL	MEANING
Reactants	Left side of equation
Products	Right side of equation
$\longrightarrow$, $\rightleftharpoons$	Separates the products from the reactants
(g)	Gas or gas as a product
(l)	Liquid
(s)	Solid or solid as a product precipitating or coming out of a solution
(aq)	Aqueous solution (dissolved in water)
Δ above or below arrow ($\xrightarrow{\Delta}$ or $\xrightarrow[\Delta]{}$)	Heat needed for reaction to start or go to completion
Symbol above or below arrow ($\xrightarrow{Pt}$ or $\xrightarrow[Pt]{}$)	Catalyst

9.3 Guidelines for Balancing Chemical Equations

Just as not all symbols appear in all equations, so there are no absolute rules for how to balance equations. (Remember, though, that it *is* a rule that you *must* balance equations.) You should, however, find the following guidelines generally applicable for most of the simple equations you will encounter in this chapter. Also remember that you must balance the number of atoms or moles of atoms of each element. Hence, there must be the *same number* of atoms or moles of atoms of each element on each side of the equation. We call this process "balancing by inspection." This expression refers to the fact that no mathematical process is involved. Rather, we look at (inspect) the equation, work with the equation according to the guidelines, and balance it.

To help you understand this process and the guidelines, we will use them to balance an equation for the reaction between aqueous solutions of calcium hydroxide and phosphoric acid to give solid calcium phosphate and liquid water.

Guideline 1. Write the correct formulas for the reactants and the products, with the reactants on the left and the products on the right separated by $\longrightarrow$ or $\rightleftharpoons$. Separate each reactant and each product from every other by a + sign. *Once you have written a correct formula do not change it during the subsequent balancing operation.* Instead, place numbers, called coefficients, *in front* of the formula to obtain a balanced equation.

> **Study Hint:** Review Chapter 7 on nomenclature to be sure you write the correct formulas.

The example equation becomes

$$Ca(OH)_2(aq) + H_3PO_4(aq) \longrightarrow Ca_3(PO_4)_2(s) + H_2O(l)$$

Guideline 2. Begin the balancing process by selecting a *specific element* to balance. Generally, you should select an element from the *compound* containing the *most atoms*, and you should select the element present in the largest number in that compound. The selected element should *not* be an element in a polyatomic ion, and it should *not* be H or O. Balance the number of atoms of this element by placing a coefficient *in front* of the appropriate formula containing the selected element. For example, if you place a 3 in front of the formula NaCl (3 NaCl), this means that 3 formula units of sodium chloride are required in the reaction. If you place no number in front of a formula, the coefficient is assumed to be a 1. *Under no circumstances should you change the original (correct) formula of a compound to balance the equation.*

Selecting the Ca in $Ca_3(PO_4)_2$ as our element, the example equation now becomes

$$3 \, Ca(OH)_2(aq) + H_3PO_4(aq) \longrightarrow Ca_3(PO_4)_2(s) + H_2O(l)$$

Note that there are now 3 Ca atoms on each side of the equation.

Guideline 3. Next, balance the *polyatomic ions* that remain the *same* on both sides of the equation. You may balance these polyatomic ions as a single unit. In a rare case, you may have to adjust the coefficient you placed in step 2 to do this. If this happens, be sure to repeat step 2 to ensure that the original selected element is still balanced.

The PO_4 group is the polyatomic ion PO_4^{3-}. Balancing this ion, the example equation now becomes

$$3\ Ca(OH)_2(aq) + 2\ H_3PO_4(aq) \longrightarrow Ca_3(PO_4)_2(s) + H_2O(l)$$

Note that by placing a 2 in front of H_3PO_4, we have 2 PO_4^{3-} ions on each side of the equation.

Guideline 4. Balance the *H atoms* and then the *O atoms*. If they appear in the polyatomic ion and were balanced in step 3, you need not consider them again.

We balance the H atoms by placing a 6 in front of H_2O. The example equation now becomes

$$3\ Ca(OH)_2(aq) + 2\ H_3PO_4(aq) \longrightarrow Ca_3(PO_4)_2(s) + 6\ H_2O(l)$$

Note that there are now 12 H atoms on each side of the equation: 6 each on the left in $3\ Ca(OH)_2$ and $2\ H_3PO_4$, and 12 on the right in $6\ H_2O$. The O atoms are balanced now, too.

Guideline 5. Check all coefficients to see that they are *whole numbers* and in the *lowest possible ratio*. If the coefficients are fractions, then you must multiply all coefficients by some number so as to make them all, including the fraction, whole numbers. If a coefficient like $\frac{5}{2}$ or $2\frac{1}{2}$ exists, then you must multiply *all* coefficients by 2. The $\frac{5}{2}$ or $2\frac{1}{2}$ is then 5, a whole number. You must reduce the coefficients to the lowest possible ratios. If the coefficients are 6, 9 $\longrightarrow$ 3, 12, you can reduce them all by dividing *each* by 3 to give the lowest possible ratio of 2, 3 $\longrightarrow$ 1, 4.

Since there are no fractional coefficients in our example, this guideline does not apply.

$$3\ Ca(OH)_2(aq) + 2\ H_3PO_4(aq) \longrightarrow Ca_3(PO_4)_2(s) + 6\ H_2O(l)$$

Guideline 6. Check each atom or polyatomic ion with a ✓ above the atom or ion on both sides of the equation to ensure that the equation is balanced. As you become more proficient in balancing equations, this will not be necessary; but for the first few equations that you balance, we feel that you should check each atom or ion. These ✓ are not part of the final form of the equation, but we use them as a teaching device to make sure that you balance each atom or ion.

In our example,

$$3\ \overset{✓}{Ca}(\overset{✓}{O}\overset{✓}{H})_2(aq) + 2\ \overset{✓}{H}_3\overset{✓}{PO}_4(aq) \longrightarrow \overset{✓}{Ca}_3(\overset{✓}{PO}_4)_2(s) + 6\ \overset{✓}{H}_2\overset{✓}{O}(l) \qquad \text{(balanced)}$$

These guidelines are summarized as follows:

Guideline 1. Write the correct formulas.

Guideline 2. Start with a specific element in a compound with the most atoms.

Guideline 3. Balance the polyatomic ions.

Guideline 4. Balance the H atoms and then the O atoms.

Guideline 5. Check the coefficients to see that they are whole numbers and are in the lowest possible ratio.

Guideline 6. Check off each atom or polyatomic ion with a ✓.

9.4 Examples Involving Balancing Equations

Now let us apply these guidelines in balancing equations by inspection.

EXAMPLE 9.1   Balance each of the following equations by inspection:

a. $Fe(s) + HCl(aq) \longrightarrow FeCl_2(aq) + H_2(g)$

b. $Al(OH)_3(s) + H_3PO_4(aq) \longrightarrow AlPO_4(s) + H_2O(l)$

c. $C_4H_{10}(g) + O_2(g) \xrightarrow{\Delta} CO_2(g) + H_2O(g)$
   butane

SOLUTION

a. We need not consider guideline 1 because the formulas have already been given. Considering guideline 2, the compound with the greatest number of atoms besides hydrogen is $FeCl_2$, and the element to start with is Cl since there are 2 Cl atoms in $FeCl_2$. To balance the Cl atoms, place a 2 in front of the HCl and write 2 HCl. *The formula of HCl does not change to balance the Cl atoms.* The equation now appears as

$$Fe(s) + 2\,HCl(aq) \longrightarrow FeCl_2(aq) + H_2(g) \quad \text{(balanced)}$$

Guideline 3 is not applicable because no polyatomic ions are present. For guideline 4, the H atoms are balanced and no O atoms are present. In guideline 5, all coefficients are whole numbers and in the lowest possible ratio. Check off each atom as in guideline 6; the final balanced equation is

$$\checkmark \quad \checkmark\checkmark \qquad \checkmark\checkmark \qquad \checkmark$$
$$Fe(s) + 2\,HCl(aq) \longrightarrow FeCl_2(aq) + H_2(g) \quad \text{(balanced)} \quad \textit{Answer}$$

b. We need not consider guideline 1 because the formulas have already been given. The starting element is aluminum since you should not generally start with a polyatomic ion, hydrogen, or oxygen (guideline 2). The aluminum atoms are already balanced, so move on to the phosphate polyatomic ion (guideline 3). This, too, is already balanced. To balance the hydrogens (guideline 4), there are 6 H atoms on the left side, so place a 3 in front of H_2O to get 6 atoms on the right side of the equation. The oxygen atoms are also balanced by this action.

$$Al(OH)_3(s) + H_3PO_4(aq) \longrightarrow AlPO_4(s) + 3\,H_2O(l) \quad \text{(balanced)}$$

The coefficients are whole numbers in the lowest possible ratio (guideline 5). Check off each atom according to guideline 6 to give the final balanced equation.

$$\checkmark \;\checkmark\checkmark \qquad \checkmark \;\checkmark \qquad \checkmark \;\checkmark \qquad \checkmark\checkmark$$
$$Al(OH)_3(s) + H_3PO_4(aq) \longrightarrow AlPO_4(s) + 3\,H_2O(l) \quad \text{(balanced)} \quad \textit{Answer}$$

c. We need not consider guideline 1 because the formulas have already been given. Considering guideline 2, the compound with the greatest number of atoms is C_4H_{10}, and the element we start with is C since there are 4 C atoms in C_4H_{10} (We will balance the H atoms in guideline 4). To balance the C atoms, place a 4 in front of CO_2 and write 4 CO_2.

$$C_4H_{10}(g) + O_2(g) \xrightarrow{\Delta} 4\,CO_2(g) + H_2O(g) \quad \text{(unbalanced)}$$
   butane

Guideline 3 is not applicable because there are no polyatomic ions. Thus, we should consider guideline 4 and balance the H atoms by placing a 5 in front of H_2O to give 5 H_2O:

$$C_4H_{10}(g) + O_2(g) \xrightarrow{\Delta} 4\,CO_2(g) + 5\,H_2O(g) \quad \text{(unbalanced)}$$

Study Hint: *Do not change the formula of HCl.* The following equation is **wrong**:

$$Fe(s) + H_2Cl_2(aq) \longrightarrow$$
$$FeCl_2(aq) + H_2(g)$$

Place a 2 in *front* of HCl as shown in the balanced equation to the left.

The result is a total of 13 O atoms in the products (8 O from 4 CO_2 and 5 O from 5 H_2O, and so we must place $\frac{13}{2}$ or $6\frac{1}{2}$ in front of O_2 to obtain 13 O atoms in the reactants. The equation now appears as

$$C_4H_{10}(g) + \tfrac{13}{2}O_2(g) \xrightarrow{\Delta} 4\,CO_2(g) + 5\,H_2O(g)$$

Following guideline 5, we make the coefficients whole numbers by multiplying *all* the coefficients by 2 ($\frac{13}{2} \times 2 = 13$) and note that the coefficients are in the lowest possible ratio. We check off each atom as in guideline 6. The final balanced equation is

$$2\,C_4H_{10}(g) + 13\,O_2(g) \xrightarrow{\Delta} 8\,CO_2(g) + 10\,H_2O(g) \quad \text{(balanced)} \qquad Answer$$

> **Study Hint:** Checking off each atom provides a double check and assures you that you multiplied *each* coefficient by 2.

Word equations are another form of chemical equations without the coefficients. A *word equation* expresses a chemical equation in words instead of symbols and formulas. To write and balance chemical equations from word equations, we need only apply the guidelines from Section 9.3 with special emphasis on guideline 1. We must write the correct formulas for the elements or compounds from the names. Here we apply the nomenclature you learned in Chapter 7.

EXAMPLE 9.2 Change the following word equations into chemical equations and complete and balance them by inspection:

a. Solid potassium nitrate is heated to give solid potassium nitrite and oxygen gas.
b. Hydrochloric acid solution is added to solid limestone (calcium carbonate), producing carbon dioxide gas and calcium chloride solution.

SOLUTION
a. Following guideline 1, first write the equation.

$$KNO_3(s) \xrightarrow{\Delta} KNO_2(s) + O_2(s) \quad \text{(unbalanced)}$$

Before you continue reading, try to balance this equation yourself.

The correct coefficients are $2 \longrightarrow 2 + 1$. Did you get the answer? Now, let us see where this answer comes from. Because both K and N atoms are present as only one atom on each side and are already balanced, begin with oxygen since there is more than one O atom present (guideline 2). Place a $\frac{1}{2}$ in front of O_2, which gives 3 O atoms on the right side of the equation. These atoms are balanced by the 3 O atoms on the left side of the equation in KNO_3.

$$KNO_3(s) \xrightarrow{\Delta} KNO_2(s) + \tfrac{1}{2}O_2(g)$$

Because the polyatomic ion NO_3^- does not appear as NO_3^- on the right side of the equation, guideline 3 is not applicable. The oxygen atoms are already balanced (guideline 4), but the coefficients are not whole numbers (guideline 5). Hence, we must multiply all the coefficients by 2 to make them whole numbers and verify that they are in the lowest possible ratio. Checking off each atom according to guideline 6, we have the following balanced equation:

$$2\,KNO_3(s) \xrightarrow{\Delta} 2\,KNO_2(s) + O_2(g) \quad \text{(balanced)} \qquad Answer$$

b. Following guideline 1, first write the equation:

$$HCl(aq) + CaCO_3(s) \longrightarrow CO_2(g) + H_2O(l) + CaCl_2(aq) \quad \text{(unbalanced)}$$

The starting point in balancing the equation is $CaCl_2$ because there are 2 Cl atoms (guideline 2). Place a 2 in front of the HCl to balance the Cl atoms. Next, check each of the other elements in the equation to make sure they are balanced (guidelines 3 to 5). Check each atom (guideline 6) to give the final balanced equation (see Figure 9.2).

$$2\ HCl(aq) + CaCO_3(s) \longrightarrow CO_2(g) + H_2O(l) + CaCl_2(aq) \quad \text{(balanced)} \quad \textit{Answer}$$

FIGURE 9.2
Limestone or marble (calcium carbonate) reacts with dilute hydrochloric acid solution to produce bubbles of carbon dioxide (CO_2) gas.

| EXAMPLE 9.3 | Octane (C_8H_{18}) is one component of gasoline. The incomplete combustion (burning) of gasoline in automobiles (see Figure 9.3) produces carbon monoxide. Carbon monoxide is harmful to animals and human beings. Hemoglobin in red blood cells has a greater affinity for carbon monoxide than for oxygen. Thus, the hemoglobin is "tied up" by the carbon monoxide and is not able to carry oxygen. Carbon monoxide hence robs the tissues of the oxygen required for survival. Write the word equation and the chemical equation for the burning of liquid octane in a limited amount of oxygen gas to produce carbon monoxide gas and gaseous water and then balance by inspection. |

SOLUTION   Following guideline 1, first write the equation:

$$O_2(g) + C_8H_{18}(l) \longrightarrow CO(g) + H_2O(g) \quad \text{(unbalanced)}$$

Balance the carbon atoms (guideline 2) by placing an 8 in front of CO to give 8 CO. Guideline 3 is not applicable because there are no polyatomic ions. Balance the hydrogen atoms (guideline 4) by placing a 9 in front of the H_2O. This step requires 17 O atoms on the reactants side (8 O in 8 CO and 9 O in H_2O) and is accomplished by placing a $\frac{17}{2}$ or $8\frac{1}{2}$ in front of O_2 on the left side of the equation. The following equation results:

$$\tfrac{17}{2} O_2(g) + C_8H_{18}(l) \longrightarrow 8\ CO(g) + 9\ H_2O(g)$$

FIGURE 9.3
Older cars (pre-1975) burn gasoline less efficiently and produce a large portion of the air pollutants carbon monoxide (CO) and nitrogen oxides (NO and NO_2).

The coefficients are not whole numbers (guideline 5); hence, we must multiply *all* coefficients by 2. Checking each atom (guideline 6), we obtain the following balanced equation:

$$17\ O_2(g) + 2\ C_8H_{18}(l) \longrightarrow 16\ CO(g) + 18\ H_2O(g)\ \text{(balanced)}\quad Answer$$

Study Exercise 9.1
Balance each of the following chemical equations by inspection:

a. $I_2O_7(s) + H_2O\ (l) \longrightarrow HIO_4(aq)$ $(1 + 1 \rightarrow 2)$

b. $Ca_3(PO_4)_2(s) + H_3PO_4(aq) \longrightarrow Ca(H_2PO_4)_2(aq)$ $(1 + 4 \rightarrow 3)$

Study Exercise 9.2
Change the following word equations into chemical equations and complete and balance them by inspection:

a. silver + hydrogen sulfide gas $\longrightarrow$ silver sulfide + hydrogen gas
$$[2\,Ag + H_2S(g) \rightarrow Ag_2S + H_2(g)]$$
This is the equation for the tarnishing of silverware, that is, when it turns black upon exposure to air. The air contains minute amounts of sulfides as pollutants.

b. sodium dichromate + ammonium chloride $\longrightarrow$
chromium(III) oxide + sodium chloride + nitrogen gas + water
$$[Na_2Cr_2O_7 + 2\,NH_4Cl \rightarrow Cr_2O_3 + 2\,NaCl + N_2(g) + 4\,H_2O]$$

You *and Chemistry*

Work Problems 4 through 7.

9.5 The Five Simple Types of Chemical Equations

While the preceding examples have provided you with some excellent practice in balancing equations, they lack a major factor present in real-life chemistry: chemists must first predict how two or more substances will react with one another. Later in this text (Chapter 16), we will consider complex oxidation–reduction reactions that require special balancing methods. At this time, we will deal with only five simple types of reactions:

> **Study Hint:** Knowing the five general types of chemical reactions will help you learn many reactions later in this text.

1. Combination reactions
2. Decomposition reactions
3. Single-replacement reactions
4. Double-replacement reactions
5. Neutralization reactions.

9.6 Combination Reactions

Combination reaction Type of chemical reaction in which two or more substances (either elements or compounds) react to produce one substance (always a compound).

In chemical terms, a **combination reaction** (also known as a *synthesis reaction*) occurs when two or more substances (either elements or compounds) react to produce one substance (always a compound). In general terms, we can write this equation as

$$A + Z \longrightarrow AZ$$

where A and Z are elements or compounds and AZ is a compound.

Among the different types of combination reactions are the following:

1. metal + oxygen $\xrightarrow{\Delta}$ metal oxide

$2\,Mg(s) + O_2(g) \xrightarrow{\Delta} 2\,MgO(s)$

2. nonmetal + oxygen $\xrightarrow{\Delta}$ nonmetal oxide

$C(s) + O_2(g) \xrightarrow{\Delta} CO_2(g)$ (oxygen in excess)

3. metal + nonmetal $\longrightarrow$ salt (MX, general formula; see Section 7.6)

$2\,Na(s) + Cl_2(g) \longrightarrow 2\,NaCl(s)$

This reaction is dramatically shown in Figure 9.4.

4. water + metal oxide $\longrightarrow$ base   (MOH, general formula;

$H_2O(l) + MgO(s) \longrightarrow Mg(OH)_2(s)$ see Section 7.6)

5. water + nonmetal oxide $\longrightarrow$ oxyacid (HX or HXO, general formula;

$H_2O(l) + SO_3(s) \longrightarrow H_2SO_4(l)$ see Section 7.6)

This reaction is used in the industrial preparation of sulfuric acid which is used in automobile batteries.

The reactions involving oxygen gas, such as those in cases 1 and 2, are sometimes called **combustion reactions** because combustion (burning) is the reaction between oxygen and many substances. Metal oxides that react with water to form a base are called **basic oxides** or **basic anhydrides**. Nonmetal oxides that react with water to form an acid are called **acid oxides** or **acid anhydrides**.

For *combination reactions* you will be asked only to balance the equations and classify the type of reaction, not to predict the products of the reaction.

Study Exercise 9.3

Balance the following combination reaction equations:

a. $Ga_2O_3(s) + H_2O(l) \longrightarrow Ga(OH)_3(s)$ $(1 + 3 \rightarrow 2)$

b. $P_4O_{10}(s) + H_2O(l) \longrightarrow H_3PO_4(aq)$ $(1 + 6 \rightarrow 4)$

Combustion reaction   Type of combination reaction in which oxygen reacts with a substance; also called burning.

Basic oxide (basic anhydride) Metal oxide that reacts with water in a combination reaction to form a base.

Acid oxides (acid anhydrides)   Nonmetal oxide that reacts with water in a combination reaction to form an acid.

Work Problem 8.

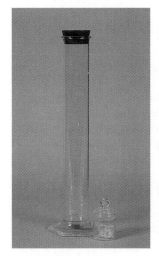

FIGURE 9.4
Combination reaction. Sodium and chlorine react violently to give sodium chloride.

9.7 Decomposition Reactions

Decomposition reactions
Type of chemical reaction in which one substance (always a compound) breaks down to form two or more substances (either elements or compounds).

In **decomposition reactions**, one substance undergoes a reaction to form two or more substances. The substance broken down is always a compound, and the products may be elements or compounds. Heat is often necessary for this process. A general equation represents this reaction:

$$AZ \longrightarrow A + Z$$

where A and Z are elements or compounds. It is not always easy to predict the products of a decomposition reaction. Therefore, for *decomposition reactions* you will be asked only to balance the equation and classify the type of reaction, not to predict the products of the reaction. Many compounds decompose upon heating, and often the reactions are unique to that compound. Two general classes of decomposition reactions are given below with some examples.

1. Metal carbonates and hydrogen carbonates (bicarbonates) often decompose to produce carbon dioxide gas.

$$2\,NaHCO_3(s) \xrightarrow{\Delta} Na_2CO_3(s) + CO_2(g) + H_2O(g)$$

This reaction describes how baking soda works. Baking soda is just sodium hydrogen carbonate or bicarbonate ($NaHCO_3$). Heating it in the oven produces carbon dioxide (CO_2) gas, which expands on heating and makes the cake "rise." Baking soda is also used to put out fires (see Figure 7.6) because carbon dioxide and water are given off on heating. The carbon dioxide helps put out the fire by smothering the flames and excluding oxygen from the air.

When limestone (calcium carbonate, $CaCO_3$) is heated, carbon dioxide is one of the products.

$$CaCO_3(s) \xrightarrow{\Delta} CaO(s) + CO_2(g)$$

The production of carbon dioxide is tested by inserting a glowing splint of wood into the test tube. The glowing splint goes out because carbon dioxide does not support burning.

2. Some compounds decompose to produce oxygen gas.

$$2\,HgO(s) \xrightarrow{\Delta} 2\,Hg(l) + O_2(g)$$

The red mercury(II) oxide, which when heated forms droplets of mercury along the edge of the test tube and oxygen, which supports burning, is evolved. The production of oxygen is tested by inserting a glowing splint into the test tube. The glowing splint immediately catches fire and burns (see Figure 9.5).

$$2\,KClO_3(s) \xrightarrow{\Delta} 2KCl(s) + 3\,O_2(g)$$

This is the laboratory method for the preparation of oxygen.

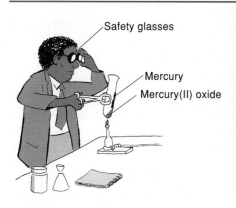

FIGURE 9.5
Decomposition reaction. Heating mercury(II) oxide (orange solid) to produce mercury (silver droplets) and oxygen gas.

Study Exercise 9.4
Balance the following decomposition reaction equations:

a. $KHCO_3(s) \xrightarrow{\Delta} K_2CO_3(s) + H_2O(g) + CO_2(g)$

$$(2 \longrightarrow 1 + 1 + 1)$$

b. $CaSO_4 \cdot 2H_2O(s) \xrightarrow{\Delta} CaSO_4(s) + H_2O(l)$

The water in $CaSO_4 \cdot 2H_2O$ is held as part of the solid.
Gentle heating releases the water. $(1 \longrightarrow 1 + 2)$ Work Problem 9.

9.8   Single-Replacement Reactions. The Electromotive or Activity Series

> **Study Hint:** This type of reaction is similar to a man "cutting in" on a couple at a dance. The first man *replaces* the man from the couple, just as a metal *replaces* a metal ion from a salt.

In **single-replacement reactions**, an element and a compound react, and the element replaces another element in the compound. Single-replacement reactions are also called *replacement*, *substitution*, or *displacement* reactions. Two general types of single-replacement reactions will be discussed in this text:

1. A metal (A) replacing a metal ion in its salt or acid. B can be a metal ion or a hydrogen ion.

$$A + BZ \longrightarrow AZ + B$$

2. A nonmetal (X) replacing a nonmetal ion in its salt or acid. B can be a metal ion or a hydrogen ion.

$$X + BZ \longrightarrow BX + Z$$

In the first case, the replacement depends on the two metals involved, A and B. The metals can be arranged in a series called the **electromotive** or **activity series**, which is shown in Figure 9.6a and on the inside back cover of this text. Each element in the series displaces any element below it from a salt or acid of the second

Single-replacement reactions
Type of chemical reaction in which an element and a compound react and the element replaces another element in the compound.

Electromotive (activity) series   Arrangement of metals in order of descending reactivity; thus, each element in the series displaces any element below it from a salt or acid of the second element.

(a) Li   (b) F₂
    K   Cl₂
    Ba   Br₂
    Ca   I₂
    Na
    Mg
    Al
    Zn
    Fe
    Cd
    Ni
    Sn
    Pb
    (H)
    Cu
    Hg
    Ag
    Au

FIGURE 9.6
Electromotive or activity series. (a) Activity series for metals. (b) Activity series for the halogens.

> **S**tudy Hint: The electromotive series is like a "pecking order." A species that is higher in the series is stronger and will win.

element. For example, zinc displaces copper(II) ions from a copper(II) salt such as copper(II) sulfate, $CuSO_4$.

Although hydrogen is not a metal, it is included in this series:

Li, K, Ba, Ca, Na, Mg, Al, Zn, Fe, Cd, Ni, Sn, Pb, (H), Cu, Hg, Ag, Au

In general all metals above hydrogen displace hydrogen ions from an acid. The most reactive metals (Li, K, Ba, Ca, and Na) replace *one* hydrogen from water itself to form the metal hydroxide and hydrogen gas. You will be asked to *complete* and *balance* single-replacement reaction equations using this series and to classify the type of reaction.

In the second type of single-replacement reaction, when a nonmetal displaces another nonmetal from its salt or acid, the reaction depends on the two nonmetals involved, X and Z. A series similar to the electromotive or activity series exists for the halogen nonmetals—F_2, Cl_2, Br_2, and I_2—as shown in Figure 9.6b. Bromine replaces iodide ions from an aqueous solution of an iodide salt, chlorine displaces bromide ion or iodide ion, and fluorine replaces any of the three halogen ions. This series follows the decrease in nonmetallic properties in the halogen family according to the periodic table (see Section 5.3, number 4).

Some examples of completing and balancing single-replacement reactions are given next.

EXAMPLE 9.4   Complete and balance the equation for the reaction between zinc metal and aqueous copper(II) sulfate (see Figure 9.7).

SOLUTION   Write the formulas of the reactants (see Chapter 7). Zinc is higher in the electromotive or activity series than copper, and so zinc displaces copper(II) ions from its salt. Following guideline 1, complete and write the equation.

$$Zn(s) + CuSO_4(aq) \longrightarrow ZnSO_4(aq) + Cu(s) \quad \textit{Answer}$$

The equation is balanced as written.

EXAMPLE 9.5   Complete and balance the equation for the reaction between magnesium metal and aqueous hydrochloric acid.

SOLUTION   Write the formulas of the reactants (see Chapter 7). Magnesium is higher in the electromotive or activity series than hydrogen, and so magnesium displaces hydrogen from an acid. Following guideline 1, complete and write the equation.

$$Mg(s) + HCl(aq) \longrightarrow MgCl_2(aq) + H_2(s) \text{ (unbalanced)}$$

FIGURE 9.7
Single-replacement reaction. When a clean strip of zinc is inserted in a solution of aqueous copper(II) sulfate, copper is deposited onto the zinc.

We can write the formula for magnesium chloride from a knowledge of the ionic charges on magnesium $(+2)$ and chloride (-1). Hydrogen is written as a diatomic gas (see Section 6.4). Balancing the equation according to the guidelines produces the final, balanced equation:

$$Mg(s) + 2\,HCl(aq) \longrightarrow MgCl_2(aq) + H_2(g) \qquad Answer$$

EXAMPLE 9.6   Complete and balance the equation for the reaction between chlorine gas and aqueous sodium bromide (see Figure 9.8).

SOLUTION   Chlorine is higher in the halogen series than bromine, and so chlorine displaces bromide from its salt. Following guideline 1, complete and write the equation:

$$Cl_2(g) + NaBr(aq) \longrightarrow NaCl(aq) + Br_2(aq) \quad \text{(unbalanced)}$$

We can write the formula for sodium chloride from a knowledge of the ionic charges on sodium $(+1)$ and chloride (-1). Both chlorine and bromine are diatomic. Balancing the equation according to the guidelines produces the final, balanced equation:

$$Cl_2(g) + 2\,NaBr(aq) \longrightarrow 2\,NaCl(aq) + Br_2(aq) \qquad Answer$$

This is one of the processes by which bromine is prepared industrially.

EXAMPLE 9.7   Complete and balance the equation for the reaction between solid potassium metal and water.

SOLUTION   Write the formulas of the reactants (see Chapter 7). Potassium is higher in the electromotive or activity series than hydrogen and can replace *one* hydrogen atom from water to form the metal hydroxide and hydrogen gas. Writing water as H—OH, the

FIGURE 9.8
Preparation of bromine by a single-replacement reaction. Mixing two aqueous solutions of chlorine and sodium bromide gives the reddish-brown color of bromine in water.

replacement of *one* hydrogen atom from water by potassium simplifies the understanding of the equation.

$$K(s) + H\!-\!OH(l) \longrightarrow KOH(aq) + H_2(g)$$

We can write the formula for potassium hydroxide from a knowledge of the ionic charges on potassium $(+1)$ and hydroxide (-1). Hydrogen is a *diatomic* gas. Balancing the equation according to the guidelines produces the final, balanced equation.

$$2\,K(s) + 2\,H\!-\!OH(l) \longrightarrow 2\,KOH(aq) + H_2(g) \qquad Answer$$

Study Exercise 9.5

Predict the products using the electromotive (activity) series and the periodic table for the following single-replacement reaction equations and balance them.

a. $Al(s) + SnCl_2(aq) \longrightarrow$

   $[2\,Al(s) + 3\,SnCl_2(aq) \longrightarrow 2\,AlCl_3(aq) + 3\,Sn(s)]$

b. $Ba(s) + H_2O(l) \longrightarrow$

   (*Hint:* Write water as $H\!-\!OH$.)

Work Problem 10.

   $[Ba(s) + 2\,HOH(l) \longrightarrow Ba(OH)_2(aq) + H_2(g)]$

9.9 Double-Replacement Reactions. Rules for the Solubility of Inorganic Substances in Water

Double-replacement reactions   Type of chemical reaction in which two compounds react and the cation of one compound exchanges places with the cation of the other compound.

In **double-replacement reactions**, *two* compounds are involved in a reaction, with the positive ion (cation) of one compound *exchanging* with the positive ion (cation) of the other compound. In other words, the two positive ions exchange negative ions (anions) or partners. A double-replacement reaction is also called a *metathesis* (meaning "a change in state, substance, or form") or *double-decomposition* reaction. The following general equation represents this reaction:

$$AX + BZ \longrightarrow AZ + BX$$

In double-replacement reactions there are four separate particles—A, X, B, and Z—while in single-replacement reactions there are only three, A, B, and Z. In double-replacement reactions the particles are ions, while in single-replacement reactions A is *not* an ion, but a free metal or nonmetal. Single-replacement reactions depend on the electromotive or activity series, but double-replacement reactions do not.

Double-replacement reactions will generally occur if one of the following three conditions is satisfied:

S tudy Hint: A double-replacement reaction is similar to two couples at a dance *exchanging partners* and dancing some more.

1. An insoluble or slightly soluble product (precipitate) is formed.

2. A weakly ionized species is produced as a product. The most common species of this type is water.

3. A gas is formed as a product.

The most common type of double-replacement reaction belongs to the first of these three classes. A **precipitate**, a solid appearing in solution, is produced during the reaction because one of the products is insoluble (or only slightly soluble) in water (see Figure 9.9 and see picture at the beginning of this chapter). Chemists indicate this precipitate in an equation by including an (*s*) next to the compound, as in AgCl(*s*). To recognize that a precipitate will form, you must be able to interpret and use the following rules. These rules are listed below and on the inside back cover of this text. You will not need to memorize them as they will be given to you on exams and quizzes unless indicated otherwise by your instructor. But, you must be able to *use* these rules.

Precipitate   Any solid that appears in a solution in the course of a chemical reaction.

FIGURE 9.9
Double-replacement reaction. Formation of silver chromate (dark-red) precipitate from solutions of silver nitrate and sodium chromate.

Rules for the Solubility of Inorganic Substances in Water

1. Nearly all *nitrates* (NO_3^-) and *acetates* ($C_2H_3O_2^-$) are *soluble*.
2. All *chlorides* (Cl^-) are *soluble*, except AgCl, Hg_2Cl_2, and $PbCl_2$. ($PbCl_2$ is soluble in hot water.)
3. All *sulfates* (SO_4^{2-}) are *soluble*, except $BaSO_4$, $SrSO_4$, and $PbSO_4$. ($CaSO_4$ and Ag_2SO_4 are only slightly soluble.)
4. Most of the *alkali metal* (group IA (1), Li, Na, K, and so on) salts and *ammonium* (NH_4^+) salts are *soluble*.
5. All the common *acids* are soluble.
6. All *oxides* (O^{2-}) and *hydroxides* (OH^-) are *insoluble*, except those of the alkali metals and certain alkaline earth metals (group IIA (2), Ca, Sr, Ba, Ra). [$Ca(OH)_2$ is only moderately soluble.]
7. All *sulfides* (S^{2-}) are *insoluble*, except those of the alkali metals, alkaline earth metals, and ammonium sulfide.
8. All *phosphates* (PO_4^{3-}) and *carbonates* (CO_3^{2-}) are *insoluble*, except those of the alkali metals and ammonium salts.

These rules will be quite useful in writing double-replacement equations. They also help us understand the actions of some chemicals used in everyday life. For example, vinegar (about 5% acetic acid) is often used to remove water spots on glass caused by the presence of calcium, magnesium, and iron salts in hard water. Vinegar is used because the acetic acid in it reacts with certain calcium, magnesium, and iron salts to form a new salt, an acetate salt, which is soluble in water (rule 1) and hence can be easily removed with water.

Consider the following double-replacement reactions:

✔ A salt and an acid react to form a new acid and a precipitate.

$$AgNO_3(aq) + HCl(aq) \longrightarrow HNO_3(aq) + AgCl(s)$$

Silver chloride is insoluble in water (rule 2). Figure 9.10 illustrates this reaction.

✔ A salt and a base react to form a new salt and a new base, one of which is insoluble and precipitates.

$$Ni(NO_3)_2(aq) + 2\,NaOH(aq) \longrightarrow Ni(OH)_2(s) + 2\,NaNO_3(aq)$$

FIGURE 9.10
A double-replacement reaction. Formation of silver chloride precipitate from solutions of silver nitrate and hydrochloric acid (Courtesy Dr. E. R. Degginger)

Nickel(II) hydroxide is insoluble in water (rule 6), and sodium nitrate is soluble (rules 1 and 4).

✔ Two salts react to form two new salts, one soluble and the other a precipitate.

$$2\,NaCl(aq) + Pb(NO_3)_2(aq) \xrightarrow{\text{cold}} PbCl_2(s) + 2\,NaNO_3(aq)$$

Lead (II) chloride is insoluble in cold water (rule 2), and sodium nitrate is soluble (rules 1 and 4).

✔ A metal carbonate or hydrogen carbonate (bicarbonate) and an acid react to form a salt, water, and carbon dioxide gas.

$$CaCO_3(s) + 2\,HCl(aq) \longrightarrow CaCl_2(aq) + H_2O(l) + CO_2(g)$$

$$CaCO_3(s) + 2\,NaHSO_4(aq) \longrightarrow CaSO_4(aq) + Na_2SO_4(aq) + H_2O(l) + CO_2(g)$$

Two moles of hydrogen ions (H^+) react with the carbonate anion (CO_3^{2-}) to form carbonic acid, H_2CO_3. Carbonic acid is unstable and decomposes to form water and carbon dioxide (CO_2).

The first reaction in the last category describes the action of the antacid, Tums, which contains calcium carbonate. The calcium carbonate neutralizes some acid (HCl) in the stomach. Calcium carbonate also is the solid that forms in a toilet bowl. Most toilet bowl cleaners contain sodium hydrogen sulfate ($NaHSO_4$), which dissolves the calcium carbonate in the bowl, as shown in the second reaction.

The insolubility of calcium carbonate is responsible for important natural formations (see Figure 9.11) as well as toilet bowl deposits. Limestone results from the precipitation of calcium carbonate from bodies of water. Coral reefs are made largely of calcium carbonate deposited by the coral as it grows. Stalactites and stalagmites in caves are made of precipitated calcium carbonate resulting from the evaporation of springs containing calcium carbonate. Calcium carbonate may also precipitate in lakes where freshwater springs containing calcium ions meet lake water containing carbonate ions. The resulting structures, called tufa, are distinctive and unusual.

You will be asked to *complete* and *balance* double-replacement reaction equations using the rules for solubility and the periodic table and to classify this type of reaction. Consider the following examples.

EXAMPLE 9.8 Complete and balance the equation for the reaction between barium nitrate and potassium sulfate in aqueous solution. [Indicate any precipitate by (*s*) and any gas by (*g*).]

SOLUTION Write the formulas of the reactants (see Chapter 7). Complete the equation and look for water, a gas, or an insoluble compound in the products.

$$Ba(NO_3)_2(aq) + K_2SO_4(aq) \longrightarrow BaSO_4(s) + KNO_3(aq) \quad \text{(unbalanced)}$$

(a)

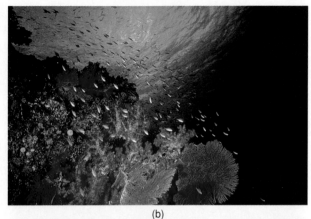

(b)

(c)

(d)

FIGURE 9.11
Natural structures made of calcium carbonate. (a) Limestone rock is a common and important geological structure. (b) Coral reefs are built by the coral as it grows and provides an important marine habitat. (c) Stalactites (top) and stalagmites (bottom) give caves a unique look. (d) Tufa, like these at Mono Lake, California, grow only underwater where springs containing calcium ions enter a lake. The structures are revealed only when lake levels drop.

Barium sulfate ($BaSO_4$) is insoluble in water (rule 3), and so a double-replacement reaction does occur. The other product, potassium nitrate, is soluble (rules 1 and 4). The balanced equation is

$$Ba(NO_3)_2(aq) + K_2SO_4(aq) \longrightarrow BaSO_4(s) + 2\,KNO_3(aq) \qquad \textit{Answer}$$

EXAMPLE 9.9 Complete and balance the equation for the reaction between nitric acid and sodium hydrogen carbonate in aqueous solution. [Indicate any precipitate by (s) and any gas by (g).]

SOLUTION Write the formulas of the reactants (see Chapter 7). Complete the equation and look for water, a gas, or an insoluble compound in the products.

$$HNO_3(aq) + NaHCO_3(aq) \longrightarrow H_2O(l) + CO_2(g) + NaNO_3(aq)$$
$$\text{(unbalanced)}$$

Carbonic acid (H_2CO_3) is formed from a hydrogen ion (H^+) and the hydrogen carbonate ion (HCO_3^-), and so water and carbon dioxide gas are products. The other product, sodium nitrate, is soluble (rules 1 and 4). The balanced equation is

$$HNO_3(aq) + NaHCO_3(aq) \longrightarrow H_2O(l) + CO_2(g) + NaNO_3(aq) \quad \textit{Answer}$$

Study Exercise 9.6
Using the rules for the solubility of inorganic substances in water and the periodic table, predict the products for the following double-replacement reaction equations and balance them. [Indicate any precipitate by (s) and any gas by (g).]

a. $CdSO_4(aq) + NaOH(aq) \longrightarrow$

$$[CdSO_4(aq) + 2\,NaOH(aq) \longrightarrow Cd(OH)_2(s) + Na_2SO_4(aq)]$$

b. $KCl(aq) + AgNO_3(aq) \longrightarrow$

Work Problem 11.

$$[KCl(aq) + AgNO_3(aq) \longrightarrow AgCl(s) + KNO_3(aq)]$$

9.10 Neutralization Reactions

Neutralization reaction
Type of chemical reaction in which an acid or acid oxide reacts with a base or basic oxide, usually producing water as one of its products.

A **neutralization reaction** is one in which an acid or an acid oxide reacts with a base or basic oxide. In most of these reactions, water is one of the products. The formation of water acts as the driving force behind the neutralization. Heat is also given off in its formation.

A general equation represents this reaction:

$$\overset{\frown}{HX + MOH} \longrightarrow MX + HOH$$

where HX is an acid and MOH is a base.

As you may have already noted, neutralization reactions are just a special type of double-replacement reaction. There are four general types of neutralization reactions.

1. acid + base $\longrightarrow$ salt + water

$$HCl(aq) + NaOH(aq) \longrightarrow NaCl(aq) + H_2O(l)$$

$$H_2SO_4(aq) + Ba(OH)_2(aq) \longrightarrow BaSO_4(s) + 2\,H_2O(l)$$

2. metal oxide (basic oxide) + acid $\longrightarrow$ salt + water

$$ZnO(s) + 2\,HCl(aq) \longrightarrow ZnCl_2(aq) + H_2O(l)$$

$$CaO(s) + 2\,HNO_3(aq) \longrightarrow Ca(NO_3)_2(aq) + H_2O(l)$$

Study Hint: From equations 3 and 4 you should note the following: CO_2 forms the CO_3^{2-} ion, SO_2 forms the SO_3^{2-} ion, and SO_3 forms the SO_4^{2-} ion. The oxidation number of the nonmetal in the nonmetal oxide is the same as in the ion. For example, C is +4 in CO_2 and +4 in CO_3^{2-} (see Section 6.3 for calculating oxidation numbers). For S, it is +4 in SO_2 and SO_3^{2-}, and +6 in SO_3 and SO_4^{2-}.

3. nonmetal oxide (acid oxide) + base $\longrightarrow$ salt + water

$$CO_2(g) + 2\,LiOH(aq) \longrightarrow Li_2CO_3(aq) + H_2O(l)$$

$$SO_2(g) + 2\,NaOH(aq) \longrightarrow Na_2SO_3(aq) + H_2O(l)$$

Apollo space capsule filters containing lithium hydroxide were used to absorb carbon dioxide from the cabin atmosphere in keeping with the first equation.

4. nonmetal oxide (acid oxide) + metal oxide (basic oxide) $\longrightarrow$ salt

$$BaO(s) + SO_3(g) \longrightarrow BaSO_4(s)$$

$$NaO(s) + CO_2(g) \longrightarrow Na_2CO_3(s)$$

You will be asked to *complete* and *balance* neutralization reaction equations using the rules for solubility and the periodic table and to classify this type of reaction. Consider the following examples.

EXAMPLE 9.10 Complete and balance the equation for the reaction between strontium hydroxide and sulfuric acid in aqueous solution. [Indicate any precipitate by (*s*) and any gas by (*g*).]

SOLUTION   Write the formulas of the reactants (see Chapter 7). Complete the equation and look for water, a gas, or an insoluble compound in the products.

$$Sr(OH)_2(aq) + H_2SO_4(aq) \longrightarrow SrSO_4(s) + H_2O(l) \quad \text{(unbalanced)}$$

Strontium sulfate ($SrSO_4$) is insoluble in water (rule 3), and water is formed so a neutralization reaction does occur. The balanced equation is

$$Sr(OH)_2(aq) + H_2SO_4(aq) \longrightarrow SrSO_4(s) + 2 H_2O(l) \quad \textit{Answer}$$

EXAMPLE 9.11 Complete and balance the equation for the reaction between copper(II) oxide solid and aqueous hydrochloric acid. [Indicate any precipitate by (*s*) and any gas by (*g*).]

SOLUTION   Write the formula of the reactants (Chapter 7). Complete the equation and look for water, a gas, or an insoluble compound in the products.

$$CuO(s) + HCl(aq) \longrightarrow CuCl_2(aq) + H_2O(l) \quad \text{(unbalanced)}$$

Copper(II) chloride ($CuCl_2$) is soluble in water (rule 2), but water is formed, so a neutralization reaction does occur. The balanced equation is

$$CuO(s) + 2 HCl(aq) \longrightarrow CuCl_2(aq) + H_2O(l) \quad \textit{Answer}$$

Study Exercise 9.7
Using the rules for the solubility of inorganic substances in water and the periodic table, predict the products for the following neutralization reaction equations and balance them. [Indicate any precipitate by (*s*) and any gas by (*g*).]

a. $Ba(OH)_2(aq) + H_3PO_4(aq) \longrightarrow$

$$[3 Ba(OH)_2(aq) + 2 H_3PO_4(aq) \longrightarrow Ba_3(PO_4)_2(s) + 6 H_2O(l)]$$

b. $CaO(s) + CO_2(g) \longrightarrow$ $\qquad [CaO(s) + CO_2(g) \longrightarrow CaCO_3(s)]$ Work Problem 12.

Table 9.2 summarizes the general reactions for the five simple types of chemical reactions discussed in this chapter.

Study Exercise 9.8

Balance the following reaction equations and classify each as (i) combination reaction, (ii) decomposition reaction, (iii) single-replacement reaction, (iv) double-replacement reaction, or (v) neutralization reaction.

a. $N_2O_3(g) + H_2O(l) \longrightarrow HNO_2(aq)$

$(1 + 1 \longrightarrow 2, \text{combination})$

b. $HgO(s) + HCl(aq) \longrightarrow HgCl_2(aq) + H_2O(l)$

Work Problems 13 through 17.

$(1 + 2 \longrightarrow 1 + 1, \text{neutralization})$

CHEMISTRY OF THE ATMOSPHERE

Pollutants in the Atmosphere

The essay following Chapter 2 discussed a number of trace gases in the atmosphere, including carbon dioxide (CO_2), sulfur dioxide (SO_2), dinitrogen monoxide (N_2O), nitrogen monoxide (NO), nitrogen dioxide (NO_2), methane (CH_4), and chlorofluorocarbons (CFCs). Large amounts of these gases are released into the atmosphere every day by human activities. In later essays we will discuss the serious problems arising from the buildup of such gases. First, however, we need to consider the sources of these trace gases. Only then can we see what steps are needed to reduce the amounts of these gases released.

Carbon dioxide (CO_2): Green plants convert carbon dioxide to sugar by photosynthesis and regenerate some of that carbon dioxide by living and growing, a process called *respiration*. These processes are an essential part of life on earth, as the figure shows. The plants are then eaten by plant-eating animals, and the plant-eating animals are in turn eaten by meat-eating animals. The death of these living organisms then returns some of the carbon to the atmosphere as carbon dioxide. Some of the dead matter is also converted to coal, oil, natural gas, or rocks. Humans release carbon dioxide to the atmosphere by burning massive amounts of fuels such as oil, gas, coal, and wood. For example,

Coal: $C(s) + O_2(g) \longrightarrow CO_2(g)$

Gasoline (C_8H_{18}):

$2\,C_8H_{18}(l) + 25\,O_2(g) \longrightarrow$
$16\,CO_2(g) + 18\,H_2O(g)$

Carbon monoxide (CO) can also be generated in these processes.

Sulfur dioxide (SO_2): There are a few natural sources of sulfur dioxide (volcanoes, hot springs), but most sulfur dioxide in the atmosphere results from burning coal, oil, or natural gas that contains significant amounts of sulfur. Combustion of the fuel converts the sulfur, which may be present as elemental sulfur (shown below), hydrogen sulfide (H_2S), or other sulfur-containing compounds, to sulfur dioxide.

$S(s) + O_2(g) \longrightarrow SO_2(g)$

Additional sulfur dioxide is generated during the preparation of metals from their sulfide ores (see The Element MERCURY: The Mad Hatter, Chapter 12)

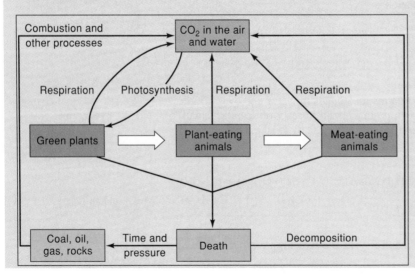

TABLE 9.2	Summary of the Five Simple Types of Chemical Reactions
Combination reaction	$A + Z \longrightarrow AZ$
Decomposition reaction	$AZ \longrightarrow A + Z$
Single-replacement reaction	$A + BZ \longrightarrow AZ + B$
	$X + BZ \longrightarrow BX + Z$
Double-replacement reaction	$AX + BZ \longrightarrow AZ + BX$
Neutralization reaction[a]	$HX + MOH \longrightarrow MX + HOH$

[a] HX is an acid, and MOH is a base. An acid oxide (nonmetal oxide) may be substituted for the acid, and a basic oxide (metal oxide) may be substituted for the base in a neutralization reaction. A nonmetal oxide plus a metal oxide forms a salt but no water.

Dinitrogen monoxide: This pollutant is generated primarily through the burning of forests, agricultural waste, and jungles in South America. Often, land in the tropics is cleared by simply burning everything. Such burning contributes carbon dioxide *and* dinitrogen monoxide to the atmosphere.

Nitrogen monoxide (NO) *and nitrogen dioxide* (NO_2): This mixture of nitrogen oxides, often referred to as NO_x, is a major component of automobile exhaust. When air is used as a source of oxygen in gasoline engines, the nitrogen (N_2) and oxygen (O_2) in air react with each other as they pass through the hot chambers of the engine to produce a mixture of nitrogen monoxide and nitrogen dioxide.

$$N_2(g) + O_2(g) \longrightarrow 2\,NO(g)$$

$$2\,NO(g) + O_2(g) \longrightarrow 2\,NO_2(g)$$

Methane (CH_4): Most methane is generated in trash dumps, peat bogs, and in the guts of cows, people, and other animals. The increasing population of the earth generates more trash and requires more food (cows!). The amount of methane in the atmosphere has been rising over the past 10 to 15 years. In 1978, about 1 out of every 660,000 gas molecules in the atmosphere was methane. In

The automobile is one of the biggest sources of atmospheric pollution.

1990, about one out of every 590,000 gas molecules in the air was methane.

Chlorofluorocarbons (CFCs): These materials are introduced to the environment solely by the activities of humans. Chlorofluorocarbons are chemicals containing carbon, fluorine, chlorine, and sometimes bromine or hydrogen. They are used as refrigerants and fire retardants, as solvents in the electronics industry, and to blow foam insulation into attics and walls. Some of the more important chlorofluorocarbons include the following:

CFC-11
(blowing agent)

CFC-12
(refrigerant)

Halon 1301
(fire retardant)

CFC-113
(solvent)

Summary

A *chemical equation* is a shorthand way of expressing a chemical change (reaction) in symbols and formulas. Chemical equations need to be balanced because of the law of conservation of mass, which requires that the number of atoms of each element be the same on both sides of the equation (Sections 9.1 and 9.2).

The simple guidelines for writing and balancing chemical equations (Section 9.3) require that you know the names, formulas, and charges on the cations (Table 6.1), anions (Table 6.2), and polyatomic ions (Table 6.4). These guidelines allow us to balance a variety of chemical equations (Section 9.4).

There are five basic types of simple chemical reactions: decomposition reactions, combination reactions, single-replacement reactions, double-replacement reactions, and neutralization reactions (Section 9.5).

In a combination (synthesis) reaction, two or more substances (elements or compounds) react to produce one substance (a compound). Some combination reactions involving oxygen gas are called combustion reactions (Section 9.6).

In a decomposition reaction, one substance (a compound) undergoes a reaction to form two or more substances (elements or compounds). This process often requires heat (Section 9.7).

In a single-replacement reaction, an element and a compound react, and the element replaces another element in the compound. Such reactions generally occur either (1) when a metal replaces a metal ion in its salt or a hydrogen ion in an acid, or (2) when a nonmetal replaces a nonmetal in its salt or acid. The electromotive (activity) or halogen series helps us to predict such reactions (Section 9.8).

In a double-replacement reaction, two compounds exchange positive ions. Such reactions generally occur either (1) when a precipitate is formed, (2) when a weakly ionized species is produced as a product, or (3) when a gas is formed as a product. The rules for the solubility of inorganic substances in water can help us to predict such reactions (Section 9.9).

In a neutralization reaction, an acid or an acid oxide reacts with a base or basic oxide, generally producing water and heat (Section 9.10).

Exercises

1. Define or explain the following terms (the number in parentheses refers to the section in the text where the term is mentioned):

    a. chemical reaction (9.1)   b. chemical equation (9.1)

    c. reactants (9.2)   d. products (9.2)

    e. catalyst (9.2)   f. combination reaction (9.6)

    g. combustion reaction (9.6)   h. basic oxide (basic anhydride) (9.6)

    i. acid oxide (acid anhydride) (9.6)   j. decomposition reaction (9.7)

    k. single-replacement reaction (9.8)   l. electromotive (activity) series (9.8)

    m. double-replacement reaction (9.9) n. precipitate (9.9)

    o. neutralization reaction (9.10)

2. Explain the meaning of the following symbols or terms in chemical equations:

    a. $\longrightarrow$ b. $\rightleftharpoons$ c. $=$ d. ✓

    e. (g) f. (l) g. (s) h. (aq)

3. Distinguish between:

   a. products and reactants

   b. $\xrightarrow{\Delta}$ and $\xrightarrow{MnO_2}$

 ## Problems

Balancing Equations (See Section 9.1 through 9.4)

4. Balance each of the following chemical equations by inspection:

   a. $BaCl_2(aq) + (NH_4)_2CO_3(aq) \longrightarrow BaCO_3(s) + NH_4Cl(aq)$

   b. $KClO_3(s) \xrightarrow{\Delta} KCl(s) + O_2(g)$

   c. $Al(OH)_3(s) + NaOH(aq) \longrightarrow NaAlO_2(aq) + H_2O(l)$

   d. $Fe(OH)_3(s) + H_2SO_4(aq) \longrightarrow Fe(SO_4)_3(aq) + H_2O(l)$

   e. $Na(s) + H_2O(l) \longrightarrow NaOH(aq) + H_2(g)$

   f. $Mg(s) + N_2(g) \xrightarrow{\Delta} Mg_3N_2(s)$

   g. $Mg(s) + O_2(g) \xrightarrow{\Delta} MgO(s)$

   h. $AgNO_3(aq) + CuCl_2(aq) \longrightarrow AgCl(s) + Cu(NO_3)_2(aq)$

   i. $C_2H_6O(l) + O_2(g) \xrightarrow{\Delta} CO_2(g) + H_2O(l)$

   j. $FeCl_2(aq) + Na_3PO_4(aq) \longrightarrow Fe_3(PO_4)_2(s) + NaCl(aq)$

5. Balance each of the following chemical equations by inspection:

   a. $CaC_2(s) + H_2O(l) \longrightarrow C_2H_2(g) + Ca(OH)_2(aq)$

   b. $MnO_2(s) + Al(s) \longrightarrow Al_2O_3(s) + Mn(s)$

   c. $CaCO_3(s) + H_3PO_4(aq) \longrightarrow Ca_3(PO_4)_2(s) + CO_2(g) + H_2O(l)$

   d. $Al(s) + H_2SO_4(aq) \longrightarrow Al_2(SO_4)_3(s) + H_2(g)$

   e. $P_4O_{10}(s) + H_2O(l) \longrightarrow H_3PO_4(aq)$

   f. $C_3H_8(g) + O_2(g) \longrightarrow CO_2(g) + H_2O(l)$

   g. $Na_2O(s) + P_4O_{10}(s) \longrightarrow Na_3PO_4(s)$

   h. $PCl_5(s) + H_2O(l) \longrightarrow H_3PO_4(aq) + HCl(g)$

   i. $Sb_2O_3(s) + NaOH(aq) \longrightarrow NaSbO_2(aq) + H_2O(l)$

   j. $TiCl_4(l) + H_2O(l) \longrightarrow TiO_2(s) + HCl(g)$

6. Change the following word equations into chemical equations and complete and balance them by inspection:

   a. sodium chloride + lead(II) nitrate $\longrightarrow$
                       lead(II) chloride + sodium nitrate

   b. ferric oxide + hydrochloric acid $\longrightarrow$ ferric chloride + water

   c. sodium hydrogen carbonate + phosphoric acid $\longrightarrow$
                sodium phosphate + carbon dioxide + water

   d. mercury + oxygen $\longrightarrow$ mercury(II) oxide

   e. calcium iodide + sulfuric acid $\longrightarrow$
                    hydrogen iodide + calcium sulfate

   f. barium nitrate + sulfuric acid $\longrightarrow$ barium sulfate + nitric acid

g. magnesium cyanide + hydrochloric acid $\longrightarrow$
hydrogen cyanide + magnesium chloride

h. iron(II) sulfide + hydrobromic acid $\longrightarrow$
iron(II) bromide + hydrogen sulfide

i. sodium hydrogen sulfite + sulfuric acid $\longrightarrow$
sodium sulfate + sulfur dioxide + water

j. aluminum sulfate + sodium hydroxide $\longrightarrow$
aluminum hydroxide + sodium sulfate

7. Change the following word equations into chemical equations and complete and balance them by inspection:

a. iron + chlorine $\longrightarrow$ iron(III) chloride

b. potassium nitrate $\xrightarrow{\Delta}$ potassium nitrite + oxygen

c. barium + water $\longrightarrow$ barium hydroxide + hydrogen

d. sodium hydroxide + sulfuric acid $\longrightarrow$
sodium hydrogen sulfate + water

e. ammonium sulfide + mercuric bromide $\longrightarrow$
ammonium bromide + mercuric sulfide

f. zinc hydroxide + sulfuric acid $\longrightarrow$ zinc sulfate + water

g. tin(II) oxide + hydrochloric acid $\longrightarrow$ tin(II) chloride + water

h. strontium sulfite + acetic acid $\longrightarrow$
strontium acetate + water + sulfur dioxide

i. hydrogen bromide + calcium hydroxide $\longrightarrow$
calcium bromide + water

j. bismuth(III) sulfide + oxygen $\longrightarrow$
bismuth(III) oxide + sulfur dioxide

Balancing Combination Reactions (Section 9.6)

8. Balance the following combination reaction equations:

a. $Al(s) + O_2(g) \xrightarrow{\Delta} Al_2O_3(s)$

b. $C(s) + O_2(g) \xrightarrow{\Delta} CO_2(g)$

c. $Si(s) + O_2(g) \xrightarrow{\Delta} SiO_2(s)$

d. $Mg(s) + S(s) \xrightarrow{\Delta} MgS(s)$

e. $Al(s) + N_2(g) \xrightarrow{\Delta} AlN(s)$

f. $Na_2O(s) + H_2O(l) \longrightarrow NaOH(aq)$

g. $Al_2O_3(s) + H_2O(l) \longrightarrow Al(OH)_3(s)$

h. $BaO(s) + SO_3(g) \longrightarrow BaSO_4(s)$

i. $CaO(s) + SO_3(g) \longrightarrow CaSO_4(s)$

j. $N_2O_5(g) + H_2O(l) \longrightarrow HNO_3(aq)$

Balancing Decomposition Reactions (Section 9.7)

9. Balance the following decomposition reaction equations:

a. $NaNO_3(s) \xrightarrow{\Delta} NaNO_2(s) + O_2(g)$

b. $C_{12}H_{22}O_{11}(s)$ (sucrose, table sugar) $\xrightarrow{\Delta} C(s) + H_2O(l)$

c. $Na_2CO_3 \cdot H_2O(s) \longrightarrow Na_2CO_3(s) + H_2O(l)$

The water in the $Na_2CO_3 \cdot H_2O$ is held as part of the solid. Gentle heating releases the water.

d. $MgCO_3(s) \xrightarrow{\Delta} MgO(s) + CO_2(g)$

e. $NaHCO_3(s) \xrightarrow{\Delta} Na_2CO_3(s) + CO_2(g) + H_2O(l)$

f. $KClO_3(s) \xrightarrow{\Delta} KCl(s) + O_2(g)$

g. $Ca(HCO_3)_2(s) \xrightarrow{\Delta} CaCO_3(s) + H_2O(l) + CO_2(g)$

h. $NaCl(l) \xrightarrow[\text{current}]{\text{direct electric}} Na(l) + Cl_2(g)$

i. $CoSO_4 \cdot 7H_2O(s) \xrightarrow{\Delta} CoSO_4(s) + H_2O(l)$

The water in the $CoSO_4 \cdot 7H_2O$ is held as part of the solid. Gentle heating releases the water.

j. $Ba(HCO_3)_2(s) \xrightarrow{\Delta} BaCO_3(s) + H_2O(l) + CO_2(s)$

Completing and Balancing Single-Replacement Reactions [Section 9.8, Electromotive (Activity) Series, and the Periodic Table]

10. Using the electromotive (activity) series or halogen series and the periodic table, predict the products for the following single-replacement reaction equations and balance them.

a. $Al(s) + HCl(aq) \longrightarrow$

b. $Cd(s) + CuSO_4(aq) \longrightarrow$

c. $Al(s) + HC_2H_3O_2(aq) \longrightarrow$

d. $Cl_2(g) + CaI_2(aq) \longrightarrow$

e. $Al(s) + SnCl_2(aq) \longrightarrow$

f. $Cu(s) + FeCl_2(aq) \longrightarrow$

Hint: Is Cu above Fe in the activity series?

g. $Pb(s) + HgBr_2(aq) \longrightarrow$

h. $Al(s) + HgBr_2(aq) \longrightarrow$

i. $Ba(s) + HCl(aq) \longrightarrow$

j. $Br_2(aq) + NaI(aq) \longrightarrow$

Completing and Balancing Double-Replacement Reactions (Section 9.9, Rules for the Solubility of Inorganic Substances in Water, and the Periodic Table)

11. Using the rules for the solubility of inorganic substances in water and the periodic table, predict the products for the following double-replacement reaction equations and balance them. [Indicate any precipitate by (s) and any gas by (g).]

a. $AgNO_3(aq) + H_2S(g) \longrightarrow$

b. $FeCl_3(aq) + NaOH(aq) \longrightarrow$

c. $Na_2CO_3(aq) + HC_2H_3O_2(aq) \longrightarrow$

d. $Pb(NO_3)_2(aq) + K_2CrO_4(aq) \longrightarrow$

e. $SnCl_2(aq) + H_2S(g) \longrightarrow$

f. $BaCl_2(aq) + (NH_4)_2CO_3(aq) \longrightarrow$

g. $FeCO_3(s) + H_2SO_4(aq) \longrightarrow$

h. $Bi(NO_3)_3(aq) + H_2S(g) \longrightarrow$

i. $Pb(NO_3)_2(aq) + H_2S(g) \longrightarrow$

j. $Ba(NO_3)_2(aq) + NaOH(aq) \longrightarrow$

Completing and Balancing Neutralization Reactions (Section 9.10, Rules for the Solubility of Inorganic Substances in Water, and the Periodic Table)

12. Using the rules for the solubility of inorganic substance in water and the periodic table, predict the products for the following neutralization reaction equations and balance them. [Indicate any precipitate by (s) and any gas by (g).]

   a. $Zn(OH)_2(s) + HNO_3(aq) \longrightarrow$

   b. $Fe_2O_3(s) + H_3PO_4(aq) \longrightarrow$

   c. $SO_3(g) + Fe(OH)_3(s) \longrightarrow$

   d. $BaO(s) + HCl(aq) \longrightarrow$

   e. $H_2SO_4(aq) + NaOH(aq) \longrightarrow$

   f. $Al(OH)_3(s) + H_2SO_4(aq) \longrightarrow$

   g. $CO_2(g) + NaOH(aq) \longrightarrow$

   h. $Ca(OH)_2(s) + HC_2H_3O_2(aq) \longrightarrow$

   i. $HNO_3(aq) + Sr(OH)_2(aq) \longrightarrow$

   j. $SO_2(g) + NaOH(aq) \longrightarrow$

Identifying the Five Simple Reaction Types (Sections 9.5 through 9.10)

13. Balance the following reaction equations and classify each as a (i) combination reaction, (ii) decomposition reaction, (iii) single-replacement reaction, (iv) double-replacement reaction, or (v) neutralization reaction.

   a. $Ca(s) + O_2(g) \xrightarrow{\Delta} CaO(s)$

   b. $HgO(s) \xrightarrow{\Delta} Hg(l) + O_2(g)$

   c. $Cd(s) + H_2SO_4(aq) \longrightarrow CdSO_4(aq) + H_2(g)$

   d. $C(s) + O_2(g) \xrightarrow{\Delta} CO(g)$

   e. $Pb(NO_3)_2(aq) + HCl(aq) \xrightarrow{\text{cold}} PbCl_2(s) + HNO_3(aq)$

   f. $Zn(OH)_2(s) + H_2SO_4(aq) \longrightarrow ZnSO_4(aq) + H_2O(l)$

   g. $CdSO_4(aq) + KOH(aq) \longrightarrow Cd(OH)_2(s) + K_2SO_4(aq)$

   h. $KNO_3(s) \xrightarrow{\Delta} KNO_2(s) + O_2(g)$

   i. $BaO(s) + H_2SO_4(aq) \longrightarrow BaSO_4(s) + H_2O(l)$

   j. $Br_2(aq) + MgI_2(aq) \longrightarrow MgBr_2(aq) + I_2(aq)$

14. Balance the following reaction equations and classify each as a (i) combination reaction, (ii) decomposition reaction, (iii) single-replacement reaction, (iv) double-replacement reaction, or (v) neutralization reaction.

   a. $Pb(s) + HCl(aq) \xrightarrow{\Delta} PbCl_2(aq) + H_2(g)$

   b. $ZnCO_3(s) + H_3PO_4(aq) \longrightarrow Zn_3(PO_4)_2(s) + H_2O(l) + CO_2(g)$

   c. $Fe(OH)_3(s) + H_3PO_4(aq) \longrightarrow FePO_4(s) + H_2O(l)$

d. $S(s) + O_2(g) \xrightarrow{\Delta} SO_3(g)$

e. $Bi(NO_3)_3(aq) + NaOH\ (aq) \longrightarrow Bi(OH)_3(s) + NaNO_3(aq)$

f. $Na(s) + H_2O(l) \longrightarrow NaOH(aq) + H_2(g)$

g. $CO_2(g) + Ca(OH)_2(aq) \longrightarrow CaCO_3(s) + H_2O(l)$

h. $H_2O(l) \xrightarrow{\text{direct electric current}} H_2(g) + O_2(g)$

i. $SrCO_3(s) \xrightarrow{\Delta} SrO(s) + CO_2(g)$

j. $Al(s) + Cl_2(g) \longrightarrow AlCl_3(s)$

15. Balance the following reaction equations and classify each as a (i) combination reaction, (ii) decomposition reaction, (iii) single-replacement reaction, (iv) double-replacement reaction, or (v) neutralization reaction.

   a. $CdCO_3(s) \xrightarrow{\Delta} CdO(s) + CO_2(g)$

   b. $Na_2O(s) + HNO_2(aq) \longrightarrow NaNO_2(aq) + H_2O(l)$

   c. $CaO(s) + H_2O(l) \longrightarrow Ca(OH)_2(aq)$

   d. $Li_2O(s) + H_2O(l) \longrightarrow LiOH(aq)$

   e. $Pb(C_2H_3O_2)_2(aq) + K_2SO_4\ (aq) \longrightarrow PbSO_4(s) + KC_2H_3O_2(aq)$

   f. $C_6H_{12}O_6(s)$ (dextrose) $\xrightarrow{\Delta} C(s) + H_2O(l)$

   g. $SO_2(g) + KOH(aq) \longrightarrow K_2SO_3(aq) + H_2O(l)$

   h. $MnSO_4(aq) + (NH_4)_2S(aq) \longrightarrow MnS(s) + (NH_4)_2SO_4(aq)$

   i. $Ca(s) + H_2O(l) \longrightarrow Ca(OH)_2(aq) + H_2(g)$

   j. $Zn(s) + HCl(aq) \longrightarrow ZnCl_2(aq) + H_2(g)$

16. Balance the following reaction equations and classify each as a (i) combination reaction, (ii) decomposition reaction, (iii) single-replacement reaction, (iv) double-replacement reaction, or (v) neutralization reaction.

   a. $CaCO_3(s) + HCl(aq) \longrightarrow CaCl_2(aq) + H_2O(l) + CO_2(g)$

   b. $Al(OH)_3(s) + HCl(aq) \longrightarrow AlCl_3(aq) + H_2O(l)$

   c. $Fe(s) + CuCl_2(aq) \longrightarrow FeCl_2(aq) + Cu(s)$

   d. $KHCO_3(s) \xrightarrow{\Delta} K_2CO_3(s) + H_2O(l) + CO_2(g)$

   e. $Pb(NO_3)_2(aq) + H_2S(g) \longrightarrow PbS(s) + HNO_3(aq)$

   f. $Cl_2(g) + KI(aq) \longrightarrow KCl(aq) + I_2(aq)$

   g. $SO_2(g) + H_2O(l) \longrightarrow H_2SO_3(aq)$

   h. $PbCO_3(s) \xrightarrow{\Delta} PbO(s) + CO_2(g)$

   i. $P_4O_{10}(s) + H_2O(l) \longrightarrow H_3PO_4(aq)$

   j. $KOH(aq) + CO_2(g) \longrightarrow K_2CO_3(aq) + H_2O(l)$

17. Balance the following reaction equations and classify each as a (i) combination reaction, (ii) decomposition reaction, (iii) single-replacement reaction, (iv) double-replacement reaction, or (v) neutralization reaction.

   a. $MgO(s) + HCl(aq) \longrightarrow MgCl_2(aq) + H_2O(l)$

   b. $SO_3(g) + H_2O(l) \longrightarrow H_2SO_4(aq)$

   c. $BaCO_3(s) + HNO_3(aq) \longrightarrow Ba(NO_3)_2(aq) + H_2O(l) + CO_2(g)$

   d. $Cl_2(g) + NaBr(aq) \longrightarrow NaCl(aq) + Br_2(aq)$

e. $H_2O_2(l) \xrightarrow{\Delta} H_2O(l) + O_2(g)$

f. $FeSO_4(aq) + (NH_4)_2S(aq) \longrightarrow FeS(s) + (NH_4)_2SO_4(aq)$

g. $Na(s) + O_2(g) \longrightarrow Na_2O(s)$

h. $Zn(s) + NiCl_2(aq) \longrightarrow ZnCl_2(aq) + Ni(s)$

i. $Pb(OH)_2(s) + HNO_3(aq) \longrightarrow Pb(NO_3)_2(aq) + H_2O(l)$

j. $NaClO_3(s) \xrightarrow{\Delta} NaCl(s) + O_2(g)$

General Problems

18. (1) Complete and balance the following reaction equations. Indicate any precipitate by (s) and any gas by (g). (2) Classify the following reactions as (i) combination, (ii) decomposition, (iii) single-replacement, (iv) double-replacement, or (v) neutralization.

a. $Cd(s) + HCl(aq) \longrightarrow$

b. $BaCl_2(aq) + H_2SO_4(aq) \longrightarrow$

c. $CaCO_3(s) + HNO_3(aq) \longrightarrow$

d. $Cu(s) + MgCl_2(aq) \longrightarrow$

19. (1) Complete and balance the following reaction equations. Indicate any precipitate by (s) and any gas by (g). (2) Classify the following reactions as (i) combination, (ii) decomposition, (iii) single-replacement, (iv) double-replacement, or (v) neutralization.

a. aluminum + lead(II) chloride $\longrightarrow$

b. iron(II) sulfide + hydrochloric acid $\longrightarrow$

c. barium chloride + sodium carbonate $\longrightarrow$

d. cadmium oxide + hydrochloric acid $\longrightarrow$

20. The U.S. Department of Agriculture is studying methods for keeping meat fresh longer in supermarket display cases. One of their proposals is to have a packet of powder inserted inside the meat package. This powder contains citric acid ($H_3C_6H_5O_7$) and sodium hydrogen carbonate. As moisture builds up in the package, carbon dioxide gas is produced and escapes through the package pores. The carbon dioxide controls most of the microorganisms responsible for meat spoilage and hence keeps the meat fresh longer. Complete and balance the reaction equation described. (*Hint:* Citric acid has *three* hydrogens that come off as H^+.)

21. The Acropolis in Athens, Greece, is slowly deteriorating. It is composed of marble ($CaCO_3$), which slowly reacts with sulfuric acid from air pollution to form a salt, which is washed away, destroying this famous historical site. The sulfuric acid is formed from the air pollutant sulfur trioxide and water. Complete and balance the two reaction equations described (Figure 9.12a and b).

22. In a train derailment, one box car contained 55-gal drums of phosphorus trichloride. Some of these drums broke and reacted with water on the ground and in the moist air to give a dense "white gas," which was noticed over a 12-square-mile area. Complete and balance the reaction equation described.

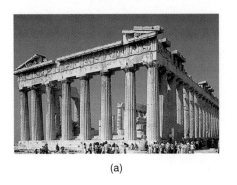

(a)

(b)

FIGURE 9.12
Deterioration of the Acropolis and other buildings. (a) Before serious air pollution. (b) After considerable air pollution.

23. Titanium diboride (TiB_2) has been developed as an extremely hard, wear-resistant coating for materials that must survive extremely erosive (eroding) environments. It is expected to be used as a coating on valves in coal liquefaction reactors. Titanium diboride is prepared by allowing titanium tetrachloride, boron trichloride, and hydrogen to react at atmospheric pressure to yield the diboride and hydrogen chloride. Write a balanced equation for this reaction.

24. In downtown Denver, Colorado, 20,000 gal of nitric acid was spilled from a tank car in a rail yard. The accident was brought under control within a few hours by blowing soda ash (sodium carbonate) onto the spilled nitric acid. Complete and balance the equation for the reaction described.

25. In the refining of magnesium, waste material is produced that contains 2% magnesium nitride (Mg_3N_2). This waste material reacts with water in the atmosphere to produce ammonia gas and magnesium hydroxide. Complete and balance the equation for the reaction described. Recently, this waste has been used as a fertilizer. It has been applied to peas, beans, and corn in Oregon and to winter wheat, barley, and oats in Idaho with no toxicity and an increase in crop yield over control plots.

26. Rust (Fe_2O_3) can be removed from tools by soaking them in vinegar (5% acetic acid). Write a balanced equation for this reaction.

✓ Chapter Quiz 9

You may use the periodic table, the electromotive (activity) series, and the rules for the solubility of inorganic substances in water.

1. Balance the following equations:

   a. $Mg(s) + N_2(g) \xrightarrow{\Delta} Mg_3N_2(s)$

   b. $C_2H_6(g) + O_2(g) \xrightarrow{\Delta} CO_2(g) + H_2O(g)$

2. Change the following word equation into a chemical equation and complete and balance it by inspection.

   aluminum sulfate + sodium hydroxide ⟶

   aluminum hydroxide + sodium sulfate

3. Classify each of the following balanced reaction equations as a (i) combination reaction, (ii) decomposition reaction, (iii) single-replacement reaction, (iv) double-replacement reaction, or (v) neutralization reaction.

   a. $2 Na(s) + Cl_2(g) \longrightarrow 2 NaCl(s)$

   b. $Mg(s) + CuSO_4(aq) \longrightarrow MgSO_4(aq) + Cu(s)$

   c. $2 KClO_3(s) \xrightarrow{\Delta} 2 KCl(s) + 3 O_2(g)$

   d. $Mn(NO_3)_2(aq) + Na_2S(aq) \longrightarrow MnS(s) + 2 NaNO_3(aq)$

4. Complete and balance the following single-replacement, double-replacement, or neutralization reactions. Indicate any precipitates by (s) and any gases by (g).

   a. $Ba(OH)_2(aq) + HC_2H_3O_2(aq) \longrightarrow$

   b. $Mg(s) + HCl(aq) \longrightarrow$

   c. $CaCl_2(aq) + H_3PO_4(aq) \longrightarrow$

   d. $Pb(NO_3)_2(aq) + HCl(aq) \xrightarrow{cold}$

Carbon Dioxide (Symbol: CO₂)

The Compound CARBON DIOXIDE: The "Life" of the Party

Carbon dioxide is found in carbonated beverages. It is also part of the carbon cycle and life on our planet.

Name: Carbon dioxide (CO_2) is occasionally called carbonic acid gas or carbonic anhydride. Solid CO_2 is frequently called *dry ice*.

Appearance: Carbon dioxide is a colorless, odorless gas under normal conditions of temperature and pressure. In the solid form, it is a cold ($-78°C$), white solid that passes directly from the solid phase into the gas phase (sublimes).

Occurrence: Carbon dioxide can be found all over the earth. It comprises about 0.03% of the atmosphere by volume. Carbon dioxide also dissolves in water (oceans, streams, and lakes) to produce a dilute solution of carbonic acid (H_2CO_3):

$$CO_2(g) + H_2O(l) \rightleftharpoons H_2CO_3(aq)$$

Carbon dioxide also occurs in the earth as carbonate salts, typically limestone (calcite, $CaCO_3$) or dolomite [$CaMg(CO_3)_2$]. The carbon dioxide tied up in such rocks is easily released by heating (Section 9.7) or adding acid (Section 9.9):

$$CaCO_3(s) \xrightarrow{\Delta} CaO(s) + CO_2(g)$$

$$CaCO_3(s) + 2\,HCl(aq) \longrightarrow$$
$$CaCl_2(aq) + H_2O(l) + CO_2(g)$$

Source: Most commercially available carbon dioxide is produced by one of four methods: (1) the reaction between methane (natural gas, CH_4) and water to give CO_2 and H_2, hydrogen gas; (2) the burning of coal in air to give CO_2; (3) heating limestone ($CaCO_3$); and (4) the fermentation of sugar ($C_{12}H_{22}O_{11}$) by yeast to give ethyl alcohol (C_2H_6O) and carbon dioxide (CO_2).

(1) $CH_4(g) + 2\,H_2O(g) \xrightarrow{\Delta} CO_2(g) + 4\,H_2(g)$

(2) $C(s) + O_2(g) \xrightarrow{\Delta} CO_2(g)$

(3) $CaCO_3(s) \xrightarrow{\Delta} CaO(s) + CO_2(g)$

(4) $C_{12}H_{22}O_{11}(s) + H_2O(l) \xrightarrow{yeast} 4\,CO_2(g) + 4\,C_2H_6O(l)$

Its Role in Our World: Carbon dioxide has many uses in our society. Large amounts of CO_2 are consumed during the industrial production of hydrogen gas (H_2), methanol (CH_3OH), urea (CH_4N_2O, a fertilizer and used in the preparation of certain plastics), and ammonia (NH_3). On a more personal level, CO_2 is used in the production of *carbonated beverages*, and dry ice helps *ice cream* vendors keep their product from melting.

But carbon dioxide plays a far more important role in our world than this. It is part of the carbon cycle, and life on our planet as discussed in the essay CHEMISTRY OF THE ATMOSPHERE: Pollutants in the Atmosphere (before the summary in this chapter).

Unusual Facts: Carbon dioxide actually has a slightly acid taste. The taste is due to the carbonic acid (H_2CO_3) that forms in your mouth as the CO_2 dissolves in your saliva. You can "taste" CO_2 by drinking *carbonated mineral water*. Humans cannot breathe air containing more than 5 to 10% CO_2 without losing consciousness, and prolonged exposure can result in death.

CHAPTER 10

Calculations Involving Chemical Equations. Stoichiometry

The modern automobile uses the combustion of gasoline (a chemical reaction) to obtain its power. Gasoline is a complex mixture of volatile, low-molecular-mass hydrocarbon molecules. The number of carbon atoms in these molecules ranges from 6 to 12. An equation for the complete combustion of a typical gasoline component is

$$2\ C_8H_{18} \text{ (an octane)} + 25\ O_2 \longrightarrow 16\ CO_2 \text{ (gas)} + 18\ H_2O \text{ (vapor)} + \text{heat}$$

GOALS FOR CHAPTER 10

1. To be able to obtain information on quantities of reactants and products from a balanced chemical equation (Section 10.1).
2. To understand the three steps in solving stoichiometry problems with emphasis on the *mole* and to identify the three types of stoichiometry problems (Sections 10.2 and 10.3).
3. To solve *mass–mass stoichiometry problems* of the following categories: mass to mass, mass to moles, moles to mass, moles to moles, and limiting reagent. To understand the difference between theoretical yield and actual yield and to calculate percent yield (Section 10.4).
4. To solve *mass–volume stoichiometry problems* of the following categories: mass to volume, volume to mass, moles to volume, volume to moles, and limiting reagent (Section 10.5).
5. To solve *volume–volume stoichiometry problems* of the following categories: volume to volume and limiting reagent (Section 10.6).
6. To understand the meaning of the term *heat of reaction*, to identify *exothermic* and *endothermic reactions*, and to calculate the amount of heat produced or needed in a given reaction (Section 10.7).

Countdown

You may use the periodic table and the rules for the solubility of inorganic substances in water.

ELEMENT	ATOMIC MASS UNITS (AMU)
Cr	52.0
Cl	35.5

5. Write the correct formulas for the following compounds (Sections 7.3, 7.4, and 7.6).
   a. calcium chloride $(CaCl_2)$
   b. chromium(III) oxide (Cr_2O_3)
   c. calcium phosphate $[Ca_3(PO_4)_2]$
   d. chromium(III) hydroxide $[Cr(OH)_3]$

4. Calculate the number of moles in each of the following quantities (Sections 8.2 and 8.3).
   a. 12.7 g Cr   (0.244 mol)
   b. 3.75 L (STP) Cl_2 gas   (0.167 mol)

3. Calculate the number of grams in each of the following quantities (Section 8.2).
   a. 0.172 mol Cr   (8.94 g)
   b. 0.245 mol Cl_2 gas   (17.4 g)

2. Calculate the number of liters at STP that the following gases would occupy (Sections 8.2 and 8.3).
   a. 0.205 mol Cl_2 gas   (4.59 L)
   b. 10.4 g Cl_2 gas   (3.28 L)

1. Predict the products and balance the following chemical reaction equations. Indicate any precipitate by (*s*) (Sections 9.3, 9.4, and 9.9).
   a. $Pb(C_2H_3O_2)_2(aq) + Na_2SO_4(aq) \longrightarrow$
   $[Pb(C_2H_3O_2)_2(aq) + Na_2SO_4(aq) \longrightarrow$
   $PbSO_4(s) + 2\ NaC_2H_3O_2]$
   b. $AsCl_3(aq) + H_2S(aq) \longrightarrow$
   $[2\ AsCl_3(aq) + 3\ H_2S(aq) \longrightarrow$
   $As_2S_3(s) + 6\ HCl]$

S uppose you are the president of a chemical company like the person shown in Figure 10.1. Chemists in the laboratories of your company believe they can combine two inexpensive chemicals—let's call them cheapium and thriftium—to form a new compound—let's call it dearium—that can be sold for a high price (that is, they propose to buy cheap and sell dear). Before you go out and buy any cheapium or thriftium, you need to know how much of each will produce how much dearium. Fortunately, the chemists in your company can give you these answers (at least on a theoretical level) by using stoichiometry.

Stoichiometry (pronounced stoi'kēom'i trē)—measurement of the relative quantities of chemical reactants and products in a chemical reaction—is just an extension of what you already know. In Chapter 8, you learned to calculate formula and molecular masses (Section 8.1), moles (Section 8.2), and molar volumes of gases (Section 8.3). In Chapter 9 you learned to balance and complete certain types of chemical reaction equations. In this chapter, we will use molar information from balanced chemical reaction equations to calculate the amounts of material or energy produced or needed in these chemical reactions.

Stoichiometry Measurement of the relative quantities of chemical reactants and products in a chemical reaction.

10.1   Information Obtained from a Balanced Chemical Equation

A completed and *balanced* chemical equation gives more information than simply which substances are reactants and which are products. It also gives the relative quantities involved, and it is very useful in carrying out certain calculations. Consider the reaction between ethane gas and oxygen to produce carbon dioxide and water.

$$2\ C_2H_6(g)\ +\ 7\ O_2(g)\ \xrightarrow{\ \Delta\ }\ 4\ CO_2(g)\ +\ 6\ H_2O(g)$$
ethane

This balanced chemical equation gives the following information:

✓ *Reactants and products.* C_2H_6 (ethane) reacts with O_2 (oxygen) when sufficient heat (Δ) is applied to produce CO_2 (carbon dioxide) and H_2O (gaseous water).

✓ *Molecules of reactants and products.* 2 molecules of C_2H_6 need 7 molecules of O_2 to react and produce 4 molecules of CO_2 and 6 molecules of H_2O.

FIGURE 10.1
Ms. Jones ponders the question, "How much of D can we get?"

✓ *Moles of reactants and products.* 2 mol of C_2H_6 molecules need 7 mol of O_2 molecules to react and produce 4 mol of CO_2 molecules and 6 mol of H_2O molecules.

✓ *Relative masses of reactants and products.* 60.0 g of C_2H_6

$$\left(2 \text{ mol } C_2H_6 \times \frac{30.0 \text{ g } C_2H_6}{1 \text{ mol } C_2H_6} = 60.0 \text{ g } C_2H_6 \right)$$

Study Hint: The mole relations (coefficients) are regarded as *exact* values and are not considered in determining significant digits.

and 224 g of O_2

$$\left(7 \text{ mol } O_2 \times \frac{32.0 \text{ g } O_2}{1 \text{ mol } O_2} = 224 \text{ g } O_2 \right)$$

react and produce 176 g of CO_2

$$\left(4 \text{ mol } CO_2 \times \frac{44.0 \text{ g } CO_2}{1 \text{ mol } CO_2} = 176 \text{ g } CO_2 \right)$$

and 108 g of H_2O

$$\left(6 \text{ mol } H_2O \times \frac{18.0 \text{ g } H_2O}{1 \text{ mol } H_2O} = 108 \text{ g } H_2O \right)$$

Note that the sum of the reactant masses (60.0 g + 224 g = 284 g) is equal to the sum of the product masses (176 g + 108 g = 284 g), obeying the law of conservation of mass (see Section 3.6).

✓ *Volume of gases.* 2 volumes of C_2H_6 need 7 volumes of O_2 to react and produce 4 volumes of CO_2 and 6 volumes of H_2O if all volumes are measured as *gases* at the same temperature and pressure.

10.2 The Mole Method of Solving Stoichiometry Problems

Using the information in a balanced chemical equation, we can solve stoichiometry problems in many different ways. The method we consider the best is the *mole method*, which is an application of our general method of problem solving, *dimensional analysis*. Three basic steps are involved in working problems by the *mole method*:

Study Hint: Before you read on, study Examples 8.6, 8.9, 8.15, and 8.20. You must know (1) how to calculate moles given grams or liters of a gas at STP, and (2) how to calculate grams or liters of a gas at STP given moles. Prior to step I and after step III, you may be required to make an additional calculation to convert from or to some mass measurement other than grams.

Step I: Calculate *moles* of elementary units (atoms, formula units, molecules, or ions) of the element, compound, or ion from the mass or volume (if gases) of the known substance or substances in the problem.

Step II: Using the coefficients of the substances in the balanced chemical equation, calculate *moles* of the unknown quantities in the problem.

Step III: From *moles* of the unknown quantities calculated, determine the mass or volume (for gases) of these unknowns in the units requested by the problem.

As you can see, the key to this method is the *mole*. Think **MOLES!**

Figure 10.2 shows the application of these basic steps in diagram form.

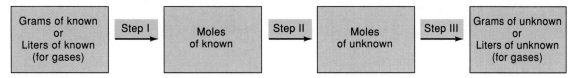

FIGURE 10.2
The three basic steps in solving stoichiometry problems. (Prior to step I and after step III, you may be required to make an additional calculation to convert to or from some mass measurement other than grams.)

10.3 Types of Stoichiometry Problems

The mole method works well with all three types of stoichiometry problems:

1. Mass–mass
2. Mass–volume or volume–mass
3. Volume–volume

We will now apply the three basic steps to these three types of stoichiometry problems.

10.4 Mass–Mass Stoichiometry Problems

In mass–mass problems, the quantities of both the known and unknown are given or asked for in mass units. Whether the known quantity is expressed in grams or in moles affects which of the steps we need to apply.

Mass to Mass Examples

We will first consider some *mass to mass examples*, in which the known is expressed in mass units as grams and the unknown is asked for in mass units as grams. These examples involve all three basic steps. Remember, however, that the chemical equation must be balanced *before* you begin the calculation.

EXAMPLE 10.1 Calculate the number of grams of oxygen required to burn 72.0 g of C_2H_6 to CO_2 and H_2O. The equation for the reaction is

$$2\,C_2H_6(g) + 7\,O_2(g) \xrightarrow{\Delta} 4\,CO_2(g) + 6\,H_2O(g)$$
ethane

SOLUTION The chemical equation is balanced, and so we can proceed to calculate the molar masses of the substances involved in the calculation, which are C_2H_6 and O_2:

$$\text{molar mass of } C_2H_6 = 30.0 \text{ g}$$

$$\text{molar mass of } O_2 = 32.0 \text{ g}$$

Organize the data:

Known: 72.0 g of C_2H_6

Unknown: grams of O_2 required

Follow the mole method:

Step I: Calculate the moles of C_2H_6 molecules given. One mole of C_2H_6 has a molar mass of 30.0 g, and so

$$72.0 \text{ g } C_2H_6 \times \frac{1 \text{ mol } C_2H_6}{30.0 \text{ g } C_2H_6} = \frac{72.0}{30.0} \text{ mol } C_2H_6 \text{ given}$$

Step II: Calculate the moles of oxygen molecules needed. From the balanced equation, the relation of C_2H_6 to O_2 is given as 2 mol of C_2H_6 to 7 mol of O_2. Therefore,

$$\frac{72.0}{30.0} \text{ mol } C_2H_6 \times \frac{7 \text{ mol } O_2}{2 \text{ mol } C_2H_6} = \frac{72.0}{30.0} \times \frac{7}{2} \text{ mol } O_2 \text{ needed}$$

Step III: Calculate the grams of oxygen needed. One mole of O_2 has a molar mass of 32.0 g O_2, and so

$$\frac{72.0}{30.0} \times \frac{7}{2} \text{ mol } O_2 \times \frac{32.0 \text{ g } O_2}{1 \text{ mol } O_2} = \frac{72.0}{30.0} \times \frac{7}{2} \times 32.0 \text{ g } O_2$$

$$= 268.9 \text{ g } O_2 = 269 \text{ g } O_2 \qquad \textit{Answer}$$

We express our answer to three significant digits (269 g of O_2) because our given quantity (72.0 g of C_2H_6) was expressed to three significant digits. We can write this solution in a linear manner as follows:

$$72.0 \text{ g } C_2H_6 \times \underbrace{\frac{1 \text{ mol } C_2H_6}{30.0 \text{ g } C_2H_6}}_{\text{Step I}} \times \underbrace{\frac{7 \text{ mol } O_2}{2 \text{ mol } C_2H_6}}_{\text{Step II}} \times \underbrace{\frac{32.0 \text{ g } O_2}{1 \text{ mol } O_2}}_{\text{Step III}} = 269 \text{ g } O_2 \qquad \textit{Answer}$$

Figure 10.3
Manganese(IV) oxide and hydrochloric acid react to produce chlorine (yellow-green gas). To do this reaction safely, it must be done in a fume hood.

EXAMPLE 10.2

Calculate the number of grams of chlorine molecules produced by the reaction of 22.1 g of manganese (IV) oxide with excess hydrochloric acid (see Figure 10.3).

$$MnO_2(s) + 4 HCl(aq) \longrightarrow MnCl_2(aq) + Cl_2(g) + 2 H_2O(l)$$

SOLUTION

Step I: Calculate the moles of manganese(IV) oxide. One mole of MnO_2 has a molar mass of 87.0 g.

$$22.1 \text{ g } MnO_2 \times \frac{1 \text{ mol } MnO_2}{87.0 \text{ g } MnO_2} = \frac{22.1}{87.0} \text{ mol } MnO_2$$

Step II: Calculate the moles of chlorine molecules produced. From the balanced equation the relation of MnO_2 to Cl_2 is given as 1 mol MnO_2 to 1 mol Cl_2.

$$\frac{22.1}{87.0} \text{ mol } MnO_2 \times \frac{1 \text{ mol } Cl_2}{1 \text{ mol } MnO_2} = \frac{22.1}{87.0} \times \frac{1}{1} \text{ mol } Cl_2 \text{ produced}$$

Step III: Calculate the grams of chlorine molecules produced. One mole of Cl_2 has a molar mass of 71.0 g.

$$\frac{22.1}{87.0} \times \frac{1}{1} \text{ mol } Cl_2 \times \frac{71.0 \text{ g } Cl_2}{1 \text{ mol } Cl_2} = \frac{22.1}{87.0} \times \frac{1}{1} \times 71.0 \text{ g } Cl_2$$

$$= 18.04 \text{ g } Cl_2 = 18.0 \text{ g } Cl_2 \quad \textit{Answer}$$

We express the answer to three significant digits (18.0 g Cl_2) because our given quantity (22.1 g of MnO_2) was expressed to three significant digits. This solution may also be expressed as follows:

$$\underbrace{22.1 \text{ g } MnO_2 \times \frac{1 \text{ mol } MnO_2}{87.0 \text{ g } MnO_2}}_{\text{Step I}} \times \underbrace{\frac{1 \text{ mol } Cl_2}{1 \text{ mol } MnO_2}}_{\text{Step II}} \times \underbrace{\frac{71.0 \text{ g } Cl_2}{1 \text{ mol } Cl_2}}_{\text{Step III}} = 18.0 \text{ g } Cl_2 \quad \textit{Answer}$$

Study Exercise 10.1

Calculate the number of grams of chromium that can be produced from the reaction of 45.6 g of chromium(III) oxide with excess aluminum according to the following balanced chemical reaction equation:

$$Cr_2O_3(s) + 2 Al(s) \longrightarrow 2 Cr(s) + Al_2O_3(s) \tag{31.2 g}$$

Work Problems 3 through 10.

Mass to Moles and Moles to Mass Examples

Sometimes the information to be determined *must be* expressed in moles and we do not need to calculate the mass in grams. At other times the information *is* expressed in moles and we do not need to calculate it. In these cases, step I or III may be eliminated. The following problems illustrate how one step may be eliminated.

EXAMPLE 10.3 Calculate the number of moles of oxygen produced by heating 1.65 g of potassium chlorate.

$$KClO_3(s) \xrightarrow{\Delta} KCl(s) + O_2(g) \quad \text{(unbalanced)}$$

SOLUTION We must balance the chemical equation before we can make any calculations. The balanced chemical reaction equation is

$$2 KClO_3(s) \xrightarrow{\Delta} 2 KCl(s) + 3 O_2(g)$$

The known quantity, 1.65 g of $KClO_3$, is given in grams. Therefore, we need step I to calculate moles of $KClO_3$. Step II converts moles of $KClO_3$ to moles of O_2, and hence we do *not* need step III. The molar mass of $KClO_3$ is 122.6 g, as calculated from the atomic masses.

The solution is expressed as follows:

$$\underbrace{1.65 \text{ g } KClO_3 \times \frac{1 \text{ mol } KClO_3}{122.6 \text{ g } KClO_3}}_{\text{Step I}} \times \underbrace{\frac{3 \text{ mol } O_2}{2 \text{ mol } KClO_3}}_{\text{Step II}} = 0.0202 \text{ mol } O_2 \quad \textit{Answer}$$

EXAMPLE 10.4 Calculate the number of grams of O_2 produced by heating 0.105 mol of $KClO_3$.

$$KClO_3(s) \xrightarrow{\Delta} KCl(s) + O_2(g) \quad \text{(unbalanced)}$$

SOLUTION   The balanced equation is

$$2 KClO_3(s) \xrightarrow{\Delta} 2 KCl(s) + 3 O_2(g)$$

The known quantity, 0.105 mol $KClO_3$, is given in moles. Therefore, we do not need step I. Step II converts moles of $KClO_3$ to moles of O_2. We are asked for grams of O_2, and so we will need step III. The molar mass of O_2 is 32.0 g.

The solution is expressed as follows:

$$0.105 \; \text{mol } KClO_3 \times \underbrace{\frac{3 \; \text{mol } O_2}{2 \; \text{mol } KClO_3}}_{\text{Step II}} \times \underbrace{\frac{32.0 \; \text{g } O_2}{1 \; \text{mol } O_2}}_{\text{Step III}} = 5.04 \; \text{g } O_2 \quad \textit{Answer}$$

Study Exercise 10.2
Calculate the number of moles of chromium that can be produced from the reaction of 28.5 g of chromium(III) oxide with excess aluminum according to the following balanced chemical reaction equation:

Work Problems 11 through 14.

$$Cr_2O_3(s) + 2 Al(s) \longrightarrow 2 Cr(s) + Al_2O_3(s) \qquad \text{(0.375 mol)}$$

Moles to Moles Examples

EXAMPLE 10.5 Consider the following balanced equation:

$$2 Na(s) + 2 H_2O(l) \longrightarrow 2 NaOH(aq) + H_2(g)$$

If 0.15 mol of Na atoms reacts with water, calculate the number of moles of H_2 molecules produced.

SOLUTION   We must carry the known only through step II to get the answer; hence, we may eliminate *both* steps I and III.

$$0.15 \; \text{mol } Na \times \underbrace{\frac{1 \; \text{mol } H_2}{2 \; \text{mol } Na}}_{\text{Step II}} = 0.075 \; \text{mol } H_2 \quad \textit{Answer}$$

> **Study Hint:** As you may have noticed from these examples, step I, step III, and *both* steps I and III may be eliminated. Step II can *never* be eliminated.

Study Exercise 10.3
Calculate the number of moles of chromium that can be produced from the reaction of 0.225 mol of chromium(III) oxide with excess aluminum according to the followed balanced chemical reaction equation:

Work Problems 15 through 18.

$$Cr_2O_3(s) + 2 Al(s) \longrightarrow 2 Cr(s) + Al_2O_3(s) \qquad \text{(0.450 mol)}$$

Limiting Reagent Examples

Suppose you want to use a candy machine that requires 1 quarter and 2 dimes for the purchase of 1 candy bar. If you have 7 quarters and 10 dimes, how many candy bars can you buy? Seven quarters allows you to purchase a maximum of 7 candy bars.

$$7 \text{ quarters} \times \frac{1 \text{ candy bar}}{1 \text{ quarter}} = 7 \text{ candy bars}$$

Ten dimes allow you to purchase a maximum of only 5 candy bars.

$$10 \text{ dimes} \times \frac{1 \text{ candy bar}}{2 \text{ dimes}} = 5 \text{ candy bars}$$

Thus, you may purchase only 5 candy bars and will have 2 quarters left over, as Figure 10.4 shows.

In much the same way, chemists can have "leftover" reactants when a chemical reaction between two elements or compounds uses up all of one reactant before it does another. In such cases, chemists refer to the reactant that is *completely used up* as the **limiting reagent** because the amount of that reagent limits the amount of a new compound that can be formed. The other (leftover) reactant is called the **excess reagent**. Chemical manufacturers make the *cheaper* reagent the *excess* reagent so as to be sure to use up the expensive reagent. The entire production is usually recycled in a continuous process.

Limiting reagent   Reactant in a chemical reaction that is completely used up in the reaction, so called because the amount of this reactant limits the amount of new compounds that can be formed.

Excess reagent   Reactant in a chemical reaction that is *not* completely used up in the reaction, so called because when the last trace of the new compound is formed, some of this reactant is left over.

EXAMPLE 10.6 A 50.0-g sample of calcium carbonate is allowed to react with 35.0 g of H_3PO_4. (a) How many grams of calcium phosphate can be produced? (b) Calculate the number of moles of excess reagent at the end of the reaction.

$$3 \, CaCO_3(s) \; + \; 2 \, H_3PO_4(aq) \; \longrightarrow \; Ca_3(PO_4)_2(s) \; + \; 3 \, CO_2(g) \; + \; 3 \, H_2O(l)$$

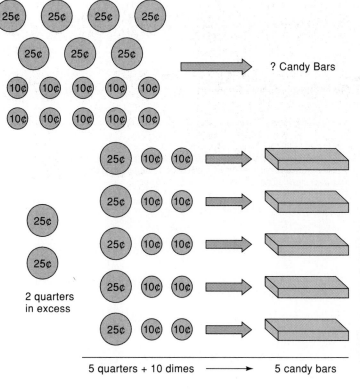

FIGURE 10.4
Limiting reagent analogy: You have 7 quarters and 10 dimes. How many candy bars can you buy from a machine that needs 1 quarter and 2 dimes for each candy bar?

Study Hint: An even simpler analogy is the gasoline in your car's gas tank. If it is 20 miles from your home to college and you know you have less than a gallon of gas in the tank, then the gasoline may *limit* your getting to your chemistry class on time!

> **Study Hint:** You may ask, How do I know this is a limiting reagent problem? The answer is that there are *two* given reactant quantities.

SOLUTION

a. We can calculate the molar masses of the substances involved using the atomic mass units: $CaCO_3 = 100.1$ g, $H_3PO_4 = 98.0$ g, and $Ca_3(PO_4)_2 = 31\overline{0}$ g. The question is, Which of the reactants, $CaCO_3$ or H_3PO_4, is the limiting reagent?

1. Calculate the moles of each used in step I.

$$50.0 \text{ g } CaCO_3 \times \frac{1 \text{ mol } CaCO_3}{100.1 \text{ g } CaCO_3} = 0.500 \text{ mol } CaCO_3$$

$$35.0 \text{ g } H_3PO_4 \times \frac{1 \text{ mol } H_3PO_4}{98.0 \text{ g } H_3PO_4} = 0.357 \text{ mol } H_3PO_4$$

2. Calculate the moles of product that can be produced from each reactant as in step II.

$$0.500 \text{ mol } CaCO_3 \times \frac{1 \text{ mol } Ca_3(PO_4)_2}{3 \text{ mol } CaCO_3} = 0.167 \text{ mol } Ca_3(PO_4)_2$$

$$0.357 \text{ mol } H_3PO_4 \times \frac{1 \text{ mol } Ca_3(PO_4)_2}{2 \text{ mol } H_3PO_4} = 0.178 \text{ mol } Ca_3(PO_4)_2$$

3. *The reactant that gives the **least** number of moles of the product is the **limiting** reagent.* Hence, in this example, $CaCO_3$ is the *limiting reagent* [0.167 mol versus 0.178 mol of $Ca_3(PO_4)_2$] and H_3PO_4 is the *excess reagent*. Thus, $CaCO_3$ is the reactant we use to solve part (a) of this example. Using $CaCO_3$, the number of grams of $Ca_3(PO_4)_2$ that can be produced is

$$0.167 \text{ mol } Ca_3(PO_4)_2 \times \frac{31\overline{0} \text{ g } Ca_3(PO_4)_2}{1 \text{ mol } Ca_3(PO_4)_2} = 51.8 \text{ g } Ca_3(PO_4)_2 \qquad Answer *$$

b. The amount of excess H_3PO_4 is equal to 0.357 mol of H_3PO_4 present at the start of the reaction (see step I) *minus* the amount consumed in the reaction between the H_3PO_4 and the limiting reagent ($CaCO_3$). The amount consumed is

$$0.500 \text{ mol } CaCO_3 \times \frac{2 \text{ mol } H_3PO_4}{3 \text{ mol } CaCO_3} = 0.333 \text{ } H_3PO_4$$

and the amount in excess is

$$0.357 \text{ mol } H_3PO_4 - 0.333 \text{ mol } H_3PO_4 = 0.024 \text{ mol } H_3PO_4 \text{ in excess} \qquad Answer$$

---

*An alternate acceptable solution is to use *both* given quantities of the reactants in *all* three steps. The *least* amount of product is the answer. The reactant that gives the least amount of product is the *limiting reagent*. The calculations are as follows:

$$50.0 \text{ g } CaCO_3 \times \frac{1 \text{ mol } CaCO_3}{100.1 \text{ g } CaCO_3} \times \frac{1 \text{ mol } Ca_3(PO_4)_2}{3 \text{ mol } CaCO_3}$$

$$\times \frac{31\overline{0} \text{ g } Ca_3(PO_4)_2}{1 \text{ mol } Ca_3(PO_4)_2} = 51.6 \text{ g } Ca_3(PO_4)_2 \qquad Answer$$

The limiting reagent is $CaCO_3$. (The difference between this answer and the above answer is in the rounding off of the above answer.)

$$35.0 \text{ g } H_3PO_4 \times \frac{1 \text{ mol } H_3PO_4}{98.0 \text{ g } H_3PO_4} \times \frac{1 \text{ } Ca_3(PO_4)_2}{2 \text{ mol } H_3PO_4}$$

$$\times \frac{31\overline{0} \text{ g } Ca_3(PO_4)_2}{1 \text{ mol } Ca_3(PO_4)_2} = 55.4 \text{ g } Ca_3(PO_4)_2$$

Percent Yields

The amount of product that we just calculated in Example 10.5 is called a theoretical yield. The **theoretical yield** is the amount of product expected if all the limiting reagent forms product and none of it is left. This assumes that none of the product is lost in isolation and purification. However, in real life, the theoretical yield is seldom the **actual yield**, the amount of product obtained. In organic reactions particularly, side reactions occur, giving minor products in addition to the major one. Also, some of the product is lost in the process of isolation and purification and in transferring it from one container to another. (In the chemical industry, this loss in isolation and purification is often minimized by a continuous process in which the materials lost in isolation and purification are recycled.) Thus, chemical manufacturers are generally concerned with the **percent yield**, the percent of the theoretical yield actually obtained. The percent yield is calculated as

$$\% \text{ yield} = \frac{\text{actual yield}}{\text{theoretical yield}} \times 100$$

Theoretical yield   Amount of product expected if all the limiting reagent forms product with none of it left. This assumes that none of the product is lost in isolation and purification.

Actual yield   Amount of product obtained in an actual chemical reaction; it is always less than the theoretical yield because of losses in the isolation and purification stages and the production of minor by-products.

Percent yield   Percent of the theoretical yield actually obtained; it is expressed as the actual yield divided by the theoretical yield, with the result multiplied by 100.

EXAMPLE 10.7 If 48.7 g of $Ca_3(PO_4)_2$ is actually obtained in Example 10.6, what is the percent yield?

SOLUTION

$$\frac{48.7 \text{ g } Ca_3(PO_4)_2 = \text{actual yield}}{51.8 \text{ g } Ca_3(PO_4)_2 = \text{theoretical yield}} \times 100 = 94.0\% \qquad \textit{Answer}$$

> **S**tudy Hint: The difference between a theoretical yield and an actual yield is similar to the difference between the amount you earn from a job and the amount you have left after you pay taxes!

EXAMPLE 10.8 A 20.2-g sample of calcium carbonate is allowed to react with 13.2 g of hydrochloric acid. Calculate (a) the number of grams of calcium chloride that can be produced, (b) the number of moles of excess reagent remaining at the end of the reaction, and (c) the percent yield if 18.3 g of calcium chloride is actually obtained (see Figure 10.5).

$$CaCO_3(s) + 2\,HCl(aq) \longrightarrow CaCl_2(aq) + CO_2(g) + H_2O(l)$$

SOLUTION
a. We can calculate the molar masses of the substances using the atomic mass units: $CaCO_3$ = 100.1 g, HCl = 36.5 g, and $CaCl_2$ = 111.1 g. Now, we must determine whether $CaCO_3$ or HCl is the limiting reagent.

1. Calculate the moles of each used in step I.

$$20.2 \text{ g } CaCO_3 \times \frac{1 \text{ mol } CaCO_3}{100.1 \text{ g } CaCO_3} = 0.202 \text{ mol } CaCO_3$$

$$13.2 \text{ g } HCl \times \frac{1 \text{ mol } HCl}{36.5 \text{ g } HCl} = 0.362 \text{ mol } HCl$$

FIGURE 10.5
Calcium carbonate and hydrochloric acid react to produce carbon dioxide gas.

2. Calculate the moles of product that can be produced from each reactant as in step II.

$$0.202 \text{ mol CaCO}_3 \times \frac{1 \text{ mol CaCl}_2}{1 \text{ mol CaCO}_3} = 0.202 \text{ mol CaCl}_2$$

$$0.362 \text{ mol HCl} \times \frac{1 \text{ mol CaCl}_2}{2 \text{ mol HCl}} = 0.181 \text{ mol CaCl}_2$$

3. Since the reactant that gives the *least* number of moles of product is the limiting reagent, HCl is the limiting reagent (0.181 mol versus 0.202 mol of $CaCl_2$) and $CaCO_3$ is the excess reagent. Using HCl, the number of grams of $CaCl_2$ that can be produced is

$$0.181 \text{ mol CaCl}_2 \times \frac{111.1 \text{ g CaCl}_2}{1 \text{ mol CaCl}_2} = 20.1 \text{ g CaCl}_2 \qquad Answer$$

b. The amount of excess $CaCO_3$ is equal to 0.202 mol of $CaCO_3$ present at the start of the reaction (step I) *minus* the amount consumed in the reaction between $CaCO_3$ and the limiting reagent (HCl). The amount consumed is

$$0.362 \text{ mol HCl} \times \frac{1 \text{ mol CaCO}_3}{2 \text{ mol HCl}} = 0.181 \text{ mol CaCO}_3$$

and the amount in excess is

$$0.202 \text{ mol CaCO}_3 - 0.181 \text{ mol CaCO}_3 = 0.021 \text{ mol CaCO}_3 \text{ in excess} \qquad Answer$$

c. Finally, we can calculate the percent yield from the theoretical yield and the actual yield as

$$\frac{18.3 \text{ g CaCl}_2}{20.1 \text{ g CaCl}_2} \times 100 = 91.0\% \qquad Answer$$

EXAMPLE 10.9   The industrial preparation of ethylene glycol, used as an automobile antifreeze and in the preparation of a polyester fiber, Dacron, involves the following reaction:

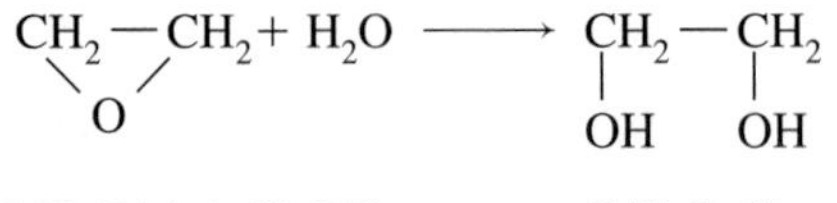

$$C_2H_4O(g) + H_2O(l) \longrightarrow C_2H_6O_2(l)$$

ethylene oxide   ethylene glycol

If 166 g of ethylene oxide is allowed to react with 75.0 g of water, calculate (a) the theoretical yield of ethylene glycol in grams, (b) the number of moles of excess reagent remaining at the end of the reaction, and (c) the percent yield if 215 g of ethylene glycol is actually obtained.

SOLUTION

a. We can calculate the molar masses of the substances using the atomic mass units: $C_2H_4O = 44.0$ g, $H_2O = 18.0$ g, and $C_2H_6O_2 = 62.0$ g. Now, we must determine whether ethylene oxide (C_2H_4O) or water (H_2O) is the limiting reagent.

1. Calculate the moles of each used in step I.

$$166 \text{ g } C_2H_4O \times \frac{1 \text{ mol } C_2H_4O}{44.0 \text{ g } C_2H_4O} = 3.7 \text{ mol } C_2H_4O$$

$$75.0 \text{ g } H_2O \times \frac{1 \text{ mol } H_2O}{18.0 \text{ g } H_2O} = 4.17 \text{ mol } H_2O$$

2. Calculate the moles of product that can be produced from each reactant as in step II.

$$3.77 \text{ mol } C_2H_4O \times \frac{1 \text{ mol } C_2H_6O_2}{1 \text{ mol } C_2H_4O} = 3.77 \text{ mol } C_2H_6O_2$$

$$4.17 \text{ mol } H_2O \times \frac{1 \text{ mol } C_2H_6O_2}{1 \text{ mol } H_2O} = 4.17 \text{ mol } C_2H_6O_2$$

3. Since the reactant that gives the *least* number of moles of product is the limiting reagent, ethylene oxide (C_2H_4O) is the limiting reagent (3.77 mol versus 4.17 mol of $C_2H_6O_2$, ethylene glycol) and water is the excess reagent. We should have expected this answer because water is considerably cheaper than any organic substance, such as ethylene oxide, that is derived from oil. Water is cheaper than oil! Using ethylene oxide, the number of grams of ethylene glycol that can be produced is

$$3.77 \text{ mol } C_2H_6O_2 \times \frac{62.0 \text{ g } C_2H_6O_2}{1 \text{ mol } C_2H_6O_2} = 234 \text{ g } C_2H_6O_2 \qquad Answer$$

b. The amount of excess water is equal to the 4.17 mol of H_2O at the start of the reaction (see step I) *minus* the amount consumed in the reaction between H_2O and the limiting reagent ethylene oxide. The amount consumed is

$$3.77 \text{ mol } C_2H_4O \times \frac{1 \text{ mol } H_2O}{1 \text{ mol } C_2H_4O} = 3.77 \text{ mol } H_2O$$

and so the amount in excess is

$$4.17 \text{ mol } H_2O - 3.77 \text{ mol } H_2O = 0.40 \text{ mol } H_2O \text{ in excess} \qquad Answer$$

c. Finally, we can calculate the percent yield from the theoretical yield and the actual yield as

$$\frac{215 \text{ g } C_2H_6O_2}{234 \text{ g } C_2H_6O_2} \times 100 = 91.9\% \qquad Answer$$

Study Exercise 10.4
A 16.5-g sample of chromium(III) oxide is allowed to react with 8.20 g of aluminum. Calculate (a) the number of grams of chromium that can be produced, (b) the number of moles of excess reagent remaining at the end of the reaction, and (c) the percent yield if 10.9 g of chromium is actually obtained.

$$Cr_2O_3(s) + 2 Al(s) \longrightarrow 2 Cr(s) + Al_2O_3(s)$$

[(a) 11.3 g; (b) 0.086 mol; (c) 96.5%]   Work Problems 19 through 22.

10.5 Mass–Volume Stoichiometry Problems

Next, let us consider mass–volume stoichiometry problems. Like mass–mass problems, mass–volume problems do not always require that we use steps I and/or III of the mole method for solving stoichiometry problems. Also, like mass–mass problems, mass–volume problems sometimes involve limiting reagents. Unlike mass–mass problems, however, in mass–volume problems, *either* the known *or* the unknown is a *gas*. The known may be given in mass units and we will be asked to calculate the unknown in volume units (if it is a gas), or the known may be given in volume units (if it is a gas) and we will be asked to calculate the unknown in mass units. In either case, we need to apply the molar volume: 22.4 L per mol of any gas at STP [0°C and 76$\overline{0}$ mm Hg (torr)], discussed in Section 8.3. (Volumes of gases at non-STP conditions are discussed in Chapter 11.)

Mass to Volume and Volume to Mass Examples

EXAMPLE 10.10   Calculate the volume in liters of oxygen gas measured at 0°C and 76$\overline{0}$ mm Hg that can be obtained by heating 28.0 g of potassium nitrate.

$$KNO_3(s) \xrightarrow{\Delta} KNO_2(s) + O_2(g) \quad \text{(unbalanced)}$$

SOLUTION   First, we must balance the equation.

$$2\,KNO_3(s) \xrightarrow{\Delta} 2\,KNO_2(s) + O_2(g)$$

Then we can calculate the molar mass of KNO_3 as 101.1 g from the atomic mass units. The conditions of 0°C and 76$\overline{0}$ mm Hg are STP conditions. Hence, in step III we must use the relation 1 mol of O_2 *molecules* at STP occupies 22.4 L.

$$28.0 \text{ g } KNO_3 \times \frac{1 \text{ mol } KNO_3}{101.1 \text{ g } KNO_3} \times \frac{1 \text{ mol } O_2}{2 \text{ mol } KNO_3} \times \frac{22.4 \text{ L } O_2 \text{ STP}}{1 \text{ mol } O_2} = 3.10 \text{ L } O_2 \text{ STP} \quad \textit{Answer}$$

$$\underbrace{\hspace{5cm}}_{\text{Step I}} \quad \underbrace{\hspace{2cm}}_{\text{Step II}} \quad \underbrace{\hspace{2cm}}_{\text{Step III}}$$

Study Exercise 10.5
Calculate the number of liters of oxygen gas (STP) needed to react with 8.95 g of chromium to prepare chromium(III) oxide, a green oxide used as a pigment. The balanced chemical reaction equation is

Work Problems 23 through 26. $4\,Cr(s) + 3\,O_2(g) \xrightarrow{\Delta} 2\,Cr_2O_3(s)$ (2.89 L)

Moles to Volume and Volume to Moles Examples

EXAMPLE 10.11   Calculate the number of liters of O_2 (at STP) produced by heating 0.480 mol of $KClO_3$.

$$KClO_3(s) \xrightarrow{\Delta} KCl(s) + O_2(s) \quad \text{(unbalanced)}$$

SOLUTION   First, balance the equation.

$$2\,KClO_3(s) \xrightarrow{\Delta} 2\,KCl(s) + 3\,O_2(s)$$

The conditions given are STP. Hence, we use the relation 1 mol O_2 molecules at STP occupies 22.4 L. (Note that step I is not needed.)

$$0.480 \text{ mol KClO}_3 \times \frac{3 \text{ mol } O_2}{2 \text{ mol KClO}_3} \times \frac{22.4 \text{ L } O_2 \text{ STP}}{1 \text{ mol } O_2} = 16.1 \text{ L } O_2 \text{ STP} \quad \textit{Answer}$$

$$\underbrace{\qquad\qquad\qquad}_{\text{Step II}} \qquad \underbrace{\qquad\qquad\qquad\qquad\qquad}_{\text{Step III}}$$

EXAMPLE 10.12 Calculate the number of moles of Cu produced if 4200 mL of H_2 measured at 0°C and $76\overline{0}$ torr reacts with an excess of CuO.

$$CuO(s) + H_2(g) \xrightarrow{\Delta} Cu(s) + H_2O(l)$$

SOLUTION The conditions 0°C and $76\overline{0}$ torr are STP conditions. Hence, in step I, we use the relation 1 mol of a gas at STP occupies 22.4 L. (Note that step III is not needed.)

$$4200 \text{ mL } H_2 \text{ STP} \times \frac{1 \text{ L}}{1000 \text{ mL}} \times \frac{1 \text{ mol } H_2 \text{ STP}}{22.4 \text{ L } H_2 \text{ STP}} \times \frac{1 \text{ mol Cu}}{1 \text{ mol } H_2 \text{ STP}} = 0.19 \text{ mol Cu} \quad \textit{Answer}$$

$$\underbrace{\qquad\qquad\qquad\qquad\qquad\qquad}_{\text{Step I}} \qquad \underbrace{\qquad\qquad\qquad}_{\text{Step II}}$$

Study Exercise 10.6

Calculate the number of liters of oxygen gas (STP) needed to react with 0.0650 mol of chromium to prepare chromium(III) oxide. The balanced chemical reaction equation is

$$4 \text{ Cr}(s) + 3 O_2(g) \xrightarrow{\Delta} 2 \text{ Cr}_2O_3(s) \qquad\qquad (1.09 \text{ L}) \qquad \text{Work Problems 27 and 28.}$$

Limiting Reagent Example

EXAMPLE 10.13 A 28.0-g sample of zinc reacts with 75.0 g of sulfuric acid. (a) How many liters of hydrogen measured at STP can be produced? (b) Calculate the number of moles of excess reagent remaining at the end of the reaction (see Figure 10.6).

$$Zn(s) + H_2SO_4(aq) \longrightarrow ZnSO_4(aq) + H_2(g)$$

SOLUTION

a. Determine the limiting reagent. The molar masses of Zn and H_2SO_4 are 65.4 g and 98.1 g, respectively.

1. Calculate the moles of each used in step I.

$$28.0 \text{ g Zn} \times \frac{1 \text{ mol Zn}}{65.4 \text{ g Zn}} = 0.428 \text{ mol Zn}$$

$$75.0 \text{ g } H_2SO_4 \times \frac{1 \text{ mol } H_2SO_4}{98.1 \text{ g } H_2SO_4} = 0.765 \text{ mol } H_2SO_4$$

(a)

(b)

FIGURE 10.6
Zinc and an acid: (a) before addition of the acid (b) after the addition of the acid showing the hydrogen gas.

2. Calculate the moles of product that can be obtained from each reactant as in step II.

$$0.428 \ \cancel{\text{mol Zn}} \times \frac{1 \ \text{mol H}_2}{1 \ \cancel{\text{mol Zn}}} = 0.428 \ \text{mol H}_2$$

$$0.765 \ \cancel{\text{mol H}_2\text{SO}_4} \times \frac{1 \ \text{mol H}_2}{1 \ \cancel{\text{mol H}_2\text{SO}_4}} = 0.765 \ \text{mol H}_2$$

3. The reactant that gives the least number of moles of product is the limiting reagent, in this case Zn.

$$0.428 \ \cancel{\text{mol H}_2} \times \frac{22.4 \ \text{L H}_2 \ \text{STP}}{1 \ \cancel{\text{mol H}_2}} = 9.59 \ \text{L H}_2 \ \text{STP} \qquad \textit{Answer}$$

b. The amount of excess H_2SO_4 is equal to 0.765 mol of H_2SO_4 present at the start of the reaction (see step I) *minus* the amount consumed by the reaction between H_2SO_4 and the limiting reagent (Zn). The amount consumed is

$$0.428 \ \cancel{\text{mol Zn}} \times \frac{1 \ \text{mol H}_2\text{SO}_4}{1 \ \cancel{\text{mol Zn}}} = 0.428 \ \text{mol H}_2\text{SO}_4$$

and the amount in excess is

$$0.765 \ \text{mol H}_2\text{SO}_4 - 0.428 \ \text{mol H}_2\text{SO}_4 = 0.337 \ \text{mol H}_2\text{SO}_4 \qquad \textit{Answer}$$

Study Exercise 10.7
A 3.50-g sample of chromium reacts with 1.25 L of oxygen gas (STP). (a) How many grams of chromium(III) oxide can be produced? (b) Calculate the number of moles of excess reagent remaining at the end of the reaction. The balanced chemical reaction equation is

Work Problems 29 and 30.

$$4 \ \text{Cr}(s) + 3 \ \text{O}_2(g) \xrightarrow{\Delta} 2 \ \text{Cr}_2\text{O}_3(s) \qquad\qquad \text{[(a) 5.11 g; (b) 0.0053 mol]}$$

10.6 Volume–Volume Stoichiometry Problems

Gay-Lussac's law of combining volumes   Principle stating that whenever gases react or are formed, their volumes are in small whole-number ratios, provided these are measured at the same temperature and pressure. The ratio of volumes for such a reaction is directly proportional to the values of the coefficients in the balanced equation.

About the time that Dalton was developing his atomic theory (see Section 4.2), the French chemist and physicist Joseph Louis Gay-Lussac (gā′lü·sak′) (1778–1850) was studying the chemical combination of gases. He found experimentally that when gases reacted, the ratios of their volumes were small whole numbers. This was true as long as the volumes were measured at the same temperature and pressure. His results are stated in **Gay-Lussac's law of combining volumes**: Whenever gases react or are formed, their volumes are in small whole-number ratios, provided these are measured at the same temperature and pressure. The ratio of volumes for such a reaction is directly proportional to the values of the coefficients in the balanced equation. This is the same principle we applied in mass–mass problems except that here we use volumes instead of moles and *all substances are gases* and measured at the *same temperature and pressure*. The condition that all the gases must be at the same temperature and pressure is extremely critical since, as we will see in Chapter 11, the volume occupied by a gas is affected by changes in the temperature and pressure. Volume–volume stoichiometry problems are based on Gay-Lussac's law of combining volumes.

For example, in the reaction

$$CH_4(g) + 2\,O_2(g) \xrightarrow{\Delta} CO_2(g) + 2\,H_2O(g)$$

all compounds are in the gaseous state and at the same temperature and pressure. One (1) volume of CH_4 gas (methane) reacts with two (2) volumes of O_2 gas to form one (1) volume of CO_2 gas and two (2) volumes of H_2O vapor. If we measure these volumes all at STP and assume that they *all* remain gases at STP, we can state that 1 mol (22.4 L) of CH_4 gas reacts with 2 mol (44.8 L) O_2 to form 1 mol (22.4 L) of CO_2 gas and 2 mol (44.8 L) of H_2O vapor. This relationship is illustrated in Figure 10.7.

Solving volume–volume stoichiometry problems is both similar to and different from solving mass–mass and mass–volume problems. Like those earlier problem types, volume–volume problems can (although they do not always) involve limiting reagents. But unlike mass–mass and mass–volume problems, volume–volume problems never use step I or III of the mole method. Step II alone is sufficient.

Volume to Volume Example

EXAMPLE 10.14	Calculate the volume of O_2 in liters required and the volume of CO_2 and H_2O in liters formed from the complete combustion of 1.50 L of C_2H_6, all volumes being measured at $4\overline{0}0$ °C and $76\overline{0}$ mm Hg pressure.

> **Study Hint:** Note that the ratio of *volumes* is always the same, 1:2:1:2 for CH_4:O_2:CO_2:H_2O, respectively.

$$2\,C_2H_6(g) + 7\,O_2(g) \xrightarrow{\Delta} 4\,CO_2(g) + 6\,H_2O(g)$$
ethane

SOLUTION All these substances are gases measured at the same temperature and pressure, and so their volumes are related to their coefficients in the balanced equation.

$$1.50\,\text{L C}_2\text{H}_6 \times \frac{7\,\text{L O}_2}{2\,\text{L C}_2\text{H}_6} = 5.25\,\text{L O}_2 \qquad Answer$$

Step II

$$1.50\,\text{L C}_2\text{H}_6 \times \frac{4\,\text{L CO}_2}{2\,\text{L C}_2\text{H}_6} = 3.00\,\text{L CO}_2 \qquad Answer$$

Step II

$$1.50\,\text{L C}_2\text{H}_6 \times \frac{6\,\text{L H}_2\text{O}(g)}{2\,\text{L C}_2\text{H}_6} = 4.50\,\text{L H}_2\text{O}(g) \qquad Answer$$

Step II

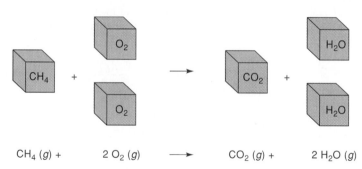

$$CH_4\,(g) + \qquad 2\,O_2\,(g) \longrightarrow CO_2\,(g) + \qquad 2\,H_2O\,(g)$$

FIGURE 10.7
Gay-Lussac's law of combining volumes: 1 volume of methane (CH_4) reacts with 2 volumes of oxygen gas (O_2) to give 1 volume of carbon dioxide (CO_2) and 2 volumes of water vapor (H_2O). All substances are gases, and the volumes are measured at the same temperature and pressure.

Study Exercise 10.8

Calculate the number of liters of carbon dioxide gas that can be produced from 4.85 L of oxygen gas when both gases are measured at the same temperature and pressure.

Work Problems 31 through 34.

$$2 \, CO(g) \; + \; O_2(g) \xrightarrow{\Delta} 2 \, CO_2(g)$$ (9.70 L)

Limiting Reagent Example

EXAMPLE 10.15 Methyl alcohol (methanol, CH_3OH) is being examined as an alternative to gasoline as a fuel for automobiles. One commercial preparation of methyl alcohol involves the reaction of carbon monoxide with hydrogen at 350° to 400°C and 3000 lb/in.^{2} pressure in the presence of metallic oxides such as a chromium(III) oxide–zinc oxide mixture.

$$CO(g) \; + \; 2 \, H_2(g) \xrightarrow[\Delta, \, P]{Cr_2O_3 - ZnO} CH_3OH(g)$$
$$\underset{\text{alcohol}}{\underset{\text{methyl}}{}}$$

If 60.0 L of CO is allowed to react with 80.0 L of H_2 in a sealed container, calculate (a) the number of liters of $CH_3OH(g)$ that can be produced, and (b) the number of liters of both CO and H_2 that would remain at the end of the reaction, assuming that all volumes are measured at the same temperature and pressure.

SOLUTION

a. The first part involves determining the limiting reagent. All these substances are gases at the same temperature and pressure, and so their volumes are related to their coefficients in the balanced equation. Thus, we can calculate directly the volume in liters of $CH_3OH(g)$ that can be produced. Using 60.0 L of CO, we have

$$60.0 \; \cancel{L \, CO(g)} \times \frac{1 \, L \, CH_3OH(g)}{1 \; \cancel{L \, CO(g)}} = 60.0 \, L \, CH_3OH(g)$$

Using 80.0 L of H_2, we have

$$80.0 \; \cancel{L \, H_2(g)} \times \frac{1 \, L \, CH_3OH(g)}{2 \; \cancel{L \, H_2(g)}} = 40.0 \, L \, CH_3OH(g) \qquad \textit{Answer}$$

The reactant that gives the least amount of product, 40.0 L $CH_3OH(g)$, is the *limiting reagent*, in this case H_2.

b. If H_2 is the limiting reagent, then CO is the excess reagent. The amount of excess CO is equal to 60.0 L of CO present at the start of the reaction *minus* the amount that is consumed by the reaction between CO and the limiting reagent (H_2). The amount consumed is

$$80.0 \; \cancel{L \, H_2(g)} \times \frac{1 \, L \, CO(g)}{2 \; \cancel{L \, H_2(g)}} = 40.0 \, L \, CO$$

and

$$60.0 \, L \, CO \; - \; 40.0 \, L \, CO \; = \; 20.0 \, L \, CO \text{ in excess}$$

$$0 \, L \, H_2 \qquad \textit{Answers}$$

Study Exercise 10.9

Consider the following balanced chemical reaction equation:

$$2\,CO(g) + O_2(g) \xrightarrow{\Delta} 2\,CO_2(g)$$

(a) Calculate the number of liters of carbon dioxide gas that can be produced when 6.00 L of carbon monoxide gas reacts with 5.00 L of oxygen gas. All gases are measured at the same temperature and pressure. (b) Calculate the number of liters of excess reagent remaining at the end of the reaction.

[(a) 6.00 L; (b) 2.00 L]   Work Problems 35 and 36.

10.7 Heats in Chemical Reactions

Besides the mass–mass, mass–volume, and volume–volume relations just outlined, energy relationships are also important in chemical reactions. The energy involved is usually observed as heat and is expressed as the heat of reaction. The **heat of reaction** is the number of calories or joules of heat energy given off (evolved) or absorbed in a particular chemical reaction for a given amount of reactants and/or products. In **exothermic reactions**, heat energy is *given off* (evolved), and in **endothermic reactions**, heat energy is *absorbed*. Heat energy is given off as a *product* in *exothermic* reactions, while heat energy is absorbed and serves as a *reactant* in *endothermic* reactions.

An example of an exothermic change is the combination of 2 mol of hydrogen gas with 1 mol of oxygen, forming 2 mol of water (liquid) and *evolving* 5.73×10^5 J (1.37×10^5 cal) of heat energy at 25°C. Thus, for this exothermic reaction, the heat of reaction is 5.73×10^5 J (1.37×10^5 cal) for the formation of 2 mol of liquid water or 2.87×10^5 J (6.85×10^4 cal) for 1 mol of liquid water. Note that the heat appears as a *product* on the *right* side of the reaction equation.

$$2\,H_2(g) + O_2(g) \longrightarrow 2\,H_2O(l) + 5.73 \times 10^5\,J\,(1.37 \times 10^5\,cal)\ \text{at }25°C$$

Two common examples of exothermic reactions that you may already have discovered in the laboratory are the preparation of dilute solutions of acids (by adding concentrated sulfuric acid to water) and bases (by adding sodium hydroxide pellets to water). In both cases you may have noticed that the flask got warm. Also, when sulfuric acid and sodium hydroxide react in a neutralization reaction (see Section 9.10), an exothermic reaction occurs. The flask gets warm—heat is evolved.

An example of an endothermic reaction is the combination of 1 mol of hydrogen gas with 1 mol of iodine gas, forming 2 mol of gaseous hydrogen iodide and absorbing 5.19×10^4 J (1.24×10^4 cal) of heat energy at 25°C. Thus, for this endothermic process, the heat of reaction is 5.19×10^4 J (1.24×10^4 cal) absorbed during the formation of 2 mol of gaseous hydrogen iodide, or 2.60×10^4 J (6.20×10^3 cal) absorbed for 1 mol of HI.

$$5.19 \times 10^4\,J\,(1.24 \times 10^4\,cal) + H_2(g) + I_2(g) \rightleftharpoons 2\,HI(g)\ \text{at }25°C$$

Note that heat energy is a reactant and appears on the *left* side of the reaction equation. The $\rightleftharpoons$ that separates the reactants and products indicates that this reaction is reversible. A **reversible reaction** is never complete because the products of the reaction also react to reform the original reactant. No matter how long you wait, some starting material will always remain. We will discuss these types of reactions in more detail in Chapter 17.

Heat of reaction Number of calories or joules of heat energy given off (evolved) or absorbed in a particular chemical reaction for a given amount of reactants or products.

Exothermic reaction Any chemical reaction in which heat energy is given off (evolved).

Endothermic reaction Any chemical reaction in which heat energy is absorbed.

Reversible reaction Chemical reaction that is never complete because the products of the reaction also react to reform the original reactants. When the reaction is finished, there are *both* reactants and products present.

Study Hint: We do not have to actively apply heat to an endothermic process. It may occur by absorbing heat from the water or air, which explains why the cold pack in Figure 10.8 gets quite cold.

Some very simple processes can be endothermic. For example, when certain salts dissolve in water, heat energy is absorbed and the resulting solutions become quite cold. Potassium iodide (KI) and ammonium nitrate (NH_4NO_3) are examples of this kind of salt. Figure 10.8 illustrates a practical application of this process.

We can use this heat energy in stoichiometric calculations. The quantity of heat energy, either exothermic or endothermic, is related to the moles of reactants or products in the balanced equation. Thus, we simply use the heat of the reaction as we used moles in step II of our three basic steps.

EXAMPLE 10.16 Natural gas (CH_4) burns in air to produce carbon dioxide, water vapor, and heat energy. Calculate the number of kilocalories of heat energy produced by the burning of 25.0 g of natural gas according to the following balanced chemical reaction equation:

$$CH_4(g) \ + \ 2\,O_2(g) \ \longrightarrow \ CO_2(g) \ + \ 2\,H_2O(g) \ + \ 213 \text{ kcal at } 25°C$$

The molar mass of CH_4 is 16.0 g. The relationship between methane and the heat of reaction is 1 mol of CH_4 to 213 kcal. We therefore solve this problem by using steps I and II of the mole method.

$$25.0 \text{ g } CH_4 \ \times \ \underbrace{\frac{1 \text{ mol } CH_4}{16.0 \text{ g } CH_4}}_{\text{Step I}} \ \times \ \underbrace{\frac{213 \text{ kcal}}{1 \text{ mol } CH_4}}_{\text{Step II}} \ = \ 333 \text{ kcal} \qquad \textit{Answer}$$

FIGURE 10.8
Endothermic process. When a chemical cold pack is twisted, an inner membrane is broken to allow mixing of water with ammonium nitrate (NH_4NO_3). This dissolving is an endothermic process. As heat is absorbed from the water, the pack grows cold.

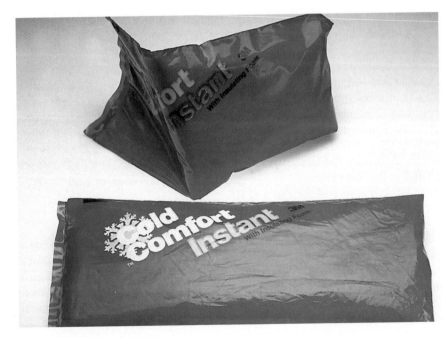

EXAMPLE 10.17 | Calculate the number of grams of hydrogen gas that must be burned to produce 3.1×10^6 J of heat energy according to the following balanced chemical reaction equation:

$$2\,H_2(g) + O_2(g) \longrightarrow 2\,H_2O(l) + 5.73 \times 10^5 \text{ J at } 25°C$$

SOLUTION   The molar mass of H_2 is 2.0 g. The relationship between hydrogen and the heat of reaction is 2 mol of H_2 to 5.73×10^5 J. We therefore solve this problem by using steps II and III of the mole method:

$$3.1 \times 10^6 \text{ J} \times \underbrace{\frac{2 \text{ mol } H_2}{5.73 \times 10^5 \text{ J}}}_{\text{Step II}} \times \underbrace{\frac{2.0 \text{ g } H_2}{1 \text{ mol } H_2}}_{\text{Step III}} = 22 \text{ g } H_2 \qquad \textit{Answer}$$

Study Exercise 10.10
Calculate the number of kilojoules of heat energy produced by the burning of 8.75 g of carbon monoxide according to the following balanced chemical reaction equation:

$$2\,CO(g) + O_2(g) \xrightarrow{\Delta} 2\,CO_2(g) + 566 \text{ kJ}$$ (88.4 kJ)   Work Problems 37 through 39.

 Summary

Stoichiometry **is the measurement of the relative quantities of chemical reactants and products in a chemical reaction. A stoichiometric calculation requires that you have a** *balanced chemical equation.* **The coefficients in the balanced chemical equation describe the molar relationships among the reactants and the products (Section 10.1).**

A three-step method based on the *mole* **and the** *dimensional analysis* **method of problem solving is useful for doing many types of stoichiometric calculations (Section 10.2). Stoichiometry problems fall into three basic types: mass–mass problems, mass–volume problems, and volume–volume problems (Section 10.3).**

Problems relating the masses of reactants and products (mass–mass problems) include *limiting reagent* **problems, which raise issues of** *theoretical yield, actual yield,* **and** *percent yield.* **Limiting reagent problems involve situations where one substance is "used up" and another is "leftover" when a reaction is completed. The theoretical yield is the amount of product obtained when we assume that all the limiting reagent forms products and that none of the product is lost. The actual yield takes into account real-world effects that produce somewhat different results. The percent yield is the percent of the theoretical yield actually obtained (Section 10.4).**

In mass–volume problems, one of the reactants or products is a gas. Such problems relate the mass of one reactant or product to the volume of another reactant or product. Limiting reagents are a factor in many mass–volume problems (Section 10.5).

Similarly, volume–volume problems relate the volumes of reactants or products to each other. In such problems, the coefficients in the balanced equation describe the volume relationships among gaseous reactants and products *for gases measured at the same temperature and pressure.* **Limiting reagents are also a factor in many volume–volume problems (Section 10.6).**

The *heat of reaction* **is the amount of energy involved in a chemical reaction. In** *exothermic* **reactions, heat energy is released as the products form, whereas in**

endothermic reactions, heat energy is absorbed as the reaction progresses. The amount of heat involved in a reaction can be related to the amounts of reactants consumed or products formed (Section 10.7).

 # Exercises

1. Define or explain the following terms (the number in parentheses refers to the section in the text where the term is mentioned):

   a. stoichiometry (Introduction)

   b. limiting reagent (10.4)

   c. excess reagent (10.4)

   d. theoretical yield (10.4)

   e. actual yield (10.4)

   f. percent yield (10.4)

   g. Gay-Lussac's law of combining volumes (10.6)

   h. heat of reaction (10.7)

   i. exothermic reaction (10.7)

   j. endothermic reaction (10.7)

   k. reversible reaction (10.7)

2. Distinguish between:

   a. theoretical and actual yields

   b. limiting and excess reagents

   c. exothermic and endothermic reactions

 # Problems

(*Hints:* Check each chemical reaction equation to make sure it is balanced; if it is not, balance it. For those questions in which an equation is not given, see if you can write one. See Sections 9.8 through 9.10 on how to complete chemical reaction equations.)

Mass–Mass Stoichiometry Problems (See Section 10.4)

Mass to Mass Problems

3. Calculate the number of grams of zinc chloride that can be prepared from 26.5 g of zinc.

$$Zn(s) + 2\,HCl(aq) \longrightarrow ZnCl_2(aq) + H_2(g)$$

4. Calculate the number of grams of hydrogen that can be produced from 5.40 g of aluminum.

$$2\,Al(s) + 6\,NaOH(aq) \longrightarrow 2\,Na_3AlO_3(aq) + 3\,H_2(g)$$

5. How many grams of silver chloride can be prepared from 6.20 g of silver nitrate?

$$AgNO_3(aq) + NaCl(aq) \longrightarrow AgCl(s) + NaNO_3(aq)$$

6. How many kilograms of iron(III) oxide can be obtained by heating 865 g of iron(II) sulfide with excess oxygen gas?

$$4\,FeS(s) + 7\,O_2(g) \longrightarrow 2\,Fe_2O_3(s) + 4\,SO_2(g)$$

7. Sodium hydroxide (5.00 g) is neutralized with sulfuric acid. How many grams of sodium sulfate can be formed?

$$2\,NaOH(aq) + H_2SO_4(aq) \longrightarrow Na_2SO_4(aq) + 2\,H_2O(l)$$

8. How many kilograms of hydrogen sulfide can be prepared by treating 625 g of iron(II) sulfide with an excess of hydrochloric acid?

$$FeS(s) + HCl(aq) \longrightarrow FeCl_2(aq) + H_2S(g) \quad \text{(unbalanced)}$$

9. Calculate the number of grams of potassium nitrate necessary to produce 2.10 g of oxygen.

$$KNO_3(s) \xrightarrow{\Delta} KNO_2(s) + O_2(g) \quad \text{(unbalanced)}$$

10. Calculate the number of grams of oxygen that can be produced by heating 3.50 g of potassium chlorate.

$$KClO_3(s) \xrightarrow{\Delta} KCl(s) + O_2(g) \quad \text{(unbalanced)}$$

Mass to Moles and Moles to Mass Problems

11. Calculate the number of moles of barium sulfate that can be prepared from 42.0 g of barium chloride.

$$BaCl_2(aq) + Na_2SO_4(aq) \longrightarrow BaSO_4(s) + 2\,NaCl(aq)$$

12. Calculate the number of moles of calcium chloride that would be necessary to prepare 67.0 g of calcium phosphate.

$$3\,CaCl_2(aq) + 2\,Na_3PO_4(aq) \longrightarrow Ca_3(PO_4)_2(s) + 6\,NaCl(aq)$$

13. Calculate the number of grams of carbon dioxide produced from the burning of 1.25 mol of propane (C_3H_8).

$$C_3H_8(g) + 5\,O_2(g) \xrightarrow{\Delta} 3\,CO_2(g) + 4\,H_2O(g)$$

14. Calculate the number of grams of water that can be produced from the burning of 0.650 mol of ethane (C_2H_6).

$$C_2H_6(g) + O_2(g) \xrightarrow{\Delta} CO_2(g) + H_2O(g) \quad \text{(unbalanced)}$$

Moles to Moles Problems

15. Sodium chloride (0.325 mol) is allowed to react with an excess of sulfuric acid. How many moles of hydrogen chloride can be formed?

$$2\,NaCl(aq) + H_2SO_4(aq) \longrightarrow Na_2SO_4(aq) + 2\,HCl(g)$$

16. If 0.350 mol of barium nitrate is allowed to react with an excess of phosphoric acid, how many moles of barium phosphate will be formed?

$$3\,Ba(NO_3)_2(aq) + 2\,H_3PO_4(aq) \longrightarrow Ba_3(PO_4)_2(s) + 6\,HNO_3(aq)$$

17. How many moles of hydrogen molecules can be produced by the reaction of 2.10 mol of sodium atoms with water?

$$2\,Na(s) + 2\,H_2O(l) \longrightarrow 2\,NaOH(aq) + H_2(g)$$

18. How many moles of HI are necessary to produce 0.250 mol of iodine according to the following balanced chemical reaction equation?

$$10\,HI(aq) + 2\,KMnO_4(aq) + 3\,H_2SO_4(aq) \longrightarrow$$
$$5\,I_2(s) + MnSO_4(aq) + K_2SO_4(aq) + 8\,H_2O(l)$$

Limiting Reagent Problems

19. A 36.0-g sample of calcium hydroxide is allowed to react with a 40.5-g sample of phosphoric acid.

   a. How many grams of calcium phosphate can be produced?

   b. If 45.2 g of calcium phosphate is actually obtained, what is the percent yield?

   $$3\ Ca(OH)_2(s)\ +\ 2\ H_3PO_4(aq)\ \longrightarrow\ Ca_3(PO_4)_2(s)\ +\ 6\ H_2O(l)$$

20. Copper(II) sulfide (0.600 mol) is treated with 1.40 mol of nitric acid.

   a. How many moles of copper(II) nitrate can be produced?

   b. If 0.500 mol of copper(II) nitrate is actually obtained, what is the percent yield?

   c. Calculate the number of moles of excess reagent remaining at the end of the reaction.

   $$3\ CuS(s)\ +\ 8\ HNO_3(aq)\ \longrightarrow$$
   $$3\ Cu(NO_3)_2(aq)\ +\ 3\ S(s)\ +\ 2\ NO(g)\ +\ 4\ H_2O(l)$$

21. A 1.4-g sample of magnesium is treated with 8.1 g of sulfuric acid.

   a. How many grams of hydrogen can be produced?

   b. If 0.060 g of hydrogen is actually obtained, what is the percent yield?

   c. Calculate the number of moles of excess reagent remaining at the end of the reaction.

   $$Mg(s)\ +\ H_2SO_4(aq)\ \longrightarrow\ MgSO_4(aq)\ +\ H_2(g)$$

22. Iron(II) hydroxide (0.320 mol) is treated with 0.250 mol of phosphoric acid.

   a. How many grams of iron(II) phosphate can be produced?

   b. If 34.0 g of iron(II) phosphate is actually obtained, what is the percent yield?

   c. Calculate the number of moles of excess reagent remaining at the end of the reaction. (See Section 9.10 on how to complete and balance this equation.)

Mass–Volume Stoichiometry Problems (See Section 10.5)

Mass to Volume and Volume to Mass Problems

23. How many liters of hydrogen sulfide measured at STP can be prepared from 4.00 g of iron(II) sulfide?

   $$FeS(s)\ +\ 2\ HCl(aq)\ \longrightarrow\ FeCl_2(aq)\ +\ H_2S(g)$$

24. Calculate the number of liters of hydrogen gas at STP that can be produced by the reaction of 5.40 g of magnesium with excess hydrochloric acid.

   $$Mg(s)\ +\ 2\ HCl(aq)\ \longrightarrow\ MgCl_2(aq)\ +\ H_2(g)$$

25. How many liters of oxygen measured at STP can be obtained by heating 0.700 g of potassium chlorate?

   $$KClO_3(s)\ \xrightarrow{\Delta}\ KCl(s)\ +\ O_2(g)\quad\text{(unbalanced)}$$

26. Calculate the number of grams of magnesium nitride necessary to produce 2.45 L of ammonia gas at STP. How many moles of magnesium hydroxide can also be formed?

   $$Mg_3N_2(s)\ +\ 6\ H_2O(l)\ \longrightarrow\ 3\ Mg(OH)_2(aq)\ +\ 2\ NH_3(g)$$

Moles to Volume and Volume to Moles Problems

27. Calculate the number of liters of hydrogen, measured at STP, that can be produced from the reaction of 0.275 mol of aluminum according to the following unbalanced chemical reaction equation:

$$Al(s) + NaOH(aq) + H_2O(l) \longrightarrow NaAlO_2(aq) + H_2(g) \quad \text{(unbalanced)}$$

28. How many moles of potassium chlorate can be produced from 1.65 L of chlorine gas at STP?

$$3\,Cl_2(g) + 6\,KOH(aq) \longrightarrow 5\,KCl(aq) + KClO_3(aq) + 3\,H_2O(l)$$

Limiting Reagent Problems

29. A 46.0-g sample of iron is allowed to react with 66.0 g of sulfuric acid.

  a. How many liters of hydrogen measured at STP can be produced?

  b. Calculate the number of moles of excess reagent remaining at the end of the reaction.

  The balanced chemical reaction equation is

$$Fe(s) + H_2SO_4(aq) \longrightarrow FeSO_4(aq) + H_2(g)$$

30. A 68.0-g sample of bismuth(III) nitrate is treated with 8.00 L of hydrogen sulfide at STP.

  a. How many grams of bismuth(III) sulfide can be produced?

  b. Calculate the number of moles of excess reagent remaining at the end of the reaction. (See Section 9.9 on how to complete and balance this equation.)

Volume–Volume Stoichiometry Problems (See Section 10.6)

Volume to Volume Problems

31. Calculate the number of liters of nitrogen gas that will react in the production of 3.50 L of gaseous ammonia, both gases being measured at the same temperature and pressure.

$$N_2(g) + 3\,H_2(g) \longrightarrow 2\,NH_3(g)$$

32. Calculate the number of liters of ammonia gas measured at STP that can be formed from 6.00 L of hydrogen (measured at STP)? (See Problem 31 for the balanced chemical reaction equation.)

33. Calculate the number of liters of gaseous nitrogen dioxide measured at STP that can be prepared from 4.25 L of gaseous nitrogen monoxide measured at STP?

$$NO(g) + O_2(g) \longrightarrow NO_2(g) \quad \text{(unbalanced)}$$

34. Calculate the number of liters of gaseous oxygen needed to yield 5.25 L of gaseous nitrogen dioxide according to the unbalanced chemical reaction equation in Problem 33, both gases being measured at the same temperature and pressure.

Limiting Reagent Problems

35. If 4.25 L of gaseous oxygen reacts with 3.10 L of gaseous nitrogen monoxide to form gaseous nitrogen dioxide, calculate:

  a. The number of liters of nitrogen dioxide that can be produced.

  b. The number of liters of excess reagent that will remain at the end of the reaction. All gases are measured at the same temperature and pressure. (See Problem 33 for the equation.)

36. Consider the following unbalanced chemical reaction equation:

$$CO(g) + O_2(g) \xrightarrow{\Delta} CO_2(g) \quad \text{(unbalanced)}$$

    a. Calculate the number of liters of carbon dioxide gas produced if 8.25 L of carbon monoxide gas reacts with 4.25 L of oxygen gas. All gases are measured at the same temperature and pressure.

    b. Calculate the number of liters of excess reagent remaining at the end of the reaction.

Heats in Chemical Reactions (See Section 10.7)

37. Consider the following balanced chemical reaction equation:

$$H_2(g) + F_2(g) \longrightarrow H_2F_2(g) + 1.284 \times 10^5 \text{ cal}$$

    a. Is the reaction exothermic or endothermic?

    b. Calculate the number of kilocalories of heat energy produced in the reaction of 3.70 g of fluorine gas with sufficient hydrogen gas.

38. Consider the following balanced chemical reaction equation:

$$11.0 \text{ kcal} + O_2(g) + 2 F_2(g) \longrightarrow 2 OF_2(g)$$

    a. Is the reaction exothermic or endothermic?

    b. Calculate the number of grams of fluorine gas needed for the reaction with 1.80 kcal of heat energy and sufficient oxygen gas.

39. Consider the following balanced reaction equation:

$$C(s) + O_2(g) \xrightarrow{\Delta} CO_2(g) + 394 \text{ kJ}$$

    a. Is the reaction exothermic or endothermic?

    b. Calculate the number of kilojoules of heat energy produced in the reaction of 13.2 g of carbon with sufficient oxygen gas.

General Problems

40. Methane gas (CH_4) burns in oxygen to produce carbon dioxide gas and water vapor.

    a. Write the balanced chemical equation for this reaction.

    b. Calculate the number of moles of hydrogen atoms in 9.00 g of methane.

    c. Calculate the number of moles of oxygen needed to completely burn 6.25 mol of methane.

    d. Calculate the number of grams of oxygen needed to completely burn 8.00 g of methane.

    e. Calculate the number of liters of carbon dioxide gas at STP that can be produced from 12.0 g of methane.

    f. Calculate the number of liters of oxygen gas required to produce 5.60 L of carbon dioxide gas, both gases being measured at the same temperature and pressure.

    g. Calculate the number of grams of carbon dioxide that can be produced from 13.2 g of methane.

    h. Calculate the percent yield if 31.3 g of carbon dioxide is actually obtained; see part (g).

41. A 30.0-g sample of iron is dissolved in concentrated hydrochloric acid (sp gr 1.18 and 35.0% by mass HCl). How many milliliters of concentrated hydrochloric acid are necessary to dissolve the iron?

$$Fe(s) \ + \ 2\,HCl(aq) \ \longrightarrow \ FeCl_2(aq) \ + \ H_2(g)$$

Hint: 35.0% by mass HCl means 35.0 g of pure HCl in $1\overline{0}0$ g of concentrated hydrochloric acid.

42. A 47.1-g sample of copper is dissolved in concentrated nitric acid (sp gr 1.42 and 68.0% by mass HNO_3). How many milliliters of nitric acid are necessary to dissolve the copper?

$$Cu(s) \ + \ 4\,HNO_3(aq) \ \longrightarrow \ Cu(NO_3)_2(aq) \ + \ 2\,NO_2(g) \ + \ 2\,H_2O(l)$$

Hint: 68.0% by mass HNO_3 means 68.0 g of pure HNO_3 in $1\overline{0}0$ g of concentrated nitric acid.

43. If 1.5 g of cadmium is allowed to react with 4.9 mL of 20.0% hydrochloric acid (sp gr 1.10), how many grams of hydrogen can be produced? If 0.020 g of hydrogen is actually obtained, what is the percent yield? (See Section 9.8 on how to complete and balance this equation.)

44. A 0.10-mol sample of iron atoms is allowed to react with 180 mL of 5.0% hydrochloric acid (sp gr 1.02; see Section 2.9).

$$Fe(s) \ + \ 2\,HCl(aq) \ \longrightarrow \ FeCl_2(aq) \ + \ H_2(g)$$

a. How many grams of hydrogen can be produced?

b. If 0.18 g of hydrogen is actually obtained, what is the percent yield?

Hint: 5.0% hydrochloric acid solution means 5.0 g of pure HCl in $1\overline{0}0$ g of HCl solution.

45. One of the components of the fuel mixture on the Apollo lunar module involved the reaction between hydrazine, $N_2H_4(l)$, and dinitrogen tetroxide, $N_2O_4(g)$. The balanced chemical reaction equation for this reaction is

$$2\,N_2H_4(l) \ + \ N_2O_4(g) \ \longrightarrow \ 3\,N_2(g) \ + \ 4\,H_2O(g)$$

What volume in liters of nitrogen gas, measured at STP, will result from the reaction between $15\overline{0}0$ kg of hydrazine and $10\overline{0}0$ kg of dinitrogen tetroxide?

✓ Chapter Quiz 10

ELEMENT	ATOMIC MASS UNITS (AMU)
C	12.0
O	16.0
H	1.0

1. Calculate the number of moles of hydrochloric acid required to produce 6.20 g of carbon dioxide gas according to the following balanced chemical reaction equation.

$$CaCO_3(s) \ + \ 2\,HCl(aq) \ \longrightarrow \ CaCl_2(aq) \ + \ H_2O(l) \ + \ CO_2(g)$$

2. Calculate the number of liters of carbon dioxide produced at STP if 0.510 mol of nitric acid reacts with potassium carbonate according to the following balanced chemical reaction equation.

$$K_2CO_3(aq) + 2\,HNO_3(aq) \longrightarrow 2\,KNO_3(aq) + H_2O(l) + CO_2(g)$$

3. Calculate the number of kilocalories of heat energy produced by burning 14.0 g of methane (CH_4) according to the following balanced chemical reaction equation.

$$CH_4(g) + 2\,O_2(g) \xrightarrow{\Delta} CO_2(g) + 2\,H_2O(g) + 213\ kcal\ (at\ 25°C)$$

4. Consider the following balanced chemical reaction equation.

$$CH_4(g) + 2\,O_2(g) \xrightarrow{\Delta} CO_2(g) + 2\,H_2O(g)$$

   a. If 28.6 g of CH_4 reacts with 57.6 g of O_2, calculate the number of grams of CO_2 that can be produced.

   b. If 32.1 g of CO_2 is actually obtained, calculate the percent yield.

   c. Calculate the number of moles of excess reagent remaining at the end of the reaction.

Chromium (Symbol: Cr)

The chrome plating on your car protects and beautifies some of the trim.

Heating of ammonium dichromate resembles a volcano. The products are chromium(III) oxide, nitrogen, and water vapor. See if you can write and balance the chemical equation.

The Element CHROMIUM: Emeralds, Rubies, and School Buses

Name: The name derives from the Greek *chroma*, meaning "color." It was first isolated from the mineral *crocite* (Siberian red lead), $PbCrO_4 \cdot PbO$, by the French chemist Vauquelin. Many chromium compounds exhibit bright colors.

Appearance: Chromium is a shiny, brittle, gray metal.

Occurrence: The only common ore from which chromium is extracted is *chromite*, $FeCr_2O_4$, a mixture of iron(II) oxide and chromium(III) oxide (Cr_2O_3). Chromium(III) ions are essential to human life, but chromium(VI) ions are toxic and may cause cancer.

Source: Most chromium is obtained industrially by treating chromium(III) oxide (Cr_2O_3) with aluminum or silicon:

$$Cr_2O_3 + 2\,Al \longrightarrow 2\,Cr + Al_2O_3$$
$$2\,Cr_2O_3 + 3\,Si \longrightarrow 4\,Cr + SiO_2$$

Common Uses: By far the most important use of chromium is in the preparation of steel alloys. Adding chromium to iron steel makes the steel much more resistant to corrosion and abrasion. Chromium–nickel steels are commonly referred to as stainless steel. Nickel–chromium alloys are used as hot-wire heating elements.

Another common use for chromium is in electroplating a thin layer of chromium metal onto other metals, particularly steel. This process not only improves the corrosion and wear resistance of the metal but also is quite attractive. Chrome plating is used extensively in making car and bicycle accessories, tools, and furniture and cabinet trim.

Chromium-containing compounds are useful in tanning leather [$Cr_2(SO_4)_3$] and as pigments for ceramics and cloth. Important chromium pigments include chrome yellow ($PbCrO_4$), chrome orange ($PbCrO_4 \cdot PbO$), and chromic-oxide green (Cr_2O_3).

Unusual Facts: The colors of the gems emerald (green) and ruby (red) are a result of the substitution of a small number of Cr^{3+} ions in place of Al^{3+} ions in ordinary minerals. The result is a valuable gemstone. The color of school buses in the United States is chrome yellow, and the yellow lines painted on streets contain the wear-resistant pigment lead silicochromate ($PbCrO_4 \cdot SiO_2$).

CHAPTER 11

Gases

Hot air balloons achieve their buoyancy from air inside the balloon heated by a propane burner. The density of the heated air is lower than that of the surrounding air.

GOALS FOR CHAPTER 11

1. To identify the six general characteristics of gases and relate them to the kinetic theory (Section 11.1).

2. To identify various units for measuring the pressure of gases (Section 11.2).

3. To solve gas problems relating volume and pressure at constant temperature (Boyle's law) (Section 11.3).

4. To solve gas problems relating volume and temperature at constant pressure (Charles's law) (Section 11.4).

5. To solve gas problems relating pressure and temperature at constant volume (Gay-Lussac's law) (Section 11.5).

6. To solve gas problems relating volume, temperature, and pressure (combined gas laws) (Section 11.6).

7. To solve gas problems involving mixtures of gases (Dalton's law of partial pressures) (Section 11.7).

8. To solve gas problems relating pressure, volume, amount, and temperature (ideal-gas equation) (Section 11.8).

9. To solve molecular mass and molar mass, density, and stoichiometry gas problems using the gas laws (Section 11.9).

Countdown

ELEMENT	ATOMIC MASS UNIT (AMU)
O	16.0
Cu	63.5
S	32.1

5. Convert the following temperatures to kelvins (K) (Section 2.8, Temperature Conversions).
   a. $20°C$ (293 K)   **b.** $-35°C$ (238 K)

4. Calculate the molecular mass and molar mass of a gas if 2.30 L measured at STP has a mass of 3.29 g (Section 8.3, Molecular Mass and Molar Mass).
   (32.0 amu, 32.0 g)

3. Calculate the density of oxygen gas at STP (Section 6.5, More About Covalent Bonds, and Section 8.3, Density).
   (1.43 g/L)

2. Solid copper(I) sulfide reacts with excess oxygen gas and heat to produce copper metal and sulfur dioxide gas.
   a. Write a balanced chemical equation for this reaction (Sections 7.2, 7.3, Metals with Variable Ionic Charges, Sections 9.3, and 9.4).
   b. Calculate the volume in liters of sulfur dioxide gas produced at STP with 0.565 mol of copper(I) sulfide and excess oxygen gas (Section 10.5, Mass to Volume and Volume to Mass Problems).
   [**a.** $Cu_2S(s) + O_2(g) \xrightarrow{\Delta} 2\,Cu(s) + SO_2(g)$;
   b. 12.7 L]

1. Solid copper(I) oxide reacts with excess solid carbon and heat to produce copper metal and carbon dioxide gas.
   a. Write a balanced chemical equation for this reaction (Sections 7.2, 7.3, Metals with Variable Ionic Charges, Sections 9.3, and 9.4).
   b. Calculate the mass in grams of copper metal formed when 3.15 L of carbon dioxide gas at STP is released from the reaction of copper(I) oxide with carbon (Section 10.5, Mass to Volume and Volume to Mass Problems).
   [**a.** $2\,Cu_2O(s) + C(s) \xrightarrow{\Delta} 4\,Cu(s) + CO_2(g)$
   b. 35.7 g]

Take a deep breath. You have just inhaled a mixture of gases—life-giving oxygen, to be sure, but also nitrogen, carbon dioxide, and traces of argon, neon, helium, methane (CH_4), carbon monoxide, and water vapor.

In Chapter 3, we noted that gases are one of three physical states of matter. In previous chapters, we learned how to write formulas for many gases and to balance equations involving these gases. Thus far, however, we have dealt with gases largely in a fixed state—assuming constant temperature and pressure. In real life, of course, temperature and pressure change frequently. Research over the last three centuries has enabled chemists to formulate a series of laws regarding gases and their response to such changes. In this chapter, we will consider these laws and their implications.

11.1 Characteristics of Ideal Gases under the Kinetic Theory

You may wonder why we begin our examination of physical states of matter with gases instead of with the solids or liquids that are so much more visible in our world. We start with **gases** for two reasons: (1) They are the simplest of the states, and (2) scientists have learned more about gases than about solids or liquids.

The first thing we need to know about gases is that they share five general characteristics:

1. *Expansion.* Gases expand indefinitely and uniformly to fill all the space in which they are placed.
2. *Indefinite shape or volume.* A given sample of gas has no definite shape or volume but adapts to fit the vessel in which it is placed.
3. *Compressibility.* Gases can be highly compressed. For example, a very large volume of oxygen gas can be compressed into pressurized tanks.
4. *Low density.* The densities of gases are much lower than the densities of solids or liquids. Hence, gas densities are measured in g/L in the metric system, rather than in g/mL as solids and liquids are (see Section 2.2).
5. *Mixing.* Two or more nonreacting gases normally mix completely and uniformly when placed in contact with each other. Consider two examples. (a) A room is filled with air. We are able to breathe in all areas of the room at all times because the gases in the air mix. If this did not happen, there would be sufficient oxygen only in certain parts of the room. (b) The gas company takes advantage of this property to facilitate the detection of leaks in natural gas lines. Natural gas is an odorless mixture of gases (mostly methane, CH_4). The gas company add traces of a foul-smelling gas (C_2H_6S) to natural gas. The C_2H_6S quickly diffuses into room air and is detected when a leak is present in the line.

These characteristics are derived from application of the kinetic theory, which also applies (although in different ways) to solids and liquids. In essence, the **kinetic theory** states that heat and motion are related, that particles of all matter are in *motion* to some degree, and that *heat* is an indication of this motion. The kinetic theory when applied to gases is called the *kinetic molecular theory of gases* and makes the following five assumptions:

Gas   One of three states of matter; it is characterized by (1) infinite and uniform expansion, (2) indefinite shape or volume, (3) compressibility, (4) low density, and (5) complete and rapid mixing in other gases.

Kinetic theory   Hypothesis that heat and motion are related, that particles of all matter are in motion to some degree, and that heat is an indication of this motion; this theory helps to explain the different behaviors and properties of various states of matter.

1. Gases are composed of *molecules*.* The *distance* between these molecules is very *great* compared to the size of the molecules themselves, and the total volume of the molecules is only a small fraction of the entire space occupied by the gas. Therefore, in considering the volume of a gas, we are considering primarily empty space. This assumption explains why gases can be highly compressed and have low density.

2. *No attractive forces* exist between molecules in a gas. This is what keeps a gas from spontaneously becoming a liquid.

3. The molecules in gases are in a state of constant, *rapid motion*, colliding with each other and with the walls of the container in a perfectly random manner, much like the motion of a small "bumper car" at an amusement park. This assumption explains why different gases normally mix completely. The collisions between gas molecules and the walls of the container account for the pressure exerted by the gas.

4. All of these molecular collisions are *perfectly elastic*. That is, like bumper cars, molecules in a gas are not damaged by their collisions but instead continue to move and collide again and again. As a result, the system *as a whole* experiences no loss of kinetic energy, the energy derived from the motion of a particle (see Section 3.5).

5. The *average kinetic energy* per molecule of a gas is proportional to the *temperature* in kelvins, and the *average kinetic energy per molecule is the same at a given temperature for all gases*. The molecules in a gas possess a range of kinetic energies; some molecules have more energy (are "hotter") than the average kinetic energy, and some molecules have less (are "colder"). Theoretically, at 0 K, molecular motion ceases and the kinetic energy of any particle becomes zero.

> **Study Hint:** The average kinetic energy of a gas does *not* depend on the molecular mass or molar mass of the gas.

Ideal gases   Gases that conform to the basic assumptions of the kinetic theory; they are composed of molecules that have no attractive forces for one another and are in constant rapid motion, colliding with one another in a perfectly elastic fashion, and have an average kinetic energy per molecule that is proportional to the temperature in kelvins.

Real gases   Gases such as hydrogen, oxygen, and nitrogen that behave as ideal gases under moderate conditions of temperature and pressure but deviate from these properties if the temperature is very low or the pressure is very high.

Gases that conform to these assumptions are called **ideal gases**, as opposed to **real gases** such as hydrogen, oxygen, nitrogen, and others. An ideal gas is considered to have the following characteristics:

✔ The volume of the molecules is negligible compared to the volume occupied by the gas.

✔ The individual molecules exert no attractive or repulsive forces on each other.

✔ All collisions are perfectly elastic.

Under moderate conditions of temperature and pressure, real gases essentially behave as ideal gases, but if the *temperature* is very *low* or the *pressure* is very *high*, then the properties of real gases deviate considerably from those of ideal gases. By avoiding extremely low temperatures and high pressures, however, we can assume that real gases behave like ideal gases and apply the basic gas laws and the equations discussed later in this chapter to real gas systems.

---

*When we think of molecules of elemental gases, we usually think of the *diatomic* gases such as N_2, O_2, F_2, Cl_2, and H_2. The noble gases exist as *monatomic* gases such as helium (He), neon (Ne), argon (Ar), krypton (Kr), xenon (Xe), and radon (Rn).

11.2 Pressure of Gases

Before we can understand the laws governing changes in gases with changes in temperature or pressure, we need to know more about pressure. **Pressure** is defined as force per unit area. For example, you may fill your car tires to 32 psi [pounds (of force) per square inch].

Pressure is the result of gas molecules colliding with the wall of the container (such as an inner tube). The more frequent the collisions, the greater the pressure of the gas. The less frequent the collisions, the lower the pressure of the gas.

All gases exert at least a small amount of pressure. Even gases in the atmosphere (primarily nitrogen, oxygen, and a small amount of argon, plus smaller amounts of other gases) exert pressure on their "container" (the atmosphere). We measure atmospheric pressure with a mercury *barometer*, which was first devised in 1643 by Evangelista Torricelli (tŏr·rĕ·chĕl′le, 1608–1647), an Italian mathematician and physicist. His barometer consisted of a long glass tube which was sealed at one end, filled with mercury, and inverted with the open end in a dish of mercury (see Figure 11.1). At sea level, the mercury level dropped to a height of 76.0 cm in the tube.

Repeated experiments with tubes of different diameters revealed an important phenomenon: *regardless of the diameter of the tube, the mercury level dropped to 76.0 cm.* To understand why, suppose we were to use a sealed tube with a diameter of 1.13 cm (a cross-sectional *area* of 1.00 cm^2). The *volume* of mercury in the tube at sea level would be 76.0 cm $\times$ 1.00 cm^2 = 76.0 cm^3. At 0°C, the density of mercury is 13.6 g/cm^3. Hence, the mass of the mercury is

$$76.0 \ \cancel{cm^3} \times \frac{13.6 \text{ g}}{1 \ \cancel{cm^3}} = 1030 \text{ g}$$

The weight of an object is the mass times the attraction due to the earth's gravity, which we will call A. Thus, the weight of the mercury is

$$\text{weight} = (\text{attraction from gravity}) \times 1030 \text{ g} = (A \times 1030 \text{ g})$$

Pressure Force per unit area, whether expressed as pounds per square inch (psi), centimeters of mercury (cm Hg), millimeters of mercury, (mm Hg), torr, inches of mercury (in. Hg) atmospheres (atm), pascals (Pa), or millibars (mbar).

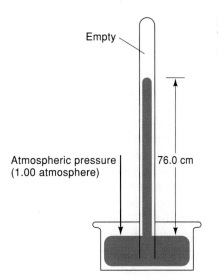

FIGURE 11.1
Torricelli's mercury barometer.

Empty

Atmospheric pressure
(1.00 atmosphere)

76.0 cm

Because the pressure is defined as the force per unit area and the force is simply the weight of the mercury, the pressure at sea level in the barometer is

$$\text{pressure} = \frac{\text{force}}{\text{unit area}} = \frac{A \times 1030 \text{ g}}{1.00 \text{ cm}^2} = (A \times 1030 \text{ g})/\text{cm}^2$$

If the cross-sectional area had been 2.00 cm^2, the weight of the mercury would have been twice as great, but so would the cross-sectional area, and the pressure would still have been the same:

$$\text{pressure} = \frac{\text{force}}{\text{unit area}} = \frac{2(A \times 1030 \text{ g})}{2.00 \text{ cm}^2} = (A \times 1030 \text{ g})/\text{cm}^2$$

This pressure [$(A \times 1030 \text{ g})/\text{cm}^2$], which at sea level supports a column of mercury at a height of 76.0 cm at 0°C, is called **standard pressure**. Standard pressure may be expressed in many other units:

Standard pressure   Pressure that supports a column of mercury at a height of 76.0 cm at 0°C at sea level: 14.7 psi, 76.0 cm of mercury, 76$\overline{0}$ mm of mercury, 76$\overline{0}$ torr, 29.9 in. of mercury, 1.00 atm, 1.013 × 10^5 Pa, or 1013 mbar.

✔ pounds per square inch, 14.7 psi
✔ centimeters of mercury, 76.0 cm Hg
✔ millimeters of mercury, 76$\overline{0}$ mm Hg
✔ torr (1 torr = 1 mm of mercury), 76$\overline{0}$ torr
✔ inches of mercury, 29.9 in. of mercury
✔ atmospheres, 1.00 atm
✔ pascals, 1.013 × 10^5 Pa
✔ millibars, 1013 mbar

The *pascal* [named in honor of Blaise Pascal (1623–1662), a French scientist] is the pressure unit recommended by the International System of Units (SI). Weather reports in many countries besides the United States report atmospheric pressure in kilopascals (kPa) with standard pressure equal to 101.3 kPa. In this book, however, we will generally use the *torr* (named in honor or Torricelli), the *centimeter* of mercury, the *millimeter* of mercury, or the *atmosphere* as the pressure unit. We can readily convert to atmospheres from torr, and vice versa, by knowing that 1 atm = 76$\overline{0}$ torr. Convert 63$\overline{0}$ torr to atmospheres as follows:

$$63\overline{0} \text{ torr} \times \frac{1 \text{ atm}}{76\overline{0} \text{ torr}} = 0.829 \text{ atm}$$

While standard pressure is a useful measure in many cases, much of the world—and thus many gases—are found at higher elevations. Atmospheric pressure decreases as altitude increases (approximately 25 torr/1000 ft). Thus, at an altitude of 1 mile, the pressure is approximately 630 torr. You may have experienced this decrease in pressure on an airplane or when you go to the mountains. You relieve this uncomfortable feeling in your ears by yawning. Yawning equalizes the pressure on your eardrums by opening a tube from the middle portion of your ears to your mouth.

Atmospheric pressure also varies with weather conditions, as you may have noted on TV weather reports. When there is considerable moisture in the atmosphere, the pressure may be low because moist air has a lower density than dry air. Moist air can create a low-pressure area over an entire region. On the other hand, a high-pressure area results when the air in a region contains very little moisture.

11.3 Boyle's Law: The Effect of Pressure Change on the Volume of a Gas at Constant Temperature

Scientists can predict the exact effects of pressure changes by using a law formulated by the British physicist and chemist Robert Boyle (1627–1691). His experiments on the change in volume of a given amount of gas with the pressure of the gas at constant temperature are the basis of **Boyle's law**. According to this law, *at constant temperature, the volume of a fixed mass of a given gas is **inversely** proportional* to the pressure it exerts*. For example, if the pressure on a given gas is doubled, the volume will be halved; if the pressure is halved, the volume will be doubled, as Figure 11.2 shows. Thus, as pressure decreases, volume increases. As pressure increases, volume decreases. Boyle's law may be expressed mathematically as

$$V \propto \frac{1}{P} \text{ (temperature constant, fixed mass)}$$

where volume (V) is inversely ($1/P$) proportional ($\propto$) to the pressure (P). By introducing a proportionality constant (k) for the constant temperature and fixed mass, we can write the equation

$$V = k \times \frac{1}{P}$$

Boyle's law   Principle that, at constant temperature, the volume of a fixed mass of a given gas is *inversely* proportional to the pressure; thus, if the pressure is doubled, the volume will be halved.

Study Hint: As you squeeze something (higher pressure), it gets smaller (its volume decreases), right?

2 volumes   1 volume   1/2 volume

FIGURE 11.2
A demonstration of Boyle's law. Temperature is constant. (As the volume is decreased, the frequency of collisions is increased, resulting in an increase in pressure.)

**Inversely proportional* means that an *increase* in one variable results in a *decrease* in the other variable, or a *decrease* in one variable results in an *increase* in the other variable.

We can then express the equation as

$$PV = k \tag{11.1}$$

That is, the product of the pressure and volume of a gas is equal to a constant at constant temperature. Because k is constant, we can equate gases with different conditions of pressure and volume as long as they have the same mass and are at a constant temperature:

$$P_{new} \times V_{new} = k = P_{old} \times V_{old} \tag{11.2}$$

From this equation, we can determine the new pressure

$$P_{new} = P_{old} \times \frac{V_{old}}{V_{new}} = P_{old} \times V_{ratio} \tag{11.3}$$

in which the new pressure is equal to the old pressure times a volume ratio, V_{ratio}. We will determine the volume ratio by considering the effect on P of the given volume change, as illustrated in Example 11.1. This method has the advantage that you *do not* need to *memorize* Equation (11.3), but rather, you can *reason out* the nature of V_{ratio} from the problem statement and an understanding of Boyle's law.

EXAMPLE 11.1 The volume of a gas is 17.4 L, measured at standard pressure. Calculate the pressure in millimeters of mercury if the volume is changed to 20.4 L and the temperature remains constant.

SOLUTION

$P_{old} = 76\overline{0}$ mm Hg $V_{old} = 17.4$ L | volume increases
$P_{new} = ?$ $V_{new} = 20.4$ L   ↓ pressure decreases T is constant

$$P_{new} = P_{old} \times V_{ratio}$$

We can now figure out the volume ratio (V_{ratio}). The volume has increased; therefore, the new pressure must be less, and we must write the volume ratio so that the new pressure will be less. The volume ratio must be less than 1; hence $\dfrac{17.4\ L}{20.4\ L}$.

Study Hint: In evaluating these and other ratios, make sure that *both* the numerator and the denominator have the same units!

$$P_{new} = 76\overline{0} \text{ mm Hg} \times \underbrace{\frac{17.4\ L}{20.4\ L}}_{\substack{\text{volume ratio} \\ \text{less than 1}}} = 648 \text{ mm Hg} \qquad Answer$$

We can also solve problems where a pressure change causes a change in the volume of a gas sample. The new volume can be calculated as

$$V_{new} = V_{old} \times \frac{P_{old}}{P_{new}} = V_{old} \times P_{ratio} \tag{11.4}$$

Similarly, the new volume is equal to the old volume times a pressure ratio (P_{ratio}). We can use this ratio to determine the pressure ratio by the same type of reasoning we used before, as Example 11.2 illustrates.

EXAMPLE 11.2	A sample of gas occupies a volume of 73.5 mL at a pressure of $71\overline{0}$ torr and a temperature of 30°C. What will the volume be in milliliters at standard pressure and 30°C?

SOLUTION

V_{old} = 73.5 mL P_{old} = $71\overline{0}$ torr   pressure increases,

V_{new} = ? P_{new} = $76\overline{0}$ torr   volume decreases T is constant

$$V_{new} = V_{old} \times P_{ratio}$$

We can now figure out the pressure ratio (P_{ratio}). The pressure has increased from $71\overline{0}$ to $76\overline{0}$ torr; therefore, the new volume must be less, and we must write the pressure ratio so that the new volume will be less. The pressure ratio must be less than 1; hence, $\dfrac{71\overline{0} \text{ torr}}{76\overline{0} \text{ torr}}$. Substituting into Equation (11.4) we have

$$V_{new} = 73.5 \text{ mL} \times \underbrace{\frac{71\overline{0} \text{ torr}}{76\overline{0} \text{ torr}}}_{\substack{\text{pressure ratio} \\ \text{less than 1}}} = 68.7 \text{ mL} \qquad Answer$$

> **Study Hint:** To reflect a *decrease* in volume, the ratio of the pressures must be *less than* 1. To reflect an *increase* in volume, the ratio of the pressures must be *greater than* 1.

Study Exercise 11.1

A sample of a gas occupies a volume of 75.0 mL at a pressure of 725 mm Hg and a temperature of 25°C. Calculate the volume in milliliters at 685 mm Hg and 25°C.

(79.4 mL)   Work Problems 8 through 11.

11.4   Charles's Law: The Effect of Temperature Change on the Volume of a Gas at Constant Pressure

Boyle's law applies when temperature is constant. But sometimes it is temperature, not pressure, that changes. To predict volume changes in such cases, we rely on a law developed by Jacques Charles (1746–1823), a French physicist. His experiments showed that the volume of a gas increases by $\frac{1}{273}$ of its value at 0°C for every degree rise in temperature (see Table 11.1).

TABLE **11.1**	**Relation of Temperature to Volume**	
TEMPERATURE (°C)	VOLUME (mL)	TEMPERATURE (K)
273	546	546
100	373	373
10	283	283
1	274	274
0	273	273
−1	272	272
−10	263	263
−100	173	173
−273	0 (theoretical)	0

Although the volume of a gas changes uniformly with changes in temperature, the volume is not directly proportional to the Celsius temperature. Instead, **Charles's law** states that *at constant pressure, the volume of a fixed mass of a given gas is **directly** proportional* to the Kelvin temperature*, as Figure 11.3 illustrates. That is, at 0 K (equivalent to −273°C) the volume of a gas is, in theory, zero. However, because gases form liquids and solids on cooling, this zero value is only theoretical. Figure 11.3 gives these data in a graph showing the theoretical 0-mL volume. To convert from degrees Celsius to kelvins, we need only add 273 to degrees Celsius as we did in Chapter 2. (In this text, we will use 273 instead of the more precise 273.15 to simplify calculations.)

$$K = °C + 273 \tag{11.5}$$

Charles's law Principle that, at constant pressure the volume of a fixed mass of a given gas is *directly* proportional to the temperature in kelvins; thus, if the Kelvin temperature doubles, so does the volume.

Note that the direct relationship between volume and Kelvin temperature means that if the Kelvin temperature is doubled at constant pressure, the volume will be doubled; if the Kelvin temperature is halved, the volume will be halved. That is, any increase in temperature causes an increase in volume. Any decrease in temperature causes a decrease in volume as depicted in Figure 11.4.

We can express Charles's law mathematically as

$$V \propto T \text{ (pressure constant, fixed mass)}$$

where V is volume and T is temperature in kelvins. As before, we introduce a proportionality constant (**k**) for the constant pressure and fixed mass and write the equation

$$V = \mathbf{k}T$$

We can then express the equation as

$$\frac{V}{T} = \mathbf{k} \tag{11.6}$$

Study Hint: The lid on a pot may jiggle or come off as steam expands with greater heat.

FIGURE 11.3
Graph relating temperature to volume of a gas. Zero-milliliter volume is only theoretical because gases form liquids and solids on cooling.

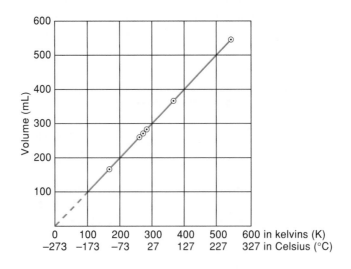

**Directly proportional* means that an *increase* in one variable results in an *increase* in the other variable, or a *decrease* in one variable results in a *decrease* in the other variable.

FIGURE 11.4
A demonstration of Charles's law (temperature in kelvins). Pressure is constant. (As the temperature is increased, the kinetic energy of the molecules is increased. This results in an increase in the volume in order to keep the pressure constant.)

1/2 volume   1 volume   2 volumes

That is, the volume divided by the Kelvin temperature is equal to a constant at constant pressure. Because **k** is a constant, we can equate different conditions of temperature and volume for the same mass of a gas at constant pressure.

$$\frac{V_{\text{new}}}{T_{\text{new}}} = \mathbf{k} = \frac{V_{\text{old}}}{T_{\text{old}}} \tag{11.7}$$

From this equation, we can determine the new temperature in *kelvins*.

$$T_{\text{new}} = T_{\text{old}} \times \frac{V_{\text{new}}}{V_{\text{old}}} = T_{\text{old}} \times V_{\text{ratio}} \tag{11.8}$$

in which the new temperature is equal to the old temperature times a volume ratio (V_{ratio}). We will determine the volume ratio by considering the effect on T of the given volume change as illustrated in Example 11.3. Once again, you *do not* need to *memorize* Equation (11.8). You can *reason* out the nature of V_{ratio} from the problem statement and an understanding of Charles's law.

EXAMPLE 11.3   A gas occupies a volume of 4.50 L at 27°C. At what temperature in degrees Celsius will the volume be 6.00 L if the pressure remains constant?

SOLUTION

$V_{\text{old}} = 4.50\text{ L}$	volume increases	$t_{\text{old}} = 27°C$	$T_{\text{old}} = 3\overline{00}\text{ K}$	
$V_{\text{new}} = 6.00\text{ L}$	temperature increases	$t_{\text{new}} = ?$	$T_{\text{new}} = ?$	P is constant

$$T_{\text{new}} = T_{\text{old}} \times V_{\text{ratio}}$$

Study Hint: The most important point to remember about Charles's law is that the temperature *must* be expressed in *kelvins*.

We can now figure out the volume ratio, V_{ratio}. The volume increases, and so we must write the volume factor so that the new temperature will be larger. The volume ratio must be greater than 1; hence, $\dfrac{6.00 \cancel{L}}{4.50 \cancel{L}}$.

$$T_{new} = 30\overline{0} \text{ K} \times \underbrace{\dfrac{6.00 \cancel{L}}{4.50 \cancel{L}}}_{\substack{\text{volume ratio} \\ \text{greater than 1}}} = 40\overline{0} \text{ K}$$

Finally, convert this Kelvin temperature to degrees Celsius by subtracting the constant, 273.

$$40\overline{0} \text{ K} = (40\overline{0} - 273)°\text{C} = 127°\text{C} \quad \textit{Answer}$$

Now, let us consider the case where a temperature change causes a change in the volume of a gas sample. Suppose you decide to have a backyard party and for the party you blow up balloons. You blow up the balloons in the house where the temperature is 26°C (about 79°F). Then, you take them outside in the backyard where it is cooler, say 18°C (about 64°F). What happens to the size of the balloons? They get smaller according to Charles's law.

We can express this mathematically. From Equation (11.7) we can express the new temperature as a temperature ratio (T_{ratio}) times the old volume.

$$V_{new} = V_{old} \times \dfrac{T_{new}}{T_{old}} = V_{old} \times T_{ratio} \qquad (11.9)$$

Similarly, the new volume is equal to the old volume times a temperature ratio (T_{ratio}). We can determine the temperature ratio by the same type of reasoning we used before. Now, suppose the volume of the balloons was 3.00 L in the house; what would the volume in liters of the balloons be outside in the backyard? To determine the T_{ratio} we must first convert the temperature in degrees Celsius to kelvins.

$$26°\text{C} + 273 = 299 \text{ K} \qquad 18°\text{C} + 273 = 291 \text{ K}$$

The new volume will be smaller, and so the temperature ratio must be less than 1. Hence,

$$3.00 \text{ L} \times \dfrac{291 \text{ K}}{299 \text{ K}} = 2.92 \text{ L}$$

Now let us consider some more examples.

EXAMPLE 11.4 A sample of gas occupies a volume of $16\overline{0}$ mL at 1.00 atm and 27°C. Calculate its volume at 0°C and 1.00 atm.

SOLUTION

$V_{old} = 16\overline{0} \text{ mL}$ $t_{old} = 27°\text{C}$ $T_{old} = 27 + 273 = 30\overline{0} \text{ K}$ │ temperature decreases

$V_{new} = ?$ $t_{new} = 0°\text{C}$ $T_{new} = 0 + 273 = 273 \text{ K}$ │ volume decreases

P = constant

$$V_{new} = V_{old} \times T_{ratio}$$

The temperature has decreased from $30\overline{0}$ to 273 K; hence, the new volume will be less. The temperature ratio must be less than 1, hence $\dfrac{273\,K}{30\overline{0}\,K}$.

$$V_{\text{new}} = 16\overline{0}\,\text{mL} \times \underbrace{\dfrac{273\,K}{30\overline{0}\,K}}_{\substack{\text{temperature ratio} \\ \text{less than 1}}} = 146\,\text{mL} \qquad \textit{Answer}$$

Study Hint: To reflect a *decrease* in volume, the ratio of temperatures must be *less than* 1. To reflect an *increase* in volume, the ratio of temperatures must be *greater than* 1.

Study Exercise 11.2
A gas occupies a volume of 2.50 L at $10\overline{0}°C$. What will its volume be at $15\overline{0}\ °C$ if the pressure remains constant?

(2.84 L)   Work Problems 12 through 15.

11.5 Gay-Lussac's Law: The Effect of Temperature Change on the Pressure of a Gas at Constant Volume

Following up on Charles's work was his fellow Frenchman Joseph Gay-Lussac (1778–1850), a physicist whose experiments led him to formulate **Gay-Lussac's law**. Under this law, *at constant volume, the pressure of a fixed mass of a given gas is* **directly** *proportional to the Kelvin temperature*. Thus, if the Kelvin temperature is doubled at constant volume, the pressure will be doubled; if the Kelvin temperature is halved, the pressure will be halved, as Figure 11.5 shows. That is, any increase in pressure increases temperature, while any decrease in pressure decreases temperature.

Gay-Lussac's law Principle that, at constant volume, the pressure of a fixed mass of a given gas is *directly* proportional to the temperature in kelvins. Thus, if the temperature doubles, so will the pressure.

FIGURE 11.5
A demonstration of Gay-Lussac's law (temperature in kelvins). Volume is constant. (As the temperature is increased, the kinetic energy of the molecules is increased and the frequency and force of collisions is increased, resulting in an increase in pressure.)

We can express this statement mathematically as

$$P \propto T \text{ (volume constant, fixed mass)}$$

where P is pressure and T is temperature in kelvins. As before, we introduce a proportionality constant ($\mathbf{k}$) for the constant volume and fixed mass and write

$$P = \mathbf{k} \, T$$

We can then express the equation as

$$\frac{P}{T} = \mathbf{k} \tag{11.10}$$

The use of a constant also allows us to equate gases of equal volume under different conditions of pressure and temperature.

$$\frac{P_{\text{new}}}{T_{\text{new}}} = \mathbf{k} = \frac{P_{\text{old}}}{T_{\text{old}}} \tag{11.11}$$

Hence, solving for P_{new} and T_{new}, the equations are

$$P_{\text{new}} = P_{\text{old}} \times \frac{T_{\text{new}}}{T_{\text{old}}} = P_{\text{old}} \times T_{\text{ratio}} \tag{11.12}$$

and

$$T_{\text{new}} = T_{\text{old}} \times \frac{P_{\text{new}}}{P_{\text{old}}} = T_{\text{old}} \times P_{\text{ratio}} \tag{11.13}$$

Consider Example 11.5.

EXAMPLE 11.5 The temperature of 1 L of a gas originally at STP is changed to $22\overline{0}°C$ at constant volume. Calculate the final pressure of the gas in torr.

SOLUTION

$P_{\text{old}} = 76\overline{0}$ torr	$t_{\text{old}} = 0°C$	$T_{\text{old}} = 0 + 273 = 273$ K	temperature increases
			V is constant
$P_{\text{new}} = ?$	$t_{\text{new}} = 22\overline{0}°C$	$T_{\text{new}} = 22\overline{0} + 273 = 493$ K	pressure increases

$$P_{\text{new}} = P_{\text{old}} \times T_{\text{ratio}}$$

Study Hint: To reflect a *decrease* in pressure, the ratio of temperatures must be *less than* 1. To reflect an *increase* in pressure, the ratio of temperatures must be *greater than* 1. Again the temperature *must* be expressed in kelvins.

Because the temperature increases, the pressure increases, and we must write the temperature ratio so that the new pressure will be greater. To reflect this increase, we must make the temperature ratio greater than 1, hence $\dfrac{493 \text{ K}}{273 \text{ K}}$.

$$P_{\text{new}} = 76\overline{0} \text{ torr} \times \frac{493 \text{ K}}{273 \text{ K}} = 1370 \text{ torr (to three significant digits)} \qquad \textit{Answer}$$

Study Exercise 11.3
The gas in a cylinder has a volume of 125 mL at $2\overline{0}°C$ (room temperature) and 1.00 atm. At what temperature in degrees Celsius will the pressure be 0.970 atm if the volume remains constant?

$$(11°C)$$

Work Problems 16 through 19.

11.6 The Combined Gas Laws

Boyle's and Charles's laws can be combined into one mathematical expression:

$$\frac{P_{\text{new}} V_{\text{new}}}{T_{\text{new}}} = \frac{P_{\text{old}} V_{\text{old}}}{T_{\text{old}}} \text{ (fixed mass)} \qquad (11.14)*$$

Solving Equation (11.14) for V_{new}, P_{new}, and T_{new} gives

$$V_{\text{new}} = V_{\text{old}} \times \frac{P_{\text{old}}}{P_{\text{new}}} \times \frac{T_{\text{new}}}{T_{\text{old}}} = V_{\text{old}} \times P_{\text{ratio}} \times T_{\text{ratio}} \qquad (11.15)$$

$$P_{\text{new}} = P_{\text{old}} \times \frac{V_{\text{old}}}{V_{\text{new}}} \times \frac{T_{\text{new}}}{T_{\text{old}}} = P_{\text{old}} \times V_{\text{ratio}} \times T_{\text{ratio}} \qquad (11.16)$$

$$T_{\text{new}} = T_{\text{old}} \times \frac{V_{\text{new}}}{V_{\text{old}}} \times \frac{P_{\text{new}}}{P_{\text{old}}} = T_{\text{old}} \times V_{\text{ratio}} \times P_{\text{ratio}} \qquad (11.17)$$

With all the variables involved, these equations may seem intimidating, but they are actually easy to use. Just remember to solve for one variable at a *time*, just as we did in each of the previous three sections. For example, in the case of Equation (11.15), we need to consider first the effect of the pressure change and then the effect of the temperature change on the volume:

1. If the pressure increases, the pressure ratio must be less than 1 because increasing the pressure decreases the old volume. If the pressure decreases, the pressure ratio must be greater than 1 because decreasing the pressure increases the old volume.

2. If the temperature increases, the ratio of the Kelvin temperatures must be greater than 1 because increasing the temperature increases the old volume. If the temperature decreases, the ratio of the Kelvin temperatures must be less than 1 because decreasing the temperature decreases the old volume.

By applying similar reasoning to Equations (11.16) and (11.17), we can draw similar conclusions for pressure and temperature and solve the following examples:

> **Study Hint:** Consider each variable *separately*. It's just like working two simpler problems of the type we have already covered!

EXAMPLE 11.6   A certain gas occupies $5\overline{00}$ mL at $76\overline{0}$ mm Hg and 0°C. What volume in milliliters does it occupy at 10.0 atm and $1\overline{00}$ °C?

---

*Equation 11.14 can be written as

$$\frac{P_2 V_2}{T_2} = \frac{P_1 V_1}{T_1}$$

where P_2, V_2, and T_2 are the *new* conditions and P_1, V_1, and T_1 are the *old* conditions. By using algebra, equations similar to (11.15) through (11.17) can be obtained, substituting subscript 2 for "new" and subscript 1 for "old." By substituting the conditions in these equations, the various gas law problems can be solved. If a certain condition is constant, then that condition is canceled on both sides of the equation.

SOLUTION

$$V_{old} = 5\overline{00} \text{ mL} \quad P_{old} = 76\overline{0} \text{ mm Hg} = 1.00 \text{ atm} \quad \Big| \text{ pressure increases}$$
$$V_{new} = ? \qquad P_{new} = 10.0 \text{ atm} \qquad \Big\downarrow \text{ volume decreases}$$
$$T_{old} = 0 + 273 = 273 \text{ K} \quad \Big| \text{ temperature increases}$$
$$T_{new} = 1\overline{00} + 273 = 373 \text{ K} \;\Big\downarrow \text{ volume increases}$$
$$V_{new} = V_{old} \times P_{ratio} \times T_{ratio}$$

Because the units of P_{old} must be the same as those of P_{new}, we must express both pressures in the same units. The pressure ratio should make the new volume less $\left(\dfrac{1.00 \text{ atm}}{10.0 \text{ atm}}\right)$; the temperature ratio should make the new volume greater $\left(\dfrac{373 \text{ K}}{273 \text{ K}}\right)$. The result is a new volume that is less because of the magnitude of the pressure ratio.

$$V_{new} = 5\overline{00} \text{ mL} \times \frac{1.00 \text{ atm}}{10.0 \text{ atm}} \times \frac{373 \text{ K}}{273 \text{ K}} = 68.3 \text{ mL} \qquad \textit{Answer}$$

Note that we consider the effect of one factor to be *independent* of any other factor, and that in each case, we consider the effect of each factor on the old volume.

Actually, what we do when we consider these factors independently is to consider the effect of a change in pressure ($P_{old} \longrightarrow P_{new}$) at constant temperature (T_{old}), followed by a change in temperature ($T_{old} \longrightarrow T_{new}$) at constant pressure ($P_{new}$). The following diagram illustrates the process.

$V = 5\overline{00}$ mL	$\xrightarrow{\substack{\text{Pressure} \\ \text{factor}}}$	50.0 mL	$\xrightarrow{\substack{\text{New pressure} \\ \text{(constant)}}}$	68.3 mL
$P = 1.0$ atm	$\left(\frac{1}{10}\right)$	10.0 atm	$\left(\frac{10}{10} = 1\right)$	10.0 atm
$T = 273$ K	$\xrightarrow{\substack{\text{Old temperature} \\ \text{(constant)}}}$	273 K	$\xrightarrow{\substack{\text{Temperature} \\ \text{factor}}}$	373 K
	$\left(\frac{273}{273} = 1\right)$		$\left(\frac{373}{273}\right)$	

Note that the effect of each factor is considered *independently*, and the final volume depends on correctly applying *both* factors.

EXAMPLE 11.7 A certain gas occupies 20.0 L at $5\overline{0}°$C and $78\overline{0}$ torr. Under what pressure in torr does this gas occupy 75.0 L at 0°C?

SOLUTION

$$V_{old} = 20.0 \text{ L} \quad \Big| \text{ volume increases} \qquad P_{old} = 78\overline{0} \text{ torr}$$
$$V_{new} = 75.0 \text{ L} \;\Big\downarrow \text{ pressure decreases} \qquad P_{new} = ?$$
$$T_{old} = 5\overline{0} + 273 = 323 \text{ K} \quad \Big| \text{ temperature decreases}$$
$$T_{new} = 0 + 273 = 273 \text{ K} \;\Big\downarrow \text{ pressure decreases}$$
$$P_{new} = P_{old} \times V_{ratio} \times T_{ratio}$$

The volume increases (from 20.0 L to 75.0 L), and so the pressure decreases and the ratio of volumes must be less than 1. A decrease in pressure will also result from the decrease in temperature (50° to 0°C), and so the ratio of Kelvin temperatures must also be less than 1.

$$P_{new} = 78\overline{0} \text{ torr} \times \frac{273 \text{ K}}{323 \text{ K}} \times \frac{20.0 \text{ L}}{75.0 \text{ L}} = 176 \text{ torr} \qquad Answer$$

Study Exercise 11.4
A certain gas occupies 20.0 L at $5\overline{0}$°C and $78\overline{0}$ torr. Calculate its volume in liters at STP.

(17.3 L)

> **Study Hint:** In this particular problem, you will notice that if you had not changed 0°C to 273 K, the answer would have been 0 since in place of the 273 K you would have had 0°C. The temperature *must* be expressed in *kelvins*.

Work Problems 20 through 28.

11.7 Dalton's Law of Partial Pressures

Up to this point, we have been examining the relationships among the pressure, temperature, and volume of a sample of a single gaseous substance. Most real-life situations involve *mixtures of gases*. For example, the air we breathe is a mixture of many gases, as is automobile exhaust. During an operation, the patient breathes a mixture of gases that includes nitrogen, oxygen, and an anesthetic agent like *halothane* (see Problem 51). Thus, we need a way to work with the pressures, volumes, and temperatures of mixtures of gases. John Dalton, whose atomic theory we referred to in Section 4.2, was also keenly interested in meteorology. This interest led him to study gases, and in 1801 he announced his conclusions, known as **Dalton's law of partial pressures**. This law states that *each gas in a mixture of gases exerts a partial pressure equal to the pressure it would exert if it were the only gas present in the same volume. The total pressure of the mixture is then the **sum** of the partial pressures of all the gases present.* For example, if two gases, such as oxygen and nitrogen, are present in a 1-L flask, and the pressure of the oxygen is $25\overline{0}$ torr and that of the nitrogen is $30\overline{0}$ torr, then the total pressure is $55\overline{0}$ torr, as Figure 11.6 shows.

We can express Dalton's law of partial pressures mathematically as

$$P_{total} = P_1 + P_2 + P_3 \cdots \qquad (11.18)$$

in which P_1, P_2, and P_3 are the partial pressures of the individual gases in the mixture.

Dalton's law of partial pressures   Principle that each gas in a mixture of gases exerts a partial pressure equal to the pressure it would exert if it were the only gas present in the same volume; hence, the total pressure of the mixture is the *sum* of the partial pressures of all the gases present.

EXAMPLE 11.8   A 1.00-L cylinder with a movable piston at 27°C contains a mixture of three gases, A, B, and C, at partial pressures of $30\overline{0}$ torr for A, $25\overline{0}$ torr for B, and 425 torr for C. (a) Calculate the total pressure in torr of the mixture. (b) Calculate the volume in liters at STP occupied by the gases remaining if gas A is selectively removed.

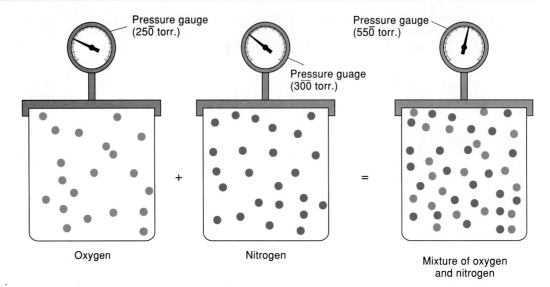

FIGURE 11.6
Dalton's law of partial pressures. The partial pressure of oxygen gas (⚬) is $25\overline{0}$ torr, and the partial pressure of nitrogen gas (●) is $30\overline{0}$ torr. When these amounts are mixed, the total pressure of both gases equals $25\overline{0}$ torr $+$ $30\overline{0}$ torr $=$ $55\overline{0}$ torr, assuming all temperatures are constant and equal.

SOLUTION

a. From Dalton's law of partial pressures, the total pressure of the mixture is equal to the sum of the individual pressures of each gas. So if the pressure of A is P_1, the pressure of B is P_2, and the pressure of C is P_3, then

$$P_{total} = P_1 + P_2 + P_3$$
$$= 30\overline{0} \text{ torr} + 25\overline{0} \text{ torr} + 425 \text{ torr}$$
$$= 975 \text{ torr}$$

b. If gas A is selectively removed at the same temperature and volume (27°C and 1.00 L), leaving only gases B and C in the flask, the pressure will decrease by P_1 or $30\overline{0}$ torr and the new pressure will be $P_{total} - P_1 = 975 \text{ torr} - 30\overline{0} \text{ torr} = 675 \text{ torr}$. Now we are confronted with the problem of calculating a new volume at STP for a gas mixture originally occupying 1.00 L at 27°C and 675 torr.

$$V_{old} = 1.00 \text{ L} \quad P_{old} = 675 \text{ torr} \quad | \quad \text{pressure increases}$$
$$V_{new} = ? \quad\quad P_{new} = 76\overline{0} \text{ torr} \quad \downarrow \quad \text{volume decreases}$$
$$T_{old} = 27 + 273 = 30\overline{0} \text{ K} \quad | \quad \text{temperature decreases}$$
$$T_{new} = 0 + 273 = 273 \text{ K} \quad \downarrow \quad \text{volume decreases}$$
$$V_{new} = V_{old} \times P_{ratio} \times T_{ratio}$$
$$= 1.00 \text{ L} \times \frac{675 \text{ torr}}{76\overline{0} \text{ torr}} \times \frac{273 \text{ K}}{30\overline{0} \text{ K}} = 0.808 \text{ L} \quad\quad \textit{Answer}$$

In the real world, we can see the effects of Dalton's law of partial pressures in the red blood cell count of people living at different altitudes. The percent of oxygen and nitrogen in the atmosphere is constant, but the *partial pressures* of the gases vary with altitude. At sea level the atmospheric pressure is $76\overline{0}$ torr, and the sum of the partial pressures of the other gases excluding oxygen (primarily nitrogen with small amounts of argon and carbon dioxide) is $60\overline{0}$ torr. As a result, the partial pressure for oxygen is 160 torr ($76\overline{0}$ torr $-$ $60\overline{0}$ torr). At mile-high altitude, the altitude of many cities in the Rocky Mountain region, the atmospheric pressure is approximately 630 torr, and the sum of the partial pressures of the other gases is $50\overline{0}$ torr. Hence, the partial pressure of oxygen is only 130 torr ($63\overline{0}$ torr $-$ $50\overline{0}$ torr).

Our bodies need a given amount of oxygen for normal metabolic processes, regardless of the altitude at which we live. The amount of oxygen carried by each red blood cell depends on the partial pressure of oxygen. The lower the partial pressure of oxygen is the less oxygen each cell can carry. Thus, at higher altitudes we need more red blood cells to compensate for the reduced amount of oxygen that each cell can carry. If we live at a high elevation, our bodies increase their production of red blood cells to generate the necessary additional red blood cells. The number of red blood cells in a normal healthy person at sea level is approximately 4.2 million per cubic millimeter of blood. At mile-high altitude, it is approximately 5.4 million. In going from low to high altitude, the body must *acclimate* to the reduced partial pressure of oxygen by increasing the number of red blood cells. For some people, acclimating may take several days, making the person sleepy because of a mild lack of oxygen. Other people may observe no apparent effect. Factors that affect this acclimation are age and general physical condition.

A more direct application of Dalton's law of partial pressures is the collection of a gas over water, as shown in Figure 11.7. The gas contains a certain amount of water vapor (gaseous water), the pressure of which is a constant value *at any given temperature* (if enough time is allowed to permit equilibrium conditions to be established). The total pressure at which the volume of the "wet" gas is measured is equal to the sum of the gas pressure and the water vapor pressure at the temperature at which the gas is collected and measured, or, mathematically,

$$P_{\text{total}} = P_{\text{gas}} + P_{\text{water}} \tag{11.19}$$

The vapor pressure of water varies with temperature, but it has a constant and predictable value at any given temperature. Thus, it is easy to calculate the pressure of the dry gas. Just subtract the known equilibrium vapor pressure of water at the given temperature from the total pressure of the "wet" gas mixture.

$$P_{\text{gas}} = P_{\text{total}} - P_{\text{water}} \tag{11.20}$$

Study Hint: "Wet" gas is not really wet; it just contains a little gaseous water (water vapor). The air on a very humid day contains water vapor and can feel quite wet.

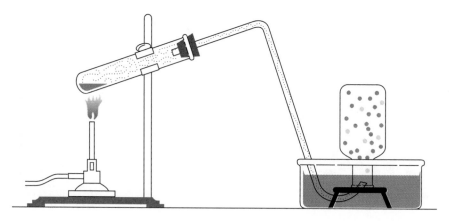

FIGURE 11.7
Collection of a gas over water. The water vapor is shown in the collection as ⚬, and the gas as •. This type of apparatus is used in the laboratory for generating oxygen gas (O_2) by heating potassium chlorate ($KClO_3$). A catalyst of manganese(IV) oxide or manganese dioxide (MnO_2) is often used.

Appendix V gives the vapor pressure of water at various temperatures. Consider the following example.

EXAMPLE 11.9   The volume of a certain gas, collected over water, is $15\overline{0}$ mL at $3\overline{0}$°C and 720.0 torr. Calculate the volume in milliliters of the dry gas at STP.

SOLUTION   The first step in the calculation is to determine the pressure of the dry gas at the initial volume ($15\overline{0}$ mL) and temperature ($3\overline{0}$°C). The pressure of the wet gas (720.0 torr) is equal to the sum of the pressure of the dry gas and the vapor pressure of water at the initial temperature. From Appendix V, the vapor pressure of water at $3\overline{0}$°C is 31.8 torr. The pressure of the dry gas is therefore equal to

$$P_{total} - P_{water} = 720.0 \text{ torr} - 31.8 \text{ torr} = 688.2 \text{ torr}$$

Thus, if the water vapor were removed, that is, if the gas were dry, the pressure of the gas would have measured 688.2 torr in a volume of $15\overline{0}$ mL at $3\overline{0}$°C, as Figure 11.8 shows. With these data, the next step is to work a combined gas law problem to calculate the volume of the dry gas at STP.

$$
\begin{array}{lll}
V_{old} = 15\overline{0} \text{ mL} & P_{old} = 688.2 \text{ torr} & \text{pressure increases} \\
V_{new} = ? & P_{new} = 76\overline{0} \text{ torr} & \text{volume decreases} \\
T_{old} = 3\overline{0} + 273 = 303 \text{ K} & & \text{temperature decreases} \\
T_{new} = 0 + 273 = 273 \text{ K} & & \text{volume decreases} \\
\end{array}
$$

$$V_{new} = V_{old} \times P_{ratio} \times T_{ratio}$$

$$= 15\overline{0} \text{ mL} \times \frac{688.2 \text{ torr}}{76\overline{0} \text{ torr}} \times \frac{273 \text{ K}}{303 \text{ K}} = 122 \text{ mL} \qquad \textit{Answer}$$

Study Exercise 11.5
The volume of a certain gas, collected over water, is 175 mL at 27°C and 635.0 mm Hg. Calculate the volume in milliliters of the *dry* gas at STP (see Appendix V).

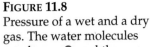

Work Problems 29 through 32.

(127 mL)

FIGURE 11.8
Pressure of a wet and a dry gas. The water molecules are shown ● and the gas molecules are shown as ●.

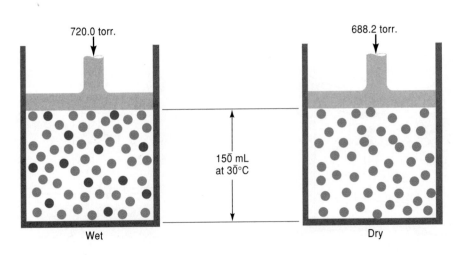

11.8 Ideal-Gas Equation

In the gas laws of Boyle (see Section 11.3), Charles (see Section 11.4), and Gay-Lussac (see Section 11.5), the mass of gas is fixed and one of the three variables—temperature, pressure, or volume—is also constant. Using a new equation, the **ideal-gas equation**, we can vary not only the temperature, pressure, and volume but also the mass of the gas. We can state the ideal-gas equation mathematically as

$$PV = nRT \qquad (11.21)$$

where P is pressure, V is volume, n is the amount of gas in moles, T is temperature, and R is the universal gas constant. We can obtain the numerical value of R by substituting known values of P, V, n, and T in the expression $R = PV/nT$. Because we know that at STP [0°C (273 K) and 1.00 atm] 1 mol ($n = 1.00$) of an ideal gas occupies 22.4 L, we can evaluate R as

$$R = \frac{1.00 \text{ atm} \times 22.4 \text{ L}}{1.00 \text{ mol} \times 273 \text{ K}} = 0.0821 \frac{\text{atm·L}}{\text{mol·K}}$$

We must know the ideal-gas equation and the numerical value of R and its *units* to work problems involving the ideal-gas equation with the four variables n (moles), T (temperature), P (pressure), and V (volume).*

Ideal-gas equation   Formula that allows scientists to vary not only the temperature, pressure, and volume of a gas but also its mass; it is expressed mathematically as $PV = nRT$, where P is pressure, V is volume, n is the amount of gas in moles, T is temperature, and R is the universal gas constant.

---

*Using the ideal-gas equation, we can develop Boyle's, Charles's, and Gay-Lussac's laws. From the ideal-gas equation ($PV = nRT$), holding the mass and temperature constant; and equating the constant mass and constant temperature with the universal gas constant to a new constant (k), we obtain Boyle's law.

(1) $PV = nRT$
(2) $k = nRT$
(3) $PV = k$ [Boyle's law, Equation (11.1)]

Again from the ideal-gas equation, holding the mass and pressure constant and equating the constant mass and constant pressure along with the universal gas constant to a new constant (k), we obtain Charles's law.

(1) $PV = nRT$

(2) $\dfrac{V}{T} = \dfrac{nR}{P}$

(3) $k = \dfrac{nR}{P}$

(4) $\dfrac{V}{T} = k$ [Charles's law, Equation (11.6)]

Returning again to the ideal-gas equation, holding the mass and volume constant and equating the constant mass and constant volume along with the universal gas constant to a new constant k we obtain Gay-Lussac's law.

(1) $PV = nRT$

(2) $\dfrac{P}{T} = \dfrac{nR}{V}$

(3) $k = \dfrac{nR}{V}$

(4) $\dfrac{P}{T} = k$ [Gay - Lussac's law, Equation (11.10)]

As you can see, the ideal-gas equation involves all three gas laws with their variables (volume, pressure, and temperature) in addition to the moles variable.

Consider the following examples.

EXAMPLE 11.10 Calculate the volume in liters of 2.15 mol of oxygen gas at 27°C and 1.25 atm.

SOLUTION Using the ideal-gas equation, $PV = nRT$, and solving for V (volume), we obtain

$$V = \frac{nRT}{P}$$

> **Study Hint:** The units should *always* work out this neatly in this type of problem. If not, then you have done something wrong!

Substituting the values for n (2.15 mol), R (0.0821 atm·L/mol·K), T (27°C + 273 = $3\overline{0}0$ K), and P (1.25 atm), we obtain

$$V = \frac{2.15 \text{ mol} \times 0.0821 \dfrac{\text{atm·L}}{\text{mol·K}} \times 3\overline{0}0 \text{ K}}{1.25 \text{ atm}} = 42.4 \text{ L} \qquad Answer$$

EXAMPLE 11.11 Calculate the pressure in torr of 0.652 mol of oxygen gas occupying a 10.0-L cylinder at $3\overline{0}$°C.

SOLUTION Using the ideal-gas equation, $PV = nRT$, and solving for P (pressure), we obtain

$$P = \frac{nRT}{V}$$

Substituting the values for n (0.652 mol), R (0.0821 atm·L/mol·K), T ($3\overline{0}$°C + 273 = 303 K), and V (10.0 L), we obtain

$$P = \frac{0.652 \text{ mol} \times 0.0821 \dfrac{\text{atm·L}}{\text{mol·K}} \times 303 \text{ K}}{10.0 \text{ L}} = 1.62 \text{ atm}$$

From the equation, we find the pressure in *atmospheres*, but the answer is to be expressed in *torr*. Because $7\overline{6}0$ torr = 1 atm (Section 11.2), we convert atmospheres to torr as follows:

$$1.62 \text{ atm} \times \frac{7\overline{6}0 \text{ torr}}{1 \text{ atm}} = 1230 \text{ torr} \quad \text{(to three significant digits)} \qquad Answer$$

EXAMPLE 11.12 Calculate the number of grams of oxygen gas in a 5.25-L cylinder at 27°C and 1.30 atm.

SOLUTION Using the ideal-gas equation, $PV = nRT$, and solving for n (moles), we obtain

$$n = \frac{PV}{RT}$$

> **Study Hint:** Note that V is expressed in liters (L), P is expressed in atmospheres (atm), and T is in kelvins (K), so the units will all cancel correctly with the units of R.

Substituting the values for P (1.30 atm), V (5.25 L), R (0.0821 atm·L/mol·K), and T (27°C + 273 = $3\overline{0}0$ K), we obtain

$$n = \frac{1.30 \text{ atm} \times 5.25 \text{ L}}{0.0821 \dfrac{\text{atm·L}}{\text{mol·K}} \times 3\overline{0}0 \text{ K}} = 0.277 \text{ mol } O_2 \text{ gas}$$

Converting the 0.277 mol of oxygen gas to grams using the molar mass of 32.0 g for oxygen, we calculate the number of grams of oxygen gas as follows:

$$0.277 \, \text{mol O}_2^- \times \frac{32.0 \text{ g O}_2}{1 \, \text{mol O}_2^-} = 8.86 \text{ g O}_2 \qquad \textit{Answer}$$

Study Exercise 11.6
Calculate the volume in milliliters of 0.0230 mol of nitrogen gas at 27°C and 1.15 atm.

(493 mL)   Work Problems 33 through 37.

11.9 Problems Related to Gas Laws

Various types of problems can be related to the gas laws. All these problems involve calculating the *new volume when changes occur in temperature and pressure.* We will consider three basic types of problems. You may encounter variations of these basic problem types; but if you follow the basic principles outlined here, you should be able to solve any similar problem.

Molecular Mass and Molar Mass Problems

In Section 8.3, we calculated the *molecular mass* and *molar mass* of a gas by solving for g/mol, which is numerically equal to the molecular mass in amu and the molar mass in grams. This calculation involved using a given mass of gas, its volume at STP, and the molar volume, 22.4 L/mol of any gas at STP [0°C and $76\overline{0}$ mm Hg (torr)].

Generally, in the actual performance of an experiment designed to determine the molecular mass and molar mass of a gas, it is difficult to measure the volume of the gas specifically at STP. Therefore, scientists measure the volume at some convenient temperature and pressure (wet or dry). They then convert this volume, using the gas laws, *to what it would have been if the gas were dry and at STP.*

Consider the following examples.

EXAMPLE 11.13   Calculate the molecular mass and molar mass of a certain gas if $60\overline{0}$ mL of the gas measured at $3\overline{0}$°C and $63\overline{0}$ torr has a mass of 0.600 g.

SOLUTION   *First,* correct the volume ($60\overline{0}$ mL) at $3\overline{0}$°C and $63\overline{0}$ torr to STP conditions so that we can use the molar volume of a gas in the next step:

$$V_{\text{old}} = 60\overline{0} \text{ mL} \quad P_{\text{old}} = 63\overline{0} \text{ torr} \quad \bigg| \quad \text{pressure increases}$$
$$V_{\text{new}} = ? \qquad\qquad P_{\text{new}} = 76\overline{0} \text{ torr} \quad \bigg\downarrow \quad \text{volume decreases}$$
$$T_{\text{old}} = 3\overline{0} + 273 = 303 \text{ K} \quad \bigg| \quad \text{temperature decreases}$$
$$T_{\text{new}} = 0 + 273 = 273 \text{ K} \quad \bigg\downarrow \quad \text{volume decreases}$$
$$V_{\text{new}} = V_{\text{old}} \times P_{\text{ratio}} \times T_{\text{ratio}}$$

$$= 60\overline{0} \text{ mL} \times \frac{63\overline{0} \text{ torr}}{76\overline{0} \text{ torr}} \times \frac{273 \text{ K}}{303 \text{ K}} = 448 \text{ mL (at STP)}$$

Then calculate the molecular mass and molar mass:

$$\frac{0.600 \text{ g}}{448 \text{ mL (STP)}} \times \frac{1000 \text{ mL}}{1 \text{ L}} \times \frac{22.4 \text{ L (STP)}}{1 \text{ mol}} = 30.0 \text{ g/mol}$$

molecular mass = 30.0 amu   *Answer*

molar mass = 30.0 g   *Answer* *

EXAMPLE 11.14 Calculate the molecular mass and molar mass of a certain gas if $45\overline{0}$ mL of the gas collected over water and measured at $3\overline{0}°$C and 720.0 torr has a mass of 0.515 g.

SOLUTION   *First*, correct the volume ($45\overline{0}$ mL) at $3\overline{0}°$C and 720.0 torr to *dry* STP conditions so that we can use the molar volume.

$V_{old} = 45\overline{0}$ mL $P_{old} = 720.0 \text{ torr} - 31.8 \text{ torr} = 688.2 \text{ torr}$ pressure
$\qquad\qquad\qquad\qquad$ (see Appendix V for vapor pressure   increases
$\qquad\qquad\qquad\qquad$ of water

$V_{new} = ?$ $P_{new} = 76\overline{0}$ torr   volume
$\qquad\qquad\qquad\qquad\qquad\qquad\qquad\qquad$ decreases

$T_{old} = 3\overline{0} + 273 = 303$ K   temperature decreases
$T_{new} = 0 + 273 = 273$ K   volume decreases
$V_{new} = V_{old} \times P_{ratio} = T_{ratio}$

$$= 45\overline{0} \text{ mL} \times \frac{688.2 \text{ torr}}{76\overline{0} \text{ torr}} \times \frac{273 \text{ K}}{303 \text{ K}} = 367 \text{ mL (}dry \text{ gas at STP)}$$

Then calculate the molecular mass and molar mass:

$$\frac{0.515 \text{ g}}{367 \text{ mL (STP)}} \times \frac{1000 \text{ mL}}{1 \text{ L}} \times \frac{22.4 \text{ L (STP)}}{1 \text{ mol}} = 31.4 \text{ g/mol}$$

molecular mass = 31.4 amu   *Answer*

molar mass = 31.4 g   *Answer* *

---

*Molecular mass and *molar mass problems* may be solved using the ideal-gas equation, $PV = nRT$, by substituting, $g/\text{(m.m.)}$ (g for grams; m.m. for molecular mass of the gas) for n in the ideal-gas equation.

(1) $PV = \dfrac{gRT}{\text{m.m.}}$

(2) $\text{m.m.} = \dfrac{gRT}{PV}$

An alternate solution to Example 11.13 illustrates application of this equation. By substituting the values of Example 11.13 into equation (2) above, we obtain the value for the molecular mass. (Note that V is expressed in liters and P in atmospheres.)

$$\text{m.m.} = \frac{0.600 \text{ g} \times 0.0821 \dfrac{\text{L·atm}}{\text{mol·K}} \times 303 \text{ K}}{63\overline{0} \text{ torr} \times \dfrac{1 \text{ atm}}{76\overline{0} \text{ torr}} \times 60\overline{0} \text{ mL} \times \dfrac{1 \text{ L}}{1000 \text{ mL}}}$$

$$= 30.0 \text{ g/mol, } 30.0 \text{ amu, } 30.0 \text{ g} \qquad Answers$$

See if you can solve this problem using the ideal-gas equation (see footnote on page 298, *Answers* 31.4 amu, 31.4 g). (*Hint:* Be sure to correct for the vapor pressure of water.)

Study Exercise 11.7
Calculate the molecular mass and molar mass of a certain gas if 455 mL of the gas collected over water and measured at 27°C and 720.0 mm Hg has a mass of 0.472 g (see Appendix V).

(28.0 amu, 28.0 g)   Work Problems 38 through 40.

Density Problems

In Section 8.3, we also calculated the density of a gas at STP conditions. The density of the gas need not be expressed only at STP conditions but may also be calculated for any temperature and pressure. To do this, we need only to calculate the volume at the new temperature and pressure.

An application of the density of gases is used in the special effects of movie or theater productions. Carbon dioxide gas sinks when it is released from dry ice (solid CO_2) (see Figure 11.9). This is because the carbon dioxide gas is more dense [about 1.96 g/L (STP) (44.0 g/mol $\times$ 1 mol/22.4 L)] than air (with a density of about 1.3 g/L). When the special effect of dense fog is required in a movie or theater production, dry ice is dropped into hot water. The carbon dioxide gas sinks to the ground simulating fog.

Consider the following example.

EXAMPLE 11.15   Calculate the density of sulfur dioxide gas in g/L at $64\overline{0}$ torr and $3\overline{0}$°C.

FIGURE 11.9
Carbon dioxide gas is more dense than air. Dry ice (solid CO_2) is used in movie or theater productions for special effects. This is a scene from the classic movie *The Wolf Man* with Lon Chaney, Jr., produced in 1941.

SOLUTION

1. Solve for the density of the gas at STP. The molar mass of sulfur dioxide (SO_2) is 64.1 g (see table of approximate atomic masses on the inside back cover of this text). Hence, we can calculate the density of the gas at STP as

$$\frac{64.1 \text{ g } SO_2}{1 \text{ mol } SO_2} \times \frac{1 \text{ mol } SO_2}{22.4 \text{ L } SO_2 \text{ (STP)}} = 2.86 \text{ g/L (at STP)}$$

2. Correct *1.00 L* at STP to $64\overline{0}$ torr and $3\overline{0}°C$.

$$
\begin{aligned}
V_{old} &= 1.0\overline{0} \text{ L} \qquad P_{old} = 76\overline{0} \text{ torr} \quad | \quad \text{pressure decreases} \\
V_{new} &= ? \qquad\qquad P_{new} = 64\overline{0} \text{ torr} \quad \downarrow \quad \text{volume increases} \\
T_{old} &= 0 + 273 = 273 \text{ K} \quad | \quad \text{temperature increases;} \\
T_{new} &= 3\overline{0} + 273 = 303 \text{ K} \quad \downarrow \quad \text{volume increases} \\
V_{new} &= V_{old} \times P_{ratio} \times T_{ratio} \\
&= 1.00 \text{ L} \times \frac{76\overline{0} \text{ torr}}{64\overline{0} \text{ torr}} \times \frac{303 \text{ K}}{273 \text{ K}} = 1.32 \text{ L}
\end{aligned}
$$

3. Calculate the density at $64\overline{0}$ torr and $3\overline{0}°C$, knowing that the 2.86 g of the gas that occupied 1.00 L at STP will now occupy a volume of 1.32 L at $64\overline{0}$ torr and $3\overline{0}°C$.

$$\frac{2.86 \text{ g}}{1.32 \text{ L}} = 2.17 \text{ g/L} \qquad Answer *$$

Study Exercise 11.8

Calculate the density of methane (CH_4) gas in g/L at $68\overline{0}$ mm Hg and $1\overline{0}°C$.

(0.616 g/L)

Work Problems 41 through 43.

Stoichiometry Problems

As a final use of the gas laws, let's apply them to mass–volume stoichiometry problems. In Section 10.5, we expressed the volume of a gas at STP conditions. By applying the gas laws, we can express this volume at any conditions we desire.

Density problems may also be solved using the ideal-gas equation by recalling that the density of gases is measured in g/L. Hence, solving for g/V equal to the density (D), we obtain

(1) $PV = nRT$

(2) $PV = \dfrac{gRT}{\text{m.m.}} \qquad n = \dfrac{g}{\text{m.m.}}$

(3) $D = \dfrac{g}{V} = \dfrac{P(\text{m.m.})}{RT}$

where m.m. = molecular mass. By substituting the values of Example 11.15 into equation (3), we obtain the value for the density.

$$D = \frac{64\overline{0} \text{ torr} \times \dfrac{1 \text{ atm}}{76\overline{0} \text{ torr}} \times \dfrac{64.1 \text{ g}}{\text{mol}}}{0.0821 \dfrac{\text{L} \cdot \text{atm}}{\text{mol} \cdot \text{K}} \times 303 \text{ K}}$$

$$= 2.17 \text{ g/L} \qquad Answer$$

EXAMPLE 11.16	Calculate the volume of oxygen in liters measured at 35°C and 63$\overline{0}$ torr that can be obtained by heating 10.0 g of potassium chlorate.

$$KClO_3(s) \xrightarrow{\Delta} KCl(s) + O_2(g) \quad \text{(unbalanced)}$$

SOLUTION

1. Balance the equation.

$$2\ KClO_3(s) \xrightarrow{\Delta} 2\ KCl(s) + 3\ O_2(g)$$

2. Calculate the volume of oxygen gas at STP. The molar mass of $KClO_3$ is 122.6 g.

$$10.0\ \text{g KClO}_3 \times \frac{1\ \text{mol KClO}_3}{122.6\ \text{g KClO}_3} \times \frac{3\ \text{mol O}_2}{2\ \text{mol KClO}_3} \times \frac{22.4\ \text{L O}_2\ \text{at STP}}{1\ \text{mol O}_2} = 2.74\ \text{L O}_2\ \text{at STP}$$

Step I   Step II   Step III

3. Correcting 2.74 L of O_2 to 35°C and 63$\overline{0}$ torr, we have

V_{old} = 2.74 L P_{old} = 76$\overline{0}$ torr | pressure decreases

V_{new} = ? P_{new} = 63$\overline{0}$ torr ↓ volume increases

T_{old} = 0 + 273 = 273 K | temperature increases

T_{new} = 35 + 273 = 308 K ↓ volume increases

V_{new} = $V_{\text{old}} \times P_{\text{ratio}} \times T_{\text{ratio}}$

$$= 2.74\ \text{L} \times \frac{76\overline{0}\ \text{torr}}{63\overline{0}\ \text{torr}} \times \frac{308\ \text{K}}{273\ \text{K}} = 3.73\ \text{L O}_2\ \text{at 35°C and 63}\overline{0}\ \text{torr} \quad \textit{Answer} *$$

EXAMPLE 11.17	Calculate the number of moles of potassium chlorate required to produce 4.13 L of oxygen gas at 3$\overline{0}$°C and 68$\overline{0}$ torr.

$$KClO_3(s) \xrightarrow{\Delta} KCl(s) + O_2(g) \quad \text{(unbalanced)}$$

Stoichiometry problems related to gas laws may also be solved by using the ideal-gas equation, but first the number of moles of the *gas* must be calculated. The alternate solution to Example 11.16 is as follows:

1. Calculate the moles of oxygen gas.

$$10.0\ \text{g KClO}_3 \times \frac{1\ \text{mol KClO}_3}{122.6\ \text{g KClO}_3} \times \frac{3\ \text{mol O}_2}{2\ \text{mol KClO}_3} = 0.122\ \text{mol O}_2$$

2. Using the ideal-gas equation to calculate the volume of the *gas* at 35°C (308 K) and 63$\overline{0}$ torr.

$$V = \frac{nRT}{P}$$

$$= \frac{0.122\ \text{mol} \times 0.0821\ \dfrac{\text{L·atm}}{\text{mol·K}} \times 308\ \text{K}}{63\overline{0}\ \text{torr} \times \dfrac{1\ \text{atm}}{76\overline{0}\ \text{torr}}}$$

$$= 3.72\ \text{L} \qquad \textit{Answer} \quad \text{(the difference is due to rounding off)}$$

Study Hint: This example differs from the previous one (Example 11.16) in that you are given the volume of a gas at non-STP conditions and not the mass of the reactant. You just then start with what you are given—volume of a gas at non-STP—and convert it to volume of a gas at STP.

SOLUTION

1. Balance the equation.

$$2\ KClO_3(s) \xrightarrow{\Delta} 2\ KCl(s) + 3\ O_2(g)$$

2. Calculate the volume of oxygen gas at STP.

$V_{old} = 4.13\ L$ $T_{old} = 3\overline{0}°C + 273 = 303\ K$ | temperature decreases

$V_{new} = ?$ $T_{new} = 0 + 273 = 273\ K$ ↓ volume decreases

$P_{old} = 68\overline{0}\ torr$ | pressure increases

$P_{new} = 76\overline{0}\ torr$ ↓ volume decreases

$$V_{new} = V_{old} \times T_{ratio} \times P_{ratio}$$

$$= 4.13\ L \times \frac{273\ K}{303\ K} \times \frac{68\overline{0}\ torr}{76\overline{0}\ torr} = 3.33\ L\ O_2\ at\ STP$$

3. Calculate the moles of $KClO_3$ needed to produce 3.30 L O_2 at STP.

$$3.33\ L\ O_2\ (STP) \times \frac{1\ mol\ O_2}{22.4\ L\ O_2\ (STP)} \times \frac{2\ mol\ KClO_3}{3\ mol\ O_2} = 0.0991\ mol\ KClO_3 \quad \textit{Answer}$$

See if you can solve this problem using the ideal-gas equation (see footnote on page 301). [*Hint:* After correcting the volume of oxygen gas to STP conditions (3.33 L), solve for moles of oxygen gas using the ideal-gas equation (0.149 mol), then for moles of $KClO_3$]. *Answer:* 0.0993 mol (the difference is due to rounding off)

Study Exercise 11.9

Calculate the volume of carbon dioxide gas in liters measured at $3\overline{0}°C$ and 0.900 atm that can be obtained when 5.60 g of copper is produced according to the following unbalanced chemical reaction equation:

Work Problems 44 through 46.

$$Cu_2O(s) + C(s) \xrightarrow{\Delta} Cu(s) + CO_2(g) \quad (unbalanced)$$

$$(0.609\ L)$$

CHEMISTRY OF THE ATMOSPHERE

The Greenhouse Effect and Global Warming

For many years, scientific journals have carried articles on the possibility of global warming by the *greenhouse effect*. The popular press over the last 10 years has begun to describe the problem to the public at large. While a detailed understanding of all the factors involved in the greenhouse effect is still far away, you can gain an appreciation of the basic problem. You may want to review the essay at the end of Chapter 6, "Sunlight, Energy for the Earth."

The general idea behind the greenhouse effect is summarized in the figure on the next page. In the first step (1), short- and middle-wavelength light from the sun arrives at the earth. A portion of this light is absorbed at the earth's surface and heats the earth. Slightly less than one-half of the original light from the sun is actually absorbed by the earth. The rest is absorbed by the upper atmosphere or reflected by the atmosphere or the earth's surface.

In the second step (2) of the process, some of the energy absorbed by the earth in the first step evaporates surface water. Because the earth's temperature is lower than that of the sun, the earth reemits the energy it absorbs, but in the form of light energy that contains very little short-wavelength light energy and a large proportion of long-wavelength infrared light. This light energy can be absorbed by certain trace gases in the atmosphere, or

The massive clearing of forests by burning contributes to the buildup of carbon dioxide in the atmosphere.

it can pass out of the atmosphere unabsorbed. Because this light generally has a long wavelength, much of it is absorbed in the atmosphere.

In the third step (3) of the process the trace atmospheric gases that absorbed the energy in step 2 reemit this energy in all directions. Thus, much of the energy that was reemitted at the earth's surface and absorbed by the trace gases is redirected back toward the earth. This action slows down the process by which the earth and its atmosphere cool and warms them slightly. *As long as the amounts of trace gases are reasonably constant, temperature will stabilize at some level.* In fact, this is what makes the earth inhabitable.

As you can see, the role of the trace gases in this process is most important. Nitrogen (N_2) and oxygen (O_2) do *not* play significant roles in trapping this energy because they *do not absorb* much

long-wavelength light. The problem lies with gases like methane (CH_4), dinitrogen monoxide (N_2O), carbon dioxide (CO_2), and chlorofluorocarbons (CFCs) which *do absorb* these wavelengths and play a significant role in trapping this energy.

$$O = C = O$$

carbon dioxide

$$H - \overset{\displaystyle H}{\underset{\displaystyle H}{\vert}}{\overset{\vert}{\underset{\vert}{C}}} - H$$

methane

$$N = N = O$$

dinitrogen monoxide

$$F - \overset{\displaystyle Cl}{\underset{\displaystyle Cl}{\vert}}{\overset{\vert}{\underset{\vert}{C}}} - F$$

CFC-12
(a chlorofluorocarbon)

As human activity *increases* the amounts of these gases, the temperature rises. It is this *extra* warming due to the *increased* amounts of trace gases that is referred to as *global warming*. If the amounts of these gases get large enough, the earth and its atmosphere may warm to temperatures high enough to cause dramatic shifts in weather patterns.

This process is somewhat analogous to you throwing a yo-yo toward the floor. You act as the sun, and the floor acts as the earth. If someone were to place a wooden paddle between you and the floor, then the yo-yo could not return to you and would be trapped between the paddle and the floor. The wooden paddle acts like the trace gases (methane, dinitrogen monoxide, carbon dioxide, and chlorofluorocarbons).

Most of the increases in these trace gases are a result of either the industrialization of our society or the increasing world population. While CO_2 can be removed from the atmosphere by green plants or absorption into the ocean, the other three major trace gases (CH_4, N_2O, and chlorofluorocarbons) all remain in the atmosphere for many years.

The natural variability in global temperatures makes it difficult to know whether the earth is

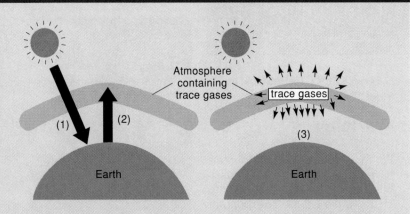

(1) Short-wavelength light from the sun is absorbed by the earth. (2) This energy is reemitted as long-wavelength light, which is absorbed by trace gases in the atmosphere. (3) The trace gases reemit this absorbed energy in all directions, much of it back toward earth.

really warming up. Although the annual increases are expected to be quite small (less than 0.1°C), many scientists estimate that the overall effect could be as large as 2° to 6°C over the next 100 years unless we reduce the amounts of these gases that we release into the atmosphere. This increase may look small, but a global warming of only 2°C could prove disastrous. Significant melting of the polar ice masses would raise the world's sea level enough to flood major coastal population centers. Prime agricultural areas could lose their rainfall and become deserts. Entire species of plant life could become extinct because of

climate changes. At best, our industrial society would endure major disruptions. At worst, food supplies could be affected and worldwide famine might result.

While scientists agree that these things *may* happen, there is as yet no general agreement on how soon they might happen, how severe the effects will be, or if global warming has even begun. There are too many poorly understood factors involved at this time. Even if scientists can tell us what is happening, avoiding the consequences of global warming will require sweeping changes in existing human behavior patterns.

 # Summary

In this chapter we examined the characteristics and behavior of gases: expansion, indefinite shape or volume, compression, low density, and mixing. This behavior is based on the kinetic theory, which states that heat and motion are related and that particles of all matter are in motion to some degree. Heat is an indication of this motion. The kinetic theory explains the characteristics and properties of gases as well as of matter in general (Section 11.1).

The pressure of a gas is the force per unit area exerted by the gas on the walls of its container. Common measurements of pressure include pounds per square inch, torr, centimeters of mercury, and atmospheres (Section 11.2).

Relationships among pressure, temperature, and volume for a given gas sample are defined by several gas laws. According to Boyle's law, at constant temperature the volume is inversely proportional to the pressure (Section 11.3). According to Charles's law, at constant pressure the volume is directly proportional to the temperature in kelvins (Section 11.4). According to Gay-Lussac's law, at constant volume the pressure is directly proportional to the temperature in kelvins (Section 11.5). These three simple gas laws can be merged into the combined gas laws to work a variety of problems (Section 11.6).

Dalton's law of partial pressures is used to deal with mixtures of gases like those we encounter in daily life. The total pressure of a mixture of gases is the sum of the pressures that the individual gases would exert if they were present in the container alone (Section 11.7).

The ideal-gas equation, $PV = nRT$, relates the pressure (P), volume (V), temperature (T, in kelvins), and *amount* of gas (n, in moles) using the universal gas constant, R (Section 11.8).

All the gas laws can be used to solve a variety of problems, including the calculation of molecular masses and molar masses, gas densities, and stoichiometry problems involving gases (Section 11.9).

Exercises

1. Define or explain the following terms (the number in parentheses refers to the section in the text where the term is mentioned):

   a. gas (11.1)

   b. kinetic theory (11.1)

   c. ideal gases (11.1)

   d. real gases (11.1)

   e. pressure (11.2)

   f. standard pressure (11.2)

   g. Boyle's law (11.3)

   h. Charles's law (11.4)

   i. Gay-Lussac's law (11.5)

   j. Dalton's law of partial pressures (11.7)

   k. ideal-gas equation (11.8)

2. Distinguish between:

   a. a real gas and an ideal gas

   b. psi and cm of mercury pressure

   c. force and pressure

   d. torr and centimeters of mercury pressure

3. Identify the following mathematical expressions as being related to Boyle's law, Charles's law, Gay-Lussac's law, or Dalton's law of partial pressures:

   a. $V \propto T$

   b. $P_t = P_1 + P_2 + P_3$

   c. $V \propto \dfrac{1}{P}$

   d. $P \propto T$

4. List five general characteristics of gases.

5. In your own words, list the five assumptions, as you understand them, of the kinetic molecular theory of gases.

6. Under what general conditions of temperature and pressure does the behavior of real gases appreciably deviate from that of ideal gases?

7. Standard atmospheric pressure can be expressed in many ways as shown in Section 11.2. Use dimensional analysis to show that 14.7 lb/in.^{2} and 1.013×10^5 Pa [1 Pa = 1 newton (N)/m^2] are equivalent. *Hint:* 1 in. = 2.54 cm and 1 N = 0.225 lb. The newton is the SI unit of force.

✓ Problems

The table of approximate atomic masses on the inside back cover of this text contains the atomic masses, and Appendix V gives the vapor pressure of water at various temperatures.

Boyle's Law (See Section 11.3)

8. A sample of gas has a volume of 285 mL when measured at 25°C and $76\overline{0}$ mm Hg. What volume in milliliters will it occupy at 25°C and 195 mm Hg?

9. The volume of a given mass of gas is 325 mL at $1\overline{0}$°C and $38\overline{0}$ torr. What will its volume be in milliliters measured at $1\overline{0}$°C and 2.00 atm?

10. What final pressure in torr must be applied to a sample of gas having a volume of $19\overline{0}$ mL at $2\overline{0}$°C and $75\overline{0}$ torr pressure to permit its expansion to a volume of $60\overline{0}$ mL at $2\overline{0}$°C?

11. The volume of a gas is 10.1 L at 10.0 atm and 273 K. Calculate the pressure in atm of the gas if its volume is changed to $50\overline{0}$ mL while the temperature remains constant.

Charles's Law (See Section 11.4)

12. A sample of gas occupies 185 mL at $1\overline{0}$°C and $75\overline{0}$ mm Hg. What volume in milliliters will the gas have at $2\overline{0}$°C and $75\overline{0}$ mm Hg?

13. A gas occupies a volume of 87.0 mL at 27°C and $74\overline{0}$ torr. What volume in milliliters will the gas have at 5°C and $74\overline{0}$ torr?

14. A gas occupies a volume of $13\overline{0}$ mL at 27°C and $63\overline{0}$ torr. At what temperature in degrees Celsius would the volume be 80.0 mL at $63\overline{0}$ torr?

15. The volume of a gas is $20\overline{0}$ mL at $3\overline{0}$°C. At what temperature in degrees fahrenheit would the volume be $26\overline{0}$ mL, assuming the pressure remains constant?

Gay-Lussac's Law (See Section 11.5)

16. A sample of gas occupies 10.0 L at $11\overline{0}$ torr and $3\overline{0}$°C. Calculate its pressure in torr if the temperature is changed to 127°C while the volume remains constant.

17. The temperature of $20\overline{0}$ mL of a gas originally at STP is changed to −35°C at constant volume. Calculate the final pressure of the gas in torr.

18. A gas occupies a volume of 50.0 mL at $3\overline{0}$°C and $63\overline{0}$ mm Hg. At what temperature in °C would the pressure be $77\overline{0}$ mm Hg if the volume remains constant?

19. A sample of gas occupies a volume of 5.00 L at $7\overline{0}0$ torr and 27°C. At what temperature in °C will the pressure be $62\overline{0}$ torr if the volume remains constant?

Combined Gas Laws (See Section 11.6)

20. A certain gas occupies a volume of 495 mL at 27°C and $74\overline{0}$ torr. What volume in milliliters will it occupy at STP?

21. A certain gas has a volume of 205 mL at $2\overline{0}$°C and 1.00 atm. Calculate its volume in milliliters at $6\overline{0}$°C and $6\overline{0}0$ mm Hg.

22. A gas has a volume of 265 mL at 25°C and $6\overline{0}0$ mm Hg. Calculate its volume in milliliters at STP.

23. A sample of a gas has a volume of 5.10 L at 27°C and 635 mm Hg. Its volume and temperature are changed to 2.10 L and $1\overline{0}0$°C, respectively. Calculate the pressure in mm Hg at these conditions.

24. A gas measures $31\overline{0}$ mL at STP. Calculate its pressure in atmospheres if the volume is changed to $45\overline{0}$ mL and the temperature to $6\overline{0}$°C.

25. A given sample of gas has a volume of 4.40 L at $6\overline{0}$°C and 1.00 atm pressure. Calculate its pressure in atmospheres if the volume is changed to 5.00 L and the temperature to $3\overline{0}$°C.

26. A gas measures $15\overline{0}$ mL at STP. Calculate its temperature in degrees Celsius if the volume is changed to $32\overline{0}$ mL and the pressure to $95\overline{0}$ torr.

27. A gas has a volume of 125 mL at 57°C and $64\overline{0}$ torr. Calculate its temperature in degrees Celsius if the volume is increased to 325 mL and the pressure is decreased to $59\overline{0}$ torr.

28. A gas has a volume of 2.50 L at 27°C and 1.00 atm. Calculate its temperature in degrees Celsius if the volume is decreased to 1.90 L and the pressure is decreased to 0.870 atm.

Dalton's Law of Partial Pressures (See Section 11.7)

29. A mixture of gases has the following partial pressures for the component gases at $2\overline{0}$°C in a volume of 2.00 L: oxygen, $18\overline{0}$ torr; nitrogen, $32\overline{0}$ torr; hydrogen, 246 torr.

    a. Calculate the total pressure in torr of the mixture.

    b. Calculate the volume in liters at STP occupied by the gases remaining if the hydrogen gas is selectively removed.

30. A mixture of gases has the following partial pressures for the component gases at $5\overline{0}$°C in a volume of $45\overline{0}$ mL: helium, $12\overline{0}$ torr; argon, $18\overline{0}$ torr; krypton, $6\overline{0}$ torr; xenon, 25 torr.

    a. Calculate the total pressure in torr of the mixture.

    b. Calculate the volume in milliliters at STP occupied by the gases remaining if the krypton is selectively removed.

31. The volume of oxygen collected over water is 165 mL at 25°C and 600.0 torr. Calculate the dry volume in milliliters of oxygen at STP.

32. The volume of nitrogen collected over water is 255 mL at 25°C and 700.0 torr. Calculate the dry volume in milliliters of nitrogen at STP.

Ideal-Gas Equation (See Section 11.8)

33. Calculate the volume in milliliters of 0.0270 mol of nitrogen gas at $3\overline{0}$°C and 1.10 atm.

34. Calculate the pressure in atmospheres of 16.8 g of nitrogen gas occupying a 10.0-L cylinder at 35°C.

35. Calculate the temperature in degrees Celsius of 0.310 mol of nitrogen gas occupying a 10.0-L cylinder at 0.950 atm.

36. Calculate the number of moles of oxygen gas (O_2) in a 4.25-L cylinder at $3\overline{0}$°C and 0.900 atm.

37. Calculate the number of grams of oxygen gas (O_2) in a 6.00-L cylinder at 27°C and $8\overline{00}$ torr. (*Hint:* Convert the pressure in torr to atmospheres.)

Related to Gas Laws: Molecular Mass and Molar Mass Problems (See Section 11.9)

38. If 485 mL of a gas measured at $3\overline{0}$°C and $6\overline{00}$ torr has a mass of 0.384 g, what are the molecular mass and molar mass?

39. A volume of 0.972 L of a gas measured at $5\overline{0}$°C and $7\overline{00}$ torr has a mass of 0.525 g. Calculate its molecular mass and molar mass.

40. Calculate the molecular mass and molar mass of a gas if 255 mL of the gas collected over water and measured at 27°C and 685.0 torr has a mass of 0.288 g.

Related to Gas Laws: Density Problems (See Section 11.9)

41. Calculate the density of carbon dioxide gas in g/L at 35°C and 3.00 atm.

42. Calculate the density of methane (CH_4) gas in g/L at −45°C and $3\overline{00}$ torr.

43. Calculate the density of carbon monoxide gas at $1\overline{0}$°C and 0.900 atm.

Related to Gas Laws: Stoichiometry Problems (See Section 11.9)

44. Calculate the number of milliliters of hydrogen gas at 27°C and $64\overline{0}$ torr produced by the reaction of 0.520 g of magnesium with excess hydrochloric acid.

$$Mg(s) + HCl(aq) \longrightarrow MgCl_2(aq) + H_2(g) \quad \text{(unbalanced)}$$

45. Calculate the number of moles of potassium nitrate required to produce 4.25 L of oxygen gas at $3\overline{0}$°C and $71\overline{0}$ mm Hg.

$$KNO_3(s) \xrightarrow{\Delta} KNO_2(s) + O_2(g) \quad \text{(unbalanced)}$$

46. Calculate the number of grams of potassium chlorate required to produce 385 mL of oxygen gas at 27°C and $65\overline{0}$ mm Hg.

$$KClO_3(s) \xrightarrow{\Delta} KCl(s) + O_2(g) \quad \text{(unbalanced)}$$

General Problems

47. Calculate the molecular mass and molar mass of a gas having a density of 2.20 g/L at 27°C and $46\overline{0}$ mm Hg.

48. Calculate the number of oxygen molecules in 4.50 L of oxygen gas measured at 27°C and $8\overline{00}$ torr. (*Hint:* See Sections 8.2 and 8.3).

49. A hydrocarbon has a density of 2.30 g/L at 27°C and $50\overline{0}$ torr. Its composition is 83.7% carbon and 16.3% hydrogen. What are its molecular mass and its molecular formula? (*Hint:* See Section 8.5.)

50. A hydrocarbon has a density of 0.681 g/L at 37°C and 438 torr. Its composition is 80.0% carbon and 20.0% hydrogen. What are its molecular mass and its molecular formula? (*Hint:* See Section 8.5.)

51. Halothane (fluothane) is a nonflammable, nonirritating general anesthetic, and in many instances it is superior to ethyl ether. At 57°C and $64\overline{0}$ torr, 0.529 g of the gas occupies a volume of 86.4 mL. Its composition is 12.2% carbon, 0.5% hydrogen, 40.5% bromine, 18.0% chlorine, and 28.9% fluorine. Calculate the molecular mass and molecular formula for halothane. (*Hint:* See Section 8.5.)

52. Given the ideal-gas equation, $PV = nRT$, calculate the value of R in the units $Pa \cdot m^3/(mol \cdot K)$. (*Hint:* Standard pressure $= 1.013 \times 10^5$ Pa; 1 m^3 = 1000 L.)

53. The pressure in a tire is equal to the total pressure that is exerted by the vehicle on the ground. Consider a bicycle ridden by one of the authors (GWD). The bicycle weighs $3\overline{0}$ lb, and the tire pressure is $6\overline{0}$ pounds per square inch (psi). The surface area of each tire that is in contact with the road is equivalent to a rectangle, 1.0 in. by 1.7 in. Estimate the weight in pounds of the author to two significant digits.

54. Marble or limestone ($CaCO_3$) reacts with aqueous hydrochloric acid to give carbon dioxide. Calculate the number of milliliters of carbon dioxide at 18°C and 715 torr that can be generated from 87.0 g of limestone.

✓ Chapter Quiz 11

ELEMENT	ATOMIC MASS UNITS (AMU)
Mg	24.3
N	14.0

1. A certain gas occupies 275 mL at 1.00 atm and 27°C. What volume in milliliters will it occupy at 10.0 atm and $10\overline{0}$ °C?

2. Exactly 525 mL of oxygen gas is collected over water at 27°C and 627 torr. Calculate the volume of *dry* oxygen gas at STP. The vapor pressure of water at 27°C is 27 torr.

3. Calculate the molecular mass and molar mass of a gas if 1.20 g has a volume of $55\overline{0}$ mL at 27°C and $70\overline{0}$ torr pressure.

4. Calculate the density of nitrogen gas at 27.0°C and 1.25 atm pressure.

5. Calculate the volume in liters of nitrogen needed at 645 torr and 27°C to react with 12.5 g of magnesium according to the following unbalanced equation:

$$Mg(s) + N_2(g) \xrightarrow{\Delta} Mg_3N_2(s) \quad \text{(unbalanced)}$$

Oxygen (Symbol: O)

The Element OXYGEN: Chemistry and Life on Earth

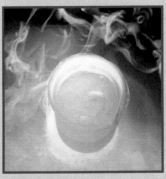

Oxygen is normally a gas at room temperature, but it liquefies at −183°C (1 atm) to form a pale blue liquid.

Name: From the Greek *oxy-* and the French *-gen*, meaning "acid producing." Many simple acids (Chapters 7 and 15) contain oxygen in the combined form.

Appearance: Colorless, odorless and tasteless gas. Can be liquefied to produce a pale-blue liquid and solidified to produce a light-blue solid.

Occurrence: Oxygen is found as a diatomic gas (O_2) and also as ozone (O_3). Oxygen makes up 21% of dry atmospheric air as O_2 and 47% of the earth's crust, mostly in the form of oxides and silicates. The concentration of ozone in the atmosphere varies with location, altitude, and season, but it is typically quite small (about $1 \times 10^{-6}\%$).

Source: Obtained commercially by the large-scale liquefaction and distillation of air.

Common Uses: Over half of all commercially produced oxygen is used in the steel industry. The production of steel-grade iron for various steel alloys and stainless steels requires large amounts of oxygen gas.

Oxygen is used in the production of many important industrial chemicals, such as sulfuric acid (battery acid).

Medical uses of oxygen include treating people with breathing problems and surgery patients under anesthesia.

Oxygen is used as a fuel or an oxidizer for the propulsion of space vehicles. America's space shuttle is a prime example of this application.

Unusual Facts: While oxygen is an absolute necessity for life on earth, there are hazards associated with it. Combustion of flammable materials proceeds explosively in the presence of pure oxygen. One such tragic consequence was the fire that destroyed an Apollo space capsule and killed three astronauts during a test years ago (January 1967).

CHAPTER 12

Liquids and Solids

Liquids and solids can be found all around you in your daily lives.

GOALS FOR CHAPTER 12

1. To identify the six general characteristics of liquids and relate them to the kinetic theory (Section 12.1).
2. To describe the processes of condensation and evaporation in terms of the behavior of molecules (Section 12.2).
3. To define vapor pressure and to explain how dynamic equilibrium is reached between a liquid and its vapor (Section 12.3).
4. To define boiling point and how it is affected by reduced or increased pressure and to calculate energy changes from the liquid to the gaseous state and the reverse (Section 12.4).
5. To describe how distillation is used to purify a liquid (Section 12.5).
6. To define surface tension and viscosity of a liquid and the attractive forces involved (Section 12.6).
7. To identify the six general characteristics of solids and relate them to the kinetic theory (Section 12.7).
8. To define crystalline solids and amorphous solids and to distinguish between them (Section 12.8).
9. To define melting and freezing points and how they are affected by increased pressure, to describe the processes of fusion and solidification in terms of behavior of molecules, and to calculate energy changes from the solid to the liquid state and the reverse (Section 12.9).
10. To define sublimation and deposition in terms of the behavior of molecules (Section 12.10).
11. To calculate the heat evolved or required during the changes among the three physical states of matter: solid, liquid, and gas (Section 12.11).

Countdown

ELEMENT	ATOMIC MASS UNITS (AMU)
C	12.0
H	1.0
Hg	200.6
N	14.0
O	16.0

5. Perform the indicated mathematical operations and express your answer to the proper number of significant digits (Sections 2.5 and 2.6).

   a. $\begin{array}{r} 289.2 \\ 47.63 \\ \underline{32.3} \end{array}$ (369.1)

   b. $\begin{array}{r} 2{,}4\underline{0}0 \\ 6{,}0\underline{0}0 \\ \underline{28{,}200} \end{array}$ (36,600)

   c. $1.00 \times 35.0(45 - 25) =$ (7$\overline{0}$0)

   d. $\dfrac{28{,}500}{(36 - 21)} =$ (1900)

4. Calculate the number of kilocalories in 9250 cal (Sections 2.2, 2.8, Metric Conversion, and 3.5).

   (9.25 kcal)

3. Calculate the number of calories of heat required to raise the temperature of 25.0 g of liquid mercury from 35.0° to 85.0°C. The specific heat of liquid mercury is 0.0331 cal/(g·°C) (Section 3.5).

   (41.4 cal)

2. Calculate the molecular or formula mass and molar mass for the following (Sections 8.1 and 8.2):

   a. ethane (C_2H_6) (30.0 amu, 30.0 g)
   b. mercury(II) nitrate [$Hg(NO_3)_2$]

   (324.6 amu, 324.6 g)

1. Calculate the number of grams in the following (Section 8.2):

   a. 1.50 mol of ethane (C_2H_6) (45.0 g)
   b. 0.385 mol of mercury(II) nitrate [$Hg(NO_3)_2$]

   (125 g)

L iquids and solids can be found all around us in our daily lives. By far the most important liquid and solid is water. Wars have been fought over "water rights." In the winter we worry about ice on the roads. We are mostly water (60%), and we require water to live (about 2.7 qt daily). In this chapter, we will consider the nature of liquids and solids and some of the chemical laws that enable scientists to predict their behavior.

12.1 The Liquid State

In Chapter 11, we noted that gases share some characteristics that can be explained by the kinetic theory. This same theory applies equally to solids and liquids. In these substances, too, the particles are in motion to a greater or lesser degree, with heat being a reflection of that motion. The differences among the various physical states of matter, then, are differences in the degree of this *motion*.

For example, remember from Section 11.1 that *ideal gases* consist of molecules or atoms that occupy only a minute portion of the total volume of the gas, do not attract or repel one another, and are perfectly elastic in their collisions. In contrast, the molecules in **liquids** are closer together, exert some attractive or repulsive forces on one another, and are not perfectly elastic. The result is the following six general characteristics of liquids:

Liquid One of three states of matter, it is characterized by (1) limited expansion, (2) lack of characteristic shape, (3) maintenance of volume, (4) slight compressibility, (5) high density, and (6) mixing in other liquids (usually slow).

1. *Limited expansion.* Liquids do not expand infinitely as gases do.
2. *Shape.* Liquids have no characteristic shape and take on the shapes of the containers they occupy. Just consider what happens when you spill a glass of milk!
3. *Volume.* Liquids maintain their volumes no matter what size container they occupy.
4. *Compressibility.* Liquids are only slightly compressible for a given temperature or pressure change. This lack of compressibility is evident in the brake fluid of the hydraulic braking system of an automobile. If the fluid could be compressed considerably, the pressure applied to the brake pedal would compress the fluid and the brakes would not stop the car. Instead, the pressure from your foot is transferred through the brake fluid in the system to the brake drum.
5. *High density.* Liquids have considerably higher densities than gases. As we saw in Section 2.2, chemists usually measure the density of a gas in g/L, but they measure liquid densities in g/mL. For example, water in the liquid state at $10\overline{0}°C$ and $76\overline{0}$ torr has a density of 0.958 g/mL, but water in the gaseous state under the same conditions has a density of only 0.598 g/L (0.000598 g/mL). Thus, liquid water is more dense than water vapor by a factor of 1600.
6. *Mixing.* Liquid molecules, like gas molecules, are in constant motion. Unlike gas molecules, however, a molecule in a liquid can move only a short distance before it strikes another molecule, slowing its motion. Hence, one liquid mixes throughout another liquid in which it is soluble, but this mixing is much slower in liquids than in gases, as is quite evident if you try to mix honey and water.

12.2 Condensation and Evaporation

Despite the differences between solids and liquids, the two are closely linked. Unlike *ideal* gases, *real* gases do involve some attraction between molecules. If the molecules of real gases are kept far enough apart (that is, if the pressure remains low enough) and if the temperature remains high enough, real gases act just like ideal gases because the kinetic energy of the molecules negates the weak attraction involved. However, if the *temperature* of such a real gas is *lowered* and the *pressure* is *increased*, these attractive forces can partially overcome the kinetic energies of the molecules, and the real gas will deviate from the behavior predicted by the gas laws.

Real gases can be liquefied when these attractive forces overcome the kinetic energy of the molecules enough to keep the molecules confined to a relatively small volume. We can achieve this effect by decreasing the temperature (lowering the kinetic energy of the molecules) and/or by applying pressure so that the molecules are brought closer together, allowing the attractive forces to become more significant, as Figure 12.1 shows. The result is **condensation**, the conversion of a gas to a liquid.

A reversal of this process leads liquids to become gases via *evaporation*. **Evaporation** is the actual escape of molecules from the surface of a liquid to form a vapor in the surrounding space above the liquid. It may occur because temperature increases (as when we bring water to a boil on the stove) or because pressure is decreased (as when water molecules have a chance to disperse into the broader volume of the atmosphere).

In any sample of a liquid, the kinetic energies of some of the molecules are *above* average, while energies of others are *below* average. "Hotter" molecules—those with *higher* kinetic energy—can more readily overcome the attractive forces in the liquid and *escape* from its surface, as Figure 12.2 illustrates.

Because the molecules of higher kinetic energy escape, the average kinetic energy of the molecules left in the liquid is *lowered*, assuming that no heat is supplied from some outside source. Lowering the average kinetic energy results in a temper-

Condensation Conversion of vapor (gas) molecules to a liquid; this reverse of the evaporation process is an exothermic change of state.

Evaporation Escape of molecules from the surface of a liquid to form a vapor in the surrounding space above the liquid; this reverse of the condensation process is an endothermic change of state.

Liquid   Gas

FIGURE 12.1
Molecules in the liquid state are closer together than in the gaseous state. This change of state from gas to liquid is accomplished by decreasing the temperature and/or applying pressure, allowing the attractive forces to become more significant.

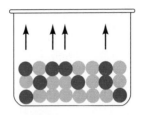

FIGURE 12.2
Evaporation of a liquid. The molecules on the surface of the liquid that have higher kinetic energies can escape into the space above the liquid surface. (Degree of kinetic energy is shown in shades of color.)

> **S**tudy Hint: A dynamic equilibrium always consists of two *opposing* processes. In the case of vapor pressure, the processes are *evaporation* and *condensation*.

Dynamic equilibrium   A situation in which the rate of a forward process is equal to the rate of the reverse process occurring simultaneously, as when the rate of evaporation is equal to the rate of condensation for a liquid in a closed container.

Vapor pressure   In a closed container, the pressure exerted by a vapor in dynamic equilibrium with its liquid state.

Boiling point   Temperature at which the vapor pressure of a liquid is equal to the external pressure above the surface of the liquid.

ature drop in the liquid. This cooling effect is evident when you first come out of a swimming pool on a hot day and feel quite cool because water evaporates from the surface of your body. Evaporation can even freeze tissue. Ethyl chloride is a local anesthetic used in medicine to reduce pain near the surface of the skin. When sprayed on the skin, ethyl chloride evaporates so fast that it chills the underlying tissue and results in a temporary loss of feeling in that region, allowing the doctor to make minor incisions in the skin.

The ease of escape of a molecule from the surface of a liquid is related to the strength of the attractive forces between the molecules in the liquid. For example, gasoline and alcohol (ethyl alcohol) evaporate faster than water. The attractive molecular forces in gasoline and alcohol are weaker than those in water. As the *temperature increases*, molecules escape more readily because more have sufficient energy to leave the surface of the liquid and overcome the attractive forces in the liquid.

12.3   Vapor Pressure

When a quantity of liquid is placed in a *closed* container, the molecules of the liquid that are escaping from the surface are *not* removed from the immediate space above the liquid. Thus, the system arrives at a steady state where the *rate of evaporation is equal to the rate of condensation*. When the rate of molecules leaving from the surface (evaporation) is equal to the rate of molecules reentering the liquid (condensation), a **dynamic equilibrium** like that shown in Figure 12.3 is established. Once a dynamic equilibrium has been established, the concentration of the molecules in the space above the liquid remains constant (at constant temperature) and the vapor exerts a *definite constant pressure*. This equilibrium pressure at any fixed temperature is called the **vapor pressure** of the liquid.

If you increase the temperature of a liquid, more molecules in the liquid will have higher kinetic energies and will break away from the surface of the liquid. That is, the concentration of vapor molecules, and thus the vapor pressure of the liquid, will increase before equilibrium is again established since the pressure is directly proportional to the concentration of the vapor molecules.

In Section 11.7, we used this vapor pressure at various temperatures in calculations involving Dalton's law of partial pressures. Appendix V gives the vapor pressure of water expressed in mm Hg (torr), atmospheres, and pascals. However, the vapor pressures of various liquids differ considerably depending on the liquid. As we mentioned earlier in this chapter, gasoline and alcohol evaporate faster than water. Hence, we would expect gasoline and alcohol to have higher vapor pressures than water at a given temperature. This behavior is what Figure 12.4 shows for water, alcohol, and *heptane*, a component of gasoline whose vapor pressure is similar to that of gasoline.

12.4   Boiling Point. Heat of Vaporization or Condensation

When the *vapor pressure* of a liquid *equals* the external pressure on the liquid, bubbles of vapor form rapidly throughout the liquid and the liquid boils. The **boiling point** of a liquid is the temperature at which the vapor pressure of the liquid is *equal* to the external pressure above the surface of the liquid. Because the atmospheric pressure varies with altitude and atmospheric conditions, scientists use standard pressure [$76\overline{0}$ mm Hg (torr) or 1 atm] in reporting boiling points. When the external pressure is $76\overline{0}$ mm Hg (torr), this temperature is called the *normal boiling point*. Therefore, the *normal boiling point* of a liquid is the temperature at which the vapor pressure of the liquid is $76\overline{0}$ mm Hg (torr). As Figure 12.4 shows, the normal boiling point of alcohol (ethyl alcohol) is about 78°C; of heptane, 98°C; and of water, $10\overline{0}$°C.

Study Exercise 12.1

Using the graph in Figure 12.4, determine the following:

a. the boiling point of water at $30\overline{0}$ mm Hg (torr).   (75°C)

b. the boiling point of alcohol (ethyl alcohol) at $25\overline{0}$ mm Hg (torr).   (55°C)

This relationship explains why the boiling point is lower at higher altitudes, where the atmospheric pressure is less. At mile-high altitude, the atmospheric pressure is approximately 630 mm Hg, and so the boiling points of alcohol (ethyl alcohol), heptane, and water are about 76°C, 92°C, and 95°C, respectively. Increased atmospheric pressure occurs below sea level, as in Death Valley, California, and in a pressure cooker, and raises the boiling point. Because temperature—not boiling point—is the crucial factor in cooking foods, however, when cooking at a lower temperature, we must increase cooking time. At sea level, where water boils at $10\overline{0}$°C, it takes about 10 min to hard-boil an egg. At Pike's Peak in Colorado (14,110 ft), where water boils at about 86°C, it takes a little over 20 min to hard-boil an egg!

Evaporation and boiling involve a loss of energy by the liquid, and heat must be *supplied* if the temperature is to remain constant. Thus, evaporation is an **endothermic change of state** requiring the *absorption* of energy. In contrast, condensation is an **exothermic change of state** requiring the *release* of heat. The exothermic property of condensation is evident from a steam burn. The steam condenses on the cooler skin, and in the process heat is released, which "cooks" the tissue and produces blisters.

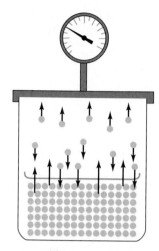

FIGURE 12.3
The pressure exerted by the gaseous molecules is the *vapor pressure* of the liquid at this fixed temperature. This establishes a *dynamic equilibrium* in a closed container. The rate of evaporation, ↑, equals the rate of condensation, ↓.

Work Problem 13.

Endothermic change of state
Any change in the state of matter in which heat energy is absorbed.

Exothermic change of state
Any change in the state of matter in which heat is released.

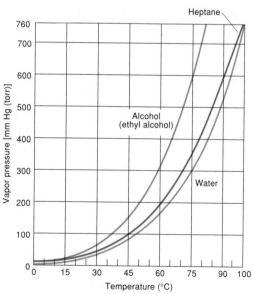

FIGURE 12.4
Vapor pressure changes with temperature for alcohol, heptane, and water.

FIGURE 12.5
Heat of vaporization ($\rightarrow$)
and condensation ($\leftarrow$).

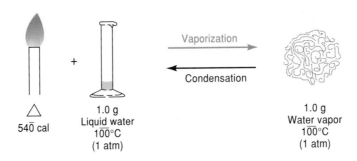

How much heat is absorbed or released in these changes? The quantity of heat required to evaporate 1 g of a given liquid at its boiling point and at constant pressure is called the **heat of vaporization** of the liquid and is shown in Figure 12.5. The quantity of heat that must be released in order to condense 1 g of a gas to a liquid at its boiling point and at constant pressure is called the **heat of condensation** and has the *same numerical value* as the heat of vaporization. Table 12.1 lists the heats of vaporization or condensation of various liquids in various units.

Consider a problem involving these quantities.

Heat of vaporization Quantity of heat required to evaporate 1 g of a given liquid at its boiling point and at constant pressure.

Heat of condensation Quantity of heat released in order to condense 1 g of a gas to a liquid at its boiling point and at constant pressure.

EXAMPLE 12.1 Calculate the quantity of heat energy in kilocalories needed to vaporize 28.0 g of liquid water to steam at $1\overline{00}°C$.

SOLUTION From Table 12.1, the heat of vaporization of water is $54\overline{0}$ cal/g; we calculate the amount of heat energy in kilocalories as follows:

$$28.0 \text{ g} \times \frac{54\overline{0} \text{ cal}}{1 \text{ g}} \times \frac{1 \text{ kcal}}{1000 \text{ cal}} = 15.1 \text{ kcal} \qquad \textit{Answer}$$

Study Hint: The amount of heat is the *same* whether condensation (exothermic) or vaporization (endothermic) is occurring. This is like climbing stairs: the number of steps involved in going up or down a flight of stairs is the same; only the *direction* (up or down) of the process is different.

TABLE 12.1 **Heats of Vaporization or Condensation of Various Liquids at Their Normal Boiling Point and 1 Atm Pressure**

LIQUID	NORMAL BOILING POINT (°C)	HEAT OF VAPORIZATION OR CONDENSATION	
		cal/g	J/kg
Water	$1\overline{00}$	$54\overline{0}$	2.26×10^6
Alcohol	78.3	204	8.54×10^5
Heptane	98.4	76.5	3.20×10^5
Carbon tetrachloride	76.7	52.1	2.18×10^5
Benzene	80.1	94.1	3.94×10^5
Sodium chloride[a]	1465	698	2.92×10^6

[a] The extremely high values for the boiling point and heat of vaporization or condensation for sodium chloride over those values for the other liquids are due to the relatively strong attractive forces found in sodium chloride. These strong attractive forces are related to the ionic bonding in sodium chloride.

EXAMPLE 12.2 Calculate the quantity of heat energy in joules needed to vaporize 0.105 mol of liquid water to steam at $10\overline{0}$°C.

SOLUTION From Table 12.1, the heat of vaporization is 2.26×10^6 J/kg. We must convert the mole to grams and then kilograms to use this factor. Using the molar mass of water as 18.0 g, the calculation is as follows:

$$0.105 \, \cancel{mol} \times \frac{18.0 \, \cancel{g}}{1 \, \cancel{mol}} \times \frac{1 \, \cancel{kg}}{1000 \, \cancel{g}} \times \frac{2.26 \times 10^6 \, J}{1 \, \cancel{kg}} = 4270 \, J \qquad \textit{Answer}$$

ALTERNATE SOLUTION From Table 12.1, the heat of vaporization is also $54\overline{0}$ cal/g. Using this factor and the conversion of calories to joules (**4.184 J = 1 cal**; see Section 3.5), the calculation is as follows:

$$0.105 \, \cancel{mol} \times \frac{18.0 \, \cancel{g}}{1 \, \cancel{mol}} \times \frac{54\overline{0} \, \cancel{cal}}{1 \, \cancel{g}} \times \frac{4.184 \, J}{1 \, \cancel{cal}} = 4270 \, J \qquad \textit{Answer}$$

Study Exercise 12.2
Calculate the quantity of heat in kilocalories needed to vaporize 0.235 mol of liquid water to steam at $10\overline{0}$°C.

(2.28 kcal) Work Problems 14 through 17.

12.5 Distillation

Chemists often take advantage of vaporization and condensation to *distill* liquids. In **distillation**, we heat a liquid to its boiling point and then cool the vapors in a condenser to form the purified liquid. Figure 12.6 shows a simple distillation apparatus.

Distillation Purifying of a liquid by heating it to the boiling point and cooling the vapors in a condenser.

FIGURE 12.6
Distillation of water.

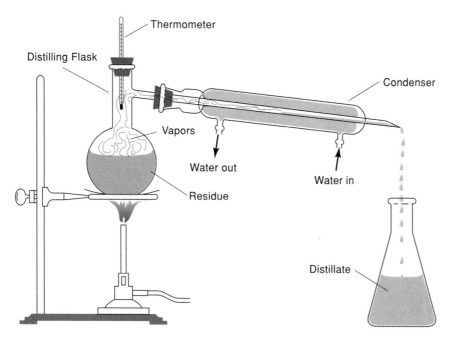

Thermometer

Distilling Flask

Condenser

Vapors

Water out

Residue

Water in

Distillate

A common application of distillation is to separate water from dissolved salts and other nonvolatile impurities. Colorless vapors appear in the distilling flask above the impure water and are condensed to give clear, colorless droplets of water called the *distillate*. The impurities remain in the distilling flask and are called the *residue*. Distilled water is prepared in this manner. The residue contains salts composed of calcium, magnesium, or iron with hydrogen carbonate (bicarbonate), carbonate, or sulfate.

The petroleum industry uses *fractional distillations* (many simple distillations) to refine gasoline, which is a mixture of carbon–hydrogen containing compounds.

12.6 Surface Tension and Viscosity

Surface tension   Property of a liquid that tends to draw the surface molecules into the body of the liquid and hence to reduce the surface to a minimum.

Distillation is possible because of the relatively free flowing nature of liquids. This characteristic, in turn, depends on the limited attraction between molecules in a liquid. Note, however, that this attraction, while limited, does exist. Two properties of this attraction are the surface tension and viscosity of liquids. **Surface tension** is the property of a liquid that tends to draw the surface molecules into the body of the liquid and hence to reduce the surface to a minimum. For example, mercury, because of its high surface tension, forms droplets on a glass surface, but water, whose surface tension is appreciably lower than that of mercury, tends to spread out on the surface. This property of a liquid can be explained by differences in attractive forces among molecules in different liquids. Any molecule in the body of a liquid is surrounded by molecules and is attracted equally in all directions by neighboring molecules. But a molecule at the *surface* of the liquid is attracted by other molecules *beneath* it and *not* above it. The result is an unbalanced force downward, tending to draw the surface molecules *into the body of the liquid* and to reduce the surface to a minimum.

FIGURE 12.7
Reading the volume of water in a graduated cylinder. Read at the bottom of the meniscus. The volume of water in this cylinder is 15 mL.

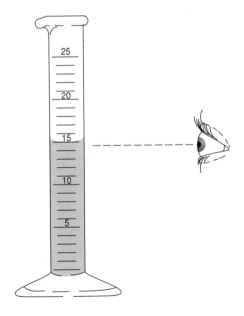

Some substances have greater surface tension than others because the attractive forces in these substances are greater (mercury versus water). Alcohol is often used to prepare an area for medical treatment because it has a low surface tension and can easily penetrate into a wound to cleanse the area. One reason for the cleansing action of soap solution is that it lowers the surface tension of water, allowing the solution to penetrate into the skin creases to clean out grease that holds dirt in place.

As temperature increases, the average kinetic energy of the molecules increases, and this increase in energy tends to overcome the attractive forces among molecules. As a result, the surface tension decreases. As you are aware, you can wash your hands more efficiently in hot water than in cold, in part because of the decreased surface tension in hot water.

In reading the volume of a liquid in a graduated cylinder or by other means for measuring volumes, you are told to read at the bottom of the *meniscus*, that is, the bottom of the curved surface of the liquid. For water the surface curves upward along the walls of the cylinder (see Figure 12.7). The surface tension of the liquid is one factor that causes this kind of behavior in liquids.

Another important property of liquids is **viscosity**, the resistance of a liquid to flow. Some liquids, such as honey, resist flow. Other liquids, such as water, flow more readily. Honey has a high viscosity; water has a more intermediate viscosity. A liquid like gasoline has an even lower viscosity than water because it flows more readily (see Figure 12.8). This property can be explained by the attractive forces among molecules. The greater these attractive forces, the more viscous the liquid. As temperature increases, the average kinetic energy of molecules increases, which breaks the attractive forces between molecules and decreases the viscosity of a substance.

Motor oil grades are based on the viscosity of the oil. Lower-viscosity oils (10W) are used in winter when it is cold, and higher-viscosity oils (40W) are used in summer when it is hot. Multiviscosity oils (10W-40) may be used year-round depending on the engine; the viscosity varies with the temperature and any additives.

Viscosity   Property of a liquid that describes the resistance of a liquid to flow; highly viscous liquids like honey flow very slowly.

(a)   (b)   (c)

FIGURE 12.8
Viscosity. (a) Honey has a high viscosity. (b) Water has an intermediate viscosity. (c) Gasoline has a low viscosity.

12.7 The Solid State

Now that we understand something about the makeup and behavior of gases and liquids, we are able to predict a good deal about solids. As you might have expected from our earlier descriptions of gases and liquids, **solids** have particles that are much closer together than particles in liquids. Particles in solids are also subject to strong attractive or repulsive forces among themselves. As a result, the particles in a solid have a relatively *fixed position* with respect to one another, as Figure 12.9 shows.

Finally, like gases and liquids, solids have six general characteristics deriving from the kinetic theory:

Solid   One of three states of matter; it is characterized by (1) lack of expansion, (2) definite shape, (3) constant volume, (4) lack of compressibility, (5) high density, and (6) severely limited mixing.

1. *No expansion (at constant temperature).* Like liquids, solids do not show infinite expansion as gases do although water, on freezing, does expand slightly.

2. *Shape.* Solids in general have definite shapes. They are relatively rigid and do not flow like gases and liquids do, except under extreme pressure. Thus, they do not take the shape of the container they occupy.

3. *Volume.* Solids maintain their volumes, as do liquids.

4. *Compressibility.* Solids are practically incompressible, as the particles are very close to each other because of their strong attractive forces.

5. *High density.* Solids, like liquids, have relatively high densities.

6. *Mixing.* Solids mix or diffuse extremely slowly except under extreme pressure. The particles in solids have essentially permanent positions because of the strong attractive forces between them. Hence, the motion of solid particles is generally very slow.

Table 12.2 summarizes the characteristics of all three physical states of matter.

> **Study Hint:** Of the three physical states of matter, particles in the gaseous state have the greatest freedom to move. The particles in liquids move less, and the particles in solids move the least.

FIGURE 12.9
Particles in the solid state are held in a relatively fixed position because attractive forces are stronger in the solid state than in the liquid state. A solid does not take the shape of the container it occupies, but a liquid does.

Solid

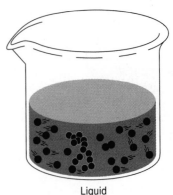

Liquid

TABLE **12.2**	Summary of the Properties of Solids, Liquids, and Gases		
PROPERTY	GASES	LIQUIDS	SOLIDS
Particle positions	Random	Limited mobility	Fixed
Expansion	Infinite	Very limited	Very limited
Shape	No definite shape	No definite shape	Definite shape
Volume	Occupies any volume	Maintains volume	Maintains volume
Compressibility	Very compressible	Slightly compressible	Incompressible
Density	Low	High	High
Mixing	Rapid	Slow	Extremely slow

12.8 The Shape of Solids

Before we consider the effect of these properties, the characteristic of shape merits further consideration. Solids can be conveniently divided into *crystalline* and *amorphous* solids, which differ from each other in structure. A **crystalline solid** consists of particles arranged in a definite geometric shape or form that is distinctive for a given solid. Examples of crystalline solids include sodium chloride, diamond, and quartz (a crystalline form of silica, silicon dioxide). Figure 12.10 shows photographs of crystalline structures. An **amorphous solid** consists of particles arranged in an irregular manner and thus lacks the order found in crystals. Examples of amorphous solids include glass and many plastics.

Amorphous solids differ from crystalline solids in the way they melt. If we monitor the temperature of a pure crystalline solid as it melts, we will find that it remains constant. Amorphous solids do not have such well-defined melting temperatures; they soften and melt over a *temperature range* and do not have a characteristic "melting point" (Figure 12.11).

Crystalline solid Any solid that consists of particles arranged in a definite geometric shape or form that is distinctive for that solid.

Amorphous solid Any solid that consists of particles arranged in an irregular manner and thus lacks the regular structure of a crystalline solid.

(a)   (b)

FIGURE 12.10
Crystalline solids: (a) sodium chloride; (b) quartz (a crystalline form of silica, silicon dioxide).

FIGURE **12.11**
(a) Diamond is an example of a crystalline solid. (Courtesy Fred J. Maroon/Photo Researchers) (b) Sealing wax is an example of an amorphous solid. (Courtesy Richard Megna/Fundamental Photographs)

(a)

(b)

12.9 Melting or Freezing Point. Heat of Fusion or Solidification

As you have certainly seen in your experience, if you cool water sufficiently, it turns into a solid—ice. And, of course, if you heat the ice, you can find a point where it turns into liquid water. Let's look at these simple processes from a molecular standpoint.

As a liquid cools, the kinetic energy of the particles decreases. At the *freezing point*, the particles become oriented in a definite geometric pattern characteristic of the substance, and crystals or irregular particles of the solid begin to appear. Removal of heat energy is necessary to allow the freer liquid particles to be "tied down" in the solid if crystallization is to continue. This temperature (**freezing point**) of the mixture of *solid* and *liquid* in dynamic equilibrium remains *constant* until *all* the liquid has solidified. Freezing, then, is an *exothermic change of state*.

If the solid is heated, the kinetic energy of some of the particles in the crystal will become sufficient to match the attractive forces in the crystal, and the solid will begin to melt at the solid's **melting point**. Because melting is an *endothermic change of state*, if melting is to continue, more heat must be applied to free the particles of the solid from each other, and the solid will change to the liquid at *constant* temperature. Therefore, the *freezing point is equal to the melting point* and is the temperature at which the liquid and solid forms are in dynamic equilibrium with each other:

$$\text{solid} \underset{\text{freezing}}{\overset{\text{melting}}{\rightleftharpoons}} \text{liquid} \tag{12.1}$$

where $\rightleftharpoons$ indicates an equilibrium (goes in both directions).

In Section 12.4, we observed that pressure considerably alters the boiling point of a liquid. That is, reduced pressure lowers the boiling point and increased pressure raises the boiling point. Only *large* pressures affect the melting or freezing point. For most substances, *increased pressure raises* the melting point. The exception is ice, where the melting point decreases with increased pressure because water's volume decreases in going from the solid to the liquid state. Other sub-

Freezing point Temperature at which the particles of a liquid begin to form crystals or irregular particles of a solid.

Melting point Temperature at which the kinetic energy of some of the particles in a solid matches the attractive forces in the solid and the solid begins to liquefy.

Study Hint: The two opposing processes in this equilibrium are melting and freezing. If we remove heat from the system, freezing predominates; if we add heat to the system, melting predominates.

stances occupy a *smaller* volume in the solid state. For ice, an increase of 1 atm pressure *decreases* the melting point by 0.0075°C.

This slight decrease in melting point with an increase in pressure is important in ice skating. The ice underneath the blades is at a temperature above its new melting point and therefore melts. The liquid layer provides the lubricant that enables the skater to skate across the ice. Another factor also important in melting the ice is the heat created by the friction between the ice and the blade. After the skate has passed over the ice, the pressure and friction decrease and the water refreezes. The skater is actually skating on liquid water. In fact, many speed skaters prefer ice that has small pools of water on it. If the ice temperature is too low, the pressure and friction are insufficient to melt the ice and skating becomes difficult (see Figure 12.12).

Consider an example.

EXAMPLE 12.3 Calculate the melting point of ice if a pressure of 25 atm is exerted by an ice skater on the ice.

SOLUTION A pressure of 1 atm lowers the melting point of ice 0.0075°C. Therefore, the calculation is as follows:

$$25 \text{ atm} \times \frac{0.0075°C}{1 \text{ atm}} = \begin{array}{l} 0.19° \text{ lowering; hence the melting} \\ \text{point of ice is } -0.19°C. \qquad \textit{Answer} \end{array}$$

Study Exercise 12.3
Calculate the melting point of ice if a pressure of 9600 mm Hg is exerted on the ice.

$(-0.095°C)$ Work Problems 18 and 19.

Like evaporation and boiling, conversion of a solid to a liquid at constant temperature requires heat (is endothermic). This heat is the **heat of fusion** of a substance and is the quantity of heat necessary to convert 1 g of a solid to the liquid at the *melting point* of the substance, as Figure 12.13 shows.

Heat of fusion Quantity of heat required to convert 1 g of a solid to liquid at the melting point of the substance; thus, fusion is an endothermic change of state.

FIGURE 12.12
An increase in pressure decreases the melting point of ice. This is applied in ice skating, especially speed skating. The skater is Dan Jansen (USA).

FIGURE 12.13
Heat of fusion (melting, →)
and solidification (freezing,
←).

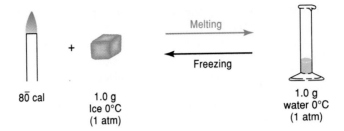

The reverse of the fusion process is solidification (crystallization), an exothermal process. The **heat of solidification (crystallization)** is the quantity of heat released by 1 g of a liquid as it becomes a solid at the melting point of the substance. The heat of solidification has the *same numerical value* as the heat of fusion. Table 12.3 lists the heats in cal/g and J/kg.

Consider problems involving these quantities.

Heat of solidification (crystallization) Quantity of heat released by 1 g of a liquid as it becomes a solid at the melting point of the substance; it is an exothermic change of state.

> **S**tudy Hint: Melting, liquefying, and fusing are the same process.

EXAMPLE 12.4 Calculate the quantity of heat energy in kilocalories evolved when 75.0 g of liquid water forms ice at 0°C.

SOLUTION From Table 12.3, the heat of solidification for water is $8\overline{0}$ cal/g. We calculate the amount of heat energy in kilocalories as follows:

$$75.0 \text{ g} \times \frac{8\overline{0} \text{ cal}}{1 \text{ g}} \times \frac{1 \text{ kcal}}{1000 \text{ cal}} = 6.0 \text{ kcal} \quad \textit{Answer}$$

(two significant digits because of the $8\overline{0}$ cal/g)

TABLE 12.3	Heats of Fusion or Solidification of Various Solids at Their Melting Points		

| | | HEAT OF FUSION OR SOLIDIFICATION | |
SOLID	MELTING POINT (°C)	cal/g	J/kg
Water	0	$8\overline{0}$	3.35×10^5
Alcohol	−117	24.9	1.04×10^5
Heptane	−91	33.7	1.41×10^5
Carbon tetrachloride	−23	5.09	2.13×10^4
Benzene	6	30.1	1.26×10^5
Sodium chloride[a]	804	124	5.19×10^5

[a] The high values for sodium chloride are again due to the strong attractive forces in the substance because of its ionic bonding. (See the footnote to Table 12.1).

EXAMPLE 12.5 Calculate the quantity of heat energy in kilojoules evolved when 0.510 mol of liquid benzene (C_6H_6) is converted to solid benzene at its melting point of 6°C.

SOLUTION From Table 12.3, the heat of solidification of benzene is 1.26×10^5 J/kg. We must convert the mole to grams and then to kilograms to use this factor. Using the molar mass of benzene (C_6H_6) as 78.0 g, the calculation is as follows:

$$0.510 \,\cancel{mol} \times \frac{78.0 \,\cancel{g}}{1 \,\cancel{mol}} \times \frac{1 \,\cancel{kg}}{1000 \,\cancel{g}} \times \frac{1.26 \times 10^5 \,\cancel{J}}{1 \,\cancel{kg}} \times \frac{1 \text{ kJ}}{1000 \,\cancel{J}} = 5.01 \text{ kJ} \quad Answer$$

ALTERNATE SOLUTION From Table 12.3, the heat of solidification for benzene is also 30.1 cal/g. Using this factor and the conversion of calories to joules (**4.184 J = 1 cal**; see Section 3.5), the calculation is as follows:

$$0.510 \,\cancel{mol} \times \frac{78.0 \,\cancel{g}}{1 \,\cancel{mol}} \times \frac{30.1 \,\cancel{cal}}{1 \,\cancel{g}} \times \frac{4.184 \,\cancel{J}}{1 \,\cancel{cal}} \times \frac{1 \text{ kJ}}{1000 \,\cancel{J}} = 5.01 \text{ kJ} \quad Answer$$

Study Exercise 12.4
Calculate the quantity of heat energy in kilocalories needed to convert 0.970 mol of ice to liquid water at 0°C.

(1.4 kcal)

Work Problems 20 through 23.

> **Study Hint:** The two opposing processes in this equilibrium are sublimation and deposition. If we remove heat from the system, deposition predominates; if we add heat to the system, sublimation predominates.

Sublimation Direct conversion of a solid to a vapor without passing through the liquid state; it is an endothermic change of state.

12.10 Sublimation

Substances in the solid state, like those in the liquid state, exhibit a definite vapor pressure at a given temperature and can pass directly from the solid state into the gaseous state. These particles in the solid state have obtained sufficient energy of motion to break away from the relatively fixed position in the solid state and can then pass directly into the gaseous state. This process, called **sublimation**, is defined as the direct conversion of a *solid* to the *vapor* without passing through the liquid state (see Figure 12.14). Since gas molecules have larger kinetic energies than molecules in the solid state, sublimation must be an endothermic process. Mothballs (*p*-dichlorobenzene), iodine, dry ice (solid carbon dioxide, Figure 12.15), and ice are examples of solids that can sublime. Wet clothes dry at subzero temperatures even though they are frozen because ice can sublime. This process oc-

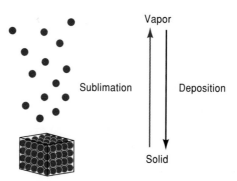

FIGURE 12.14
Sublimation. The solid is converted directly to the vapor state without passing through the liquid state. Deposition is the reverse of sublimation.

FIGURE 12.15
Solid carbon dioxide sublimes directly into the gas phase. Notice that no liquid can be detected.

Deposition   Direct conversion of a vapor to a solid without passing through the liquid state; it is an exothermic change of state.

curs best with a wind blowing the water vapor away and in relatively dry climates where the humidity is low. Freeze-dried foods are prepared by such a process.

The reverse of sublimation, **deposition**, is also shown in Figure 12.14. Heat energy is evolved in the process, and so deposition is an exothermic process. Ice or snow forming in clouds is an example of deposition.

12.11 Heat Energy Transformations in the Three Physical States of Matter

Now that we are well acquainted with all three physical states of matter, we can consider the actual heat transformations accompanying changes between states. Figure 12.16 summarizes these transformations and also shows the energy changes involved.

Figure 12.17 illustrates these changes by plotting heat energy added versus temperature. The process can be broken down into five separate steps.

1. As heat energy is added to a given solid, the temperature rises according to the *specific heat* of the solid. The kinetic energy of the particles in the solid increases, causing the particles to move more rapidly. As more heat energy is added, the kinetic energy continues to increase up to the melting point.

FIGURE 12.16
Summary of the changes of state. Endothermic changes are shown as →, while exothermic changes are shown as ←.

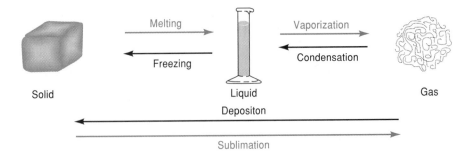

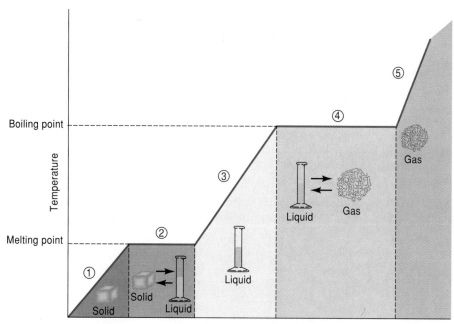

2. At the melting point, the solid begins to melt. The temperature remains constant until all the solid is converted to liquid. During this process, the solid and liquid states are in dynamic equilibrium and the addition of heat energy (*heat of fusion*) serves only to change the substance from a solid to a liquid, breaking the attractive forces in the solid state, and does *not* raise the temperature. Note that if heat energy is removed from the equilibrium mixture of a solid and liquid, the temperature will again remain constant and liquid will be converted to solid. Remember, solidification is an exothermic process. The heat given off by the solidification process just matches the heat energy removed from the system, and so the temperature *does not fall.*

3. After all the solid has been converted to liquid, further addition of heat energy increases the kinetic energy of the particles, and the temperature rises in accord with the specific heat of the liquid. This process continues until the temperature reaches the boiling point.

4. At the boiling point, the liquid begins to boil. The temperature remains constant until all the liquid is converted to gas. The heat energy put into the system just balances the energy required by the endothermic vaporization process. During this process, the liquid and gaseous states are in dynamic equilibrium, and the addition of heat energy (*heat of vaporization*) only changes the substance from a liquid to a gas, breaking the attractive forces in the liquid state to give the gaseous state with almost no attractive forces in the gaseous state. The heat energy does *not* raise the temperature. If heat energy is removed from the equilibrium mixture of a liquid and gas, the temperature will also remain constant and gas will be converted to liquid (condensation is exothermic).

5. After all the liquid has been converted to gas, further heating increases the kinetic energy of the gas molecules, and the temperature rises according to the specific heat of the gas.

Now let us consider how to determine the quantity of heat energy required for a given transformation.

EXAMPLE 12.6 Calculate the quantity of heat energy in calories required to convert $12\overline{0}$ g of ice at exactly 0°C to steam at $1\overline{00}$°C.

SOLUTION While a diagram like the following is not required as part of the solution, it often makes it easier to see what steps are involved.

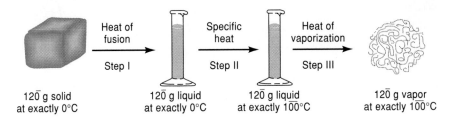

$12\overline{0}$ g solid at exactly 0°C	$12\overline{0}$ g liquid at exactly 0°C	$12\overline{0}$ g liquid at exactly $1\overline{00}$°C	$12\overline{0}$ g vapor at exactly $1\overline{00}$°C

From Table 12.3, the heat of fusion of water is $8\overline{0}$ cal/g. Thus, the energy required for step I is

$$12\overline{0}\,g \times \frac{8\overline{0}\,\text{cal}}{1\,g} = 9600\,\text{cal}$$

From Table 3.4, the specific heat of water is 1.00 cal/(g·°C). Thus, the energy required for step II is

$$\frac{1.00\,\text{cal}}{1\,g\cdot°C} \times 12\overline{0}\,g \times (1\overline{00} - 0)°C = 12,\overline{0}00\,\text{cal}$$

(Review Section 3.5 for the calculation in step II.) From Table 12.1, the heat of vaporization of water is $54\overline{0}$ cal/g. Hence, the energy required for step III is

$$12\overline{0}\,g \times \frac{54\overline{0}\,\text{cal}}{1\,g} = 64,800\,\text{cal}$$

The total heat energy requirement of steps I through III is

$$
\begin{array}{rl}
Step\ \ I: & 9,600\ \text{cal} \\
Step\ \ II: & 12,\overline{0}00\ \text{cal} \\
Step\ \ III: & \underline{64,800\ \text{cal}} \\
\text{total heat energy required} = & 86,400\ \text{cal} \qquad \textit{Answer}
\end{array}
$$

Note that the quantity of heat energy required in step I (9600 cal) and step II (12,$\overline{0}$00 cal) is nearly equal and relatively small, but that over five times this quantity of heat energy is required in step III (64,800 cal) to convert the liquid water (1$\overline{00}$°C) to vapor (1$\overline{00}$°C).

The process can be reversed; that is, heat energy is given off on cooling. Consider the following example.

> **Study Hint:** In problems involving energy changes, you should always break the problem down into simple individual steps.

EXAMPLE 12.7   Calculate the quantity of heat energy in calories liberated when $45\overline{0}$ g of steam at $10\overline{0}°C$ is changed to liquid water at $2\overline{0}°C$.

SOLUTION   Diagram the change as follows:

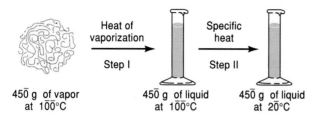

$45\overline{0}$ g of vapor $45\overline{0}$ g of liquid $45\overline{0}$ g of liquid
at $10\overline{0}°C$ at $10\overline{0}°C$ at $2\overline{0}°C$

Step I: $45\overline{0} \, \cancel{g} \times \dfrac{54\overline{0} \, \text{cal}}{1 \, \cancel{g}} = 243{,}000$ cal evolved

Step II: $\dfrac{1.00 \, \text{cal}}{1 \, \text{g·}1°C} \times 45\overline{0} \, \cancel{g} \times (10\overline{0} - 2\overline{0})°\cancel{C} = 36{,}000$ cal evolved

The total of steps I and II is

$$\begin{array}{rl} \textit{Step I}: & 243{,}000 \text{ cal} \\ \textit{Step II}: & \underline{36{,}000 \text{ cal}} \\ \text{total heat energy liberated} = & 279{,}000 \text{ cal} \quad \textit{Answer} \end{array}$$

Of the two steps, the greater quantity of heat energy is liberated in step I, which again emphasizes the scalding property of steam when it condenses on a cooler surface.

EXAMPLE 12.8   Calculate the quantity of heat energy in kilojoules required to convert 20.0 g of ice at exactly 0°C to liquid water at $4\overline{0}°C$.

SOLUTION   Diagram the change as follows:

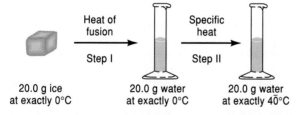

20.0 g ice   20.0 g water   20.0 g water
at exactly 0°C   at exactly 0°C   at exactly $4\overline{0}°C$

Step I: Using the heat of fusion of water as 3.35×10^5 J/kg (Table 12.3), the calculation is as follows:

$$20.0 \, \cancel{g} \times \dfrac{1 \, \cancel{kg}}{1000 \, \cancel{g}} \times \dfrac{3.35 \times 10^5 \, \cancel{J}}{1 \, \cancel{kg}} \times \dfrac{1 \, \text{kJ}}{1000 \, \cancel{J}} = 6.70 \, \text{kJ}$$

Step II: Using the specific heat of water as 4.18×10^3 J/(kg·K) (Table 3.4) and the temperature change of $4\overline{0}$ K [313 K ($4\overline{0}°C$) $-$ 273 K (0°C)], the calculation is as follows:

$$\dfrac{4.18 \times 10^3 \, \cancel{J}}{\cancel{kg}·\cancel{K}} \times 20.0 \, \cancel{g} \times \dfrac{1 \, \cancel{kg}}{1000 \, \cancel{g}} \times 4\overline{0} \, \cancel{K} \times \dfrac{1 \, \text{kJ}}{1000 \, \cancel{J}} = 3.3 \, \text{kJ}$$

The total of steps I and II is

$$
\begin{array}{lr}
\textit{Step I}: & \text{6.70 kJ} \\
\textit{Step II}: & \underline{\text{3.3 \ kJ}}
\end{array}
$$

total heat energy required $=$ 10.00 kJ , 10.0 kJ (to the tenths place) *Answer*

ALTERNATE SOLUTION

Step I: Using the heat of fusion of water as $8\overline{0}$ cal/g (Table 12.3) and converting to kilo-joules (**4.184 J = 1 cal**; see Section 3.5), the calculation is as follows:

$$
20.0 \text{ g} \times \frac{8\overline{0} \text{ cal}}{1 \text{ g}} \times \frac{4.184 \text{ J}}{1 \text{ cal}} \times \frac{1 \text{ kJ}}{1000 \text{ J}} = \text{6.7 kJ (to two significant digits because of } 8\overline{0} \text{ cal/g)}
$$

Step II: Using the specific heat of water as 1.00 cal/(g·°C) and converting to kilojoules, the calculation is as follows:

$$
\frac{1.00 \text{ cal}}{\text{g·°C}} \times 20.0 \text{ g} \times 4\overline{0}\text{°C} \times \frac{4.184 \text{ J}}{1 \text{ cal}} \times \frac{1 \text{ kJ}}{1000 \text{ J}} = \text{3.3 kJ}
$$

The total of steps I and II is

$$
\begin{array}{lr}
\textit{Step I}: & \text{6.7 kJ} \\
\textit{Step II}: & \underline{\text{3.3 kJ}}
\end{array}
$$

total heat energy required $=$ 10.0 kJ   *Answer*

Study Exercise 12.5

Calculate the quantity of heat energy in kilocalories required to convert 30.0 g of ice at exactly 0°C to steam at $10\overline{0}$°C.

Work Problems 24 through 28.

(21.6 kcal)

 Summary

The kinetic theory explains six general characteristics of liquids: limited expansion, lack of shape, constant volume, limited compressibility, high density, and limited mixing (Section 12.1).

Real gases condense to form liquids when attractive forces between the molecules overcome the kinetic energy of these molecules, at high pressure or low temperature. A reversal of these conditions causes liquids to evaporate to form gases (Section 12.2). Vapor pressure is the equilibrium pressure at which the rate of evaporation of a pure liquid is equal to the rate of condensation of the vapor in a closed container (Section 12.3).

When the vapor pressure of a liquid equals the external pressure on the liquid, the liquid's boiling point is reached. Boiling and evaporation are endothermic changes of state that require the absorption of energy. Condensation is an exothermic change of state that requires the release of heat. The heat of vaporization required to vaporize a liquid at the boiling point and the heat of condensation released when a gas liquefies have the same numerical value for a given amount of substance (Section 12.4). Chemists use vaporization and condensation to distill (purify) liquids (Section 12.5).

Liquids differ in their surface tension, the tendency to draw surface molecules into the body of the liquid and hence to reduce the surface to a minimum, and in their viscosity, or resistance to flow (Section 12.6).

The kinetic theory also explains six general characteristics of solids: lack of expansion, definite shape, constant volume, lack of compressibility, high density, and very limited mixing (Section 12.7). Solids may be either crystalline or amorphous (Section 12.8).

Solids are subject to exothermic changes of state such as freezing and endothermic changes of state such as melting. Solids liquefy when heat is added (heat of fusion), and heat is released when liquids solidify (heat of solidification) (Section 12.9).

In some cases, solids convert directly to the gas state without passing through the liquid state, an endothermic process called sublimation. The reverse of sublimation, deposition, is an exothermic process (Section 12.10). In general, however, substances pass from the gaseous to the liquid and on to the solid state as more and more heat is removed and reverse the process as more and more heat is added (Section 12.11).

 # Exercises

1. Define or explain the following terms (the number in parentheses refers to the section in the text where the term is mentioned):

   a. liquid (12.1)

   b. condensation (12.2)

   c. evaporation (12.2)

   d. dynamic equilibrium (12.3)

   e. vapor pressure (12.3)

   f. boiling point (12.4)

   g. endothermic change of state (12.4)

   h. exothermic change of state (12.4)

   i. heat of vaporization (12.4)

   j. heat of condensation (12.4)

   k. distillation (12.5)

   l. surface tension (12.6)

   m. viscosity (12.6)

   n. solid (12.7)

   o. crystalline solid (12.8)

   p. amorphous solid (12.8)

   q. freezing point (12.9)

   r. melting point (12.9)

   s. heat of fusion (12.9)

   t. heat of solidification (crystallization) (12.9)

   u. sublimation (12.10)

   v. deposition (12.10)

2. Distinguish between:

   a. evaporation and condensation

   b. boiling point and vapor pressure

   c. a crystalline solid and an amorphous solid

   d. sublimation and evaporation

3. List the characteristics of gases, liquids, and solids.

Evaporation (See Sections 12.2 and 12.4)

4. Upon evaporating, rubbing alcohol (isopropyl alcohol) gives a cooling sensation to the skin. Why?

Boiling Point (See Section 12.4)

5. Potatoes, when cooked in boiling water in the mountains, never appear to be completely cooked. Explain.

6. Explain why your hands get cleaner when you wash with soap and *hot* water than they do when you use cold water.

7. Experiments show that the viscosity of castor oil decreases as the temperature increases. Explain this observation in terms of the kinetic theory.

8. Which physical state of matter exhibits the strongest attractive forces between its particles? The weakest?

Melting Point (See Section 12.9)

9. In outdoor speed ice skating, the skate blade is very thin. Explain.

10. Mercury thermometers are of no use below −39°C; alcohol (ethyl alcohol) thermometers are used in extremely cold regions when the temperature dips to below −39°C. Why? (*Hint:* In *Handbook of Chemistry and Physics* or other suitable references, look up the physical properties of the two liquids.)

Sublimation (See Section 12.10)

11. Dry ice is usually stored in a covered box. Explain.

12. A chemistry student placed an odorous solid in an open beaker in his laboratory desk to dry until the next laboratory period. To his astonishment, when he returned to the laboratory the next week, he found only half the solid remaining. Explain.

 # Problems

Refer to Tables 12.1 and 12.3 for additional data.

Boiling Point (See Section 12.4)

13. Use the graph in Figure 12.4 to determine the following:
    a. The boiling point of heptane at $6\overline{0}0$ mm Hg (torr).
    b. The boiling point of water on top of Mt. Whitney, California (14,495 ft); atmospheric pressure is approximately 435 mm Hg (torr).
    c. The boiling point of water on top of Mt. Everest, Nepal–Tibet border (29,028 ft); atmospheric pressure is approximately 240 mm Hg (torr).
    d. The boiling point of heptane on top of Mt. Everest.

Heat of Vaporization or Condensation (See Section 12.4)

14. Calculate the quantity of heat energy in kilocalories evolved when 16.0 g of steam condenses to form liquid water at $1\overline{0}0$°C.

15. Calculate the quantity of heat energy in kilojoules required to vaporize 135 g of liquid water to steam at $1\overline{0}0$°C.

16. Calculate the quantity of heat energy in kilojoules required to vaporize 0.128 mol of liquid water to steam at $1\overline{0}0$°C.

17. Calculate the quantity of heat energy in calories required to vaporize 0.625 mol of carbon tetrachloride (CCl_4) at its boiling point (76.7°C). Repeat the calculation for the same amount of sodium chloride at its boiling point (1465°C). Explain the large difference in calculated values.

Melting Point (See Section 12.9)

18. Calculate the melting point of ice if a pressure of 17 atm is exerted by a young woman skating on the ice.

19. In ice skating, a skate blade can exert a pressure of approximately $3\overline{0}$ atm by a hefty man on the surface of the ice. What is the melting point of ice at this pressure?

Heat of Fusion or Solidification (Crystallization) (See Section 12.9)

20. Calculate the quantity of heat energy in kilocalories evolved when 55.0 g of liquid water form ice at 0°C.

21. Calculate the quantity of heat energy in kilocalories required to melt 0.750 mol of ice to liquid water at 0°C.

22. Calculate the quantity of heat energy in kilojoules required to melt 0.0996 mol ice to liquid water at 0°C.

23. Calculate the quantity of heat energy in kilocalories required to liquify 0.435 mol of solid benzene (C_6H_6) at its melting point (6°C).

Changes of State (See Section 12.11)

24. Calculate the quantity of heat energy in kilocalories required to convert 25.0 g of ice at exactly 0°C to steam at exactly $10\overline{0}$°C. [Specific heat of water $=$ 1.00 cal/(g·°C).]

25. Calculate the quantity of heat energy in kilojoules required to convert 35.0 g of ice at exactly 0°C to steam at exactly $10\overline{0}$°C. [Specific heat of water $=$ 4.18×10^3 J/(kg·K).]

26. Calculate the quantity of heat energy in kilocalories required to convert 1.10 mol of ice at exactly 0°C to steam at $10\overline{0}$°C. [Specific heat of water $=$ 1.00 cal/(g·°C).]

27. Calculate the quantity of heat energy in kilojoules evolved when 0.120 kg of steam at $10\overline{0}$°C condenses to form liquid water at $6\overline{0}$°C. [Specific heat of water $=$ 4.18×10^3 J/(kg·K).]

28. Calculate the quantity of heat energy to the nearest calorie required to convert 15.00 g of ice at −8.0°C to steam at 105.0°C. [Specific heats: ice $=$ 0.500 cal/(g·°C); water $=$ 1.00 cal/(g·°C); steam $=$ 0.480 cal/(g·°C).]

General Problems

29. Calculate the mass of water in grams that can be heated from exactly 0°C to 15°C by the heat evolved on cooling 1.00 kg of water from $10\overline{0}$° to 25°C.

30. Seeding a hurricane involves dropping silver iodide crystals from a plane just outside the eye of the storm. The seed acts to condense the supercooled water droplets (between −20° and 0°C) to ice, giving off heat. This heat energy then increases the temperature of the surrounding air and decreases the pressure at the edge of the eye. The intent is to reduce the extremely high wind velocity in the area. In the seeding of a certain hurricane, silver iodide was released along a line approximately 18.0 mi long at $33,\overline{0}00$ ft and scattered to approximately $18,\overline{0}00$ ft.

    a. Considering this "curtain of silver iodide" to be on the average approximately 5.00 mi wide because of the wind, what volume of air in cubic meters was seeded? (1 in. $=$ 2.54 cm, 5,280 ft $=$ 1 mi).

    b. "On the average," this volume of air contained 2.00 g of water droplets per cubic meter. Assuming that all these droplets at 0°C were converted to ice at 0°C, calculate the number of kilocalories of heat energy released in the seeding process.

✓ Chapter Quiz 12

1. Calculate the quantity of heat energy in kilocalories evolved when 65.0 g of steam condenses to form liquid water at $1\overline{00}°C$. (Heat of condensation = $54\overline{0}$ cal/g.)

2. Calculate the quantity of heat energy in kilojoules required to convert 0.220 mol of liquid water to steam at $1\overline{00}°C$. (Heat of vaporization = 2.26×10^6 J/kg.)

3. Calculate the quantity of heat in calories evolved when 0.138 kg of liquid water forms ice at 0°C. (Heat of solidification = $8\overline{0}$ cal/g.)

4. Calculate the quantity of heat energy in kilojoules required to convert 75.0 g of ice at exactly 0°C to steam at $1\overline{00}°C$. (Heat of fusion = 3.35×10^5 J/kg; specific heat of water = 4.18×10^3 J/(kg·K); heat of vaporization = 2.26×10^6 J/kg.)

Mercury (Symbol: Hg)

The Element MERCURY: The Mad Hatter

Mercury is a mobile silver liquid at ordinary temperatures. It is one of only four elements that is liquid at room temperature.

The Mad Hatter at the tea party after the design by Sir John Tenniel for the first edition, 1865, of Lewis Carroll's Alice's Adventures in Wonderland.

Name: Symbol derives from *hydrargyrum* (Latin), meaning "liquid silver." Named after the Roman god Mercury, the messenger of the gods, who was clever and fast. Also called quicksilver.

Appearance: Bright, silver, metallic liquid.

Occurrence: Mercury occurs in small amounts in the earth's crust combined with other elements such as oxygen and sulfur.

Source: Obtained commercially by heating cinnabar [mercury(II) sulfide] in oxygen to 600°C. The products of this process are metallic mercury and sulfur dioxide gas.

$$HgS(s) \ + \ O_2(g) \ \xrightarrow{\Delta} \ Hg(l) \ + \ SO_2(g)$$

Common Uses: Mercury is used extensively in the electronics industry because it is one of the few liquid materials that is an excellent conductor of electricity.

Mercury is used in batteries for hearing aids, calculators, watches, and pacemakers.

Mercury is used in fluorescent lamps for home, industrial, and street lighting. These lamps produce a characteristic yellow light.

Mercury amalgams (solutions of metals or metal salts in mercury) are used in a variety of applications, including dental fillings.

Mercury is used in a variety of measuring instruments, including thermometers (see Section 2.3) and barometers (see Section 11.2) because it has a high density and remains liquid over a large range of temperatures.

Unusual Facts: Mercury and its compounds are quite toxic (poisonous), affecting mainly the nervous system. For years, mercury was used to extract gold and silver, and mercury compounds were used as pesticides, antiseptics, and explosives. Today these functions are largely carried out in less hazardous ways. Felt used to be made by softening organic fibers with mercury salts, and hat makers often exhibited the effects of constant exposure to mercury-containing substances. The Mad Hatter in *Alice's Adventure in Wonderland* is based on this historical premise.

CHAPTER 13

Water

The launch of the space shuttle Discovery. The combustion of liquid hydrogen and liquid oxygen to form water and energy is used to launch rockets into space (NASA).

GOALS FOR CHAPTER 13

1. To list various physical properties of water (Section 13.1).

2. To describe the structure of water, to explain why the H—O—H bond angle is 105°, and to explain how this bond angle makes water a polar compound (Sections 13.2 and 13.3).

3. To describe a hydrogen bond and how it affects the physical properties of water (Section 13.4).

4. To complete and balance chemical reaction equations in which water is a product (Section 13.5) or a reactant (Section 13.6).

5. To define the term *hydrate*, to explain the mechanisms by which water is bonded within a hydrate, to calculate the percent water in a hydrate and the formula of a hydrate, and to define the terms *efflorescent substances, hygroscopic substances,* and *deliquescent substances* (Section 13.7).

6. To explain how an ion-exchange column can be used to convert hard water to soft water and to explain how dangerous impurities are removed from water (Section 13.8).

7. To describe the structure, preparation, properties, and uses of hydrogen peroxide (Section 13.9).

Countdown

You may use the periodic table and the rules for the solubility of inorganic substances in water.

ELEMENT	ATOMIC MASS UNITS (amu)
Ba	137.3
Cl	35.5
C	12.0
O	16.0
H	1.0
S	32.1

5. Write Lewis structures and structural formulas for the following molecules (Section 6.7):

   a. CBr_4

   $$\left(\begin{array}{c} \ddot{\ddot{Br}} \\ \ddot{\ddot{Br}} \overset{\times\times}{\underset{\cdot\cdot}{C}} \ddot{\ddot{Br}} \\ \ddot{\ddot{Br}} \end{array} \right) \quad \left(\begin{array}{c} Br \\ | \\ Br-C-Br \\ | \\ Br \end{array} \right)$$

   b. PBr_3

   $$\left(\begin{array}{c} \ddot{\ddot{Br}} \overset{\times\times}{\underset{\cdot\cdot}{P}} \ddot{\ddot{Br}} \\ \ddot{\ddot{Br}} \end{array} \right) \quad \left(\begin{array}{c} Br-P-Br \\ | \\ Br \end{array} \right)$$

4. Calculate the following (Sections 7.3, 7.4, and 8.2.):
   a. moles of barium chloride in 70.5 g of barium chloride
   (0.338 mol)
   b. moles of barium carbonate in 85.8 g of barium carbonate. (0.435 mol)

3. Calculate the percent hydrogen in the following acids (Sections 7.6 and 8.4):
   a. sulfuric acid (2.0%)
   b. acetic acid (6.7%)

2. Balance the following chemical reaction equations (Sections 9.3 and 9.4):
   a. $PbO_2(s) \xrightarrow{\Delta} PbO(s) + O_2(g)$
   $$(2 \longrightarrow 2 + 1)$$
   b. $C_6H_{12}O_6(s) + O_2(g) \xrightarrow{\Delta} CO_2(g) + H_2O(g)$
   $$(1 + 6 \longrightarrow 6 + 6)$$

1. Predict the products using the rules for the solubility of inorganic substances in water and the periodic table for the following neutralization reactions and balance the chemical equations. [Indicate any precipitate by (s).] (Section 9.10).
   a. $Pb(OH)_2(s) + HC_2H_3O_2(aq) \longrightarrow$
   $$[Pb(OH)_2(s) + 2\,HC_2H_3O_2(aq) \longrightarrow$$
   $$Pb(C_2H_3O_2)_2(aq) + 2\,H_2O(l)]$$
   b. $Pb(OH)_2(s) + H_3PO_4(aq) \longrightarrow$
   $$[3\,Pb(OH)_2(s) + 2\,H_3PO_4(aq) \longrightarrow$$
   $$Pb_3(PO_4)_2(s) + 6\,H_2O(l)]$$

D id you know?

✔ Water is the most abundant substance on earth.

✔ In solid (ice) and liquid form, water covers 71% of the earth's surface.

✔ Your body is composed of 60% water.

✔ You and all other living organisms require water to live.

✔ Human beings require about 2.7 qt of water daily.

✔ The average American gets about 1.6 qt of water each day from foods and 1.1 qt from beverages.

Because water is so much a part of human life, and because it is an integral part of so many chemical compounds that make up living and nonliving objects in our world, in this chapter we will consider this precious fluid, its characteristics, bonding, formation, reactions, and purification. In addition, we will examine some of the most common substances containing water.

13.1 Physical Properties of Water

In studying the previous few chapters, we learned a great deal about the properties and behavior of water. This section attempts to put it all together so that we feel comfortable using information about water as we continue our study of chemistry.

Appearance

Pure water is colorless, odorless, and tasteless. Any variation in these properties of water is due to impurities dissolved in it. An example of such an impurity is hydrogen sulfide, which is often dissolved in sulfur spring water, giving it the odor of rotten eggs.

Density

At 4°C, the density of liquid water is 1.00000 g/mL, which is its maximum density. This value is used as the standard in calculations involving specific gravity (see Section 2.9). At 0°C, the density of liquid water is 0.99987 g/mL, while the density of ice is 0.917 g/mL. Because ice is *less dense* than liquid water at the same temperature, it floats on water. The volume of floating ice exposed to the air is approximately 8%. About 92% is below the surface (see Figure 13.1).

Ice and water at 0°C have a lower density than water at temperatures above 4°C. As a result, lakes and ponds freeze from the *top down*. If ice were denser than water, as most solids are denser than their corresponding liquids, lakes and ponds would freeze from the bottom up. They would be completely frozen in the winter, and fish could not live in them.

The expansion of water on freezing also accounts for the weathering of rocks. As water freezes in cracks, it expands and thus assists in the rocks' erosion.

Table 13.1 summarizes some of the physical properties of water.

13.2 The Structure of the Water Molecule

Many of the properties of water that we take for granted are a direct result of its molecular structure. Thus, to truly understand the properties of water, we must understand its structure. In other words, water is much like an automobile. We can use it and appreciate its qualities without understanding its inner workings. But we cannot know *why* it behaves as it does without "looking under the hood."

In Example 6.2 (Section 6.7), we constructed the following Lewis structure for the water molecule:

$$H : \overset{..}{\underset{..}{O}} :$$
$$H$$

TABLE 13.1	Some Physical Properties of Water
PROPERTY	VALUE
Appearance	Colorless, odorless, tasteless
Normal boiling point	$10\overline{0}°C$
Melting point	$0°C$
Density at 4°C	1.00000 g/mL
Specific heat of liquid	1.00 cal/(g·°C), 4.18×10^3 J/(kg·K)
Heat of fusion	$8\overline{0}$ cal/g, 3.35×10^5 J/kg
Heat of vaporization	$54\overline{0}$ cal/g, 2.26×10^6 J/kg

This structure suggests that each water molecule consists of a central oxygen atom bonded to two hydrogen atoms by covalent bonds containing two electrons each. In addition, each oxygen atom has two unshared pairs of electrons.

Despite the positions of the O and H atoms in this model, we noted that the H—O—H bond angle in water is 105°. In Section 6.8 we described the water molecule as *bent*. The reason for this is that the oxygen atom has *four* groups of electrons around it, two pairs of bonding electrons (O—H, single bonds), and two pairs of unshared electrons. Each pair of electrons repels all the other pairs because of the repulsion of like charges (Coulomb's law; see Section 6.6).

To minimize these repulsive interactions the four pairs of electrons get as far away from each other as they can. This, then, forms a tetrahedron with the oxygen atom at the center (see Section 6.8, Special Cases of Tetrahedral: Bent and Pyramidal Molecules) and the electron pairs at the four corners of the tetrahedron, as shown in Figure 13.2a, creating a bond angle of 109.5°. In this way, the electrons pairs remain close to the oxygen atom while maximizing the distance between them. The repulsive interaction between unshared pairs of electrons in water is somewhat greater than the repulsive interaction between the electrons pairs involved in the hydrogen–oxygen covalent bonds. The result is the slightly smaller bond angle of 105° in the water molecules as shown in Figure 13.2b.

13.3 Polarity of Water

We noted in Section 6.5 that oxygen is more electronegative than hydrogen. This difference in the electronegativities of oxygen and hydrogen results in an unequal sharing of electrons in the formation of the covalent bonds between oxygen and

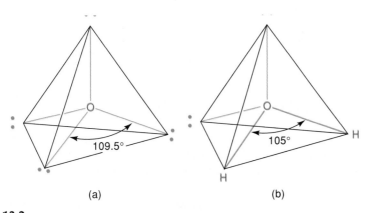

(a)   (b)

FIGURE 13.2
The structure of an individual water molecule. (a) The four electron pairs are as far away from each other as they can be. The oxygen atom is at the center of a tetrahedron, creating a bond angle of 109.5°. The electron pairs remain close to the oxygen atom, while maximizing the distance among themselves. (b) The structure of the water molecule showing the covalent bonds (blue) and the unshared pairs of electrons (red). The repulsive interaction between the unshared pairs of electrons is slightly more than the repulsive interaction between the electron pairs in the oxygen–hydrogen covalent bond. The result is a slightly smaller bond angle of 105° in the water molecule.

Polar bond Type of chemical bond, also known as a polar covalent bond, formed by the *unequal* sharing of electrons between two atoms whose electronegativities differ.

Dipole moment Property of a molecule where the presence of polar bonds generates a center of positive charge and a center of negative charge that do not coincide; such molecules have regions that are always positively or negatively charged.

> **Study Hint:** A dipole moment is something like chocolate chip ice cream in which the chips are not uniformly distributed throughout the ice cream. Some parts of the ice cream (molecule) have more chips (charge) than others.

Work Problem 8.

hyrdogen, with the hydrogen being partially positive (δ^+) and the oxygen partially negative (δ^-), as Figure 13.3a shows. This imbalance produces a **polar bond** (polar covalent bond). A molecule with such polar bonds has a center of positive charge (center of the partially positive sites) and a center of negative charge (center of the partially negative sites) within the molecule. A molecule in which these centers of positive and negative charge do not coincide, but are separated by a finite distance, is said to have a **dipole moment**. Such molecules have regions that are always positively or negatively charged. As Figure 13.3b shows, an arrow symbol ($\longmapsto$) indicates this moment, with the head of the arrow pointing toward the negative center. Compounds having molecules possessing *dipole moments*, like water, are referred to as *polar*.

If the hydrogen–oxygen bond angle were 180°C (linear), then *no* net dipole moment would exist for the compound because the center of the positive charges would coincide with the center of negative charge on the oxygen atom.

$$\text{H} \underset{\text{(linear)}}{\text{—O—H}} \quad \text{dipole moment} = 0$$

Therefore, the *shape* of a molecule is one of the important factors in determining its dipole moment. The presence of a dipole moment in water indicates that the angle between the hydrogens is not 180°.

Study Exercise 13.1

Represent the *net dipole moment*, if any, for the following molecules. [*Hint:* Write the Lewis structure and structural formula (Section 6.7), consider the electronegativities of the elements (Figure 6.11), and then decide on the shape of the molecule, that is, linear, trigonal planar, tetrahedral, bent, or pyramidal (Section 6.8)].

a. HF   (H—F)
  $\longmapsto$

b. H_2S

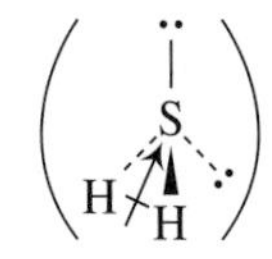

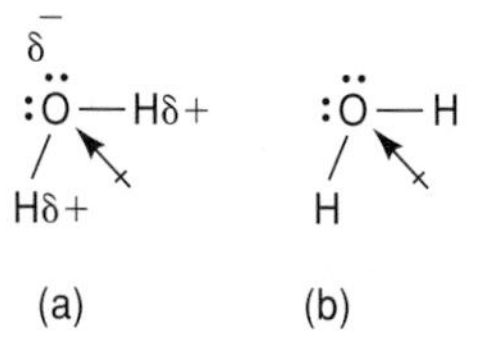

FIGURE 13.3
Polarity of water: (a) unequal sharing of electrons in the water molecule with the hydrogen partially positive (δ^+), and the oxygen partially negative (δ^-); (b) *net dipole moment* in a water molecule.

13.4 Hydrogen Bonding in Water

As Table 13.2 shows, water has some extraordinary properties when compared with the hydrogen compounds of the other elements (S, Se, Te) in the same group as oxygen, group VIA (16). The melting and boiling points of these other hydrogen compounds are plotted in Figure 13.4. Note the steady decrease in both properties from H_2Te to H_2Se to H_2S. If we were to predict the melting and boiling points (shown as open circles in Figure 13.4) of water from the trends shown by H_2S, H_2Se, and H_2Te, we would get values of $-106°C$ (melting) and $-81°C$ (boiling), not the true values of 0°C and 100°C, respectively. Also from Table 13.2, note that the heats of fusion and vaporization for water are high.

Why, then, does water have comparatively high melting and boiling points and heats of fusion and vaporization? These properties result from a special type of bond that forms between water molecules to a much greater extent than between the molecules of the other hydrogen compounds. As we noted earlier in this chapter, water is polar. The partial negative charge (δ^-) on the oxygen atom from the un-

TABLE 13.2	**Some Physical Properties of Hydrogen Compounds of Group VIA (16)**			
PROPERTY	H_2O	H_2S	H_2Se	H_2Te
Molar mass (g)	18.0	34.1	81.0	129.6
Melting point (°C)	0	−86	−66	−49
Boiling point (°C)	100	−61	−41	−2
Heat of fusion				
(kcal/mol)	1.44	0.57	0.60	1.0
(kJ/mol)	6.02	2.38	2.51	4.18
Heat of vaporization				
(kcal/mol)	9.72	4.46	4.62	5.55
(kJ/mol)	40.7	18.7	19.3	23.2

shared pairs of electrons in *one* molecule of water attracts a partial positive charge (δ^+) on the hydrogen atom of another molecule of water.

This attraction results in the formation of a weak chemical bond called a *hydrogen bond*. A **hydrogen bond** results when a hydrogen atom bonded to a highly electronegative atom (X) becomes partially bonded to another electronegative atom. *Electronegative* atoms, such as F, O, and N (compare the order of electronegativities, Figure 6.11), are generally involved in such bonding. In the case of water, these hydrogen bonds hold *two* or *more* molecules together in chains or clusters. This linkage is depicted with a dashed line (-----), as shown below and in Figures 13.5 and 13.6.

Hydrogen bond Type of weak chemical bond formed when a hydrogen atom bonded to a highly electronegative atom (F, O, or N) also bonds partially to another electronegative atom.

$$X—H\text{-----}X—H$$
$$\uparrow$$
$$\text{hydrogen bond}$$

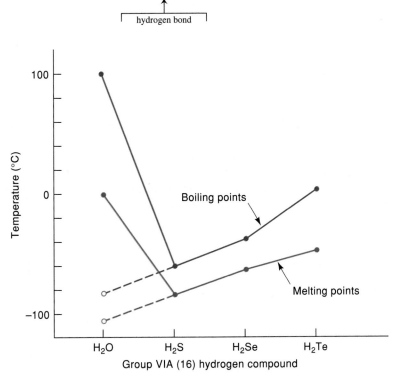

FIGURE 13.4
Prediction of the melting and boiling points of water. The melting (red) and boiling (blue) points of the group VIA (16) hydrogen compounds are plotted as solid circles. The estimated melting and boiling points of water are shown as open circles and are predicted by extrapolating (broken lines) the data of H_2S, H_2Se, and H_2Te.

FIGURE 13.5
Hydrogen bonding in
water: (a) Lewis formula;
(b) ball-and-stick model.

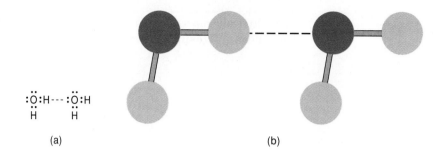

:Ö:H---:Ö:H
 H   H

(a)   (b)

A hydrogen bond is much weaker than a covalent bond. For example, the bond energy of a hydrogen bond in ice is 6.4 kcal (27 kJ) per mole. In contrast, the bond energy of a truly covalent bond between oxygen and hydrogen is 119 kcal (498 kJ) per mole. The average hydrogen bond energy is about 5 kcal (20 kJ) per mole.

In a hydrogen bond the hydrogen atom is not positioned at the exact center between the two electronegative oxygen atoms in the two water molecules. The hydrogen bond distance (177 pm) is longer than the hydrogen–oxygen covalent bond distance (99 pm), as Figures 13.5 and 13.6 show. The shorter H—O bond is the strong covalent bond; the longer H—O bond is the weaker hydrogen bond.

In all cases, in going from solid to liquid to gas, heat energy must be supplied to overcome the attractive forces in the solid or liquid. The attractive forces include hydrogen bonds, which although weaker than covalent bonds, are stronger than most attractive forces found between molecules that lack hydrogen bonds. Therefore, as ice begins to melt to its liquid state at 0°C, some, but not all, hydrogen bonds break. Breaking these bonds requires more energy than breaking the attractions between molecules that lack hydrogen bonds, in which the attractive forces are weaker. This higher energy requirement accounts for the high melting point and high heat of fusion for water.

In raising the temperature of liquid water from 0° to 100°C, the number of hydrogen bonds decreases. In changing from its liquid state to a gas (steam) at $1\overline{0}0$°C (1 atm), nearly all the hydrogen bonds in water break. Again, breaking these bonds requires more energy than breaking the attractions between molecules in compounds without hydrogen bonds. As a result, water has a high boiling point and a high heat of vaporization (see Figure 13.7).

The hydrogen bonding of water molecules also helps to explain the unusual freezing behavior of water. In Chapter 12, we noted that the molecules in solids are generally closer together than those in liquids. Yet you have no doubt noticed how water expands when it freezes (a disaster if you leave soda pop in the freezer to cool

FIGURE 13.6
Water and ice contain many
hydrogen bonds. Notice
that the "hydrogen bond"
distance, (a) (177 pm), is
greater than the
hydrogen–oxygen covalent
bond distance, **b** (99 pm).

---H—Ö:---H—Ö:---H—Ö:---H—Ö:---
     \   \   \   \
     H   H   H   H

    |← (a) →|← (b) →|

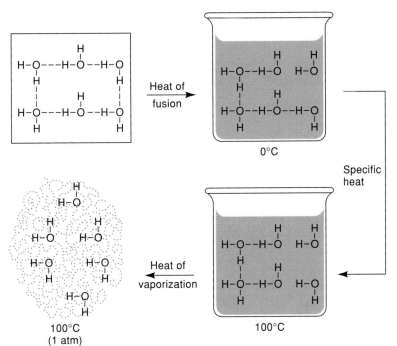

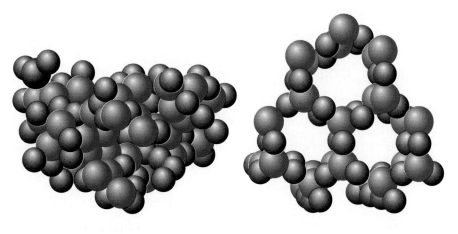

FIGURE 13.7
Transformation of ice to liquid water to gaseous steam, showing hydrogen bonding. Notice the fewer hydrogen bonds in the series of transformations. (Shown only in two dimensions and with a 90° bond angle in water for convenience.)

for too long). The formation of hydrogen bonds between water molecules forces them to orient in a very specific way, as Figure 13.8 shows. This orientation creates a great deal of empty space in the structure of ice. Hence, the volume increases during the change from liquid (water) to solid (ice).

The hydrogen compounds of the other elements in group VIA (16; S, Se, and Te) do *not* appear to form hydrogen bonds appreciably. These elements are *not* as electronegative as oxygen. Hence, the hydrogen derivatives have considerably lower melting points, boiling points, heats of fusion, and heats of vaporization.

FIGURE 13.8
The density of ice is less than the density of water. (a) In liquid water, the molecules are close together. (b) In ice, hydrogen bonding forces the molecules to orient so that empty space is generated, giving a decreased density for the solid.

Study Exercise 13.2

Explain the apparent irregularity in the series of melting points shown for the compounds NH_3, PH_3, and AsH_3. The electronegativities are as follows: N = 3.0, P = 2.1, As = 2.0. You may use the periodic table.

COMPOUND	MELTING POINT (°C)
NH_3	−78
PH_3	−133
AsH_3	−117

(The melting point of NH_3 is higher than that of PH_3 and AsH_3. Because of the high electronegativity of N, NH_3 forms hydrogen bonds (H_2N — H---NH_3). P and As have low electronegativities and do not hydrogen bond. The energy required to separate the molecules of NH_3 includes energy to break the hydrogen bonds. PH_3 and AsH_3 do not have hydrogen bonds to break and thus separate with less energy needed.)

Work Problem 9.

13.5 Important Reactions That Produce Water

As noted at the start of this chapter, water is very common in our lives and on our planet. Thus, it may not surprise you that water is a product of many chemical reactions, particularly combustion reactions, combination reactions, and neutralization reactions. In fact, many of the biochemical reactions of the body include water as either reactant or product.

Combustion

Many *organic* (carbon-containing) compounds burn in oxygen (air contains approximately 20% oxygen), forming water and carbon dioxide on complete combustion. An example is the exothermic reaction of methane (natural gas), which burns in air to produce water and carbon dioxide, thus releasing heat energy that warms many homes.

$$CH_4(g) + 2\,O_2(g) \xrightarrow{\Delta} 2\,H_2O(g) + CO_2(g) + energy \,(=213 \text{ kcal at } 25°C) \quad (13.1)$$

The complete combustion of kerosene to produce carbon dioxide, water, and energy is another example:

$$2\,C_{12}H_{26}(l) + 37\,O_2(g) \text{ or } (l) \xrightarrow{\Delta} 26\,H_2O(g) + 24\,CO_2(g) + energy \quad (13.2)$$

This reaction was used in the first stage of the Saturn V rocket to launch the astronauts to the moon in the early Apollo space missions. The oxidizer was pure liquid oxygen instead of air.

Another example of combustion producing water is the metabolism of foods. The complete combustion of sucrose (sugar) yields carbon dioxide, water, and energy:

$$C_{12}H_{22}O_{11}(s) + 12\,O_2(g) \xrightarrow{\Delta} 11\,H_2O(g) + 12\,CO_2(g) + energy \quad (13.3)$$

This energy keeps the body warm and maintains body temperature. In the preceding reaction, one tablespoon of sugar (12 g) yields 45 kcal (Cal).

Combination of Hydrogen and Oxygen

Under appropriate conditions, hydrogen gas may combine with oxygen gas to form water:

$$2 H_2(g) + O_2(g) \longrightarrow 2 H_2O(g) + \text{energy} \qquad (13.4)$$

If pure hydrogen is burned in oxygen or air, the hydrogen will burn smoothly where the hydrogen gas makes contact with the oxygen gas and will form a very hot, colorless flame. As shown above, water is the only product in such a reaction. But if a 2:1 mixture of hydrogen and oxygen gas is allowed to mix thoroughly *before* being ignited, a violent explosion will occur (see Figure 13.9 and the photograph in the essay, the Element HYDROGEN: Lighter Than Air, at the end of this chapter).

The National Aeronautics and Space Administration (NASA) utilizes this reaction of hydrogen and oxygen when it combines liquid hydrogen and liquid oxygen in the initial stage of the space shuttle launch. The tragic accident of the space shuttle *Challenger* in 1987 was caused by the failure of the giant O-rings made of synthetic rubber that seal each joint between the booster-rocket segments. Most likely because of the low temperature (36°F) at launch time, these O-rings did not seal the joint where two segments of the booster are connected. The hot gases escaping from this leak eventually caused the tanks holding the liquid hydrogen and oxygen to rupture. The resulting catastrophic explosion destroyed the rockets and the *Challenger*, killing all seven people on board.

Product of Neutralization Reactions

In nearly all neutralization reactions (see Section 9.10), water is one of the products. Consider various types of neutralization reaction equations with water as one of the products.

$$KOH(aq) + HCl(aq) \longrightarrow KCl(aq) + H_2O(l) \qquad (13.5)$$

$$BaO(s) + 2 HCl(aq) \longrightarrow BaCl_2(aq) + H_2O(l) \qquad (13.6)$$

$$Ba(OH)_2(aq) + CO_2(g) \longrightarrow BaCO_3(s) + H_2O(l) \qquad (13.7)$$

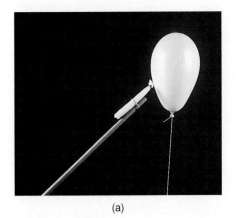

(a)

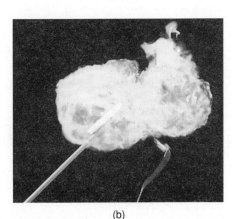

(b)

FIGURE 13.9
The combustion of hydrogen and oxygen. A violent explosion occurs when the balloon containing hydrogen and oxygen (a) is touched by a flame (b).

13.6 Reactions of Water

Just as important as the reactions that produce water are the reactions in which water participates. Two of the most fundamental reactions include the electrolysis of water and the reactions between water and certain metals.

Electrolysis

Electrolysis Application of a direct electric current to a substance to produce a chemical change.

As noted in Section 3.3, the application of a direct electric current (**electrolysis**) to water containing a little sulfuric acid or some other ionic substance decomposes the water and releases hydrogen and oxygen gases:

$$2\,H_2O(l) \xrightarrow[\text{electric current}]{\text{direct}} 2\,H_2(g) + O_2(g) \qquad (13.8)$$

This reaction is just the reverse of that for the combination of hydrogen and oxygen. Is this decomposition reaction exothermic or endothermic? [See Equation (13.4)].

When a lead storage battery (see Section 16.7) is charged with a battery charger, a slight amount of electrolysis also occurs. Small amounts of hydrogen gas and oxygen gas are given off. Any nearby flame or spark can ignite these gases and produce an explosion, as shown in Equation (13.4). This is the reason why batteries and battery chargers carry a warning against smoking or lighting a flame.

Reaction with Certain Metals

In Section 9.8, we introduced the electromotive or activity series and stated that certain metals in the series displace any of those following it from an aqueous solution of its salt or acid. The first five metals in the series—*lithium, potassium, barium, calcium,* and *sodium*—are very active and can replace *one* hydrogen atom from water to form the *metal hydroxide* and hydrogen gas (see list in the margin):

Study Hint: Writing water as H—OH makes it easier to see the replacement of *one* hydrogen atom in water.

$$Ba(s) + 2\,H\!-\!OH(l) \longrightarrow Ba(OH)_2(aq) + H_2(g) \qquad (13.9)$$
$$\text{(excess)}$$

The next five metals—*magnesium, aluminum, zinc, iron,* and *cadmium*—displace hydrogen atoms from *steam* under proper conditions and form the *metal oxide* and hydrogen gas:

$$Mg(s) + H_2O(g) \xrightarrow{\Delta} MgO(s) + H_2(g) \qquad (13.10)$$

The other metals in the electromotive or activity series are not normally active enough to react with water, either cold or hot.

Study Exercise 13.3

Complete and balance the following chemical reaction equations; indicate any precipitate by (s) and any gas by (g). (You may use the periodic table.)

a. $CH_4O(l) + O_2(g) \xrightarrow{\Delta}$
methyl alcohol (excess)

$$[2\,CH_4O(l) + 3\,O_2(g) \xrightarrow{\Delta} 2\,CO_2(g) + 4\,H_2O(g)]$$

b. $Li(s) + H\!-\!OH(l) \longrightarrow$

$$[2\,Li(s) + 2\,H\!-\!OH(l) \longrightarrow 2\,LiOH(aq) + H_2(g)]$$

Li
K
Ba
Ca
Na
—
Mg
Al
Zn
Fe
Cd
—
Ni
Sn
Pb
(H)
Cu
Hg
Ag
Au

Work Problems 10 and 11.

13.7 Hydrates

Now that we have discussed the processes that produce water, let's look at a process where water *trapped* in a compound is allowed to get out. **Hydrates**, crystalline compounds that contain chemically bound water in *definite* proportions, are a group of substances that, when heated, decompose to yield water. For example, Epsom salts ($MgSO_4 \cdot 7\ H_2O$) is more properly magnesium sulfate *hepta*hydrate. The Greek prefixes (see Table 7.1) indicate the number of water molecules; for example, *2* H_2O is *di*hydrate, and *3* H_2O is *tri*hydrate.

The water may be held to the salt by any one of three possible mechanisms:

1. Coordinate covalent bonding (Section 6.6) with the metal cation

2. Hydrogen bonding (Section 13.4) with the anion

3. Entrapment in the crystalline structure of the hydrate, like a stone or rock in a cement sidewalk.

The type of bonding that holds the water to the salt is directly related to the structure of the compound.

A given hydrate may have various combinations of these types of attachments to the water molecules. For example, the hydrate cobalt(II) sulfate heptahydrate, $CoSO_4 \cdot 7\ H_2O$, has *six* water molecules attached by coordinate covalent bonding to the metal cation (mechanism 1) and *one* water molecule attached by hydrogen bonding to the anion (mechanism 2), as Figure 13.10 shows.

We can calculate the percent of water in a hydrate from the formula of the hydrate by using a method similar to that used in Chapter 8 (see Section 8.4) to calculate the percent composition of compounds given the formula, as the following example shows.

Hydrates Crystalline compounds that contain chemically bound water in definite proportions.

Study Hint: Epsom salt is not a chemical curiosity; you can buy it in a drugstore. It is used externally to bathe inflamed skin and internally as a strong laxative.

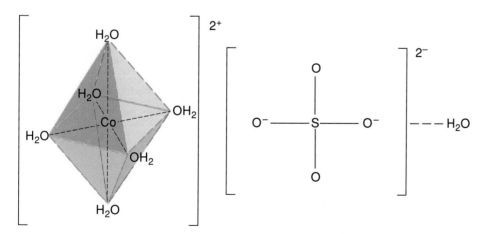

FIGURE 13.10
Cobalt(II) sulfate heptahydrate. Bonding of the water molecules. *Six* water molecules are attached by coordinate covalent bonding with the cobalt ion, and *one* is attached by hydrogen bonding to the sulfate ion. [Notice the octahedron (eight plane surfaces) formed by the six water molecules about the cobalt ion.]

Work Problem 12.

Study Hint: The molar mass (219.1 g) and 108.0 g could have been used to give the *same* results. Both amu and grams give the percent of water.

EXAMPLE 13.1 Calculate the percent of water in calcium chloride hexahydrate.

SOLUTION Calculate the formula mass of $CaCl_2 \cdot 6 H_2O$ as 219.1 amu.

$$1 \times 40.1 = 40.1 \text{ amu}$$
$$2 \times 35.5 = 71.0 \text{ amu}$$
$$6 \times 18.0 = 108.0 \text{ amu (molecular mass } H_2O = 18.0 \text{ amu)}$$
$$\text{formula mass } CaCl_2 \cdot 6 H_2O = \overline{219.1 \text{ amu}}$$

Calculate the percent of water in the hydrate by dividing the *total* contribution of the water in amu by the formula mass in amu of the hydrate and multiplying by 100.

$$\frac{108.0 \text{ amu}}{219.1 \text{ amu}} \times 100 = 49.3\% \text{ water} \qquad Answer$$

Study Exercise 13.4

Calculate the percent of water in sodium bromide dihydrate. (25.9%)

Figure 13.11 shows the difference in appearance of a hydrate and the anhydrous salt (no water present). The hydrate [copper(II) sulfate pentahydrate] is blue, and the anhydrous salt [copper(II) sulfate] is white. Some hydrates, but not all, show this dramatic color change in going from the hydrate to the anhydrous salt.

Knowing the percent of the anhydrous salt and the percent of water in a given hydrate, we can calculate the formula of the hydrate. We use a method similar to that used in Chapter 8 (see Section 8.5) to calculate the formula of a compound given the percent composition. Consider the following example.

EXAMPLE 13.2 Calculate the formula of the hydrate of barium chloride containing 14.7% water.

SOLUTION If 14.7% of the hydrate is water, then 85.3% $(100.0 - 14.7 = 85.3)$ must be barium chloride. Hence, in $1\overline{00}$ g of the hydrate there are 14.7 g of water and 85.3 g of barium chloride. The molar masses of H_2O and $BaCl_2$ are 18.0 g and 208.3 g, respectively. Therefore, the first step is to calculate the moles of H_2O and of $BaCl_2$ in $1\overline{00}$ g of the hydrate.

$$14.7 \text{ g } H_2O \times \frac{1 \text{ mol } H_2O}{18.0 \text{ g } H_2O} = 0.817 \text{ mol } H_2O$$

$$85.3 \text{ g } BaCl_2 \times \frac{1 \text{ mol } BaCl_2}{208.3 \text{ g } BaCl_2} = 0.410 \text{ mol } BaCl_2$$

FIGURE 13.11
Copper(II) sulfate pentahydrate (blue) appears very different than the anhydrous copper(II) sulfate salt (white).

The second step is to express these relationships in small whole numbers by dividing by the smallest value, as follows:

$$For\ BaCl_2: \quad \frac{0.410}{0.410} = 1.00$$

$$For\ H_2O: \quad \frac{0.817}{0.410} = 1.99, \quad \text{approximately } 2$$

Hence, the hydrate has 2 mol of water for every mole of barium chloride, and the formula is $BaCl_2 \cdot 2\ H_2O$. *Answer*

Study Exercise 13.5
Calculate the formula of the hydrate of sodium iodide containing 19.4% water.

$(NaI \cdot 2\ H_2O)$ Work Problems 13 and 14.

A number of substances, such as efflorescent, hygroscopic, and deliquescent substances, are related to hydrates.

Efflorescent Substances

Although many hydrates decompose when *heated* to produce water, some release water when simply *exposed* to the atmosphere. These hydrates are called **efflorescent substances**. The process is called *efflorescence*. The term "efflorescence" should not be confused with "effervescence," which means "to bubble"—Alka Seltzer effervesces in water.

Efflorescent substance Any hydrate that releases water when simply exposed to the atmosphere.

(a)

(b)

(c)

FIGURE 13.12
Efflorescence. The crystal on the left is a fresh sodium acetate trihydrate crystal grown from a saturated solution of the salt. The crystal in the middle is a crystal after exposure to the arid New Mexico atmosphere for a few *minutes*. The crystal on the right is a similar crystal after exposure to the New Mexico atmosphere for several *hours*. This crystal has lost its three waters of hydration. Notice the change of the fresh crystal as it becomes opaque as a result of losing its water of hydration.

Examples of *efflorescent substances* include washing soda ($Na_2CO_3 \cdot 10\ H_2O$), which forms $Na_2CO_3 \cdot H_2O$ upon standing in the open air, and sodium acetate trihydrate ($NaC_2H_3O_2 \cdot 3\ H_2O$), which forms the anhydrous salt, ($NaC_2H_3O_2$), as Figure 13.12 illustrates. For efflorescence to occur, however, the vapor pressure of water in the atmosphere must be *less* than the vapor pressure of the hydrate. If the vapor pressure in the atmosphere is *greater* than the vapor pressure of the hydrate, efflorescence does not occur.

Hygroscopic Substances

Hygroscopic substance Any substance that readily absorbs moisture from the air.

In contrast to efflorescent substances, **hygroscopic substances** readily *absorb* moisture from the air. Many finely divided substances are slightly hygroscopic. For example, sugar absorbs moisture from the air on humid days and forms a cake. A small packet containing a hygroscopic substance is placed in many medicine bottles containing tablets or capsules to prevent water absorption by the drug. On the packet is printed "Do not eat."

Deliquescent Substances

Deliquescent substance Any hygroscopic substance that continues to absorb water from the air to the point where it becomes a solution.

A hygroscopic substance that continues to absorb water and eventually forms a *solution* is a **deliquescent substance**. Hence, all deliquescent substances are hygroscopic, although not all hygroscopic substances are deliquescent. An example of a deliquescent substance is anhydrous calcium chloride, which first absorbs water to form the solid hexahydrate and then absorbs more water to form a *solution*. Another example is sodium hydroxide, which readily absorbs water to form a solution. You may have already noticed this in the laboratory. If someone spilled sodium hydroxide pellets and you came along later, you may have placed your laboratory book or your hand in the sodium hydroxide *solution* formed there!

13.8 Purification of Water

Thus far, we have dealt only with pure water. Yet pure water is a relatively rare occurrence. Nearly all the water you encounter contains something more than just hydrogen and oxygen atoms. Impurities in water may or may not be obvious. You can immediately detect the salts in seawater. But water from the purest of wells can contain salts of calcium, magnesium, and/or iron picked up as water trickles over rocks. You may be aware of these salts only because their presence creates "hard water" ("soft water" has very few calcium or magnesium salts). Hard-water salts use up some amount of soap before a soapy lather appears. The hard-water cations react with the anion from the soap to form an insoluble "soap." This insoluble soap forms the "ring around the bathtub" and leaves a deposit on your skin. To prevent waste of soap through this reaction, hard water is frequently "softened" with water softeners or agents that remove these undesirable cations. Such purification involves one of two general processes: (1) distillation or (2) ion exchange.

We have already seen how scientists use distillation to purify water (Section 12.5).

In the ion-exchange process, ions from the salts dissolved in water are replaced by less troublesome ions from the ion-exchanging unit. One type of ion exchanger commonly found in commercial water softeners uses zeolite, a hydrated sodium aluminum silicate. Zeolite *exchanges* its sodium ions for calcium, magnesium, or iron ions from the salts in hard water. The hard water containing the salts seeps through the zeolite layers in the exchanger tank, and the cations of the salts are exchanged for sodium ions. Note that this is a double-replacement reaction:

$$\underset{\substack{\text{(in tank, Z =} \\ \text{aluminum silicate} \\ \text{portion of the} \\ \text{zeolite)}}}{Na_2Z(s)} + \underset{\text{(from hard water)}}{CaSO_4(aq)} \longrightarrow \underset{\text{(in water)}}{Na_2SO_4(aq)} + \underset{\text{(in tank)}}{CaZ(s)} \qquad (13.11)$$

The sodium salts are soluble and do *not* precipitate the soap, nor do they interfere with lather formation.

Another type of ion exchanger removes both cations—such as calcium, magnesium, and iron—and anions—such as hydrogen carbonate, carbonate, and sulfate—by exchanging them for hydrogen (H^+) ions and hydroxide (OH^-) ions, respectively. The exchanger consists of two types of resins: One of the resins exchanges the cations for hydrogen ions (H^+); the other exchanges the anions for hydroxide ions (OH^-). This exchange results in the formation of water according to the following equation:

$$H^+ + OH^- \longrightarrow H_2O \qquad (13.12)$$

Water purified by this method is called *demineralized water* because all the mineral salts have been removed.

Not all impurities are as harmless as the salts in well water tend to be. Each year, billions of dollars are spent around the world to remove dangerous impurities, such as the bacteria caused by dumping sewage into a water supply. Such purification generally follows two steps:

1. *Settling out and filtering.* The water is allowed to stand in a reservoir to permit suspended particles, such as mud and silt, to settle. After settling, the water is filtered through sand and gravel beds to remove suspended material, such as silt, and also some bacteria. Prior to the filtering process, chemicals such as lime and aluminum sulfate may be added to aid in the filtering. Lime (calcium oxide) reacts with the water to form calcium hydroxide, which in turn reacts with the aluminum sulfate to form aluminum hydroxide. Aluminum hydroxide is an insoluble gelatinous precipitate that aids in removing some bacteria from the water by retaining them in the gelatinous precipitate. The following chemical equations illustrate the reactions involved in the addition of chemicals prior to the filtering process:

$$CaO(s) + H_2O(l) \longrightarrow Ca_2(OH)_2(aq) \qquad (13.13)$$

$$3\,Ca(OH)_2(aq) + Al_2(SO_4)_3(s) \longrightarrow 3\,CaSO_4(aq) + 2\,Al(OH)_3(s) \qquad (13.14)$$

2. *Chlorination.* Chlorine is added to the water to kill any harmful bacteria that may have remained through the filtering process. Chlorinated lime (a mixture of calcium hypochlorite, calcium chloride, and calcium hydroxide) is often used in place of gaseous chlorine.

Other contaminants that make water unsafe for human use call for other purification processes. The Du Pont Company has developed a process for adding powdered activated carbon to acidic wastewater containing a wide variety of organic (carbon-containing) compounds. Highly contaminated water requires very specialized treatment before it is safe for drinking purposes. Such processes generally increase the cost of the purification.

To conserve water, some communities are reusing wastewater. This water is usually purified by gaseous oxygen and is used for watering golf courses and large yards. Reusing one resource helps solve an environmental problem caused by our using a different resource. The destruction of large forested areas is decreasing the earth's capacity to convert carbon dioxide to oxygen. Growing grass or green plants increases the oxygen supply and decreases the carbon dioxide supply in the air by the process of photosynthesis, thereby helping slow carbon dioxide increase (thought to contribute to global warming; see CHEMISTRY OF THE ATMOSPHERE, The Greenhouse Effect and Global Warming, in Chapter 11). A chemical reaction equation for photosynthesis is

$$6\,CO_2(g) \; + \; 6\,H_2O(l) \; \xrightarrow[\text{sunlight}]{\text{chlorophyll,}} \; \underset{\text{glucose, a sugar}}{C_6H_{12}O_6(s)} \; + \; 6\,O_2(g)$$

Substances may be added to the water, depending on the amount of these substances present from the natural sources. For example, small amounts [0.1 to 1 part per million (ppm); see Section 14.7] of *sodium fluoride* may be added to help prevent tooth decay. In the United States natural fluoride exists in water in the range of 0.05 to 8 ppm, with the least in the Northeast and the most in the Southwest.

13.9 Hydrogen Peroxide

Not all combinations of hydrogen and oxygen atoms form water. When two hydrogen atoms bond to *two* oxygen atoms, the result is hydrogen peroxide (H_2O_2). Although hydrogen peroxide is less common than water, it is still an important compound. However, because hydrogen peroxide has a different molecular formula, it is a *different compound*, with a different structure and different properties than water.

Structure

In hydrogen peroxide, the two oxygen atoms each share a covalent bond with a hydrogen atom. In addition, they share one covalent bond with each other, as shown by the following Lewis and structural formulas:

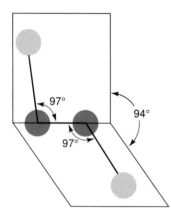

FIGURE 13.13
Ball-and-stick model of the structure of hydrogen peroxide.

The hydrogen peroxide molecule has a H—O—O bond angle of approximately 97°. The other hydrogen atom extends from these three atoms at approximately 94°, as Figure 13.13 shows. Hydrogen peroxide, like water, has a dipole moment and forms hydrogen bonds. As a result, hydrogen peroxide has a high boiling point and is quite polar, forming hydrogen bonds.

Preparation

Hydrogen peroxide can be prepared in the laboratory by treating barium peroxide (BaO_2) with an aqueous solution of sulfuric acid:

$$BaO_2(s) + H_2SO_4(aq) \longrightarrow BaSO_4(s) + H_2O_2(aq) \qquad (13.15)$$

The insoluble barium sulfate is removed by filtration, and the resulting solution of hydrogen peroxide may be concentrated by carefully distilling it to 30% at atmospheric pressure. Distilling the solution at reduced pressure gives 98% hydrogen peroxide.

Properties

Pure hydrogen peroxide is a pale-blue, odorless, oily liquid that readily decomposes to form water and oxygen:

$$2\ H_2O_2(aq) \xrightarrow[\text{or catalyst}]{\text{heat, light,}} 2\ H_2O(l) + O_2(g) \qquad (13.16)$$

You can see the release of oxygen as bubbles when hydrogen peroxide is used as an antiseptic and comes into contact with a wound.

Because its decomposition may be accelerated by heat or light, hydrogen peroxide sold in drugstores is stored in brown bottles (which helps cut down on light absorption) and kept in a relatively cool place. A stabilizer is often added to a solution of hydrogen peroxide to prevent decomposition. Other substances, such as silver, carbon, manganese(IV) oxide, saliva, dirt, and an enzyme in the blood, often act as catalysts in the decomposition of hydrogen peroxide.

Uses

Relatively dilute solutions of hydrogen peroxide are used principally as antiseptics and bleaching agents. A 3% solution is used as an antiseptic, and slightly higher concentrations (6% to 30%) are used as bleaching agents for cloth, silk, straw, flour, and hair. A 90% solution has been used as fuel in the propulsion of rockets.

A solution of half 3% hydrogen peroxide solution and half warm water has been used as a gargle for a sore throat. In the process of gargling, foam builds up in the mouth. What do you think produces this foam? See Equation (13.16).

Study Exercise 13.6
Write Lewis structures and structural formulas for the following molecules. (You may use the periodic table.)

a. H_2S $\qquad \left(\begin{matrix} H \!:\!\overset{\times\times}{\underset{\bullet\bullet}{S}}\!\times \\ H \end{matrix} \qquad \begin{matrix} H - S \\ \qquad\ \ H \end{matrix} \right)$

b. H_2S_2 $\qquad \left(\begin{matrix} H \\ \overset{\times\bullet\ \ \bullet\circ\circ}{\underset{\times\times\ \ \bullet\circ}{\times S \bullet S \circ}} \\ H \end{matrix} \qquad \begin{matrix} H \\ \ \ \diagdown \\ \quad S - S \\ \qquad\qquad \diagdown \\ \qquad\qquad\ H \end{matrix} \right)$

Study Exercise 13.7
Complete and balance the following chemical reaction equations; indicate any precipitate by (s) and any gas by (g). You may use the periodic table.

a. $H_2O_2(aq) \xrightarrow{MnO_2}$ $\qquad\qquad [2\ H_2O_2(aq) \xrightarrow{MnO_2} 2\ H_2O(l) + O_2(g)]$

b. $SrO_2(s) + H_2SO_4(aq) \longrightarrow$
$\qquad\qquad [SrO_2(s) + H_2SO_4(aq) \longrightarrow SrSO_4(s) + H_2O_2(aq)]$ Work Problems 15 and 16.

CHEMISTRY OF THE ATMOSPHERE

Ozone Depletion in the Stratosphere

A layer of ozone-enriched air is found in the stratosphere (see figure) with its maximum amount about 25 km above the earth's surface. This ozone layer absorbs most of the short-wavelength light from sunlight and protects living systems from being damaged by this high-energy light.

Ozone is formed in the stratosphere by the action of ultraviolet (short-wavelength) light on oxygen gas by the following process: (1) oxygen molecules absorb ultraviolet light and break apart into two oxygen atoms; (2) each of these oxygen atoms in turn reacts with another oxygen molecule to give a molecule of ozone:

(1) $O_2(g) + \text{light} \longrightarrow 2\ O(g)$

(2) $O(g) + O_2(g) \longrightarrow O_3(g)$

The amount of ozone in the stratosphere is not large; there are only about one or two molecules of ozone for every 1 million other gas molecules. However, this small amount of ozone is enough to absorb much of the ultraviolet light from the sun.

In 1974, two American chemists, F. S. Rowland and M. J. Molina, suggested that chlorofluorocarbons (CFCs, see Chemistry of the Atmosphere, Chapter 9) could partially destroy this layer of ozone. Chlorofluorocarbons are so stable that they remain in the atmosphere for 80 to 120 years from the time they are released. Two important examples are CFC-11 ($CFCl_3$) and CFC-12 (CF_2Cl_2). Since these compounds do not readily decompose, they are

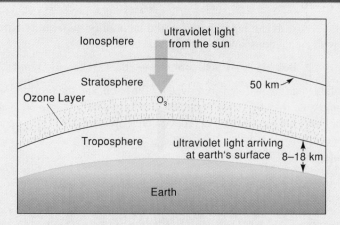

The layer of ozone in the stratosphere, 25 km from the earth's surface, absorbs much of the harmful ultraviolet light from the sun.

mixed into the atmosphere by wind processes over the years.

About the only way chlorofluorocarbons decompose in the atmosphere is by encountering high-energy ultraviolet light (short wavelength). This encounter occurs only when chlorofluorocarbons reach the stratosphere, where the ultraviolet light from the sun has not yet been screened out by the protective ozone blanket. Although most chlorofluorocarbons do not reach this altitude, those that do are broken down by ultraviolet light into species that convert ozone back to oxygen molecules:

$$2\ O_3(g) \longrightarrow 3\ O_2(g)$$

The resulting decrease in the amount of ozone in the stratosphere allows increasing amounts of ultraviolet light to reach the earth's surface. The essay following Chapter 17 discusses this process in more detail.

The destruction of the ozone layer has been very dramatic over the Arctic and Antarctica. The popular press has done a great deal to describe the overall problem of ozone depletion since ozone holes as large as the continent of

Antarctica have appeared over the poles during their respective spring seasons. Special circumstances of temperature and moisture make the destruction of ozone over Antarctica in September and over the Arctic in March especially efficient. The ozone losses over the rest of the globe are far less dramatic. Still, the observations at the polar regions indicate that we do have a serious problem.

Thus far, scientists have not reached agreement on the extent of the problem. Predictions of global ozone reduction by the year 2075 range from 2% to 10%. Yet even a 2% reduction in the amount of stratospheric ozone could translate into 3 to 15 million additional cases of skin cancer and 550,000 to 2.8 million additional cases of cataracts in the United States alone. This increase in skin cancer has already occurred in the Southern Hemisphere, including Chile and Australia. Exposure of plants and crops to increasing amounts of ultraviolet light might also have negative effects on our ability to grow or produce food.

The ozone depletion problem is rapidly becoming as much a political question as a scientific one. The cost of developing alternatives to chlorofluorocarbons and the price of changing existing industrial processes create economic hardships. Developing countries do not want the ozone depletion problem to keep them from developing the modern technology that is already available in developed countries. Some developed countries do not wish to slow their own growth by allocating economic and physical resources to CFC control. Thus, international discussions on how quickly to phase out CFCs involve economic and resource development issues as well as technological ones. In all likelihood, CFCs will be phased out by the beginning of the twenty-first century and acceptable substitutes will have been developed. But even if we were to stop using

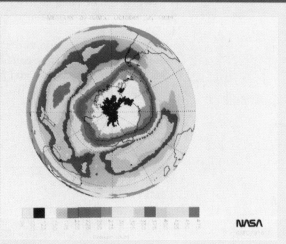

On October 2, 1994, Russian Meteor-3 satellite detected ozone losses of over 60% since 1975 over Antarctica (dark area in center).

CFCs tomorrow, there are enough already in the atmosphere to cause the earth significant problems for the next 100 to 150 years. The biggest question of all then remains: How significant will the ozone depletion problem become and how will it affect life on earth into the next century?

Summary

In this chapter we described the physical and chemical properties of water and the related compound hydrogen peroxide. Water is a very common material, and we are very familiar with many of its properties (Section 13.1). The structure of the water molecule is "bent," with an H — O — H bond angle of 105°. This structure results from the placement of the four electron pairs about the central oxygen (Section 13.2).

A water molecule has a *dipole moment* as a result of the 105° H — O — H bond angle (Section 13.3). The polarity of water and the high electronegativity of oxygen make water molecules capable of bonding to each other through *hydrogen bonds*. While hydrogen bonds are not as strong as covalent or ionic bonds, they are strong enough to make water molecules harder to pull apart. As a result, water has an unusually high melting point, boiling point, heat of fusion, and heat of vaporization (Section 13.4).

Water is produced in the combustion of organic (carbon-containing) compounds, in the combination of oxygen and hydrogen, and in neutralization reactions (Section 13.5). Water can be broken down into hydrogen gas and oxygen gas by *electrolysis*, and water reacts with metals above nickel in the activity series to liberate hydrogen gas (Section 13.6). Water can also be obtained by heating *hydrates*, which are crystalline salts that contain chemically bound water in definite proportions. Efflorescent, hygroscopic, and deliquescent substances are related to hydrates. Efflorescent substances give off water when simply exposed to the atmosphere. Hygroscopic and deliquescent substances absorb water from the air (Section 13.7).

Water can be purified in a number of ways, including filtration, chlorination, and passage through ion-exchange systems to remove unwanted inorganic ions (Section 13.8).

In hydrogen peroxide, two hydrogen atoms bond to two oxygen atoms. In the laboratory, hydrogen peroxide can be prepared by treating barium peroxide with sulfuric acid. Although its structure and properties differ somewhat from those of water, both have dipole moments, form hydrogen bonds, and have high boiling points (Section 13.9).

 Exercises

1. Define or explain the following terms (the number in parentheses refers to the section in the text where the term is mentioned):

   a. polar bond (13.3)

   b. dipole moment (13.3)

   c. hydrogen bond (13.4)

   d. electrolysis (13.6)

   e. hydrates (13.7)

   f. efflorescent substance (13.7)

   g. hygroscopic substance (13.7)

   h. deliquescent substance (13.7)

2. Distinguish between:

   a. soft water and hard water

   b. H and O found in $CaCl_2 \cdot 2\,H_2O$ and H and O found in $C_{12}H_{22}O_{11}$

   c. hydrate and hygroscopic substance

   d. efflorescence and effervescence

Physical Properties of Water (See Section 13.1)

3. Would ice ($d^{0°} = 0.917$ g/mL) float when placed in the following liquids? Explain.

   a. ethyl alcohol ($d^{0°} = 0.806$ g/mL)

   b. benzene ($d^{0°} = 0.899$ g/mL)

   c. glycerine ($d^{0°} = 1.26$ g/mL)

4. Discuss the consequences to the earth if, on going from liquid water to solid ice, the volume of the solid did not expand.

Hydrates (See Section 13.7)

5. List the three possible mechanisms by which water combines with a salt to form a hydrate.

Purification of Water (See Section 13.8)

6. Explain how a mixture of lime and aluminum sulfate aids the filtering process in water purification.

7. Explain briefly the operation of a commercial water softener.

Problems

Polarity of Molecules (See Section 13.3)

8. Represent the *net dipole monent*, if any, for the following molecules. [*Hint:* Write the Lewis structure (Section 6.7), consider the electronegativities of the elements (Figure 6.11), and then decide on the shape of the molecule, that is, linear, trigonal planar, tetrahedral, bent, or pyramidal (Section 6.8)].

   a. H_2O **b.** CO_2 **c.** HCl   **d.** BrCl

Hydrogen Bonding in Water (See Section 13.4)

9. Explain the apparent irregularity in the series of boiling points shown for the compounds HF, HCl, and HBr. The electronegativities are as follows: F = 4.0, Cl = 3.0, and Br = 2.8. You may use the periodic table.

Compound	Boiling Point (°C at 1 atm)
HF	19.7
HCl	−85.0
HBr	−66.8

Important Reactions That Produce Water and Reactions of Water (See Sections 13.5 and 13.6)

10. Complete and balance the following chemical reaction equations; indicate any precipitate by (s) and any gas by (g). (You may use the periodic table.)

   a. $C_6H_{12}O_6(s) + O_2(g) \xrightarrow{\Delta}$
     glucose, (excess)
     dextrose

   b. $H_2(g) + O_2(g) \xrightarrow{\text{"spark"}}$

   c. $Ca(OH)_2(aq) + H_2SO_4(aq) \longrightarrow$

   d. $BaO(s) + HCl(aq) \longrightarrow$

   e. $CO_2(g) + KOH(aq) \longrightarrow$

   f. $H_2O(l) \xrightarrow[\text{electric current}]{\text{direct}}$

   g. $K(s) + HOH(l) \longrightarrow$
     (excess)

   h. $Na(s) + HOH(l) \longrightarrow$
     (excess)

   i. $Mg(s) + H_2O(g) \xrightarrow{\Delta}$

   j. $Al(s) + H_2O(g) \xrightarrow{\Delta}$

11. Complete and balance the following chemical reaction equations; indicate any precipitate by (s) and any gas by (g). (You may use the periodic table.)

   a. $C_2H_6(g) + O_2(g) \xrightarrow{\Delta}$
     ethane (excess)

   b. $C_2H_6O(g) + O_2(g) \xrightarrow{\Delta}$
     ethyl alcohol (excess)

   c. $Ca(OH)_2(aq) + CO_2(g) \longrightarrow$

   d. $KOH(aq) + SO_2(g) \longrightarrow$

   e. $KOH(aq) + SO_3(g) \longrightarrow$

   f. $Zn(OH)_2(s) + HCl(aq) \longrightarrow$

   g. $ZnO(s) + HCl(aq) \longrightarrow$

   h. $Ca(s) + HOH(l) \longrightarrow$
     (excess)

   i. $Li(s) + HOH(l) \longrightarrow$
     (excess)

   j. $Zn(s) + H_2O(g) \xrightarrow{\Delta}$

Hydrates (See Section 13.7)

12. Calculate the percent of water in the following hydrates:

   a. magnesium sulfate heptahydrate

   b. calcium sulfate dihydrate

  c. zinc sulfate heptahydrate

  d. cobalt(II) sulfate heptahydrate

  e. calcium chloride dihydrate

13. Calculate the formula for the following hydrates:

  a. a hydrate of copper(II) sulfate containing 10.1% water

  b. a hydrate of copper(II) sulfate containing 25.3% water

  c. a hydrate of copper(II) sulfate containing 36.1% water

  d. a hydrate of sodium carbonate containing 14.5% water

  e. a hydrate of sodium carbonate containing 63.0% water

14. A student heated 1.75 g of a hydrate of calcium chloride and then found it to have a mass of 0.89 g. Calculate the formula of the hydrate.

Lewis Structures and Structural Formulas (See Sections 13.3 and 13.9)

15. Write the Lewis structures and structural formulas for the following molecules. (You may use the periodic table.)

  a. H_2O **b.** H_2O_2

Hydrogen Peroxide (See Section 13.9)

16. Complete and balance the following chemical reaction equations; indicate any precipitate by (s) and any gas by (g). (You may use the periodic table.)

  a. $H_2O_2(aq) \xrightarrow{\text{light}}$

  b. $BaO_2(s) + H_2SO_4(aq) \longrightarrow$

General Problems

17. Calculate the number of grams of water required to produce 4.75 g of oxygen by electrolysis.

18. Calculate the volume of oxygen in liters produced at STP from the electrolytic decomposition of 7.00 g of water.

19. Calculate the volume in liters of the hydrogen produced from the electrolysis of 4.00 g of water if it is collected over water at 627 torr and 27°C. (See Appendix V for additional data.)

20. In the decomposition of hydrogen peroxide, how many milliliters of oxygen can be produced at STP from 3.00 g of a 6.00% hydrogen peroxide solution? (*Hint:* A 6.00% hydrogen peroxide solution contains 6.00 g of pure hydrogen peroxide in $10\overline{0}$ g of solution.)

21. The isotope ^2_1H of hydrogen is also known as deuterium, with the symbol D. "Heavy water" is water (D_2O) with deuterium instead of the normal isotope, ^1_1H. One way of producing deuterium gas (D_2) is to pass electricity through D_2O and collect D_2 over the reaction mixture. Calculate the volume of D_2O that would have to react to produce 1.00 L of D_2 gas at 635 torr and 29°C. (D or ^2_1H = 2.0 amu, O = 16.0 amu, density of D_2O = 1.11 g/mL.)

22. Plaster of paris is a traditional building material used in the preparation of wall plaster, wallboard, moldings, tiles, and blocks. Calcium sulfate hemihydrate is mixed with water to form a thick liquid slurry that slowly sets to a hard solid. The final product is calcium sulfate dihydrate. (Hemihydrate means that there is one molecule of water for every two formula units of calcium sulfate.) Write a balanced chemical equation for this process. To which one of the five simple chemical reaction types (see Section 9.5) does it belong?

✓ Chapter Quiz 13

COMPOUND	BOILING POINT ($°C$ at 1 atm)
NH_3	-33.4
PH_3	-87.7
AsH_3	-62.5

You may use the periodic table and the rules for the solubility of inorganic substances in water.

1. Explain the apparent irregularity in the series of boiling points shown for the compounds NH_3, PH_3, and AsH_3. The electronegativities are as follows: $N = 3.0, P = 2.1$, and $As = 2.0$.

2. Complete and balance the following chemical reaction equations; indicate any precipitate by (s) and any gas by (g).

    a. $BaO_2(s) + H_2SO_4(aq) \longrightarrow$

    b. $CO_2(g) + NaOH(aq) \longrightarrow$

    c. $Ca(OH)_2(aq) + H_3PO_4(aq) \longrightarrow$

    d. $Al(s) + H_2O(g) \xrightarrow{\Delta}$

3. Calculate the percent water in $CrPO_4 \cdot 6 H_2O$ [chromium (III) phosphate hexahydrate]. ($Cr = 52.0$ amu, $P = 31.0$ amu, $O = 16.0, H = 1.0$ amu.)

4. Calculate the formula of a hydrate of lithium bromide (LiBr) containing 29.3% water. ($Li = 6.9$ amu, $Br = 79.9$ amu, $H = 1.0$ amu, $O = 16.0$ amu.)

Hydrogen (Symbol: H)

The Element HYDROGEN: Lighter than Air

Hydrogen gas is lighter than air, and so it was used to fill dirigibles (rigid framework blimps) in the first part of the century. Hydrogen is also very combustible, which was vividly illustrated by the explosion of the Hindenburg, *the largest dirigible ever built, in 1937 at Lakehurst, New Jersey. Understandably, modern blimps (nonrigid or semirigid framework) use helium as the flotation gas.*

Name: Hydrogen is found in water and many other compounds such as methane (CH_4), kerosene ($C_{12}H_{26}$), and sucrose ($C_{12}H_{22}O_{11}$), all mentioned in Section 13.5. Its name derives from the Greek *hydro-* meaning "water," and the French *-gene*, meaning "to form." The element was named by the French chemist and physicist Antoine Laurent Lavoisier (1743–1794) (see Section 3.6, Figure 3.13).

Appearance: Hydrogen occurs as a colorless and odorless diatomic gas, H_2. Hydrogen is the lightest element and has a density of only 0.08988 g/L.

Occurrence: Hydrogen comprises only 0.1% of the earth's crust, but it is 10.8% of the oceans. On the earth, hydrogen gas (H_2) occurs in the free (uncombined) state only to a very small extent. Hydrogen is most commonly found as a component of water. However, hydrogen (as uncombined hydrogen atoms in free space or in stars) is the most common element in the universe and accounts for 80% of the mass and 94% of the molecules in the universe.

Source: Hydrogen gas is obtained from water when an electric current is passed through water. It is also obtained when steam reacts with various carbon-containing compounds such as methane gas (CH_4) and carbon monoxide.

Its Role in Our World: Hydrogen gas is a very important material in the chemical industry. It is used in the production of a number of chemicals such as ammonia (NH_3) and methanol (methyl alcohol, CH_4O). Ammonia is used as a fertilizer, and methanol as an additive to gasoline to increase the octane number and decrease pollution.

Hydrogen is used to remove sulfur-containing pollutants from coal, gas, and oil.

Hydrogen is also used at high temperature to prepare a number of metallic elements, such as iron (Fe), chromium (Cr), and nickel (Ni), from their oxides.

Hydrogen can help make things hot or cool things off. Hydrogen gas is used in high-temperature welding torches (approximately 2000°C) such as the oxyhydrogen torch. Liquid hydrogen is used as a low-temperature coolant (−252°C or 21 K).

Unusual Facts: Hydrogen gas has been proposed as the "fuel of the future" because it burns cleanly and completely in air to give water and *no pollutants*. Scientists have suggested the use of hydrogen in the internal combustion engine. The technological problems that must be solved for hydrogen to serve as a fuel in the future include the following: (1) finding a way to turn water into hydrogen and oxygen inexpensively, probably by the use of sunlight, and (2) finding a way to *safely* and efficiently store and transport large quantities of hydrogen gas.

CHAPTER 14

Solutions and Colloids

A saxophone. Did you know that brass is really a solution of zinc (30 to 40%) dissolved in copper (60 to 70%)?

GOALS FOR CHAPTER 14

1. To distinguish between solute and solvent and to explain how molecules and ions dissolve in water (Section 14.1).

2. To describe the most common combinations of different states of matter that can form a solution (Section 14.2).

3. To discuss three factors that affect the rate of dissolution of various substances in other substances and three factors that affect the *rate* of solubility (Section 14.3).

4. To explain the differences among saturated, unsaturated, and supersaturated solutions in terms of their composition and preparation in the laboratory (Section 14.4).

5. To list five ways of expressing the concentration of solutions (Section 14.5).

6. To calculate the concentration of a solution in percent by mass and to use percent by mass in various types of calculations (Section 14.6).

7. To calculate the concentration of a solution in parts per million and to use parts per million in various types of calculations (Section 14.7).

8. To calculate the concentration of a solution in molarity and to use molarity in various types of calculations including dilution of stock solutions (Section 14.8).

9. To calculate the concentration of a solution in normality and to use normality in various types of calculations including conversion to molarity and from molarity to normality (Section 14.9).

10. To calculate the concentration of a solution in molality and to use molality in various types of calculations (Section 14.10).

11. To calculate the boiling point elevation and freezing point depression of solutions and to calculate the molecular mass and molar mass of an unknown solute (Section 14.11).

12. To distinguish among solutions, colloids, and suspensions and to discuss properties peculiar to colloids (Section 14.12).

Countdown

You may use the periodic table.

ELEMENT	ATOMIC MASS UNITS (amu)
C	12.0
H	1.0
O	16.0
Na	23.0
S	32.1

5. Using the periodic table, classify the following compounds as essentially ionic or covalent (Section 6.10, Types of Bonding).
   a. $BaCl_2$ (ionic)   b. NO_2 (covalent)
   c. Na_2SO_4 (ionic)   d. CoF_2 (ionic)

4. Write the correct formula for the following compounds (Sections 7.2 through 7.4 and 7.6).
   a. sodium sulfide (Na_2S)
   b. nitrogen trichloride (NCl_3)
   c. cobalt(II) phosphate [$Co_3(PO_4)_2$]
   d. dichromic acid ($H_2Cr_2O_7$)

3. Calculate the molecular or formula mass and molar mass for the following (Sections 8.1 and 8.2).
   a. glycerol ($C_3H_8O_3$)   (92.0 amu, 92.0 g)
   b. sodium sulfate (Na_2SO_4) (142.1 amu, 142.1 g)

2. Calculate the number of moles in each of the following quantities (Section 8.2).
   a. 43.6 g glycerol ($C_3H_8O_3$)   (0.474 mol)
   b. 565 mg sodium sulfate (Na_2SO_4)
   (0.00398 mol)

1. Calculate the number of grams in each of the following quantities (Section 8.2).
   a. 0.164 mol glycerol ($C_3H_8O_3$)   (15.1 g)
   b. 0.00875 mol sodium sulfate (Na_2SO_4)
   (1.24 g)

I t's just one of those days. You stumble to the breakfast table and blearily stir your coffee, only to grimace when you find that you've put salt, not sugar into your morning brew. It's a bitter cold morning, and you curse your stupidity when the car won't start because you haven't had antifreeze put in. Then, between classes, you feed money to the vending machine but it refuses to give you the can of soda pop you craved—or your money back. That afternoon, you head to the dentist and get more bad news: you'll need to have two cavities filled.

You might not think of it in these terms, but you've been done in by a series of solutions. As you will see in this chapter, coffee with sugar, antifreeze, soda pop—even a dental filling—are all solutions.

Over the course of the last three chapters, we have considered the three physical states of matter—solid, liquid, and gas—in some detail. In this chapter, we will consider how substances in these different states sometimes combine to form solutions, including the factors that affect such combining and ways of describing these combinations.

14.1 Solutions

As we noted in Chapter 3, by definition a **solution** is homogeneous throughout. It is composed of two or more pure substances, and its composition can be *varied*, usually *within certain limits*. For example, a solution of sugar dissolved in water could contain 1 g, 2 g, or 3 g of sugar in 100 g of water and still be a "sugar solution." A solution consists of a solute dissolved in a solvent. The term **solute** usually refers to the component present in lesser quantity than the solvent. The term **solvent** refers to the component present in greater quantity than the solute. For example, in a 5.00% sugar solution in water, sugar is the solute and water is the solvent. This solution is called an *aqueous* solution because the solvent is water.

A covalently bonded substance typically disperses in the solvent as individual molecules, while an ionic substance usually dissolves as individual ions among the solvent molecules. Often, the molecules or ions associate or attach themselves to one or more solvent molecules. As Figure 14.1a shows, in an aqueous sugar solution, sugar molecules leave the solid mass of sugar at the bottom of the container and disperse themselves into the bulk of the water. As they do so, they become loosely tied to a number of water (solvent) molecules by hydrogen bonding. These dissolved molecules are considered to be *hydrated* (solvated) by water molecules. More molecules continue to dissolve in the water and become hydrated until either all the sugar present dissolves or the water can hold no more sugar. Most solutes have a limit to the amount which will dissolve in a given solvent.

In Chapter 9 (Section 9.9) we gave rules for the solubility of inorganic substances in water. These rules are also found on the inside back cover of this book. Referring to this table, you will notice that sodium chloride (NaCl) is soluble in water (rules 2 and 4). We will now consider how sodium chloride dissolves in water.

In an aqueous sodium chloride solution, the individual sodium ions (Na^+) and chloride ions (Cl^-) leave the solid mass of sodium chloride at the bottom of the container and disperse into the water. Each ion becomes surrounded by water molecules as shown in Figure 14.1b. Note that the water molecules surround a sodium ion with their negative dipole ends (oxygen atoms) toward the ion, whereas the positive ends of their dipoles (hydrogen atoms) surround the negative chloride ions.

Solution Homogeneous mixture involving two or more pure substances; its composition usually can be varied within certain limits.

Solute Substance dissolved in a solution; it is usually present in lesser quantity than the solvent.

Solvent Dissolving substance in a solution; it is usually present in greater quantity than the solute.

Study Hint: The energy released on hydration of each ion is small, but the sum of interactions over many ions gives enough energy to overcome strong ionic attractions in the solid.

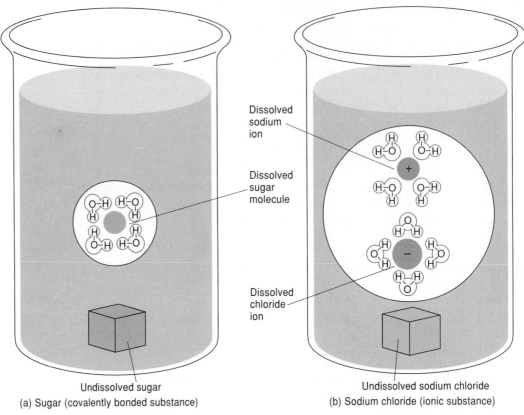

FIGURE 14.1

The process of dissolving: (a) sugar dissolving in water to form hydrated sugar molecules [$C_{12}H_{22}O_{11}(H_2O)_x$]; (b) sodium chloride dissolving in water to form hydrated sodium ions [$Na^+(H_2O)_x$] (relatively negative oxygen atoms of water attracted to positive sodium ions) and hydrated chloride ions [$Cl^-(H_2O)_y$] (relatively positive hydrogen atoms of water attracted to negative chloride ions).

The energy released in formation of the hydrated ions overcomes the strong ionic forces of attraction between the sodium cations and the chloride anions in solid crystalline NaCl.

14.2 Types of Solutions

Although aqueous solutions are common, they are far from being the only types of solutions. Solvents can be water or any liquid. Solutes can be solids, liquids, or gases. The most common types of solutions are a gas in a liquid, a liquid in a liquid, and a solid in a liquid. Table 14.1 gives some examples of common solutions.

Solutions of liquids dissolved in liquids are common (vinegar, wine, gasoline). When two liquids are combined, they may (1) mix completely, forming a solution; (2) mix partially, forming two solutions; or (3) not mix at all and form a heterogeneous mixture.

TABLE 14.1		Some Types of Solutions
SOLUTE	**SOLVENT**	**EXAMPLE**
Gas	Liquid	Carbonated beverages (carbon dioxide in aqueous solution)
Liquid	Liquid	Antifreeze in a car radiator (ethylene glycol in water)
Liquid	Solid	Dental fillings (mercury in silver)
Solid	Liquid	Sugar in water
Solid	Solid	Brass (zinc in copper)

In the *first* case, where the two liquids completely mix in all proportions to form a solution, the two liquids are called **miscible liquids**. An alcohol solution in water and a solution of water and ethylene glycol (a common antifreeze) are examples of pairs of *miscible liquids*.

In the *second* case, one liquid partially dissolves in the other. The result is two solutions appearing as two layers with the solution having the greater density becoming the bottom layer. Two such liquids are considered to be *partially miscible*. If ethylene glycol is mixed with chloroform, two solutions result. The bottom solution consists of a small amount of ethylene glycol dissolved in chloroform; the top solution is a small amount of chloroform dissolved in ethylene glycol. A distinct boundary between the two layers is visible.

In the *third* case, the two liquids do not mix appreciably; they form two separate layers and are called **immiscible liquids**. Two such immiscible liquids are gasoline and water. The less dense gasoline floats as a separate layer on top of the water. Very little gasoline dissolves in the water, and very little water dissolves in the gasoline. Oil and water act like gasoline and water and are immiscible liquids (see Figure 14.2). One of the ways to clean up oil slicks is to get the oil and water to mix and become miscible liquids. An agent must be added to the immiscible oil–water to promote this.

Miscible liquids Two liquids that mix completely in all proportions to form a solution.

Immiscible liquids Two liquids that do not mix, but form separate layers; neither liquid dissolves appreciably in the other.

14.3 Factors Affecting Solubility and Rate of Solution

Not every substance is soluble in every other substance. As we just mentioned, oil and water are immiscible. The oil is not soluble in the water. Some solutes are harder to dissolve than others. In this section we will consider two sets of factors:

1. Factors that affect the actual *solubility* (how much?) of a given solute in a solvent

2. Factors that affect the *rate* (how fast?) at which a given solute dissolves in a given solvent.

The actual solubility of a solute in a given solvent depends on three factors: (1) the properties of the solute and the solvent, (2) temperature, and (3) pressure. The *rate* at which a given solute dissolves in a given solvent depends on three factors: (1) particle size of a solute, (2) rate of stirring, and (3) temperature.

FIGURE 14.2
Oil and water are immiscible, and so the less dense oil floats on top of the water.

Properties of Solute and Solvent

In Chapter 13, we noted that water is considered a *polar* compound because its molecules possess dipole moments (see Section 13.3). Many ionic compounds, such as sodium chloride, are generally soluble in water because their ions are at-

FIGURE 14.3
Carbon tetrachloride is a nonpolar molecule. (a) Structure of carbon tetrachloride, showing zero dipole moment because the centers of positive and negative charge coincide on the carbon atom as a result of the tetrahedral structure. (b) Ball-and-stick model of carbon tetrachloride. (Laima Druskis)

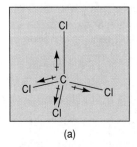

(a)

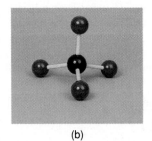

(b)

tracted to one of the poles of the dipole in water, or because they are held by hydrogen bonds, or both.

In contrast, ionic compounds (see Section 6.10, Types of Bonding) generally are not soluble in nonpolar compounds. To understand why, consider one such nonpolar compound, carbon tetrachloride (CCl_4). In carbon tetrachloride the center of negative charge coincides with the center of positive charge because of the tetrahedral arrangement of the molecule, as Figure 14.3 shows. Hence, we cannot dissolve sodium chloride in carbon tetrachloride because the oppositely charged ions of this ionic compound (Na^+, Cl^-) attract each other more than they attract the nonpolar carbon tetrachloride molecule.

However, covalent compounds are generally soluble in carbon tetrachloride and insoluble in water because the water molecules are bound by hydrogen bonds more firmly to each other than they could be to the nonionic, nonpolar molecules. This solubility explains why household "spot" removers composed of tetrachloroethylene (which acts like carbon tetrachloride) are effective solvents for relatively nonpolar organic materials such as fats. Similar substances are used in dry cleaning to dissolve the fatty grease binding dirt to the fabric.

In most cases, *like dissolves like*. Thus, *ionic compounds (polar) are generally soluble in polar solvents, and covalent compounds (nonpolar or weakly polar) are generally soluble in nonpolar solvents.*

Temperature

The solubility of a gas in water decreases with an increase in temperature. For example, at 0°C, 4.89 mL of oxygen can dissolve in $1\overline{0}0$ mL of water. But at 50°C, only 2.46 mL of oxygen can dissolve in $1\overline{0}0$ mL of water. People who fish in lakes

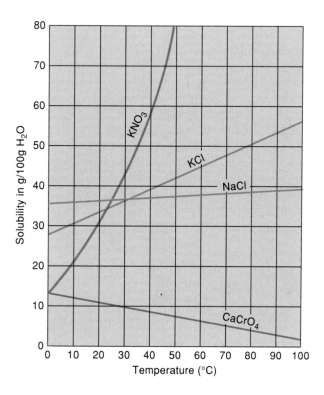

FIGURE 14.4
Relation between the solubility of various salts and temperature.

know that fish move away from the shore as the day grows warmer. They may not be aware, however, that the sun heats the water near the shore, decreasing the solubility of oxygen in the water. The fish move into deeper, cooler water for a better oxygen supply. You may also have observed the decrease in solubility of a gas with an increase in temperature when you heat a glass container and bubbles of gas form along the side of the container before the water boils. The water contains dissolved gases, such as oxygen, nitrogen, and carbon dioxide, that escape from the water when it is heated.

The solubility of a solid in a liquid *usually increases* with an *increase* in *temperature*. There are, however, a number of exceptions. Figure 14.4 illustrates the variable effect of increasing temperature on solubility. Note that as the temperature increases, the solubilities of KCl and KNO_3 increase, but that of NaCl remains relatively constant and that of $CaCrO_4$ decreases. At $3\overline{0}°C$, $1\overline{0}$ g of $CaCrO_4$ can dissolve in $1\overline{00}$ g of water, but at $8\overline{0}°C$ only about 4 g of $CaCrO_4$ can dissolve in $1\overline{00}$ g of water.

Study Exercise 14.1
From Figure 14.4, determine the solubility of the following in grams per $1\overline{00}$ g of water.

Work Problem 10.

a. KCl at 25°C (35 g)
b. KCl at $6\overline{0}°C$ (45 g)

Pressure

Solutions involving only liquids and solids are not influenced to any great extent by pressure. However, solutions of gases in liquids are appreciably affected by pressure. William Henry (1774–1836), an English chemist and physician, carried out experiments on the solubility of gases in a liquid as a function of pressure. From his experiments, he formulated what we now call **Henry's law**. This law states that the solubility of a gas in a liquid is *directly proportional to the partial pressure* (see Section 11.7) *of the gas above the liquid.* For example, if the partial pressure of the gas above the liquid is *doubled*, the *solubility* of the gas in the liquid will also be *doubled*. Conversely, if the partial *pressure is reduced by one-half*, then the *solubility* of the gas in the liquid will be *reduced by one-half* and some of the gas will escape from the liquid. When you open a carbonated beverage, the release of the pressure allows the excess carbon dioxide gas to escape and bubbles and foam rise. If it does not bubble and foam, we say it is "flat."

Henry's law Principle that the solubility of a gas in a liquid is directly proportional to the partial pressure of the gas above the liquid.

Consider the following example involving Henry's law:

EXAMPLE 14.1 Calculate the solubility in g/L of a certain gas in water at a partial pressure of 3.50 atm and 0°C. The solubility is 0.530 g/L at a *total* pressure of 1.000 atm and 0°C.

SOLUTION The total pressure is the sum of the partial pressures of water and the gas, according to Dalton's law of partial pressures (see Section 11.7).

$$P_{total} = P_{gas} + P_{water}$$

P_{water} at 0°C is 0.0061 atm (see Appendix V)

$$P_{gas} = 1.000 \text{ atm} - 0.0061 \text{ atm}$$

$$= 0.994 \text{ atm}$$

Now solve the problem using the same general method we used for solving gas law problems in Chapter 11:

solubility$_{old}$ = 0.530 g/L P_{old} = 0.994 atm | pressure increases
solubility$_{new}$ = ? P_{new} = 3.50 atm   ↓ solubility increases

$$= \text{solubility}_{old} \times \text{pressure ratio of new to old}$$

$$\text{solubility}_{new} = 0.530 \text{ g/L} \times \frac{3.50 \text{ atm}}{0.994 \text{ atm}} = 1.87 \text{ g/L} \qquad Answer$$

Study Exercise 14.2

Calculate the solubility in g/L of carbon dioxide gas in water at a partial pressure of 1.50 atm and 5°C. The solubility is 2.77 g/L at a *total* pressure of 1.000 atm and 5°C (see Appendix V).

(4.19 g/L)   Work Problems 11 and 12.

Particle Size

Now that we have some understanding of why substances do and do not act as solutes or solvents for other substances, we can explore some factors that affect the *rate of dissolution* (dissolving) for soluble substance combinations. Because smaller solute particles have a greater *total* surface exposed to the solvent, they dissolve faster than larger solute particles. This factor explains why a solid lump of sugar dissolves much more slowly than the same amount of sugar in small granular form. The small sugar granules have a greater *total* surface area exposed to the solvent molecules than the lump of sugar does. Stirring also tends to break up larger particles.

Rate of Stirring

Stirring can also increase the rate of dissolution. Stirring increases the rate of direct contact between undissolved solute particles and solvent molecules not yet bound to solute particles. This is the reason you stir your coffee or iced tea after you add sugar.

Temperature

Assuming that one substance is soluble in another, an increase in temperature results in an increase in the *rate* of dissolution. This increase in rate is related to the increase in kinetic energy of the solute, solvent, and solution. With increased kinetic energy of the solute, the particles more readily break away from the other solute particles. With increased kinetic energy of the solvent, the solvent molecules can interact more often with the solute particles. Hence, the rate of dissolution increases. As the solution is being prepared, the higher temperature increases the kinetic energy of the entire solution, resulting in a faster motion of all particles, both solute and solvent. The effect of temperature on rate is quite evident if you try to add sugar to hot and iced tea. Sugar dissolves readily in the hot tea, but even repeated stirring may not dissolve the same amount of sugar in ice-cold tea.

14.4 Saturated, Unsaturated, and Supersaturated Solutions

Concentration Measure of how much solute is contained in a given amount of solvent or solution.

Still another factor that affects how readily one substance dissolves in another is how much of a given solute is already contained in the solvent. No matter how fine the particles of sugar, how hot the tea, or how fast you stir, at some point the water in the tea simply will not hold any more sugar molecules in the solution. The excess sugar falls to the bottom of the cup because you have exceeded the maximum *concentration* for a sugar solution in tea. The **concentration** of a solution indicates how much solute is contained in a given amount of solvent or solution. The concentration of any solution can be described as *saturated, unsaturated,* or *supersaturated*.

Saturated Solutions

Saturated solution Solution that contains just as much solute as can be dissolved in the solvent by ordinary means. Dissolved solute is in dynamic equilibrium with any undissolved solute. That is, the rate of dissolution (dissolving) of any undissolved solute is equal to the rate of crystallization of dissolved solute.

A **saturated solution** is a solution that contains just as much solute as can be dissolved in the solvent by ordinary means. Dissolved solute is in dynamic equilibrium ($\rightleftharpoons$; see Section 12.3) with any dissolved solute. That is, the rate of dissolution (dissolving) of any undissolved solute is equal to the rate of crystallization of dissolved solute:

$$\text{undissolved solute} \xrightleftharpoons[\text{rate of crystallization}]{\text{rate of dissolution}} \text{dissolved solute}$$

A cup of tea that contains just as much sugar as it can maintain in solution—no more and no less—is said to be saturated. Note, though, that saturation is not a result but an ongoing process. Even though a saturated solution does not appear to be changing, the *two processes of dissolution and crystallization are still occurring* whenever excess solute is present. How much of a solute must be present for a solution to be saturated varies with temperature and other factors.

Study Hint: Recall that a dynamic equilibrium always consists of two *opposing* processes. The two processes in this equilibrium are *dissolution (dissolving)* and *crystallization*.

In the laboratory, we can prepare a saturated solution by adding an *excess* of solute to a given amount of solvent and allowing sufficient time for the maximum amount of solute to dissolve. As Figure 14.5a shows, the excess solute sinks to the bottom. The solution formed *above* the undissolved solute is a saturated solution. Filtering off the excess solute leaves us with a saturated solution.

For example, to prepare a saturated solution of potassium chloride (solubility at $2\overline{0}°C$, 34 g of solute per $1\overline{00}$ g of water) in $1\overline{00}$ g of water at $2\overline{0}°C$, we add more than 34 g of potassium chloride to $1\overline{00}$ g of water. This quantity of potassium chloride ensures the presence of excess solute. Next, we warm the water and potassium chloride and stir until saturation or near-saturation is achieved at the elevated temperature. Then we cool the solution at $2\overline{0}°C$ to allow the excess solute to crystallize. After filtering, this solution is saturated at $2\overline{0}°C$.

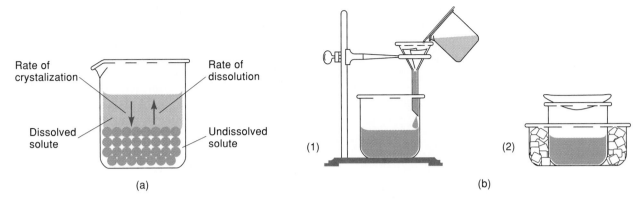

FIGURE 14.5
Saturated and supersaturated solutions. (a) *Saturated* solution before filtering. (b) Preparing a *supersaturated* solution: (1) filtering excess solute from a hot saturated solution and (2) cooling the hot saturated filtrate to form a supersaturated solution if no crystallization occurs.

Unsaturated Solutions

Not all solutions are saturated. An **unsaturated solution** is one in which the concentration of solute is *less* than that of a saturated (equilibrium) solution under the *same conditions*. In an unsaturated solution, the rate of dissolution of undissolved solute is *greater* than the rate of crystallization of dissolved solute, so that, in time, no undissolved solute remains. Such is the case in a cup of hot tea with one teaspoon of sugar.

In the laboratory, you can prepare an unsaturated solution by *diluting* (adding more solvent to) a saturated solution or by using less solute than was dissolved in forming a given saturated solution. In the case of an unsaturated solution of potassium chloride, *less* than 34 g of potassium chloride per $1\overline{00}$ g of water at $2\overline{0}°C$ is present. A solution consisting of $2\overline{0}$ g of potassium chloride in $1\overline{00}$ g of water is thus an unsaturated solution at $2\overline{0}°C$.

Unsaturated solution Solution in which the concentration of solute is less than that in a saturated solution under the same conditions.

Supersaturated Solutions

Finally, some solutions are *supersaturated*. A **supersaturated solution** is one in which the concentration of solute is actually *greater* than that possible in a saturated (equilibrium) solution under the *same conditions*. Such a solution is unstable and will revert to a saturated solution if a "seed" crystal of the solute is added, the excess solute crystallizing out of solution. Honey is an example of a supersaturated solution of sugar.

In the laboratory, we can prepare a supersaturated solution by filtering the undissolved solute from a saturated solution at an *elevated* temperature and then allowing the solution to cool. As Figure 14.5b shows, if crystallization does *not* occur during the cooling process, a supersaturated solution will be obtained at the *lower* temperature. However, the addition of one seed crystal of solute to this solution usually starts crystallization of the excess solute, which continues until a saturated solution is obtained at the lower temperature.

Supersaturated solution Solution in which the concentration of solute is greater than that in a saturated solution under the same conditions. This solution is unstable and reverts to a saturated solution if a seed crystal of solute is added; the excess solute crystallizes out of solution.

FIGURE 14.6
Supersaturated solutions:
(a) a supersaturated solution of sodium acetate; (b)
after adding a crystal of
sodium acetate, excess
sodium acetate crystallizes
out of solution.

(a)   (b)

Note that supersaturated solutions are somewhat unusual. Sodium acetate regularly forms such solutions (see Figure 14.6), but in the case of most solids dissolved in liquids, crystallization of the excess solute occurs on cooling. In the case
of a solution of potassium chloride saturated at $2\overline{0}°C$ (34 g/$1\overline{00}$ g of water), cooling
to 0°C results in the crystallization of 6 g of potassium chloride since the solubility
at 0°C is only 28 g in $1\overline{00}$ g of water.

Purification by Crystallization

Crystallization Method of
purifying solids in which a
sample is dissolved in a hot
solvent, filtered, and cooled to
induce crystallization of the
purified solid.

Chemists prepare saturated solutions, and in some cases supersaturated solutions,
in order to purify a particular solid by **crystallization**. Suppose that a given solid
contains both water-soluble impurities and water-insoluble impurities and that this
solid is soluble in hot water but relatively insoluble in cold water. The solid may be
purified by a two-step process. *First*, dissolve it in sufficient hot water to form a
saturated (or nearly saturated) solution and filter out the water-insoluble impurities.
Second, cool the filtrate or let it stand until it cools to room temperature so that the
solid crystallizes out of solution and the water-soluble impurities remain in solution; then filter out the purified solid. Sometimes, upon cooling, a supersaturated
solution results and a seed crystal must be added to induce crystallization. (Since
not all the impurities may be removed in one such operation, several *re*crystallizations may be needed to achieve a desired state of purity.)

Figure 14.7 outlines this process of purification by crystallization of a solid. Industrially, this process is carried out on a large scale with automated equipment in
the preparation of medicines such as aspirin (acetylsalicylic acid).

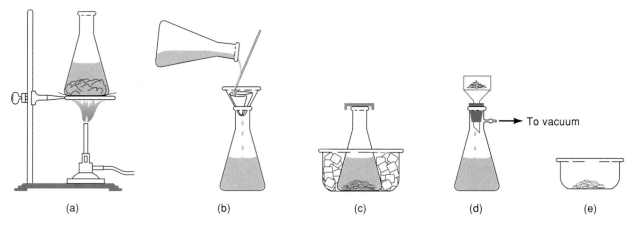

FIGURE 14.7
Purification of a solid by crystallization: (a) dissolving the solid in water to form a solution; (b) filtering the solution to remove water-insoluble impurities; (c) cooling the filtrate to form crystals of purified solid with the water-soluble impurities staying in the solution; (d) filtering the crystals on a Büchner funnel; (e) allowing the crystals to dry.

14.5 Concentration of Solutions

Although saturation is important in purification, knowing whether a solution is unsaturated, saturated, or supersaturated is not enough. Saturation occurs under different conditions of temperature and other factors. Moreover, most solutions are not saturated or supersaturated but unsaturated, which raises the question of just how unsaturated they are.

Scientists often use the terms "concentrated" (abbreviated as "conc" or "concd") and "dilute" (abbreviated as "dil") to express concentration, but these terms are very qualitative. Concentrated hydrochloric acid contains approximately 37 g of hydrogen chloride per $10\overline{0}$ g of solution, but concentrated nitric acid has approximately 72 g of nitric acid per $10\overline{0}$ g of solution. Dilute solutions are less concentrated, but beyond this little more can be said concerning them. A dilute solution of hydrochloric acid, for example, can contain 1.00 g, 5.00 g, or 10.0 g of hydrogen chloride per $10\overline{0}$ g of solution, depending on the particular purpose intended for the acid.

The need to express concentrations precisely is of great concern to scientists. A weak concentration of a solution may have one effect, while a stronger concentration may yield a very different result. For example, hospitals often give injured patients a dilute saline (solution) to help stabilize them. But a concentrated saline solution would quickly kill them.

In the next five sections, we will learn the most common quantitative methods for expressing the concentration of solutions:

1. Percent by mass
2. Parts per million
3. Molarity
4. Normality
5. Molality

Each method has an advantage over the other methods *depending on the eventual use of the solution*. For example, if we want to know the mass of salt in a given mass of ocean water, it is more convenient to express the concentration in percent by mass.

14.6 Percent by Mass

The concentration of a solution may be expressed as parts by mass of solute per $10\overline{0}$ parts by mass of *solution*. This method is known as the **percent by mass** of a solute in a solution:

Percent by mass Measure of the concentration of a solution expressed as the parts by mass of solute per $10\overline{0}$ parts by mass of solution; it is calculated as the mass of solute divided by the mass of solution, with the result multiplied by 100.

$$\% \text{ by mass } = \frac{\text{mass of solute}}{\text{mass of } solution} \times 100 \qquad (14.1)$$

Note that the mass of *solution* is equal to the mass of the *solute* plus the mass of the *solvent*. For example, a 20.0% solution of sodium sulfate contains 20.0 g of sodium sulfate in $10\overline{0}$ g of solution (80.0 g of water), as Figure 14.8 shows.

Consider the following examples involving percent by mass.

> **S**tudy Hint: When you use percent by mass, be sure that the mass units for the solute and the solvent are the same.

EXAMPLE 14.2 Calculate the percent by mass of sodium chloride if 19.0 g of sodium chloride is dissolved in enough water to make 175 g of solution.

SOLUTION The mass of the solution is 175 g. Therefore,

$$\frac{19.0 \text{ g NaCl}}{175 \text{ g solution}} \times 100 = 10.9 \text{ parts NaCl per } 10\overline{0} \text{ parts of solution}$$

$$= 10.9\% \text{ NaCl} \qquad Answer$$

EXAMPLE 14.3 Calculate the percent by mass of sodium chloride if 16.0 g of sodium chloride is dissolved in 80.0 g of water.

SOLUTION The mass of the solution (16.0 g + 80.0 g) is 96.0 g. Therefore,

$$\frac{16.0 \text{ g NaCl}}{96.0 \text{ g solution}} \times 100 = 16.7\% \text{ NaCl} \qquad Answer$$

FIGURE 14.8
A 20.0% by mass aqueous solution of sodium sulfate (Na_2SO_4).

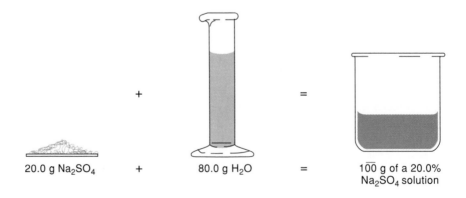

20.0 g Na_2SO_4 +   80.0 g H_2O = $10\overline{0}$ g of a 20.0% Na_2SO_4 solution

Study Exercise 14.3
Calculate the percent by mass of potassium chloride if 6.30 g of potassium chloride is dissolved in 52.5 g of water.

(10.7%)   Work Problems 13 and 14.

EXAMPLE 14.4   Calculate the number of grams of sugar ($C_{12}H_{22}O_{11}$) that must be dissolved in 825 g of water to prepare a 20.0% sugar solution.

SOLUTION   In this solution, there are 20.0 g of sugar for every 80.0 g of water (100.0 g of solution − 20.0 g of sugar = 80.0 g of water); thus, we can calculate the number of grams of sugar needed for 825 g of water as

$$825 \text{ g } H_2O \times \frac{20.0 \text{ g sugar}}{80.0 \text{ g } H_2O} = 206 \text{ g sugar needed for 825 g water} \qquad Answer$$

EXAMPLE 14.5   Calculate the number of grams of water that must be added to 10.0 g of phenol to prepare a 2.00% aqueous phenol solution.

SOLUTION   In a 2.00% phenol solution, there are 2.00 g of phenol for every 98.0 g of water (100.0 g of solution − 2.00 g of phenol = 98.0 g of water); thus, we can calculate the number of grams of water needed for 10.0 g of phenol as

$$10.0 \text{ g phenol} \times \frac{98.0 \text{ g water}}{2.00 \text{ g phenol}} = 49\overline{0} \text{ g water needed for 10.0 g phenol} \qquad Answer$$

EXAMPLE 14.6   Calculate the number of grams of solution necessary to provide 12.5 g of sugar ($C_{12}H_{22}O_{11}$) from a 10.0% sugar solution.

SOLUTION   In a 10.0% sugar solution, there are 10.0 g of sugar for 100.0 g of solution; thus, we can calculate the number of grams of solution for 12.5 g of sugar as

$$12.5 \text{ g sugar} \times \frac{100.0 \text{ g solution}}{10.0 \text{ g sugar}} = 125 \text{ g solution} \qquad Answer$$

Study Exercise 14.4
Calculate the number of grams of water that must be added to 8.50 g of potassium chloride to prepare a 12.0% aqueous potassium chloride solution.

(62.3 g)   Work Problems 15 through 17.

In addition to percent by mass, it is occasionally convenient to express concentration as *percent by volume*, although percent by volume varies with temperature and hence is less reliable. Percent by volume expresses concentration as parts by volume of the solute per $10\overline{0}$ parts by volume of solution. This method is generally used to express the concentration of alcohol in alcoholic beverages. A wine may be 12.5% by volume, and so there is 12.5 mL of alcohol in $10\overline{0}$ mL of the wine. This 12.5% by volume concentration of alcohol in wine is equivalent to only 10.0% by mass of alcohol. In chemical usage, concentrations expressed as percent are *always* understood to mean *percent by mass* unless specified otherwise.

14.7 Parts per Million

Parts per million (ppm)
Measure of the concentration of a solution expressed as parts by mass of solute per 1 million parts by mass of solution; it is calculated as the mass of solute divided by the mass of solution, with the result multiplied by 1,000,000.

In percent by mass the concentration is expressed as parts by mass of solute per $\overline{100}$ parts by mass of solution. In **parts per million (ppm)** the concentration is expressed as parts by mass of solute per *1,000,000* parts by mass of solution.

$$\text{parts per million (ppm)} = \frac{\text{mass of solute}}{\text{mass of solution}} \times 1{,}000{,}000 \qquad (14.2)$$

This method is used for very dilute solutions, such as in water analysis or in biological preparations. In special cases where even smaller amounts of solute are being measured, a concentration might be expressed in *parts per billion* (parts by mass of solute per 1,000,000,000 parts by mass of solution). In these very dilute solutions the amount of solute is so small that the density of the solution is so near to that of water that it is assumed to be 1.00 g/mL.

Consider the following examples involving parts per million.

EXAMPLE 14.7 A water sample contains 3.5 mg of fluoride (F^-) ions in 825 mL of the water sample. Calculate the parts per million of the fluoride ion in the sample.

SOLUTION

$$\frac{3.5 \text{ mg } F^-}{825 \text{ mL sample}} \times \frac{1 \text{ mL sample}}{1.00 \text{ g sample}} \times \frac{1 \text{ g sample}}{1000 \text{ mg sample}} \times 1{,}000{,}000 = 4.2 \text{ ppm} \quad \textit{Answer}$$

Note that as long as the mass units of the solute and the solution are the same, we get the same answer:

$$\frac{3.5 \text{ mg } F^-}{825 \text{ mL sample}} \times \frac{1 \text{ mL sample}}{1.00 \text{ g sample}} \times \frac{1 \text{ g } F^-}{1000 \text{ mg } F^-} \times 1{,}000{,}000 = 4.2 \text{ ppm} \quad \textit{Answer}$$

Study Exercise 14.5
Calculate the parts per million of an aqueous solution containing 355 mg of sodium (Na^+) ions in $75\overline{0}$ mL of a water sample. (Assume that the density of the very dilute sample is 1.00 g/mL.)

Work Problems 18 and 19.

(473 ppm)

EXAMPLE 14.8 Calculate the milligrams of fluoride (F^-) in 1.25 L of a water sample having 4.0 ppm fluoride ion.

SOLUTION Assigning units to the 4.0 ppm as 4.0 g F^- per 1,000,000 g sample, the solution is

$$1.25 \text{ L sample} \times \frac{1000 \text{ mL sample}}{1 \text{ L sample}} \times \frac{1.00 \text{ g sample}}{1 \text{ mL sample}} \times \frac{4.0 \text{ g F}^-}{1,000,000 \text{ g sample}}$$

$$\times \frac{1000 \text{ mg F}^-}{1 \text{ g F}^-} = 5.0 \text{ mg F}^- \qquad Answer$$

We could have also assigned units to the 4.0 ppm as 4.0 mg F^- per 1,000,000 mg sample as long as both units are the same. The solution is

$$1.25 \text{ L sample} \times \frac{1000 \text{ mL sample}}{1 \text{ L sample}} \times \frac{1.00 \text{ g sample}}{1 \text{ mL sample}} \times \frac{1000 \text{ mg sample}}{1 \text{ g sample}}$$

$$\times \frac{4.0 \text{ mg F}^-}{1,000,000 \text{ mg sample}} = 5.0 \text{ mg F}^- \qquad Answer$$

Study Exercise 14.6

Calculate the number of milligrams of sodium (Na^+) ions in 1.50 L of a water sample having 285 ppm sodium ion.

(428 mg)   Work Problem 20.

14.8 Molarity

When using volumetric equipment (graduated cylinders, burets, and the like) to measure a quantity of a solution, scientists often find it helpful to express concentration as molar concentration or molarity. **Molarity (M)** is the number of moles of solute per *liter* of *solution*:

$$M = \text{molarity} = \frac{\text{moles of solute}}{\text{liter of } solution} \qquad (14.3)$$

Molarity (M) Measure of the concentration of a solution expressed as the number of moles of solute per liter of solution; it is calculated as the moles of solute divided by the liters of solution.

Thus, from the volume measured, a simple calculation gives the mass of solute used.

For example, to prepare 1 L of a 1-molar aqueous solution of sodium sulfate, 1 mol of sodium sulfate (142.1 g) is dissolved in water. Enough water is then added to bring the volume of the solution to 1 L in a volumetric flask, as Figure 14.9 shows. An important point to note here is that no information is stated as to the amount of solvent added, only that the solution is made to bring the total volume *to* 1 L.

Consider the following examples involving molarity.

EXAMPLE 14.9   **a.** Calculate the molarity of an aqueous sodium chloride solution containing 284 g of sodium chloride in 2.20 L of solution.

b. Calculate the molarity of the chloride ion in the solution.

SOLUTION

a. The formula of sodium chloride is NaCl, and the molar mass of NaCl is 58.5 g; therefore, calculate the molarity as

$$\frac{284 \text{ g NaCl}}{2.20 \text{ L solution}} \times \frac{1 \text{ mol NaCl}}{58.5 \text{ g NaCl}} = \frac{2.21 \text{ mol NaCl}}{1.00 \text{ L solution}} = 2.21 \text{ } M \qquad Answer$$

FIGURE 14.9
A 1.00 *M* aqueous solution
of sodium sulfate (Na_2SO_4).

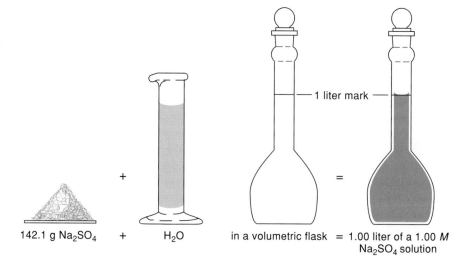

142.1 g Na_2SO_4 + H_2O in a volumetric flask = 1.00 liter of a 1.00 *M* Na_2SO_4 solution

1 liter mark

b. One mole of sodium chloride forms 1 mol of sodium ions and 1 mol of chloride ions according to the following balanced chemical equation:

$$NaCl(aq) \longrightarrow Na^+(aq) + Cl^-(aq)$$

In a 2.21 *M* sodium chloride solution there are 2.21 mol of sodium chloride per liter of solution. Hence, 2.21 mol of NaCl can form 2.21 mol of Na^+ and 2.21 mol of Cl^-.

$$2.21 \text{ mol NaCl} \times \frac{1 \text{ mol Cl}^-}{1 \text{ mol NaCl}} = 2.21 \text{ mol Cl}^-$$

The molarity of the chloride ion is 2.21 *M*. *Answer*

Study Exercise 14.7
a. Calculate the molarity of an aqueous potassium chloride solution containing 125 g of potassium chloride in 1.50 L of solution. (1.12 *M*)
b. Calculate the molarity of the potassium ion. (1.12 *M*)

Work Problems 21 and 22.

EXAMPLE 14.10 **a.** Calculate the number of grams of sodium chloride (NaCl) necessary to prepare $23\overline{0}$ mL of a 2.00 *M* aqueous sodium chloride solution.
b. Explain how this solution is prepared.

SOLUTION
a. The formula of sodium chloride is NaCl, and the molar mass of NaCl is 58.5 g. In a 2.00 *M* NaCl solution, there are 2.00 mol of NaCl per 1.00 L of solution. Calculate the number of grams of NaCl necessary for preparing $23\overline{0}$ mL of a 2.00 *M* solution as

$$23\overline{0} \text{ mL solution} \times \frac{1 \text{ L solution}}{1000 \text{ mL solution}} \times \frac{2.00 \text{ mol NaCl}}{1 \text{ L solution}} \times \frac{58.5 \text{ g NaCl}}{1 \text{ mol NaCl}}$$
$$= 26.9 \text{ g NaCl needed} \quad \textit{Answer}$$

b. The sodium chloride (26.9 g) is dissolved in sufficient water to make the total volume of the solution equal to $23\overline{0}$ mL. *Answer*

EXAMPLE 14.11   Calculate the number of liters of a 6.00 M sodium hydroxide solution required to provide $41\overline{0}$ g of sodium hydroxide.

SOLUTION   The formula of sodium hydroxide is NaOH, and the molar mass of NaOH is 40.0 g. In a 6.00 M NaOH solution, there are 6.00 mol of NaOH per 1.00 L of solution. Calculate the number of liters of 6.00 M solution necessary to provide $41\overline{0}$ g of NaOH as

$$41\overline{0}\text{ g NaOH} \times \frac{1\text{ mol NaOH}}{40.0\text{ g NaOH}} \times \frac{1.00\text{ L solution}}{6.00\text{ mol NaOH}} = 1.71\text{ L solution} \qquad Answer$$

Study Exercise 14.8
Calculate the number of grams of potassium chloride necessary to prepare $65\overline{0}$ mL of a 1.50 M aqueous potassium chloride solution.

(72.7 g)   Work Problems 23 and 24.

Dilution of Molar Solutions

Solutions in a laboratory are often available in a concentrated form and then diluted to the desired concentration with water. These concentrated solutions are called *stock solutions*. You may have previously used a stock solution without recognizing it. If you have ever used a bleach solution (5.25% sodium hypochlorite) in your laundry, the bleach solution is the stock solution. Usually, 1 cup of bleach solution (stock solution) is *diluted* in the washing machine wash water before the clothes are added, or 1 cup of bleach is *diluted* with 1 qt of water and then added to the wash water. In chemistry we do the same thing except that we have to calculate the exact amount of stock solution to add to the water to make the desired concentration. Because *only water* is added, the number of moles of solute remains the same in the diluted solution as in the volume of stock solution removed for dilution. We, therefore, need to calculate the moles of solute in the diluted solution. Consider the following example.

EXAMPLE 14.12   Calculate the number of milliliters of 12.0 M concentrated hydrochloric acid that must be added to water to prepare 1.00 L of a 5.00 M hydrochloric acid solution.

SOLUTION
(1) Calculate the moles of hydrochloric acid in the diluted solution using 5.00 M (5.00 mol HCl per 1 L solution) and 1.00 L of solution.

$$1.00\text{ L solution} \times \frac{5.00\text{ mol HCl}}{1\text{ L solution}} = 5.00\text{ mol HCl needed in the diluted solution}$$

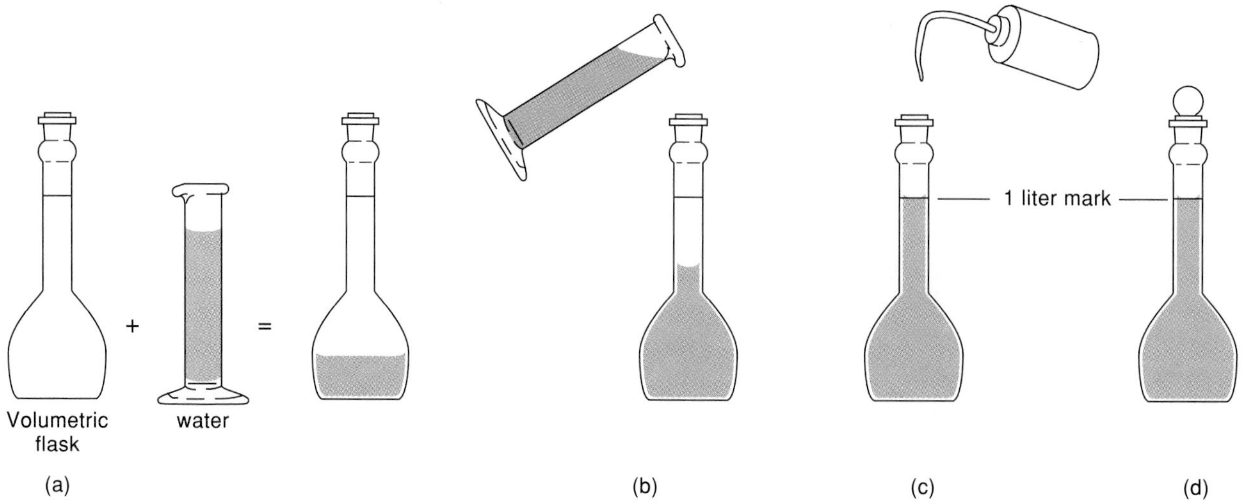

FIGURE 14.10
Preparation of 1.00 L of a 5.00 *M* hydrochloric acid solution. (a) Water (about 400 mL) is added to a volumetric flask. (b) Then 417 mL of 12.0 *M* hydrochloric acid solution is added, and the flask is swirled. (c) More water is added with a wash bottle until the volume of the solution reaches the 1-L mark on the volumetric flask. (d) The cap is placed on the volumetric flask, and the flask is turned upside-down a number of times to mix the solution.

(2) The 5.00 mol HCl needed is obtained from the stock solution consisting of 12.0 *M* (12.0 mol HCl per 1 L solution). The number of milliliters of the stock solution is calculated as

$$5.00 \cancel{\text{ mol HCl}} \times \frac{1 \cancel{\text{ L solution}}}{12.0 \cancel{\text{ mol HCl}}} \times \frac{1000 \text{ mL solution}}{1 \cancel{\text{ L solution}}}$$

$$= 417 \text{ mL of } 12.0 \text{ } M \text{ HCl} \qquad \textit{Answer *}$$

Now how is this diluted solution prepared? The 417 mL of the 12.0 *M* hydrochloric acid stock solution is added to a 1-L volumetric flask containing some water (about 400 mL). The reason that water is present in the volumetric flask is that heat is given off (exothermic; see Section 10.7) in the dilution process and

---

*An alternate solution to these problems is to solve them using the formula

$$\boxed{V_1 \times M_1 = V_2 \times M_2}$$ (14.4)

where V_1 is the volume in liters at a concentration of M_1 in mol/L and V_2 is the volume in liters at a concentration of M_2 in mol/L. Solving this equation using these units gives

$$\cancel{\text{liters}_1} \times \frac{\text{mol}_1}{\cancel{\text{liter}_1}} = \cancel{\text{liters}_2} \times \frac{\text{mol}_2}{\cancel{\text{liter}_2}}$$

$$\text{mol}_1 = \text{mol}_2$$

Therefore, the moles of solute in the two solutions remain the same since *only water* was added to make the diluted solution from the stock solution.

some of this heat is absorbed by the water. *Always add acid to water. NEVER* add water to acid, as the flask may break because of the excess heat. The volumetric flask with the water and stock solution acid is swirled, and then more water is added with a wash bottle to bring the *total* volume of the solution to the 1-L mark on the volumetric flask. The cap is placed on the volumetric flask, and the flask is turned upside-down a number of times to mix the solution. This then gives 1 L of a 5.00 *M* hydrochloric acid solution. See Figure 14.10.

We can also calculate the molar concentration of a diluted solution from the volume and concentration of the stock solution and the volume of the diluted solution. Again, since only water is being added, the number of moles of solute remains the same in the diluted solution as in the volume of the stock solution used.

Consider the following example.

EXAMPLE 14.13 Calculate the molarity of a hydrochloric acid solution prepared by using 125 mL of 12.0 *M* hydrochloric acid stock solution and diluting it to $5\overline{0}0$ mL with water.

SOLUTION
(1) Calculate the moles of hydrochloric acid in 125 mL of the 12.0 *M* solution (12.0 mol HCl per 1 L solution) as

$$125 \text{ mL solution} \times \frac{1 \text{ L solution}}{1000 \text{ mL solution}} \times \frac{12.0 \text{ mol HCl}}{1 \text{ L solution}} = 1.50 \text{ mol HCl}$$

(2) The 1.50 mol HCl is then diluted with water to $5\overline{0}0$ mL. The molarity in moles of HCl per liter of solution is calculated as

$$\frac{1.50 \text{ mol HCl}}{5\overline{0}0 \text{ mL solution}} \times \frac{1000 \text{ mL solution}}{1 \text{ L solution}} = 3.00 \text{ M} \qquad Answer$$

> **Study Hint:** The moles of solute is always calculated first. It is calculated from the volume of the diluted stock solution *or* the volume of the stock solution and the molarity of the respective solutions. The volume and the molarity are written next to each other in the problem.

Study Exercise 14.9
a. Calculate the number of milliliters of 17.4 *M* concentrated acetic acid that must be added to water to prepare 1.00 L of a 6.00 *M* acetic acid solution.

(345 mL)

b. Explain how this diluted solution is prepared.
[The 345 mL is added to a 1-L volumetric flask containing some water (about 400 mL) and swirled. Then, more water is added until the volume reaches the 1-L mark on the volumetric flask. The cap is added to the flask, and the flask is turned upside-down a number of times to mix the solution.]

Work Problems 25 and 26.

14.9 Normality

When a solution involves a reaction of acids with bases, scientists often prefer to express concentration in terms of *normality*. **Normality (*N*)** is the number of equivalents (eq) of solute per liter of *solution*:

> **Normality (*N*)** Measure of the concentration of a solution expressed as the number of equivalents of solute per liter of solution; it is calculated as the equivalents of solute divided by the liters of solution.

$$\boxed{N = \text{normality} = \frac{\text{equivalents of solute}}{\text{liter of } solution}} \qquad (14.5)$$

The normality definition is similar to that for molarity except that it considers equivalents instead of moles of solute per liter. But what is an equivalent? In this

Equivalent of an acid Quantity of a substance that reacts to yield 1 mol (6.02×10^{23}) of hydrogen ions (H^+).

Equivalent of a base Quantity of a substance that reacts with 1 mol (6.02×10^{23}) of hydrogen ions (H^+) or supplies 1 mol (6.02×10^{23}) of hydroxide ions (OH^-).

> **Study Hint:** To use normality, we *must* know how many hydrogen ions or hydroxide ions are replaced or react.

text, we will use two definitions: *One **equivalent of an acid** is that quantity that reacts to yield 1 mol (6.02×10^{23}) of hydrogen ions. One **equivalent of a base** is that quantity that reacts with 1 mol (6.02×10^{23}) of hydrogen ions or supplies 1 mol (6.02×10^{23}) of hydroxide ions.* Thus, *one equivalent of any acid combines exactly with one equivalent of any base.*

To determine how much 1 eq of an acid is, we must divide the molar mass of the acid by the number of moles of hydrogen ion per mole of acid *used in the reaction*. To determine how much 1 eq of a base is, we divide the molar mass of the base by the number of moles of hydrogen ion that combine with 1 mol of the base or the number of moles of hydroxide ion per mole of base *used in the reaction*.

Consider the case of acids. An example of a reaction in which *two* H^+ are replaced on sulfuric acid is

$$2\,NaOH(aq) + H_2SO_4(aq) \longrightarrow Na_2SO_4(aq) + 2\,H_2O(l)$$

(both H^+ are replaced)   (14.6)

Therefore, in 1 eq of H_2SO_4 there are

$$\frac{98.1\ g}{2} = 49.0\ g$$

The replacement of only *one* H^+ from sulfuric acid is

$$1\,NaOH(aq) + H_2SO_4(aq) \longrightarrow NaHSO_4(aq) + H_2O(l)$$

(only one H^+ is replaced)   (14.7)

Therefore, in *one* equivalent of H_2SO_4 there are

$$\frac{98.1\ g}{1} = 98.1\ g$$

Consider the case of bases. An example of a reaction in which *2 mol* H^+ ion reacts with the base is

$$Ca(OH)_2(aq) + 2\,HCl(aq) \longrightarrow CaCl_2(aq) + 2\,H_2O(l)$$

(both OH^- are replaced)   (14.8)

Therefore, in *one* equivalent of $Ca(OH)_2$ there are

$$\frac{74.1\ g}{2} = 37.0\ g$$

Bases usually react completely with an acid to replace all the hydroxide ions. Replacement of less than the total number of hydroxides in a base is not common and will not be considered here.

EXAMPLE 14.14 Calculate the normality of an aqueous sulfuric acid solution containing 275 g of sulfuric acid in 1.20 L of solution in reactions that replace both hydrogen ions.

SOLUTION The formula of sulfuric acid is H_2SO_4, and the molar mass of H_2SO_4 is 98.1 g. Therefore, 1 eq of H_2SO_4 is equal to 49.0 g (98.1 g/2) since both hydrogen ions in H_2SO_4 are replaced in the reaction. We can calculate the normality as

$$\frac{275\ g\ H_2SO_4}{1.20\ L\ solution} \times \frac{1\ eq\ H_2SO_4}{49.0\ g\ H_2SO_4} = \frac{4.68\ eq\ H_2SO_4}{1\ L\ solution} = 4.68\ N \qquad Answer$$

Study Exercise 14.10
Calculate the normality of an aqueous solution of calcium hydroxide containing 2.25 g of calcium hydroxide in 1.50 L of solution in reactions that replace *both* hydroxide ions.

(0.0405 *N*)   Work Problems 27 and 28.

EXAMPLE 14.15   Calculate the number of grams of sulfuric acid necessary to prepare $52\overline{0}$ mL of 0.100 *N* aqueous sulfuric acid solution in reactions that replace *both* hydrogen ions.

SOLUTION   The formula of sulfuric acid is H_2SO_4, and the molar mass of H_2SO_4 is 98.1 g. Therefore, 1 eq of H_2SO_4 is equal to 49.0 g (98.1 g/2) since both hydrogen ions of H_2SO_4 are replaced in the reaction. In a 0.100 *N* H_2SO_4 solution there is 0.100 eq of H_2SO_4 in 1.00 L of solution. Therefore, the number of grams of H_2SO_4 necessary for preparing $52\overline{0}$ mL of 0.100 *N* H_2SO_4 solution is

$$52\overline{0} \text{ mL solution} \times \frac{1 \text{ L solution}}{1000 \text{ mL solution}} \times \frac{0.100 \text{ eq } H_2SO_4}{1 \text{ L solution}} \times \frac{49.0 \text{ g } H_2SO_4}{1 \text{ eq } H_2SO_4}$$

$$= 2.55 \text{ g } H_2SO_4 \qquad Answer$$

EXAMPLE 14.16   Calculate the number of milliliters of sulfuric acid solution required to provide $12\overline{0}$ g of sulfuric acid from a 2.00 *N* aqueous sulfuric acid solution in a reaction that replaces *both* hydrogen ions.

SOLUTION   A 2.00 *N* sulfuric acid solution contains 2.00 eq of H_2SO_4 in each liter of solution. Since 2 mol of hydrogen ions is replaced per mole of H_2SO_4, 1 eq of H_2SO_4 is equal to 49.0 g (98.1 g/2). Therefore, we can calculate the number of milliliters of solution required as

$$12\overline{0} \text{ g } H_2SO_4 \times \frac{1 \text{ eq } H_2SO_4}{49.0 \text{ g } H_2SO_4} \times \frac{1 \text{ L solution}}{2.00 \text{ eq } H_2SO_4} \times \frac{1000 \text{ mL solution}}{1 \text{ L solution}}$$

$$= 1220 \text{ mL solution (to three significant digits)} \qquad Answer$$

Study Exercise 14.11
Calculate the number of milliliters of calcium hydroxide solution required to provide 0.650 g of calcium hydroxide from a 0.0300 *N* calcium hydroxide solution in reactions that replace *both* hydroxide ions.

(586 mL)   Work Problems 29 and 30.

Relation of Molarity to Normality

As previously mentioned, the definition of *normality* is similar to the definition of *molarity*. In normality we use equivalents; in molarity we use moles. The two definitions are related. *Normality* is always a *multiple of molarity* because there are 1, 2, 3, and so on, equivalents *per mole*. Consider some examples. A 1 *M* hydrochloric acid (HCl) solution is 1 *N* because there is *one equivalent per mole* of HCl. A 1 *M* sulfuric acid (H_2SO_4) solution is a 2 N solution in which both hydrogen ions are replaced because there are *two equivalents per mole* of sulfuric acid.

$$\frac{1.00 \text{ mol } H_2SO_4}{1 \text{ L solution}} \times \frac{2 \text{ eq } H_2SO_4}{1 \text{ mol } H_2SO_4} = \frac{2 \text{ eq } H_2SO_4}{1 \text{ L solution}} = 2.00 \text{ } N \text{ } H_2SO_4$$

There is always *one* mole of the acid or base with a whole number of equivalents to this one mole.

In reactions of calcium hydroxide [$Ca(OH)_2$] with *both* hydroxide ions replaced, 1 N calcium hydroxide is 0.500 M because there are *two* equivalents of calcium hydroxide in *one* mole of calcium hydroxide.

$$\frac{1.00 \text{ eq } Ca(OH)_2}{1 \text{ L solution}} \times \frac{1 \text{ mol } Ca(OH)_2}{2 \text{ eq } Ca(OH)_2} = \frac{0.500 \text{ mol } Ca(OH)_2}{1 \text{ L solution}} = 0.500 \text{ } M \text{ } Ca(OH)_2$$

EXAMPLE 14.17 Calculate the normality of a 3.50 M sulfuric acid solution in reactions that replace *both* hydrogen ions.

SOLUTION   The formula of sulfuric acid is H_2SO_4. Since both hydrogen ions are replaced, there are 2 eq H_2SO_4 per 1 mol H_2SO_4. Solve for equivalents per liter using 3.50 mol H_2SO_4 per 1 L solution as

$$\frac{3.50 \text{ mol } H_2SO_4}{1 \text{ L solution}} \times \frac{2 \text{ eq } H_2SO_4}{1 \text{ mol } H_2SO_4} = \frac{7.00 \text{ eq } H_2SO_4}{1 \text{ L solution}} = 7.00 \text{ } N \text{ } H_2SO_4 \quad \textit{Answer}$$

Study Exercise 14.12
Calculate the molarity of a 8.50 N phosphoric acid solution in reactions that replace all *three* hydrogen ions.

Work Problems 31 and 32.

(2.83 M)

14.10 Molality

At times it is more convenient to measure the mass of the solvent rather than the volume of the solution. In such cases, the molarity or normality is not useful. Instead, chemists measure the concentration in terms of *molality*.

Molality (m)   Measure of the concentration of a solution expressed as the number of moles of solute per kilogram of solvent; it is calculated as the moles of solute divided by the kilograms of solvent.

Molality (m) is the number of moles of solute per kilogram of *solvent*:

$$\boxed{m = \text{molality} = \frac{\text{moles of solute}}{\text{kilogram of } \textit{solvent}}} \qquad (14.9)$$

Note that to express a concentration in units of molality we must know the *mass* of the solute and the mass (not the volume) of the *solvent*. Consider Figure 14.11, which shows the preparation of a 1-molal aqueous solution of sodium sulfate. The solute in this solution is 1 mol of sodium sulfate (142.1 g). The solvent is 1.000 kg ($1\overline{000}$ g) of water.

In working problems involving molality, factors relating mass of solute to mass of *solvent* and mass of solute to mass of *solution* are frequently useful. The mass units for the solute in these factors may be expressed in grams or moles, and the mass units for the solvent or solution may be expressed in grams or kilograms. The best choice of factor and units depends on the application.

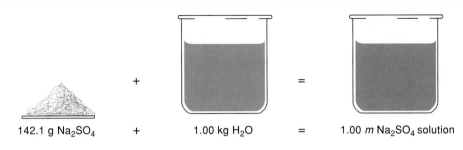

142.1 g Na₂SO₄ + 1.00 kg H₂O = 1.00 *m* Na₂SO₄ solution

FIGURE 14.11
A 1.00 *m* aqueous solution
of sodium sulfate (Na₂SO₄).

Consider the following examples involving molality.

EXAMPLE 14.18 Calculate the molality of a glycerol ($C_3H_8O_3$) solution containing 32.7 g of glycerol in $1\overline{0}0$ g of water.

SOLUTION The molar mass of glycerol ($C_3H_8O_3$) is 92.0 g. The molality of the solution must express the concentration of glycerol as mol/kg of water. Calculate the molality as

$$\frac{32.7 \text{ g } C_3H_8O_3}{1\overline{0}0 \text{ g } H_2O} \times \frac{1 \text{ mol } C_3H_8O_3}{92.0 \text{ g } C_3H_8O_3} \times \frac{1000 \text{ g } H_2O}{1 \text{ kg } H_2O} = \frac{3.55 \text{ mol } C_3H_8O_3}{1 \text{ kg } H_2O}$$

$$= 3.55 \text{ } m \qquad Answer$$

Study Exercise 14.13
Calculate the molality of a sugar ($C_{12}H_{22}O_{11}$) solution containing 96.0 g of sugar in $2\overline{0}0$ g of water.

$$(1.40 \text{ } m) \qquad \text{Work Problems 33 and 34.}$$

EXAMPLE 14.19 Calculate the number of grams of glycerol ($C_3H_8O_3$) necessary to prepare $52\overline{0}$ g of a 2.00 *m* solution of glycerol in water.

SOLUTION The molar mass of glycerol is 92.0 g. A 2.00 *m* glycerol solution contains 2.00 mol (184 g) of glycerol in 1.000 kg ($1\overline{0}00$ g) of water. The *total* mass of this solution is

$1184 \text{ g} \left(2.00 \text{ mol} \times \dfrac{92.0 \text{ g}}{1 \text{ mol}} + 1\overline{0}00 \text{ g of water} \right)$. Calculate the mass of the glycerol nec-

essary for $52\overline{0}$ g of a 2.00 *m* solution as

$$52\overline{0} \text{ g solution} \times \frac{184 \text{ g glycerol}}{1184 \text{ g solution}} = 80.8 \text{ g glycerol} \qquad Answer$$

> **Study Hint:** In Example 14.19 the $52\overline{0}$ g refers to the *solution*—not the water. That is the reason we must calculate the total mass of the 2.00 *m* solution. You must read the problem very carefully.

EXAMPLE 14.20 Calculate the number of grams of water that must be added to 5.80 g of glycerol ($C_3H_8O_3$) to prepare a 0.100 *m* glycerol solution.

SOLUTION The molar mass of glycerol ($C_3H_8O_3$) is 92.0 g. We are asked to find the number of grams of water required to prepare the solution. Since the mass of the solvent is in the denominator of the molality definition, we use the inverse (kg H_2O/mol $C_3H_8O_3$). Calculate the number of grams of water as

$$5.80 \text{ g } C_3H_8O_3 \times \frac{1 \text{ mol } C_3H_8O_3}{92.0 \text{ g } C_3H_8O_3} \times \frac{1 \text{ kg } H_2O}{0.100 \text{ mol } C_3H_8O_3} \times \frac{1000 \text{ g } H_2O}{1 \text{ kg } H_2O}$$

$$= 63\overline{0} \text{ g } H_2O \qquad Answer$$

TABLE 14.2	Expressing Concentrations of Solutions

$$\text{percent by mass} = \frac{\text{mass of solute}}{\text{mass of } solution} \times 100$$

$$\text{parts per million (ppm)} = \frac{\text{mass of solute}}{\text{mass of } solution} \times 1{,}000{,}000$$

$$\text{molarity } (M) = \frac{\text{moles of solute}}{\text{liter of } solution}$$

$$\text{normality } (N) = \frac{\text{equivalents of solute}}{\text{liter of } solution}$$

$$\text{molality } (m) = \frac{\text{moles of solute}}{\text{kilogram of } solvent}$$

Study Exercise 14.14
Calculate the number of grams of sugar ($C_{12}H_{22}O_{11}$) necessary to prepare $6\overline{0}0$ g of a 1.00 m solution of sugar in water.

(153 g)

Study Exercise 14.15
Calculate the number of grams of water that must be added to $18\overline{0}$ g of sugar ($C_{12}H_{22}O_{11}$) to prepare a 1.20 m solution of sugar in water.

Work Problems 35 and 36.

(439 g)

Table 14.2 reviews the five different types of solutions discussed in this chapter.

14.11 Colligative Properties of Solutions

Colligative properties Properties of a solution that depend only on the number of particles of solute present in a solution and not on the actual identity of these solute particles.

Colligative properties of solutions are properties that depend only on the *number* of particles of solute present in the solution and not on the actual identity of these solute particles. Both sucrose ($C_{12}H_{22}O_{11}$) and urea (CH_4N_2O) change the boiling and freezing points of a given solvent in an identical fashion but have completely different properties, such as solubility and density. We will examine two of these colligative properties: (1) boiling point elevation and (2) freezing point depression.

Before we consider these properties, however, we need to know the types of solutes with which we will be concerned. We will perform calculations *only* for solutes that are *not* ionic (nonionic) and do *not* evaporate (nonvolatile) from the solution. Thus, solutes like sugar ($C_{12}H_{22}O_{11}$), glucose or dextrose ($C_6H_{12}O_6$), ethylene glycol ($C_2H_6O_2$), urea (CH_4N_2O), and nitrobenzene ($C_6H_5NO_2$, a liquid that does behave well) will serve as examples.

Boiling Point Elevation

If we were to measure the boiling point of seawater or of a sugar solution, we would find that these solutions boil at temperatures somewhat *higher* than the boiling

TABLE 14.3	**Molal Boiling Point Elevation and Freezing Point Depression Constants for Water and Benzene**			
SOLVENT	NORMAL BOILING POINT (°C)	K_b (°C/m)	FREEZING POINT (°C)	K_f (°C/m)
Water	100.00	0.52	0.00	1.86
Benzene	80.10	2.53	5.50	5.12

point of pure water. For example, a solution containing 1.00 mol of sucrose (table sugar) in $10\overline{0}0$ g of water boils at 100.52°C rather than at the normal 100.00°C observed for pure water at 1 atm. The dissolved solute raises the boiling point of the solution.

We can easily calculate this elevation of the boiling point for a "well-behaved" solute like sugar. The elevation is related in a simple way to the molality of the sugar solution under consideration and the *molal boiling point elevation constant* of the solvent (K_b). Table 14.3 gives this constant for the solvents water and benzene. The units of K_b are degrees Celsius per molal solution. Thus, water has a K_b of 0.52°C/m, and so a 1.00 m solution boils at 100.52°C (at 1 atm), 0.52°C *above* the normal boiling point of pure water (100.00°C). The equation for calculating the boiling point elevation is

> **Study Hint:** Remember that boiling points are *elevated*, and so the boiling point of the solution has to be *higher* than the boiling point of the pure solvent.

$$\text{change in } T_b = \text{molality } (m) \cdot K_b \qquad (14.10)$$

EXAMPLE 14.21   Calculate the boiling point (at 1 atm) in degrees Celsius of the following solutions:

a. a 2.50 m aqueous sugar ($C_{12}H_{22}O_{11}$) solution
b. a solution containing 9.75 g of urea (CH_4N_2O) in $25\overline{0}$ g of water

SOLUTION
a. Calculate the boiling point elevation (the value of K_b for water is in Table 14.3) as

$$\text{change in } T_b = \text{molality } (m) \cdot K_b = 2.50 \, m \times \frac{0.52°C}{1 \, m} = 1.3°C$$

$$T_b = 100.00°C + 1.3°C = 101.3°C \qquad Answer$$

> **Study Hint:** The boiling point of the solution (101.3°C) is *higher* than the boiling point of pure water (100.0°C).

b. Calculate the molality of the urea (CH_4N_2O, molar mass = 60.0 g) solution as

$$\frac{9.75 \text{ g } CH_4N_2O}{25\overline{0} \text{ g water}} \times \frac{1 \text{ mol } CH_4N_2O}{60.0 \text{ g } CH_4N_2O} \times \frac{1000 \text{ g water}}{1 \text{ kg water}} = \frac{0.650 \text{ mol } CH_4N_2O}{1 \text{ kg water}}$$

$$= 0.650 \, m$$

Calculate the boiling point elevation (the value of K_b for water is in Table 14.3) as

$$\text{change in } T_b = \text{molality } (m) \cdot K_b = 0.650 \, m \times \frac{0.52°C}{1 \, m} = 0.34°C$$

$$T_b = 100.00°C + 0.34°C = 100.34°C \qquad Answer$$

Study Exercise 14.16
Calculate the boiling point (at 1 atm) in degrees Celsius of a solution prepared by dissolving 24.6 g of glucose ($C_6H_{12}O_6$, molar mass $= 18\overline{0}$ g) in $25\overline{0}$ g of water.

(100.28°C)

Freezing Point Depression

Seawater does not freeze solid when the temperature is at 0°C. Neither does a sugar solution. Instead, they freeze at a somewhat *lower* temperature than pure water. Thus, a solution of 1.00 mol of sucrose (table sugar) in $10\overline{00}$ g of water freezes at −1.86°C rather than at the normal 0.00°C observed for pure water. The dissolved solute lowers the freezing point of the solution.

Like the boiling point elevation, we can readily calculate the freezing point depression for a well-behaved solute like sugar. The drop in the freezing point is related to the molality of the sugar solution under consideration and the *molal freezing point depression constant* of the solvent (K_f). Table 14.3 gives this constant for the solvents water and benzene. The units of K_f are degrees Celsius per molal solution. Thus, water has a K_f of 1.86°C/*m*, and so a 1.00 *m* solution freezes at −1.86°C, or −1.86°C *below* the normal freezing point of pure water (0.00°C). The equation for calculating the freezing point depression is

$$\boxed{\text{change in } T_f \ = \ \text{molality } (m) \cdot K_f} \qquad (14.11)$$

The principle of freezing point depression is applied in melting ice on highways or sidewalks. A salt, such as sodium chloride or calcium chloride, is used. The freezing point of the mixture (ice–salt) is lower than that of pure ice, and the ice melts if the outside temperature is not lower than the freezing point of the mixture. The principles of freezing point depression and boiling point elevation also help protect the engine in your car. Serious damage can occur to your engine if the solution in the radiator freezes (recall that water expands on freezing; see Section 13.4) or turns into steam. Adding ethylene glycol to the radiator water raises the boiling point and lowers the freezing point. As a result, the radiator solution has a lower freezing point and does not freeze solid in the winter. Raising the boiling point helps prevent the solution from boiling away on hot summer days.

EXAMPLE 14.22 Calculate the freezing point in degrees Celsius of the following solutions:

a. a solution containing 4.27 g of sugar ($C_{12}H_{22}O_{11}$) in 50.0 g of water.
b. a solution containing 9.75 g of nitrobenzene ($C_6H_5NO_2$) in 175 g of *benzene*.

SOLUTION
a. Before we can calculate the freezing point depression, we must determine the molality of the sugar solution. The molar mass of sugar ($C_{12}H_{22}O_{11}$) is 342 g; calculate the molality of this solution as

$$\frac{4.27 \text{ g } C_{12}H_{22}O_{11}}{50.0 \text{ g } H_2O} \times \frac{1 \text{ mol } C_{12}H_{22}O_{11}}{342 \text{ g } C_{12}H_{22}O_{11}} \times \frac{1000 \text{ g } H_2O}{1 \text{ kg } H_2O} = \frac{0.250 \text{ mol } C_{12}H_{22}O_{11}}{1 \text{ kg } H_2O}$$

$$= 0.250 \ m$$

Calculate the freezing point depression (the value of K_f for water is in Table 14.3) as

$$\text{change in } T_f = \text{molality } (m) \cdot K_f = 0.250 \, m \times \frac{1.86°C}{1 \, m} = 0.465°C$$

$$T_f = 0.00°C - 0.465°C = -0.46°C \qquad Answer$$

b. Calculate the molality of the nitrobenzene ($C_6H_5NO_2$, molar mass = 123.0 g) solution as

$$\frac{9.75 \text{ g } C_6H_5NO_2}{175 \text{ g benzene}} \times \frac{1 \text{ mol } C_6H_5NO_2}{123.0 \text{ g } C_6H_5NO_2} \times \frac{1000 \text{ g benzene}}{1 \text{ kg benzene}} = \frac{0.453 \text{ mol } C_6H_5NO_2}{1 \text{ kg benzene}}$$

$$= 0.453 \, m$$

Study Hint: The solvent in Example 14.22b is *benzene* and not water. The freezing point of the solution (3.18°C) is *lower* than the freezing point of pure benzene (5.50°C).

Calculate the freezing point depression (the value of K_f for *benzene* is in Table 14.3) as

$$\text{change in } T_f = \text{molality } (m) \cdot K_f = 0.453 \, m \times \frac{5.12°C}{1 \, m} = 2.32°C$$

$$T_f = 5.50°C - 2.32°C = 3.18°C \qquad Answer$$

Study Exercise 14.17
Calculate the freezing point in degrees Celsius for a solution prepared by dissolving 35.0 g of glucose ($C_6H_{12}O_6$, molar mass = $18\overline{0}$ g) in $20\overline{0}$ g of water.

$$(-1.81°C)$$

Work Problems 37 and 38.

Determination of Molecular Mass and Molar Mass

You might wonder if the colligative properties are useful for anything besides melting the ice on city roads and highways. The following example shows how the freezing point depression can be used to determine the molecular mass and molar mass of a solid, sometimes a difficult measurement to make.

EXAMPLE 14.23 Calculate the molecular mass and molar mass of an unknown solid from the following data:

a. 9.75 g of the unknown was dissolved in 125 g of water, and the resulting solution had a freezing point of −1.25°C.

b. 5.65 g of the unknown was dissolved in 125 g of *benzene*, and the resulting solution had a freezing point of 4.25°C.

SOLUTION
a. Calculate the change in $T_f = 1.25°C$ [freezing point of water − freezing point of solution = 0.00°C − (−1.25°C) = 1.25°C]. Calculate the molality of the solution (the value of K_f for water is given in Table 14.3).

$$\text{molality } (m) = \frac{\text{change in } T_f}{K_f} = 1.25°C \times \frac{1 \, m}{1.86°C} = 0.672 \, m$$

$$= \frac{0.672 \text{ mol unknown}}{1 \text{ kg water}}$$

Using the unknown solution data and the molality (0.672 mol of unknown per kilogram of water), calculate the molecular mass and molar mass by solving for *g/mol*.

$$\frac{9.75 \text{ g unknown}}{125 \text{ g water}} \times \frac{1000 \text{ g water}}{1 \text{ kg water}} \times \frac{1 \text{ kg water}}{0.672 \text{ mol unknown}}$$

$$= 116 \text{ g/mol, } 116 \text{ amu} \qquad \textit{Answer}$$

$$116 \text{ g} \qquad \textit{Answer}$$

b. Calculate the change in $T_f = 1.25°C$ (freezing point of *benzene* $-$ freezing point of solution $= 5.50°C - 4.25°C = 1.25°C$). Calculate the molality of the solution (the value of K_f for *benzene* is given in Table 14.3).

$$\text{molality } (m) = \frac{\text{change in } T_f}{K_f} = 1.25°C \times \frac{1 \text{ m}}{5.12°C} = 0.244 \text{ m}$$

$$= \frac{0.244 \text{ mol unknown}}{1 \text{ kg benzene}}$$

Using the unknown solution data and the molality (0.244 mol of unknown per kilogram of benzene), calculate the molecular mass and molar mass by solving for *g/mol*.

$$\frac{5.65 \text{ g unknown}}{125 \text{ g benzene}} \times \frac{1000 \text{ g benzene}}{1 \text{ kg benzene}} \times \frac{1 \text{ kg benzene}}{0.244 \text{ mol unknown}}$$

$$= 185 \text{ g/mol, } 185 \text{ amu} \qquad \textit{Answer}$$

$$185 \text{ g} \qquad \textit{Answer}$$

Study Exercise 14.18

Calculate the molecular mass and molar mass of an unknown solid if 18.5 g of the unknown solid was dissolved in $15\overline{0}$ g of water and the resulting solution had a freezing point of $-1.55°C$.

Work Problem 39.

(148 amu, 148 g)

14.12   Colloids and Suspensions

Colloid   A mixture in which the particles are dispersed without appreciable bonding to the solvent molecules and do not settle out on standing.

Dispersed particles (dispersed phase)   The colloidal particles in a colloid, comparable to the solute in a solution with a diameter range of 1000 to 200,000 pm.

Dispersing medium (dispersing phase)   The substance in a colloid in which the colloidal particles are distributed, comparable to the solvent in a solution.

Colloids bridge the gap between matter in a solution and matter dispersed in a suspension. In a *solution*, the particles are homogeneously dispersed and *do not* settle out on standing because they may be partially bound to the solvent molecules (Section 14.1). In a *suspension*, the particles are *not* bound to solvent molecules and *do* settle out on standing. In **colloids**, the dispersed particles are *not* appreciably bound to solvent molecules but *do not* settle out on standing. In this section we will discuss properties and examples of colloids and relate colloids to solutions.

In discussing solutions we spoke of solutes and solvents, but in discussing colloids we use the terms "dispersed particles" (dispersed phase) and "dispers*ing* medium" (dispers*ing* phase). **Dispersed particles** (dispersed phase) are the colloidal particles, comparable to the solute in a solution. The **dispers*ing* medium** (dispers*ing* phase) is the substance in which the colloidal particles are distributed, comparable to the solvent in a solution. Milk is an example of a colloid; butterfat constitutes the dispersed particles, and water is the dispersing medium.

In a solution, the particles of solute and solvent are molecules or ions with one or more solvent molecules bound to each solute particle. In colloids, the suspended

FIGURE 14.12
A walk in the fog. Fog and mist are examples of colloids (liquid water dispersed in a gas–liquid aerosol). (Dr. E. R. Degginger)

particles are large molecules or aggregates of molecules, *larger* than those found in *solutions* but *smaller* than those found in *suspensions*, and on the average with less than one molecule of dispersing medium bound per molecule of dispersed particles. The colloidal particles are in various shapes and range from 1000 to 200,000 pm (see Section 2.2) in diameter. They can consist of one huge molecule, such as a starch molecule, or aggregates of *many* molecules.

Like solutions, colloids may exist in any one of the three physical states of matter. These different types of colloids have different names: a *foam* is a gas dispersed in a liquid or solid; a *liquid aerosol* is a liquid dispersed in a gas (see Figure 14.12); a *solid aerosol* or a *smoke* is a solid dispersed in a gas; an *emulsion* is a liquid dispersed in a liquid; a *gel* is a liquid dispersed in a solid; and a *sol* is a solid dispersed in a liquid or solid. Table 14.4 lists these various types of colloids and gives examples of each.

TABLE 14.4		Types of Colloids	
DISPERSED PARTICLES	DISPERSING MEDIUM	NAME	EXAMPLE(S)
Gas	Liquid	Foam	Foaming shaving cream
Gas	Solid	Solid foam	Styrofoam
Liquid	Gas	Liquid aerosol	Fog; mist; clouds
Liquid	Liquid	Emulsion	Milk; some pharmaceutical preparations such as liniments and mineral oil emulsion in water; mayonnaise; butter; hand cream
Liquid	Solid	Gel	Jell-O; certain hair gels
Solid	Gas	Solid aerosol or smoke	Dust; smoke
Solid	Liquid	Sol	Latex paints
Solid	Solid	Solid sol	Ruby-colored glass and the gem, opal

Oil and water mixed together with a special agent can form a colloid of a liquid dispersed in a liquid—an emulsion. This is one way of cleaning up an oil slick.

Besides particle size, colloids have other identifying properties. Properties peculiar to colloids are (1) optical effect, (2) motion effect, (3) electric charge effect, and (4) adsorption effect.

Optical Effect

When a relatively narrow beam of light is passed through a colloid such as *dust* particles in the air (Table 14.4, solid in gas), the dust particles scatter the light and appear in the beam as bright, tiny specks of light (see Figure 14.13). In a solution, the appearance is different. The scattering of light in a colloid occurs because larger colloid particles reflect the light, producing a *visible* beam of light. Smaller solute particles in a solution cause no observable reflection. Hence, a beam of light passing through a solution is *invisible*. This optical effect is named the *Tyndall effect*, after British physicist John Tyndall (1820–1893) who critically investigated it in 1869.

Motion Effect

If a colloid is viewed with a special microscope, the dispersed colloidal particle appears to move in a zigzag, random motion through the dispersing medium. The special microscope consists of a regular microscope focused on the colloid with a strong light source at right angles to the colloid. It is viewed against a black background. What is actually seen are reflections from the colloidal particles. The erratic, random motion (a "jittering dance") of the dispersed colloidal particles in the dispersing medium occurs because the medium bombards the dispersed colloidal particles, creating a zigzag, random motion, as Figure 14.14 shows. Because this

(a)

(b)

FIGURE 14.13
(a) Optical effect (Tyndall effect) of colloids. (b) Sunlight appearing through clouds showing the path of light rays. (Courtesy of Yva Momatiuk/Photo Researchers).

reflection of light is due to the colloidal particle size, no such motion is seen in a solution, even though the solute and solvent particles are in continuous random motion. This motion effect of colloids is one reason that colloidal particles do not settle out on standing. The constant bombardment by the dispersing medium particles on the dispersed colloidal particles keeps the dispersed colloidal particles suspended indefinitely. This effect, named after the English botanist Robert Brown (1773–1858) who first studied it with cytoplasmic granules in pollen grains in the summer of 1827, is called *Brownian movement*.

Electric Charge Effect

A dispersed colloidal particle can *adsorb* (or bind) electrically charged particles (ions) on its *surface*. The charged species adsorbed on the surface of a given kind of

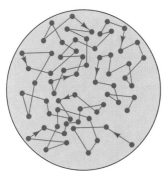

FIGURE 14.14
Motion effect (Brownian movement) of colloids.

colloidal particle may be either positive or negative. Colloidal particles of a given kind *all* have the *same* sign of excess charge. Because like charges repel each other, the dispersed colloidal particles repel each other. In addition to the random motion of the particles (Brownian movement), this electric charge effect also prevents the dispersed colloidal particles from coagulating and precipitating.

In colloidal dispersions of hydrated metal oxides, such as iron(III) hydroxide, the adsorbed ions are positive and the colloidal particles acquire a positive charge. Conversely, in the case of certain sulfides, such as arsenic(III) sulfide sol, the colloidal particles adsorb negative ions and thus acquire a negative charge. Dirt in river beds adsorbs negative ions and acquires a negative charge.

If a colloid of one charge comes in contact with a colloid of another charge or an ion of opposite charge, the dispersed colloidal particles generally precipitate from the dispersing medium. Such is the case in the mixing of the two colloids iron(III) hydroxide and arsenic(III) sulfide. When rivers water with negatively charged particles meets the ocean with positively charged particles (ionic salts), dirt is deposited, forming a fertile river delta.

The coagulation effect of an electric charge on colloids is used in removing suspended particles from the effluent gases in industrial smokestacks. The device used in this operation is called a *Cottrel precipitator*, and smaller models of it are used as electronic air cleaners in homes. Airborne colloidal particles (dust, pollen, and so on) are effectively removed from the atmosphere in the home as these airborne particles pass through an "ionizing" section of the device where they are given an intense electric charge. Next, the air carries them into the collecting section where they are attracted to metal plates under the influence of a powerful electric field, as Figure 14.15 shows. The particles cling to these metal plates, and the clean air is returned to the ventilation system.

Adsorption Effect

Colloids, because of their great surface area, have great adsorbing power. The property of adsorption is not limited to colloids, for other substances—such as charcoal, filter paper (cellulose), aluminum oxide, and bentonite—also have great adsorbing power. Charcoal has been used as an adsorbing agent in the brewing of beer in three different ways. In the first way, it is used to purify the water and to remove any odors or taste in the water. In the second way, it is used to clarify the beer and to adsorb the

FIGURE 14.15
Electronic air cleaner for homes (✶ are large airborne particles; dots are smaller airborne particles).

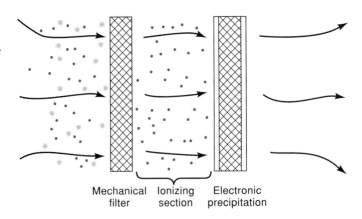

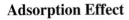

Mechanical filter   Ionizing section   Electronic precipitation

cloudy colloids formed during the fermentation process. And, in the third way, it is used to purify the carbon dioxide added to carbonate the beer. The carbon dioxide gives the beer a tang, and its presence causes the beer to form a "head" and be foamy.

 # Summary

A *solution*, by definition, is homogeneous throughout and is composed of two or more pure substances, the composition of which can usually be varied within certain limits. The substance present in lesser quantity is called the *solute*, and the substance present in greater quantity is called the *solvent*. The solute typically exists in solution as individual molecules or ions that are hydrated (solvated) by solvent molecules (Section 14.1). Solutions can be composed of a gas dissolved in a liquid, a liquid dissolved in a liquid or a solid, or a solid dissolved in a liquid or a solid (Section 14.2). The solubility of a solute in a solvent depends on the nature of the two materials, the temperature, and (for a gaseous solute) the pressure. The rate at which a solute dissolves depends on the temperature, the particle size, and the amount of stirring (Section 14.3).

The *concentration* of a solution tells us *how much* solute is dissolved in a given amount of solvent. The concentration of a solution can be qualitatively described as *saturated, unsaturated,* or *supersaturated* (Section 14.4). More quantitative expressions of concentration (Section 14.5) include percent by mass (Section 14.6), parts per million (Section 14.7), molarity (Section 14.8), normality (Section 14.9), and molality (Section 14.10). Molar solutions are usually available in concentrated form called *stock solutions*. These stock solutions are often diluted (Section 14.8). Molar and normal solutions can easily be converted from one concentration to another (Section 14.9).

The *colligative properties* of a solution are those properties that change as a result of addition of the solute. The boiling point of a solution is higher than the boiling point of the pure solvent. Similarly, the freezing point of a solution is lower than the freezing point of the pure solvent. For solutions containing simple solutes, the change in boiling or freezing point can be calculated from the molality of the solution. The molecular mass and molar mass of a pure substance can be calculated from the freezing point depression (Section 14.11).

Colloids are intermediate between solution and suspensions. Properties of colloids differ from those of solutions. Many common examples of colloids are foaming shaving cream, fog, milk, butter, hand cream, dust, smoke, and latex paint (Section 14.12).

Exercises

1. Define or explain the following terms (the number in parentheses refers to the section in the text where the term is mentioned):

   a. solution (14.1)

   b. solute (14.1)

   c. solvent (14.1)

   d. miscible liquids (14.2)

   e. immiscible liquids (14.2)

   f. Henry's law (14.3)

   g. concentration (14.4)

   h. saturated solution (14.4)

   i. unsaturated solution (14.4)

   j. supersaturated solution (14.4)

   k. crystallization (14.4)

   l. percent by mass (14.6)

   m. parts per million (14.7)

   n. molarity (14.8)

   o. normality (14.9)

   p. equivalent of an acid (14.9)

q. equivalent of a base (14.9)   **r.** molality (14.10)

s. colligative properties (14.11)   **t.** colloid (14.12)

u. dispersed particles (14.12)   **v.** dispersing medium (14.12)

2. Distinguish between:

   a. solubility and rate of solution

   b. unsaturated and saturated solutions

   c. saturated and supersaturated solutions

   d. miscible and immiscible

   e. dispersed particles and dispersing medium

Types of Solutions (See Section 14.2)

3. Classify the following solutions according to the physical states of the solute and solvent.

   a. salt in water

   b. ammonia in water

   c. instant coffee in hot water

   d. antifreeze added to the cooling system of your car

   e. glucose (a sugar) in the blood

   f. alcohol in water

   g. chlorine added to water for purification

   h. oxygen dissolved in water

   i. sulfur dioxide, a pollutant, in the fluid surrounding mucous membranes, as in your nose and throat

   j. sugar in water

Factors Affecting Solubility and Rate (See Section 14.3)

4. Classify the following compounds as to whether they are more soluble in water or in carbon tetrachloride.

   a. KCl   **b.** NaF   **c.** gasoline   **d.** $Pb(NO_3)_2$

Saturated, Unsaturated, and Supersaturated Solutions (See Section 14.4)

5. Describe in your own words the process of purifying a solid by crystallization.

6. How would you determine if a given solution were unsaturated, saturated, or supersaturated? Explain.

Colloids and Suspensions (See Section 14.12)

7. Classify the following colloids according to possible types and give the name of the colloid:

   a. gold in water   **b.** soap foam

   c. clouds   **d.** mayonnaise

8. Two solids, **A** and **B**, are dispersed in pure water. When a narrow beam of light is passed through **A** in aqueous medium, no visible light path is observed, but with **B** in aqueous medium, the path of the beam of light is visible. Which one of the two, **A** or **B**, forms a colloid and which one forms a solution?

9. Approximately 1000 years ago, a river delta was formed where the city of New Orleans is now located. This delta consists of deposited silt formerly held in colloidal dispersion in the Mississippi River. Explain why this silt from the Mississippi River was deposited at the point where the river empties into the salt-laden Gulf of Mexico.

✓ Problems

Factors Affecting Solubility and Rate (See Section 14.3)

10. From Figure 14.4, determine the solubility of the following in grams per $1\overline{00}$ g of water:

    a. KNO_3 at 41°C

    b. $CaCrO_4$ at 75°C

    c. NaCl at $2\overline{0}$°C

    d. KCl at $2\overline{0}$°C

    e. KCl at 75°C

    f. KNO_3 at 35°C

11. Calculate the solubility in g/L of oxygen gas in water at a partial pressure of 3.00 atm and 5°C. The solubility is 0.0607 g of oxygen per liter of water at 5°C and a *total* pressure of 1.000 atm. (See Appendix V.)

12. Calculate the solubility in g/L of nitrogen gas in water at a partial pressure of $76\overline{00}$ torr and 0°C. The solubility is 0.0292 g of nitrogen per liter of water at 0°C and a *total* pressure of $76\overline{0}$ torr. (See Appendix V.)

Percent by Mass (See Section 14.6)

13. Calculate the percent of solute in each of the following aqueous solutions:

    a. 7.25 g of sodium chloride in 95.0 g of solution

    b. 25.2 g of potassium carbonate in 100.0 g of water

    c. 3.88 g of calcium chloride in 90.0 g of water

14. Calculate the percent of solute in each of the following solutions:

    a. 13.0 g of sodium chloride in sufficient water to make $11\overline{0}$ g of solution

    b. 12.4 g of barium chloride in 80.7 g of water

    c. 0.155 g of phenol (C_6H_6O) in 15.000 g of glycerol

15. Calculate the grams of solute that must be dissolved in

    a. $35\overline{0}$ g of water to prepare a 15.0% potassium sulfate solution

    b. 15.0 g of water to prepare a 12.0% sodium chloride solution

    c. 275 g of water to prepare a 10.0% aqueous potassium nitrate solution

16. Calculate the grams of water that must be added to

    a. 16.0 g of sugar ($C_{12}H_{22}O_{11}$) to prepare a 20.0% sugar solution

    b. 4.00 g of potassium iodide to prepare a 1.90% potassium iodide solution

    c. 6.00 g of potassium nitrate to prepare a 7.50% aqueous potassium nitrate solution

17. Calculate the number of grams of solution necessary to provide the following:

    a. 68.3 g of sodium chloride from a 15.0% aqueous sodium chloride solution

    b. 1.20 g of sodium hydrogen carbonate from a 6.00% aqueous sodium hydrogen carbonate solution

    c. 5.00 g of potassium nitrate from a 10.0% aqueous potassium nitrate solution

Parts per Million (See Section 14.7)

18. Calculate the parts per million of the solute in each of the following aqueous solutions. (Assume that the density of the very dilute sample is 1.00 g/mL.)

    a. 128 mg of sodium (Na^+) ions in $55\overline{0}$ mL of a water sample

    b. 172 mg of potassium (K^+) ions in $85\overline{0}$ mL of a water sample

    c. 2.5 mg of aluminum ions (Al^{3+}) in 1.5 L of ocean water (*Hint:* The concentration of aluminum is independent of other ions in the water.)

19. Calculate the parts per million of the solute in each of the following aqueous solutions. (Assume that the density of the very dilute water sample is 1.00 g/mL.)

    a. 225 mg of sodium chloride (NaCl) in $30\overline{0}$ mL of a water sample

    b. 6.5 mg of potassium (K^+) in $5\overline{0}$ mL of a water sample

    c. 2.7×10^{-3} mg of gold (Au) in 450 mL of ocean water

20. Calculate the milligrams of solute dissolved in the following aqueous solutions. (Assume that the density of the very dilute water sample is 1.00 g/mL.)

    a. 5.50 L of a water sample having 15 ppm strontium (Sr^{2+}) ions

    b. 9.80 L of ocean water having 65 ppm bromide (Br^-) ions

    c. 15.0 L of ocean water having 3.0×10^{-4} ppm silver (Ag)

Molarity (See Section 14.8)

21. Calculate the molarity of each of the following aqueous solutions:

    a. 82.5 g of ethyl alcohol (C_2H_6O) in $45\overline{0}$ mL of solution

    b. 2.65 g of sodium chloride in 40.0 mL of solution; also, calculate the molarity of the chloride ion

    c. 20.8 g of sugar ($C_{12}H_{22}O_{11}$) in 275 mL of solution

22. Calculate the molarity of each of the following aqueous solutions:

    a. 27.0 g of sodium bromide in $85\overline{0}$ mL of solution; also, calculate the molarity of the bromide ion

    b. 12.0 g of calcium chloride in $64\overline{0}$ mL of solution; also, calculate the molarity of the chloride ion

    c. 15.0 g of barium bromide in 1150 mL of solution; also, calculate the molarity of the bromide ion

23. Calculate the grams of solute necessary to prepare the following aqueous solutions. Explain how each solution would be prepared.

    a. $45\overline{0}$ mL of a 0.110 M sodium hydroxide solution

    b. $25\overline{0}$ mL of a 0.220 M calcium chloride solution

    c. $10\overline{0}$ mL of a 0.155 M sodium sulfate solution

24. Calculate the milliliters of aqueous solution required to provide the following:

    a. 5.50 g of sodium bromide from a 0.100 M solution

    b. 7.65 g of calcium chloride from a 1.40 M solution

    c. 1.20 mol of sulfuric acid from a 6.00 M solution

25. Calculate the number of milliliters of a 15.4 M nitric acid stock solution needed to prepare the following diluted nitric acid solutions. Explain how the diluted solutions are prepared.

   a. 1.00 L of a 6.00 M nitric acid solution

   b. 1.00 L of a 8.00 M nitric acid solution

   c. $50\overline{0}$ mL of a 3.00 M nitric acid solution

26. Calculate the molarity of $50\overline{0}$ mL of a sulfuric acid solution prepared from the following sulfuric acid stock solutions:

   a. 10.0 mL of 17.8 M sulfuric acid

   b. 25.0 mL of 17.8 M sulfuric acid

   c. 45.0 mL of 17.8 M sulfuric acid

Normality (See Section 14.9)

27. Calculate the normality of each of the following aqueous solutions:

   a. 9.50 g of sodium hydroxide in $45\overline{0}$ mL of solution

   b. 2.10 g of barium hydroxide in $50\overline{0}$ mL of solution in reactions that replace *both* hydroxide ions

   c. 65.5 g of phosphoric acid in $25\overline{0}$ mL of solution in reactions that replace all *three* hydrogen ions

28. Calculate the normality of each of the following aqueous solutions:

   a. 18.2 g of sulfuric acid in $75\overline{0}$ mL of solution in reactions that replace *both* hydrogen ions

   b. 14.1 g of potassium hydroxide in 625 mL of solution

   c. 0.900 g of calcium hydroxide in $83\overline{0}$ mL of solution in reactions that replace *both* hydroxide ions

29. Calculate the grams of solute necessary to prepare the following aqueous solutions:

   a. $35\overline{0}$ mL of a 0.0100 N sulfuric acid solution in reactions that replace *both* hydrogen ions

   b. 145 mL of a 0.800 N phosphoric acid solution in reactions that replace all *three* hydrogen ions

   c. $25\overline{0}$ mL of a 0.0200 N calcium hydroxide solution in reactions that replace *both* hydroxide ions.

30. Calculate the milliliters of aqueous solution required to provide the following:

   a. 75.0 g of sulfuric acid (H_2SO_4) from a 4.00 N solution in reactions that replace *both* hydrogen ions

   b. 1.85 g of barium hydroxide from a 0.0400 N solution in reactions that replace *both* hydroxide ions

   c. 0.500 g of calcium hydroxide from a 0.0350 N solution in reactions that relace *both* hydroxide ions

31. Calculate the normality of the following solutions:

   a. 1.50 M sulfuric acid solution in reactions that replace *both* hydrogen ions

   b. 2.50 M phosphoric acid solution in reactions that replace all *three* hydrogen ions

   c. 0.0150 M calcium hydroxide solution in reactions in which 2 mol hydrogen ions replaces *both* hydroxide ions

32. Calculate the molarity of the following solutions:

   a. 3.00 N sulfuric acid solution in reactions that replace *both* hydrogen ions

   b. 5.00 N chromic acid solution in reactions that replace *both* hydrogen ions

   c. 0.100 N barium hydroxide solution in reactions that replace *both* hydroxide ions

Molality (See Section 14.10)

33. Calculate the molality of each of the following solutions:

   a. $17\overline{0}$ g of ethyl alcohol (C_2H_6O) in $80\overline{0}$ g of water

   b. 3.50 g of sulfuric acid in 12.0 g of water

   c. 2.60 g of glucose ($C_6H_{12}O_6$) in $11\overline{0}$ g of water

34. Calculate the molality of each of the following solutions:

   a. 15.0 g of ethylene glycol ($C_2H_6O_2$, antifreeze) in 485 g of water

   b. 28.0 g of calcium chloride in $62\overline{0}$ g of water

   c. 2.40 mol of sugar ($C_{12}H_{22}O_{11}$) in $86\overline{0}$ g of water

35. Calculate the grams of solute necessary to prepare the following aqueous solutions:

   a. $45\overline{0}$ g of a 0.400 m solution of ethyl alcohol (C_2H_6O)

   b. $70\overline{0}$ g of a 0.500 m solution of sulfuric acid (H_2SO_4)

   c. 425 g of a 3.20 m solution of ethylene glycol ($C_2H_6O_2$)

36. Calculate the grams of water that must be added to

   a. 85.0 g of glucose ($C_6H_{12}O_6$) in the preparation of a 2.00 m solution

   b. 95.0 g of sugar ($C_{12}H_{22}O_{11}$) in the preparation of an 8.00 m solution

   c. 4.10 mol of sulfuric acid in the preparation of a 12.0 m solution

Colligative Properties of Solutions
(See Section 14.12 and Table 14.3 for additional data)

37. Calculate the boiling point (at 1 atm) and freezing point in degrees Celsius of the following solutions:

   a. 1.85 m aqueous glucose ($C_6H_{12}O_6$) solution

   b. glucose solution containing 5.35 g of glucose ($C_6H_{12}O_6$) in 75.0 g of water

   c. glycerol solution containing 7.65 g of glycerol ($C_3H_8O_3$) in 125 g of water

38. Calculate the boiling point (at 1 atm) and freezing point in degrees Celsius of the following solutions:

   a. 0.800 m aqueous urea (CH_4N_2O) solution

   b. urea solution containing 4.35 g of urea (CH_4N_2O) in $11\overline{0}$ g of water

   c. urea solution containing 8.50 g of urea (CH_4N_2O) in $20\overline{0}$ g of water

39. Calculate the molecular mass and molar mass of an unknown solid from the following data:

   a. 8.50 g of an unknown was dissolved in 135 g of water, and the resulting solution had a freezing point of $-1.25°C$

   b. 7.30 g of another unknown was dissolved in $2\overline{0}0$ g of water, and the resulting solution had a freezing point of $-1.65°C$

   c. 4.32 g of an unknown was dissolved in 75.0 g of *benzene*, and the resulting solution had a freezing point of 4.00°C

General Problems

40. A nonvolatile, nonionic unknown gave, on analysis, 40.0% carbon, 6.7% hydrogen, and 53.3% oxygen. A solution of this compound (10.2 g in $1\overline{0}0$ g of water) had a freezing point of $-1.05°C$. Calculate the molecular formula of this compound. (*Hint:* See Section 8.5.)

41. Urea (CH_4N_2O) is found in the urine and is a very common way for the body to excrete nitrogen wastes. It is typically reported as grams of nitrogen rather than grams of urea. A patient's chart indicated that a 1230-mL urine sample from one 24-h period contained 13.7 g of nitrogen. Calculate the molarity of urea in the urine sample.

42. Calculate the boiling point and freezing point in degrees Celsius of the fluid in a car radiator. It consists of 2.67 kg of ethylene glycol ($C_2H_6O_2$) in 7.45 kg of water.

✓ Chapter Quiz 14A (Sections 14.1 to 14.8)

You may use the periodic table.

ELEMENT	ATOMIC MASS UNITS (amu)
K	39.1
Cl	35.5

1. Calculate the grams of water that must be added to 17.0 g of potassium chloride in the preparation of a 20.0% potassium chloride solution.

2. Calculate the milligrams of potassium ions (K^+) in 3.0 L of a water sample containing $2\overline{0}0$ ppm of potassium ions. (Assume that the density of the very dilute water sample is 1.00 g/mL.)

3. Calculate the molarity of a solution of potassium chloride containing 45.0 g of potassium chloride in $4\overline{0}0$ mL of solution.

4. Calculate the grams of potassium chloride needed to prepare $6\overline{0}0$ mL of a 0.500 *M* potassium chloride solution.

5. Calculate the number of milliliters of 17.4 *M* concentrated acetic acid that must be added to water to prepare $5\overline{0}0$ mL of a 3.00 *M* acetic acid solution.

✓ Chapter Quiz 14B
(Sections 14.9 to 14.12)

ELEMENT	ATOMIC MASS UNITS (AMU)
H	1.0
S	32.1
O	16.0
C	12.0

1. Calculate the normality of a sulfuric acid solution containing 32.5 g sulfuric acid in $85\overline{0}$ mL of solution in reactions that replace *both* hydrogen ions.

2. Calculate the normality of a 4.00 *M* phosphoric acid solution in reactions that replace all *three* hydrogen ions.

3. Calculate the grams of sugar ($C_{12}H_{22}O_{11}$) necessary to prepare $75\overline{0}$ g of a 0.500 *m* sugar solution.

4. Calculate the freezing point of a sugar solution containing 8.00 g of sugar ($C_{12}H_{22}O_{11}$) in 100.0 g of water. (K_f for water $=$ 1.86°C/*m*, freezing point of water $=$ 0.00°C.)

5. To the left of each number place a letter corresponding to the appropriate number. Only one letter per number may be used, and a given letter may be used only once.

_____ 1. Particle size in colloids   A. Sol

_____ 2. Liquid in gas   B. Foam

_____ 3. Liquid in liquid   C. Liquid aerosol

_____ 4. Optical effect   D. Emulsion

    E. Less than 1000 pm in diameter

    F. 1000 to 200,000 pm in diameter

    G. Greater than 200,000 pm in diameter

    H. Brownian movement

    I. Tyndall effect

Soap
(Formula: $R\text{-}CO_2^-Na^+$)

Soap was invented for little children!

The Compound SOAP:
Chemistry Cleans Up!

Name: The word "soap" probably derives from *saipon*, from the Germanic family of languages in northern Europe. The German *seifen*, the French *savon*, the Italian *sapone*, and the Spanish *jabon* all came from *saipon*.

Appearance: Soap is a soft, waxy solid that dissolves to some extent in water.

Occurrence: Credit is generally given to the Phoenicians for inventing soap about 2600 years ago. However, it is believed that soap was first prepared 5000 years ago from wood ashes and animal fat. Until the early 1700s soap was prepared in small batches in homes or as a cottage industry for home and local use. At that time it began to be prepared on a larger scale as an industry. All this time, soap was made one batch at a time. It wasn't until 1938 that a large-scale continuous process was perfected by the Procter and Gamble Company and soap could be made efficiently on an industrial scale. A number of other continuous processes have been developed since 1938.

Source: Soap was historically made by heating animal fats (triglycerides) with soda ash (Na_2CO_3). A more efficient process involves treating animal fats with a sodium or potassium hydroxide solution to give soap and glycerol (glycerine). A typical example is

triglycerides (fats) + 3 NaOH $\longrightarrow$ glycerol + 3 soap

$$R = -\overset{\displaystyle H}{\underset{\displaystyle H}{C}} - \overset{\displaystyle H}{\underset{\displaystyle H}{C}} - \cdots - \overset{\displaystyle H}{\underset{\displaystyle H}{C}} - \overset{\displaystyle H}{\underset{\displaystyle H}{C}} - H$$

variable number of CH_2 units

Most soaps involve R groups that contain a total of 11, 13, 15, or 17 carbon atoms. Soft soaps involve potassium hydroxide (KOH) instead of sodium hydroxide (NaOH), and thus the soap contains potassium ions (K^+) instead of sodium ions (Na^+).

Its Role in Our World:

Soap can be envisioned as a long, skinny molecule with a $-CH_2-CH_2-\cdots-CH_2-CH_3$ portion or "tail" (shown as 〜〜〜〜) and a negatively charged "head" (shown as ⊖). The charged head likes to be in water since water molecules can hydrate (solvate, Section 14.1) the negative charge. The long, skinny tail does not like to be polar (Section 14.3, Properties of Solute and Solvent) and does not hydrate well. In fact, the tails prefer to be in contact with grease or oil if possible.

charged head—

likes water ⊖〜〜〜〜〜〜 uncharged tail— does not like water

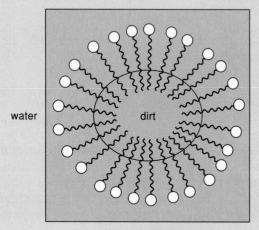

micelle with dirt inside suspended in water

The structure of soap molecules enables soap to *suspend dirt* and *grease* in a *water solution*. It does this by forming little aggregates (called *micelles*) of soap around each bit of greasy dirt; the tails of the soap molecules surround the greasy dirt, and the charged heads maintain contact with the water solution that surrounds the micelle. In this way the greasy dirt is "dissolved" in the water solution.

Detergents also form micelles and suspend greasy dirt, the only difference being that the charged head has a different structure in detergents. Detergents are more important in laundry and dishwashing formulations, while soap is most useful for cleaning faces and bodies and little children!

Unusual Facts:

Animal fat has gotten a bad reputation in terms of healthy eating habits. However, animal fats (beef and lamb fat, lard) are the most common fats used for making soap. Other fats and oils used in soap production include coconut oil, palm oil, and palm kernel oil.

The afternoon "soaps" on TV got their name from their original radio sponsors in the 1940s—soap companies such as Procter and Gamble.

CHAPTER 15

Acids, Bases, and Ionic Equations

The reaction between citric acid ($C_6H_8O_7$) and sodium hydrogen carbonate (baking soda) to produce carbon dioxide gas is the same reaction that produces the fizz in Alka-Seltzer.

GOALS FOR CHAPTER 15

1. To identify an acid and a base using both the Arrhenius and the Brønsted–Lowry definitions, an amphoteric substance, a conjugate acid, and a conjugate base (Section 15.1).

2. To identify strong and weak acids and strong and weak bases and describe the differences between strong and weak acids and strong and weak bases (Section 15.2).

3. To describe the ionization of water, the dissociation of ionic substances, and the ionization of covalent substances (Section 15.3).

4. To describe the three general types of reactions of acids and/or bases and to calculate the concentration of an unknown acid or base in a titration experiment (Section 15.4).

5. To define pH and pOH and to use pH and pOH in calculations involving solutions of acids or bases (Section 15.5).

6. To explain the differences among the terms electrolytes, nonelectrolytes, strong electrolytes, and weak electrolytes (Section 15.6).

7. To complete and write balanced chemical ionic equations and net ionic equations (Section 15.7 and 15.8).

Countdown

ELEMENT	ATOMIC MASS UNITS (amu)
H	1.0
P	31.0
O	16.0

You may use the periodic table, the rules for the solubility of inorganic substances in water, and the electromotive or activity series.

5. Write the correct formula for the following compounds: (Sections 7.3 through 7.6).
  a. magnesium iodide (MgI_2)
  b. magnesium iodate [$Mg(IO_3)_2$]
  c. magnesium phosphate [$Mg_3(PO_4)_2$]
  d. tin(II) phosphate [$Sn_3(PO_4)_2$]
  e. phosphoric acid [H_3PO_4]

4. Classify each of the following compounds as (1) an acid, (2) a base, or (3) a salt (Section 7.6).
  a. $Sn_3(PO_4)_2$ [(3), salt] **b.** H_3PO_4 [(1), acid]
  c. $Mg(OH)_2$ [(2), base] **d.** $Ga(OH)_3$ [(2), base]

3. Complete and balance the following chemical reaction equations. Indicate any precipitate by (s) and any gas by (g) (Sections 9.3, 9.4, and 9.8 through 9.10).
  a. $HNO_3(aq) + Mg(OH)_2(s) \longrightarrow$
  $[2\,HNO_3(aq) + Mg(OH)_2(s) \longrightarrow Mg(NO_3)_2(aq)$
  $+ 2\,H_2O(l)]$
  b. $Mg(s) + HNO_3(aq) \longrightarrow$
  $[Mg(s) + 2\,HNO_3(aq) \longrightarrow Mg(NO_3)_2(aq) + H_2(g)]$
  c. $MgCO_3(s) + HCl(aq) \longrightarrow$
  $[MgCO_3(s) + 2\,HCl(aq) \longrightarrow MgCl_2(aq)$
  $+ H_2O(l) + CO_2(g)]$
  d. $Mg(OH)_2(s) + HCl(aq) \longrightarrow$
  $[Mg(OH)_2(s) + 2\,HCl(aq) \longrightarrow MgCl_2(aq)$
  $+ 2\,H_2O(l)]$

2. Calculate the molarity of an aqueous phosphoric acid solution containing 175 g of phosphoric acid in 1.25 L of solution (Section 14.8).   (1.43 *M*)

1. Calculate the normality of an aqueous phosphoric acid solution containing 195 g of phosphoric acid in 1.50 L of solution in reactions that replace all *three* hydrogen ions (Section 14.9).   (3.98 *N*)

Acids and bases are very common substances in your life. Much of the food you eat is acidic, but your blood is slightly basic. If the acid–base balance in the blood changes slightly in either direction, death may result. Table 15.1 lists some common substances and their acid or base ingredient.

In Chapter 7, we introduced the terms "acid" and "base" in discussing nomenclature. In Chapter 9, we considered the reaction of an acid with a base as an example of a neutralization reaction. In this chapter, we will define an acid and a base more explicitly and examine their properties more fully. After we have considered acids and bases and their properties, we will discuss how to write *ionic equations*.

15.1 Properties and Definitions of Acids and Bases

As early as the seventeenth century, chemists began to recognize that the substances we call acids and bases have certain distinctive properties. For example, *acids*

✔ *Taste sour when dissolved in water.* Citrus fruits (lemons, limes, and the like) have a sour taste because of the citric acid in them. Dill pickles are sour because vinegar (acetic acid) is present. Sour milk contains lactic acid.

✔ *Turn blue litmus paper red.* Litmus is an indicator. **Indicators** are compounds that change color in the presence of certain chemicals, such as acids and bases. Litmus paper is the best known indicator.

✔ *Neutralize bases.* Acids react with bases as described in Section 9.11.

Indicators   Compounds that change color in the presence of certain chemicals, such as acids and bases. Litmus paper is the best known indicator.

TABLE 15.1	Some Common Substances and Their Acid or Base Ingredient
SUBSTANCE	INGREDIENT
Acids	
Battery acid	Sulfuric acid
Carbonate water	Carbonic acid
Eye wash	Boric acid
Food preservative	Benzoic acid
Lemon, lime, tomato	Citric acid
Rust remover	Phosphoric acid
Sour milk	Lactic acid
Stomach acid	Hydrochloric acid
Vinegar	Acetic acid
Vitamin C	Ascorbic acid
Bases	
Drain cleaner	Sodium hydroxide
Milk of magnesia	Magnesium hydroxide
Mortar and plaster	Calcium hydroxide
Window cleaner	Ammonia solution

You and Chemistry

In contrast, *bases*

✔ *Taste bitter when dissolved in water.* Milk of magnesia [$Mg(OH)_2$] has a relatively bitter taste, which is sometimes masked with a mint flavor.

✔ *Feel soapy or slick when dissolved in water.*

✔ *Turn red litmus paper blue.*

✔ *Neutralize acids.*

Although these properties are useful in helping to characterize substances as acidic or basic, chemists generally prefer to define these substances according to their chemical properties.

Arrhenius Definition

One of the first to attempt such a definition was Svante August Arrhenius (1859–1927), a Swedish chemist and physicist. In 1884, Arrhenius proposed definitions of acids and bases that were linked to the *ions* they release when they are dissolved in water. Arrhenius defined an **acid** as a substance that releases *hydrogen ions* (H^+) when dissolved in water. According to Arrhenius's definition, common examples of acids include nitric acid (HNO_3) and acetic acid ($HC_2H_3O_2$). In contrast, Arrhenius defined a **base** as a substance that releases *hydroxide ions* (OH^-) when dissolved in water. Common examples of such bases include sodium hydroxide ($NaOH$) and calcium hydroxide [$Ca(OH)_2$].

According to Arrhenius's definition, bases are ionic compounds composed of a cation and one or more hydroxide ions. When the base dissolves in water, the ions are released to produce hydroxide ions in solution. In contrast, acids are covalent compounds that do not contain hydrogen ions as such. Acids release hydrogen ions into the solution by breaking a covalent bond between a hydrogen and some other atom. The hydrogen ion then associates with water molecules to form what is called an *aqueous hydrogen ion* or a *hydronium ion*. Such an ion is typically designated as $H^+(aq)$ or H_3O^+. These terms are used to describe the hydrogen ions produced by an acid in aqueous solution. The opposite natures of acids and bases explain why these substances neutralize one another.

Acid (Arrhenius definition) Any substance that releases hydrogen ions (H^+) when dissolved in water.

Base (Arrhenius definition) Any substance that releases hydroxide ions (OH^-) when dissolved in water.

Study Exercise 15.1
Classify each of the following substances as an acid or a base according to the Arrhenius theory.

a. $H_2C_2O_4(aq)$ (acid) **b.** $Cu(OH)_2(aq)$ (base)

Work Problem 6.

Brønsted–Lowry Definition

The Arrhenius definitions of acids and bases are limited to solutions prepared with water as the solvent. In 1923, Johannes N. Brønsted (1879–1947), a Danish chemist, and Thomas M. Lowry (1874–1936), an English chemist, independently proposed more general definitions of acids and bases. The Brønsted–Lowry concept defines an **acid** as any substance that can donate a hydrogen ion or *proton* to another substance, and a **base** as any substance that can accept a hydrogen ion or *proton* from another substance. In simple terms, an *acid* is a *proton donor* and a *base* is a *proton acceptor*. Any substance that is an Arrhenius acid or base is also a Brønsted–Lowry acid or base. However, the Brønsted–Lowry definitions hold

Acid (Brønsted–Lowry definition) Any substance that can donate a proton (H^+) to some other substance.

Base (Brønsted–Lowry definition) Any substance that can accept a proton (H^+) from another substance.

regardless of the solvent used to prepare the solution of acid or base. According to the Brønsted–Lowry concept, *ions* as well as *uncharged molecules* may also be acids or bases.

A few examples may help to illustrate this concept. In the following equations, the HCl and HNO_3 molecules behave as Brønsted–Lowry acids by donating protons to a water molecule, which acts as a base when it accepts a proton.

$$HCl(g) + H_2O(l) \longrightarrow H_3O^+(aq) + Cl^-(aq) \qquad (15.1)$$

$$HNO_3(l) + H_2O(l) \longrightarrow H_3O^+(aq) + NO_3^-(aq) \qquad (15.2)$$

Water does not always act as a base:

$$NH_3(g) + H_2O(l) \rightleftharpoons NH_4^+(aq) + OH^-(aq) \qquad (15.3)$$

Here the water is behaving as a Brønsted–Lowry acid as it donates a proton to a molecule of ammonia (NH_3) in a reaction that proceeds from left to right. If we consider the *reverse reaction* (a reaction that proceeds from right to left), then the ammonium ion acts as an acid and the hydroxide ion acts as a base.

A double arrow ($\rightleftharpoons$) indicates that not all the reactants proceed to give products. In Equation (15.3), the top arrow is shorter than the bottom arrow because there are more reactant molecules than product molecules when the reaction is at equilibrium. If the top arrow were longer than the bottom arrow, it would mean that there are more product molecules than reactant molecules when the reaction is at equilibrium.

Some substances such as water are capable of behaving as a Brønsted–Lowry acid *or* base. These substances are called *amphoteric substances* (*amphi-* means "of both kinds"). An **amphoteric substance** is any substance that can act as an acid or a base, depending on the nature of the solution. Water behaves as a base (proton acceptor) toward hydrogen chloride [Equation (15.1)] and as an acid toward ammonia [Equation (15.3)]. Certain ions, hydrogen sulfate (HSO_4^-) and hydrogen carbonate (HCO_3^-), are also amphoteric substances since they can both donate and accept a proton.

In any acid–base or proton-transfer reaction, both an acid and a base are on the reactants and products side of the equation. For example,

$$\underset{\text{acid 1}}{HC_2H_3O_2(aq)} + \underset{\text{base 1}}{H_2O(l)} \rightleftharpoons \underset{\text{acid 2}}{H_3O^+(aq)} + \underset{\text{base 2}}{C_2H_3O_2^-(aq)} \qquad (15.4)$$

Amphoteric substance   Any substance that can act as an acid or base, depending on the nature of the solution; water is the best known amphoteric substance.

We give special names to the acid (acid 2) and base (base 2) on the products side. Acid 2 is called the *conjugate acid*. A **conjugate acid** is the species formed when a proton (H^+) is added to a base. The base and conjugate acid for this reaction are H_2O and H_3O^+, respectively. The pair is called a base–conjugate acid pair. Base 2 is called the *conjugate base*. A **conjugate base** is the species formed when a proton (H^+) is removed from an acid. The acid and conjugate base for this reaction are $HC_2H_3O_2$ and $C_2H_3O_2^-$, respectively. The pair is called an acid–conjugate base pair.

Conjugate acid   Species formed when a proton (H^+) is added to a base.

Conjugate base   Species formed when a proton (H^+) is removed from an acid.

Study Exercise 15.2

Classify each of the following as an acid, a base, or an amphoteric substance according to the Brønsted–Lowry theory.

a. $H_2PO_4^-$ (amphoteric)   **b.** PO_4^{3-} (base)

Study Exercise 15.3

From the following reaction equation, identify the conjugate acid and conjugate base.

$$HCl(aq) + H_2O(l) \longrightarrow H_3O^+(aq) + Cl^-(aq)$$

(conjugate acid, H_3O^+; conjugate base, Cl^-)   Work Problems 7 and 8.

15.2   The Strengths of Acids and Bases

Different acids can have very different abilities to donate a proton to another molecule or ion. Just as some people are stronger than others, some acids are stronger than other acids. For example, consider the differences between hydrochloric acid (HCl) and acetic acid ($HC_2H_3O_2$). Experiments reveal that virtually all the HCl molecules in a solution of hydrochloric acid have donated a proton to a water molecule. In contrast, very few of the acetic acid molecules have accomplished this transfer in an acetic acid solution. Arrows of unequal size or just in one direction are used to indicate these facts in Equations 15.5 and 15.6:

$$HCl(g) + H_2O(l) \longrightarrow H_3O^+(aq) + Cl^-(aq) \qquad (15.5)$$

$$HC_2H_3O_2(aq) + H_2O(l) \rightleftharpoons H_3O^+(aq) + C_2H_3O_2{}^-(aq) \qquad (15.6)$$

Hydrochloric acid is a *strong acid* because it is very effective in accomplishing the transfer of a proton to water, while acetic acid is a *weak acid* because it is much less able to do the job.

Likewise, there are also strong and weak bases. *Strong bases* dissolve in water to give solutions with high concentrations of OH^-. *Weak bases* dissolve in water and generate only modest concentrations of hydroxide ion in solution.

$$NaOH(s) \xrightarrow{H_2O} Na^+(aq) + OH^-(aq) \qquad (15.7)$$

$$NH_3(aq) + H_2O(l) \rightleftharpoons NH_4{}^+(aq) + OH^-(aq) \qquad (15.8)$$

The most common strong bases are soluble hydroxide compounds like sodium hydroxide and potassium hydroxide. The most important class of weak bases includes nitrogen-containing compounds similar to ammonia. The transfer of a proton from water to the ammonia molecule generates the hydroxide ion as shown above. The equilibrium arrows indicate that only small amounts of the product ions are formed. Thus, ammonia is a weak base.

How can you tell if a common compound is a strong or a weak acid or a strong or a weak base? There are only a few common strong acids (six) and strong bases (eight). *Almost all of the rest of the acids or bases are relatively weak.* Table 15.2 lists the strong acids and bases.

Study Exercise 15.4

Classify each of the following substances as a strong acid, a weak acid, a strong base, or a weak base.

a. oxalic acid solution ($H_2C_2O_4$)   (weak acid)
b. barium hydroxide solution [$Ba(OH)_2$]   (strong base)   Work Problems 9 and 10.

TABLE 15.2	Summary of Strong Acids and Bases

STRONG ACIDS	STRONG BASES
H_2SO_4 (sulfuric acid)	Group IA (1) hydroxides
HCl (hydrochloric acid)	LiOH (lithium hydroxide)
HBr (hydrobromic acid)	NaOH (sodium hydroxide)
HI (hydriodic acid)	KOH (potassium hydroxide)
HNO_3 (nitric acid)	RbOH (rubidium hydroxide)
$HClO_4$ (perchloric acid)	CsOH (cesium hydroxide)
	Group IIA (2) hydroxides
	(*only* following)
	$Ca(OH)_2$ (calcium hydroxide)
	$Sr(OH)_2$ (strontium hydroxide)
	$Ba(OH)_2$ (barium hydroxide)

15.3 Ion Formation in Aqueous Solutions

We have defined strong and weak acids and bases in terms of their behavior in water. Water is an excellent solvent for the formation of ions. Many processes that produce ions in solution are acid–base processes. Indeed, the acid–base chemistry that occurs in our bodies is in an aqueous system. Proton-transfer reactions in aqueous solutions are also the best understood of the acid–base processes. Thus, the formation of ions in aqueous solutions merits a closer look.

Ionization of Water

As we previously mentioned, water can act as a Brønsted–Lowry acid or base and thus is an *amphoteric substance*. We might expect that water can react with itself to form H_3O^+ and OH^-, as shown in the following equation:

| water acting as a Brønsted– Lowry base | water acting as a Brønsted– Lowry acid | aqueous hydrogen ion or hydronium ion | hydroxide ion |

$$(15.9)$$

In this *self-ionization* reaction, water reacts with itself to create aqueous hydrogen or hydronium ions and hydroxide ions. How significant is this ionization reaction? As suggested by the unequal arrows in the equation, the effect is small, but it is measurable. Experiments that measure the concentrations of ions in pure water show that at 25°C there are 1×10^{-7} mol/L of hydrogen ions and 1×10^{-7} mol/L of hydroxide ions. Further examination reveals that in any aqueous solution the product of the concentration of hydrogen ions in mol/L and the concentration of hydroxide ions in mol/L is equal to a *constant*, which we will call K_w. The constant K_w is called the *ion product constant* for water, and it has a value of 1×10^{-14}

(mol^2/L^2) at 25°C. The constant has no units, but we have placed units next to the value for the constant for your convenience in solving problems. These "convenience units" will be in *parentheses*. For any aqueous solution at 25°C,

$$[H^+][OH^-] = K_w = 1 \times 10^{-14} \, (mol^2/L^2) \tag{15.10}$$

In this equation, the brackets represent the concentration in mol/L of the substance whose formula is enclosed in the brackets. A change in the hydrogen ion concentration results in a corresponding change in the hydroxide ion concentration such that the product of these concentrations in any aqueous solution at 25°C is always equal to the constant K_w, $1 \times 10^{-14} \, (mol^2/L^2)$.

If a solution has a *hydrogen ion* concentration *larger* (that is, a smaller negative exponent) than 10^{-7} mol/L (that is, there are more hydrogen than hydroxide ions), the solution is termed *acidic*. If the *hydroxide ion* concentration is *larger* than 10^{-7} mol/L or the *hydrogen ion* concentration is *less* (has a larger negative exponent), the solution is called *basic*. If the hydrogen ion concentration in mol/L $[H^+]$ and the hydroxide ion concentration in mol/L $[OH^-]$ are equal to 1×10^{-7} mol/L, the solution is neutral. That is,

$$[H^+] > 1 \times 10^{-7} \text{ mol/L} \qquad \textit{acidic} \text{ solution}$$

$$\left.\begin{array}{c} [OH^-] > 1 \times 10^{-7} \text{ mol/L} \\ \text{or} \\ [H^+] < 1 \times 10^{-7} \text{ mol/L} \end{array}\right\} \qquad \textit{basic} \text{ solution}$$

$$[H^+] = [OH^-] = 1 \times 10^{-7} \text{ mol/L} \qquad \textit{neutral} \text{ solution}$$

> **Study Hint:** Notice that whenever the $[H^+]$ is larger than $1.0 \times 10^{-7} \, M$ the $[OH^-]$ must be less than $1.0 \times 10^{-7} \, M$, and vice versa.

Carbonated soft drinks have a hydrogen ion concentration of approximately 1×10^{-4} mol/L; therefore, they are acidic. Milk of magnesia, an antacid and laxative, has a hydrogen ion concentration of approximately 1×10^{-10} mol/L; therefore, it is basic.

Dissociation and Ionization

In addition to self-ionization, water also plays a role in the ionization of many other compounds. As we have seen before, some substances dissolve in water to form ions. This process of ion formation from compounds can occur in two fundamentally different ways. One involves the formation of ions from *ionic* compounds, while the other involves the formation of ions from *covalent* compounds.

An ionic compound is a solid that is composed of positive and negative ions called cations and anions, respectively. The ions *already exist* as a result of the formation of an ionic bond as described in Section 6.4. When such a substance dissolves in water, the cations and anions are simply pulled apart from their close association in the crystalline state to give separate ions in solution that are hydrated by the highly polar water molecules. This process, **dissociation**, refers to the separation of ions in *ionic compounds* by the action of the solvent. For example,

$$(Na^+OH^-)(s) \xrightarrow{\text{in } H_2O(l)} Na^+(aq) + OH^-(aq) \tag{15.11}$$

$$(Na^+Cl^-)(s) \xrightarrow{\text{in } H_2O(l)} Na^+(aq) + Cl^-(aq) \tag{15.12}$$

Dissociation Process in which the ions in ionic substances separate when acted on by a solvent.

FIGURE 15.1
Ionization in weak and strong acids. (a) Hydrochloric acid, a strongly ionizing acid; (b) hydrofluoric acid, a weakly ionizing acid.

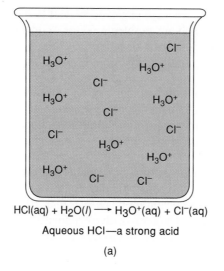

$HCl(aq) + H_2O(l) \longrightarrow H_3O^+(aq) + Cl^-(aq)$

Aqueous HCl—a strong acid

(a)

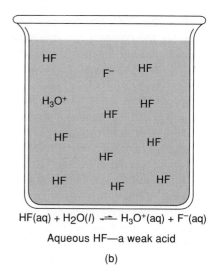

$HF(aq) + H_2O(l) \rightleftharpoons H_3O^+(aq) + F^-(aq)$

Aqueous HF—a weak acid

(b)

Ionization   Process in which covalent compounds dissolve to produce ions in aqueous solution.

Study Hint: Note the difference between ionization and dissociation. *Ionization* is the formation of ions from a *covalent* compound (HCl); *dissociation* is the separation of ions from ionic compounds (crystalline solids) *composed* of ions (Na^+, Cl^-).

The ions in sodium hydroxide and sodium chloride already exist, and the dissociation process disperses these ions throughout the solution.

In contrast, **ionization** is the process by which *covalent compounds* dissolve to produce ions in aqueous solution. During ionization, a covalent bond is broken and a cation and an anion are created. These ions are then hydrated by water molecules. An example of ionization is the reaction that occurs when gaseous hydrogen chloride (HCl) is dissolved in water to give hydrochloric acid:

$$H-Cl(g) + H_2O(l) \longrightarrow H_3O^+(aq) + Cl^-(aq) \qquad (15.13)$$

Note that there are no ions in gaseous HCl. Ions form as the H—Cl covalent bond breaks and a proton (H^+) is transferred to a molecule of water.

Not all covalently bonded substances undergo ionization. Strong acids ionize extensively when dissolved in water, but weak acids ionize only slightly (as indicated by the arrows of unequal size), as when acetic acid and hydrofluoric acid (both weak acids) dissolve in water. Relatively few acetic acid and hydrofluoric acid molecules actually react to form ions:

$$H-C_2H_3O_2(aq) + H_2O(l) \rightleftharpoons H_3O^+(aq) + C_2H_3O_2^-(aq) \qquad (15.14)$$

$$H-F(aq) + H_2O(l) \rightleftharpoons H_3O^+(aq) + F^-(aq) \qquad (15.15)$$

Figure 15.1 summarizes the difference between a solution of a strongly ionizing acid (HCl, a strong acid) and a weakly ionizing acid (HF, a weak acid).

15.4   Reactions of Acids and Bases

Acids and bases react with other substances in three general ways. We are familiar with all these reactions already: (1) reactions of acids with metals, (2) reactions of bases with a salt to form insoluble compounds, and (3) reactions of bases with acids (neutralization reactions).

Reactions of Acids with Metals

The reaction between a metal and an acid is a single-replacement reaction (see Section 9.8). Acids react with *all* metals that appear *above* hydrogen in the electromotive or activity series to liberate (release) hydrogen gas. For example,

$$Mg(s) + 2\,HCl(aq) \longrightarrow MgCl_2(aq) + H_2(g) \qquad (15.16)$$

$$2\,Al(s) + 6\,HCl(aq) \longrightarrow 2\,AlCl_3(aq) + 3\,H_2(g) \qquad (15.17)$$

The first reaction is illustrated in Figure 15.2. The second reaction suggests why aluminum foil begins to get holes in it when it is used to cover pans of lasagna during baking. The tomatoes contain a number of acidic substances, most notably citric acid ($H_3C_6H_5O_7$). The citric acid slowly reacts with the aluminum foil at elevated temperatures to dissolve the aluminum and create holes in the foil!

In contrast, acids do *not* react with the metals that lie below hydrogen in the activity series (Cu, Hg, Ag, and Au). As a result, copper pipe is preferred to iron pipe for plumbing use.

FIGURE 15.2
Hydrogen gas is liberated when magnesium metal comes into contact with hydrochloric acid.

Study Exercise 15.5
Complete and balance the following equations for the reaction between an acid and a metal:

a. acetic acid + zinc metal $\longrightarrow$
$$[2\,HC_2H_3O_2(aq) + Zn(s) \longrightarrow Zn(C_2H_3O_2)_2(aq) + H_2(g)]$$
b. acetic acid + aluminum metal $\longrightarrow$
$$[6\,HC_2H_3O_2(aq) + 2\,Al(s) \longrightarrow 2\,Al(C_2H_3O_2)_3(aq) + 3\,H_2(g)]$$

Work Problem 11.

Reactions of Bases to Form Insoluble Compounds

The reaction between a base and a metal cation to form an insoluble compound is a double-replacement reaction (see Section 9.9). Applying the rules for the solubility of inorganic substances in water (see Section 9.9 or the inside back cover of this text) shows when such a reaction will take place.

$$Al(NO_3)_3(aq) + 3\,NaOH(aq) \longrightarrow Al(OH)_3(s) + 3\,NaNO_3(aq) \qquad (15.18)$$

$$Fe(NO_3)_3(aq) + 3\,KOH(aq) \longrightarrow Fe(OH)_3(s) + 3\,KNO_3(aq) \qquad (15.19)$$

The reddish rust that emerges from seldom used iron water pipes is iron(III) hydroxide that results from the reaction between iron(III) ions and hydroxide ions (see Figure 15.3).

FIGURE 15.3
Iron(III) hydroxide precipitates out of solution when base is added to solutions of iron(III) salts.

Study Exercise 15.6
Complete and balance the following equations for the reaction between a base and a salt to give an insoluble compound:

a. sodium hydroxide + copper(II) acetate $\longrightarrow$
$$[2\,NaOH(aq) + Cu(C_2H_3O_2)_2(aq) \longrightarrow Cu(OH)_2(s) + 2\,NaC_2H_3O_2(aq)]$$
b. potassium hydroxide + chromium(III) acetate $\longrightarrow$
$$[3\,KOH(aq) + Cr(C_2H_3O_2)_3(aq) \longrightarrow Cr(OH)_3(s) + 3\,KC_2H_3O_2(aq)]$$

Work Problem 12.

Neutralization Reactions and Titrations

By far the most common reaction of acids and bases is the neutralization reaction (see Section 9.10). Neutralization of an acid with a base results in the transfer of a proton from the acid to the base. Such proton transfer or acid–base reactions are among the most common chemical processes. In its simplest form, a strong acid transfers a proton to a hydroxide ion to give a salt and water, as follows:

$$\underset{\substack{\text{strong} \\ \text{acid}}}{HCl(aq)} + \underset{\substack{\text{strong} \\ \text{base}}}{NaOH(aq)} \longrightarrow \underset{\text{salt}}{NaCl(aq)} + \underset{\text{water}}{H_2O(l)} \qquad (15.20)$$

In addition, weak acids (such as acetic acid, $HC_2H_3O_2$) react with strong bases, while weak bases (such as ammonia, NH_3) react with strong acids:

$$\underset{\substack{\text{weak} \\ \text{acid}}}{HC_2H_3O_2(aq)} + \underset{\substack{\text{strong} \\ \text{base}}}{NaOH(aq)} \longrightarrow \underset{\text{salt}}{NaC_2H_3O_2(aq)} + \underset{\text{water}}{H_2O(l)} \qquad (15.21)$$

$$\underset{\substack{\text{strong} \\ \text{acid}}}{HCl(g \text{ or } aq)} + \underset{\substack{\text{weak} \\ \text{base}}}{NH_3(g \text{ or } aq)} \longrightarrow \underset{\text{salt}}{NH_4Cl(s) \text{ or } NH_4{}^+(aq)} + Cl^-(aq) \qquad (15.22)$$

Water is not always a *product* in a neutralization reaction. In the reaction between ammonia and hydrochloric acid above, the *ammonia* is the *proton acceptor* and the product is the salt, ammonium chloride (NH_4Cl) (see Figure 15.4).

Study Exercise 15.7
Complete and balance the following equations for the reaction between an acid and a base:

a. potassium hydroxide + hydrobromic acid $\longrightarrow$

$$[KOH(aq) + HBr(aq) \longrightarrow KBr(aq) + H_2O(l)]$$

b. potassium hydroxide + phosphoric acid $\longrightarrow$
(*Hint:* All three hydrogens on phosphoric acid react.)

$$[3\,KOH(aq) + H_3PO_4(aq) \longrightarrow K_3PO_4(aq) + 3\,H_2O(l)]$$

Work Problem 13.

Titration Procedure for determining the concentration of an acid or base by adding a base or an acid of *known* concentration.

We can utilize the stoichiometric relationship between an acid and a base in a neutralization reaction in a process called *titration*. **Titration** is a procedure for determining the concentration of an acid or a base by adding a base or an acid of known concentration. For example, if we add just enough of a base of known concentration to *exactly* neutralize a given quantity of the original acid, we can use stoichiometry to determine the concentration of the original acid. The point at which

FIGURE 15.4
Acid–base reactions can occur in the gas phase. Gaseous HCl reacts with gaseous NH_3 to form a white cloud of solid NH_4Cl.

| | APPROXIMATE | | |
| TABLE 15.3 | **Color Changes of Indicators** | | |

INDICATOR	APPROXIMATE pH AT WHICH COLOR CHANGES[a]	COLOR IN ACID	COLOR IN BASE
Phenolphthalein	9	Colorless	Red
Methyl orange	4	Red	Yellow
Methyl red	5	Red	Yellow
Bromthymol blue	7	Yellow	Blue

[a] The choice of an indicator depends on the pH (see Section 15.5) of the aqueous solution of the salt formed when the acid or base is neutralized at the equivalence point.

we have exactly neutralized the original substance is called the **equivalence point**. The equivalence point is usually determined by a change in color of an appropriately selected indicator (see Section 15.1, usually phenolphthalein, methyl orange, methyl red, or bromthymol blue; see Table 15.3). The point at which the indicator in an acid–base titration changes color is called the **end point**. The end point is often the same as the equivalence point but not always. At the equivalence point, *the number of moles of added base or acid is exactly enough to neutralize the original acid or base*. See Figure 15.5 for an actual titration.

Equivalence point   Point at which an acid or base is exactly neutralized in the titration process.

End point   Point at which indicator in an acid–base titration changes color. The end point is often the same as the equivalence point, but not always.

(a)   (b)   (c)

FIGURE 15.5
Titration of a sodium hydroxide solution of unknown concentration. The procedure is as follows: (a) measure an exact amount of sodium hydroxide solution of unknown concentration into an Erlenmeyer flask; (b) add to this solution one or two drops of phenolphthalein (an indicator) to give a pink solution; (c) from a buret add hydrochloric acid solution of known concentration until the pink color fades and a nearly colorless solution appears at the end point. (Dr. E. R. Degginger).

Consider the following examples.

EXAMPLE 15.1 In the titration of 30.00 mL of sodium hydroxide solution of unknown concentration, 45.20 mL of 0.100 M hydrochloric acid was required to neutralize the sodium hydroxide solution to a phenolphthalein end point. Calculate the molarity of the sodium hydroxide solution.

SOLUTION As with stoichiometry problems, the first thing we must know is the balanced equation:

$$HCl(aq) + NaOH(aq) \longrightarrow NaCl(aq) + H_2O(l)$$

Next, apply the stoichiometry procedure (Chapter 10) and the units of molarity (Section 14.8) to calculate the moles of sodium hydroxide neutralized with 45.20 mL of 0.100 M hydrochloric acid solution:

$$45.20 \text{ mL solution} \times \frac{1 \text{ L}}{1000 \text{ mL}} \times \frac{0.100 \text{ mol HCl}}{1 \text{ L solution}} \times \underset{\substack{| \\ \text{determined from the} \\ \text{balanced equation}}}{\frac{1 \text{ mol NaOH}}{1 \text{ mol HCl}}} = 0.00452 \text{ mol NaOH}$$

Finally, calculate the concentration of the sodium hydroxide solution in mol/L from the 30.00 mL of sodium hydroxide solution that was used:

$$\frac{0.00452 \text{ mol NaOH}}{30.00 \text{ mL solution}} \times \frac{1000 \text{ mL}}{1 \text{ L}} = \frac{0.151 \text{ mol NaOH}}{1 \text{ L solution}} = 0.151 \ M \qquad \textit{Answer}$$

EXAMPLE 15.2 Household ammonia is a dilute solution of ammonia in water. In the titration of 2.00 mL of household ammonia, 34.90 mL of 0.110 M hydrochloric acid solution was required to neutralize this solution to a methyl red end point. Calculate (a) the molarity and (b) the percent (density of solution = 0.985 g/mL) of the dilute ammonia solution.

SOLUTION
a. Calculate the molarity of the NH_3. The equation is

$$NH_3(aq) + HCl(aq) \longrightarrow NH_4Cl(aq)$$

The moles of NH_3 are

$$34.90 \text{ mL solution} \times \frac{1 \text{ L}}{1000 \text{ mL}} \times \frac{0.110 \text{ mol HCl}}{1 \text{ L solution}} \times \underset{\substack{| \\ \text{determined from the} \\ \text{balanced equation}}}{\frac{1 \text{ mol NH}_3}{1 \text{ mol HCl}}} = 0.00384 \text{ mol NH}_3$$

and the molarity of the NH_3 solution is

$$\frac{0.00384 \text{ mol NH}_3}{2.00 \text{ mL solution}} \times \frac{1000 \text{ mL}}{1 \text{ L}} = \frac{1.92 \text{ mol NH}_3}{1 \text{ L solution}} = 1.92 \ M \qquad \textit{Answer}$$

b. Calculate the percent of ammonia using the density of the solution as 0.985 g/mL and the molar mass of ammonia as 17.0 g.

$$\frac{1.92 \text{ mol NH}_3}{1 \text{ L solution}} \times \frac{17.0 \text{ g NH}_3}{1 \text{ mol NH}_3} \times \frac{1 \text{ L}}{1000 \text{ mL}} \times \frac{1 \text{ mL solution}}{0.985 \text{ g solution}} \times 100$$

$$= 3.31\% \text{ NH}_3 \qquad \textit{Answer}$$

EXAMPLE 15.3   Pure sodium carbonate is used as a standard in determining the molarity of an acid. If 0.875 g of pure sodium carbonate was dissolved in water and the solution was titrated with 35.60 mL of hydrochloric acid to a methyl orange end point, calculate the molarity of the hydrochloric acid solution. (*Hint:* This neutralization takes the carbonate ion to carbon dioxide.)

SOLUTION   Equation:

$$\text{Na}_2\text{CO}_3(s) + 2 \text{ HCl}(aq) \longrightarrow 2 \text{ NaCl}(aq) + \text{CO}_2(g) + \text{H}_2\text{O}(l)$$

Moles of HCl (the molar mass of Na_2CO_3 is 106.0 g):

$$0.875 \text{ g Na}_2\text{CO}_3 \times \frac{1 \text{ mol Na}_2\text{CO}_3}{106.0 \text{ g Na}_2\text{CO}_3} \times \frac{2 \text{ mol HCl}}{1 \text{ mol Na}_2\text{CO}_3} = 0.0165 \text{ mol HCl}$$

determined from the balanced equation

Molarity of HCl:

$$\frac{0.0165 \text{ mol HCl}}{35.60 \text{ mL solution}} \times \frac{1000 \text{ mL}}{1 \text{ L}} = \frac{0.463 \text{ mol HCl}}{1 \text{ L solution}} = 0.463 \, M \qquad \textit{Answer}$$

Study Exercise 15.8

In the titration of a potassium hydroxide solution of unknown concentration, 30.20 mL of a 0.100 M hydrochloric acid solution was required to neutralize 25.00 mL of potassium hydroxide to a phenolphthalein end point. Calculate the molarity of the potassium hydroxide solution.   (0.121 M)   Work Problems 14 through 20.

An alternative approach is to use the definitions of normality and equivalents to solve such problems. In acid–base titrations the number of equivalents of acid is equal to the number of equivalents of base.

$$\text{equivalents (eq) A} = \text{equivalents (eq) B} \qquad (15.23)$$

This statement follows from our definition of *one* equivalent (see Section 14.9) of acid or base supplying 1 mol (6.02×10^{23}) of hydrogen ions (if an acid) or 1 mol (6.02×10^{23}) of hydroxide ions (if a base). Thus, *one equivalent of any acid combines exactly with one equivalent of any base.*

Since the number of equivalents of an acid or base is the volume of the solution (in liters) times the normality (in equivalents per liter), we can rewrite this equation as

$$\boxed{V_{\text{acid}} \cdot N_{\text{acid}} = V_{\text{base}} \cdot N_{\text{base}}} \qquad (15.24)$$

where V_{acid} and V_{base} are the volumes (in liters) of the acid and base, respectively, and N_{acid} and N_{base} are the normalities of the acid and base solutions, respectively. We can now use this expression to solve titration problems involving acids and bases.

EXAMPLE 15.4 In the titration of 34.50 mL of sodium hydroxide solution of unknown concentration, 27.50 mL of 0.100 N sulfuric acid solution was required to neutralize the sodium hydroxide in reactions where both hydrogen ions of the sulfuric acid react. Calculate (a) the normality and (b) the molarity of the sodium hydroxide solution.

SOLUTION
a. The volumes of acid (V_{acid}) and base (V_{base}) are 27.50 mL (0.02750 L) and 34.50 mL (0.03450 L), respectively. We can calculate the normality of the base from the normality of the acid (N_{acid} = 0.100 N) as shown:

$$V_{acid} \cdot N_{acid} = V_{base} \cdot N_{base}$$

$$(0.02750 \text{ L})\frac{0.100 \text{ eq}}{1.000 \text{ L}} = (0.3450 \text{ L})(N_{base})$$

$$N_{base} = \frac{(0.02750)(0.100) \text{ eq}}{0.03450 \text{ L}}$$

$$= 0.0797 \text{ eq/L} = 0.0797 \ N \qquad Answer$$

b. In sodium hydroxide (NaOH) there is only one hydroxide ion. Therefore, there is only one equivalent per mole of sodium hydroxide (see Section 14.9, relation of molarity to normality). The calculation of molarity is

$$\frac{0.0797 \text{ eq NaOH}}{1 \text{ L solution}} \times \frac{1 \text{ mol NaOH}}{1 \text{ eq NaOH}} = \frac{0.0797 \text{ mol NaOH}}{1 \text{ L solution}} = 0.0797 \ M \qquad Answer$$

Study Exercise 15.9
In the titration of 20.00 mL of potassium hydroxide solution of unknown concentration, 28.50 mL of a 0.110 N hydrochloric acid solution was required to neutralize the potassium hydroxide solution. Calculate (a) the normality and (b) the molarity of the potassium hydroxide solution.   [(a) 0.157 N; (b) 0.157 M]

Work Problems 21 and 22.

15.5 pH and pOH

If we wish to deal more quantitatively with solutions of acids and bases, we will need to know the actual concentrations of [H$^+$] and [OH$^-$] in these solutions. These concentrations can be very small, and so chemists often express these concentrations using "p" numbers, which are based on logarithms and defined for any concentration C by the following equation:

$$pC = -\log C = (-1) \times \log C$$

pH Quantitative way of expressing the acidic or basic nature of solutions using the negative logarithmic values of their hydrogen ion (H$^+$) concentrations; a substance with pH < 7 is acidic, one with pH > 7 is basic, and one with pH = 7 is neutral.

Hence, the formula for the "p" of the hydrogen ion concentration, [H$^+$], is

$$\boxed{\text{pH} = -\log[\text{H}^+] = (-1) \times \log[\text{H}^+]} \qquad (15.25)$$

Likewise, the formula for the "p" of the hydroxide ion concentration, [OH⁻], is

$$pOH = -log[OH^-] = (-1) \times log[OH^-]$$ (15.26)

Logarithms are just the exponent (see Section 2.7) portion of a number expressed as a power of 10. That is, the logarithm of 10^6 is 6, and the logarithm of 10^{-12} is -12. (Appendix II will help you use your calculator to calculate the logarithms of numbers.)

The general range for the hydrogen ion concentration in acidic and basic solutions falls between $1.0\ M$ and $1.0 \times 10^{-14}\ M$, which translates into a pH range of 0 to 14 for most common solutions. Solutions with a pH below 7 are acidic, while solutions with a pH above 7 are basic. The *lower* the pH number, the *higher* the hydrogen ion concentration. A solution is *neutral* when the hydrogen ion concentration and the hydroxide concentration are equal, which corresponds to a pH of 7:

0	⟷	7.00	⟷	14
acidic		*neutral*		*basic*

Another relationship that comes in handy is one that relates pH and pOH to each other. We can derive such a relationship from the K_w expression in the following way:

K_w expression:

$$K_w = [H^+][OH^-] = 1.0 \times 10^{-14}$$

Take the logarithms of both sides:

$$log\{[H^+][OH^-]\} = log(1.0 \times 10^{-14})$$

Simplify:

$$log[H^+] + log[OH^-] = -14.00$$

Multiply by -1:

$$-log[H^+] - log[OH^-] = 14.00$$

And finally:

$$pH + pOH = 14.00$$ (15.27)

Thus, if we know the pH, we can calculate the pOH very easily, and vice versa.

Table 15.4 summarizes the relationships among [H⁺], [OH⁻], pH, and pOH in aqueous solutions and presents a number of common substances and their pH values. Notice that the pH of your blood is slightly *basic*. Moreover, it has a very narrow pH range—a mere 0.2 pH unit (7.3 to 7.5). If the pH of the blood goes much below 7.3, *acidosis* occurs; if it falls below 7.0, death may ensue. If the pH goes above 7.5, *alkalosis* occurs; if it goes above 7.8, death may result. The pH of the blood is maintained within this narrow range by *buffers*, solutions of substances that prevent a rapid change in the pH. Buffer solutions will be described in more detail in Section 17.5. Three main types of buffers in the blood are

1. Carbonic acid (H_2CO_3) and sodium hydrogen carbonate ($NaHCO_3$)
2. Sodium dihydrogen phosphate (NaH_2PO_4) and disodium hydrogen phosphate (Na_2HPO_4)
3. Certain proteins

Now that we know something about pH and its applications, we can calculate pH and pOH given the hydrogen ion concentration in mol/L, as the following examples illustrate.

pOH Quantitative way of expressing the acidic or basic nature of solutions using the negative logarithmic values of their hydroxide ion (OH⁻) concentrations; a substance with pOH < 7 is basic, one with pOH > 7 is acidic, and one with pOH = 7 is neutral.

You and Chemistry

TABLE **15.4**		Relationship among [H⁺], [OH⁻], pH, and pOH and the pH of Some Common Examples				

	$[H^+]$ (mol/L)	$[OH^-]$ (mol/L)	pH[a]	pOH	ACID OR BASE STRENGTH	COMMON EXAMPLES (APPROXIMATE pH RANGE)
Acidic	10^0 (1)	10^{-14}	0	14	Strongly acidic	1 *M* HCl (0)
	10^{-1}	10^{-13}	1	13		Gastric juice (1–3)
	10^{-2}	10^{-12}	2	12		Limes (1.8–2.0)
						Soft drinks (2.0–4.0)
						Lemons (2.2–2.4)
	10^{-3}	10^{-11}	3	11	Weakly acidic	Dill pickles (3.2–3.6)
	10^{-4}	10^{-10}	4	10		Acid rain (below 5.65)
	10^{-5}	10^{-9}	5	9		Urine (4.5–8.0)
	10^{-6}	10^{-8}	6	8		Sour milk (6.0–6.2)
						Milk (6.5–6.7)
						Saliva (6.5–7.5)
	10^{-7}	**10^{-7}**	**7**	**7**	**Neutral**	Blood (7.3–7.5)
Basic	10^{-8}	10^{-6}	8	6	Weakly basic	Intestines (6–8)
	10^{-9}	10^{-5}	9	5		
	10^{-10}	10^{-4}	10	4		Milk of magnesia (9.9–10.1)
	10^{-11}	10^{-3}	11	3		Household ammonia (11.5–12.0)
	10^{-12}	10^{-2}	12	2	Strongly basic	
	10^{-13}	10^{-1}	13	1		
	10^{-14}	10^0 (1)	14	0		1 *M* NaOH (14)

[a] The *normal* pH range is from 0 to 14, although solutions with negative pH (to −2) exist, as do solutions having a pH greater than 14 (to 16).

EXAMPLE 15.5	Gatorade, a popular antithirst drink, has a hydrogen ion concentration of 8.0×10^{-4} mol/L. Calculate (a) its pH and (b) its pOH.

SOLUTION

$$pH = -\log [H^+] = -\log [8.0 \times 10^{-4}]$$

Use your calculator to determine the logarithm of 8.0×10^{-4} as -3.0969 (see Appendix II).

Study Hint: The logarithm of a number with *two* significant digits has *two* digits to the *right* of the decimal point.

$$pH = -(-3.0969) = 3.10 \qquad \textit{Answer (a)}$$

$$pOH = 14.00 - pH = 14.00 - 3.10 = 10.90 \qquad \textit{Answer (b)}$$

EXAMPLE 15.6 A commercial tomato juice has a hydrogen ion concentration of 2.5×10^{-5} mol/L. Calculate (a) its pH and (b) its pOH.

SOLUTION

$$pH = -\log [H^+] = -\log [2.5 \times 10^{-5}]$$

Use your calculator to determine the logarithm of 2.5×10^{-5} as -4.6021 (see Appendix II).

$$pH = -(-4.6021) = 4.60 \quad \textit{Answer (a)}$$

$$pOH = 14.00 - pH = 14.00 - 4.60 = 9.40 \quad \textit{Answer (b)}$$

Study Exercise 15.10
Calculate the pH and pOH of the following solutions:
a. hydrogen ion concentration is 2.7×10^{-3} mol/L   (2.57, 11.43)
b. hydrogen ion concentration is 9.5×10^{-8} mol/L   (7.02, 6.98)   Work Problems 23 through 26.

The pH of a solution can be determined directly with a pH meter, as Figure 15.6 shows, where the pH is read either to the tenths' or hundredths' place. For the purpose of this text, we will report pH values to the nearest hundredths' place.

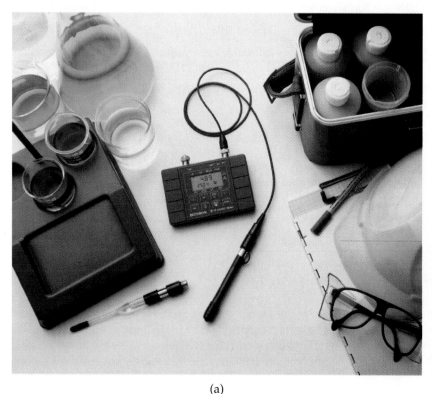

(a)

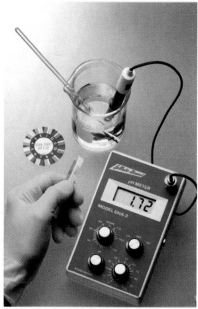

(b)

FIGURE 15.6
Modern pH meters make pH measurements with a high degree of accuracy both (a) in the laboratory and (b) in the field.

The pH of a solution as determined with a pH meter may be used to calculate the hydrogen ion or hydroxide ion concentrations of the solution, as Examples 15.7 and 15.8 illustrate.

Study Hint: The antilogarithm of a number with *two* digits to the *right* of the decimal point has *two* significant digits.

EXAMPLE 15.7 A urine sample has a pH of 5.40. What is the hydrogen ion concentration in the sample?

SOLUTION The pH of the urine sample is 5.40; thus, log $[H^+]$ = -5.40 and

$$[H^+] = 10^{-5.40}$$

Use your calculator to determine the antilogarithm of -5.40 as 3.98×10^{-6} (see Appendix II).

$$[H^+] = 3.98 \times 10^{-6} \, mol/L = 4.0 \times 10^{-6} \, mol/L \qquad Answer$$

EXAMPLE 15.8 The pH of the rain in Pitlochry, Scotland, on April 10, 1974, was 2.40. Calculate the hydrogen ion concentration in the rain.

SOLUTION The pH of the rain sample is 2.40; thus, log $[H^+]$ = -2.40 and

$$[H^+] = 10^{-2.40}$$

Use your calculator to determine the antilogarithm of -2.40 as 3.98×10^{-3} (see Appendix II).

$$[H^+] = 3.98 \times 10^{-3} \, mol/L = 4.0 \times 10^{-3} \, mol/L \qquad Answer$$

Study Exercise 15.11
Calculate the hydrogen ion concentration in mol/L for each of the following solutions:

a. a solution whose pH is 3.40 $\qquad$ (4.0×10^{-4} mol/L)
b. a solution whose pOH is 6.80 $\qquad$ (6.3×10^{-8} mol/L)

Work Problems 27 through 30.

15.6 Solutions of Electrolytes and Nonelectrolytes

In addition to differences in pH, solutions of acids and bases differ from many other solutions in their ability to conduct electricity. The underlying nature of such solutions was first recognized by Svante Arrhenius (see Section 15.1), who would later develop his theory concerning acids and bases. At the time he presented his theory, Arrhenius was only 25 years old and his elder colleagues were skeptical of the new theory. Scientists at the time knew that aqueous solutions of acids, bases, and salts, as well as melted salts, act as electrolytes. *Electrolytes* are substances that in aqueous solution conduct an electric current. In contrast, *nonelectrolytes*, as we might expect, are substances that in aqueous solution do *not* conduct an electric current.

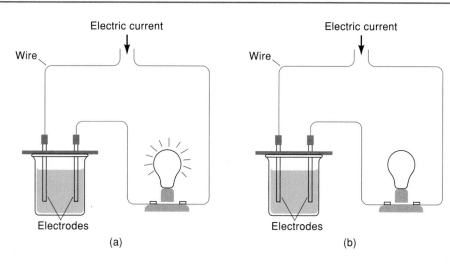

FIGURE 15.7
Apparatus for determining conduction of an electric current in an aqueous solution of a substance: (a) electrolyte; (b) nonelectrolyte.

Until Arrhenius, however, no one had been able to explain *why* electrolytes could conduct an electric current. Arrhenius successfully argued that aqueous solutions of electrolytes contain ions, while solutions of nonelectrolytes do not.

To determine whether a substance is an electrolyte or a nonelectrolyte, chemists prepare an aqueous solution of the substance and then test the solution with two electrodes connected to a source of electric current with a standard light bulb in the circuit, as Figure 15.7 shows. If the *bulb glows*, the substance is an *electrolyte*. If the bulb does *not glow*, the substance is a *non*electrolyte. This experiment *should not be attempted unless it is supervised by a qualified individual.*

We can therefore define **electrolytes** as substances that produce *ions* in aqueous solution and solutions that *conduct* an electric current. **Nonelectrolytes** then are substances that *do not* produce *ions* in aqueous solution and solutions that *do not* conduct an electric current.

Common *electrolytes* include salts, acids, and bases. Some common *nonelectrolytes* are sugar (sucrose, $C_{12}H_{22}O_{11}$), ethyl alcohol (C_2H_6O), glycerine ($C_3H_8O_3$), and urea (CH_4N_2O). Solutions of these nonelectrolytes do *not* conduct an electric current and exist as molecules, not ions, in solution. Pure water is shown to be essentially a nonelectrolyte because a standard light bulb *does not glow* in the electrode test. There do not appear to be sufficient ions present in pure water to conduct an electric current. Tap water, however, contains dissolved salts and can conduct electricity, which is why it is dangerous to touch electrical objects when taking a bath.

Electrolytes may be further subdivided into strong and weak electrolytes based on the degree of dissociation or ionization that is occurring. For *strong electrolytes*, such as ionic salts and strong acids and bases, a standard light bulb glows brightly because there are many ions in solution. For *weak electrolytes*, such as weak acid and weak bases, a standard bulb has only a dull glow because only a few ions exist in solution. Figure 15.8 summarizes the process for determining whether a substance is a nonelectrolyte, a strong electrolyte, or a weak electrolyte.

Table 15.5 summarizes the strong and weak electrolytes. Study it carefully. Knowing the examples of nonelectrolytes and strong and weak electrolytes in this table will help you write ionic equations in the next part of this chapter.

Electrolytes Substances that produce *ions* in aqueous solution and solutions that *conduct* an electric current.

Nonelectrolytes Substances that *do not* produce *ions* in aqueous solution and solutions that *do not* conduct an electric current.

FIGURE 15.8
A substance can be classed as a nonelectrolyte, a strong electrolyte, or a weak electrolyte by testing an aqueous solution of the substance with the apparatus shown in Figure 15.7.

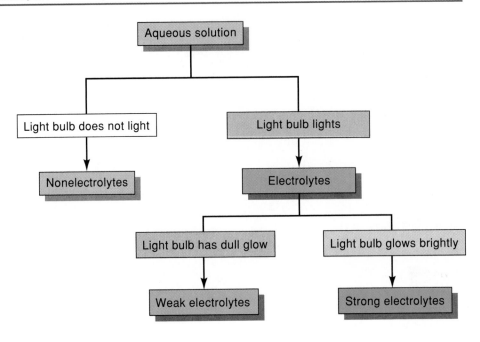

TABLE 15.5	Summary of Strong Electrolytes, Weak Electrolytes, and Nonelectrolytes	
STRONG ELECTROLYTES	WEAK ELECTROLYTES	NONELECTROLYTES
Strong acids	*Weak acids*	$C_{12}H_{22}O_{11}$ (sugar or sucrose)
H_2SO_4 (sulfuric acid)	$HC_2H_3O_2$ (acetic acid)	C_2H_6O (ethyl alcohol)
HCl (hydrochloric acid)	Most other acids	$C_3H_8O_3$ (glycerine)
HBr (hydrobromic acid)		H_2O (water)
HI (hydriodic acid)		CH_4N_2O (urea)
HNO_3 (nitric acid)		
$HClO_4$ (perchloric acid)		
Strong base	*Weak bases*	
Group IA (1) hydroxides	NH_3	
LiOH (lithium hydroxide)	Most other bases	
NaOH (sodium hydroxide)		
KOH (potassium hydroxide)		
RbOH (rubidium hydroxide)		
CsOH (cesium hydroxide)		
Group IIA (2) (hydroxides)		
(*only* following)		
$Ca(OH)_2$ (calcium hydroxide)		
$Sr(OH)_2$ (strontium hydroxide)		
$Ba(OH)_2$ (barium hydroxide)		
Ionic salts		

15.7 Guidelines for Writing Ionic Equations

The ionic nature of acids and bases calls for a special means of expressing these substances and their reactions. In our discussion of chemical equations (Section 9.1), we stated that equations may be written in two general ways: as *complete* (or *molecular*) *equations* and as *ionic equations*. We considered complete equations in detail in Chapter 9. Now we will consider ionic equations. **Ionic equations** express a chemical reaction involving compounds that exist mostly in ionic form in aqueous solution. For *ionic compounds*, the reacting particles are actually *ions*. Hence, in ionic equations the ions are written as they *actually* exist in the solution. Ionic equations thus give a better representation of a chemical change in an aqueous solution than complete equations do.

In the discussion on balancing complete equations (Section 9.3), we presented a few guidelines to help you balance equations by inspection. We suggest the following seven guidelines for writing ionic equations:

1. Complete and balance an equation in the form in which it is given to you. If it is given in the complete form, complete the equation in the complete form, balance it, and *then* change it to the ionic form. If it is given in the ionic form, complete it and balance it in the ionic form. For discussion on how to complete chemical equations, see Chapter 9.

2. Write the formulas for the following compounds as *complete formulas*:
   a. *Nonelectrolytes* such as those listed in Table 15.5.
   b. *Weak acids* and *weak bases* such as those listed in Table 15.5.
   c. *Solids* and *insoluble salts* (precipitates) from aqueous solutions (see solubility rules in Section 9.9 or on the inside back cover of this text) such as $CaCO_3(s)$ and $AgCl(s)$. These insoluble salts contain ions but are written as *complete formulas* because the ions are *not* bound to solvent molecules.
   d. *Gases*, such as H_2, N_2, and O_2, written as the diatomic gas, $H_2(g)$, and so on.

3. Write the formulas for the following compounds as *ionic formulas*:
   a. *Strong acid* and *strong bases* such as those listed in Table 15.5.
   b. *Soluble salts* (see solubility rules in Section 9.9 or on the inside back cover of this text).

4. When you write compounds in ionic form, use subscripts only to express polyatomic ions. For example, write **1** mol of sulfuric acid (H_2SO_4) in ionic form as $2\,H^+ + SO_4^{2-}$ (use a subscript 4 because SO_4^{2-} is a polyatomic ion). Write **3** mol of sodium sulfate (3 Na_2SO_4) as $6\,Na^+ + 3\,SO_4^{2-}$.

5. Check (✔) each ion (monatomic or polyatomic) and each atom to make sure it is balanced on both sides of the equation. The net charge on each side of the equation must be the *same*.

6. The *net ionic equation* shows only those ions that have actually undergone a chemical change. Cross out the ions appearing on *both sides* of the equation that have *not* undergone a change; do not include these ions in the net ionic equation. Include these unaltered ions in the *total ionic equation* but *not* in the net ionic equation. They are called "spectator ions" because they just sit back and enjoy the show.

Ionic equations   Express chemical reactions involving compounds that exist mostly in ionic form in aqueous solution.

Study Hint: Table 15.5 is *very* important to you. You must memorize examples of strong and weak acids and bases and nonelectrolytes to be able to write compounds as *complete* and *ionic* formulas. Notice that *soluble* salts are written as *ionic formulas*, while *insoluble* salts (precipitates) are written as *complete formulas*. The solubility rules will be given to you.

7. Finally, check ($\checkmark$) the net ionic equation for ions, atoms, and charge and to see that the coefficients are in the lowest possible integral ratio.

15.8 Examples of Ionic Equations

Now let us apply these guidelines to writing ionic equations.

EXAMPLE 15.9 Complete and balance the following chemical equation in ionic form:

SOLUTION

$$AgNO_3(aq) + HCl(aq) \longrightarrow$$

Completing and balancing the equation according to guideline 1 gives

$$AgNO_3(aq) + HCl(aq) \longrightarrow AgCl(s) + HNO_3(aq)$$

Consider each of the reactants and products and decide whether they should be written as *complete* or *ionic* formulas.

FORMULA	IDENTIFICATION	CONCLUSION
$AgNO_3$	Soluble salt (see solubility rules on inside back cover)	Ionic formula
HCl	Strong acid	Ionic formula
AgCl	Insoluble salt (see solubility rules)	Complete formula
HNO_3	Strong acid	Ionic formula

Write the total ionic equation by applying guidelines 2, 3, and 4. All compounds here are written as ionic formulas except AgCl which is an insoluble salt (see Table 15.5 and the solubility rules). The total ionic equation is

$$Ag^+(aq) + NO_3^-(aq) + H^+(aq) + Cl^-(aq) \longrightarrow AgCl(s) + H^+(aq) + NO_3^-(aq)$$

Check off each ion, atom, and charge according to guideline 5. The total ionic equation is

$$\overset{\checkmark}{Ag^+}(aq) + \overset{\checkmark}{NO_3^-}(aq) + \overset{\checkmark}{H^+}(aq) + \overset{\checkmark}{Cl^-}(aq) \longrightarrow \overset{\checkmark\checkmark}{AgCl}(s) + \overset{\checkmark}{H^+}(aq) + \overset{\checkmark}{NO_3^-}(aq)$$

Charges: $+1 \quad + (-1) \quad + (+1) \quad + (-1) = 0 \quad = 0 + (+1) \quad + (-1) = 0$

Write the net ionic equation by crossing out ions that appear on both sides of the equation, according to guideline 6. Check the final net ionic equation for ions, atoms, charge, and lowest possible ratio of coefficients (guideline 7).

$$Ag^+(aq) + \cancel{NO_3^-}(aq) + \cancel{H^+}(aq) + Cl^-(aq) \longrightarrow AgCl(s) + \cancel{H^+}(aq) + \cancel{NO_3^-}(aq)$$

The net ionic equation is

$$\overset{\checkmark}{Ag^+}(aq) + \overset{\checkmark}{Cl^-}(aq) \longrightarrow \overset{\checkmark\checkmark}{AgCl}(s) \qquad Answer$$

Charges: $+1 \quad + (-1) = 0 \quad = 0$

From the net ionic equation, this reaction is the reaction of any soluble ionic silver salt with a soluble strongly ionic chloride compound.

Study Hint: Another example is the reaction between silver acetate and potassium chloride. See if you get the *same* net ionic equation.

EXAMPLE 15.10   Complete and balance the following chemical equation in ionic form:

$$NaOH(aq) + H_2SO_4(aq) \longrightarrow$$

SOLUTION   Completing and balancing the equation according to guideline 1 gives

$$2\,NaOH(aq) + H_2SO_4(aq) \longrightarrow Na_2SO_4(aq) + 2\,H_2O(l)$$

Consider each of the reactants and products and decide whether they should be written as *complete* or *ionic* formulas.

FORMULA	IDENTIFICATION	CONCLUSION
NaOH	Strong base	Ionic formula
H_2SO_4	Strong acid	Ionic formula
Na_2SO_4	Soluble salt (see solubility rules)	Ionic formula
H_2O	Nonelectrolyte	Complete formula

Write the total ionic equation by applying guidelines 2 through 4. All compounds here are written as ionic formulas except H_2O (a nonelectrolyte; see Table 15.5). The total ionic equation is

$$2\,Na^+(aq) + 2\,OH^-(aq) + 2\,H^+(aq) + SO_4{}^{2-}(aq) \longrightarrow 2\,Na^+(aq) + SO_4{}^{2-}(aq) + 2\,H_2O(l)$$

Check each ion, atom, and charge according to guideline 5.

$$2\,Na^+(aq) + 2\,OH^-(aq) + 2\,H^+(aq) + SO_4{}^{2-}(aq) \longrightarrow 2\,Na^+(aq) + SO_4{}^{2-}(aq) + 2\,H_2O(l)$$

Charges:
$$2(+1) \quad + 2(-1) \quad + 2(+1) \quad + (-2) = 0 \quad = 2(+1) + (-2) \quad +0 = 0$$

Crossing out the ions that appear on both sides of the equation according to guideline 6 gives the net ionic equation. Check the net ionic equation for ions, atoms, charge, and lowest possible ratio of coefficients (guideline 7):

$$\cancel{2\,Na^+(aq)} + 2\,OH^-(aq) + 2\,H^+(aq) + \cancel{SO_4{}^{2-}(aq)} \longrightarrow \cancel{2\,Na^+(aq)} + \cancel{SO_4{}^{2-}(aq)} + 2\,H_2O(l)$$

$$2\,OH^-(aq) + 2\,H^+(aq) \longrightarrow 2\,H_2O(l)$$

Divide both sides of the equation by 2; the net ionic equaton is

$$OH^-(aq) + H^+(aq) \longrightarrow H_2O(l) \quad \textit{Answer}$$
$$\text{Charges:}\ -1 \quad + (+1) \quad = 0 = 0$$

This reaction is a *neutralization* reaction (Section 9.10), and as a net ionic equation, it is simply the reaction of a hydroxide ion with a hydrogen ion to form water. This then is the reaction of any *strong acid* with any *strong base*.

> **Study Hint:** Another example is the reaction between hydrochloric acid and barium hydroxide. See if you get the *same* net ionic equation.

EXAMPLE 15.11   Complete and balance the following chemical equation in ionic form:

$$Al(s) + H_2SO_4(aq) \longrightarrow$$

SOLUTION   This is a single-replacement reaction involving the electromotive or activity series (see Section 9.8). According to guideline 1, completing and balancing the equation gives

$$2\,Al(s) + 3\,H_2SO_4(aq) \longrightarrow Al_2(SO_4)_3(aq) + 3\,H_2(g)$$

Consider each of the reactants and products and decide whether they should be written as *complete* or *ionic* formulas.

FORMULA	IDENTIFICATION	CONCLUSION
Al	Solid metal	Complete formula
H_2SO_4	Strong acid	Ionic formula
$Al_2(SO_4)_3$	Soluble salt (see solubility rules)	Ionic formula
H_2	Gas	Complete formula

Write the total ionic equation by applying guidelines 2 through 4. All substances here are written as ionic formulas except Al, a solid and a free metal (*not* an ion), and H_2, a gas. Checking (guideline 5) gives the following total ionic equation:

$$2\,Al(s) + 6\,H^+(aq) + 3\,SO_4^{2-}(aq) \longrightarrow 2\,Al^{3+}(aq) + 3\,SO_4^{2-}(aq) + 3\,H_2(g)$$

Charges: 0   +6(+1)   +3(−2)   = 0 = 2(+3)   + 3(−2)   + 0 = 0

Crossing out the ions that appear on both sides of the equation and checking again as in guideline 6, we have the following net ionic equation:

$$2\,Al(s) + 6\,H^+(aq) + 3\,\cancel{SO_4^{2-}}(aq) \longrightarrow 2\,Al^{3+}(aq) + 3\,\cancel{SO_4^{2-}}(aq) + 3\,H_2(g)$$

> **Study Hint:** Neither Al nor H^+ can be crossed out because they appear in the products as Al^{3+} and H_2, respectively.

The net ionic equation is

$$2\,Al(s) + 6\,H^+(aq) \longrightarrow 2\,Al^{3+}(aq) + 3\,H_2(g) \qquad Answer$$

Charges: 0   +6(+1) = +6 = 2(+3)   + 0 = +6

EXAMPLE 15.12 Complete and balance the following chemical equation in ionic form:

$$NH_3(aq) + H_2O(l) + Al_2(SO_4)_3(aq) \longrightarrow$$

SOLUTION   Completing and balancing the equation according to guideline 1 with $NH_3 + H_2O$ acting as $NH_4^+ + OH^-$* gives

$$6\,NH_3(aq) + 6\,H_2O(l) + Al_2(SO_4)_3(aq) \longrightarrow 3\,(NH_4)_2SO_4(aq) + 2\,Al(OH)_3(s)$$

Consider each of the reactants and products and decide whether they should be written as *complete* or *ionic* formulas.

FORMULA	IDENTIFICATION	CONCLUSION
NH_3	Weak base	Complete formula
H_2O	Nonelectrolyte	Complete formula
$Al_2(SO_4)_3$	Soluble salt (see solubility rules)	Ionic formula
$(NH_4)_2SO_4$	Soluble salt (see solubility rules)	Ionic formula
$Al(OH)_3$	Weak base and solid (see solubility rules)	Complete formula

Write the total ionic equation by applying guidelines 2 through 4. Write $NH_3(aq)$, a weak base, H_2O, a nonelectrolyte, and $Al(OH)_3$, a weak base and solid, as complete formulas with the two soluble salts [$Al_2(SO_4)_3$ and $(NH_4)_2SO_4$] as ionic formulas.

---

*The substance ammonium hydroxide (NH_4OH) does not exist. Ammonia (NH_3) in water gives ammonium ions (NH_4^+) and hydroxide ions (OH^-) but *no* NH_4OH molecules. In the laboratory you may see bottles labeled "NH_4OH, ammonium hydroxide;" which actually means "NH_3 in water, ammonia water."

✔✔   ✔✔   ✔   ✔   ✔✔   ✔   ✔ ✔✔
$6 NH_3(aq) + 6 H_2O(l) + 2 Al^{3+}(aq) + 3 SO_4^{2-}(aq) \longrightarrow 6 NH_4^+(aq) + 3 SO_4^{2-}(aq) + 2 Al(OH)_3(s)$

Charges:

0   + 0   + 2(+3)   + 3(−2)   = 0 = 6(+1)   + 3(−2)   + 0 = 0

Crossing out the ions that appear on both sides of the equation and checking again as in guideline 6, we have the following net ionic equation:

$6 NH_3(aq) + 6 H_2O(l) + 2 Al^{3+}(aq) + 3\,\cancel{SO_4^{2-}}(aq) \longrightarrow 6 NH_4^+(aq) + 3\,\cancel{SO_4^{2-}}(aq) + 2 Al(OH)_3(s)$

Divide both sides of the equation by 2:

✔✔   ✔✔   ✔   ✔✔   ✔ ✔✔
$3 NH_3(aq) + 3 H_2O(l) + Al^{3+}(aq) \longrightarrow 3 NH_4^+(aq) + Al(OH)_3(s)$ *Answer*

Charges: 0   +0   + (+3) = +3 = 3(+1)   + 0 = +3

EXAMPLE 15.13   Complete and balance the following chemical equation in ionic form:

$$Ag^+(aq) + H_2S(aq) \longrightarrow$$

SOLUTION   This is a double-replacement reaction (see Section 9.9). The anion bound to the Ag^+ is any anion that produces a soluble silver salt in water, such as acetate or nitrate. Completing and balancing the equation according to guideline 1 gives the following ionic equation:

$$2 Ag^+(aq) + H_2S(aq) \longrightarrow Ag_2S(s) + 2 H^+(aq)$$

Consider each of the reactants and products and decide whether they should be written as *complete* or *ionic* formulas.

FORMULA	IDENTIFICATION	CONCLUSION
Ag^+	Ion	Ionic formula
H_2S	Weak acid	Complete formula
Ag_2S	Insoluble salt (see solubility rules)	Complete formula
H^+	Ion	Ionic formula

Write the insoluble salt (Ag_2S) and weak acid (H_2S) as complete formulas according to guidelines 2 through 4. Check the net ionic equation for ions, atoms, and charge according to guideline 5.

✔   ✔✔   ✔ ✔   ✔
$2 Ag^+(aq) + H_2S(aq) \longrightarrow Ag_2S(s) + 2 H^+(aq)$ *Answer*

Charges: 2(+1)   + 0   = +2 = 0   + 2(+1) = +2

The net ionic equation is the same as this ionic equation because the same ions do not appear on *both* sides of the equation. This is the reaction that forms tarnish on silver (see Figure 15.9).

Study Exercise 15.12
Complete and balance the following reaction equations, writing them as total ionic equations and net ionic equations.

a. $Ba(OH)_2(aq) + HC_2H_3O_2(aq) \longrightarrow$
   [$Ba^{2+}(aq) + 2 OH^-(aq) + 2 HC_2H_3O_2(aq) \longrightarrow Ba^{2+}(aq) + 2 C_2H_3O_2^-(aq) + 2 H_2O(l)$;
        net: $OH^-(aq) + HC_2H_3O_2(aq) \longrightarrow C_2H_3O_2^-(aq) + H_2O(l)$]

b. $Pb(NO_3)_2(aq) + H_2S(aq) \longrightarrow$
        [$Pb^{2+}(aq) + 2 NO_3^-(aq) + H_2S(aq) \longrightarrow PbS(s) + 2 H^+(aq) + 2 NO_3^-(aq)$;
        net: $Pb^{2+}(aq) + H_2S(aq) \longrightarrow PbS(s) + 2 H^+(aq)$]

Study Hint: This is *not* a single-replacement reaction because neither of the reactants is an *element in the free or uncombined state*. Neither Ag^+ (an ion) nor H_2S appears in the electromotive or activity series.

You and Chemistry

Work Problems 31 and 32.

FIGURE 15.9
(a) Soluble silver salts react with hydrogen sulfide to give a black precipitate of silver sulfide (Ag$_2$S). (b) Tarnish on sterling silver is merely silver sulfide that has formed on the surface of the silverware. Polished sterling silver spoon (top); tarnished sterling silver spoon (bottom).

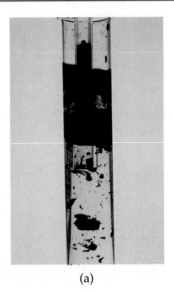

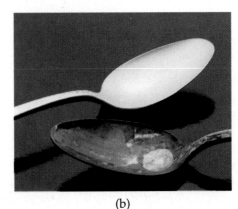

(a)   (b)

CHEMISTRY OF THE ATMOSPHERE

Acid from the Skies!

As early as the 1850s, scientists in England observed the presence of strongly acidic compounds in rain water. The term *acid rain* first appeared in a book in 1872 entitled *Air and Rain: The Beginning of Chemical Climatology*. It was not until the 1950s, however, that systematic monitoring began in northern and western Europe. By the middle of the 1960s, it had become clear that acid precipitation was affecting the forests of many northern countries and that the problem was spreading. Then, in the 1970s, studies indicated that acid precipitation was also falling on the northern United States and Canada.

Today, many scientists prefer to speak of *acid deposition*, rather than acid rain. They believe that the term "acid deposition" better describes the distribution of acidic materials over vast geographical areas by the natural processes of wind and precipitation. Regardless of its name, the root of the problem lies in the acidic materials generated by humans as they go about living their daily lives. The acidic materials may be dispersed as windblown solids or as solutions in precipitation (rain, snow, and fog). For the purposes of this discussion, we will focus on the dispersion of these acids by rainfall.

Acid rain is generally defined as rain water that has a pH of less than 5.65. Why 5.65? Because normally, pure water that has been exposed to air containing carbon dioxide (CO$_2$) has a pH of about 5.65. This pH level is due to the CO$_2$ that dissolves in the water (see Section 14.3, Henry's law) to produce a small amount of carbonic acid (H$_2$CO$_3$):

$$CO_2(g) + H_2O(g) \rightleftharpoons H_2CO_3(aq)$$

This carbonic acid is a weak acid and dissociates to produce a low but significant number of aqueous hydrogen ions in solution:

$$H_2CO_3(aq) \rightleftharpoons H^+(aq) + HCO_3^-(aq)$$

Notice that both of these equations depict equilibrium conditions where only small amounts of the products are present. Any rain that has a pH less than 5.65, then, must contain materials that are more acidic than the carbonic acid in pure water exposed to the atmosphere.

In extreme cases, acid rain can have a pH as low as 3.0 or 4.0. But the average pH of rainfall is commonly in the 4.1 to 4.5 range in the eastern United States. As you may recall, the pH scale is logarithmic, which means that rain with a pH of 4.65 is *10 times* as acidic as normal rain (pH of 5.65).

Analysis of the rain indicates that sulfuric acid (H$_2$SO$_4$) and nitric acid (HNO$_3$) account for about 94% of the acidity of low-pH rain water. These acids are the end product of chemical processes that convert atmospheric sulfur dioxide, nitrogen oxide, and nitrogen dioxide eventually to sulfuric acid and nitric acid, respectively.

SO$_2$ eventually forms H$_2$SO$_4$

NO or NO$_2$ eventually forms HNO$_3$

We will examine these processes more carefully in the essay follow-

(a)

(b)

The Black Forest in Germany has been severely damaged by acid rain. (a) Part of the countryside photographed recently; (b) the same scene photographed 14 years ago.

ing Chapter 16. The table below gives you an idea of just how much of these gases is generated by human activities.

When these acids return to earth as acid rain, they may affect the environment in many ways. Acidification of lakes and streams endangers aquatic life, including water plants, fish, and the animals that feed on them. Acidification of soils may affect tree growth in forests. Acidic rain may leach cations (especially Al^{3+}) from soil, thereby increasing the concentrations of these cations in bodies of water. These cations, in turn, may deprive plants of desirable agents or concentrate harmful materials in lakes and streams. Finally, acid precipitation may harm older buildings and monuments. For example, acid rain has been credited with the near total destruction of the marble friezes on the Acropolis in Athens, Greece (see Figure 9.12).

Ending acid rain will not be easy. The sources of the offending gases are deeply ingrained in modern society. Many industries would have to radically alter their operations, at considerable expense, to significantly reduce their emissions. Moreover, unlike the greenhouse effect and ozone depletion, the effects of acid deposition tend to be regional, being most evident in the northern United States, Canada, and areas of northern Europe. Unfortunately, the effects of acid rain are not always visible where the pollution is generated. Canada and New England suffer because of emissions that originate in the industrial parts of the Midwest, for example. Like the greenhouse effect and ozone depletion, acid rain problems involve both political and social issues and will be difficult to solve.

Contributions to Atmospheric Sulfur Dioxide, Nitrogen Oxide, and Nitrogen Dioxide by Human Activities in a Typical Year

MAJOR SOURCES	SULFUR DIOXIDE EMISSIONS (TONS)	NITROGEN OXIDES (NO AND NO_2) EMISSIONS (TONS)
Power plants that use coal, oil, or natural gas to generate electricity	22.8×10^6	13.5×10^6
Metallurgy, chemical, and petroleum industries	5.3×10^6	0.82×10^6
Use of gasoline and diesel fuel for transportation	0.91×10^6	9.4×10^6

✓ Summary

Arrhenius defined an *acid* as a substance that yields hydrogen ions (H^+) when dissolved in water, and a *base* as a substance that yields hydroxide (OH^-) ions in water. Brønsted and Lowry defined an *acid* as any substance that can donate a proton (H^+) to some other substance, and a *base* as any substance capable of accepting a proton. A substance that can act as either a Brønsted–Lowry acid or base is called an *amphoteric substance*. When any acid (for example, $HC_2H_3O_2$) loses a proton, the species ($C_2H_3O_2^-$) that is produced is called the *conjugate base*. When any base (for example, H_2O) gains a proton in a proton-transfer reaction, the new species (H_3O^+) is called the *conjugate acid* (Section 15.1).

Different acids have different abilities to donate a proton to another molecule or ion. *Strong acids* readily donate a proton to water (solvent) to produce many hydrogen ions in solution, while weak acids generate only a relatively small number of ions. Similarly, *strong bases* can produce large numbers of hydroxide ions in solution, but *weak bases* generate only a few such ions (Section 15.2).

Water is an *amphoteric substance*. In a self-ionization reaction one water molecule can actually transfer a proton to a second water molecule to produce small concentrations of aqueous hydrogen or hydronium ions and hydroxide ions in pure water. In *dissociation*, ions in *ionic* compounds (crystalline solids) are separated by the action of the solvent to form hydrated ions. In *ionization*, *covalent* compounds dissolve in water and create ions by breaking a covalent bond (Section 15.3).

Acids and bases engage in three general types of reactions: (1) reactions between an acid and a metal, (2) reactions between a base and a salt to give a precipitate, and (3) neutralization reactions. In a process called *titration*, chemists use the neutralization reaction and stoichiometry to determine the concentration of an acid or base (Section 15.4).

Precise concentrations of acids and bases are expressed in terms of the solution's pH or pOH. *pH* is defined as the negative logarithm of the hydrogen ion concentration, and *pOH* as the negative logarithm of the hydroxide ion concentration. The sum of the pH and the pOH is equal to 14 (Section 15.5).

Electrolytes are substances that produce ions in aqueous solution and form solutions that conduct an electric current. *Nonelectrolytes* are substances that do not produce ions in aqueous solution and therefore form solutions that do not conduct an electric current. Electrolytes are divided into *strong* and *weak electrolytes*, depending on their ability to conduct electricity. The more ions present in the solution, the better the conduction. Hence, strong acids and bases and ionic salts are strong electrolytes (Section 15.6).

Ionic equations express a chemical reaction in terms of ions for compounds that exist mostly in ionic form in aqueous solutions. Writing such equations requires (1) completing the equations, (2) identifying strong and weak acids and bases and nonelectrolytes, and (3) following the solubility rules of inorganic substances in water (Sections 15.7 and 15.8).

✓ Exercises

1. Define or explain the following terms (the number in parentheses refers to the section in the text where the term is mentioned):

   a. indicators (15.1)

   b. acid (Arrhenius definition) (15.1)

c. base (Arrhenius definition) (15.1)

d. acid (Brønsted–Lowry definition) (15.1)

e. base (Brønsted–Lowry definition) (15.1)

f. conjugate acid (15.1)

g. conjugate base (15.1)

h. amphoteric substance (15.1)

i. dissociation (15.3)

j. ionization (15.3)

k. titration (15.4)

l. equivalence point (15.4)

m. end point (15.4)

n. pH (15.5)

o. pOH (15.5)

p. electrolyte (15.6)

q. nonelectrolyte (15.6)

r. ionic equations (15.7)

2. Distinguish between:

a. an Arrhenius acid and a Brønsted–Lowry acid

b. an Arrhenius base and a Brønsted–Lowry base

c. hydrogen ion and hydroxide ion

d. ionization and dissociation

e. a strong electrolyte and a weak electrolyte

f. equivalence point and end point

Acids and Bases (See Section 15.1)

3. List at least three general characteristics each for an acid and for a base.

Electrolytes versus Nonelectrolytes (See Section 15.6)

4. List at least four strong electrolytes, four weak electrolytes, and four nonelectrolytes.

5. In your own words, explain why pure acetic acid does not light the standard light bulb but dilute acetic acid does.

✓ Problems

Acids and Bases (See Section 15.1)

6. Classify each of the following substances as an acid or a base according to the Arrhenius theory.

a. $HNO_3(aq)$ b. $KOH(aq)$ c. $Ca(OH)_2(aq)$

d. $HCl(aq)$ e. $H_2SO_4(aq)$ f. $NaOH(aq)$

7. Classify each of the following as an acid, a base, or amphoteric substance according to the Brønsted–Lowry theory.

   a. NH_4^+ b. HCO_3^- c. H_2O

   d. H_3O^+ e. H_2SO_4 f. HSO_4^-

8. From the following reaction equations, identify the conjugate acid and conjugate base.

   a. $H_2SO_4(aq) + H_2O(l) \longrightarrow H_3O^+(aq) + HSO_4^-(aq)$

   b. $H_2O(l) + HSO_4^-(aq) \rightleftharpoons SO_4^{2-}(aq) + H_3O^+(aq)$

The Strengths of Acids and Bases (See Section 15.2)

9. Classify each of the following substances as a strong acid, a weak acid, a strong base, or a weak base.

   a. hydrochloric acid solution (HCl)   b. ammonia solution (NH_3)

   c. hydrocyanic acid solution (HCN)   d. potassium hydroxide solution

10. Perchloric acid ($HClO_4$) is a strongly acidic substance, while hypochlorous acid (HClO) is not. Write a pair of balanced chemical equations that describe the different behaviors of these two compounds in aqueous solution. (*Hint:* To what extent does hypochlorous acid produce hydrogen ions in solution?)

Reactions of Acids and Bases (See Section 15.4)

11. Complete and balance the following chemical equations for the reaction between an acid and a metal.

    a. hydrochloric acid + zinc metal $\longrightarrow$

    b. hydrobromic acid + cadmium metal $\longrightarrow$

    c. hydrochloric acid + silver metal $\longrightarrow$

12. Complete and balance the following chemical equations for the reaction between a base and a salt to give an insoluble compound.

    a. sodium hydroxide + aluminum nitrate $\longrightarrow$

    b. barium hydroxide + iron(III) chloride $\longrightarrow$

    c. potassium hydroxide + nickel(II) bromide $\longrightarrow$

13. Complete and balance the following chemical equations for the reaction between an acid and a base.

    a. hydrochloric acid + potassium hydroxide $\longrightarrow$

    b. nitric acid + barium hydroxide $\longrightarrow$

    c. sulfuric acid + sodium hydroxide $\longrightarrow$
       (*Hint:* Both hydrogens on sulfuric acid react.)

14. If 25.00 mL of 0.100 M hydrochloric acid solution is necessary to neutralize 55.00 mL of a solution of sodium hydroxide to a phenolphthalein end point, calculate the molarity of the sodium hydroxide solution.

15. If 32.00 mL of a dilute solution of lime water (calcium hydroxide) required 12.40 mL of 0.100 M hydrochloric acid solution for neutralization to a methyl red end point, calculate the molarity of the lime water.

16. If 22.00 mL of potassium hydroxide solution required 16.40 mL of 0.150 M hydrochloric acid solution for neutralization to a bromthymol blue end point, calculate the molarity of the potassium hydroxide solution.

17. If 37.50 mL of 0.500 M sodium hydroxide solution is necessary to neutralize 25.00 mL of a hydrochloric acid solution to a phenolphthalein end point, calculate (a) the molarity and (b) the percent by mass (density of solution = 1.013 g/mL) of the hydrochloric acid solution.

18. Vinegar is a dilute solution of acetic acid. In the titration of 5.00 mL of vinegar, 37.70 mL of 0.105 M sodium hydroxide solution was required to neutralize the vinegar to a phenolphthalein end point. Calculate (a) the molarity and (b) the percent by mass (density of solution = 1.007 g/mL) of the vinegar.

19. If 0.650 g of pure sodium carbonate was dissolved in water and the solution titrated with 30.80 mL of hydrochloric acid to a methyl orange end point, calculate the molarity of the hydrochloric acid solution. (*Hint:* This process takes the carbonate ion to carbon dioxide.)

20. If 0.200 g of pure sodium hydroxide was dissolved in water and the solution titrated with 5.70 mL of hydrochloric acid solution to a phenolphthalein end point, calculate (a) the molarity and (b) the percent by mass (density of solution = 1.016 g/mL) of the hydrochloric acid solution.

21. In the titration of a 0.125 N sodium hydroxide solution, 28.50 mL of 0.155 N sulfuric acid solution was required to neutralize the sodium hydroxide in reactions that replace both hydrogen ions of the sulfuric acid. Calculate the number of milliliters of sodium hydroxide solution needed in the reaction.

22. In the titration of 24.50 mL of potassium hydroxide solution of unknown concentration, 35.70 mL of 0.110 N sulfuric acid was required to neutralize the potassium hydroxide in reactions where both hydrogen ions of the sulfuric acid react. Calculate (a) the normality and (b) the molarity of the potassium hydroxide solution.

pH and pOH (See Section 15.5)

23. Calculate the pH and pOH of the following solutions:
    a. hydrogen ion concentration is 1.0×10^{-9} mol/L
    b. household ammonia, hydrogen ion concentration of 2.0×10^{-12} mol/L
    c. commercial milk, hydrogen ion concentration of 2.0×10^{-7} mol/L
    d. sour milk, hydrogen ion concentration of 6.3×10^{-7} mol/L
    e. 7-Up, hydrogen ion concentration of 2.5×10^{-4} mol/L

24. Calculate the pH and pOH of the following solutions:
    a. hydrogen ion concentration is 2.4×10^{-6} mol/L
    b. vinegar, hydrogen ion concentration of 7.9×10^{-4} mol/L
    c. a dilute solution (0.133%) of citric acid found in various fruits, hydrogen ion concentration of 1.4×10^{-3} mol/L
    d. seawater, hydrogen ion concentration of 5.3×10^{-9} mol/L
    e. a sample of human urine with a hydrogen ion concentration of 8.6×10^{-6} mol/L

25. In general, the flavor of devil's-food cake is best if the hydrogen ion concentration is between 1.0×10^{-8} and 3.2×10^{-8} mol/L.
    a. Calculate this range on the pH scale.
    b. With a hydrogen ion concentration of less than 1.0×10^{-8} mol/L, the cake has a bitter, soapy taste. Explain.

26. The hydrogen ion concentration of soils varies considerably. In forest soils the hydrogen ion concentration is about 3.2×10^{-5} mol/L, while in desert soils the hydrogen ion concentration is about 1.0×10^{-10} mol/L. The higher hydrogen ion concentration in forest soils occurs because decomposition of organic matter results in the production of carbon dioxide. Calculate the pH and pOH for (a) forest and (b) desert soils.

27. Calculate the hydrogen ion concentration in mol/L for each of the following solutions:

    a. a solution whose pH is 6.20   **b.** a solution whose pH is 9.60

    c. a solution whose pOH is 2.50   **d.** a solution whose pOH is 12.00

    e. a solution whose pOH is 4.77

28. Calculate the hydroxide ion concentration in mol/L for each of the following solutions:

    a. a solution whose pOH is 5.00   **b.** a solution whose pOH is 9.60

    c. a solution whose pH is 5.70   **d.** a solution whose pH is 8.20

    e. a solution whose pH is 5.40

29. Rain water has a pH of 4.00. Calculate its hydrogen ion concentration in mol/L. Explain why the water is acidic instead of neutral.

30. In California, the pH of acid fog has dipped as low as 1.90. Calculate its hydrogen ion concentration in mol/L.

Ionic Equations (See Sections 15.7 and 15.8. *Hint:* First, review Chapter 9 so that you will be able to complete and balance these chemical equations.)

31. Complete and balance the following chemical equations, writing them as total ionic equations and as net ionic equations. Indicate any precipitate by (*s*) and any gas by (*g*).

    a. $BaCl_2(aq) + (NH_4)_2CO_3(aq) \longrightarrow$

    b. $Fe(NO_3)_3(aq) + NH_3(aq) + H_2O(l) \longrightarrow$
    (*Hint:* See Example 15.12.)

    c. $SrCl_2(aq) + K_2CO_3(aq) \longrightarrow$

    d. $Na_2CO_3(aq) + HCl(aq) \longrightarrow$

    e. $KCl(aq) + AgNO_3(aq) \longrightarrow$

    f. $Al(s) + HCl(aq) \longrightarrow$

    g. $CO_2(aq) + Ca(OH)_2(aq) \longrightarrow$

    h. $SrCO_3(s) + HC_2H_3O_2(aq) \longrightarrow$

    i. $Fe(s) + CuSO_4(aq) \longrightarrow$

    j. $CdCl_2(aq) + H_2S(aq) \longrightarrow$

32. Complete and balance the following chemical equations, writing them as total ionic equations and as net ionic equations. Indicate any precipitate by (*s*) and any gas by (*g*).

    a. $CuCl_2(aq) + H_2S(aq) \longrightarrow$

    b. $MgSO_4(aq) + NaOH(aq) \longrightarrow$

    c. $CaO(s) + HCl(aq) \longrightarrow$

    d. $FeCl_3(aq) + NH_3(aq) + H_2O(l) \longrightarrow$
    (*Hint:* See Example 15.12.)

e. $FeSO_4(aq) + (NH_4)_2S(aq) \longrightarrow$

f. $Al(OH)_3(s) + HCl(aq) \longrightarrow$

g. $H_3PO_4(aq) + KOH(aq) \longrightarrow$

h. $MgCl_2(aq) + Na_2CO_3(aq) \longrightarrow$

i. $Cl_2(g) + NaBr(aq) \longrightarrow$

j. $BiCl_3(aq) + H_2S(aq) \longrightarrow$

General Problems

33. If 1.50 g of 80.0% by mass potassium hydroxide was dissolved in water and the solution titrated with 10.00 mL of hydrochloric acid solution, calculate (a) the molarity and (b) the percent by mass (density of solution = 1.037 g/mL), and (c) the normality of the hydrochloric acid solution.

34. A theory used to explain how acid rain destroys forest is that the acid makes soluble the *insoluble* aluminum salts in the soil. These *soluble* aluminum salts are taken up by the tree roots and are toxic to the tree. Assuming that the insoluble salt in the soil is AlZ_3 (Z = polyatomic organic anion) and the acid rain is nitric acid (dilute aqueous solution), write a balanced chemical equation for this reaction. Write the ionic and net ionic equations for the reaction.

35. The acidity of lemon juice (see Table 15.4) is due to the presence of 5% to 8% of the weak acid citric acid ($C_6H_8O_7$). Neutralizing a 0.421-g sample of citric acid required 30.60 mL of 0.215 M sodium hydroxide. (a) How many acidic hydrogens are present in citric acid? (b) Write a balanced net ionic equation that represents the reaction described above. [*Hint:* In part (a) compare the moles of hydroxide ion to the moles of citric acid.]

36. Picric acid is used in the production of explosives and matches, as well as in a variety of other industries. It has one acidic hydrogen that may be titrated with aqueous sodium hydroxide. A 0.2690-g sample of picric acid required 12.20 mL of 0.0961 M sodium hydroxide to neutralize the acidic hydrogen. What are the molar mass and molecular mass of picric acid?

✓ Chapter Quiz 15A (Sections 15.1 and 15.4)

You may use the periodic table.

1. Classify each of the following as an acid, a base, or an amphoteric substance according to the Brønsted–Lowry theory.

    a. $H_2SO_4(aq)$ b. $HSO_4^-(aq)$ c. $SO_4^{2-}(aq)$ d. $H_2O(l)$

2. If 28.70 mL of 0.375 M sodium hydroxide is necessary to neutralize 40.00 mL of hydrochloric acid solution to a phenolphthalein end point, calculate the molarity of the hydrochloric acid solution.

3. If 0.700 g of sodium carbonate was dissolved in water and the solution titrated with 37.50 mL of hydrochloric acid solution to a methyl orange end point, calculate the molarity of the hydrochloric acid solution. (Na = 23.0 amu, C = 12.0 amu, and O = 16.0 amu) (*Hint:* This process takes the carbonate ion to carbon dioxide.)

4. In the titration of 34.50 mL of a sodium hydroxide solution of unknown concentration, 25.60 mL of 0.108 N sulfuric acid solution was required to neutralize the sodium hydroxide in a reaction where both hydrogen ions of sulfuric acid react. Calculate the normality of the sodium hydroxide solution.

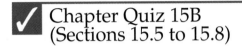

✓ Chapter Quiz 15B
(Sections 15.5 to 15.8)

You may use the periodic table, the rules for the solubility of inorganic substances in water, and the electromotive or activity series.

1. Calculate the pH and pOH for the following solutions:
   a. $[H^+] = 6.0 \times 10^{-8}$ mol/L   b. $[H^+] = 1.3 \times 10^{-7}$ mol/L

2. Calculate the hydrogen ion concentration in mol/L for each of the following solutions:
   a. pH = 4.50   b. pH = 9.60

3. Complete and balance the following chemical equations, writing them as total ionic equations and as net ionic equations. Indicate any precipitate by (s) and any gas by (g).
   a. $CdSO_4(aq) + NaOH(aq) \longrightarrow$ b. $Zn(s) + HCl(aq) \longrightarrow$
   c. $Zn^{2+}(aq) + H_2S(aq) \longrightarrow$

Compound (Formula: $CaCO_3$)	The Compound CALCIUM CARBONATE: From Chemistry Books to No Tummy Aches!

FROM

Daub Seese

Name: Calcium carbonate is the chemical name for the salt formed from calcium ions (Ca^{2+}) and carbonate ion ($CO_3{}^{2-}$).

Appearance: Calcium carbonate is a white, crystalline solid when pure but takes on other colors when trace materials are incorporated into the crystalline structure.

Occurrence: Calcium carbonate occurs in nature in a variety of forms. Limestone deposits are primarily calcium carbonate, as are marble and chalk. Geological structures like stalactites, stalagmites, coral reefs, and tufa are also deposits of solid calcium carbonate (see Figure 9.11).

TO

Source: Originally, pure calcium carbonate was prepared by allowing calcium chloride ($CaCl_2$) and sodium carbonate to react. However, the removal of all of the contaminating sodium chloride (NaCl) was difficult. Now, naturally occurring calcium carbonate deposits (limestone, chalk, and so on) are subjected to a three-step process that affords pure material. The products of the first step, lime (CaO) and carbon dioxide (CO_2), can be readily purified before they are recombined in step 3.

(1) $CaCO_3(s) \xrightarrow{\Delta} CaO(s) + CO_2(g)$

(2) $CaO(s) + H_2O(l) \longrightarrow Ca(OH)_2(aq)$

(3) $CO_2(g) + Ca(OH)_2(aq) \xrightarrow{\Delta} CaCO_3(s) + H_2O(g)$

The resulting product consists of uniformly sized crystals that are suitable for many industrial and manufacturing processes.

Uses: The most important use of calcium carbonate is in the paper industry, which consumes 75% of the calcium carbonate produced in the United States. Calcium carbonate acts as a filler in papermaking. By filling in between the fibers that constitute paper, the calcium carbonate renders the paper denser, smoother, and whiter. Such paper also receives printing ink more effectively.

Calcium carbonate is also used in both latex and alkyd paint formulations. Paints cover better and have a more uniform sheen as a result. In addition, inks, puttys, caulks, sealants, and some adhesives contain calcium carbonate as a filler and strengthening agent.

Unusual Facts: Since calcium carbonate is colorless, odorless, tasteless, and nontoxic, it appears in numerous products designed for cosmetic and pharmacological use by people. It is the active ingredient for neutralizing stomach acid in several antacids, it can be used as a mild abrasive in toothpastes, and it serves as a filler in some cosmetics. People who need supplemental calcium can use calcium carbonate for this purpose.

CHAPTER 16

Oxidation–Reduction Equations and Electrochemistry

A copper wire becomes encrusted with crystals of silver metal after immersion in a solution of silver nitrate. The bluish tinge of the solution results from the Cu^{2+} ions produced along with the silver metal in the oxidation–reduction reaction:

$$Cu(s) + 2\,Ag^+(aq) \longrightarrow Cu^{2+}(aq) + 2\,Ag(s)$$

GOALS FOR CHAPTER 16

1. To define oxidation, reduction, oxidizing agent, and reducing agent (Section 16.1).

2. To calculate the oxidation number of an element in a compound or ion (Section 16.2).

3. To balance chemical equations involving oxidation–reduction reactions using the oxidation number method and to identify the substances oxidized and reduced and the oxidizing and reducing agents (Section 16.3).

4. To balance chemical equations involving oxidation–reduction reactions using the ion electron method and to identify the substances oxidized and reduced and the oxidizing and reducing agents (Section 16.4).

5. To explain the operation of an electrolytic cell and to give examples of practical uses of such cells (Section 16.5).

6. To explain the operations of a voltaic cell including how electricity is produced in the cell (Section 16.6).

7. To explain the two basic types of batteries including the half-reactions for each and where these half-reactions occur—at the anode or cathode (Section 16.7).

8. To explain the process of corrosion including the half-reactions and where these reactions occur—at the anode or cathode (Section 16.7).

Countdown

ELEMENT	ATOMIC MASS UNITS (amu)
Mg	24.3
Cl	35.5

You may use the periodic table, the electromotive (activity) series, and the rules for the solubility of inorganic substances in water.

5. Give the abbreviation for and relative charge on each of the following subatomic particles (Section 4.3):
   a. proton   (p or p^+ and $+1$)
   b. neutron   (n or n^0 and 0)
   c. electron   (e^- and -1)

4. Diagram the structure of the following indicating the number of protons and neutrons in the nucleus and arranging the electrons in principal energy levels (Section 4.6 and 6.4):

Principal energy level

	1	2	3

a. $^{24}_{12}Mg$ $\left(\left(\begin{smallmatrix}12p\\12n\end{smallmatrix}\right)\right.$ $2e^-$ $8e^-$ $2e^-$ $\left.\right)$

b. $^{24}_{12}Mg^{2+}$ $\left(\left(\begin{smallmatrix}12p\\12n\end{smallmatrix}\right)\right.$ $2e^-$ $8e^-$ 2^+ $\left.\right)$

c. $^{35}_{17}Cl$ $\left(\left(\begin{smallmatrix}17p\\18n\end{smallmatrix}\right)\right.$ $2e^-$ $8e^-$ $7e^-$ $\left.\right)$

d. $^{35}_{17}Cl^-$ $\left(\left(\begin{smallmatrix}17p\\18n\end{smallmatrix}\right)\right.$ $2e^-$ $8e^-$ $8e^-$ 1^- $\left.\right)$

3. Predict the products using the electromotive (activity) series and the periodic table for the following single-replacement reaction equations and balance them (Section 9.8).
   a. $Mg(s) + HCl(aq) \longrightarrow$
        $[Mg(s) + 2\,HCl(aq) \longrightarrow MgCl_2 + H_2]$
   b. $Mg(s) + SnCl_2(aq) \longrightarrow$
        $[Mg(s) + SnCl_2(aq) \longrightarrow MgCl_2 + Sn]$

2. Calculate the number of kilograms of magnesium that can be produced when an electric current is passed through 4.00 kg of molten (melted) magnesium chloride according to the following balanced chemical equation (Section 10.4):

$$MgCl_2(l) \xrightarrow[\text{current}]{\text{electric}} Mg(s) + Cl_2(g) \quad (1.02\ \text{kg})$$

1. Calculate the number of liters of chlorine (STP) that can be produced when an electric current is passed through 2.50 kg of molten (melted) magnesium chloride according to the balanced chemical equation above.   (588 L)

An **oxidation–reduction (redox) reaction** is a type of chemical reaction in which one substance transfers electrons to another substance. Though you may have never heard of them, oxidation–reduction reactions are vitally important to your life. For example, the battery in your car works because of an oxidation–reduction reaction. The batteries that power your flashlight, calculator, and "boom box" also involve oxidation–reduction reactions. The bleach that keeps your clothes white, the antiseptics that keep you from infection, and the processes that develop your photos are oxidation–reduction reactions. The electroplating of silverware and jewelry and the production of computer chips also depend on oxidation–reduction reactions. Even more important, the reactions your body uses to metabolize food to energy include oxidation–reduction processes.

In this chapter, we will consider the nature of oxidation–reduction reactions and show how to balance equations involving such reactions. As we will see, these reactions sometimes require and sometimes produce electricity and are the basis for many modern products.

Oxidation–reduction (redox) reaction Type of chemical reaction in which one substance transfers electrons to another substance.

16.1 Oxidation and Reduction

In Chapter 15, we defined acid–base reactions as reactions involving the transfer of protons. In contrast, we can define *oxidation–reduction (redox) reactions* as those involving the transfer of electrons. **Oxidation**, which once referred only to reactions with oxygen, is now defined to include any chemical reaction in which a substance *loses electrons*. **Reduction** is any chemical reaction in which a substance *gains electrons*. One way of remembering these two terms is by using a mnemonic—a device designed to assist memory. There are two mnemonics you can use: (1) *LEO* the lion goes *GER* (grr . . .) (*L*oss of *E*lectrons = *O*xidation and *G*ain of *E*lectrons = *R*eduction; see Figure 16.1a), and (2) *OIL RIG* (*O*xidation *I*s *L*oss, *R*eduction *I*s *G*ain). Figure 16.1b applies this definition to a chemical example with calcium being oxidized and sulfur being reduced.

In a given chemical reaction, whenever a substance is oxidized, it loses electrons to another substance, which gains these electrons and is reduced. The substance being oxidized is called the **reducing agent (reductant)** because it produces a reduction in another substance. The substance being reduced is called the **oxidizing agent (oxidant)** because it produces oxidation in another substance.

Oxidation and reduction go hand in hand. They are analogous to the lending and borrowing of money. You cannot borrow money unless there is someone who will lend it to you. The same is true for oxidation and reduction. There can be no oxidation without reduction, and no reduction without oxidation. Electrons do not spontaneously appear; they must come from somewhere. *Oxidation always accompanies reduction, and reduction always accompanies oxidation.*

Oxidation Any chemical reaction in which a substance loses electrons.

Reduction Any chemical reaction in which a substance gains electrons.

Reducing agent (reductant) Substance in an oxidation–reduction chemical reaction that is oxidized (loses electrons).

Oxidizing agent (oxidant) Substance in an oxidation–reduction chemical reaction that is reduced (gains electrons).

16.2 Oxidation Numbers Revisited

One way to express the number of electrons lost or gained by a substance in an oxidation–reduction reaction is by assigning an oxidation number to an atom. In Chapter 6 we considered oxidation numbers, and we will now review them again.

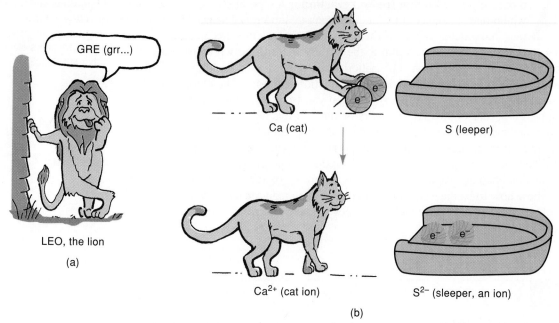

FIGURE 16.1

(a) **LEO**, the lion goes **GRE** (grr . . .). *Loss of Electrons =* Oxidation, and *Gain of Electrons =* Reduction. (b) Calcium has lost two electrons and is oxidized, while sulfur has gained two electrons and is reduced.

Oxidation number (or **oxidation state**)   A positive or negative whole number assigned to an element in a compound or ion. It is based on certain rules.

In Section 6.3 we defined an **oxidation number** (or **oxidation state**) as a positive or negative whole number assigned to an element in a compound or ion. This assignment is based on certain rules (see next page). An *oxidation number* can be zero, positive, or negative. An atom with a *zero oxidation number* is considered to have the same number of electrons as there are in an uncombined neutral atom. An atom with a *positive oxidation number* is considered to have fewer electrons than there are in an uncombined neutral atom. An atom with a *negative oxidation number* is considered to have more electrons than there are in an uncombined neutral atom.

Oxidation numbers are not inherent in a particular element. Rather, they are assigned so that chemists can use them as a method of electronic "bookkeeping." All atoms have an oxidation number of zero when they are uncombined. Their oxidation numbers change when a reaction causes them to lose or gain electrons. An atom is assigned a positive oxidation number if it loses electrons (remember, electrons are negatively charged and losing a negative charge makes an atom more positive). An atom is assigned a negative oxidation number if it gains electrons (becomes more negatively charged). Thus, we increase the oxidation number of an element when it is oxidized and decrease the oxidation number of an element when it is reduced. Figure 16.2 illustrates this relationship.

In Section 6.3 we gave certain rules for assigning oxidation numbers (ox. nos.) The following is a review of these rules which we will now use to calculate oxidation numbers.

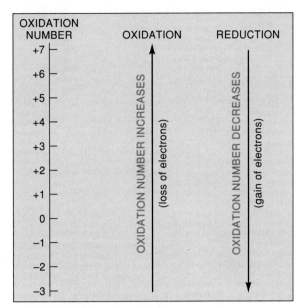

Figure 16.2
Oxidation and reduction, change in oxidation number, and electron transfer.

Rule	State of Element	Oxidation Number (ox. no.)	Example
1	Neutral compound	Sum of ox. nos. = 0	NaCl: +1(Na) + [−1(Cl)] = 0
2	Uncombined	Zero	Na^0: ox. no. = 0
3	Monatomic ion	Same as ionic charge	Na^+: ox. no. = +1
4	Polyatomic ion	Ionic charge = sum of ox. nos. on all atoms	ClO^-: +1(Cl) + [−2(O)] = −1
5	Metals combined with nonmetals	Positive	NaCl: Na^+, +1
	Nonmetals combined with metals	Negative	Cl^-, −1
6	Two nonmetals combined	More electronegative atom—negative	NO: N, +2; O, −2
7	Hydrogen	Usually +1, except hydrides −1	HCl: H, +1 NaH: H, −1
8	Oxygen	Usually −2, except peroxides −1	MgO: O, −2 H_2O_2: O, −1

EXAMPLE 16.1   Calculate the oxidation number for the element indicated in each of the following compounds or ions.

a. P in H_3PO_3

SOLUTION   The oxidation numbers (ox. no.) of H and O in the compound are +1 and −2 (see rules 7 and 8, respectively.) The sum of the oxidation numbers of all the elements in the compound must equal zero (see rule 1). Therefore,

$$3(+1) + (\text{ox. no. of P}) + 3(-2) = 0$$
$$+3 + (\text{ox. no. of P}) - 6 = 0$$
$$(\text{ox. no. of P}) - 3 = 0$$
$$\text{ox. no. of P} = +3 \quad Answer$$

b. P in P_4O_6

SOLUTION   The oxidation number of oxygen is -2 (see rule 8). The sum of the oxidation numbers of the elements must equal zero (see rule 1). There are *four* atoms of phosphorus, but we must solve for *one* atom of phosphorus. Therefore,

$$4(\text{ox. no. of P}) + 6(-2) = 0$$
$$4(\text{ox. no. of P}) - 12 = 0$$
$$4(\text{ox. no. of P}) = +12$$
$$\text{ox. no. of P} = +12/4$$
$$\text{ox. no. of P} = +3 \quad Answer$$

c. P in PO_2^-

SOLUTION   The oxidation number of oxygen is -2 (see rule 8), and the sum of the oxidation numbers of the elements in the ion must equal the charge on the ion which is -1 in this case (see rule 4). Therefore,

$$(\text{ox. no. of P}) + 2(-2) = -1$$
$$(\text{ox. no. of P}) - 4 = -1$$
$$\text{ox. no. of P} = +4 - 1$$
$$\text{ox. no. of P} = +3 \quad Answer$$

Study Exercise 16.1

Calculate the oxidation number for the element indicated in each of the following compounds or ions.

a. Cr in CrO_3 $(+6)$

b. Cr in CrO_4^{2-} $(+6)$

Work Problem 5   **c.** Cr in CrO_2^- $(+3)$

16.3   Balancing Oxidation– Reduction Equations by the Oxidation Number Method

We can balance oxidation–reduction equations by one of two methods: the *oxidation number method* or the *ion electron method*. The oxidation number method applies to both complete (or molecular) and ionic equations. We will show both types as examples.

     As we did in balancing complete equations (Section 9.3) and in writing ionic equations (Section 15.7), we will suggest guidelines to follow in balancing oxidation–reduction equations by the **oxidation number method**, a method of balancing oxidation–reduction reaction equations which uses the oxidation numbers of the elements involved in the reaction. Let's use the reaction between carbon and sulfuric acid to illustrate the guidelines.

Oxidation number method
Method of balancing oxidation–reduction reaction equations which uses the oxidation numbers of the elements involved in the reaction.

$$C(s) + H_2SO_4(aq) \xrightarrow{\Delta} CO_2(g) + SO_2(g) + H_2O(l) \quad \text{(unbalanced)}$$

1. Look over the equation and decide which elements *undergo* a *change* in *oxidation number*. At first, this step may involve calculating the oxidation number for each element in the equation, but after balancing a few oxidation–reduction equations, we should be able to recognize which elements undergo a change in oxidation number. In our example all hydrogens are $+1$ and all oxygens are -2. (Carbon has an oxidation number of zero in the free state.) Therefore, carbon changes from 0 to $+4$ and sulfur changes from $+6$ to $+4$.

2. Above each element that is oxidized or reduced, *write* its *oxidation number*. We may write the oxidation numbers below elements if we like, but we must be *consistent*. The example equation becomes

$$\overset{0}{C} + \overset{+6}{H_2SO_4} \xrightarrow{\Delta} \overset{+4}{CO_2} + \overset{+4}{SO_2} + H_2O \quad \text{(unbalanced)}$$

3. *Determine* the number of *electrons lost* or *gained* for each element undergoing a change in oxidation number. We can best do this by drawing an arrow from the element in the reactant to the element in the product and indicating the *number of electrons lost or gained* in the change above the arrows. A change in oxidation number from 0 to $+4$ involves a loss of four electrons (remember, electrons are negative). A change from $+6$ to $+4$ is a gain of two electrons. The example equation becomes

$$\overset{\text{loss of } 4e^-}{\overset{0}{C} + \overset{+6}{H_2SO_4} \xrightarrow{\Delta} \overset{+4}{CO_2} + \overset{+4}{SO_2} + H_2O} \quad \text{(unbalanced)}$$
gain of $2e^-$

> **Study Hint:** Four electrons must be *lost* to go from 0 to 4^+. Two electrons are *gained* in going from $+6$ to $+4$. *Electrons are negative!*

4. *Balance* the number of *electrons lost* and the number of electrons *gained* by placing coefficients in front of these numbers such that the *total number of electrons lost equals the total number of electrons gained*. Thus, in our example equation we must multiply the number of electrons gained (2) by 2 (in red) and the number of electrons lost (4) by 1 (in blue) so that the total number of electrons lost equals the total number of electrons gained (4).

$$\overset{1 \times (\text{loss of } 4e^-)}{\overset{0}{C} + \overset{+6}{H_2SO_4} \xrightarrow{\Delta} \overset{+4}{CO_2} + \overset{+4}{SO_2} + H_2O} \quad \text{(unbalanced)}$$
$2 \times (\text{gain of } 2e^-)$

5. *Place* these *coefficients in front* of the corresponding formulas of the reactants and complete the balancing of the equation. If we later have to double the coefficient of the oxidized or reduced substance, we must also double the coefficient of the reduced or oxidized substance, respectively. We must do this to make the *total number of electrons lost equal to the total number of electrons gained*. Our example equation becomes

$$\overset{1 \times (\text{loss of } 4e^-)}{\overset{0}{1C} + \overset{+6}{2H_2SO_4} \xrightarrow{\Delta} \overset{+4}{CO_2} + \overset{+4}{2SO_2} + 2H_2O} \quad \text{(unbalanced)}$$
$2 \times (\text{gain of } 2e^-)$

(Because the 1 is understood in front of the carbon, it can be deleted.)

6. Place a check (✔) above each atom on both sides of the equation to ensure that the equation is balanced. Also check the equation to see that the coefficients are in the lowest possible ratio. For ionic equations, the net *charges on both sides* of the equation must be *equal*. Hence, the answer to our example equation is

$$\overset{\checkmark}{C} + 2\overset{\checkmark\checkmark\checkmark}{H_2SO_4} \xrightarrow{\Delta} \overset{\checkmark\checkmark}{CO_2} + 2\overset{\checkmark\checkmark}{SO_2} + 2\overset{\checkmark\checkmark}{H_2O} \quad Answer$$

Carbon increases in oxidation number and loses electrons, and so it is oxidized and is the *reducing agent. Sulfur* (in H_2SO_4) decreases in oxidation number and gains electrons, and so H_2SO_4 is reduced and is the *oxidizing agent.*

To check your understanding of this method, work through the following examples. (In this text, because the products are often complicated, we will give the products of the oxidation–reduction reaction. Thus, you will be asked only to balance the equations. Also, to simplify writing the equation, we will normally not include the physical states in the equation.)

EXAMPLE 16.2 Balance the following equation using the oxidation number method:

$$KMnO_4 + HCl \longrightarrow MnCl_2 + Cl_2 + H_2O + KCl \quad \text{(unbalanced)}$$

SOLUTION   Apply guidelines 1 and 2 to this equation and place the oxidation numbers above the elements that change in oxidation number:

$$\overset{+7}{K}\overset{}{Mn}\overset{}{O_4} + \overset{-1}{H}\overset{}{Cl} \longrightarrow \overset{+2}{Mn}Cl_2 + \overset{0}{Cl_2} + H_2O + KCl \quad \text{(unbalanced)}$$

Following guideline 3, determine the number of electrons lost or gained by each element that has a change in oxidation number and write them above or below the arrow.

$$\begin{array}{c} \overset{\text{loss of } 1e^-}{\overbrace{}} \\ \overset{+7}{KMnO_4} + \overset{-1}{HCl} \longrightarrow \overset{+2}{MnCl_2} + \overset{0}{Cl_2} + H_2O + KCl \quad \text{(unbalanced)} \\ \underset{\text{gain of } 5e^-}{\underbrace{}} \end{array}$$

Now balance the number of electrons lost and gained (guideline 4). Quick examination seems to indicate that we need a 5 in front of the number of electrons lost, but this is not the case. Because we need an even number of Cl atoms on the right to be able to place a whole number in front of the Cl_2 molecule, we must use 10 HCl and 2 $KMnO_4$ on the left (guidelines 4 and 5):

$$\begin{array}{c} \overset{10 \times (\text{loss of } 1e^-)}{\overbrace{}} \\ \overset{+7}{2\,KMnO_4} + \overset{-1}{10\,HCl} \longrightarrow \overset{+2}{MnCl_2} + \overset{0}{Cl_2} + H_2O + KCl \quad \text{(unbalanced)} \\ \underset{2 \times (\text{gain of } 5e^-)}{\underbrace{}} \end{array}$$

Continue with guideline 5 and balance the K, Mn, O, and Cl_2 (molecule). The following equation results:

$$2\,KMnO_4 + 10\,HCl \longrightarrow 2\,MnCl_2 + 5\,Cl_2 + 8\,H_2O + 2\,KCl \quad \text{(unbalanced)}$$

At this point, note that everything is balanced but the H atoms (10 on the left and 16 on the right) and the Cl atoms (10 on the left and 16 on the right). Balance these by simply adding 6 *more* HCl molecules to the left side of the equation. The elements in these additional HCl molecules do not undergo any change in oxidation number and therefore are not involved in the reaction. The following balanced equation results:

Study Hint: To prevent confusing the terms "oxidation" and "reduction," use either one of the two mnemonics: (1) **LEO** the lion goes **GRE** (grr . . .) or (2) **OIL RIG**.

$$\checkmark\checkmark\checkmark \quad\quad \checkmark\checkmark \quad\quad \checkmark\checkmark \quad\quad \checkmark \quad\quad \checkmark\checkmark \quad\quad \checkmark\checkmark$$
$$2\,KMnO_4 \;+\; 16\,HCl \;\longrightarrow\; 2\,MnCl_2 \;+\; 5\,Cl_2 \;+\; 8\,H_2O \;+\; 2\,KCl \quad\quad \textit{Answer}$$

The *chlorine* in HCl increases in oxidation number and loses electrons, so HCl is oxidized and is the *reducing agent*. The *manganese* in $KMnO_4$ decreases in oxidation number and gains electrons, so $KMnO_4$ is reduced and is the *oxidizing agent*.

EXAMPLE 16.3 Balance the following ionic equation using the oxidation number method:

$$Ag \;+\; H^+ \;+\; NO_3^- \;\longrightarrow\; Ag^+ \;+\; NO \;+\; H_2O \quad \text{(unbalanced)}$$

SOLUTION   Applying guidelines 1 and 2, place the oxidation numbers of elements that change in oxidation number above the elements.

$$\begin{array}{cccc} 0 & +5 & +1 & +2 \\ Ag \;+\; H^+ \;+\; NO_3^- & \longrightarrow & Ag^+ \;+\; NO \;+\; H_2O \end{array} \quad \text{(unbalanced)}$$

Following guideline 3, determine the number of electrons lost or gained by each element that has a change in oxidation number and write the number above or below the arrow. Then balance the number of electrons lost and gained (guideline 4):

$$\begin{array}{c} 3 \times (\text{loss of 1e}^-) \\ \begin{array}{cccc} 0 & +5 & +1 & +2 \\ Ag \;+\; H^+ \;+\; NO_3^- & \longrightarrow & Ag^+ \;+\; NO \;+\; H_2O \end{array} \quad \text{(unbalanced)} \\ 1 \times (\text{gain of 3e}^-) \end{array}$$

Place these coefficients in front of the formulas and balance the equation by inspection (guideline 5):

$$\begin{array}{c} 3 \times (\text{loss of 1e}^-) \\ \begin{array}{cccc} 0 & +5 & +1 & +2 \\ 3\,Ag \;+\; 4\,H^+ \;+\; NO_3^- & \longrightarrow & 3\,Ag^+ \;+\; NO \;+\; 2\,H_2O \end{array} \quad \text{(balanced)} \\ 1 \times (\text{gain of 3e}^-) \end{array}$$

Check each atom on both sides of the equation and check the *charge* on both sides (guideline 6).

$$\checkmark \quad \checkmark \quad \checkmark\checkmark \quad\quad \checkmark \quad \checkmark\checkmark \quad \checkmark\checkmark$$
$$3\,Ag \;+\; 4\,H^+ \;+\; NO_3^- \;\longrightarrow\; 3\,Ag^+ \;+\; NO \;+\; 2\,H_2O \quad\quad \textit{Answer}$$
$$\text{charges:} \quad +4 \;+\; -1 \;=\; +3 \;=\; +3$$

The *silver* increases in oxidation number and loses electrons, so it is oxidized and is the *reducing agent*. The *nitrogen* in NO_3^- decreases in oxidation number and gains electrons, so NO_3^- is reduced and is the *oxidizing agent*.

EXAMPLE 16.4 Balance the following ionic equation using the oxidation number method.

$$I^- + Cr_2O_7^{2-} + H^+ \longrightarrow Cr^{3+} + I_2 + H_2O \quad \text{(unbalanced)}$$

SOLUTION Applying guidelines 1 and 2 to this ionic equation produces

$$\overset{-1}{I^-} + \overset{+6}{Cr_2O_7^{2-}} + H^+ \longrightarrow \overset{+3}{Cr^{3+}} + \overset{0}{I_2} + H_2O \quad \text{(unbalanced)}$$

Balance the total number of electrons lost and gained (guideline 3). Because there are 2 Cr atoms changing oxidation number, the total number of electrons gained by 2 Cr atoms is 6 electrons. Therefore, we must use a coefficient of 6 (guideline 4).

$$
\begin{array}{c}
\overbrace{}^{6 \times (\text{loss of } 1e^-)} \\
\overset{-1}{I^-} + \overset{+6}{Cr_2O_7^{2-}} + H^+ \longrightarrow \overset{+3}{Cr^{3+}} + \overset{0}{I_2} + H_2O \quad \text{(unbalanced)} \\
\underbrace{}_{\substack{1 \times (2 \text{ atoms} \times \text{gain} \\ \text{of } 3e^- \text{ per atom})}}
\end{array}
$$

Place these coefficients in front of the formulas and balance the equation by inspection (guideline 5):

$$
\begin{array}{c}
\overbrace{}^{6 \times (\text{loss of } 1e^-)} \\
\overset{-1}{6\,I^-} + \overset{+6}{Cr_2O_7^{2-}} + 14\,H^+ \longrightarrow \overset{+3}{2\,Cr^{3+}} + \overset{0}{3\,I_2} + 7\,H_2O \quad \text{(balanced)} \\
\underbrace{}_{\substack{1 \times (2 \text{ atoms} \times \text{gain} \\ \text{of } 3e^- \text{ per atom})}}
\end{array}
$$

Check each atom on both sides of the equation and check the *charge* on both sides (guideline 6).

$$6\,I^- + Cr_2O_7^{2-} + 14\,H^+ \longrightarrow 2\,Cr^{3+} + 3\,I_2 + 7\,H_2O \quad \text{\textit{Answer}}$$

charges: $-6 \; + \; (-2) \quad + \; (+14) = +6 = 2(+3) \; = \; +6$

The I^- increases in oxidation number and loses electrons, and so it is oxidized and is the *reducing agent*. The *chromium* in $Cr_2O_7^{2-}$ decreases in oxidation number and gains electrons, and so $Cr_2O_7^{2-}$ is reduced and is the *oxidizing agent*.

Study Exercise 16.2
(1) Balance the following equations using the oxidation number method.
(2) Indicate the substances oxidized and reduced and the oxidizing and reducing agent.

    a. $C + HNO_3 \longrightarrow CO_2 + NO_2 + H_2O$

                        [(1) $C + 4\,HNO_3 \longrightarrow CO_2 + 4\,NO_2 + 2\,H_2O$;
                        (2) C is oxidized and is the reducing agent; HNO_3 is reduced and is the oxidizing agent.]

    b. $Zn + H^+ + NO_3^- \longrightarrow Zn^{2+} + N_2O + H_2O$

                        [(1) $4\,Zn + 10\,H^+ + 2\,NO_3^- \longrightarrow 4\,Zn^{2+} + N_2O + 5\,H_2O$;
                        (2) Zn is oxidized and is the reducing agent; NO_3^- is reduced and is the oxidizing agent.]

Work Problems 6 through 9.

16.4 Balancing Oxidation–Reduction Equations by the Ion Electron Method

We can also balance oxidation–reduction equations by using the **ion electron** (or *half-reaction*) **method**. This is a method of balancing oxidation–reduction equations which separates the process into two partial reactions representing half-reactions: one describes the oxidation reaction, and the other describes the reduction reaction. The two partial equations are then added to produce a final balanced equation. Although we artificially divide the original reaction into two partial equations, these partial equations *do not take place alone; whenever oxidation occurs, so does reduction.*

As before, a few guidelines will help you balance oxidation–reduction reactions by the ion electron method. As we present the guidelines, we will look at an example equation for the reaction between the iron(II) ion and the permanganate ion to give the iron(III) ion and the manganese(II) ion.

$$Fe^{2+} + MnO_4^- \longrightarrow Fe^{3+} + Mn^{2+} \text{ in acid solution}$$

1. *Write* the equation in *net ionic form* (see Section 15.7) *without* attempting to balance it. We will typically need to refer to the solubility rules for inorganic substances to write an equation in net ionic form. This example is already in net ionic form:

   $$Fe^{2+} + MnO_4^- \longrightarrow Fe^{3+} + Mn^{2+} \text{ in acid solution}   \text{(unbalanced)}$$

2. *Determine* from the oxidation numbers the elements that undergo a *change* in *oxidation number* and then *write two partial equations:* an *oxidation half-reaction* and a *reduction half-reaction*. The example equation becomes

   $$\begin{array}{ll} (1) \quad Fe^{2+} \longrightarrow Fe^{3+} & \\ & \text{(unbalanced)} \\ (2) \quad MnO_4^- \longrightarrow Mn^{2+} & \end{array}$$

3. *Balance* the *atoms* on each side of the partial equations. In *acid* solutions, we may add H^+ ions and H_2O molecules. For each hydrogen atom (H) needed, we add a H^+ ion. For each oxygen atom (O) needed, we add a H_2O molecule, with 2 H^+ ions being shown on the other side of the partial equation. In *basic* solutions, we may add OH^- ions and H_2O molecules. For *each* hydrogen atom (H) needed, we add an H_2O molecule with an OH^- ion written on the other side of the partial equation. For *each* oxygen atom (O) needed, we add *two* OH^- ions with *one* H_2O molecule written on the other side of the partial equation. The following summarizes these additions in acid and base:

		Need	*Add*	
In acid:	H	H^+		
	O	H_2O	$\longrightarrow$	$2H^+$
In base:	H	H_2O	$\longrightarrow$	OH^-
	O	$2\,OH^-$	$\longrightarrow$	H_2O

   Now look at our example equation. The iron atoms are already balanced in equation (1). In equation (2) we need 4 O atoms on the products side, and

Ion electron method
Method of balancing oxidation reduction reaction equations which separates the process into two partial reactions representing half-reactions; one describes the oxidation reaction, and the other describes the reduction reaction.

so we add 4 H_2O to the products side and 8 H^+ to the reactants since the reaction is in acid solution.

(1) $Fe^{2+} \longrightarrow Fe^{3+}$ (unbalanced)

(2) $8\,H^+ + MnO_4^- \longrightarrow Mn^{2+} + 4\,H_2O$ (unbalanced)

4. *Balance* each partial equation *electrically* by adding electrons to the appropriate side of the equation so that the *charges* on both sides of the partial equation are *equal*. These two partial equations are defined as follows. The *oxidation* half-reaction equation, in which the reactant *loses* electrons, is written with the electrons on the *products* side; the *reduction* half-reaction equation, in which the reactant *gains* electrons, is written with the electrons on the *reactant* side.

$$\text{Oxidation:}\quad M \longrightarrow M^+ + 1e^-$$
$$\text{Reduction:}\quad X + 1e^- \longrightarrow X^-$$

We can balance our example equation electrically by adding $1e^-$ to the products of half-reaction (1) and $5e^-$ to the reactants of half-reaction (2). Equation (1) is the oxidation half-reaction because an electron is lost, and equation (2) is the reduction half-reaction because $5e^-$ are gained.

(1) *Oxidation:* $Fe^{2+} \longrightarrow Fe^{3+} + 1e^-$
　　 Charges: $+2 \;\; = \;\; +3 \; + (-1) = +2$

(2) *Reduction:* $8\,H^+ + MnO_4^- + 5e^- \longrightarrow Mn^{2+} + 4\,H_2O$
　　 Charges: $+8 \; + (-1) \;\; + (-5) = +2 = +2$

5. *Multiply* each *entire* partial equation by an appropriate number so that the *electrons lost in one partial equation* (oxidation half-reaction) *are equal to the electrons gained in the other partial equation* (reduction half-reaction). This converts the example equations to

(1) *Oxidation:* $5\,Fe^{2+} \longrightarrow 5\,Fe^{3+} + 5e^-$

(2) *Reduction:* $8\,H^+ + MnO_4^- + 5e^- \longrightarrow Mn^{2+} + 4\,H_2O$

6. *Add* the *two partial equations* and eliminate the electrons, ions, or water molecules that appear on both sides of the equation. The example becomes

$$5\,Fe^{2+} + 8\,H^+ + MnO_4^- + \cancel{5e^-} \longrightarrow 5\,Fe^{3+} + \cancel{5e^-} + Mn^2 + 4\,H_2O$$

Note how the $5e^-$ can be canceled from both sides of the equation.

7. *Place a check* (✔) above each atom on both sides of the equation to ensure that the equation is balanced. Also check the net charges on both sides of the equation to see that they are equal. Check the equation to see that the coefficients are the lowest possible ratios. Note how the charges on both sides of the example equation sum to $+17$. The balanced equation is

$$\overset{✔}{5\,Fe^{2+}} + \overset{✔}{8H^+} + \overset{✔✔}{MnO_4^-} \longrightarrow \overset{✔}{5\,Fe^{3+}} + \overset{✔}{Mn^{2+}} + \overset{✔✔}{4\,H_2O} \quad \text{(balanced)}$$

Charges: $5(+2) \; + (+8) + (-1) = +17 = 5(+3) + (+2) = +17$

In partial equation (1) the Fe^{2+} is oxidized, and so it is the *reducing agent*. In partial equation (2) the MnO_4^- is reduced, and so it is the *oxidizing agent*.

The following examples balance oxidation–reduction equations using the ion electron method.

EXAMPLE 16.5   Balance the following equation using the ion electron method.

$$Zn + HgO \longrightarrow ZnO_2{}^{2-} + Hg \quad \text{in basic solution} \quad \text{(unbalanced)}$$

SOLUTION   Excluding guideline 1, because the equation is already in ionic form, and proceeding to guideline 2, write the following two partial equations for the half-reactions:

$$(1) \quad Zn \longrightarrow ZnO_2{}^{2-}$$
$$\qquad\qquad\qquad\qquad \text{(unbalanced)}$$
$$(2) \quad HgO \longrightarrow Hg$$

Next, balance the atoms for each partial equation (guideline 3). In partial equation (1), two oxygen atoms are required in the reactants. Because the solution is basic, add 4 OH$^-$ ions to the reactants and 2 H$_2$O molecules to the products to balance the atoms. In partial equation (2), 1 oxygen atom is required in the products, so add 2 OH$^-$ ions to the products and 1 H$_2$O molecule to the reactants to balance the atoms.

$$(1) \quad Zn + 4\,OH^- \longrightarrow ZnO_2{}^{2-} + 2\,H_2O$$
$$(2) \quad HgO + H_2O \longrightarrow Hg + 2\,OH^-$$

Following guideline 4, balance the two partial equations electrically by adding electrons to the appropriate side. (Remember that a free metal has zero charge; Zn and Hg are neutral and so have zero oxidation numbers.) In partial equation (1), the electrons are lost; hence, this equation represents the *oxidation* half-reaction. In partial equation (2), electrons are gained; hence, this equation represents the *reduction* half-reaction.

$$(1) \quad \textit{Oxidation:} \quad Zn + 4\,OH^- \longrightarrow ZnO_2{}^{2-} + 2\,H_2O + 2e^-$$
$$\qquad\qquad \text{Charges:} \qquad\quad -4 = \qquad -2 \qquad + \qquad\quad (-2) = -4$$

$$(2) \quad \textit{Reduction:} \quad HgO + H_2O + 2e^- \longrightarrow Hg + 2\,OH^-$$
$$\qquad\qquad \text{Charges:} \qquad\qquad\qquad -2 = \qquad\quad -2$$

In the two partial equations, the number of electrons lost already equals the number of electrons gained, and so we do not need to apply guideline 5. Now add the two partial equations and eliminate the electrons from both sides of the equation (guideline 6).

$$Zn + 4\,OH^- + \cancel{2e^-} + HgO + H_2O \longrightarrow ZnO_2{}^{2-} + 2\,H_2O + \cancel{2e^-} + Hg + 2\,OH^-$$

Water molecules and OH$^-$ ions still appear on both sides of the equation. Eliminate the duplicates to give 2 OH$^-$ on the left side (4 OH$^-$ on the left minus 2 OH$^-$ on the right) and 1 H$_2$O on the right side (2 H$_2$O on the right minus 1 H$_2$O on the left):

$$Zn + 2\,OH^- + HgO \longrightarrow ZnO_2{}^{2-} + H_2O + Hg \quad \text{(balanced)}$$

Check each atom and the charge on both sides of the equation to obtain the final balanced equation according to guideline 7:

$$\overset{\checkmark}{}\quad\overset{\checkmark\checkmark}{}\quad\overset{\checkmark\checkmark}{}\quad\overset{\checkmark\checkmark}{}\quad\overset{\checkmark\checkmark}{}\;\overset{\checkmark}{}$$
$$Zn + 2\,OH^- + HgO \longrightarrow ZnO_2{}^{2-} + H_2O + Hg \qquad \textit{Answer}$$
$$\text{Charges: } -2 \qquad\qquad\qquad = -2$$

In partial equation (1), the Zn is oxidized, and so it is the *reducing agent*. In partial equation (2), the HgO is reduced, and so it is the *oxidizing agent*.

> **Study Hint:** Try balancing this equation by the oxidation number method. You should get the same answer both ways.

EXAMPLE 16.6 Balance the following example, using the ion electron method:

$$NaI + Fe_2(SO_4)_3 \longrightarrow I_2 + FeSO_4 + Na_2SO_4 \text{ in aqueous solution} \quad \text{(unbalanced)}$$

SOLUTION Applying guideline 1, write the equation in *net ionic* form *without* attempting to balance it. Refer to the solubility rules of inorganic substances in water on the inside back cover of the book. Note that the Na^+ and SO_4^{2-} ions can be completely removed from the equation because neither Na^+ or SO_4^{2-} changes in any way during the reaction. The following net ionic equation results:

$$Na^+ + I^- + Fe^{3+} + \cancel{SO_4^{2-}} \longrightarrow I_2 + Fe^{2+} + \cancel{SO_4^{2-}} + \cancel{Na^+} + \cancel{SO_4^{2-}}$$

Net ionic: $I^- + Fe^{3+} \longrightarrow I_2 + Fe^{2+}$ (unbalanced)

We make no attempt to balance the ions at this point. Rather we write two partial equations (guideline 2):

$$\begin{array}{l} (1)\ I^- \longrightarrow I_2 \\ (2)\ Fe^{3+} \longrightarrow Fe^{2+} \end{array} \quad \text{(unbalanced)}$$

Balance the atoms for each partial equation (guideline 3):

$$\begin{array}{l} (1)\ 2\,I^- \longrightarrow I_2 \\ (2)\ Fe^{3+} \longrightarrow Fe^{2+} \end{array} \quad \text{(unbalanced)}$$

Then balance these two equations electrically (guideline 4). In partial equation (1), electrons are lost; hence, this equation represents the *oxidation* half-reaction. In partial equation (2), electrons are gained; hence, this equation represents the *reduction* half-reaction.

(1) *Oxidation:* $2\,I^- \longrightarrow I_2 + 2e^-$
Charges: $\quad -2 \quad = \qquad\quad -2$

(2) *Reduction:* $Fe^{3+} + 1e^- \longrightarrow Fe^{2+}$
Charges: $\quad +3\ +(-1) = +2$

In the two partial equations, the number of electrons lost must be equal to the number of electrons gained (guideline 5), and so we need to multiply partial equation (2) by 2.

(1) *Oxidation:* $2\,I^- \longrightarrow I_2 + 2e^-$

(2) *Reduction:* $2\,Fe^{3+} + 2e^- \longrightarrow 2\,Fe^{2+}$

> **Study Hint:** Don't forget to multiply the *entire* partial equation by 2.

Add the two partial equations and eliminate the electrons on opposite sides of the equation (guideline 6).

$$2\,I^- + 2\,Fe^{3+} + \cancel{2e^-} \longrightarrow I_2 + \cancel{2e^-} + 2\,Fe^{2+}$$

$$2\,I^- + 2\,Fe^{3+} \longrightarrow I_2 + 2\,Fe^{2+} \quad \text{(balanced)}$$

Check each atom and the charge on both sides of the equation (guideline 7) to obtain the final balanced equation:

$$\overset{\checkmark}{2\,I^-} + \overset{\checkmark}{2\,Fe^{3+}} \longrightarrow \overset{\checkmark}{I_2} + \overset{\checkmark}{2\,Fe^{2+}} \quad \textit{Answer}$$

Charges: $-2 + 2(+3) = +4 = 2(+2) = +4$

In partial equation (1), the I^- is oxidized, and so it is the *reducing agent*. In partial equation (2), the Fe^{3+} is reduced, and so it is the *oxidizing agent*. Figure 16.3 illustrates this reaction. Note the characteristic brownish color of I_2 in aqueous solution.

FIGURE 16.3
Solution (left) of sodium iodide (NaI, Na^+, and I^-). When some filtered iron(III) sulfate solution $[Fe_2(SO_4)_3$, Fe^{3+}, and $SO_4^{2-}]$ are added to the sodium iodide solution, an oxidation–reduction reaction occurs and a reddish-yellow color (I_2) is formed (right). The balanced equation is $2\,NaI + Fe_2(SO_4)_3 \longrightarrow I_2 + 2\,FeSO_4 + Na_2SO_4$ (aqueous solution).

EXAMPLE 16.7 Balance the following equation using the ion electron method:

$$CrO_2 + ClO^- \longrightarrow CrO_4^{2-} + Cl^- \text{ in basic solution} \quad \text{(unbalanced)}$$

SOLUTION Excluding guideline 1 because the equation is already in ionic form, and proceeding to guideline 2, write the following two partial equations for the half-reactions.

$$(1)\ CrO_2 \longrightarrow CrO_4^{2-}$$
$$(2)\ ClO^- \longrightarrow Cl^- \quad \text{(unbalanced)}$$

Balance the atoms for each partial equation (guideline 3):

$$(1)\ CrO_2 + 4\,OH^- \longrightarrow CrO_4^{2-} + 2\,H_2O$$
$$(2)\ ClO^- + H_2O \longrightarrow Cl^- + 2\,OH^- \quad \text{(unbalanced)}$$

Balance these two equations electrically according to guideline 4. In partial equation (1), electrons are lost, and so this equation represents the *oxidation* half-reaction. In partial equation (2), electrons are gained; therefore, this equation represents the *reduction* half-reaction.

(1) *Oxidation:* $CrO_2 + 4\,OH^- \longrightarrow CrO_4^{2-} + 2\,H_2O + 2e^-$
  Charges: $-4 \quad = -2 \qquad + (-2) = -4$

(2) *Reduction:* $ClO^- + H_2O + 2e^- \longrightarrow Cl^- + 2\,OH^-$
  Charges: $-1 \quad + \qquad -2 = -3 = -1 + (-2) = -3$

In the two partial equations, the number of electrons lost already equals the number of electrons gained, and so we do not need to apply guideline 5. Now add the two partial equations and eliminate the electrons from both sides of the equation (guideline 6).

$$CrO_2 + 4\,OH^- + ClO^- + H_2O + \cancel{2e^-} \longrightarrow$$
$$CrO_4^{2-} + 2\,H_2O + \cancel{2e^-} + Cl^- + 2\,OH^-$$

Eliminate the duplicates to give 2 OH^- on the left side (4 OH^- on the left minus 2 OH^- on the right) and 1 H_2O on the right side (2 H_2O on the right minus 1 H_2O on the left) to give the following equation:

$$CrO_2 + 2\,OH^- + ClO^- \longrightarrow CrO_4^{2-} + H_2O + Cl^- \quad \text{(balanced)}$$

Check each atom and the charge on both sides of the equation to obtain the final balanced equation according to guideline 7:

$$\overset{\checkmark\checkmark}{CrO_2} + \overset{\checkmark\checkmark}{2\,OH^-} + \overset{\checkmark\checkmark}{ClO^-} \longrightarrow \overset{\checkmark\checkmark}{CrO_4^{2-}} + \overset{\checkmark\checkmark}{H_2O^-} + \overset{\checkmark}{Cl^-} \quad \textit{Answer}$$

Charges: $-2 \quad + (-1) = -3 = (-2) + \quad (-1) = -3$

In partial equation (1), the CrO_2 is oxidized, and so it is the *reducing agent*. In partial equation (2), the ClO^- is reduced, and so it is the *oxidizing agent*.

Study Exercise 16.3
(1) Balance the following equations using the ion electron method.
(2) Indicate the substance oxidized and reduced and the oxidizing and reducing agents.

a. $Br_2 + SO_2 \longrightarrow Br^- + SO_4^{2-}$ in acid solution

$$[(1)\ Br_2 + SO_2 + 2\,H_2O \longrightarrow 2\,Br^- + SO_4^{2-} + 4\,H^+;$$
(2) Br_2 is reduced and is the oxidizing agent; SO_2 is oxidized and is the reducing agent.]

b. $S^{2-} + MnO_4^- \longrightarrow MnO_2(s) + S$ in basic solution

$$[(1)\ 3\,S^{2-} + 2\,MnO_4^- + 4\,H_2O \longrightarrow 2\,MnO_2(s) + 3\,S + 8\,OH^-;$$
(2) S^{2-} is oxidized and is the reducing agent; MnO_4^- is reduced and is the oxidizing agent.]

Work Problems 10 through 13.

16.5 Electrochemistry and Electrolytic Cells

So far in this chapter, we have dealt with oxidation and reduction in terms of combining two chemicals. However, the transfer of electrons also produces something we use everyday: electricity. Indeed, electricity is defined as the flow of electrons, and so it is not surprising that there exists a field called **electrochemistry**, which is concerned with the relationship between electrical energy and chemical energy. Electrochemistry can involve either (1) the use of electricity to *cause* a chemical reaction or (2) the production of an electric current *by* a chemical reaction. Both of these involve oxidation and reduction.

In Section 15.6, we noted that electrolytes conduct electric current. **Electrolytic cells** use electrolytes and electrical energy to produce certain chemical reactions that would otherwise not occur. Figure 16.4 shows a simple electrolytic cell that still contains all the primary elements of such an apparatus: direct current, two electrodes, and an electrolyte solution (the X^+ and Y^- ions in the figure).

In any electrolytic cell, the direct-current source (battery or generator) withdraws electrons from one electrode and pushes them to the other. One electrode becomes negatively charged in this process and is called the **cathode**. The positive ions of the electrolyte (X^+) migrate to this cathode and accept electrons from it. Thus, *reduction occurs at the cathode* ($X^+ + e^- \longrightarrow X^0$). The other electrode, which becomes positively charged by the action of the direct-current source, is

Electrochemistry Study of the relationship between electrical and chemical energy.

Electrolytic cell Device that uses electrolytes and electrical energy to produce chemical reactions that would otherwise not occur.

Cathode Negatively charged electrode in an electrolytic cell; it is the site of reduction reactions since it attracts the positively charged ions in a solution.

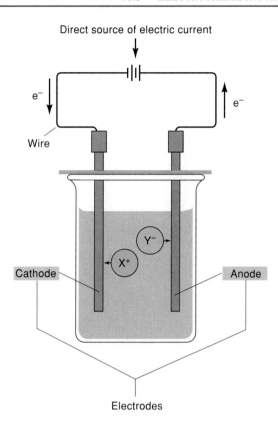

Direct source of electric current

Wire

e^-

e^-

Cathode

Anode

Y^-

X^+

Electrodes

FIGURE 16.4
Simple electrolytic cell.

called the **anode**. The negative ions of the electrolyte (Y^-) migrate to this anode and transfer electrons to the anode, thus completing the electric circuit. Hence, *oxidation occurs at the anode* ($Y^- \longrightarrow Y^0 + e^-$).

We can express this flow of electrons through the wire (external circuit) from the anode to the cathode as two half-cell reactions:

$$Cathode:\quad X^+ + 1e^- \longrightarrow \underset{\substack{\text{deposited at} \\ \text{cathode}}}{X^0} \quad \text{(reduction)} \qquad (16.1)$$

$$Anode:\quad Y^- \longrightarrow \underset{\substack{\text{deposited at} \\ \text{anode}}}{Y^0} + 1e^- \quad \text{(oxidation)} \qquad (16.2)$$

One application of an electrolytic cell is the electroplating of silver onto utensils (knives, forks, and spoons) made of other metals. As Figure 16.5 shows, the utensil acts as the cathode. Pure silver is the anode, and the electrolyte is a soluble silver salt. The following equations illustrate the two half-reactions involved:

$$Cathode:\quad \underset{\text{electrolyte}}{Ag^+} + 1e^- \longrightarrow \underset{\substack{\text{deposited} \\ \text{on utensil}}}{Ag} \quad \text{(reduction)} \qquad (16.3)$$

$$Anode:\quad \underset{\text{pure silver}}{Ag} \longrightarrow \underset{\text{electrolyte}}{Ag^+} + 1e^- \quad \text{(oxidation)} \qquad (16.4)$$

The silver ions produced at the anode (lose electrons–oxidation) migrate to the cathode (gain electrons–reduction), forming a deposit of silver on the utensil. The

Anode   Positively charged electrode in an electrolytic cell; it is the site of oxidation reactions, since it attracts the negatively charged ions in a solution.

Study Hint: At the cathode *reduction* occurs; at the *anode oxidation* occurs. A mnemonic for this is **CRAO**, the crab!

FIGURE 16.5
Electroplating a spoon. The spoon acts as the cathode, and pure silver acts as the anode. Reduction occurs at the *cathode*, and oxidation occurs at the *anode*.

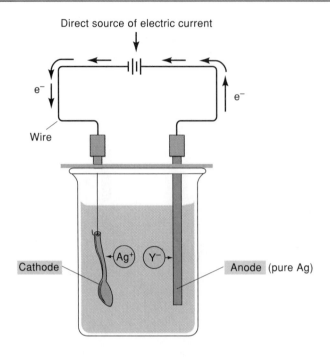

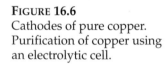

negative ions (Y^-) of the electrolyte (soluble silver salt) migrate to the anode, accompanying the production of the silver ions.

Similarly, "tin" cans can be made by electroplating a *thin* film of tin on a steel can or by dipping a steel can in molten tin. (Tin cans consist of *very little* tin.)

In addition, electrolytic cells are crucial to the industrial purification of copper and several other metals including aluminum, calcium, magnesium, potassium, and sodium. The impure copper acts as the anode, and a thin plate of pure copper acts as the cathode. Copper(II) sulfate acts as the electrolyte. The pure copper that forms on the cathode, as shown in Figure 16.6, is more than 99.9% copper. Other metals

FIGURE 16.6
Cathodes of pure copper. Purification of copper using an electrolytic cell.

found in copper ores, such as silver, gold, and platinum, are deposited as a *sludge* at the bottom of the cell. These metals are further refined, and their value makes the entire operation quite economical. Because this final step in purification requires electricity, cheap electric power must be available at or near the electrolytic cell plant site.

16.6 Electrochemistry and Voltaic (Galvanic) Cells

Unlike electrolytic cells, which use electricity, **voltaic (galvanic) cells** use oxidation–reduction reactions to *produce electrical energy*. These cells were named after two Italian physicists, Count Alessandro Volta (1745–1827) and Luigi (or Aloisio) Galvani (1737–1798), who contributed to the basic development of the cell. Like an electrolytic cell, a voltaic cell consists of two electrodes, with *reduction* occurring at the *cathode, oxidation* occurring at the *anode,* and electrons flowing through a wire *from the anode to the cathode.*

The reaction of zinc metal with copper(II) ions is a simple example of a voltaic or galvanic cell being used to produce electricity. If a piece of zinc metal is immersed in a solution of copper(II) ions, a reaction takes place immediately. The energy released as this reaction proceeds takes the form of heat.

$$Zn + Cu^{2+} \longrightarrow Zn^{2+} + Cu + heat \qquad (16.5)$$

If the reaction is carried out by *not* allowing the zinc metal to come in *direct contact* with the copper(II) ions, the energy produced will take the form of electrical energy (electricity). For example, in Figure 16.7a, a solution of zinc ions is *separated* from a solution containing copper(II) ions by a *porous partition* through which ions may diffuse. These two solutions can also be connected by a *salt bridge*, consisting of a solution of some electrolyte such as potassium sulfate, as Figure 16.7b shows.

No reaction takes place until the switch in the external circuit is closed. The electricity produced is enough to make a light bulb glow. Then Zn goes into solu-

Voltaic (galvanic) cells   Device that uses oxidation–reduction reactions to produce electricity; commonly called batteries.

Study Hint: Note the difference between an electrolytic cell and a galvanic cell. An electrolytic cell *uses* electric current to produce a reaction that would not occur otherwise; a galvanic cell *produces* electricity by an oxidation–reduction reaction.

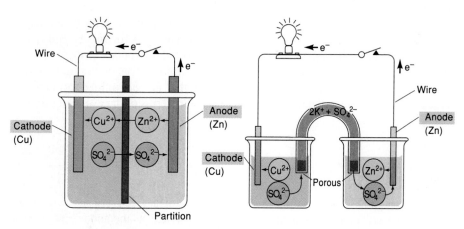

(a) Porous partition   (b) Salt bridge

FIGURE 16.7
Simple voltaic or galvanic cell: (a) a porous partition separates the two solutions of zinc sulfate and copper(II) sulfate; (b) a salt bridge of potassium sulfate connects the two solutions of zinc sulfate and copper(II) sulfate.

tion as Zn^{2+} ions, and the electrons deposited on the zinc electrode travel along the wire to the copper electrode. There, electrons are accepted by Cu^{2+} ions at the surface of the electrode (that is, copper metal is deposited on the copper electrode). At the zinc electrode, electrons are lost (*oxidation*), and so the zinc electrode is the *anode*. At the copper electrode, electrons are gained (*reduction*), and so the copper electrode is the *cathode*. The following equations give the two half-reactions:

$$\underset{\text{metal}}{Zn} \longrightarrow \underset{\text{electrolyte}}{Zn^{2+}} + 2e^- \qquad \text{(oxidation)} \qquad (16.6)$$

with *Anode:* label.

$$\textit{Cathode:} \quad \underset{\text{electrolyte}}{Cu^{2+}} + 2e^- \longrightarrow \underset{\substack{\text{metal} \\ \text{deposited}}}{Cu} \qquad \text{(reduction)} \qquad (16.7)$$

The sum of these two half-cell equations represents the overall cell reaction given in the equation.

$$Zn + Cu^{2+} \longrightarrow Zn^{2+} + Cu + \text{electricity} \qquad (16.8)$$

Because the zinc is going into solution and the copper is coming out, there should be as few zinc ions and as many copper(II) ions in solution as possible to prevent the cell reaction from slowing down because of the abundance or scarcity of certain ions. The porous partition or salt bridge allows the excess negative ions accumulating in the copper compartment to neutralize the excess positive ions accumulating in the zinc compartment.

Note the key difference between the two types of reactions between zinc and copper(II) ions. When zinc is allowed to come into contact with copper(II) ions [the first case, Equation (16.5)], the electrons are transferred *directly* from a zinc atom to a copper(II) ion in the solution. When zinc is *not* allowed to be in direct contact with copper(II) ions [the second case, Equation (16.8)], the transfer of electrons takes place through an external wire and electricity is generated.

16.7 Voltaic Cells in Practical Use

Although voltaic cells may seem rather abstract, they are very much a part of your everyday life. Dry cells, such as a flashlight battery, and lead storage batteries, like those found in your car, are voltaic cells.

Primary cell Form of voltaic cell (battery) that may not be reused after it discharges and is typically thrown away.

Secondary cell Form of voltaic cell (battery) that may be recharged and used over and over again.

There are two basic types of batteries. A **primary cell** is a battery that may not be reused after it discharges and is typically thrown away. A **secondary cell** is a battery that may be recharged and used over and over again. You are probably most familiar with primary cells, but secondary cells will find increasing use in our lives as society addresses the trash disposal problems that are beginning to plague our cities. Secondary cells make a great deal of sense in terms of the efficient use of our resources.

Primary Cells

The most popular battery is the flashlight battery or dry cell. As shown in Figure 16.8a, a *dry cell* consists of a zinc container, which acts as the anode, and a carbon

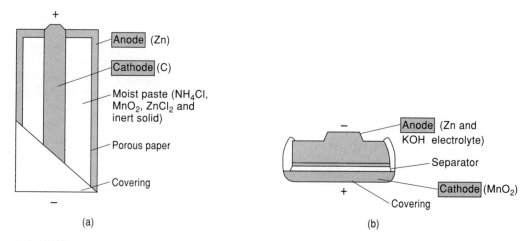

FIGURE 16.8
Cutaway view of (a) a flashlight battery or dry cell and (b) an alkaline battery.

cathode. The electrolyte is a moist paste of ammonium chloride, manganese(IV) oxide, and zinc chloride. The following two equations give the half-cell reactions:

Anode: $Zn \longrightarrow Zn^{2+} + 2e^-$ (oxidation)   (16.9)

Cathode: $2\,MnO_2 + 2\,NH_4^+ + 2e^- \longrightarrow$

$$H_2O + Mn_2O_3 + 2\,NH_3 \quad \text{(reduction)} \qquad (16.10)$$

The anode is separated from the moist paste of the electrolyte by a porous paper or plastic film which prevents the electrons given off at the anode from entering the moist paste directly. Instead, the electrons pass through an attached circuit and reenter at the carbon electrode. If the manganese(IV) oxide were absent, the cathode reaction would be

$$2\,NH_4^+ + 2e^- \longrightarrow 2\,NH_3 + H_2 \quad \text{(reduction)} \qquad (16.11)$$

The liberated hydrogen bubbles would cover and insulate the carbon cathode from the mixture of ammonium chloride and zinc chloride, and the reaction would soon stop. But the presence of the manganese(IV) oxide prevents the formation of hydrogen, making the dry cell a practical and long-lived source of direct-current electricity. The overall cell equation is

$$Zn + 2\,MnO_2 + 2\,NH_4^+ \longrightarrow Zn^{2+} + H_2O + Mn_2O_3 + 2\,NH_3 \quad (16.12)$$

This type of battery has a tendency to leak.

The alkaline battery (see Figure 16.8b) has become increasingly popular as an alternative power supply over the past two decades. The *alkaline cell* has a longer lifetime and produces a more constant current. The electrolyte paste in an alkaline cell contains KOH and is basic, in contrast to the dry cell, which contains NH_4Cl and is acidic. The half-cell reactions for the alkaline cell are

Anode: $Zn + 2\,OH^- \longrightarrow Zn(OH)_2 + 2e^-$ (oxidation)   (16.13)

Cathode: $2\,MnO_2 + H_2O + 2e^- \longrightarrow Mn_2O_3 + 2\,OH^-$ (reduction)   (16.14)

The alkaline battery is used in cameras, calculators, and watches. It does not leak.

Secondary Cells

The battery in your car, a *lead storage battery*, is a good example of a secondary cell. This battery provides the electric power needed to start the engine of the car. If this battery were not rechargeable, you would have to replace it on a regular basis as it ran down. Instead, the alternator on the engine recharges the battery as you drive by converting some of the mechanical energy from the engine to electrical energy. The alternator forces a current through the battery in the reverse direction, charging the battery. As shown in Figure 16.9, a lead storage battery consists of a lead anode, a lead(IV) oxide cathode, and an electrolyte of dilute sulfuric acid (battery acid). The half-cell reactions are given in the following two equations:

Anode: $Pb + SO_4^{2-} \longrightarrow PbSO_4 + 2e^-$ (oxidation)   (16.15)

Cathode: $PbO_2 + 4H^+ + SO_4^{2-} + 2e^- \longrightarrow PbSO_4 + 2H_2O$ (reduction) (16.16)

The concentration of the sulfate ions decreases during discharge as lead(II) sulfate forms at both electrodes. This decrease in concentration of the sulfate ion can be determined by measuring the specific gravity (see Section 2.9) of the electrolyte, dilute sulfuric acid (battery acid), using a *hydrometer* (see Figure 16.10).

The white solid that forms at the exterior terminals is lead(II) sulfate. This ionic solid can interfere with conduction and should be brushed away. Some "weekend" mechanics coat the battery terminals with a slurry of automatic transmission fluid and baking soda (sodium hydrogen carbonate). The transmission fluid maintains good electrical contact, and the baking soda neutralizes any vapors of sulfuric acid (the electrolyte) that may be near the terminals.

The overall cell reaction for the lead storage battery is

$$Pb + PbO_2 + 4H^+ + 2SO_4^{2-} \longrightarrow 2PbSO_4 + 2H_2O \qquad (16.17)$$

If they are ever to be practical, electrical cars will probably have to use secondary cells such as the lead storage battery as their power source. Development of

FIGURE 16.9
One cell of a lead storage battery.

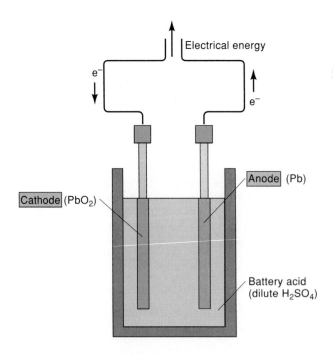

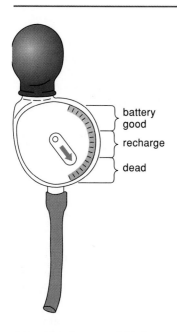

FIGURE 16.10
A hydrometer used to measure the specific gravity of dilute sulfuric acid (battery acid). The specific gravity of dilute surfuric acid in a fully charged battery is between 1.25 and 1.30. As the discharge occurs, the specific gravity decreases to less than 1.25 because of the increase in water and decrease in dissolved sulfate ions (see Equation 16.16; specific gravity of water $= 1.00$). A specific gravity around 1.20 means that the battery needs to be recharged.

battery good

recharge

dead

this technology could help to solve air pollution problems and to reduce the world's dependence on petroleum as an energy source.

A second type of secondary cell is the *nickel–cadmium (NiCad) battery*. Batteries you recharge at home in a simple recharger are nickel–cadmium cells. The anode is cadmium metal, and the cathode is nickel(IV) oxide. The half-cell reactions for this battery are

Anode: $\quad Cd + 2\,OH^- \longrightarrow Cd(OH)_2 + 2e^- \quad$ (oxidation) $\qquad$ (16.18)

Cathode: $\quad NiO_2 + 2\,H_2O + 2e^- \longrightarrow Ni(OH)_2 + 2\,OH^- \quad$ (reduction) $\qquad$ (16.19)

Figure 16.11 shows various types of common batteries, both primary and secondary cells.

FIGURE 16.11
A variety of common batteries. *Primary cells* like the dry cell (blue) or the alkaline cells (gold, silver, and hearing aid size) are discarded after they have run down. *Secondary cells* like the lead storage battery from your car and nickel–cadmium cells (black, white, and yellow) may be recharged numerous times and reused. Rechargeable batteries for everyday uses are a much more efficient way to use batteries in your home. In the long run, it's cheaper, too!

Study Exercise 16.4

A fuel cell is another type of voltaic cell; in it the reactants are continuously supplied and the products continuously removed. Such cells were used on the Apollo space missions to the moon. The fuel cell is based on the conversion of oxygen gas to hydroxide ions and hydrogen gas to water with a potassium hydroxide solution as the electrolyte. (1) Write the balanced oxidation and reduction half-reactions associated with the fuel cell, and (2) indicate whether these reactions occur at the cathode or the anode.

$$[(1) \; \textit{Reduction:} \; O_2(g) \; + \; 2\,H_2O \; + \; 4e^- \; \longrightarrow \; 4\,OH^-$$
$$\textit{Oxidation:} \; H_2(g) \; + \; 2\,OH^- \; \longrightarrow \; 2\,H_2O \; + \; 2\,e^-$$

Work Problem 14.

$$(2) \; \text{Reduction at the cathode; oxidation at the anode]}$$

16.8 Corrosion

Corrosion Process in which a metal is eaten away (oxidation) by a chemical reaction with oxygen to give a product that lacks the structural properties of the original metal.

Not all electrochemical processes are as useful to society as batteries. Indeed, if your car is in the same condition as the one shown in Figure 16.12, you may view rusting—what chemists call corrosion—as very detrimental to your lifestyle. **Corrosion** is the process in which a metal is eaten away (oxidation) by a chemical reaction with oxygen to give a product that lacks the structural properties of the original metal. The corrosion process involves contact of the metal with other substances, particularly oxygen, that react with the metal to give the metal oxide. The rusting of iron is the most common example, and it costs society billions of dollars a year to protect steel from corrosion and to replace parts that are damaged by corrosion.

Iron is oxidized by oxygen from the atmosphere to form hydrated iron(III) oxide, $Fe_2O_3 \cdot x\,H_2O$, where the value of x can vary. This iron(III) oxide is the reddish-brown material that we call *rust*. The oxidation–reduction processes in corrosion are

$$\textit{Anode:} \quad Fe \; \longrightarrow \; Fe^{2+} \; + \; 2e^- \quad \text{(oxidation)} \tag{16.20}$$

$$\textit{Cathode:} \quad O_2 \; + \; 4\,H^+ \; + \; 4e^- \; \longrightarrow \; 2\,H_2O \quad \text{(reduction)} \tag{16.21}$$

In corrosion, the electrons flow from the point where corrosion occurs (anode) to another part of the steel where oxygen and moisture come into contact with the steel (cathode). Thus, iron atoms are removed from the metal (as Fe^{2+}) at the anode

Figure 16.12
The corrosion of steel exacts an enormous financial cost to society.

as oxygen gas is reduced at the cathode. The Fe^{2+} ions formed at the anode are further oxidized to Fe^{3+} by more oxygen, and $\mathbf{Fe_2O_3 \cdot \textit{x} \ H_2O}$ forms. The result is that the steel at the anode site is eaten away and the metal loses its strength. Salts used to melt ice on highways in the winter (NaCl or $CaCl_2$) act as electrolytes to make the conduction process more efficient and thus increase the rate of corrosion.

To protect steel or iron against corrosion, a protective layer of paint or another metal [zinc (galvanized iron) or tin] is placed on the steel to minimize its contact with moisture and oxygen. Cars also are often undercoated with a layer of paint to reduce the amount of corrosion in winter.

CHEMISTRY OF THE ATMOSPHERE

Oxidation–Reduction Processes and Acid Rain

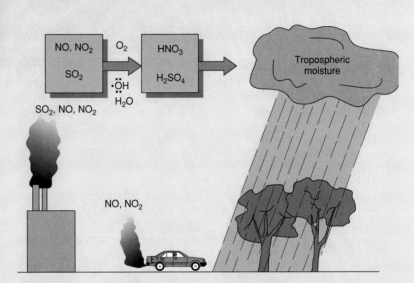

Acid rain involves the conversion of pollutants sulfur dioxide, nitrogen oxide, and nitrogen dioxide to sulfuric acid and nitric acid, followed by the return of these acidic materials to the earth dissolved in rain, snow, or fog.

In the essay at the end of Chapter 15, we introduced the basic concept of acid rain. In brief, sulfur dioxide, nitrogen oxide, and nitrogen dioxide—waste products from human activities—are released into the atmosphere where they are converted into two strongly corrosive materials: sulfuric acid (H_2SO_4) and nitric acid (HNO_3). These acids then dissolve in atmospheric water or ice crystals and return to earth as acid precipitation. On the earth's surface, the acid rain can harm plant and animal life, promote the corrosion of steel equipment, and erode buildings and stonework.

But how are sulfur dioxide, nitrogen oxide, and nitrogen dioxide transformed into sulfuric acid and

nitric acid? A quick calculation of oxidation numbers indicates that an oxidation–reduction process must be responsible, since the sulfur and nitrogen have different oxidation numbers in the waste gases and resulting acids.

While the chemistry is somewhat complex and still not fully understood, scientists believe that these oxidation processes are promoted by reactive species in the atmosphere called *radicals*. *Radicals* are fragments of molecules that do not have stable octets of electrons about each atom. Rather, one atom possesses an *unpaired electron* which renders the species unstable and highly reactive. Radicals react with stable molecules in an attempt to gain an additional electron and fill their octets. One of the most im-

OXIDATION NUMBER	REACTANT		PRODUCT	OXIDATION NUMBER
+4	SO_2	⇒	H_2SO_4	+6
+2	NO	⇒	HNO_3	+5
+4	NO_2	⇒	HNO_3	+5

portant radical species in the atmosphere is the hydroxyl radical, $\cdot\ddot{\text{O}}$—H . (*Caution:* Note that these are hydroxyl *radicals*, **not** hydroxide ions, OH^-.) Hydroxyl radicals are formed in the troposphere in small concentrations by the following process:

$$O_3(g) + \text{ultraviolet light} \longrightarrow$$
$$O(g) + O_2(g)$$

$$O(g) + H_2O(g) \longrightarrow 2\cdot\ddot{\text{O}}H(g)$$

Hydroxyl and other radicals are believed to promote the oxidation of sulfur dioxide to sulfuric acid and nitrogen oxides (NO and NO_2) to nitric acid:

$$SO_2(g) + \cdot\ddot{\text{O}}H \longrightarrow HOSO_2$$

$$HOSO_2 + O_2 \longrightarrow$$
$$SO_3 + \cdot\ddot{\text{O}}—\ddot{\text{O}}—H$$

$$SO_3 + H_2O \longrightarrow H_2SO_4$$

$$NO + \cdot\ddot{\text{O}}—\ddot{\text{O}}—H \longrightarrow$$
$$NO_2 + \cdot\ddot{\text{O}}H$$

$$NO_2 + \cdot\ddot{\text{O}}H \longrightarrow HNO_3$$

The industrial aspects of modern life have a cost. Emissions of NO, NO₂, and SO₂ from industrial plants produce acidic materials in the atmosphere that extensively damage plant and animal life hundreds of miles away.

Once this conversion is accomplished in the atmosphere, all that remains is for the acid materials to return to earth in some way. Acidifying agents may settle back to earth as "dry deposition" (droplets; see essay at the end of Chapter 15) or they may dissolve in atmospheric moisture and return in precipitation (rain, snow, or fog). The overall process is summarized in the figure on the previous page.

✓ Summary

Oxidation–reduction reactions all involve the transfer of electrons. A substance is *oxidized* (acts as reducing agent) when it loses electrons. A substance is *reduced* (acts as an oxidizing agent) when it gains electrons (Section 16.1).

To make it easier to tell whether a substance has been oxidized or reduced, chemists assign *oxidation numbers* to the elements in a compound or ion. When the oxidation number of an element increases (loss of electrons), the element and hence the compound or ion undergoes *oxidation*. When the oxidation number decreases (gain of electrons), the element and hence the compound or ion undergoes *reduction* (Section 16.2).

Equations for complete and ionic oxidation–reduction reactions can be quite complex and difficult to balance. Two methods are presented in this chapter, the *oxidation number method* (Section 16.3) and the *ion electron method* (Section 16.4). Numerous examples of each method are presented.

Electrochemistry is the study of the relationship between oxidation–reduction reactions and electricity. Among the various types of electrochemical cells in use are electrolytic and voltaic cells. *Electrolytic cells* are cells that use electric current to produce a reaction that normally does not proceed (Section 16.5). *Voltaic* (or *galvanic*) *cells* are cells that harness a reaction that wants to proceed and use it to generate an electric current (Section 16.6). In everyday life, voltaic cells serve as primary (disposable) cells, such as flashlight (dry cell) batteries, and as secondary (rechargeable) cells, such as lead storage batteries in cars (Section 16.7).

Another electrochemical process—corrosion—costs Americans and American businesses billions of dollars each year as oxygen reacts with metal to give metal oxides (Section 16.8).

 # Exercises

1. Define or explain the following terms (the number in parentheses refers to the section number in the text where the term is mentioned.)

   a. oxidation–reduction (redox) (Introduction to Chapter 16)

   b. oxidation (16.1)

   c. reduction (16.1)

   d. reducing agent (reductant) (16.1)

   e. oxidizing agent (oxidant) (16.1)

   f. oxidation number (6.3 and 16.2)

   g. oxidation number method (16.3)

   h. ion electron method (16.4)

   i. electrochemistry (16.5)

   j. electrolytic cell (16.5)

   k. cathode (16.5)

   l. anode (16.5)

   m. voltaic (galvanic) cells (16.6)

   n. primary cell (16.7)

   o. secondary cell (Section 16.7)

   p. corrosion (16.8)

2. Distinguish between:

   a. oxidation and reduction

   b. the oxidation number method and the ion electron method of balancing equations

   c. cathode and anode, in regard to oxidation and reduction

   d. electrolytic cells and voltaic (galvanic) cells

   e. primary and secondary voltaic cells

Electrolytic Cells (See Section 16.5)

3. In your own words, describe an electrolytic cell. Include a drawing and the reactions that occur at the anode and cathode.

Voltaic or Galvanic Cells (See Section 16.6)

4. In your own words, describe a voltaic or galvanic cell. Clearly point out the difference between an electrolytic cell and a voltaic or galvanic cell. Include drawings of both cells.

✓ Problems

Oxidation Numbers Revisited (See Section 16.2)

5. Calculate the oxidation number for either the element vanadium (V) or tungsten (W) as indicated in each of the following compounds or ions.

   a. VO_2

   b. V_2O_5

   c. V_2O_3

   d. VO_3^-

   e. WO_3

   f. WO_4^{2-}

Oxidation–Reduction Equations: Oxidation Number Method (See Section 16.3)

6. Balance the following oxidation–reduction equations using the oxidation number method:

   a. $HNO_3 + HI \longrightarrow NO + I_2 + H_2O$

   b. $KI + H_2SO_4(conc) \longrightarrow H_2S + H_2O + I_2 + K_2SO_4$

   c. $Cu + HNO_3(dil) \longrightarrow Cu(NO_3)_2 + NO + H_2O$

   d. $KIO_4 + KI + HCl \longrightarrow KCl + I_2 + H_2O$

   e. $HNO_3 + I_2 \longrightarrow NO_2 + H_2O + HIO_3$

   f. $Ag + H_2SO_4(conc) \longrightarrow Ag_2SO_4 + SO_2 + H_2O$

   g. $Cr_2O_7^{2-} + Fe^{2+} + H^+ \longrightarrow Cr^{3+} + Fe^{3+} + H_2O$

   h. $Cu + H^+ + SO_4^{2-} \longrightarrow Cu^{2+} + SO_2 + H_2O$

   i. $I_2 + CdS + H^+ \longrightarrow HI + S + Cd^{2+}$

   j. $Zn + Cr_2O_7^{2-} + H^+ \longrightarrow Zn^{2+} + Cr^{3+} + H_2O$

7. In the equations in Problem 6, indicate the substances oxidized and reduced and the oxidizing and reducing agents.

8. Balance the following oxidation–reduction equations using the oxidation number method:

   a. $Bi(OH)_3 + K_2SnO_2 \longrightarrow Bi + K_2SnO_3 + H_2O$

   b. $Sb + HNO_3(dil) \longrightarrow Sb_2O_5 + NO + H_2O$

   c. $Na_2TeO_3 + NaI + HCl \longrightarrow NaCl + Te + H_2O + I_2$

   d. $Mn(NO_3)_2 + NaBiO_3 + HNO_3 \longrightarrow HMnO_4 + Bi(NO_3)_3 + NaNO_3 + H_2O$

   e. $CoSO_4 + KI + KIO_3 + H_2O \longrightarrow Co(OH)_2 + K_2SO_4 + I_2$

   f. $I_2O_5 + CO \longrightarrow I_2 + CO_2$

   g. $Cr_2O_7^{2-} + H^+ + Cl^- \longrightarrow Cr^{3+} + Cl_2 + H_2O$

   h. $I^- + MnO_4^- + H^+ \longrightarrow Mn^{2+} + I_2 + H_2O$

   i. $MnO_4^- + SO_2 + H_2O \longrightarrow Mn^{2+} + SO_4^{2-} + H^+$

   j. $H_2S + Cr_2O_7^{2-} + H^+ \longrightarrow S + Cr^{3+} + H_2O$

9. In the equations in Problem 8, indicate the substances oxidized and reduced and the oxidizing and reducing agents.

Oxidation–Reduction Equations: Ion Electron Method (See Section 16.4)

10. Balance the following oxidation–reduction equations using the ion electron method:

   a. $Sn^{2+} + IO_3^- \longrightarrow Sn^{4+} + I^-$ in acid solution

   b. $AsO_2^- + MnO_4^- \longrightarrow AsO_3^- + Mn^{2+}$ in acid solution

c. $C_2O_4^{2-} + MnO_4^- \longrightarrow CO_2 + Mn^{2+}$ in acid solution

d. $Mn^{2+} + BiO_3^- \longrightarrow MnO_4^- + Bi^{3+}$ in acid solution

e. $MnO_4^- + H_2O_2 \longrightarrow Mn^{2+} + O_2$ in acid solution

f. $Fe + NO_3^- \longrightarrow Fe^{3+} + NO$ in acid solution

g. $Cl_2 \longrightarrow ClO_3^- + Cl^-$ in basic solution

h. $Cl_2 \longrightarrow ClO^- + Cl^-$ in basic solution (cold)

i. $MnO_2(s) + O_2 \longrightarrow MnO_4^{2-} + H_2O$ in basic solution

j. $PbS(s) + H_2O_2 \longrightarrow PbSO_4(s) + H_2O$ in acid solution
    (*Hint:* H_2O_2 is a nonelectrolyte.)

11. In the equations in Problem 10, indicate the substances oxidized and reduced and the oxidizing and reducing agents.

12. Balance the following oxidation–reduction equations using the ion electron method:

a. $Cr_2O_7^{2-} + C_2O_4^{2-} \longrightarrow Cr^{3+} + CO_2$ in acid solution

b. $S^{2-} + NO_3^- \longrightarrow S + NO$ in acid solution

c. $SO_3^{2-} + MnO_4^- \longrightarrow SO_4^{2-} + Mn^{2+}$ in acid solution

d. $AsO_3^{3-} + Br_2 \longrightarrow AsO_4^{3-} + Br^-$ in acid solution

e. $AsO_4^{3-} + I^- \longrightarrow AsO_3^{3-} + I_2$ in acid solution

f. $Bi(OH)_3 + SnO_2^{2-} \longrightarrow SnO_3^{2-} + Bi$ in basic solution

g. $Mn^{2+} + H_2O_2 \longrightarrow MnO_2(s) + H_2O$ in basic solution

h. $MnO_4^- + ClO_2^- \longrightarrow MnO_2(s) + ClO_4^-$ in basic solution

i. $NiS(s) + HCl + HNO_3 \longrightarrow NiCl_2 + NO(g) + S(s)$ in acid solution

j. $HSbCl_6 + H_2S(g) \longrightarrow H_3SbCl_6 + S(s)$ in acid solution
    (*Hint:* $HSbCl_6$ ionizes as H^+ and $SbCl_6^-$, and H_3SbCl_6 ionizes as $3 H^+$ and $SbCl_6^{3-}$.)

13. In the equations in Problem 12, indicate the substances oxidized and reduced and the oxidizing and reducing agents.

Voltaic Cells in Practical Use (See Section 16.7)

14. An intriguing battery called an alkaline air–zinc cell (Figure 16.13) has been used in such diverse applications as the railroad industry and hearing aids.

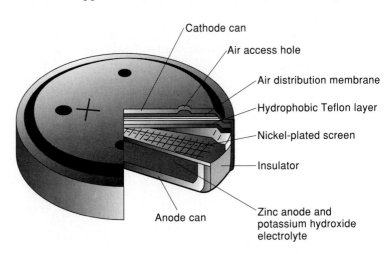

Cathode can

Air access hole

Air distribution membrane

Hydrophobic Teflon layer

Nickel-plated screen

Insulator

Anode can

Zinc anode and potassium hydroxide electrolyte

FIGURE 16.13
Alkaline air–zinc battery. Note the holes in the casing that allow atmospheric oxygen to enter the cell and participate in the oxidation–reduction reaction.

This cell is based on the conversion of zinc metal to zinc hydroxide [$Zn(OH)_2$] and oxygen gas (which diffuses from the atmosphere into the cell through holes in the cell container as Figure 16.13 shows) to hydroxide ions under alkaline (basic) conditions.

  a. Write the balanced oxidation and reduction half-reactions associated with the operation of this unusual battery.

  b. Indicate whether these reactions occurred at the cathode or the anode.

General Problems

15. Refer to Problem 6, part a.

  a. How many grams of nitrogen oxide (nitric oxide) can be produced from 66.0 g of pure nitric acid?

  b. How many liters of nitrogen oxide at STP can be produced from 66.0 g of pure nitric acid?

  c. How many liters of nitrogen oxide at 27°C and $7\overline{00}$ torr can be produced from 66.0 g of pure nitric acid?

  d. How many grams of nitrogen oxide can be produced from 45.0 mL of concentrated nitric acid? (Concentrated nitric acid is 72.0% pure nitric acid and has a density of 1.42 g/mL.)

16. In Problem 8, part a, how many grams of bismuth hydroxide are required to produce 4.10 g of precipitated bismuth metal?

17. Consider the following equation:

$$KMnO_4 + FeSO_4 + H_2SO_4 \longrightarrow MnSO_4 + Fe_2(SO_4)_3 + K_2SO_4 + H_2O$$

  a. Balance this oxidation–reduction equation by the oxidation number method.

  b. Calculate the number of moles of potassium permanganate required to oxidize 2.30 g of iron(II) sulfate.

  c. How many milliliters of a 0.100 M potassium permanganate solution would be required to oxidize 2.30 g of iron(II) sulfate?

18. In one of the tests to determine alcohol content in the breath, the police officer uses a chemical test. The chemical equation is as follows:

$$\underset{\substack{\text{ethyl}\\\text{alcohol}}}{C_2H_6O} + K_2Cr_2O_7 + H_2SO_4 \longrightarrow \underset{\substack{\text{acetic}\\\text{acid}}}{C_2H_4O_2} + Cr_2(SO_4)_3 + K_2SO_4 + H_2O$$

The basis of the test is the change of the red-orange potassium dichromate ($K_2Cr_2O_7$) to the green chromium(III) sulfate [$Cr_2(SO_4)_3$].

  a. Balance the equation by the oxidation number method.

  b. What is the oxidizing agent?

19. Iodine is generally made from sodium iodate ($NaIO_3$), which is mined in naturally occurring salt deposits found in Chile.

a. Balance the following oxidation–reduction equation in *acid solution* by the ion electron method that describes the reaction used to prepare iodine (I_2) from the sodium iodate salt.

$$IO_3^- + HSO_3^- \longrightarrow I_2 + SO_4^{2-}$$

b. Calculate the number of grams of iodine produced from 0.800 kg of sodium iodate.

20. Some toilet bowl cleaners contain hydrochloric acid (HCl). It is not to be mixed with bleach which contains sodium hypochlorite (NaClO).

a. Write a balanced half-reaction of hypochlorite ion (ClO^-) with hydrogen ion to produce chlorine gas (Cl_2).

b. Is this half-reaction an oxidation or a reduction?

 # Chapter Quiz 16

1. Calculate the oxidation numbers for the elements indicated in each of the following ions:

a. C in CO_3^{2-} b. S in $S_2O_3^{2-}$

2. Balance the following equation by the oxidation number method:

$$H_2S + HNO_3 \longrightarrow S + NO + H_2O$$

3. In question 2, what is the oxidizing agent?

4. Balance the following equations by the ion electron method.

a. $Zn + NO_3^- \longrightarrow Zn^{2+} + N_2$ in acid solution

b. $Zn + NO_3^- \longrightarrow Zn^{2+} + NH_3$ in basic solution

5. In question 4, what is the reducing agent in part (a) and in part (b)?

6. In the Edison storage battery iron metal is converted to solid iron(II) hydroxide and solid nickel(IV) oxide is converted to solid nickel(II) hydroxide with an electrolyte of potassium hydroxide solution.

a. Write the balanced oxidation and reduction half-reaction for this cell.

b. Indicate whether these reactions occurred at the cathode or the anode.

Magnesium (Symbol: Mg)

The Element MAGNESIUM: From Wheels to Flares

FROM

Magnesium is used to make alloys which were once used to make "mag wheels." Now, "mag wheels" are made from aluminum. They give a dazzling appearance to even a Ferrari!

Name: The name derives from a region called *Magnesia*, in Thessaly, Greece. *Magnesia* (Latin) refers to a number of magnesium-containing solids such as magnesium oxide (MgO) and talc (3 MgO·4 SiO$_2$·H$_2$O).

Appearance: A low-density, ductile, silver-white metal.

Occurrence: Magnesium is quite common and comprises about 2% of the earth's crust as *dolomite* [CaMg(CO$_3$)$_2$], *carnallite* [MgCl$_2$·KCl·6 H$_2$O], *epsomite* [MgSO$_4$·7 H$_2$O, epsom salts], and other minerals. It is also found in substantial concentrations in seawater. The "mining" of seawater is a new source of many elements.

Source: Magnesium is prepared by passing electric current through molten magnesium chloride to give magnesium metal and chlorine gas.

$$MgCl_2(l) \xrightarrow[\text{current}]{\text{electric}} Mg(s) + Cl_2(g)$$

TO

Magnesium burns vigorously and gives off a dazzling white light. This property makes it useful in the construction of flares and flash devices.

Common Uses: The primary use of magnesium is in the preparation of strong, durable lightweight alloys (mixtures of metals). Two common alloys include a 5% magnesium/95% aluminum alloy and a 91% magnesium/6% aluminum/3% zinc alloy.

Magnesium is used in the production of flares, fireworks, and incendiary (fire-setting) devices.

Magnesium is used in the construction of some types of dry cell batteries.

Unusual Facts: Magnesium is an essential metal in the plant world. Magnesium ions (Mg^{2+}) are incorporated in chlorophyll, a complex molecule that allows plants to convert sunlight, water, and carbon dioxide to a sugar (glucose). This process provides the basis for all life on earth.

CHAPTER 17

Reaction Rates and Chemical Equilibria

An explosion is an example of a very fast reaction in which heat and gases are liberated.

GOALS FOR CHAPTER 17

1. To define reaction rate and activation energy and to discuss the four factors that determine the rate of a reaction (Section 17.1).

2. To explain dynamic equilibrium, to write the equilibrium constant for a given reaction, and to explain the meaning of the magnitude of a given equilibrium constant (Section 17.2).

3. To use Le Châtelier's principle to explain how changes in concentration, temperature, or pressure affect a chemical equilibrium (Section 17.3).

4. To define K_a and K_b and to use them to calculate properties of solutions containing weak acids or weak bases (Section 17.4).

5. To define a buffer solution and to explain how it resists a change in pH when strong acids or strong bases are added (Section 17.5).

6. To define K_{sp}, to explain the meaning of the magnitude of a given solubility product constant, and to use it to calculate the solubility of slightly soluble electrolytes in water (Section 17.6).

Countdown

You may use the periodic table.

ELEMENT	ATOMIC MASS UNITS (amu)
N	14.0
Ag	107.9
Cl	35.5

5. Perform the following operations and express your answer to three significant digits (Section 2.5 and 2.7).

   a. $\sqrt{1.80 \times 10^{-8}}$ (1.34×10^{-4})

   b. $\sqrt{1.80 \times 10^{-9}}$ (4.24×10^{-5})

   c. Raise 2.00×10^{-3} to the second power
   (4.00×10^{-6})

   d. Raise 2.00×10^{-3} to the third power
   (8.00×10^{-9})

4. Write the correct formula for the following compounds (Sections 7.3, 7.4, 7.6, and 7.7).

   a. ammonia (NH_3)

   b. iron(II) sulfate $(FeSO_4)$

   c. potassium acetate $(KC_2H_3O_2)$

   d. calcium chromate $(CaCrO_4)$

   e. acetic acid $(HC_2H_3O_2)$

3. Calculate the number of moles in each of the following quantities. Express your answer in scientific notation (Sections 2.7 and 8.2)

   a. 1.50 g of nitrogen gas $(5.36 \times 10^{-2} \text{ mol})$

   b. 2.50 g of silver chloride $(1.74 \times 10^{-2} \text{ mol})$

2. Calculate the number of grams in each of the following quantities (Section 8.2).

   a. 8.35×10^{-2} mol of nitrogen gas   (2.34 g)

   b. 2.64×10^{-3} mol of silver chloride   (0.379 g)

1. Calculate the pH and pOH of the following solutions (Section 15.5):

   a. hydrogen ion concentration is 3.3×10^{-3} mol/L
   (2.48, 11.52)

   b. hydrogen ion concentration is 8.8×10^{-9} mol/L
   (8.06, 5.94)

A young nurse is planning for her future and invests a certain portion of her income in a mutual fund. She might ask two very different questions about her fund. One might be, "How rapidly does my fund appreciate?" Another equally important question might be, "What will my fund be worth in twenty years?" These two different questions illustrate two aspects of reaction chemistry that we will be looking at in this chapter: How fast does a reaction go? and What will the composition of the reaction mixture be when the reaction stops?

The study of how fast and by what path reactions go is called *chemical kinetics*, and knowledge about the composition of a reaction mixture when a reaction is complete relies on an understanding of *chemical equilibrium*. Both of these areas of chemistry are important to an understanding of reaction processes.

17.1 Reaction Rates

Reaction rate Rate or speed at which products are formed or reactants are consumed.

A knowledge of the speed at which a reaction goes can be essential in the real world. The length of time you bake chocolate chip cookies depends on the rate at which baking soda (sodium bicarbonate or hydrogen carbonate) decomposes at 350°F:

$$2\,NaHCO_3(s) \xrightarrow{\Delta} Na_2CO_3(s) + CO_2(g) + H_2O(l) \tag{17.1}$$

> **Study Hint:** Some people can run fast, others run more slowly, and still others do not run at all. Molecules are the same: some are very reactive, others react less rapidly, and still others react *very* slowly.

The **reaction rate** is the rate or speed at which products are formed or reactants are consumed. An explosion is an example of a fast reaction. The formation of oil from decayed organic matter is an example of a slow reaction. Through extensive experimentation, chemists have determined that the rate of a reaction depends on four factors: (1) the nature of the reactants, (2) the concentrations of the reactants, (3) the temperature, and (4) the presence of catalysts.

The Nature of the Reactants

The rate of a chemical reaction depends in part on the ease with which bonds can be made or broken. Very reactive materials can break or make bonds easily. Unreactive materials do not break or make new bonds readily. For example, the decomposition of nitrogen dioxide is complete in a few minutes.

> **Study Hint:** As the number of people in a small room increases, the chances of bumping into someone increases. The same thing is true if you increase the concentration of molecules. More collisions occur.

$$2\,NO_2(g) \longrightarrow 2\,NO(g) + O_2(g) \tag{17.2}$$

In contrast, the decomposition of a comparable amount of dinitrogen pentoxide requires over 100 min at the same temperature.

$$2\,N_2O_5(g) \longrightarrow 4\,NO_2(g) + O_2(g) \tag{17.3}$$

Concentrations of Reactants

Law of mass action Principle that for a general reaction aA + bB $\longrightarrow$ cC + dD the rate of reaction can be defined as $k[A]^x[B]^y$, where [A] and [B] are concentrations of A and B in moles per liter, respectively, k is the specific rate constant, and x and y are determined by experimentation.

Molecules must collide with each other in order to react. The more often they collide, the more chances there are for a reaction to take place. Thus, the rate of a reaction depends in part on the concentrations of the reactants. This relationship was formulated quantitatively in 1864 by two Norwegian chemists, Cato M. Guldberg (1836–1902) and Peter Waage (1833–1900). Their **law of mass action** states that for a general reaction

$$a\text{A} + b\text{B} \longrightarrow c\text{C} + d\text{D} \tag{17.4}$$

the rate of reaction can be defined as

$$\text{rate of reaction } = k[A]^x[B]^y \qquad (17.5)$$

where [A] and [B] are the concentrations of A and B in mol/L, respectively, and x and y are numbers that are determined by *experimentation*. The exponents x and y are usually positive whole numbers but may be fractional or negative numbers or 0 in complicated situations. They can be *determined* only by *experimentation* and may or may not be the same as the coefficients of the reactant(s) in the balanced equation. The constant k is the **specific rate constant** and is also determined by experimentation.

Specific rate constant Experimentally determined constant factor in the rate of reactions, expressed mathematically as k.

EXAMPLE 17.1   Consider the decomposition of nitrogen dioxide.

$$2\,NO_2(g) \longrightarrow 2\,NO(g) + O_2(g)$$
$$\text{rate } = k[NO_2]^2$$

What will be the effect on the rate of the decomposition if the concentration of NO_2 is doubled from 0.10 M to 0.20 M?

SOLUTION   The rate of the original reaction when $[NO_2] = 0.10\ M$ is

$$\text{rate } = k[0.10\ M]^2 = k(0.010\ M^2)$$

The rate of the reaction when $[NO_2] = 0.20\ M$ is

$$\text{rate } = k[0.20\ M]^2 = k(0.040\ M^2)$$

The rate of decomposition will be four times (0.040/0.010) the original rate.   *Answer*

Study Exercise 17.1
Consider the decomposition of dinitrogen pentoxide.

$$2\,N_2O_5(g) \longrightarrow 4\,NO_2(g) + O_2(g)$$
$$\text{rate } = k[N_2O_5]$$

What will be the effect on the rate of decomposition if the concentration of N_2O_5 is tripled from 0.010 M to 0.030 M?

(It will be three times the rate with 0.010 M.)   Work Problem 7.

Temperature and Activation Energy

All chemical reactions speed up as the temperature increases because at higher temperatures molecules have higher kinetic energies (see Section 3.5) and more of them have enough energy to undergo a reaction. The concept that molecules need a minimum amount of energy in order to react is fundamental to an understanding of chemical reactions. This energy requirement is called the **activation energy** of the reaction. Activation energy is similar to the energy that must be supplied by a golfer on a miniature golf course about to hit a ball on one side of a hill toward a hole on the other side. Until the golfer hits the ball hard enough for it to reach the top of the hill, the ball will not get to the hole on the other side. Unless the energy of the reactants is equal to the activation energy necessary to break a chemical bond, no reaction will occur, as Figure 17.1 shows.

Activation energy   Energy barrier that must be overcome before molecules can react.

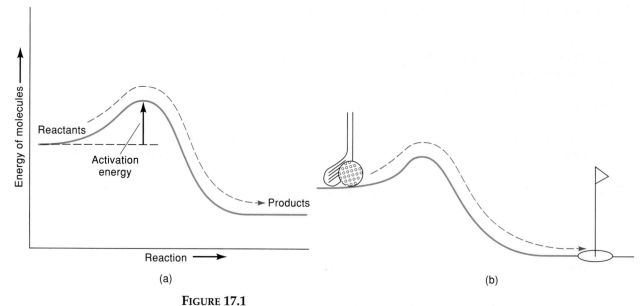

FIGURE 17.1
(a) The journey from reactants to products over an energy barrier (activation energy) is similar to (b) the progress of a golf ball over a hill and down to the hole.

Figure 17.2 depicts exothermic (heat-producing) and endothermic (heat-absorbing) reactions (see Section 10.7). In the exothermic reaction shown in Figure 17.2a, the energy of the products is *lower* than the energy of the reactants. The law of conservation of energy (see Section 3.6) requires that the difference in the energy of the products and reactants be given off as the heat of reaction. In contrast, as Fig-

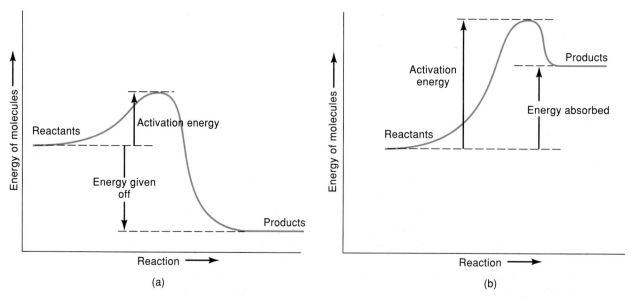

FIGURE 17.2
Energy relationships in (a) exothermic and (b) endothermic reactions.

ure 17.2b shows, the products in an endothermic reaction are *higher* in energy than the reactants, and the difference represents the heat absorbed in the reaction. Note, however, that both types of reactions require enough (activation) energy to get the reactants "over the hill." It is the energy required to get from the reactants to the "top of the hill" that we call activation energy.

Catalyst

In Section 9.2, we defined a *catalyst* as a substance that speeds up a chemical reaction but is recovered without appreciable change at the end of the reaction. Catalysts increase reaction rates by lowering the activation energies. Examples of catalysts include enzymes, chlorophyll (in photosynthesis), and manganese (IV) dioxide (in the decomposition of potassium chlorate). Small amounts of acid or base often act as catalysts in some reactions.

> **S**tudy Hint: Think of cheerleaders at a sporting event as a catalyst. The players can play without the cheerleaders, but often the shouts of encouragement give them an added burst of energy to excel.

17.2 Reversibility of Reactions and Chemical Equilibrium

So far, almost all the equations representing chemical reactions we have seen have shown one or more reactants producing one or more products. Yet, all chemical reactions are *theoretically* reversible, and most are *practically* reversible. In a reversible reaction, the reactants react to form products and the products react to form reactants until a state of **dynamic equilibrium** is reached. We have previously encountered dynamic equilibria many times. Examples include vapor pressure (Chapter 12), melting and freezing (Chapter 12), and saturated solutions (Chapter 14).

Equilibrium is indicated by a double arrow ($\rightleftharpoons$), and it exists when the rates of both the forward and reverse reactions become the same. In this manner, reactions "go both ways." Chemists refer to the reaction that proceeds from the left to the right as the *forward reaction* and to the reaction that proceeds from the right to the left as the *reverse reaction*. Thus, in the equation

$$A + B \rightleftharpoons C + D \qquad (17.6)$$

the reaction $A + B \longrightarrow C + D$ is considered the forward reaction and $C + D \longrightarrow A + B$ is considered the reverse reaction. If we start with a mixture of A and B at a given temperature and pressure, A and B will react to form C and D at a rate that depends on the starting concentrations of A and B. Initially, the reverse reaction has a rate of zero since no C and D molecules are present. However, as their concentrations increase, the rate of the reverse reaction increases. During this time, the rate of the forward reaction decreases as the concentrations of A and B decrease. Thus, during the course of a reaction, the rate of the forward reaction *decreases with time*, while the rate of the reverse reaction *increases with time*.

Eventually, the rate at which C and D molecules react to form A and B molecules *will be equal* to the rate at which A and B molecules react to form C and D molecules. *When this point is reached, the reaction is at equilibrium.* This is a dynamic equilibrium because two opposing reactions are continuously taking place. Once a reaction has achieved equilibrium, there are no further changes in the rates of the forward and reverse processes as long as no outside agent acts on the system. Figure 17.3 illustrates the process of achieving equilibrium.

Dynamic equilibrium A situation in which the rate of a forward process is equal to the rate of the reverse process occurring simultaneously, as when the rate of evaporation is equal to the rate of condensation for a liquid in a closed container.

FIGURE 17.3
Chemical equilibrium is reached when the rate of the forward reaction is equal to the rate of the reverse reaction. A catalyst (----) decreases the time required for equilibrium to be established; it does not increase the yield of the desired product or shift the position of the equilibrium in relation to reactant and product concentrations (see Section 17.3).

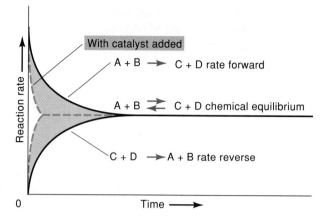

Although, under ideal conditions, any reaction is reversible, such is not the case in real life. In some cases the reverse reaction never appears to take place and the reaction essentially "goes to completion." This condition occurs when a product is removed from the system or the reverse reaction rate is so slow as to be negligible. The frying of a hamburger or an egg is an example of an irreversible reaction. Irreversible reactions generally involve one of the following substances as a product: (1) a *gas* (like that shown in Figure 17.4), (2) a *precipitate*, or (3) a *nonionized* or *partially ionized* substance such as water.

In still other cases, a reaction is only partially reversible. That is, either the forward or the reverse reaction goes further toward completion than the other. In such

FIGURE 17.4
Formation of a gas that is removed as soon as it is formed drives a reaction to completion. The reaction of a solution of sodium carbonate (Na_2CO_3) with hydrochloric acid (HCl) to form sodium chloride (NaCl), water, and the gas, carbon dioxide (CO_2). See if you can write the balanced equation and the net ionic equation for this reaction. (Courtesy Dr. E. R. Degginger)

cases, chemists sometimes use arrows of different lengths to indicate the more and the less completed reactions. For example, in the equation

$$A + B \rightleftharpoons C + D \tag{17.7}$$

the forward equation goes further toward completion than the reverse equation does. Regardless of whether they are reverisble, irreversible, or partially reversible, however, all reactions eventually reach an equilibrium at which no further changes occur without outside intervention.

Equilibrium Constant

Although two different reactions both achieve equilibrium, they may have vastly different amounts of product present at equilibrium. How can we be more quantitative in our treatment of equilibrium? The answer lies in the **equilibrium constant expression**. For the general reaction

$$aA + bB \rightleftharpoons cC + dD \tag{17.8}$$

the equilibrium constant, K, is

$$K = \frac{[C]^c[D]^d}{[A]^a[B]^b} \tag{17.9}$$

where the brackets indicate that the substances within them are expressed in concentrations of mol/L. The *powers (exponents)* are simply the *coefficients* from the balanced equation. At a *given* temperature the value of K is constant. As the temperature changes, the value of K varies.

In the equilibrium expression for an equation that contains a solid, the *solid is not considered in the equilibrium expression*. We omit the solid because, at constant temperature, its concentration (determined by its density) in mol/L is constant and will not change. This value is automatically included in K.

Figure 17.5 illustrates how colorless N_2O_4 forms red-brown NO_2 in an equilibrium reaction. The chemical equation is

$$N_2O_4(g) \rightleftharpoons 2\,NO_2(g) \tag{17.10}$$

Equilibrium constant expression For the general reaction

$$aA + bB \rightleftharpoons cC + dD$$

the equilibrium constant expression is equal to $[C]^c[D]^d$ divided by $[A]^a[B]^b$; the equilibrium constant is designated as K and is an experimentally determined constant that varies with the temperature.

(a)

(b)

(c)

FIGURE 17.5
(a) Pure N_2O_4 is colorless. (b) As the sample warms up, the reaction progresses and N_2O_4 decomposes to give NO_2 (red-brown color). (c) The reaction has achieved equilibrium, and the red-brown color does not get any darker.

Consider the following example, which illustrates how the equilibrium constant expression is formulated.

EXAMPLE 17.2   Write the expression for the equilibrium constant for each of the following reactions:

a. $CO(g) + Cl_2(g) \rightleftharpoons COCl_2(g)$
b. $H_2(g) + I_2(g) \rightleftharpoons 2 HI(g)$
c. $CaCO_3(s) \rightleftharpoons CaO(s) + CO_2(g)$

SOLUTION   Using the expression for the equilibrium constant, K is as follows for the preceding reactions:

a. $K = \dfrac{[COCl_2]}{[CO][Cl_2]}$ *Answer*

b. $K = \dfrac{[HI]^2}{[H_2][I_2]}$ *Answer*

c. $K = [CO_2]$ *Answer*

Both $CaCO_3$ and CaO are solids and hence their concentrations are constant at a given temperature. They are, therefore, not considered in the equilibrium expression because they are included in the value for the constant K.

Study Exercise 17.2
Write the expression for the equilibrium constant for the following reactions:

a. $2 NOCl(g) \rightleftharpoons 2 NO(g) + Cl_2(g)$ $\left(K = \dfrac{[NO]^2[Cl_2]}{[NOCl]^2} \right)$

Work Problems 8 and 9.

b. $2 TiCl_3(s) + 2 HCl(g) \rightleftharpoons 2 TiCl_4(g) + H_2(g)$ $\left(K = \dfrac{[TiCl_4]^2[H_2]}{[HCl]^2} \right)$

To better understand the equilibrium expression, consider the equation

$$N_2(g) + 3 H_2(g) \rightleftharpoons 2 NH_3(g) \tag{17.11}$$

which has the following equilibrium constant expression:

$$K = \frac{[NH_3]^2}{[N_2][H_2]^3} \tag{17.12}$$

The equilibrium constant describes the equilibrium composition of *any* reaction mixture containing N_2, H_2, and NH_3, regardless of the initial concentrations of the compounds. At a given temperature, K is constant.

What does the equilibrium constant expression tell us? Suppose the general equilibrium in Equation (17.9) has a value of 1.00 for K and we start with $[A] = [B] = 1.0$ mol/L. A and B react until the new concentrations of A, B, C, and D satisfy the expression

$$K = 1.00 = \frac{[C]^c[D]^d}{[A]^a[B]^b} \tag{17.13}$$

When half of A and B have reacted, [A] = [B] = [C] = [D] = 0.50 M and the reaction is at equilibrium. At this point there will be no observable changes in the reaction composition unless outside agents intervene.

The magnitude of K provides information about the *position* of the equilibrium for a reaction. If we compare two similar reactions, the reaction with the larger value of K will have more products at equilibrium than the reaction with a smaller value for K. For example, consider the ionization of the weak acids acetic acid ($HC_2H_3O_2$) and lactic acid ($HC_3H_5O_3$). These acids ionize in water to a small extent according to the following equations.

$$HC_2H_3O_2(aq) \rightleftharpoons H^+(aq) + C_2H_3O_2^-(aq) \qquad (17.14)$$
$$\text{acetic acid}$$

$$HC_3H_5O_3(aq) \rightleftharpoons H^+(aq) + C_3H_5O_3^-(aq) \qquad (17.15)$$
$$\text{lactic acid}$$

The equilibrium constants at 25°C for these reactions are 1.76×10^{-5} (mol/L) for acetic acid and 1.37×10^{-4} (mol/L) for lactic acid. The larger K value (smaller negative exponent) for lactic acid indicates that it is a stronger acid than acetic acid. Therefore, if we begin with the same concentrations and volumes of each acid, more ions are produced in the lactic acid solution than in the acetic acid solution.

The same reasoning applies for bases. The larger the value of K, the stronger the base.

Study Exercise 17.3

Listed in the following table are ionization constants for various bases (K_b) at 25°C. Arrange them in order of decreasing base strength by their K_b values.

BASE (IN APPROXIMATELY 0.1 N AQUEOUS SOLUTIONS)	K_b at 25°C (mol/L)
Papaverine	8×10^{-9}
Codeine	9×10^{-7}
Cocaine	2.6×10^{-6}
Novocain	7×10^{-6}
Caffeine	4.1×10^{-4}
Theobromine	4.8×10^{-14}

(caffeine > novocain > cocaine > codeine > papaverine > theobromine)

Work Problems 10 and 11.

If we consider the equilibrium described above for acetic acid, the equilibrium constant *would appear* to have units

$$K = \frac{[H^+][C_2H_3O_2^-]}{[HC_2H_3O_2]} = \frac{[\text{mol/L}][\text{mol/L}]}{[\text{mol/L}]} = \text{mol/L} \qquad (17.16)$$

As we did with the ion product constant for water (K_w; see Section 15.2), we will place these convenience units in parentheses following the value of the constant. This was done above in the K_a values for acetic acid and lactic acid. These convenience units are given to help you solve problems, but in reality the constant has no units.

17.3 Le Châtelier's Principle

Le Châtelier's principle
Principle that if an equilibrium system is subjected to a change in conditions, the composition of the reaction mixture will change in such a way as to try to restore the original condition.

Equilibrium position Equilibrium composition of a reaction mixture; it refers to a new composition formed in response to changes in concentration, temperature, or pressure.

In 1888, Henry Louis Le Châtelier (le·shä′ te·lyā′) (1850–1936), a French chemist, formulated a principle governing equilibrium. **Le Châtelier's principle** states that if an *equilibrium system* is subjected to a change in conditions of *concentration, temperature,* or *pressure,* the composition of the reaction mixture will change in such a way as to try to *restore* the original condition. This new equilibrium composition of the reaction mixture is referred to as a new **equilibrium position**.

In this section, we will consider the effects of each of these changes in condition separately. To give you a real-life perspective on these effects, we will also show how each change affects the commercial production of ammonia by the Haber process. In this process, developed in 1914 by Fritz Haber (1868–1934), a German chemist, nitrogen reacts with hydrogen to form ammonia. After leaving the catalyst chamber, the ammonia is liquefied and separated from the nitrogen and hydrogen gases, which are then recycled. The equation for this reaction, which we used earlier to explain the equilibrium constant, is

$$N_2(g) + 3 H_2(g) \underset{}{\overset{\text{catalyst}}{\rightleftharpoons}} 2 NH_3(g) + \text{heat energy} \qquad (17.17)$$

Concentration

Study Hint: Think of shifts in equilibrium as analogous to bailing water from a leaking boat. The rise and fall of the boat in the water is analogous to the shift of equilibrium with changes in concentration, temperature, and pressure analogous to the bailing of the water.

When the concentration of one of the substances in a system at equilibrium is increased, Le Châtelier's principle predicts that the *equilibrium will shift so as to partially use up the added substance.* Decreasing the concentration of one of the substances in a system at equilibrium causes the equilibrium to shift so as to partially replenish the substance removed. In all cases, the equilibrium constant K remains *constant.* Only the *concentration of the reactants and products vary.*

In the Haber process for ammonia, an *increase* in the concentration of either *nitrogen* or *hydrogen* shifts the equilibrium to the *right* (the products side) and decreases the concentration of the other reactant. *Increasing* the concentration of *ammonia* shifts the equilibrium to the *left* (the reactants side). Conversely, decreasing the concentration of either nitrogen or hydrogen shifts the equilibrium to the left (the reactants side), and decreasing the concentration of ammonia shifts the equilibrium to the right (the products side). Therefore, to obtain a maximum yield of ammonia by the Haber process, (1) the ammonia gas that is formed is constantly removed, (2) either hydrogen or nitrogen is used in *excess*, and (3) the excess reactant gases are recycled.

Temperature

As stated earlier (see Section 17.2, Equilibrium Constant), K is a constant for a reaction at a given temperature. However, *the equilibrium constant changes when the temperature changes.* Le Châtelier's principle can be used to understand the effect of temperature on an equilibrium constant. If the temperature of a system at equilibrium is changed, the equilibrium constant will change so as to *try* to move the system back to its original temperature.

If the reaction is *exothermic,* as in the Haber process, the heat of reaction will *act as if it is one of the products.*

$$N_2(g) \; + \; 3\,H_2(g) \; \underset{}{\overset{\text{catalyst}}{\rightleftharpoons}} \; 2\,NH_3(g) \; + \; \text{heat energy} \qquad (17.18)$$

If there is an *increase* in *temperature*, the equilibrium will shift to the *left* to use up NH_3 and reduce the amount of heat produced (heat is required to go from $2\,NH_3$ to $N_2 \; + \; 3\,H_2$). Thus, K gets *smaller* when the *temperature increases* in an *exothermic* reaction. A decrease in temperature will shift the equilibrium to produce more ammonia, and K gets *larger*. The Haber process is generally carried out at 500°C, the optimal temperature for the production of ammonia.

If the reaction is *endothermic*, the heat of reaction will *act as if it is one of the reactants*. The reaction of oxygen gas to give ozone gas serves as a good example:

$$3\,O_2(g) \; + \; \text{heat energy} \; \rightleftharpoons \; 2\,O_3(g) \qquad (17.19)$$

In such cases, an increase in temperature shifts the equilibrium to the right (forming more ozone) to use up some of the heat. Thus, K gets *larger* when the temperature *increases* in an *endothermic* reaction. A decrease in temperature shifts the oxygen–ozone equilibrium to the left (consuming ozone), and K gets *smaller*. The effects of temperature on K are summarized in Table 17.1.

Pressure

Finally, Le Châtelier's principle predicts that *increasing* the pressure by compression on a gaseous system at equilibrium will shift the equilibrium in whichever direction *decreases the volume* (or number of molecules). Decreasing the pressure by expansion will have the opposite effect. If there is no change in volume or the number of molecules in going from reactants to products, pressure will have no effect on the equilibrium. The equilibrium constant *does not change with pressure*; only the equilibrium position changes.

In the Haber process, 1 volume of nitrogen gas reacts with 3 volumes of hydrogen gas to produce 2 volumes of ammonia gas. If *pressure* is *applied*, a system at equilibrium will react to offset the strain by shifting to the *right* and forming ammonia because *4* volumes of gas are on the left side of the equation ($1 \; + \; 3 \; = \; 4$) and only *2* volumes of gas are on the right. Therefore, to maximize the amount of ammonia at equilibrium, the system should utilize high pressures of about 200 to 600 atm. (Higher pressures are not used because the cost of the equipment for handling it would offset the value of the increased yield.) At a pressure of 600 atm and a temperature of 500°C, a 42.1% yield of ammonia is obtained in the equilibrium mixture.

TABLE **17.1**	**Summary of the Effects of Temperature Change on the Equilibrium Constant (K)**		
REACTION	TYPE	TEMPERATURE CHANGE	EFFECT ON K
A + B $\rightleftharpoons$ C + D + heat energy	Exothermic	Increase	Decrease
A + B $\rightleftharpoons$ C + D + heat energy	Exothermic	Decrease	Increase
A + B + heat energy $\rightleftharpoons$ C + D	Endothermic	Increase	Increase
A + B + heat energy $\rightleftharpoons$ C + D	Endothermic	Decrease	Decrease

$N_2 + 3 H_2 \rightleftharpoons 2 NH_3 + \text{Heat energy}$	$N_2 + 3 H_2 \rightleftharpoons 2 NH_3 + \boxed{\text{Heat energy}}$
$N_2 + 3 H_2 \rightleftharpoons 2 NH_3 + \text{Heat energy}$	(b) Temperature
$N_2 + 3 H_2 \rightleftharpoons 2 NH_3 + \text{Heat energy}$	$1 N_2 + 3 H_2 \rightleftharpoons 2 NH_3 + \text{Heat energy}$
(a) Concentration	(c) Pressure

FIGURE 17.6
Summary of the effects of an increase of various factors on the equilibrium for the Haber process for the production of ammonia. (a) Concentration. (b) Temperature. (c) Pressure. (Note the relation of the *colored* symbols and/or words to the *colored* arrows and their effect on equilibrium.

Consider as another example the formation of nitrogen oxide according to the following equation:

$$N_2(g) + O_2(g) \rightleftharpoons 2 NO(g) \qquad (17.20)$$

Pressure has no effect on the equilibrium because *2* volumes of reactants (*1* volume of nitrogen and *1* volume of oxygen) react to give *2* volumes of product (nitrogen oxide). Also, as we might expect, reactions involving only solids and/or liquids are not affected by pressure changes.

Figure 17.6 summarizes the effects of concentration, temperature, and pressure changes on the Haber process.

Study Exercise 17.4
Consider the following chemical reaction:

$$2 CO(g) + O_2(g) \rightleftharpoons 2 CO_2(g) + 566 \text{ kJ}$$

Predict the effect on equilibrium with the following changes:
a. increasing the concentration of CO   (right)
b. decreasing the concentration of CO_2 (right)
c. increasing the temperature   (left)
d. increasing the pressure by compression   (right)

Work Problems 12 through 15.

An application of Le Châtelier's principle may be seen in the absorption of oral drugs. To be absorbed into the bloodstream a drug must pass through the fatty cell membrane. For this to occur, the drug should not be water-soluble but rather fat-soluble. This requires the drug to be in a nonionized, rather than ionized, form (like dissolves like; see Section 14.3). Most oral drugs are weak acids or weak bases. Now where are these weak acid or weak bases absorbed? In the acidic stomach or in the basic intestines?

Consider the case of the acid drug (HA). In the stomach the pH is low (1 to 3). Addition of acid from the stomach drives the equilibrium (Le Châtelier's principle) to the left to form the nonionized form according to the following equation.

$$\textit{Acid drug:} \quad HA \rightleftharpoons \underset{\substack{\text{acid from}\\\text{stomach}}}{H^+} + A^- \qquad (17.21)$$

The nonionized form is fat-soluble. Therefore, acid drugs are absorbed from the stomach. Examples of acid drugs are the barbiturates which produce sleep, such as phenobarbital and pentobarbital (Nembutal, yellow jackets).

Consider the case of the basic drug, B:. In the acid of the stomach, a basic drug is ionized, which drives the reaction to the right (Le Châtelier's principle) according to the following equation.

$$\textit{Basic drug:} \quad \text{B:} + \text{H}^+ \rightleftharpoons \text{B:H}^+ \tag{17.22}$$

The ionized drug is water-soluble and hence not absorbed. In the basic intestine the pH is high (6 to 8) and the proton on the protonated basic drug is removed to form the nonionized drug, driving the equilibrium to the right (Le Châtelier's principle) according to the following equation.

$$\underset{\substack{\text{base from} \\ \text{intestine}}}{\text{B:H}^+} + \text{OH}^- \rightleftharpoons \text{H}_2\text{O} + \text{B:} \tag{17.23}$$

The nonionized form is fat-soluble. Therefore, basic drugs are absorbed from the intestines. Examples of basic drugs are morphine (painkillers) and quinine (treat malaria).

The Lack of Effect by Catalysts

Earlier in this chapter, we mentioned that a catalyst acts to increase the rate of a reaction. A catalyst does *not affect the position* of equilibrium. However, a catalyst does increase the rates of both the forward and reverse reactions to the *same degree*. So, while a catalyst does *not* increase the yield of the desired product, it *decreases the time* required for equilibrium to be established. In the Haber process, the catalysts are magnetic iron oxide (Fe_3O_4) and a mixture of potassium oxide and aluminum oxide ($\text{K}_2\text{O} \cdot \text{Al}_2\text{O}_3$). These catalysts make the process economically feasible because without them the reaction would take a considerable time to reach equilibrium (see Figure 17.3).

17.4 | Weak Electrolyte Equilibria

Special equilibrium constants, called K_a and K_b, are associated with a common class of reactions—those involving *weak acids* and *weak bases*, respectively. These substances, as noted in Chapter 15, are weak electrolytes. In solution these substances are ionized to only a small degree, and the solutions contain both the *nonionized form* of the substance and the *ions* resulting from the partial ionization. For example, consider acetic acid:

$$\text{HC}_2\text{H}_3\text{O}_2(aq) \rightleftharpoons \text{H}^+(aq) + \text{C}_2\text{H}_3\text{O}_2^-(aq) \tag{17.24}$$

The equilibrium constant for this reaction is given a special symbol, K_a, since this weak electrolyte is a weak *acid*:

$$K_a = \frac{[\text{H}^+][\text{C}_2\text{H}_3\text{O}_2^-]}{[\text{HC}_2\text{H}_3\text{O}_2]} \tag{17.25}$$

In this equation, $[\text{H}^+]$ and $[\text{C}_2\text{H}_3\text{O}_2^-]$ are the concentrations (*at equilibrium*) of hydrogen ions and acetate ions expressed in mol/L. $[\text{HC}_2\text{H}_3\text{O}_2]$ is the concentration of acetic acid molecules (*at equilibrium*) in mol/L. For simplicity, the states of the reactants and products are not written. The concentration of the water is already included in the value for K_a.

A similar equilibrium is found in an aqueous ammonia solution:

$$NH_3(aq) + HOH(l) \rightleftharpoons NH_4^+(aq) + OH^-(aq) \qquad (17.26)$$

The following expression gives the equilibrium constant K_b for this weak *base*:

$$K_b = \frac{[NH_4^+][OH^-]}{[NH_3]} \qquad (17.27)$$

In this equation, $[NH_4^+]$ and $[OH^-]$ are the concentrations (*at equilibrium*) of ammonium ions and hydroxide ions expressed in mol/L. $[NH_3]$ is the concentration of aqueous ammonia (*at equilibrium*) in mol/L. Again, for simplicity, we do not write the states of the reactants and products. The concentration of the water is included in the K_b constant.

Now, let us consider some calculations involving weak electrolytes at equilibrium.

EXAMPLE 17.3 The hydrofluoric acid in 0.0400 *M* HF solution is 13.4% ionized. Calculate K_a for HF.

SOLUTION The equation representing the ionization is

$$HF(aq) \rightleftharpoons H^+(aq) + F^-(aq)$$

and the expression for K_a is

$$K_a = \frac{[H^+][F^-]}{[HF]}$$

To evaluate the K_a, determine the concentrations of H^+, F^-, and HF (*at equilibrium*) from the data given. If the HF is 13.4% ionized, the percent *nonionized* at equilibrium will be 86.6% (100.0 − 13.4), and we can calculate the concentrations of H^+, F^-, and HF in solution:

$$[H^+] = 0.0400 \text{ mol/L} \times 0.134 = 0.00536 \text{ mol/L}$$

$[F^-]$ is equal to $[H^+]$ because in the balanced equation 1 mol of HF ionizes to form 1 mol of H^+ and 1 mol of F^-; therefore, 0.00536 mol of HF ionizes to form 0.00536 mol of H^+ and 0.00536 mol of F^-.

$$[F^-] = 0.00536 \text{ mol/L}$$

$$[HF] = 0.0400 \text{ mol/L} \times 0.866 = 0.0346 \text{ mol/L } at \; equilibrium$$

Calculate the value of K_a as follows:

$$K_a = \frac{[H^+][F^-]}{[HF]} = \frac{[0.00536 \text{ mol/L}][0.00536 \text{ mol/L}]}{[0.0346 \text{ mol/L}]}$$

$$= \frac{[5.36 \times 10^{-3}][5.36 \times 10^{-3}](\text{mol}^2/L^2)}{[3.46 \times 10^{-2}](\text{mol/L})}$$

$$= 8.30 \times 10^{-4} (\text{mol/L}) \qquad Answer$$

Study Exercise 17.5
The nitrous acid in a 0.500 *M* solution is 3.00% ionized. Calculate the K_a for HNO_2.

Work Problems 16 and 17.

$$[4.64 \times 10^{-4} \text{ (mol/L)}]$$

EXAMPLE 17.4 Calculate the hydrogen ion concentration and the percent ionization at 25°C in a 0.102 M acetic acid ($HC_2H_3O_2$) solution. The K_a for $HC_2H_3O_2$ = 1.76 × 10^{-5} (mol/L) at 25°C.

SOLUTION The equation representing the ionization is

$$HC_2H_3O_2(aq) \rightleftharpoons H^+(aq) + C_2H_3O_2^-(aq)$$

and the expression for K_a is

$$K_a = \frac{[H^+][C_2H_3O_2^-]}{[HC_2H_3O_2]}$$

In the balanced equation, **1 mol** of $HC_2H_3O_2$ on *ionizing* forms **1 mol** of H^+ and **1 mol** of $C_2H_3O_2^-$. If **5 mol** of $HC_2H_3O_2$ ionizes, we will have **5 mol** of H^+ and **5 mol** of $C_2H_3O_2^-$, and if x **mol** of $HC_2H_3O_2$ ionizes, this form x **mol** of H^+ and x **mol** of $C_2H_3O_2^-$. Therefore, let x = $[H^+]$ in mol/L, and x mol/L also equals $[C_2H_3O_2^-]$. The concentration of the non-ionized $HC_2H_3O_2$ remaining *at equilibrium* is $[HC_2H_3O_2]$ = 0.102 mol/L − x mol/L. The value of 0.102 mol/L of acetic acid *initially* and the value of 0.102 mol/L − x mol/L of acetic acid at *equilibrium* may need some further explanation. This new value at equilibrium is what is left *after* some of the acetic acid has ionized. The following illustration may be helpful:

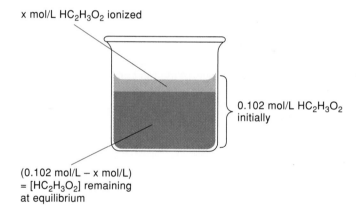

x mol/L HC₂H₃O₂ ionized

0.102 mol/L HC₂H₃O₂ initially

(0.102 mol/L − x mol/L) = [HC₂H₃O₂] remaining at equilibrium

Substituting the values for $[H^+]$, $[C_2H_3O_2^-]$, and $[HC_2H_3O_2]$ into the equilibrium expression just given, and dropping the units of x for simplicity, we have

$$1.76 \times 10^{-5} \text{ (mol/L)} = \frac{[x][x]}{[0.102 \text{ mol/L} - x]}$$

$$1.76 \times 10^{-5} \text{ (mol/L)} = \frac{[x]^2}{[0.102 \text{ mol/L} - x]}$$

To solve this equation for x, we must use the quadratic equation; however, we can simplify by assuming that x is very small in relation to the original concentration of $HC_2H_3O_2$ (0.102 M). Therefore, the quantity 0.102 mol/L − x will have a value not much different from 0.102 mol/L. Hence, we may neglect x in the *denominator* to simplify solving the equation without greatly affecting the answer. Dropping the x is analogous to taking a bucket of water from the ocean. The loss of such a small amount has a relatively negligible effect on the overall volume of water in the ocean. In some cases, such as very low concentrations and relatively large K_a values, we cannot neglect x and must solve the quadratic equation. For our purposes in this text, we shall consider only cases where we can neglect x in the denominator. Neglecting x gives

$$1.76 \times 10^{-5} \text{ (mol/L)} = \frac{[x]^2}{0.102 \text{ mol/L}}$$

and

$$x^2 = 1.76 \times 0.102 \times 10^{-5} \frac{\text{mol}^2}{\text{L}^2} = 1.80 \times 10^{-6} \frac{\text{mol}^2}{\text{L}^2}$$

$$x = \sqrt{1.80 \times 10^{-6} \frac{\text{mol}^2}{\text{L}^2}} = 1.34 \times 10^{-3} \frac{\text{mol}}{\text{L}} = [H^+] \qquad \textit{Answer}$$

Note that 1.34×10^{-3} is very small compared to 0.102, and so our approximation was reasonable. We can obtain the square root of 1.80×10^{-6} from a calculator. Calculate the percent ionization as follows:

$$\frac{\text{mol/L of acid ionized}}{\text{mol/L of acid initially}} \times 100$$

The mol/L of acid ionized is the same as the moles per liter of hydrogen ions because in the balanced equation the mole relationship is 1:1. Hence,

$$\frac{1.34 \times 10^{-3} \text{ mol/L}}{0.102 \text{ mol/L}} \times 100 = \frac{1.34 \times 10^{-3} \times 10^2}{1.02 \times 10^{-1}} = 1.31\% \qquad \textit{Answer}$$

EXAMPLE 17.5 Calculate the hydroxide ion concentration and the percent ionization at $2\bar{0}°C$ in a 0.100 M aqueous ammonia (NH_3) solution. The K_b for aqueous $NH_3 = 1.70 \times 10^{-5}$ (mol/L) at $2\bar{0}°C$.

SOLUTION The equation representing the ionization is

$$NH_3(aq) + HOH(aq) \rightleftharpoons NH_4^+(aq) + OH^-(aq)$$

and the expression for K_b is

$$K_b = \frac{[NH_4^+][OH^-]}{[NH_3]}$$

If we let $x = [OH^-]$ in mol/L at equilibrium, x in mol/L will also equal $[NH_4^+]$ because in the balanced equation 1 mol of aqueous NH_3 ionizes to form 1 mol of NH_4^+ and 1 mol of OH^-. The concentration of the nonionized NH_3 *at equilibrium* is

$$[NH_3] = 0.100 \text{ mol/L} - x \text{ mol/L}$$

since the mole relationship of NH_3 and OH^- is 1:1 in the balanced equation. Substituting these values into the equilibrium expression and dropping the units of x for simplicity, we obtain

$$1.70 \times 10^{-5} \text{ (mol/L)} = \frac{[x][x]}{[0.100 \text{ mol/L} - x]}$$

$$1.70 \times 10^{-5} \text{ (mol/L)} = \frac{[x]^2}{[0.100 \text{ mol/L} - x]}$$

Neglecting the x in 0.100 mol/L $- x$ and further solving the equation, we have

$$1.70 \times 10^{-5} \text{ (mol/L)} = \frac{[x]^2}{0.100 \text{ mol/L}}$$

$$x^2 = 0.100 \text{ mol/L} \times 1.70 \times 10^{-5} \text{ mol/L}$$

$$= 0.170 \times 10^{-5} \frac{\text{mol}^2}{\text{L}^2}$$

$$x^2 = 1.70 \times 10^{-6} \frac{\text{mol}^2}{\text{L}^2}$$

$$x = \sqrt{1.70 \times 10^{-6} \frac{\text{mol}^2}{\text{L}^2}} = 1.30 \times 10^{-3} \frac{\text{mol}}{\text{L}} \qquad Answer$$

Note that 1.30×10^{-3} is very small compared to 0.100, and so our approximation was reasonable. Calculate the percent ionization as follows:

$$\frac{1.30 \times 10^{-3} \text{ mol/L}}{0.100 \text{ mol/L}} \times 100 = \frac{1.30 \times 10^{-3} \times 10^2}{1.00 \times 10^{-1}} = 1.30\% \qquad Answer$$

Study Exercise 17.6
Calculate the hydrogen ion concentration and the percent ionization at $2\overline{0}°C$ in a 0.600 *M* hydrofluoric acid solution. The K_a for hydrofluoric acid $= 7.20 \times 10^{-4}$ (mol/L) at $2\overline{0}°C$.

$(2.08 \times 10^{-2}$ mol/L; $3.47\%)$ Work Problems 18 and 19.

17.5   Buffer Solution Equilibria

In Section 15.5, we mentioned that the pH of blood is maintained in a very narrow pH range (7.3 to 7.5) by various buffers in the blood. In this section we will consider how these buffers act.

    Normally, acids and bases cannot exist in solution without reacting with each other. But if the acid and base are a weak acid and its conjugate base (see Section 15.1), they can coexist in solution. A **buffer solution** is a solution that contains substantial amounts of both a weak acid and its conjugate base. Such a solution resists a change in its pH. When the substances involved do react, they simply regenerate themselves:

> **Buffer solution**   Solution that contains substantial amounts of both a weak acid and its conjugate base and so resists changes in its pH.

$$\underset{\substack{\text{weak} \\ \text{acid}}}{HC_2H_3O_2(aq)} + \underset{\substack{\text{conjugate} \\ \text{base}}}{C_2H_3O_2{}^-(aq)} \longrightarrow C_2H_3O_2{}^-(aq) + HC_2H_3O_2(aq) \qquad (17.28)$$

    Consider a solution containing both acetic acid and sodium acetate $(Na^+C_2H_3O_2{}^-)$. In such a solution, the following equilibrium exists:

$$HC_2H_3O_2(aq) \rightleftharpoons H^+(aq) + C_2H_3O_2{}^-(aq) \qquad (17.29)$$

If an acid such as HCl (a strong acid, H^+ and Cl^-) is introduced into this buffer solution, the acetate ion [on the right side of Equation (17.29)] will react with the H^+ to form acetic acid (a weak acid) according to the following chemical equation:

$$C_2H_3O_2{}^-(aq) + H^+(aq) + \cancel{Cl^-}(aq) \longrightarrow HC_2H_3O_2(aq) + \cancel{Cl^-}(aq) \qquad (17.30)$$

FIGURE 17.7
Buffer solution: A buffer solution contains both a weak acid ($HC_2H_3O_2$) and its conjugate base ($C_2H_3O_2^-$). Added strong acid (H^+) is converted to $HC_2H_3O_2$ by reaction with $C_2H_3O_2^-$ ion. Added strong base (OH^-) is converted to $C_2H_3O_2^-$ ion and water by reaction with $HC_2H_3O_2$.

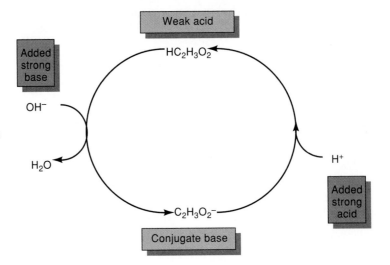

(The Cl^- is a spectator ion; see Section 15.7, rule 6).

If a base such as NaOH (a strong base, Na^+ and OH^-) is introduced into this buffer solution, the acetic acid [on the left side of Equation (17.28)] will react with the OH^- and form acetate ion (a weak base) and water according to the following chemical equation:

$$HC_2H_3O_2(aq) + \cancel{Na^+}(aq) + OH^-(aq) \longrightarrow C_2H_3O_2^-(aq) + H_2O(l) + \cancel{Na^+} \quad (17.31)$$

(The Na^+ is a spectator ion.) The buffer solution's action changes the strong acid (HCl) into a weak acid (acetic acid) and the strong base (NaOH) into a weak base (acetate ion). In both cases, the pH of the solution does not change markedly. Figure 17.7 summarizes these processes.

The three buffers in the blood are

1. Carbonic acid (H_2CO_3) and hydrogen carbonate (HCO_3^-)
2. Dihydrogen phosphate ($H_2PO_4^-$) and hydrogen phosphate (HPO_4^{2-})
3. Certain proteins.

These three buffers act in a similar manner as the acetic acid ($HC_2H_3O_2$) and acetate ion ($C_2H_3O_2^-$) buffer just described.

EXAMPLE 17.6　Given the following buffer solutions, identify the weak acid and conjugate base and write a chemical equation for the equilibrium involving these species. In addition, write a chemical equation that can neutralize some added hydrobromic acid or potassium hydroxide.

a. Solution of hydrofluoric acid and sodium fluoride
b. Solution of ammonia and ammonium chloride

SOLUTION

a. The weak acid is hydrofluoric acid (HF), and the conjugate base is F^- (recall that NaF is a strong electrolyte, Na^+ and F^-). The equilibrium is

$$HF(aq) \rightleftharpoons H^+(aq) + F^-(aq)$$

Added hydrobromic acid (HBr, H^+ and Br^-) is consumed by the reaction with F^-:

$$H^+(aq) + F^-(aq) \longrightarrow HF(aq)$$

Added potassium hydroxide (KOH, K^+ and OH^-) is consumed by the reaction with HF:

$$HF(aq) + OH^-(aq) \longrightarrow F^-(aq) + H_2O(l)$$

b. The weak acid is ammonium ion (NH_4^+), and the conjugate base is NH_3. The equilibrium is

$$NH_4^+(aq) \rightleftharpoons H^+(aq) + NH_3(aq)$$

Added hydrobromic acid (HBr, H^+ and Br^-) is consumed by the reaction with NH_3:

$$H^+(aq) + NH_3(aq) \longrightarrow NH_4^+(aq)$$

Added potassium hydroxide (KOH, K^+ and OH^-) is consumed by the reaction with NH_4^+:

$$NH_4^+(aq) + OH^-(aq) \longrightarrow NH_3(aq) + H_2O(l)$$

Figure 17.8 shows buffer solutions of a variety of pH values. These solutions are commercially available for laboratory use.

Study Exercise 17.7

Given the buffer solution of nitrous acid and sodium nitrite, identify the weak acid and the conjugate base and write the chemical equation for the equilibrium involving these species. In addition, write a chemical equation that can neutralize added hydrochloric acid or sodium hydroxide.

$$[\text{weak acid: } HNO_2; \text{ conjugate base: } NO_2^-;$$
$$HNO_2(aq) \rightleftharpoons H^+(aq) + NO_2^-(aq)$$
$$H^+(aq) + NO_2^-(aq) \longrightarrow HNO_2(aq)$$
$$HNO_2(aq) + OH^-(aq) \longrightarrow NO_2^-(aq) + H_2O(l)]$$

Work Problems 20 and 21.

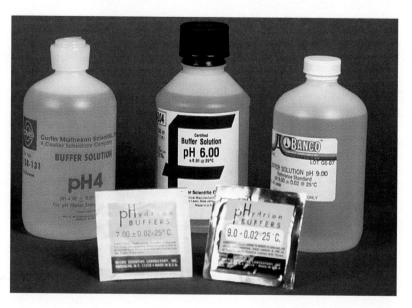

FIGURE 17.8
Buffer solutions of a variety of pH values are commercially available for laboratory use.

17.6 Solubility Equilibria

The equilibrium between a slightly soluble electrolyte and its ions in solution is a dynamic equilibrium involving constant *dissolving* of the electrolyte and *reprecipitation* of it at the surface of the solid crystals. At equilibrium, the solution is *saturated*. That is, the *rate of solution equals the rate of precipitation.*

As an example, consider what happens when the slightly soluble electrolyte Ag_2CrO_4 is added to pure water. The Ag_2CrO_4 dissolves until the solution reaches saturation. The equation for this equilibrium, pictured in Figure 17.9, is expressed as

$$Ag_2CrO_4(s) \rightleftharpoons 2\,Ag^+(aq) + CrO_4^{2-}(aq) \qquad (17.32)$$
$$\text{(in a saturated solution)}$$

The equilibrium constant for this equation can be expressed as

$$K = \frac{[Ag^+]^2[CrO_4^{2-}]}{[Ag_2CrO_4]} \qquad (17.33)$$

Solubility product constant (K_{sp}) An experimentally determined constant which describes the concentration equilibrium between a slightly soluble salt and its dissolved ions at saturation. In the expression for K_{sp}, the coefficients of the ions become exponents for the concentrations of the ions in the solution.

However, since Ag_2CrO_4 is present as a solid and its concentration in the solid phase is constant, the $[Ag_2CrO_4]$ term is incorporated into the equilibrium constant to give a new constant, the *solubility product constant*, K_{sp}.

$$K_{sp} = [Ag^+]^2[CrO_4^{2-}]$$

Here $[Ag^+]$ and $[CrO_4^{2-}]$ represent the concentrations of these ions in mol/L in a solution at equilibrium with solid Ag_2CrO_4 (at saturation).

The **solubility product constant (K_{sp})** is an experimentally determined constant which describes the concentration equilibrium between a slightly soluble salt and its dissolved ions at saturation. In the expression for K_{sp}, the coefficients of the ions become exponents for the concentrations of the ions in the solution. Figure 17.9 depicts this equilibrium.

FIGURE 17.9
(a) Solubility equilibrium for Ag_2CrO_4. (b) Precipitate of red Ag_2CrO_4 appears when soluble silver salt is added to a solution containing the chromate ion, CrO_4^{2-}.

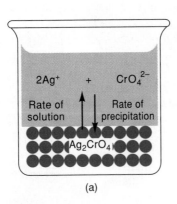

(a)

(b)

The solubility product constant K_{sp} is constant at a *given* temperature for a particular slightly soluble electrolyte. This value changes with a change in temperature, as do all equilibrium constants.

In comparing values of K_{sp}, the electrolytes must all be of the *same* type, such as all AB (AgCl) or all A_2B (Ag_2CrO_4). The larger the K_{sp} value of an electrolyte of a *given* type, the more soluble the electrolyte in water. For example, the solubility product constant for silver chloride (AgCl) is 1.56×10^{-10} (mol^2/L^2) at 25°C. That of silver bromide (AgBr) is 7.00×10^{-13} or 0.00700×10^{-10} (mol^2/L^2) at 25°C. Both are AB-type electrolytes. Therefore, because the K_{sp} value for silver chloride is larger (smaller negative exponent), the solubility of silver chloride in water is greater than that of silver bromide.

Study Exercise 17.8
Listed in the following table are solubility product constants (K_{sp}) for various slightly soluble electrolyes in water at 18°C. Arrange them in order of decreasing solubility in water.

Slightly Soluble Electrolyte	K_{sp} at 18°C (mol^2/L^2)
Lead(II) carbonate, $PbCO_3$	3.3×10^{-14}
Lead(II) chromate, $PbCrO_4$	1.8×10^{-14}
Lead(II) oxalate, PbC_2O_4	2.7×10^{-11}
Lead(II) sulfate, $PbSO_4$	1.1×10^{-8}
Lead(II) sulfide, PbS	3.4×10^{-28}

[lead(II) sulfate > lead(II) oxalate > lead(II) carbonate > lead(II) chromate > lead(II) sulfide]   Work Problem 22.

Examination of the K_{sp} values for silver chloride and silver bromide appear to have units. As we did with equilibrium constants (K_a and K_b, see Section 17.2), we will place these convenience units in parentheses following the value of the constant. These convenience units are given to help you solve problems, but in reality the constant has no units.

When the product of the concentrations of the ions, each of which is raised to its respective power, is *equal* to the value of the solubility product constant, a *saturated* solution exists. When the product of the concentrations of the ions, each raised to its respective power, is *less* than the value of the solubility product constant, an *unsaturated* solution exists. When this product is *greater* than the value of the solubility product constant, a *supersaturated* solution temporarily exists and precipitation normally occurs. Therefore, the solubility product constant gives the limit of the solubility of the electrolyte in water at a given temperature.

Consider some more K_{sp} expressions for slightly soluble electrolytes.

EXAMPLE 17.7   Write the K_{sp} expression for each of the following slightly soluble electrolytes:

a. $Ba_3(PO_4)_2$
b. $Mn(OH)_2$

SOLUTION

a. Obtain the K_{sp} for $Ba_3(PO_4)_2$:

$$Ba_3(PO_4)_s(s) \rightleftharpoons 3\,Ba^{2+}(aq) + 2\,PO_4^{3-}(aq)$$
$$\text{(in saturated solution)}$$

$$K_{sp} = [Ba^{2+}]^3[PO_4^{3-}]^2 \quad Answer$$

b. Obtain the K_{sp} for $Mn(OH)_2$:

$$Mn(OH)_s(s) \rightleftharpoons Mn^{2+}(aq) + 2\,OH^-(aq)$$
$$\text{(in saturated solution)}$$

$$K_{sp} = [Mn^{2+}][OH^-]^2 \quad Answer$$

Study Exercise 17.9

Write the K_{sp} expression for each of the following slightly soluble electrolytes.

a. Ag_2CrO_4 $(K_{sp} = [Ag^+]^2[CrO_4^{2-}])$

Work Problems 23 and 24.

b. $Fe(OH)_2$ $(K_{sp} = [Fe^{2+}][OH^-]^2)$

The following examples illustrate the application of solubility product equilibria to quantitative calculations. Calculations such as these should generally be limited to two significant digits because of experimental difficulties in measuring K_{sp} values.

EXAMPLE 17.8 The solubility of silver chloride in water at 25°C is 0.0018 g/L. Calculate the K_{sp} for AgCl at 25°C.

SOLUTION The equation for the equilibrium is

$$AgCl(s) \rightleftharpoons Ag^+(aq) + Cl^-(aq)$$
$$\text{(in saturated solution)}$$

and

$$K_{sp} = [Ag^+][Cl^-]$$

The K_{sp} expression uses the concentration of the ions in mol/L, hence, we must express the solubility in *mol/L*. The molar mass for AgCl is 143.4 g, and the concentration in mol/L (molarity) is

$$\frac{0.0018\text{ g AgCl}}{1\text{ L}} \times \frac{1\text{ mol AgCl}}{143.4\text{ g AgCl}} = 1.3 \times 10^{-5}\text{ mol AgCl/L}$$

The concentrations of both Ag^+ and Cl^- in the solution are 1.3×10^{-5} mol/L each because 1 mol of AgCl *in solution* produces 1 mol each of Ag^+ and Cl^- from the balanced equation. The solubility product constant for AgCl is, therefore,

$$K_{sp} = [Ag^+][Cl^-] = [1.3 \times 10^{-5}\text{ mol/L}][1.3 \times 10^{-5}\text{ mol/L}]$$
$$= 1.7 \times 10^{-10}\text{ (mol}^2/L^2) \quad Answer$$

Study Exercise 17.10

The solubility of silver iodide in water at 25°C is 2.8×10^{-6} g/L. Calculate the K_{sp} for AgI.

$$[1.4 \times 10^{-16}\text{ (mol}^2/L^2)]$$

Work Problems 25 and 26.

EXAMPLE 17.9   The solubility product constant for barium chromate is 2.0 $\times 10^{-10}$ (mol^2/L^2) at $2\overline{0}°C$.

a. Calculate the molarity of a saturated solution of $BaCrO_4$ at $2\overline{0}°C$.
b. What is its concentration in grams of $BaCrO_4$ per liter at $2\overline{0}°C$?
c. If the concentrations of $[Ba^{2+}]$ and $[CrO_4^{2-}]$ in a solution each reaches 1.0×10^{-4} mol/L, will precipitation occur?

SOLUTION
a. The equation for the equilibrium is

$$BaCrO_4(s) \;\rightleftharpoons\; Ba^{2+}(aq) + CrO_4^{2-}(aq)$$
$$\text{(in saturated solution)}$$

and

$$K_{sp} = 2.0 \times 10^{-10} \,(\text{mol}^2/\text{L}^2) = [Ba^{2+}][CrO_4^{2-}]$$

Let x = mol of $BaCrO_4$/L in a saturated solution. The concentrations of both Ba^{2+} and CrO_4^{2-} are equal to x mol/L because from the balanced equation 1 mol of $BaCrO_4$ *in solution* yields 1 mol of Ba^{2+} and 1 mol of CrO_4^{2-}. Hence,

$$x \text{ mol/L} = [Ba^{2+}] = [CrO_4^{2-}]$$

Express K_{sp} as

$$2.0 \times 10^{-10} \,\text{mol}^2/\text{L}^2 = [Ba^{2+}][CrO_4^{2-}] = [x][x] = [x]^2$$

$$x = \sqrt{2.0 \times 10^{-10} \frac{\text{mol}^2}{\text{L}^2}} = 1.4 \times 10^{-5} \frac{\text{mol}}{\text{L}} \qquad Answer$$

b. The molar mass of $BaCrO_4$ is 253.3 g; hence, calculate the solubility in g/L as

$$\frac{1.4 \times 10^{-5} \text{ mol BaCrO}_4}{1 \text{ L}} \times \frac{253.3 \text{ g BaCrO}_4}{1 \text{ mol BaCrO}_4} = 3.6 \times 10^{-3} \text{ g} \frac{\text{BaCrO}_4}{\text{L}} \qquad Answer$$

c. $[Ba^{2+}] = [CrO_4^{2-}] = 1.0 \times 10^{-4}$ mol/L

The product of the concentration of the ions raised to their respective powers is

$$[Ba^{2+}][CrO_4^{2-}] = (1.0 \times 10^{-4})(1.0 \times 10^{-4})(\text{mol}^2/\text{L}^2) = 1.0 \times 10^{-8} \,(\text{mol}^2/\text{L}^2)$$

This value $[1.0 \times 10^{-8} \,(\text{mol}^2/\text{L}^2)]$ is *greater* than the value of the K_{sp} $[2.0 \times 10^{-10}$ or $0.02 \times 10^{-8} \,(\text{mol}^2/\text{L}^2)]$ and so, yes, precipitation will occur.   *Answer*

Study Exercise 17.11
The solubility product constant for strontium oxalate is 5.6×10^{-8} (mol^2/L^2) at 18°C.

a. Calculate the molarity of a saturated solution of SrC_2O_4 at 18°C.
$$(2.4 \times 10^{-4} \,M)$$
b. What is its concentration in grams of SrC_2O_4 per liter at 18°C?
$$(4.2 \times 10^{-2} \,\text{g/L})$$
c. If the concentrations of Sr^{2+} and $C_2O_4^{2-}$ in a solution each reaches 1.4×10^{-5} mol/L, will precipitation occur?

(No)   Work Problems 27 and 28.

CHEMISTRY OF THE ATMOSPHERE

The Chemical Reactions Behind Ozone Depletion

The essay following Chapter 13 described the origins and beneficial effects of the layer of ozone that defines the boundary between the stratosphere and the troposphere. In that essay, we noted that ozone absorbs short-wavelength ultraviolet light and reduces the amount of this high-energy light that reaches the earth. The question that remained unanswered was *how* chlorofluorocarbons (CFCs) convert ozone back into oxygen gas.

Many scientists believe that chlorofluorocarbons *lacking hydrogen atoms* are the most significant contributors to ozone depletion. The most important examples include trichlorofluoromethane (CFC-11, CCl_3F) and dichlorodifluoromethane (CFC-12, CCl_2F_2). Chlorofluorocarbons that *do* contain hydrogen are believed to be less threatening because they may degrade in the troposphere fast enough so that they do *not* reach the stratosphere. An example of this type of chlorofluorocarbon is chlorodifluoromethane ($CHClF_2$), a refrigerant used in home air conditioners.

In contrast, consider the problem of dichlorodifluoromethane (CFC-12, CCl_2F_2). The process begins when a CCl_2F_2 molecule moves up through the troposphere and into the stratosphere. While many chlorofluorocarbon molecules never reach this altitude, these compounds are very stable

If stratospheric ozone destruction becomes significant, the beach scene will be hazardous to your health!

and some eventually are blown high enough to get above the ozone layer. Once it reaches this altitude, the molecule is no longer shielded from the heavy doses of ultraviolet light. This light has enough energy to break the carbon–chlorine covalent bond:

$$\text{F} - \overset{\displaystyle \text{Cl}}{\underset{\displaystyle \text{Cl}}{\text{C}}} - \text{F} + \text{ultraviolet light} \longrightarrow$$

$$\text{Cl}(g) + \quad \text{F} - \overset{\displaystyle \text{Cl}}{\underset{\displaystyle \cdot}{\text{C}}} - \text{F}$$

The C — Cl covalent bond is broken to give a chlorine atom (Cl(*g*) or ·C̈l:) and a ·CClF_2 radical (see *Chemistry of the Atmosphere* at the end of Chapter 16), with the dot representing the unpaired electron. Some of these chlorine atoms then fall back to earth through the stratospheric ozone layer. Here they encounter and react with ozone molecules and

oxygen atoms according to the following two reactions:

$$\text{Cl}(g) + \text{O}_3(g) \longrightarrow \text{ClO}(g) + \text{O}_2(g)$$

$$\text{ClO}(g) + \text{O}(g) \longrightarrow \text{Cl}(g) + \text{O}_2(g)$$

Reactions like these are called *chain reactions*, because a *product* [Cl(*g*)] of the final reaction in the sequence is a *reactant* for the starting reaction. Thus, one chlorine atom can react with ozone many times and consume as many as 100,000 molecules of ozone in the process. Thus, even if just a few chlorofluorocarbon molecules reach this altitude, they can conceivably convert a significant amount of ozone into oxygen gas, as the accompanying figure shows.

As noted in *Chemistry of the Atmosphere* in Chapter 13, these effects are very dramatic over the Arctic in the spring and Antarctica in the fall. The cold stratospheric temperatures and special effects of ice particles in stratospheric clouds over the poles render the destruction mechanism especially effective at these times of the year. Ozone losses as high as 95% in stratospheric clouds and 60% over the polar regions have been observed when conditions are just right. At the same time, monitors detected large increases in the amount of chlorine monoxide (ClO) in the stratosphere. After a few weeks, when somewhat warmer temperatures return and the stratospheric clouds do not form, the destruction process subsides. Wind processes then mix in gases from lower altitudes, and ozone levels return to near normal. Nevertheless, these studies show that under the right conditions the chlorine–chlorine monoxide process can be devastating to stratospheric ozone.

How much ozone has the stratosphere lost so far? Monitoring the amount of ozone worldwide is difficult since the amounts vary with the time of year, the altitude,

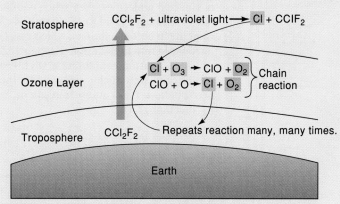

A molecule of dichlorodifluoromethane (CCl_2F_2) can move into the stratosphere and encounter high-energy ultraviolet light. This light can break the C—Cl bond to produce chlorine atoms, which in turn can convert many ozone molecules to oxygen molecules as shown.

and the latitude. Nevertheless, some studies indicate a 1 to 3% loss of ozone worldwide. The clearest demonstration of the overall process has been the appearance of increasing amounts of chlorine monoxide in the stratosphere over the past 10 years.

Even more important questions remain unanswered: (1) How soon can political and economic issues be worked out and CFC use be reduced? (2) How much damage will be done by the chlorofluorocarbons that have already been put into the atmosphere?

✓ Summary

The rate of a chemical reaction is the speed at which products are produced or reactants are consumed. The rate of a reaction depends on (1) the nature of the reactants, (2) the concentrations of the reactants, (3) the temperature, and (4) the presence of a catalyst. Reactants must collide with sufficient energy before they can react. This energy is expressed as the *activation energy* (Section 17.1).

A state of equilibrium is achieved when the composition of a reaction mixture ceases to change. All equilibria are *dynamic equilibria* and are characterized by a forward and a reverse reaction, both proceeding at the same rate. At equilibrium, some reactions have significant concentrations of reactants present. Still others go essentially to completion because the reverse reactions are very slow. These reactions are *irreversible reactions*, and they proceed because one of the products is (1) a gas, (2) a precipitate, or (3) a nonionized or only partially ionized substance, such as water. The composition of an equilibrium mixture is described by the *equilibrium constant expression* (Section 17.2).

Le Châtelier's principle states that if an equilibrium system is subjected to a change in conditions of concentration, temperature, or pressure, the system will change in a direction that tends to restore the original conditions (Section 17.3).

Because weak acids and bases are only a few percent ionized and contain some nonionized molecules, they reach equilibrium in accordance with the special equilibrium constants K_a and K_b (Section 17.4). Weak acids and their conjugate bases can also combine in buffer solutions, which resist changes in pH when strong acids or bases are introduced into the solution (Section 17.5).

In a solution containing a slightly soluble electrolyte and its ions, the dynamic equilibrium state is characterized by the constant dissolving of the electrolyte and its precipitation. This equilibrium can be expressed using the solubility product constant K_{sp} (Section 17.6).

 # Exercises

1. Define or explain the following terms (the number in parentheses refers to the section number in the text where the term is mentioned):

   a. reaction rate (17.1)

   b. law of mass action (17.1)

   c. specific rate constant (17.1)

   d. activation energy (17.1)

   e. dynamic equilibrium (17.2)

   f. equilibrium constant expression (17.2)

   g. Le Châtelier's principle (17.3)

   h. equilibrium position (17.3)

   i. buffer solution (17.5)

   j. solubility product constant (17.6)

2. Distinguish between:

   a. chemical kinetics and chemical equilibrium

   b. reaction rate and equilibrium

   c. equilibrium constant and equilibrium position

   d. K_a and K_b

   e. K_a and K_{sp}

Reaction Rates (See Section 17.1)

3. List and describe, in your own words, the four factors that influence the rate of a reaction.

4. The yield of a certain industrial preparation at equilibrium appears to be progressively decreasing in a certain chemical plant. One suggested solution to the problem is to study various catalysts that can increase the yield. What do you think about this solution to the problem?

Reversibility and Chemical Equilibrium (See Section 17.2)

5. List three classes of products that act as a driving force for a reaction to go toward completion, with the point of equilibrium shifted far toward the products side.

Le Châtelier's Principle (See Section 17.3)

6. List and describe, in your own words, the three changes in conditions that affect the equilibrium according to Le Châtelier's principle.

 # Problems

Reaction Rates (See Section 17.1)

7. Given the reaction

$$2\,NO(g) + 2\,H_2(g) \longrightarrow N_2(g) + 2\,H_2O(g)$$

with the reaction rate $= k[NO]^2\,[H_2]$. Calculate the effect on the reaction rate if the following occur:

a. The concentration of NO is increased from 0.10 M to 0.20 M with the H_2 concentration remaining the same.

b. The concentration of NO is increased from 0.10 M to 0.30 M with the H_2 concentration remaining the same.

c. The concentration of H_2 is increased from 0.10 M to 0.20 M with the NO concentration remaining the same.

Reversibility of Reactions and Chemical Equilibrium (See Section 17.2)

8. Write the expression for the equilibrium constant for each of the following reactions:

a. $CH_4(g) + Cl_2(g) \rightleftharpoons CH_3Cl(g) + HCl(g)$

b. $SO_2(g) + NO_2(g) \rightleftharpoons SO_3(g) + NO(g)$

c. $2\,NO(g) + O_2(g) \rightleftharpoons 2\,NO_2(g)$

d. $NH_4Cl(s) \rightleftharpoons NH_3(g) + HCl(g)$

e. $4\,H_2O(g) + 3\,Fe(s) \rightleftharpoons Fe_3O_4(s) + 4\,H_2(g)$

f. $3\,O_2(g) \rightleftharpoons 2\,O_3(g)$

g. $SO_2Cl_2(g) \rightleftharpoons SO_2(g) + Cl_2(g)$

h. $4\,NH_3(g) + 5\,O_2(g) \rightleftharpoons 4\,NO(g) + 6\,H_2O(g)$

i. $2\,Pb\,(NO_3)_2(s) \rightleftharpoons 2\,PbO(s) + 4\,NO_2(g) + O_2(g)$

j. $2\,H_2(g) + O_2(g) \rightleftharpoons 2\,H_2O(g)$

9. Write the expression for the equilibrium constant for each of the following reactions:

a. $N_2(g) + O_2(g) \rightleftharpoons 2\,NO(g)$

b. $2\,SO_2(g) + O_2(g) \rightleftharpoons 2\,SO_3(g)$

c. $BaSO_3(s) \rightleftharpoons BaO(s) + SO_2(g)$

d. $C(s) + H_2O(g) \rightleftharpoons CO(g) + H_2(g)$

e. $2\,NOCl(g) \rightleftharpoons 2\,NO(g) + Cl_2(g)$

f. $PCl_5(g) \rightleftharpoons PCl_3(g) + Cl_2(g)$

g. $2\,HgO(s) \rightleftharpoons 2\,Hg(g) + O_2(g)$

h. $COBr_2(g) \rightleftharpoons CO(g) + Br_2(g)$

   i. $CH_4(g) + 2\,O_2(g) \rightleftharpoons CO_2(g) + 2\,H_2O(g)$
   j. $2\,Ag(s) + Cl_2(g) \rightleftharpoons 2\,AgCl(s)$

10. Listed in the following table are ionization constants for various acids (K_a) at 25°C. Arrange them in order of decreasing acid strength by their K_a values.

ACID (IN APPROXIMATELY 0.1 N AQUEOUS SOLUTIONS)	K_a at 25°C (mol/L)
Acetic acid	1.76×10^{-5}
Barbituric acid	9.8×10^{-5}
Chloroacetic acid	1.40×10^{-3}
Formic acid	1.77×10^{-4}
Lactic acid	1.37×10^{-4}
Sulfurous acid	1.72×10^{-2}

11. Listed in the following table are ionization constants for various bases (K_b) at 25°C. Arrange them in order of decreasing base strength by their K_b values.

BASE (IN APPROXIMATELY 0.1 N AQUEOUS SOLUTION)	K_b at 25°C (mol/L)
Ammonia	1.79×10^{-5}
Codeine	9×10^{-7}
Nicotine	7×10^{-7}
Novocain	7×10^{-6}
Silver hydroxide	1.1×10^{-4}
Urea	1.5×10^{-14}

Le Châtelier's Principle (See Section 17.3)

12. One way to prepare chlorine gas is by using the Deacon process, which involves the following equilibrium:

$$4\,HCl(g) + O_2(g) \rightleftharpoons 2\,Cl_2(g) + 2\,H_2O(g)$$

In which direction will the equilibrium be shifted by each of the following changes?

   a.   increasing the concentration of HCl
   b.   decreasing the concentration of HCl
   c.   increasing the concentration of O_2
   d.   decreasing the concentration of O_2
   e.   increasing the concentration of Cl_2
   f.   decreasing the concentration of Cl_2
   g.   increasing the concentration of H_2O
   h.   decreasing the concentration of H_2O
   i.   increasing the pressure by compression
   j.   decreasing the pressure by expansion

13. Commercial preparation of hydrogen gas often involves the following equilibrium:

$$CO(g) + H_2O(g) \rightleftharpoons CO_2(g) + H_2(g)$$

In which direction will the equilibrium be shifted by each of the following changes?

a. increasing the concentration of CO

b. decreasing the concentration of CO

c. increasing the concentration of H_2O

d. decreasing the concentration of H_2O

e. increasing the concentration of CO_2

f. decreasing the concentration of CO_2

g. increasing the concentration of H_2

h. decreasing the concentration of H_2

i. increasing the pressure by compression

j. decreasing the pressure by expansion

14. Predict the effect on equilibrium of the following chemical reactions when (1) the temperature is increased, (2) the temperature is decreased, (3) the pressure is increased by compression, and (4) the pressure is decreased by expansion:

a. $2 H_2(g) + O_2(g) \rightleftharpoons 2 H_2O(g) + 115.6$ kcal

b. $H_2(g) + Cl_2(g) \rightleftharpoons 2 HCl(g) + 185$ kJ

c. $H_2(g) + I_2(g) \rightleftharpoons 2 HI(g) - 51.9$ kJ

d. $2 F_2(g) + O_2(g) \rightleftharpoons 2 OF_2(g) - 46.0$ kJ

e. $4 Al(s) + 3 O_2(g) \rightleftharpoons 2 Al_2O_3(s) + 798.2$ kcal

15. Predict the effect on equilibrium of the following reactions when (1) the temperature is increased, (2) the temperature is decreased, (3) the pressure is increased by compression, and (4) the pressure is decreased by expansion:

a. $C_6H_6(g) + 3 H_2(g) \rightleftharpoons C_6H_{12}(g) + 206$ kJ

b. $2 NO(g) \rightleftharpoons N_2(g) + O_2(g) + 43.2$ kcal

c. $2 CO(g) + O_2(g) \rightleftharpoons 2 CO_2(g) + 135.2$ kcal

d. $N_2(g) + 2 O_2(g) \rightleftharpoons 2 NO_2(g) - 16.2$ kcal

e. $C(s) + O_2(g) \rightleftharpoons CO_2(g) + 393$ kJ

Weak Electrolyte Equilibria (See Section 17.4)

16. Calculate the ionization constants for each of the following weak electrolytes from the percent ionization at the concentration given:

a. A 0.500 M solution of aqueous NH_3 is 0.600% ionized.

b. A 0.100 M solution of HF is 8.23% ionized.

c. A 0.800 M solution of HCN is 0.00300% ionized.

d. A 0.350 M solution of HA is 1.50% ionized.

17. Calculate the ionization constants for each of the following weak electrolytes from the percent ionization at the concentration given:

a. A 2.00 M solution of $HC_2H_3O_2$ is 0.300% ionized.

b. A 0.0100 M solution of formic acid ($HCHO_2$) is 13.1% ionized. (*Hint:* $HCHO_2(aq) \rightleftharpoons H^+(aq) + CHO_2^-(aq)$.)

c. A 0.0300 M solution of MOH is 5.00% ionized.

d. A 0.0200 M solution of HA is 9.00% ionized.

18. From the ionization constants of each of the following weak electrolytes, calculate the hydrogen ion concentration (for acids) or the hydroxide ion concentration (for bases) in moles per liter and the percent ionization of the weak electrolyte in each of the following solutions:

    a.   0.200 M acetic acid at 5°C. The K_a for $HC_2H_3O_2$ = 1.70 × 10^{-5} (mol/L) at 5°C.

    b.   0.500 M aqueous ammonia at $2\overline{0}$°C. The K_b for aqueous NH_3 = 1.70 × 10^{-5} (mol/L) at $2\overline{0}$°C.

    c.   2.00 M formic acid at $5\overline{0}$°C. The K_a for $HCHO_2$ = 1.65 × 10^{-4} (mol/L) at $5\overline{0}$°C. (*Hint:* $HCHO_2(aq) \rightleftharpoons H^+(aq) + CHO_2^-(aq)$.)

    d.   0.250 M MOH. The K_b for MOH = 6.40 × 10^{-7} (mol/L).

19. From the ionization constants of each of the following weak electrolytes, calculate the hydrogen ion concentration (for acids) or the hydroxide ion concentration (for bases) in mol/L and the percent ionization of the weak electrolyte in each of the following solutions:

    a.   0.100 M hypochlorous acid at 25°C. The K_a for HClO = 3.50 × 10^{-9} (mol/L) at 25°C.

    b.   0.0278 M hypoiodous acid at 25°C. The K_a for HIO = 2.30 × 10^{-11} (mol/L) at 25°C.

    c.   0.0500 M MOH. The K_b for MOH = 1.10 × 10^{-6} (mol/L).

    d.   0.300 M MOH. The K_b for MOH = 3.00 × 10^{-8} (mol/L).

Buffer Equilibria (See Section 17.5)

20. Given the following buffer solutions, identify the weak acid and conjugate base and write a chemical equation for the equilibrium involving these species. In addition, write a chemical equation that can neutralize some added hydrochloric acid or sodium hydroxide.

    a.   Solution containing acetic acid and potassium acetate

    b.   Solution containing ammonium chloride and ammonia

    c.   Solution containing hydrocyanic acid and sodium cyanide

    d.   Solution containing formic acid ($HCHO_2$) and sodium formate ($NaCHO_2$). (*Hint:* See Problem 18, part (c).)

21. According to the discussion in Section 15.5, the pH of blood is controlled by several buffers. For the blood buffers indicated below, identify the weak acid and its conjugate base and write an equation for the equilibrium involving these species. In addition, write the reactions that can serve to neutralize added H^+ or OH^- in the blood.

    a.   Carbonic acid (H_2CO_3) and hydrogen carbonate or bicarbonate (HCO_3^-)

    b.   Dihydrogen phosphate ($H_2PO_4^-$) and hydrogen phosphate (HPO_4^{2-})

Solubility Equilibria (See Section 17.6)

22. Listed in the following table are solubility product constants (K_{sp}) for various slightly soluble electrolytes in water at $2\overline{0}$°C. Arrange them in order of decreasing solubility in water.

SLIGHTLY SOLUBLE ELECTROLYTE	K_{sp} at $20°C$ (mol^2/L^2)
Silver acetate, $AgC_2H_3O_2$	4.0×10^{-3}
Silver iodide, AgI	8.5×10^{-17}
Silver bromate, $AgBrO_3$	6.0×10^{-5}
Silver bromide, $AgBr$	5.0×10^{-13}
Silver chloride, $AgCl$	1.8×10^{-10}

23. Write the expression for the solubility product constant (K_{sp}) for each of the following slightly soluble electrolytes:

   a. $AgCl$ b. CaF_2 c. $Al(OH)_3$
   d. $Pb_3(AsO_4)_2$ e. Ag_2CO_3

24. Write the expression for the solubility product constant (K_{sp}) for each of the following slightly soluble electrolytes:

   a. $AgOH$ b. BaF_2 c. $Fe(OH)_3$
   d. $Pb(IO_3)_2$ e. HgI_2

25. From the solubility of each of the following compounds in pure water at a given temperature, calculate the solubility product constant (K_{sp}) for the compound at that temperature.

   a. Silver bromide: 0.00016 g of $AgBr$/L at 25°C
   b. Barium carbonate: 9.0×10^{-5} mol of $BaCO_3$/L at 18°C
   c. Lead(II) sulfate: 1.0×10^{-4} mol of $PbSO_4$/L at 18°C
   d. AB: 0.0013 g of AB/L (molar mass AB = 85.0 g) at 25°C

26. From the solubility of each of the following compounds in pure water at a given temperature, calculate the solubility product constant (K_{sp}) for the compound at that temperature:

   a. Strontium carbonate: 0.0059 g of $SrCO_3$/L at 25°C
   b. Copper(I) iodide: 4.2×10^{-4} g of CuI/L at 18°C
   c. Barium sulfate: 3.3×10^{-3} g/L of $BaSO_4$ at 50°C
   d. AB: 0.00081 mol of AB/L at 25°C

27. From the solubility product constant for each of the following salts, (1) calculate the molarity of a saturated solution of the salt at the given temperature, and (2) calculate the solubility of the salt in g/L at the given temperature:

   a. Magnesium oxalate at 18°C; K_{sp} for MgC_2O_4 = 8.6×10^{-5} (mol^2/L^2) at 18°C
   b. Barium carbonate at 16°C; the K_{sp} for $BaCO_3$ = 7.0×10^{-9} (mol^2/L^2) at 16°C
   c. Thallium(I) bromide at 25°C; the K_{sp} for $TlBr$ = 4.0×10^{-6} (mol^2/L^2) at 25°C
   d. Strontium carbonate at 25°C; the K_{sp} for $SrCO_3$ = 1.6×10^{-9} (mol^2/L^2) at 25°C. If the concentrations of Sr^{2+} and $CO_3{}^{2-}$ in a solution each reaches 9.2×10^{-5} mol/L, will precipitation occur?

28. From the solubility product constant for each of the following salts, (1) calculate the molarity of a saturated solution of the salt at the given temperature, and (2) calculate the solubility of the salt in grams per liter at the given temperature:

   a. Silver iodide at 25°C; the K_{sp} for AgI $= 1.5 \times 10^{-16}$ (mol²/L²) at 25°C

   b. Lead(II) carbonate at 18°C; the K_{sp} for PbCO₃ $= 3.3 \times 10^{-14}$ (mol²/L²) at 18°C

   c. Calcium chromate at 18°C; the K_{sp} for CaCrO₄ $= 2.3 \times 10^{-2}$ (mol²/L²) at 18°C

   d. Barium chromate at 28°C; the K_{sp} for BaCrO₄ is 2.4×10^{-10} (mol²/L²) at 28°C. If the concentrations of Ba²⁺ and CrO₄²⁻ in a solution each reaches 3.6×10^{-6} mol/L, will precipitation occur?

General Problems

29. Calculate the pH and pOH of each of the following solutions:

   a. A 0.530 M aqueous solution of acetic acid at 5°C; the K_a for HC₂H₃O₂ $= 1.70 \times 10^{-5}$ (mol/L) at 5°C.

   b. A 0.100 M aqueous solution of aqueous ammonia at 2̄0°C; the K_b for aqueous NH₃ $= 1.70 \times 10^{-5}$ (mol/L) at 2̄0°C.

30. Calculate the pH and pOH of each of the following solutions:

   a. 0.01000 M aqueous solution of acetic acid at 2̄0°C; HC₂H₃O₂ is 4.20% ionized at 2̄0°C.

   b. A 0.500 M solution of aqueous ammonia at 27°C; aqueous NH₃ is 0.600% ionized at 27°C.

31. Swimmers often complain that the chlorine in the swimming pool "burns" their eyes when the pH is low. Explain this fact in terms of pH and chlorine formation. The ionic equation for the reaction of chlorine with water is

$$\text{Cl}_2 + \text{H}_2\text{O} \rightleftharpoons \text{HClO} + \text{H}^+ + \text{Cl}^-$$

✓ Chapter Quiz 17A
(Sections 17.1 to 17.3)

1. Listed in the following table are ionization constants for various acids (K_a) at 25°C. Arrange them in order of decreasing acid strength by their K_a values.

ACID	K_a at 25°C (mol/L)
Acetic acid	1.75×10^{-5}
Trichloroacetic acid	2.32×10^{-1}
Chloroacetic acid	1.36×10^{-3}
Dichloroacetic acid	5.53×10^{-2}

2. Write the expression for the equilibrium constant for each of the following reactions:

   a. $2\,NO(g) + O_2(g) \rightleftharpoons 2\,NO_2(g)$

   b. $N_2O_4(g) \rightleftharpoons 2\,NO_2(g)$

   c. $C(s) + CO_2(g) \rightleftharpoons 2\,CO(g)$

   d. $ZnSO_3(s) \rightleftharpoons ZnO(s) + SO_2(g)$

3. Given the following equilibrium mixture at a given temperature:

$$CuO(s) + CO(g) \rightleftharpoons Cu(s) + CO_2(g)$$

   a. Predict the effect on equilibrium when the concentration of $CO(g)$ is increased.

   b. Predict the effect on equilibrium when the concentration of $CO_2(g)$ is increased.

4. Given the following equilibrium mixture at a given temperature:

$$16.2\ kcal + N_2(g) + 2\,O_2(g) \rightleftharpoons 2\,NO_2(g)$$

   a. Predict the effect on equilibrium when the temperature is increased.

   b. Predict the effect on equilibrium when the pressure is increased by compression.

✓ Chapter Quiz 17B
(Sections 17.4 to 17.6)

1. Write the expression for the solubility product constant (K_{sp}) for each of the following slightly soluble electrolytes:

   a. PbI_2 b. $Ag_2Cr_2O_7$

2. Calculate the ionization constant (K_a) for HA if a 0.500 M solution of HA is 1.30% ionized.

3. Calculate the hydrogen ion concentration in mol/L for a 0.200 M HA solution. $K_a = 4.37 \times 10^{-7}$ mol/L for HA at 25°C.

4. Calculate the solubility product constant at 25°C for the salt AgBr if the solubility of the salt in water at 25°C is 1.6×10^{-4} g/L. (Atomic masses: Ag = 107.9 amu, Br = 79.9 amu.)

5. The K_{sp} for thallium(I) chloride (TlCl) in water is 2.7×10^{-4} at 25°C. What is the solubility of the salt in grams per liter at 25°C? (Atomic masses: Tl = 204.4 amu, Cl = 35.5 amu.)

Nitrogen (Symbol: N)

The Element NITROGEN: Our Food Depends On It

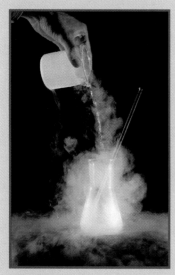

Liquid nitrogen is colorless and boils at −196°C. It is an important coolant for use in the laboratory.

Name: From the Middle English *nitre-* and the French *-gen*, meaning "produced by niter." Niter is the name for KNO_3 or saltpeter, which was use to make gunpowder.

Appearance: Colorless, odorless, and tasteless gas at room temperature. It liquefies at −196°C to give a colorless liquid that looks just like water (but is *much* colder!).

Occurrence: Nitrogen occurs in the free (uncombined) form as diatomic nitrogen (N_2) in the atmosphere. Other forms of nitrogen (ammonia, nitrite, and nitrate) can be found in soil, water, or rock formations. Most nitrogen (98%) is tied up in rocks, leaving about 2% in the atmosphere and <1% in the soil and water.

Source: Virtually all nitrogen produced for industrial applications comes from the liquefaction of air, which is 80% nitrogen gas by volume. Oxygen gas is coproduced in this process.

Common Uses: The most important use of nitrogen gas is in the production of ammonia (NH_3). Ammonia is the key chemical in the production of nitrogen fertilizers such as ammonia, ammonium nitrate (NH_4NO_3), and urea ($H_2N—CO—NH_2$). In fact, 85% of all ammonia is used in the production of fertilizers. Crop yields are typically limited by the availability of nitrogen-containing fertilizers, and so nitrogen is crucial to the world *food production* picture. Most ammonia is prepared on a large scale by the Haber process (see Section 17.3).

All living systems need nitrogen for normal biological processes. The nitrogen cycle (see the figure) describes the processes that control distribution of the various forms of nitrogen within the environment. Bacteria convert nitrogen gas into ammonia by a reduction process called nitrogen fixation (1). Ammonia (or the ammonium ion, NH_4^+) is the key species in the use of nitrogen by living organisms. Other redox processes (2) in living systems convert ammonia into nitrite and nitrate, which serve various roles in their metabolisms. Finally, plants and bacteria reduce nitrite and nitrate back to nitrogen or ammonia to complete the cycle (3).

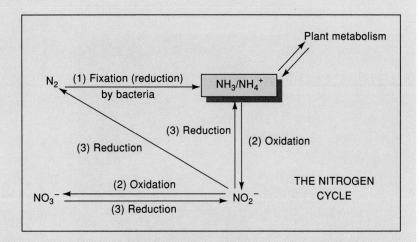

Life on earth depends on the balance among these processes.

Nitric acid is an important strong acid that is used in oxidations and the production of dyes and explosives. Ammonia, and thus nitrogen, is the raw material for the production of nitric acid (HNO_3) by the Ostwald process:

$$4 NH_3(g) + 7 O_2(g) \longrightarrow 4 NO_2(g) + 6 H_2O(l)$$

$$3 NO_2(g) + H_2O(l) \longrightarrow 2 HNO_3(aq) + NO(g)$$

Nitrogen gas is also used in the production of nitrogen-containing resins like melamine, and liquid nitrogen is increasingly used as a "cryogen" for the cooling of materials to very low temperatures. This technology could be extremely important in the development of superconducting ceramics (see The Element COPPER: Electrical Conductivity and High-Speed Trains, Chapter 8).

Unusual Facts: Typically, a mixture of 88% argon and 12% nitrogen gas is used as the "filler" gas in common incandescent light bulbs.

CHAPTER 18

Organic Chemistry

From the plastic of the frisbee to the many biochemical compounds in the grass, dog, and children, probably 99% of the chemicals in this scene are organic.

GOALS FOR CHAPTER 18

1. To define organic chemistry and to explain the classification of organic compounds (Section 18.1).

2. To identify the three basic bonding arrangements associated with carbon in organic molecules and to describe their characteristics and an example of each (Section 18.2).

3. To define alkanes and isomers and to draw condensed structural formulas for isomeric alkanes. To name and write structural formulas for alkanes and to complete and balance chemical equations for monohalogenation (chlorine or bromine) of alkanes (Section 18.3).

4. To define alkenes, to name and write structural formulas for alkenes, and to complete and balance chemical equations for halogenation (chlorine or bromine) of alkenes (Section 18.4).

5. To define alkynes, to name and write structural formulas for alkynes, and to complete and balance chemical equations for halogenation (chlorine or bromine) of alkynes (Section 18.5).

6. To define aromatic hydrocarbons, to name and write structural formulas for aromatic hydrocarbons, and to complete and balance chemical equations for halogenation (chlorine or bromine) of aromatic hydrocarbons (Section 18.6).

7. To define a functional group and to identify the functional group(s) of hydrocarbon derivatives in organic compounds (Section 18.7).

Countdown

You may use the periodic table.

5. Using the periodic table, indicate the number of valence electrons for the following elements (Section 5.3).
   a. carbon (4) **b.** chlorine (7)
   c. oxygen (6) **d.** phosphorus (5)

4. Write the electron-dot formulas for each of the following atoms (Sections 5.3 and 4.7).

   a. $^{12}_{6}C$ $(\cdot\overset{\cdot}{C}\cdot)$ **b.** $^{35}_{17}Cl$ $(:\overset{\cdot\cdot}{\underset{\cdot\cdot}{Cl}}\cdot)$

   c. $^{16}_{8}O$ $(\cdot\overset{\cdot\cdot}{\underset{\cdot\cdot}{O}}:)$ **d.** $^{31}_{15}P$ $(\cdot\overset{\cdot\cdot}{P}\cdot)$

3. Write Lewis structures and structural formulas for the following molecules (Section 6.7).

   a. CCl_4
   $$\left(\begin{array}{c} :\overset{\cdot\cdot}{Cl}: \\ :\overset{\cdot\cdot}{Cl}\cdot\overset{\cdot\cdot}{C}\cdot\overset{\cdot\cdot}{Cl}:, \\ :\overset{\cdot\cdot}{Cl}: \end{array} \quad \begin{array}{c} Cl \\ | \\ Cl-C-Cl \\ | \\ Cl \end{array} \right)$$

   b. CO_2 $\left(:\overset{\cdot\cdot}{O}: \; C \; :\overset{\cdot\cdot}{O}:, \; O{=}C{=}O \right)$

2. Write the correct name for each of the following compounds (Sections 7.2 and 7.6).
   a. CCl_4 (carbon tetrachloride)
   b. CO_2 (carbon dioxide)
   c. CO (carbon monoxide)
   d. $HC_2H_3O_2(aq)$ (acetic acid)
   e. $H_2C_2O_4(aq)$ (oxalic acid)

1. Complete and balance the following chemical reaction equations; indicate any gas by (g) (Section 13.5).
   a. $C_2H_6(g) \; + \; O_2(g) \; \xrightarrow{\Delta}$
   (excess)
   $$[2\,C_2H_6(g) \; + \; 7\,O_2(g) \; \xrightarrow{\Delta}$$
   $$4\,CO_2(g) \; + \; 6\,H_2O(g)]$$

   b. $C_2H_6O(l) \; + \; O_2(g) \; \xrightarrow{\Delta}$
   ethyl alcohol (excess)
   $$[C_2H_6O(l) \; + \; 3\,O_2(g) \; \xrightarrow{\Delta}$$
   $$2\,CO_2(g) \; + \; 3\,H_2O(g)]$$

W hat do diamonds, pencils, some golf clubs shafts, and living tissue have in common? You guessed it—they all contain carbon in one form or another. You may have noticed that we have not said much about carbon, despite the fact that it is the basis of all living—and many nonliving—things. In this chapter, we will consider a variety of substances containing carbon, their properties, and their uses.

18.1 Organic Chemistry

Scientists refer to the study of substances containing carbon as **organic chemistry** (see Section 1.2). This name arose because originally it was thought that all such compounds involved *living* matter. Today, however, chemists have been able to create many carbon-containing compounds in the laboratory without the use of any living matter.

Nevertheless, the term "organic chemistry" remains in use to distinguish these compounds from the noncarbon compounds that make up the study of *inorganic chemistry*. The number of known organic compounds, about 4 million, far exceeds the number of inorganic compounds, approximately 100,000. Some compounds that contain carbon, such as those containing the polyatomic ions cyanide (CN^-), hydrogen carbonate (HCO_3^-), and carbonate (CO_3^{2-}), are generally considered inorganic compounds. They are so classed because their properties are more similar to those of inorganic compounds than to those of organic compounds.

Organic compounds are not only the foundation of all human, animal, and plant life but are also crucial to sustaining that life. The wood that has sheltered and given heat to humans over the ages is made of organic materials. Indeed, most sources of fuel for heat and locomotion—coal, coke, peat, and petroleum—are mixtures of organic compounds. These substances are available to us today thanks to thousands of years of natural compression of dead plant and animal matter.

Moreover, coal and petroleum, and other organic substances, serve as raw materials for many other carbon compounds. Carbon is at the heart of many products we use every day: natural fibers (cotton, wool, rayons) and synthetic polyamide and polyester fibers (nylon and Dacron, respectively); vitamins (A, B_1, B_2, B_6, B_{12}, C, D, E, K); hormones (estrone, progesterone, testosterone, insulin, corticosterone, adrenalin); and drugs such as aspirin, caffeine, antihistamines, and antibiotics (penicillins, streptomycin, tetracyclines).

Organic compounds fall into two broad categories: (1) hydrocarbons and (2) derivatives of hydrocarbons. **Hydrocarbons** are organic compounds that contain only the elements carbon and hydrogen. The organic compounds in oil and coal are largely hydrocarbons. Hydrocarbons, in turn, are divided into groups based on the structure of the compounds: the *aliphatic* hydrocarbons and the *aromatic* hydrocarbons. The **aliphatic hydrocarbons** are further divided into three general groups: the alkanes, the alkenes, and the alkynes. Figure 18.1 summarizes the classification of organic compounds. We will consider each of these subclasses in this chapter.

Before we look at these subclasses, however, we need to know something about the types of bonding arrangements that characterize carbon in organic compounds.

Organic chemistry Study of the substances containing carbon.

Hydrocarbons Organic compounds that contain only the elements carbon and hydrogen.

Aliphatic hydrocarbons Hydrocarbons composed of alkanes, alkenes, and alkynes.

FIGURE 18.1
Classification of organic
compounds.

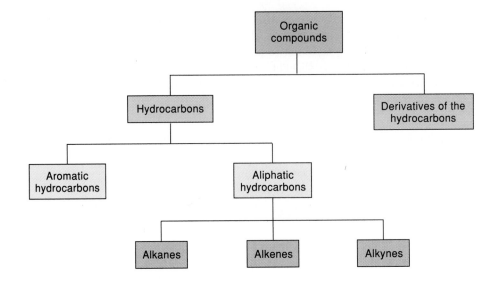

18.2 The Shapes and Structures of Organic Molecules

Chemists have studied organic substances in a systematic way for over 100 years. During that time they have discovered that carbon atoms exist in three basic bonding arrangements depending on the number of groups to which the carbon is bonded:

Category 1: Carbon atoms bonded to *four* atoms or groups of atoms

Category 2: Carbon atoms bonded to *three* atoms or groups of atoms

Category 3: Carbon atoms bonded to *two* atoms or groups of atoms.

The shape or geometry of an organic substance is determined by the number of groups bonded to the carbon atom. Let's examine four simple carbon-containing molecules that illustrate these different geometries.

Category 1: Carbon to Four Atoms. The simplest of all organic molecules is methane. A methane molecule (CH_4) is composed of four hydrogen atoms and one carbon atom. Methane is an example of a carbon atom that forms four covalent bonds to four atoms, and the Lewis structure and structural formula for methane are

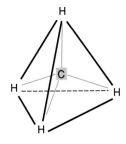

FIGURE 18.2
The structure of methane is described as *tetrahedral* because the hydrogen atoms occupy the corners of a tetrahedron. All bonds are shown in red, and all bond angles are 109.5°.

$$\begin{matrix} & & & & H \\ & H & & & | \\ H\!:\!\ddot{C}\!:\!H & \text{and} & H\!-\!C\!-\!H \\ & \ddot{H} & & & | \\ & & & & H \end{matrix}$$

The actual arrangement of the five atoms in space is a *tetrahedron* (a three-sided base pyramid; see Section 6.8) with each hydrogen atom occupying a corner of a tetrahedron and the carbon atom placed in the center, as shown in Figure 18.2. Four covalent bonds connect the C and H atoms, and the bond angles are all 109.5°C. This arrangement is referred to as *tetrahedral*. Why this geometric arrangement for methane? The four pairs of bonding electrons try to get as far apart as possible from each other to minimize the repulsion between them (like charges repel each other). The tetrahedral arrangement maximizes the bond angles and thus minimizes the repulsive force between the electron pairs. Carbon atoms in category 1 always form four single bonds and have a *tetrahedral* (bond angles of 109.5°) arrangement.

Category 2: Carbon to Three Atoms. The molecule ethylene (C_2H_4) has a double bond (see Section 6.7) between the two carbon atoms, and each carbon atom forms three covalent bonds to three atoms. The Lewis structure and structural formula of ethylene are

$$2 \text{ groups of } 2 \text{ electrons} \Bigg\} \quad \overset{H}{\underset{H}{:}} C :: C \overset{H}{\underset{H}{:}} \quad \text{and} \quad \overset{H}{\underset{H}{>}} C = C \overset{H}{\underset{H}{<}}$$

$$\underbrace{}_{\substack{1 \text{ group of} \\ 4 \text{ electrons}}}$$

The actual arrangement of the atoms in ethylene is planar, and all bond angles are 120°, as shown in Figure 18.3. By *planar*, we mean that *both* carbon atoms and *all four* hydrogen atoms lie in the same plane. The double bond and the four groups attached to them are flat. This depiction is reasonable because the three groups of bonding electrons (*two groups* of *two electrons* and *one group* of *four electrons*) that surround each carbon atom are oriented as far apart from each other as possible to minimize the repulsion between the groups of electrons. Carbon atoms in category 2 always form a double bond and two single bonds and have a *planar* (120°) arrangement.

Category 3: Carbon to Two Atoms. Two different molecules have carbon atoms which form two covalent bonds to two atoms: acetylene and carbon dioxide. Acetylene (C_2H_2) has a triple bond (see Section 6.7) between the two carbon atoms, and each carbon forms two covalent bonds. The Lewis structure and structural formula of acetylene are

$$H : C ::: C : H \quad \text{and} \quad H - C \equiv C - H$$

$$\underset{\substack{1 \text{ group of} \\ 2 \text{ electrons}}}{\Big\}} \qquad \underset{\substack{1 \text{ group of} \\ 6 \text{ electrons}}}{\Big\{}$$

The actual arrangement of the atoms in acetylene is *linear*, and all bond angles are 180°. This arrangement is reasonable because the two groups of bonding electrons (*one group* of *two electrons* and *one group* of *six electrons*) that surround each carbon atom are oriented as far apart from each other as possible to minimize the repulsion between the groups of electrons.

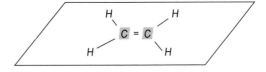

FIGURE 18.3
The structure of ethylene is *planar*, and all the bond angles are 120°.

An alternative bonding scheme for category 3 is illustrated by carbon dioxide (CO_2). Carbon forms two covalent bonds with two atoms in CO_2, but *both bonds are double bonds.* The Lewis structure and structural formula of carbon dioxide are

$$\ddot{O}::C::\ddot{O} \quad \text{and} \quad O{=}C{=}O$$

Again, the arrangement of the atoms is *linear*. Carbon atoms in category 3 form either a triple bond and a single bond *or* two double bonds, with a *linear* (180°) arrangement.

Table 18.1 summarizes the characteristics of the three basic carbon geometries and includes ball-and-stick models of methane (CH_4), ethylene (C_2H_4), acetylene (C_2H_2), and carbon dioxide (CO_2).

More complex organic molecules may contain any of these particular shapes, singly or in combination with each other. Some examples are given in the following example and in Figure 18.4.

TABLE 18.1	Summary of the Properties of the Three Basic Carbon Geometries		
GEOMETRY	NUMBER AND TYPES OF BONDS TO CARBON	BOND ANGLES	EXAMPLE
Tetrahedral	Four single	109.5°	
Planar	One double and two single	120°	
Linear	One triple and one single	180°	
Linear	Two double	180°	

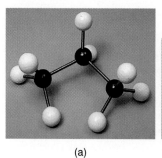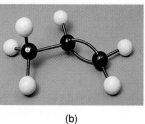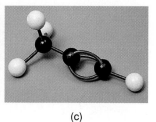

(a) (b) (c)

FIGURE 18.4
The geometry about a car-
bon atom depends on the
nature of the bonding to
that carbon atom. Ball-and-
stick model of (a) propane
(C_3H_8), (b) propene (C_3H_6),
and (c) propyne (C_3H_4).

| EXAMPLE 18.1 | Identify the geometry (tetrahedral, 109.5°; planar, 120°; lin-ear, 180°) at all carbon atoms in each of the following mole-cules: |

a. propane (C_3H_8)

$$
\begin{array}{c}
\text{H} \quad \text{H} \quad \text{H} \\
| \quad\;\; | \quad\;\; | \\
\text{H}-\text{C}-\text{C}-\text{C}-\text{H} \\
| \quad\;\; | \quad\;\; | \\
\text{H} \quad \text{H} \quad \text{H}
\end{array}
$$

b. propene (C_3H_6)

$$
\begin{array}{c}
\text{H} \qquad\quad \text{H} \\
\;\;\searrow \quad\;\; \nearrow \\
\quad\;\text{C}=\text{C} \qquad \text{H} \\
\nearrow \qquad\;\; \searrow\;\;\nearrow \\
\text{H} \qquad\qquad \text{C} \\
\qquad\qquad \nearrow \;\; \searrow \\
\qquad\quad \text{H} \qquad \text{H}
\end{array}
$$

c. propyne (C_3H_4)

$$
\begin{array}{c}
\text{H} \\
| \\
\text{H}-\text{C}\equiv\text{C}-\text{C}-\text{H} \\
| \\
\text{H}
\end{array}
$$

SOLUTION

a. propane (C_3H_8, a gas used in camping stoves, gas heaters, and propane torches)

$$
\begin{array}{c}
\text{H} \quad \text{H} \quad \text{H} \\
| \quad\;\; | \quad\;\; | \\
\text{H}-\text{C}-\text{C}-\text{C}-\text{H} \\
| \quad\;\; | \quad\;\; | \\
\text{H} \quad \text{H} \quad \text{H}
\end{array}
$$

Each carbon atom has four single bonds, and so propane belongs in category 1. The geometry about each carbon atom is *tetrahedral* (109.5°); see Figure 18.4a.   *Answer*

b. propene (C_3H_6)

$$
\begin{array}{c}
\text{H} \qquad\quad \text{H} \\
\;\;\searrow \quad\;\; \nearrow \\
\quad\;\text{C}=\text{C} \qquad \text{H} \\
\nearrow \qquad\;\; \searrow\;\;\nearrow \\
\text{H} \qquad\qquad \text{C} \\
\qquad\qquad \nearrow \;\; \searrow \\
\qquad\quad \text{H} \qquad \text{H}
\end{array}
$$

The doubly bonded carbon atoms have a double bond and two single bonds (category 2) and are *planar* (180°). The remaining carbon atom is *tetrahedral* (109.5°); see Figure 18.4b.   *Answer*

c. propyne (C_3H_4)

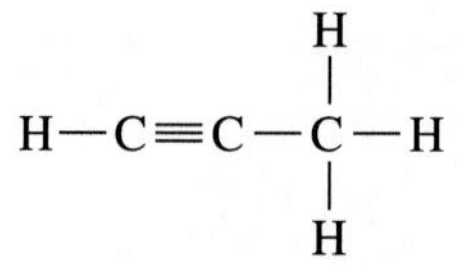

$$H-C\equiv C-\overset{\displaystyle H}{\underset{\displaystyle H}{\overset{|}{\underset{|}{C}}}}-H$$

The triply bonded carbon atoms have a triple bond and a single bond (category 3) and are *linear* (180°). The remaining carbon atom is *tetrahedral* (109.5°); see Figure 18.4c. *Answer*

Study Exercise 18.1
Identify the geometry (tetrahedral, 109.5°; planar, 120°; linear, 180°) at all carbon atoms in each of the following molecules:

a.

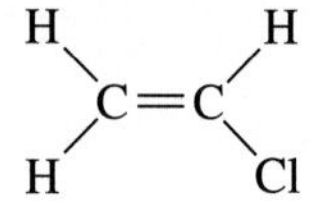

$$\overset{\displaystyle H}{\underset{\displaystyle H}{}}\!\!C=C\!\!\overset{\displaystyle H}{\underset{\displaystyle Cl}{}}$$

(Both carbons—planar, 120°)

b. 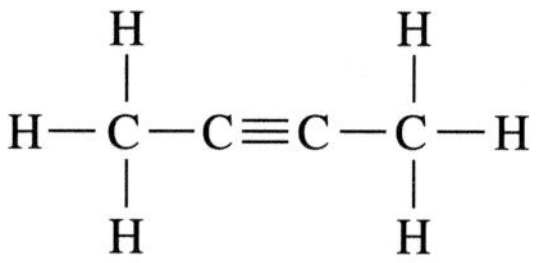 $H-\overset{\displaystyle H}{\underset{\displaystyle H}{\overset{|}{\underset{|}{C}}}}-C\equiv C-\overset{\displaystyle H}{\underset{\displaystyle H}{\overset{|}{\underset{|}{C}}}}-H$

(Triple-bonded carbons—linear, 180°; other two carbons—tetrahedral, 109.5°)

Work Problem 5.

TABLE 18.2	Alternate Formulas for Some Simple Organic Compounds	
COMPOUND	STRUCTURAL FORMULA	CONDENSED STRUCTURAL FORMULA
Methane	$H-\overset{H}{\underset{H}{\overset{|}{\underset{|}{C}}}}-H$	CH_4
Propane	$H-\overset{H}{\underset{H}{\overset{|}{\underset{|}{C}}}}-\overset{H}{\underset{H}{\overset{|}{\underset{|}{C}}}}-\overset{H}{\underset{H}{\overset{|}{\underset{|}{C}}}}-H$	$CH_3-CH_2-CH_3$ or $CH_3CH_2CH_3$
Propene	$\overset{H}{\underset{H}{}}C=C\overset{H}{\underset{\underset{H}{\overset{|}{C}}H}{}}$	$CH_2=CH-CH_3$ or $CH_2=CHCH_3$
Propyne	$H-C\equiv C-\overset{H}{\underset{H}{\overset{|}{\underset{|}{C}}}}-H$	$HC\equiv C-CH_3$ or $HC\equiv CCH_3$

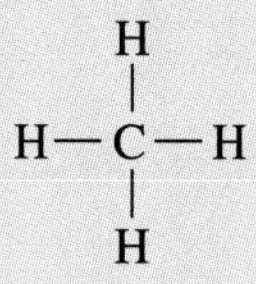

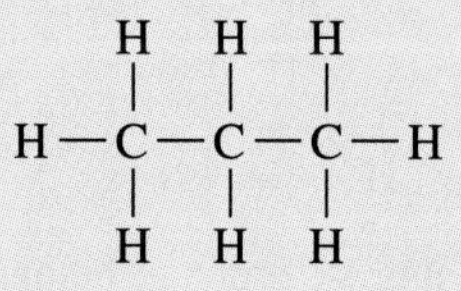

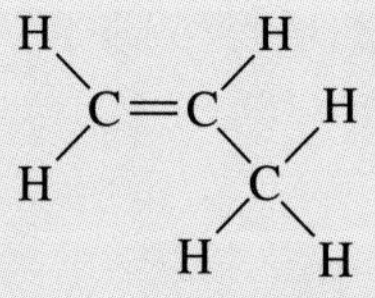

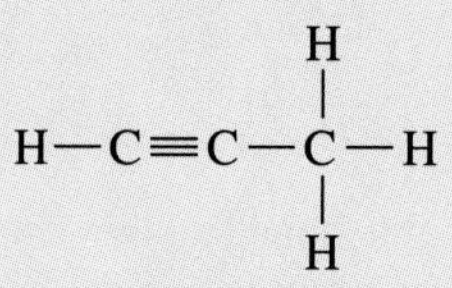

The properties of an organic molecule depend on its structure. Thus it is very important that scientists know the structure of such molecules and have ways of representing the structure. Lewis structures are useful, but they become cumbersome to use for large molecules. Organic chemists also use structural formulas and *condensed structural formulas*. In a condensed structural formula the hydrogen atoms are written collectively next to the carbon atom to which they are attached. In this text we will normally use *condensed structural formulas* for compounds. Table 18.2 illustrates a few examples.

As we can see, the structural formula for propane is written as a "straight chain." In reality, though, propane is not really a straight chain because the center carbon has tetrahedral geometry and the $C-C-C$ bond angle is approximately $109.5°$; see Figure 18.4a. The phrase "straight chain" in this case simply means that there is a continuous chain of carbon atoms without any *branches*. A branched-chain compound has one or more places where the carbon atom is bonded to three or even four other carbon atoms, thus forming a branch or branches in the chain.

Carbon atoms may also form rings by bonding carbon atoms together in a circle. Examples of compounds containing rings of carbon atoms are cyclopentane (C_5H_{10}) and cyclohexane (C_6H_{12}). Table 18.3 shows their structural formulas as condensed structural formulas and skeletal formulas, where each corner of the ring represents a $-CH_2-$. In all the formulas, each carbon has exactly four *bonds*, which are arranged in a tetrahedral manner.

18.3 Alkanes

The **alkanes** (al'kāns), also called the **saturated hydrocarbons** (paraffins), are aliphatic hydrocarbons that have the general molecular formula C_nH_{2n+2} for open-chain compounds. The simplest alkane is methane (natural gas, CH_4), which fol-

Alkanes (saturated hydrocarbons) Form of aliphatic hydrocarbons: they have the general molecular formula C_nH_{2n+2} for open-chain compounds. Each carbon has a tetrahedral structure and four single bonds.

TABLE 18.3	**Alternative Formulas for Cyclopentane and Cyclohexane**		
COMPOUND	STRUCTURAL FORMULA	CONDENSED STRUCTURAL FORMULA	SKELETAL FORMULA
Cyclopentane (C_5H_{10})			
Cyclohexane (C_6H_{12})			

lows the general molecular formula with n equal to 1. Propane, the gas commonly used in backyard gas barbecues, has the formula C_3H_8, which follows the general formula with n equal to 3. Gasoline contains a mixture of alkanes containing 5 to 10 carbon atoms plus other organic compounds to make the fuel perform better. In alkanes, all carbon atoms have four single bonds and a tetrahedral geometry (category 1). Note that the cycloalkanes have a general formula of C_nH_{2n}.

Table 18.4 shows the first 10 continuous-chain alkanes. The prefix in each of these hydrocarbons is characteristic of the number of carbon atoms in the chain and, as we will see later, these prefixes are used in naming many classes of organic compounds. The prefixes from pentane to decane are nearly the same prefixes used in inorganic nomenclature (see Table 7.1) and indicate the number of carbon atoms. You *must* memorize the names and structural formulas of the alkanes in Table 18.4.

If we examine the molecular formulas of this series, we will notice that they differ from each other by one carbon and two hydrogen atoms, a CH_2, or *methylene*, group. For example, propane (CH_3—CH_2—CH_3) differs from ethane (CH_3—CH_3) by one —CH_2— group. Butane (CH_3—CH_2—CH_2—CH_3) differs from propane (CH_3—CH_2—CH_3) also by one —CH_2— group. A series of compounds in which each compound in the series differs from the next compound by a *multiple*, such as a —CH_2—, is called a **homologous series**. Therefore, the series methane to decane is a homologous series.

Homologous series Series of compounds in which each compound in the series differs from the next compound by a multiple, such as —CH_2—.

Isomers

In the previous paragraph, we considered only those alkanes whose carbon atoms form a continuous chain and for which we write only one structural formula. Once the number of carbon atoms in the molecular formula for an alkane reaches four or more, it is possible to write more than one structural formula for a given molecular formula. To do so requires that a branch carbon be introduced. Compounds that have the *same* molecular formula but a *different* structural formula are called **isomers**. *Each isomer of a given molecular formula has physical and chemical properties*

Isomer Any compound that has the same molecular formula but a different structural formula. Each isomer has physical and chemical properties different from those of any other isomer of the same molecular formula.

TABLE 18.4	The Alkanes, Methane to Decane		
NAME[a]	MOLECULAR FORMULA	n	CONDENSED STRUCTURAL FORMULA
Methane	CH_4	1	CH_4
Ethane	C_2H_6	2	CH_3CH_3
Propane	C_3H_8	3	$CH_3CH_2CH_3$
Butane	C_4H_{10}	4	$CH_3CH_2CH_2CH_3$
Pentane	C_5H_{12}	5	$CH_3CH_2CH_2CH_2CH_3$
Hexane	C_6H_{14}	6	$CH_3CH_2CH_2CH_2CH_2CH_3$
Heptane	C_7H_{16}	7	$CH_3CH_2CH_2CH_2CH_2CH_2CH_3$
Octane	C_8H_{18}	8	$CH_3CH_2CH_2CH_2CH_2CH_2CH_2CH_3$
Nonane	C_9H_{20}	9	$CH_3CH_2CH_2CH_2CH_2CH_2CH_2CH_2CH_3$
Decane	$C_{10}H_{22}$	10	$CH_3CH_2CH_2CH_2CH_2CH_2CH_2CH_2CH_2CH_3$

[a] The hydrocarbons from butane to decane were formerly named *n*-butane to *n*-decane. The *n* stood for normal, which meant that the carbon atoms were bonded in a continuous chain.

different from those of any other isomer of the same molecular formula. For example, consider the three isomers of pentane (C_5H_{12}):

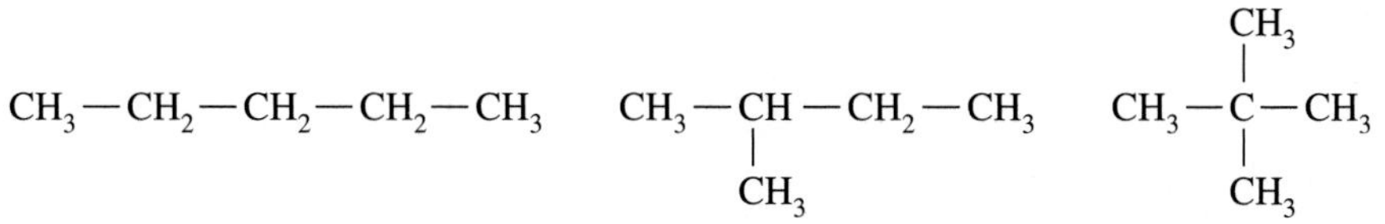

$$CH_3-CH_2-CH_2-CH_2-CH_3 \qquad CH_3-\underset{\underset{CH_3}{|}}{CH}-CH_2-CH_3 \qquad CH_3-\underset{\underset{CH_3}{|}}{\overset{\overset{CH_3}{|}}{C}}-CH_3$$

| pentane | isopentane | neopentane |

Although all of these isomers have the formula C_5H_{12}, their structural formulas differ. Moreover, these isomers differ in essential characteristics such as boiling and melting point:

ISOMER	BOILING POINT	MELTING POINT
Pentane	36°C	−130°C
Isopentane	28°C	−160°C
Neopentane	9°C	−20°C

In drawing isomers of alkanes, you may find the following four guidelines helpful:

1. Draw a carbon skeleton (no hydrogen atoms) using all the carbons in a *continuous* chain.

2. Remove *one* carbon (C) atom from the end of the chain and place it on another carbon atom so that the new skeleton differs from the previous carbon skeleton. Repeat this procedure until you exhaust all possibilities of relocating *one* carbon atom.

3. Next, if necessary, remove *two* carbon atoms from the continuous-chain skeleton in guideline 1 and relocate them either as *single* carbon atoms or as a *two-carbon* fragment on other carbons in the chain. Write all possible *different* skeletons relocating two carbon atoms. Continue this procedure, if necessary, for *three* carbon atoms until you have the number of isomers required in the problem. Check all skeletons to be sure that they are *really* all different.

4. Place H atoms on the C atoms in each skeleton in guidelines 1, 2, and 3, remembering that there are *four single bonds to each carbon atoms.*

EXAMPLE 18.2 Write condensed structural formulas for the five isomers of hexane (C_6H_{14}).

SOLUTION
Guideline 1:
(1) C—C—C—C—C—C

Guideline 2:

(2) 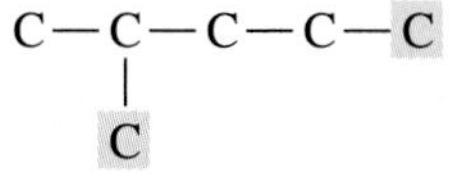 C—C—C—C—C   [If a C is placed on the end, such as
          | C—C—C—C—C this form will be the same as (1).]
          C |
                                   C

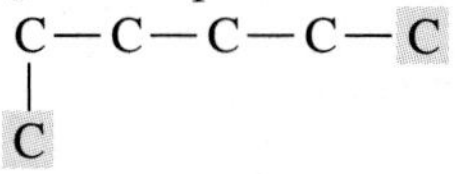

(3) 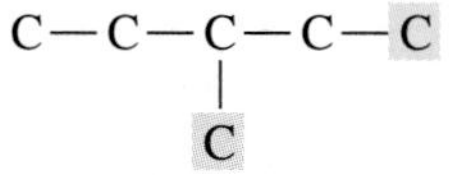 C—C—C—C—C
          |
          C

Guideline 3:

(4)
$$C-\underset{\underset{\displaystyle C}{|}}{\overset{\overset{\displaystyle C}{|}}{C}}-C-C$$
(5)
$$C-\underset{\underset{\displaystyle C}{|}}{C}-\underset{\underset{\displaystyle C}{|}}{C}-C$$

Guideline 4 (Remember that there are four single bonds to each carbon atom):

(1) $CH_3-CH_2-CH_2-CH_2-CH_2-CH_3$

(2) $CH_3-\underset{\underset{\displaystyle CH_3}{|}}{CH}-CH_2-CH_2-CH_3$

(3) $CH_3-CH_2-\underset{\underset{\displaystyle CH_3}{|}}{CH}-CH_2-CH_3$

(5) $CH_3-\underset{\underset{\displaystyle CH_3}{|}}{CH}-\underset{\underset{\displaystyle CH_3}{|}}{CH}-CH_3$

(4) $CH_3-\underset{\underset{\displaystyle CH_3}{|}}{\overset{\overset{\displaystyle CH_3}{|}}{C}}-CH_2-CH_3$

Study Exercise 18.2

Write condensed structural formulas for the two isomers of butane (C_4H_{10}).

$$[CH_3-CH_2-CH_2-CH_3 \text{ and } CH_3-\underset{\underset{\displaystyle CH_3}{|}}{CH}-CH_3]$$

(1) (2)

Work Problem 6.

Nomenclature

In naming the isomeric pentanes, we used the prefixes *iso-* and *neo-*, which chemists call *trivial prefixes*. Such trivial prefixes are awkward if used in naming isomeric hexanes and higher hydrocarbons, and so a systematic nomenclature is necessary. The International Union of Pure and Applied Chemistry (IUPAC) system developed over the years is now the preferred method of naming organic compounds. Basically, this system uses names composed of two parts. The *terminal* portion names the longest continuous chain in the molecule, the *parent chain*; the *first* portion names the *substituent groups* attached to the parent chain.

Alkyl group Hydrocarbon group obtained by removing a hydrogen atom from an alkane. They are usually represented by the symbol **R**.

Before discussing this system in more detail, we must consider the nomenclature of alkyl groups, which are frequently substituents on parent chains. **Alkyl groups** are derived by removing one hydrogen atom from an alkane and are usually represented by the symbol **R**. Generally they are named in the case of the simpler hydrocarbons by replacing the *-ane* ending of the alkane by *-yl*. In continuous chain hydrocarbons higher than ethane, "alkyl" is reserved only for the alkyl group obtained by removing a hydrogen atom from the *terminal* carbon atoms. Thus,

CH_4 is meth*ane* CH_3- is meth**yl**

CH_3-CH_3 is eth*ane* CH_3-CH_2- is eth**yl**

$CH_3-CH_2-CH_3$ is prop*ane* $CH_3-CH_2-CH_2-$ is prop**yl**

$CH_3-CH_2-CH_2-CH_3$ is but*ane* $CH_3-CH_2-CH_2-CH_2-$ is but**yl**

$CH_3-CH_2-CH_2-CH_2-CH_3$ is pent*ane* $CH_3-CH_2-CH_2-CH_2-CH_2-$ is pent**yl**

$CH_3-CH_2-CH_2-CH_2-CH_2-CH_3$ is hex*ane* $CH_3-CH_2-CH_2-CH_2-CH_2-CH_2-$ is hex**yl**

The nomenclature of the alkyl groups isomeric to propyl, butyl, pentyl, and so on, is more difficult, but a suitable trivial method of naming these groups has been developed. The groups are isopropyl, *sec*-butyl, isobutyl, and *tert*-butyl. You *must* learn the trivial names of the simple alkyl groups. Table 18.5 summarizes them for you.

To understand the use of these terms, consider the alkyl groups propyl and isopropyl. These **two** isomeric (both C_3H_7) propyl groups can be derived from propane by removing a hydrogen atom either from one of the end carbons (red H) to give the *propyl* group or from the middle carbon (blue H) to give the *isopropyl* group:

$$CH_3 — CH_2 — CH_2 — \quad or \quad CH_3CH_2CH_2 —$$

$$\mathbf{1} = propyl$$

$$CH_3 — CH — CH_2 — H$$
$$\underset{\text{propane}}{|}$$
$$H$$

$$CH_3 — CH — CH_3 \quad or \quad (CH_3)_2CH —$$
$$|$$

$$\mathbf{2} = isopropyl$$

TABLE 18.5 **Summary of Simple Alkyl Groups and Their Trivial Names**

ALKYL GROUP	TRIVIAL NAME
$CH_3 —$	methyl
$CH_3 — CH_2 —$	ethyl
$CH_3 — CH_2 — CH_2 —$	propyl
$CH_3 — CH —$ $\quad\quad\ \|$ $\quad\ \ CH_3$	isopropyl
$CH_3 — CH_2 — CH_2 — CH_2 —$	butyl
$CH_3 — CH_2 — CH —$ $\quad\quad\quad\quad\quad \|$ $\quad\quad\quad\quad CH_3$	*sec*-butyl
$\quad\quad\quad\ CH_3$ $\quad\quad\quad\ \|$ $CH_3 — CH — CH_2 —$ $\quad\quad CH_3$	isobutyl
$\quad\quad\ CH_3$ $\quad\quad\ \|$ $CH_3 — C —$ $\quad\quad\ \|$ $\quad\quad CH_3$	*tert*-butyl

Similarly, the structures of the **four** isomeric butyl groups are derived from the *two* butanes, butane and isobutane, by removing the appropriate hydrogen atoms:

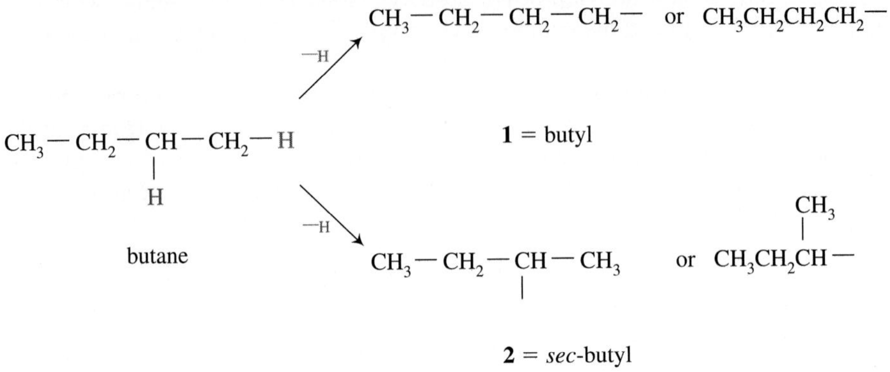

$CH_3-CH_2-CH_2-CH_2-$ or $CH_3CH_2CH_2CH_2-$

1 = butyl

$CH_3-CH_2-CH-CH_2-H$
$\quad\quad\quad\quad\quad |$
$\quad\quad\quad\quad\quad H$

butane

$CH_3-CH_2-CH-CH_3$ or CH_3CH_2CH-
$\quad\quad\quad\quad\quad |$ $\quad\quad\quad\quad\quad\quad\quad |$
$\quad\quad\quad\quad\quad\quad\quad\quad\quad\quad\quad\quad\quad\quad CH_3$

2 = *sec*-butyl

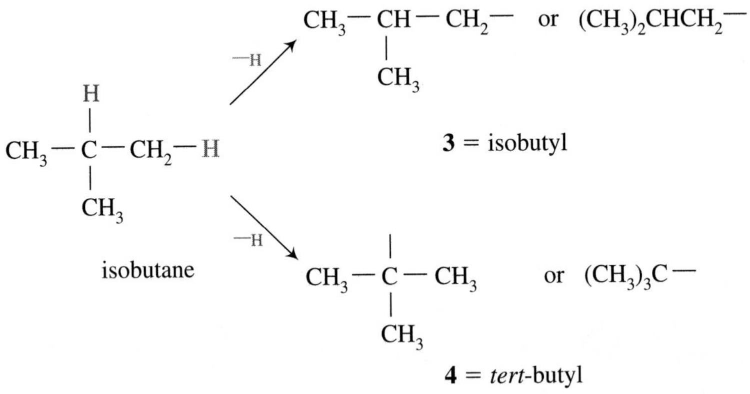

$CH_3-CH-CH_2-$ or $(CH_3)_2CHCH_2-$
$\quad\quad\quad |$
$\quad\quad\quad CH_3$

3 = isobutyl

$\quad\quad H$
$\quad\quad |$
CH_3-C-CH_2-H
$\quad\quad |$
$\quad\quad CH_3$

isobutane

$\quad\quad\quad\quad\quad |$
CH_3-C-CH_3 or $(CH_3)_3C-$
$\quad\quad\quad\quad\quad |$
$\quad\quad\quad\quad\quad CH_3$

4 = *tert*-butyl

Because two butyl groups are derived from butane and two from isobutane, we need to differentiate further between these groups in order to name them adequately. Alkyl groups can be classified into three different types depending on the type of attaching carbon atom (C):

$R-CH_2-$ primary (1°)   *One* alkyl (R) and two hydrogen atoms attached

$R\diagdown$
$\quad\;CH-$ secondary (2°)   *Two* alkyls (R) and one hydrogen atom attached
$R\diagup$

$R\diagdown$
$R-C-$ tertiary (3°)   *Three* alkyls (R) and no hydrogen atoms attached
$R\diagup$

Now we can use the following terms to describe the four different butyl groups described above.

✔ Primary butyl or butyl $CH_3CH_2CH_2CH_2-$

✔ Secondary butyl or *sec*-butyl CH_3CH_2CH-
   (read "secondary butyl") $|$
                                     CH_3

✔ Primary isobutyl or isobutyl $(CH_3)_2CHCH_2—$

✔ Tertiary butyl or *tert*-butyl $(CH_3)_3C—$
(read "tertiary butyl")

In addition to naming alkyl substituents, we need to become familiar with the IUPAC system of naming organic compounds. An IUPAC name has two parts: (1) the *terminal* part specifying the longest continuous carbon chain (the *parent* chain), and (2) the *beginning* part specifying the substituents that are attached to the parent chain. We derive the IUPAC names for the alkanes by using the following six rules:

1. The alkane hydrocarbons all have the ending -**ane**.

2. The longest continuous chain of carbons is the parent structure. For example, if the longest continuous chain of carbon atoms is **5**, the parent structure is called a *pentane*. This is the terminal portion of the IUPAC name. Consider the following example:

$$\begin{array}{ccccc} 1 & 2 & 3 & 4 & 5 \\ CH_3 — & CH — & CH — & CH_2 — & CH_3 \\ & | & | & & \\ & CH_3 & CH_3 & & \\ 5 & 4 & 3 & 2 & 1 \end{array}$$

The name of the parent structure is *pentane*. This longest continuous chain may be written as a "straight chain" but may also be bent, as follows:

$$\begin{array}{cccc} 1 & 2 & 3 & 4 \\ CH_3 — & CH — & CH — & CH_2 \\ & | & | & | \\ & CH_3 & CH_3 & CH_3 \quad 5 \end{array}$$

The name of the parent structure is still *pentane*. *If the chain is in the form of a ring, use the prefix* **cyclo-**.

3. Number the carbon atoms in this chain by starting at the end that would give the *lowest numbers* to the carbon atoms where the group or groups are attached to the parent structure. In this example, the two methyl groups (CH_3) are attached at carbons 2 and 3 (blue numbers) or 3 and 4 (red numbers). Use the blue numbering to achieve the *lowest set* of numbers.

4. Give the group attached to the parent structure, other than hydrogen, both *name* and *number*. Halogens attached to the parent structure are named *chloro-* (for Cl), *bromo-* (for Br), and *iodo-* (for I). The NO_2 group is named *nitro-*. Alkyl groups are given their accepted trivial names, see Table 18.5.

5. If more than one of the *same* group appears as a substituent in a given molecule, use the prefixes *di-*, *tri-*, *tetra-*, and *penta-* to indicate the number of times this group appears (two, three, four, or five times, respectively) and indicate the position of these groups on the numbered parent structure in *increasing numerical order. Each group* must have a *number* to indicate its position on the parent structure, even if it is the same number. In the case above, two methyl groups appear on the pentane, and

so we use a *di-* in the name *dimethyl*. We use a comma (,) between the numbers and a hyphen (-) between the number and the name. Therefore, the correct IUPAC name for the compound above is 2,3-dimethylpentane.

6. If more than *one type of group* is attached to the parent structure, we place these groups in alphabetical order in the name. However, in determining such alphabetical order, ignore prefixes such as *di-*, *tri-*, and *tetra-*, denoting the number of groups, and hyphenated prefixes such as *sec-* and *tert-*. Use the actual name of the group following such prefixes to determine alphabetical order. The prefix *iso-* is not hyphenated and is used in determination of the alphabetical sequence. In the following list of groups, the letter in **bold** print determines the alphabetical order: *sec-***b**utyl, *tert-***b**utyl, di**c**hloro, tetra**e**thyl, **i**sopropyl, tri**m**ethyl.

Study Hint: Think of the nomenclature of alkanes as a clothesline. The terminal part (parent chain) is the clothesline, and the substituents are the clothes attached to it. The clothes have both *name* and *position* (number) as to where they are attached on the clothesline.

Study Hint: Part c is a bit tricky. The longest continuous chain is *bent* and contains *six* carbons, not five!

EXAMPLE 18.3 Write the IUPAC name for each of the following compounds:

Answer

a. CH_3—CH—CH_3 2-methylpropane
           |
           CH_3

b. CH_3—CH—CH—CH_2—CH_3 2-chloro-3-methylpentane
          | | (Note the alphabetical order of the sub-
          Cl CH_3 stituents.)

c. CH_3 2,2,4-trimethylhexane
           |
   CH_3—C—CH_2—CH—CH_3
           | |
           CH_3 CH_2—CH_3

d. CH_2 cyclopropane
     ╱ ╲   (Note that each corner represents a —CH_2— .)
 CH_2—CH_2 or   △

Study Exercise 18.3
Write the IUPAC name for each of the following compounds:

a. CH_3—CH—CH_2—CH_2—CH_3
         |
         NO_2 (2-nitropentane)

b. CH_2—CH—CH_2—CH_2
    | | |
   Br   Br CH_2—CH_3 (1,2-dibromohexane)

Work Problem 7.

Now that we can name alkanes, we must also be able to write the structural formula for a compound from its name. We can do so readily by following three steps:

1. Write the chain or ring of carbon atoms for the parent structure from the terminal portion of the name.

2. Add the various groups to the correct positions of the parent structure.

3. Place the necessary hydrogen atoms on all the carbon atoms so that each carbon has *four single* covalent bonds.

EXAMPLE 18.4 Write the structural formula for each of the following compounds:

Answers

a. 2-methylpentane

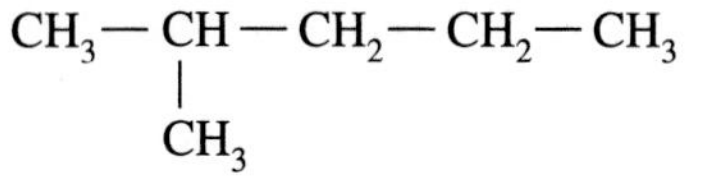

$$CH_3 - CH - CH_2 - CH_2 - CH_3$$
$$|$$
$$CH_3$$

b. 2-bromobutane

$$CH_3CH_2CHCH_3$$
$$|$$
$$Br$$

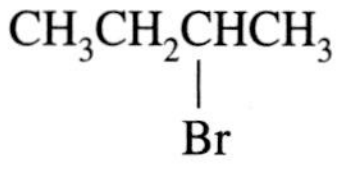

c. 3-chloro-2,4-dimethylheptane

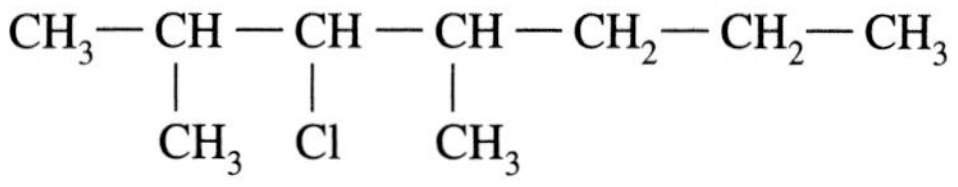

$$CH_3 - CH - CH - CH - CH_2 - CH_2 - CH_3$$
$$|\quad\quad|\quad\quad|$$
$$CH_3\ \ Cl\ \ \ CH_3$$

d. methylcyclopentane

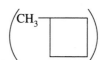   or

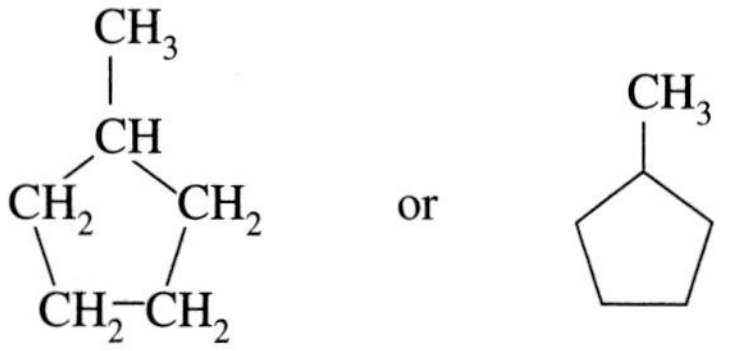

(All positions on the ring are equivalent, and so the $-CH_3$ can be attached to any one of the carbons in the cyclopentane ring. No number is needed for *one* substituent.)

Study Exercise 18.4
Write the structural formula for each of the following compounds:

a. 2,2-dichloropentane

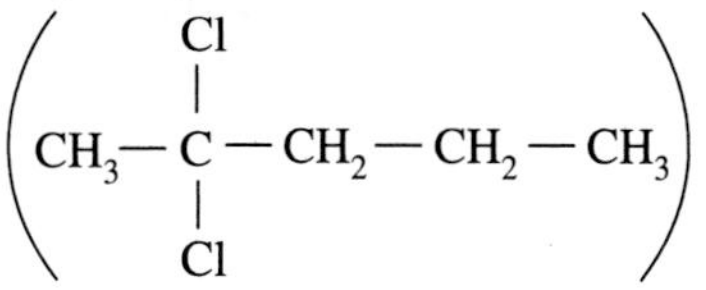

$$\left(CH_3 - \overset{\displaystyle Cl}{\underset{\displaystyle Cl}{\overset{|}{\underset{|}{C}}}} - CH_2 - CH_2 - CH_3 \right)$$

b. methylcyclobutane

Work Problem 8.

Uses and Reactions of Alkanes

Alkanes are used in the production of some common materials. For example, cyclohexane is the chief organic material used in the preparation of nylon, and hexane is used in the extraction of such common oils as soybean, peanut, and cottonseed.

The most important use of alkanes, however, is as a fuel for internal combustion engines and for heating buildings. Automobiles, trains, buses, trucks, airplanes, and motorcycles all burn fuel mixtures that are largely composed of alkanes. Much of the heat for homes and businesses is supplied by natural gas (methane) and liquefied petroleum gas (propane and the butanes).

Gasoline is composed of a mixture of alkanes that contain from 5 to 10 (mostly 7 to 9) carbon atoms per molecule. Other substances added to the gasoline make it burn more smoothly in the engine. Kerosene, used in the formulation of diesel and jet fuel, is a mixture of alkanes with 10 to 15 carbon atoms per molecule. Different formulations are used for different purposes.

Of course, combustion (see Section 13.5) is the most important reaction of alkanes since it is the basis for the use of alkanes as a fuel. The next most important reaction of alkanes is the reaction with chlorine or bromine. This process, called **halogenation**, is a *substitution reaction* in which a hydrogen atom on an alkane is replaced by a chlorine or bromine atom. The chlorination of methane serves as an example:

$$CH_4(g) \ + \ Cl_2(g) \ \xrightarrow{\Delta \text{ or} \atop \text{light}} \ CH_3{-}Cl(g) \ + \ H{-}Cl(g) \qquad (18.1)$$
$$\text{methane} \qquad\qquad\qquad \text{chloromethane} \quad\; \text{hydrogen} \atop \text{chloride}$$

Halogenation Substitution reaction in which a hydrogen atom on an alkane is replaced by a chlorine or bromine atom.

A more general equation for this process is

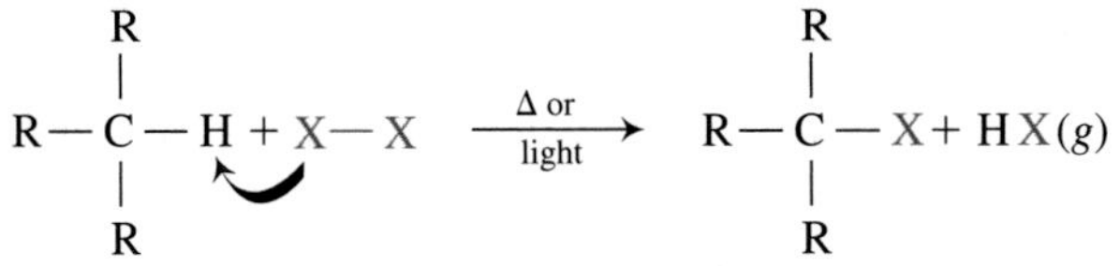

$$
\underset{\substack{|\\R}}{\overset{\substack{R\\|}}{R{-}C}}{-}H + X{-}X \ \xrightarrow{\Delta \text{ or} \atop \text{light}} \ \underset{\substack{|\\R}}{\overset{\substack{R\\|}}{R{-}C}}{-}X + HX(g) \qquad (18.2)
$$

$$(R = \text{alkyl or H}) \qquad X = Cl \text{ or } Br$$

Chloromethane is used in the production of methyl cellulose, a product you probably use everyday. Methyl cellulose is used as a thickener in cosmetics, shampoos, lotions, detergents, and a variety of other everyday materials. A by-product of the chlorination reaction above (18.1) is dichloromethane (CH_2Cl_2), which is used in paint-stripping products.

Study Exercise 18.5
Complete and balance the reaction equation for *mono*halogenation of the following chemical reaction:

$$CH_3{-}CH_3 \ + \ Cl_2 \ \xrightarrow{\Delta \text{ or} \atop \text{light}}$$

Work Problem 9.

$$[CH_3{-}CH_3 \ + \ Cl_2 \ \xrightarrow{\Delta \text{ or} \atop \text{light}} \ CH_3{-}CH_2{-}Cl \ + \ HCl(g)]$$

18.4 Alkenes

Alkenes Form of unsaturated aliphatic hydrocarbon; they have the general molecular formula C_nH_{2n} for open-chain compounds, a carbon–carbon double bond, and a planar structure.

The **alkenes** (al′kēns) are aliphatic hydrocarbons that have the general molecular formula C_nH_{2n} for open-chain compounds. This formula differs from the alkane formula of C_nH_{2n+2} by two hydrogen atoms. Thus, alkenes contain *two less hydrogens* than the alkane containing the same number of carbon atoms. As a result, all alkenes have a *double bond* (category 2). The alkenes are called **unsaturated hydrocarbons** because they contain less hydrogen than the corresponding alkane.

The simplest alkene is ethylene, which follows the general molecular formula with *n* equal to 2. As you saw in Section 18.2, a double bond is planar and all bond angles are 120°:

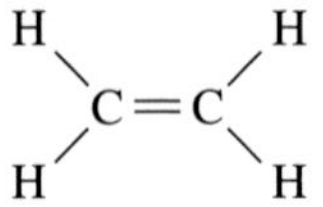

Unsaturated hydrocarbons Hydrocarbons that contain less hydrogen than the corresponding alkane. They contain a double or triple bond; for example, alkenes and alkynes.

$$\underset{H}{\overset{H}{\diagdown}}C{=}C\underset{H}{\overset{H}{\diagup}} \qquad \text{or} \qquad CH_2{=}CH_2$$

All alkenes possess a double bond with these same characteristics. The double bond may be at the end of a chain (terminal), in the middle of a chain, or at a branch point in the chain:

$$H_2C=CH-CH_2-CH_3 \qquad CH_3-CH=CH-CH_3 \qquad (CH_3)_2C=CH-CH_3$$

> **Study Hint:** Cycloalkanes also have a general formula of C_nH_{2n}. Do not be confused by this similarity. Cycloalkanes do not have a double bond.

Nomenclature

A few of the alkenes have commonly used trivial names, such as ethylene $(CH_2=CH_2)$, propylene $(CH_2=CH-CH_3)$, and isobutylene $[(CH_3)_2C=CH_2]$. These names are not used in the IUPAC system. We can name alkenes by the IUPAC system in the following way:

1. The alkene hydrocarbons all have the ending -**ene**. And so ethylene has the IUPAC name *ethene*.

2. The longest continuous chain of carbons that *contain the double bond* is the *parent* structure. For example, if the longest continuous chain of carbon atoms that contains the double bond is 5 carbon atoms, the parent structure is named as a *pent*ene. Consider the following example:

$$\overset{1}{CH_3}-\overset{2}{\underset{\underset{CH_3}{|}}{C}}=\overset{3}{CH}-\overset{4}{\underset{\underset{CH_3}{|}}{CH}}-\overset{5}{CH_3}$$

3. Number the carbon atoms in this chain so as to give the *lowest* possible number to the double bond regardless of the groups attached to the parent structure. In this example, we would get the lowest possible number for the double bond by numbering from the left side. Thus, it is a 2-pentene, not a 3-pentene.

4. Indicate the position of the double bond by placing the lower-numbered carbon atom of the double bond before the name of the parent structure. This procedure makes the pentene above a 2-pentene.

5. Give the groups attached to the parent structure, other than hydrogen, both *name* and *number*, as was done in naming the alkanes, and place these groups in alphabetical order. For the compound above, the correct IUPAC name is 2,4-dimethyl-2-pentene.

> **Study Hint:** The double bond takes precedence over the substituent(s), and its position is indicated by the lower-numbered carbon atom. For alkanes the substituent(s) takes precedence, but for alkenes the double bond does.

EXAMPLE 18.5 Write the IUPAC name for each of the following compounds:

Answer

a. $CH_2=CH-CH_2-CH_3$ 1-butene

b. $CH_3-CH=CH-CH_2-CH_3$ 2-pentene

c. 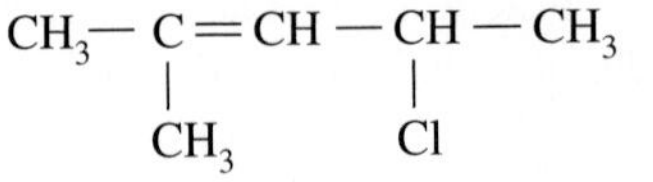   CH$_3$—C=CH—CH—CH$_3$ 4-chloro-2-methyl-2-pentene
         | |
         CH$_3$ Cl

d. 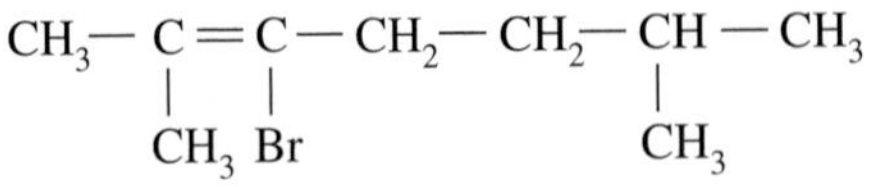   CH$_3$—C=C—CH$_2$—CH$_2$—CH—CH$_3$ 3-bromo-2,6-dimethyl-2-heptene
         | | |
         CH$_3$ Br   CH$_3$

Study Exercise 18.6
Write the IUPAC name for the following compounds:
a. 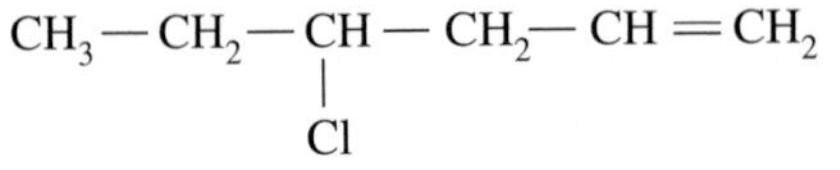   CH$_3$—CH$_2$—CH—CH$_2$—CH=CH$_2$
                 |
                 Cl
                                            (4-chloro-1-hexene)

b. 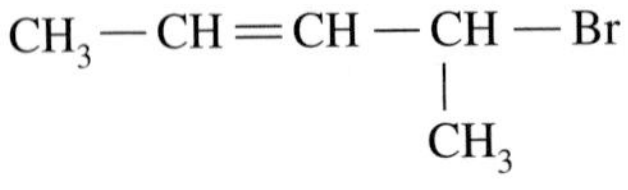   CH$_3$—CH=CH—CH—Br
                  |
                  CH$_3$
                                            (4-bromo-2-pentene)

Work Problem 10.

In writing structures of alkenes from the name, we follow the same steps as for alkanes, *except* we must add the *double bond* in the correct position of the parent structure.

EXAMPLE 18.6   Write the structural formula for each of the following compounds:

Answer

a.   propene   CH$_2$=CH—CH$_3$

b.   1-pentene   CH$_2$=CH—CH$_2$—CH$_2$—CH$_3$

c.   cyclohexene

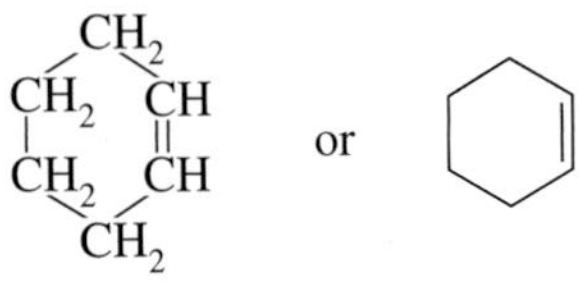

             CH$_2$
         CH$_2$ CH
          | || or
         CH$_2$ CH
             CH$_2$

(All positions on the ring are equivalent and so the double bond can be between any two of the carbon atoms in the cyclohexene ring. No number is needed in the name.)

d.   4,5,5-trimethyl-2-hexene

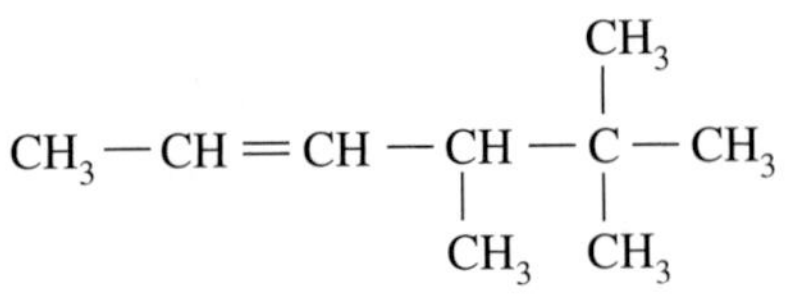

                                      CH$_3$
                                      |
         CH$_3$—CH=CH—CH—C—CH$_3$
                        | |
                        CH$_3$ CH$_3$

Study Exercise 18.7
Write the structural formula for each of the following compounds:

a.   4-nitro-2-hexene   (CH$_3$—CH=CH—CH—CH$_2$—CH$_3$
                                           |
                                           NO$_2$

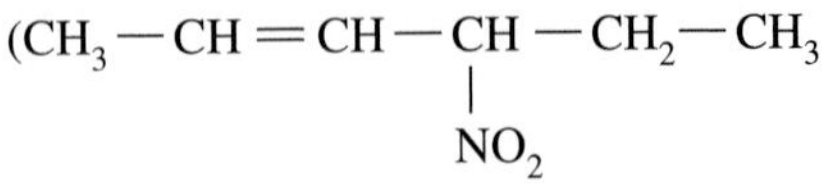

b.   cyclopentene

             CH
         CH$_2$ CH
          | | or
         CH$_2$—CH$_2$

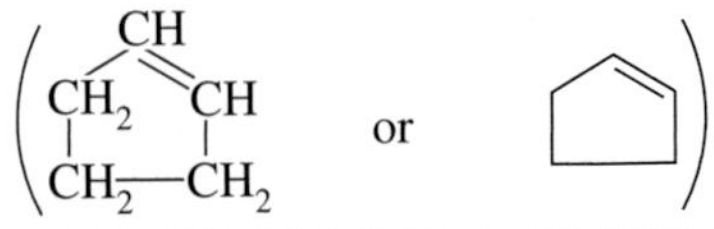

Work Problem 11.

Uses and Reactions of Alkenes

The most important use of alkenes is as a starting material or *feedstock* for some useful plastics. Thus, ethylene is important for what can be made out of it. The most common material made from ethylene is a *polymer* called *poly*ethylene. A **polymer** is a substance made up of thousands of smaller molecules (monomers) that have bonded together to form a giant molecule. (*Poly* means "many," and *mer* comes from Greek meaning "parts".) Polyethylene is a polymer made from the smaller molecule ethylene. Plastics, adhesives, and synthetic rubber are polymers.

Polymer Substance made up of thousands of smaller molecules (*monomers*) that have bonded together to form a giant molecule.

Polymers are formed from monomers which react to form long-chain molecules with high molecular masses. **Addition polymers** are polymers formed from alkene monomers that bond together by breaking one part of the double bond and forming two new single bonds.

The long chains thus formed are entirely new materials with new and often very useful properties. *Three* such polymers are polyethylene, Teflon, and polyvinyl chloride.

Addition polymers Polymers formed from alkene monomers that bond together by breaking one part of the double bond and forming two new single bonds.

Polyethylene is formed from the monomer ethylene using an organic peroxide catalyst, heat, and pressure. The equation for the reaction is

$$n\text{CH}_2\text{=}\text{CH}_2 \longrightarrow \quad \text{or} \quad \left[\begin{array}{cc} \text{H} & \text{H} \\ | & | \\ \text{C} - \text{C} \\ | & | \\ \text{H} & \text{H} \end{array}\right]_n \qquad (18.3)$$

ethylene

In the formula for polyethylene, the n denotes a large number of monomer units, giving a molecular mass for polyethylene of about 1,000,000 amu. Polyethylene is used in making plastic bags, trash cans, milk containers, squeeze bottles, and a host of other common items (see Figure 18.5).

Teflon is formed from the monomer tetrafluoroethylene using a catalyst of hydrogen peroxide and iron(III) ions. The equation for the reaction is

FIGURE 18.5
Various polyethylene materials.

$$nCF_2 = CF_2 \longrightarrow \qquad \text{or} \qquad \left[\begin{array}{cc} F & F \\ | & | \\ C - C \\ | & | \\ F & F \end{array} \right]_n \qquad (18.4)$$

tetrafluoroethylene

Teflon is resistant to most types of chemical action and can withstand high temperatures. It is used on nonstick cooking utensils, in greaseless bearings, and as an insulator (See Figure 18.6).

Polyvinyl chloride (PVC) is formed from the monomer vinyl chloride using an organic peroxide catalyst and heat. The equation for the reaction is

$$n CH_2 = CH \longrightarrow \qquad \text{or} \qquad \left[\begin{array}{c} CH_2 - CH \\ | \\ Cl \end{array} \right]_n \qquad (18.5)$$
$$\qquad | \\ \qquad Cl$$

vinyl chloride

Polyvinyl chloride has a molecular mass of about 1,500,000 amu. It is used to make vinyl tile floors, plumbing pipes, phonograph records, rainware, shower curtains, and garden hoses (see Figure 18.7).

This participation by only one of the bonds of the double bond in a reaction is characteristic of alkenes. This type of reaction is called an *addition reaction*. An **addition reaction** is a combination reaction in which an alkene and either bromine or chlorine react to give a dibromo or dichloro compound, respectively. The reaction readily occurs at room temperature.

Addition reaction Form of combination reaction in which an alkene and either bromine or chlorine reacts to give dibromo- or dichloro- compounds, respectively.

$$CH_2 = CH - CH_3 + Br_2 \longrightarrow \begin{array}{ccc} CH_2 - CH - CH_3 \\ | & | \\ Br & Br \end{array} \qquad (18.6)$$

(in CCl_4 solvent)

FIGURE 18.6
Various Teflon materials.

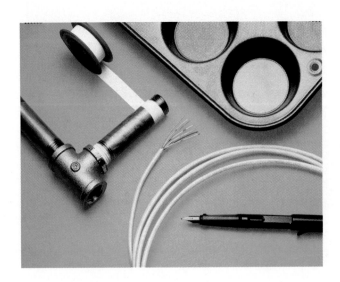

FIGURE 18.7
Various polyvinyl chloride materials.

A more general equation for this process is

$$R-CH=CH_2 + X-X \longrightarrow R-CH-CH_2 \atop \qquad\qquad\quad | \quad\; |$$
$$\qquad\qquad\qquad\qquad\qquad\qquad\qquad\; X \quad X$$

(18.7)

R = alkyl or H   X = Br or Cl

> **Study Hint:** Alkanes undergo halogenation by a *substitution reaction* to form (1) an organic halogen compound and (2) a hydrogen halide gas (HX) in the presence of heat or light. Alkenes undergo halogenation by an *addition reaction* to form a dihaloorganic compound (one product), and the reaction occurs at room temperature.

Study Exercise 18.8
Complete and balance the reaction equation for the following chemical reaction:

+ Br₂

(in CCl₄ solvent) ⟶

$$\left(\begin{array}{c} + Br_2 \\ \text{(in CCl}_4 \\ \text{solvent)} \end{array} \longrightarrow \begin{array}{c} Br \\ \quad Br \\ \end{array} \right)$$

Work Problem 12.

18.5 Alkynes

The **alkynes** (al′kīns) are aliphatic hydrocarbons that have the general molecular formula C_nH_{2n-2} for open-chain compounds. Like the alkenes, alkynes are unsaturated hydrocarbons because the number of hydrogens differs from the alkane formula of C_nH_{2n+2}. In the case of alkynes, however, *four* hydrogen atoms are missing relative to the alkane containing the same number of carbon atoms. As a result, all alkynes have a *triple bond* (category 3).

The simplest alkyne is acetylene, which follows the general molecular formula with *n* equal to 2. As we saw in Section 18.2, a triple bond is linear, and the H—C—C bond angle is 180°.

$$H-C\equiv C-H \quad \text{or} \quad HC\equiv CH$$

Alkynes Form of unsaturated aliphatic hydrocarbon; they have the general molecular formula C_nH_{2n-2} for open-chain compounds, a carbon–carbon triple bond, and a linear structure.

All alkynes possess a triple bond with these same characteristics. The triple bond may be at the end of a chain (terminal) or in the middle of a chain, but *not* at a branch point:

$$H-C\equiv C-CH_2CH_3 \qquad CH_3-C\equiv C-CH_3$$

Nomenclature

The IUPAC system uses the following five rules to name alkynes:

1. The alkyne hydrocarbons all have the ending **-yne**. And so acetylene has the IUPAC name *ethyne*.
2. The longest continuous chain of carbon atoms that *contains the triple bond* is the parent structure.
3. Number the carbon atoms in this chain so that the triple bond has the lowest possible number.
4. Indicate the position of the triple bond by placing the lower-numbered carbon atom of the triple bond before the name of the parent structure.
5. Give both *name* and *number* to the groups attached to the parent structure other than hydrogen and place them in alphabetical order.

EXAMPLE 18.7   Write the IUPAC name for each of the following compounds:

Answer

a. $HC\equiv C-CH_2-CH_3$ 1-butyne

b. 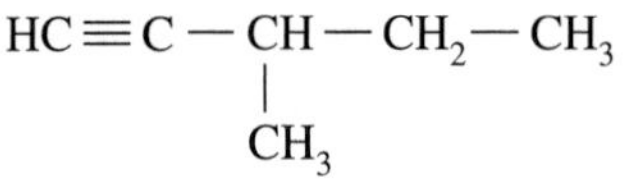 $HC\equiv C-CH-CH_2-CH_3$ 3-methyl-1-pentyne
            $\overset{\displaystyle |}{CH_3}$

c. 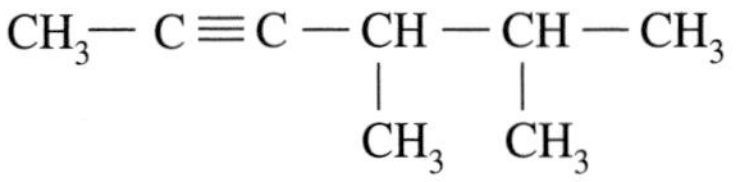 $CH_3-C\equiv C-CH-CH-CH_3$ 4,5-dimethyl-2-hexyne
                          $\overset{\displaystyle |}{CH_3}$ $\overset{\displaystyle |}{CH_3}$

d. 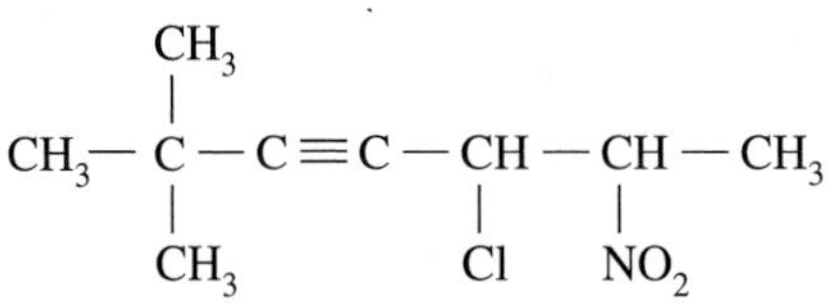 CH_3 5-chloro-2,2-dimethyl-6-nitro-3-
                   $|$ heptyne
       $CH_3-\overset{|}{\underset{|}{C}}-C\equiv C-CH-CH-CH_3$
                   CH_3 $\overset{|}{Cl}$ $\overset{|}{NO_2}$

Study Exercise 18.9
Write the IUPAC name for each of the following compounds:

a. 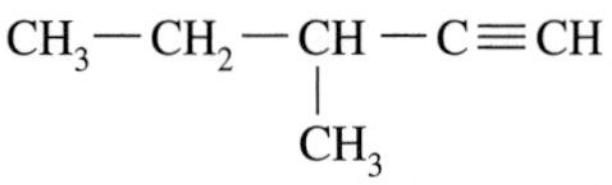 $CH_3-CH_2-CH-C\equiv CH$
                       $\overset{\displaystyle |}{CH_3}$

(3-methyl-1-pentyne)

b. 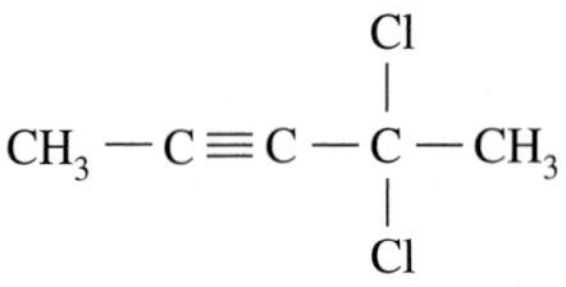 Cl
                                 $|$
       $CH_3-C\equiv C-\overset{|}{\underset{|}{C}}-CH_3$
                                 Cl

(4,4-dichloro-2-pentyne)

Work Problem 13.

To write structures of alkynes from the name of the alkyne, we just follow the same steps we took with alkenes, *except* that we must put the *triple bond* in the correct position of the parent structure.

| EXAMPLE 18.8 | Write the structural formula for each of the following compounds: |

Answer

a. 2-pentyne

$$CH_3 - C \equiv C - CH_2 - CH_3$$

b. 3-methyl-1-butyne

$$HC \equiv C - CH - CH_3$$
$$\qquad\qquad\ \ |$$
$$\qquad\qquad\ CH_3$$

c. 4,4-dimethyl-2-hexyne

$$\qquad\qquad\qquad CH_3$$
$$\qquad\qquad\qquad |$$
$$CH_3 - C \equiv C - C - CH_2 - CH_3$$
$$\qquad\qquad\qquad |$$
$$\qquad\qquad\qquad CH_3$$

d. 4-bromo-1-chloro-4,5-dimethyl-2-octyne

$$\qquad\qquad\qquad CH_3\ CH_3$$
$$\qquad\qquad\qquad |\quad\ |$$
$$CH_2 - C \equiv C - C - CH - CH_2 - CH_2 - CH_3$$
$$|\qquad\qquad\ \ |$$
$$Cl\qquad\qquad\ Br$$

Study Exercise 18.10

Write the structural formula for each of the following compounds:

a. 4-chloro-3-bromo-1-hexyne

$$(HC \equiv C - CH - CH - CH_2 - CH_3)$$
$$\qquad\qquad\quad |\qquad |$$
$$\qquad\qquad\ \ Br\quad Cl$$

b. 4,4,5-trichloro-2-octyne

$$\qquad\qquad\qquad\ Cl\ \ Cl$$
$$\qquad\qquad\qquad\ |\quad |$$
$$(CH_3 - C \equiv C - C - CH - CH_2 - CH_2 - CH_3)$$
$$\qquad\qquad\qquad\ |$$
$$\qquad\qquad\qquad\ Cl$$

Work Problem 14.

Uses and Reactions of Alkynes

Like ethylene, acetylene is an important feedstock chemical. It is used in the preparation of a variety of organic compounds needed for the synthesis of agricultural and pharmaceutical products. Acetylene also has some use as a fuel, particularly in the oxyacetylene torch. Acetylene burns in oxygen to give a very hot flame that is used for high-temperature welding:

$$2\,HC \equiv CH\ +\ 5\,O_2\ \longrightarrow\ 4\,CO_2\ +\ 2\,H_2O \qquad\qquad (18.8)$$

In alkynes we again see the participation by one or two bonds of the triple bond in a reaction. The first and second bonds in a triple bond react with bromine or chlorine, but a single bond still remains after the reaction. The bromine molecules add one at a time, and if we choose to, we can stop the reaction after one addition by limiting the amount of bromine. The reactions are *addition reactions* like those we observed for alkenes. In limited bromine we obtain a dibromo alkene.

$$H-C\equiv C-CH_3 + Br_2 \xrightarrow[\substack{(in\ CCl_4 \\ solvent)}]{} \begin{array}{c} H-C=C-CH_3 \\ | \quad | \\ Br \quad Br \end{array} \quad (18.9)$$

In excess bromine or chlorine, *2 mol* is added to break the two bonds of the triple bond.

$$H-C\equiv C-CH_3 + 2\ Br_2 \xrightarrow[\substack{(in\ CCl_4 \\ solvent)}]{} \begin{array}{c} Br\ \ Br \\ | \quad | \\ H-C-C-CH_3 \\ | \quad | \\ Br\ \ Br \end{array} \quad (18.10)$$

A more general equation for this process is:

$$R-C\equiv C-H + X-X \longrightarrow \begin{array}{c} R \quad H \\ C=C \\ X \quad X \end{array} + X-X \longrightarrow \begin{array}{c} X\ X \\ | \ | \\ R-C-C-H \\ | \ | \\ X\ X \end{array} \quad (18.11)$$

R = alkyl or H   X = Cl or Br

Study Exercise 18.11

Complete and balance the reaction equation for the following chemical reaction:

$$\begin{array}{c} CH_3-C\equiv C-CH-CH_3 + Br_2 \longrightarrow \\ | \\ Cl \end{array} \quad \substack{(excess \\ in\ CCl_4 \\ solvent)}$$

$$\left(\begin{array}{c} CH_3-C\equiv C-CH-CH_3 + 2\ Br_2 \\ | \\ Cl \end{array} \xrightarrow[\substack{(excess \\ in\ CCl_4 \\ solvent)}]{} \begin{array}{c} Br\ \ Br \\ | \quad | \\ CH_3-C-C-CH-CH_3 \\ | \quad | \quad | \\ Br\ \ Br\ Cl \end{array} \right)$$

Work Problem 15.

Aromatic hydrocarbons Hydrocarbons that have a ring of six carbon atoms (*benzene ring*) and alternating carbon–carbon double bonds within the ring.

Benzene ring Structure characteristic of aromatic hydrocarbons; it is a regular (equal-sided) hexagon with alternating double bonds around the ring.

18.6 Aromatic Hydrocarbons

The **aromatic hydrocarbons** are hydrocarbons that have a ring of six carbon atoms and alternating carbon–carbon double bonds within that ring. These compounds got their name because many such compounds—cinnamon, vanilla, and wintergreen, for example—have a pleasant smell. Not all aromatic compounds are pleasant to smell, however. Toluene, which used to be a component of airplane glue, is rather unpleasant as well as dangerous to smell for too long a period of time.

While they are not bound together by odor, aromatic hydrocarbons share a structural basis known as the **benzene ring**. The structural formula for benzene (C_6H_6) is a regular (equal-sided) hexagon with alternating double bonds around the ring. This formula can be drawn in several ways:

The structures on the left and in the middle differ only in which carbons are bonded by single bonds and which are bonded by double bonds. A double-headed arrow (⟷) between the first two structures is used to show their equivalency. To represent these two possible structures, we will use the structure on the right, which is a circle within a regular hexagon.

All the carbons in the benzene ring have three covalent bonds: a double bond and two single bonds (category 2). Thus, the benzene ring is planar, as shown by the ball-and-stick model of benzene in Figure 18.8.

Other aromatic hydrocarbons consist of benzene rings joined to each other. Two such compounds are (1) naphthalene and (2) benzo[*a*]pyrene.

napthalene   benzo[*a*]pyrene

Nomenclature

The aromatic hydrocarbons are named as derivatives of the parent, benzene; however, the IUPAC still recognizes trivial names for some of them. A few of these trivial names are

toluene   *o*-xylene   *m*-xylene   *p*-xylene

In the IUPAC system, we name aromatic compounds with **one** *substituent* group by combining the name of the substituent group with *benzene*. Because *all* hydrogen atoms are equivalent on benzene, this group may be attached at any position. The following structures are all named *chlorobenzene*, and all represent the same compound.

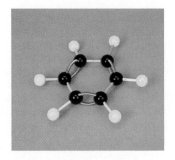

FIGURE 18.8
Ball-and-stick model of benzene (C_6H_6). Note that all the atoms lie in the same plane, making the molecule "flat."

Aromatic compounds with **two** *substituents* on the ring have three different isomers. The positions of the substituents on these compounds and the prefixes used to describe them are shown below.

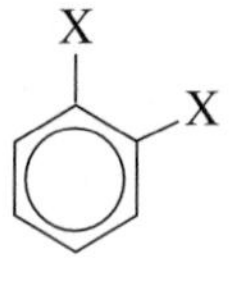

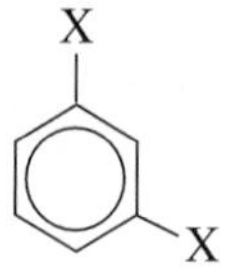

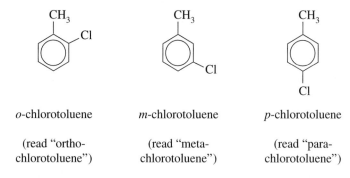

| ortho (or 1, 2-) | meta or (1, 3-) | para (or 1, 4-) |
| abbreviated *o-* | abbreviated *m-* | abbreviated *p-* |

Combining these prefixes with their appropriate abbreviations and name for the substituents plus the word benzene produces the following compound names:

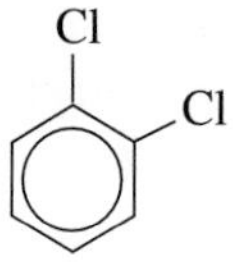

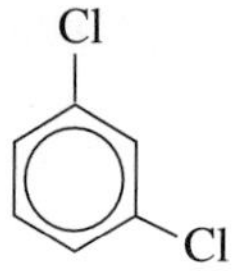

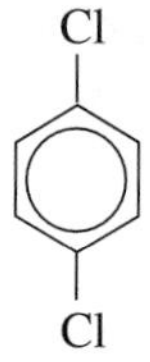

| *o*-dichlorobenzene | *m*-dichlorobenzene | *p*-dichlorobenzene |
| (read "ortho-dichlorobenzene") | (read "meta-dichlorobenzene") | (read "para-dichlorobenzene") |

Some aromatic compounds are related to toluene, not benzene, and are named as toluenes.

| *o*-chlorotoluene | *m*-chlorotoluene | *p*-chlorotoluene |
| (read "ortho-chlorotoluene") | (read "meta-chlorotoluene") | (read "para-chlorotoluene") |

For compounds with *three* or *more substituents* on the ring, we must use numbers to indicate the position on the ring and apply the following three rules:

1. The parent structure is benzene *or* toluene.

2. For *benzene*, the group name written next to the word "benzene" becomes position 1 on the ring, and we number the ring in a direction that will give the lowest possible numbers to the substituents. For *toluene*, the methyl group is at position 1 on the ring, and the ring is numbered in a direction that will give the lowest possible numbers to the substituents. Position 1 is assumed in both benzene and toluene derivatives but is not indicated in the name.

3. Place all groups in alphabetical order in the name except the group in position 1.

The following examples illustrate the application of these rules:

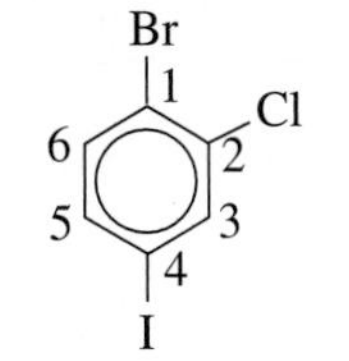

2-chloro-4-
iodobromobenzene

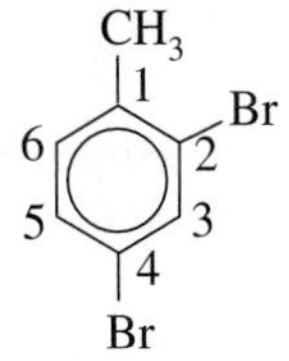

2, 4-dibromotoluene

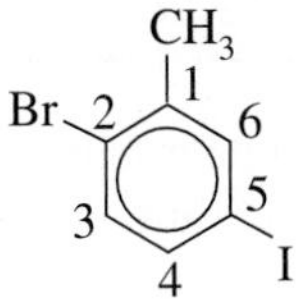

2-bromo-5-
iodotoluene

EXAMPLE 18.9 Write the IUPAC name for each of the following compounds:

Answer

a. 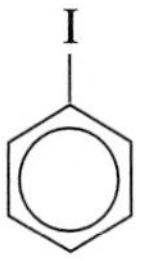 I   iodobenzene

b. 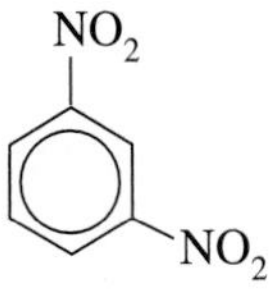 NO$_2$ *m*-dinitrobenzene

c.  Br   *p*-nitrobromobenzene

d. 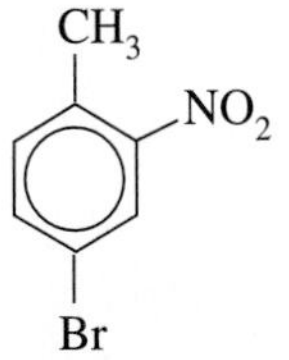 CH$_3$ 4-bromo-2-nitrotoluene

Study Exercise 18.12
Write the IUPAC name for each of the following compounds:

a. 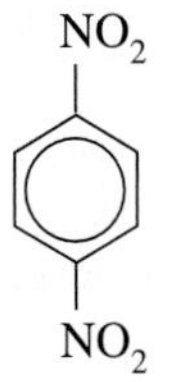 NO$_2$

(*p*-dinitrobenzene)

b.

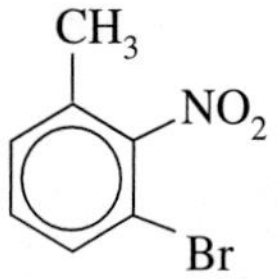

(3-bromo-2-nitrotoluene)

Work Problem 16.

In writing structures of benzene compounds from the names, draw the structure of the parent compound and then attach the various groups.

EXAMPLE 18.10 Write the structural formula for each of the following compounds:

Answer

a. nitrobenzene

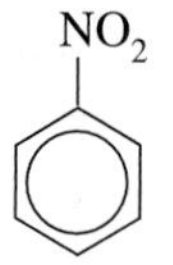

b. *o*-bromochlorobenzene

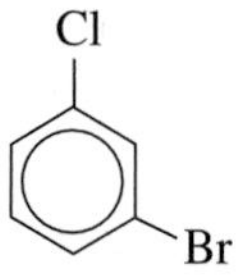

c. 2-chloro-4-nitrobromobenzene

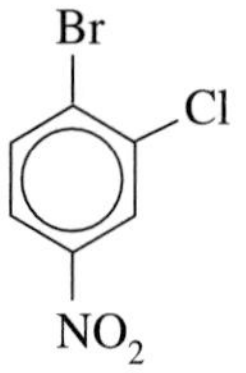

d. 2,6-dinitrotoluene

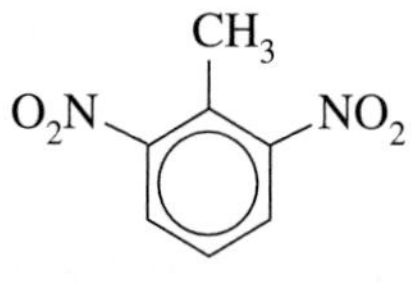

Study Exercise 18.13
Write the structural formula for each of the following compounds:

a. *m*-nitrochlorobenzene

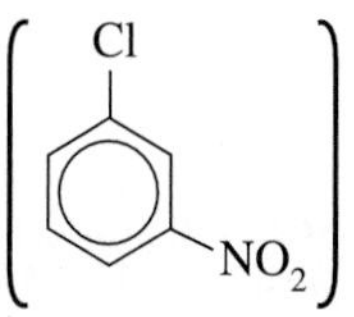

b. *o*-nitrotoluene

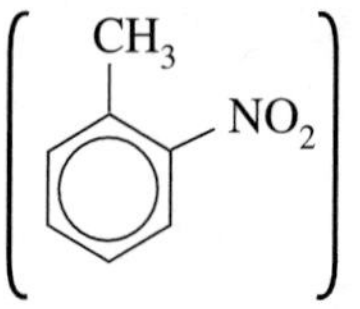

Work Problem 17.

Uses and Reactions
of Aromatic Compounds

An enormous number of compounds that contain benzene rings are used in our daily lives. Benzene and toluene are additives in unleaded gasoline. A number of polymers—the polyester Dacron, the plastics polystyrene and Bakelite, and many others—contain benzene rings in their structure. Such common pharmaceutical products as pain relievers (aspirin, Tylenol, ibuprofen), antibiotics (sulfa drugs, tetracycline, some penicillins), and asthma drugs incorporate benzene rings. Agricultural products such as pecticides, fungicides, and insecticides also contain benzene rings. Other well-known organic chemicals with benzene rings include 2,4,6-trinitrotoluene (TNT), dioxin (a highly toxic by-product of the synthesis of some pesticides), tetrahydrocannabinols (the active ingredient in marijuana), and novocaine. Naphthalene has been used as a moth repellent and insecticide, but *p*-dichlorobenzene is replacing it. Benzo[*a*]pyrene is found in cigarette smoke and automobile exhaust and is considered to be a major cause of lung cancer.

We might think that benzene would react with chlorine or bromine just like an alkene, since both compounds have double bonds. They do not, however, because benzene rings are more stable with the aromatic structure. Instead of undergoing an addition reaction, benzene rings usually undergo *substitution reactions* by bromine or chlorine (like alkanes):

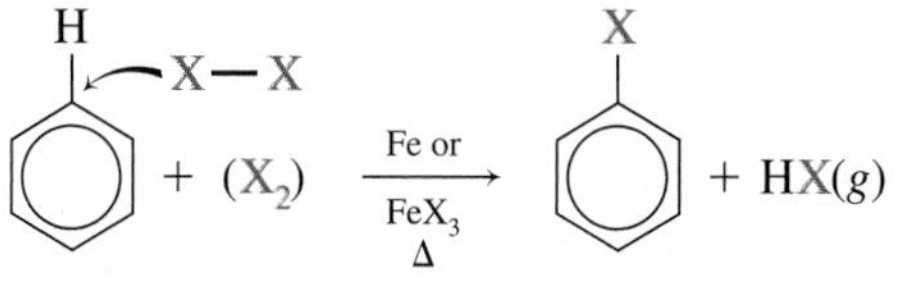

$$\text{benzene} + (X_2) \xrightarrow[\Delta]{\text{Fe or FeX}_3} \text{product} + HX(g) \qquad (18.12)$$

X = Cl, Br

In addition to the substituted benzene, a gas (HX, such as HBr or HCl) is also released. This substitution reaction generally requires heat and a catalyst in the form of powdered iron or the corresponding iron(III) halide ($FeCl_3$ or $FeBr_3$).

Study Exercise 18.14

Complete and balance the reaction equation for the following reaction:

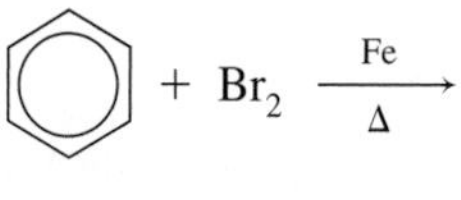

$$+ Br_2 \xrightarrow[\Delta]{\text{Fe}}$$

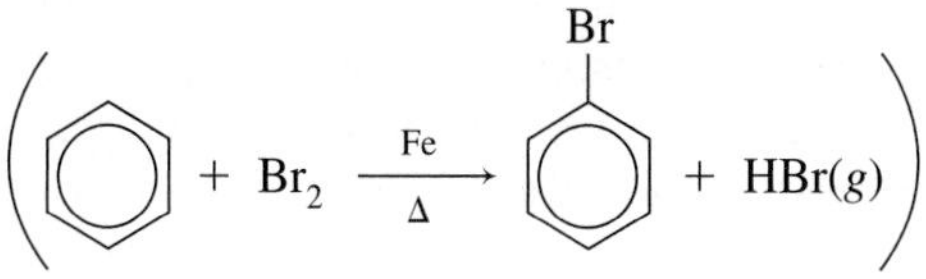

$$\left(+ Br_2 \xrightarrow[\Delta]{\text{Fe}} + HBr(g) \right)$$

Work Problem 18.

Table 18.6 summarizes the four types of hydrocarbons.

18.7 Hydrocarbon Derivatives

A wide variety of organic compounds have been derived from the hydrocarbons by substituting various *functional groups* for hydrogen atoms attached to carbon. A **functional group** is an atom or group of atoms (other than hydrogen) that is attached to a hydrocarbon chain and that confers some distinctive chemical and

Functional group   Atom or group of atoms (other than hydrogen) that is attached to a hydrocarbon chain and that confers some distinctive chemical and physical properties on the organic compound.

TABLE 18.6		Summary of the Hydrocarbons				
TYPE	NAME	FORMULA	STRUCTURE	EXAMPLE	GEOMETRY AT C	REACTION
Aliphatic (no rings)	Alkane	C_nH_{2n+2}	$-\overset{\mid}{\underset{\mid}{C}}-$	CH_3-CH_3	Tetrahedral	Substitution
	Alkene	C_nH_{2n}	$\diagdown C=C \diagup$	$CH_2=CH_2$	Planar	Addition
	Alkyne	C_nH_{2n-2}	$-C\equiv C-$	$HC\equiv CH$	Linear	Addition
Aromatic		C_6H_6 (for benzene)	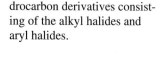	Benzene	Planar	Substitution

physical properties on the organic compound. We have already seen two functional groups. In the alkenes the carbon–carbon double bond is the functional group, and in the alkynes the functional group is the carbon–carbon triple bond. Table 18.7 lists some of the more common classes of organic compounds, a general formula for each class, and the functional group involved.

Organic Halides

Organic halides Class of hydrocarbon derivatives consisting of the alkyl halides and aryl halides.

The class of hydrocarbon derivatives known as **organic halides** consists of *alkyl halides* and *aryl halides*. Alkyl halides contain an alkyl group or substituted alkyl group attached to a halogen atom (F, Cl, Br, or I). The general formula for an alkyl halide is **R—X**, with the **X** representing any halogen atom as the functional group. The **R** may also represent an alkenyl group (carbon–carbon double bond present) or an alkynyl group (carbon–carbon triple bond present). Ethyl chloride, shown in Figure 18.9, is an alkyl halide that is used to cool and numb the skin during minor surgery.

Aryl halides are compounds having a halogen atom directly attached to an aromatic ring. The general formula for aromatic halides is **Ar—X**, with **Ar** representing the aromatic ring, and **X** the halogen atom. Chlorobenzene, shown in Figure 18.10, is an example of an aryl halide.

The insecticide dichlorodiphenyltrichloroethane (DDT) belongs to the class of organic halides. DDT is a very effective insecticide and has been responsible for saving thousands of human and animal lives. Unhappily, DDT does not readily de-

FIGURE 18.9
(a) Structural formula and
(b) ball-and-stick model of
ethyl chloride, an example
of an alkyl halide.

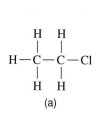

$$H-\overset{\overset{\displaystyle H}{\mid}}{\underset{\underset{\displaystyle H}{\mid}}{C}}-\overset{\overset{\displaystyle H}{\mid}}{\underset{\underset{\displaystyle H}{\mid}}{C}}-Cl$$

(a)

(b)

TABLE **18.7**	Classes of Organic Compounds Derived from Hydrocarbons

CLASS OF COMPOUND	GENERAL FORMULA	FUNCTIONAL GROUP
Organic halide	$R-X,^a Ar-X^b$	$-X^c$
Alcohol	$R-OH$	OH
Ether	$R-O-R', Ar-O-R, Ar-O-Ar'$	$-O-$

(R and R′ may be the same or different; Ar and Ar′ may be the same or different)

CLASS OF COMPOUND	GENERAL FORMULA	FUNCTIONAL GROUP
Aldehyde	$R-\overset{\displaystyle O}{\underset{\displaystyle H}{C}}$, $Ar-\overset{\displaystyle O}{\underset{\displaystyle H}{C}}$	$-\overset{\displaystyle O}{\underset{\displaystyle H}{C}}$
Ketone	$R-\overset{\displaystyle O}{\underset{\displaystyle R'}{C}}$, $Ar-\overset{\displaystyle O}{\underset{\displaystyle R}{C}}$, $Ar-\overset{\displaystyle O}{\underset{\displaystyle Ar'}{C}}$	$-\overset{\displaystyle O}{C}$
Carboxylic acid	$R-\overset{\displaystyle O}{\underset{\displaystyle OH}{C}}$, $Ar-\overset{\displaystyle O}{\underset{\displaystyle OH}{C}}$	$-\overset{\displaystyle O}{\underset{\displaystyle OH}{C}}$
Ester	$R-\overset{\displaystyle O}{\underset{\displaystyle OR}{C}}$, $Ar-\overset{\displaystyle O}{\underset{\displaystyle OR}{C}}$	$-\overset{\displaystyle O}{\underset{\displaystyle OR}{C}}$
Amine	$R-NH_2, R_2NH, R_3N$	$-NH_2, \diagup NH, \diagup N-$
Amide	$R-\overset{\displaystyle O}{\underset{\displaystyle NH_2}{C}}$, $Ar-\overset{\displaystyle O}{\underset{\displaystyle NH_2}{C}}$	$-\overset{\displaystyle O}{\underset{\displaystyle NH_2}{C}}$

[a] The symbol R designates an alkyl, alkenyl, or alkynyl group.
[b] The symbol Ar designates an aromatic group.
[c] The symbol X designates a halogen atom (F, Cl, Br, or I).

compose. Instead, it accumulates in the environment. DDT has been found to produce cancer and to have other harmful effects on laboratory animals. DDT is also transmitted through the food chain. It accumulates in insects and fish, and then fish-eating animals ingest it. For example, pelicans that ingest DDT produce eggs with softer than normal shells. These more fragile shells break prematurely, thus endangering the animal's ability to reproduce. For these reasons the Environmental Protection Agency (EPA) has banned the sale of DDT except in extreme epidemics.

Another environmental problem associated with halogenated organic compounds is the threat posed to stratospheric ozone by chlorofluorocarbons such as CFC-11 (CCl_3F) and the refrigerant CFC-12 (CCl_2F_2, see Figure 18.11). This problem was discussed in Chemistry of the Atmosphere, Chapters 13 and 17.

Organic halides are not all harmful, though. Many compounds containing halogen atoms are very useful to society. Numerous pharmaceutical and agricultural agents contain halogen atoms. The active ingredient in mothballs is *p*-dichlorobenzene. Some important cleaning solvents also contain halogen atoms.

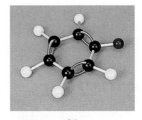

FIGURE 18.10
(a) Structural formula and (b) ball-and-stick model of chlorobenzene, an example of an aryl halide.

Alcohols

Alcohols Class of hydrocarbon derivatives in which a hydroxyl (—OH) group replaces a hydrogen atom on a carbon atom of an aliphatic hydrocarbon.

The compounds we call **alcohols** are a class of hydrocarbon derivatives in which a hydroxyl group (—OH) replaces a hydrogen atom on a carbon atom of an *aliphatic* hydrocarbon. The general formula for an alcohol is

$$\mathbf{R — O — H}$$

The hydroxyl group (—OH) is the functional group and is attached directly to an alkyl group, *not* to an aryl group. The hydroxyl group is not a hydroxide ion (OH⁻) since in alcohols the hydroxyl group is *covalently* bonded to the alkyl group.

Alcohols are related to water in that one of the hydrogen atoms of water (**H — O — H**) has been replaced by an **R** group. Just as water ionizes slightly, so alcohols ionize to yield hydrogen ions (H⁺) and act as very weak acids, even *weaker* acids than water.

Ethyl alcohol, shown in Figure 18.12 is what we mean when we say "alcohol," the inebriating component of alcoholic beverages. Ethyl alcohol is also used in pharmaceutical preparations both as a solvent in tinctures (such as tincture of iodine) and for "medicinal" purposes, as in many cough syrups. However, overconsumption of alcohol not only can leave us feeling horrible the next day but can also increase our risk of long-term health problems such as liver disease.

Gasohol, a mixture of 10% ethanol (ethyl alcohol, CH_3CH_2OH) and gasoline, is an automotive fuel that was developed to try to reduce American dependence on foreign oil. One method for preparing ethanol is by fermenting cereal grains, such as corn. It is separated from the fermentation mixture by distillation (see Section 12.5). All the water is chemically removed, and the ethanol is then mixed with gasoline. However, extremely wide use of gasohol could create another problem— a shortage of cereal grains. Another alcohol, methanol (methyl alcohol, CH_3OH), has also been used as a fuel blended into gasoline.

To make automobile engines run more efficiently, petroleum companies have always blended a variety of volatile organic compounds into gasoline. Ethanol is just one in a long line of such additives. One of the more hazardous additives has been tetraethyllead [Pb(C₂H₅)₄]. The use of leaded gasoline increases the amount of lead-containing compounds in the environment. Such compounds are very toxic to living beings, and leaded gas is currently being phased out of use. Newer and less dangerous additives include methanol, ethanol, and methyl *tert*-butyl ether.

FIGURE 18.11
Container of the refrigerant CFC-12 (Freon-12, dichlorodifluoromethane CCl_2F_2).

FIGURE 18.12
(a) Structural formula and (b) ball-and-stick model of ethyl alcohol, an example of an alcohol.

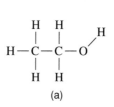

(a)

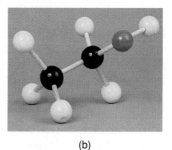

(b)

Ethers

Compounds in which *two hydrocarbon groups* [alkyl (**R**) and/or aryl (**Ar**), the same or different] are attached to an oxygen atom are called **ethers**. The general formulas for ethers are

$$\textbf{R—O—R'} \qquad \textbf{R—O—Ar} \qquad \textbf{Ar—O—Ar'}$$
$$\textbf{R = R' or R ≠ R'} \qquad\qquad\qquad \textbf{Ar = Ar' or Ar ≠ Ar'}$$

The oxygen atom (—O—) is the functional group. Ethers can be considered compounds in which both of the H atoms of water (**H—O—H**) have been replaced by an alkyl group (**R**) or an aryl group (**Ar**).

Diethyl ether, or just *ether* ($CH_3—CH_2—O—CH_2—CH_3$), shown in Figure 18.13, is the most common ether. At one time it was used as an anesthetic, but it has been largely replaced by halothane (Fluothane) because ether is highly flammable. The structural formula of halothane is

$$\begin{array}{ccc} & F & H \\ & | & | \\ F— & C—C & —Br \\ & | & | \\ & F & Cl \end{array}$$

Methyl-*tert*-butyl ether is another ether that we encounter nearly every day. This ether is added to unleaded gasoline to improve its performance and burning characteristics. In heavily polluted areas the gasoline mixture is required to contain compounds containing oxygen atoms.

$$\begin{array}{c} CH_3 \\ | \\ CH_3—O—C—CH_3 \\ | \\ CH_3 \end{array}$$

Aldehydes

Compounds whose structures consist of *one* hydrocarbon group (alkyl or aryl) attached to a carbonyl group $\left(\diagdown C{=}O \right)$ to which a *hydrogen atom* is also attached

Ethers   Class of hydrocarbon derivatives in which *two hydrocarbon groups* [alkyl (R) and/or aryl (Ar), the same or different] are attached to an oxygen atom.

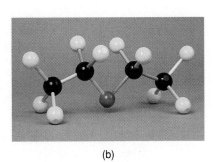

FIGURE 18.13
(a) Structural formula and (b) ball-and-stick model of diethyl ether, an example of an ether.

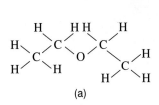
(a)

(b)

Aldehydes Class of hydrocarbon derivatives in which *one* hydrocarbon (alkyl or aryl) group is attached to a carbonyl group $\left(\begin{array}{c}\diagdown\\C=O\\\diagup\end{array}\right)$ on one side and a *hydrogen atom* on the other.

are known as **aldehydes**. A carbonyl group is a carbon with a double-bonded oxygen. Formaldehyde, the simplest aldehyde, has two hydrogen atoms attached to the carbonyl group. The general formulas for aldehydes are

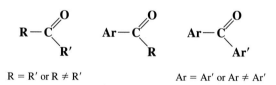

The functional group is $-C\diagfrac{\diagup O}{\diagdown H}$, the formyl group.

Acetaldehyde, shown in Figure 18.14, is an aldehyde. Acetaldehyde is one of the chief eye irritants found in smog. But not all aldehydes are noxious. A number of flavoring agents and spices rely on aldehydes for their tasty effect. Vanillin (in vanilla), cinnamaldehyde (in cinnamon), and cuminaldehyde (in cumin) all contain the aldehyde functional group.

An aldehyde plays a critical role in the chemistry of vision. Vitamin A is converted in the body to *cis*-retinal, an aldehyde. *cis*-Retinal is intimately involved in the eye's response to light and allows us to detect light and see. Interestingly, all known visual systems that occur in nature use this same chemistry to detect light.

Ketones

Ketones Class of hydrocarbon derivatives in which *two* hydrocarbon (alkyl or aryl) groups are attached to a carbonyl group $\left(\begin{array}{c}\diagdown\\C=O\\\diagup\end{array}\right)$.

Yet another class of hydrocarbon derivatives, **ketones**, consists of compounds whose structures contain *two* hydrocarbon (alkyl or aryl) groups attached to a carbonyl group $\left(\begin{array}{c}\diagdown\\C=O\\\diagup\end{array}\right)$. The general formulas for ketones are

$$R-C\diagfrac{\diagup O}{\diagdown R'} \qquad Ar-C\diagfrac{\diagup O}{\diagdown R} \qquad Ar-C\diagfrac{\diagup O}{\diagdown Ar'}$$

R = R' or R ≠ R'   Ar = Ar' or Ar ≠ Ar'

Acetone, shown in Figure 18.15, is the simplest ketone.

FIGURE 18.14
(a) Structural formula and (b) ball-and-stick model of acetaldehyde, an example of an aldehyde.

(a)   (b)

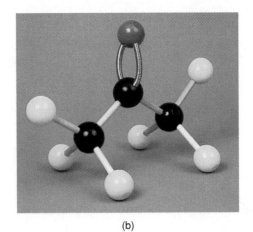

FIGURE 18.15
(a) Structural formula and (b) ball-and-stick model of acetone, an example of a ketone.

Many ketones are used as solvents. Acetone, $\left(CH_3-\overset{\overset{\displaystyle O}{\|}}{C}-CH_3\right)$, one of the most widely used solvents, mixes with water, and also dissolves many organic substances. Many important compounds contain a ketone functional group, including various hormones (such as cortisone and progesterone) and antibiotics (such as tetracycline and erythromycin).

The carbonyl group that is characteristic of aldehydes and ketones is found in many sugars, including glucose and fructose. Also notice the numerous alcohol functional groups present.

glucose

fructose

Sucrose (common table sugar) is formed by the chemical combination of a glucose unit and a fructose unit. Starch consists of a large number (1000 to 4000) of glucose molecules linked together to form a polymer. Cellulose is another polymer of glucose in which the glucose units are linked in a different way. When we eat starch, an enzyme breaks the starch down into glucose units that are used by the body. But people cannot metabolize cellulose because they lack the enzyme required to break the links between glucose molecules in cellulose. However, animals such as cows, sheep, and deer have a microorganism in their gut that does possess such an enzyme and can metabolize cellulose.

Carboxylic Acids

Carboxylic acids Class of hydrocarbon derivations in which a carboxyl group (—COOH) is attached to a hydrocarbon (alkyl or aryl) group.

Compounds that have hydrocarbon (alkyl or aryl) groups attached to a carboxyl group $\left(\begin{array}{c} O \\ \| \\ -C \\ \diagdown \\ O-H \end{array} \right)$ are called **carboxylic** (kär′bok·sil′ik) **acids**. The general formulas for carboxylic acids are

$$R-C\overset{O}{\underset{OH}{\diagup}} \quad \text{(or R—COOH)} \quad \text{and} \quad Ar-C\overset{O}{\underset{OH}{\diagup}} \quad \text{(or Ar—COOH)}$$

Carboxyl group Functional group that consists of a carbon atom that has a double bond with oxygen on one side and a single bond to a —OH group on the other side. $-C\overset{O}{\underset{OH}{\diagup}}$

The functional group is the carboxyl group. A **carboxyl group** consists of a carbon atom that has a double bond with oxygen on one side and a single bond to an OH group on the other side. The acidic hydrogen (H^+) is the H attached to the oxygen of the carboxyl group. It is not one of the hydrogens attached to the alkyl or aryl groups. Acetic acid, shown in Figure 18.16, is the most common carboxylic acid and is found in vinegar.

The carboxylic acid group is present in a variety of everyday experiences. The simplest carboxylic acid, formic acid (H—COOH), is the active irritant in ant and bee stings. Citric acid is a tricarboxylic acid (three carboxyl groups per molecule) that gives lemons, limes, and pineapples their sour taste. Penicillins have carboxylic acid groups, as do aspirin and lactic acid (found in sour milk).

Carboxylic acids play many important roles in the body. In particular, carboxylic acids are involved in the buildup and breakdown of sugars in metabolism and in the synthesis of the molecules that form membranes in the body.

Carboxylic acids are weak acids that can be readily neutralized by appropriate bases, such as sodium hydroxide or sodium bicarbonate, as follows:

$$R-C\overset{O}{\underset{O-H}{\diagup}} + NaOH \longrightarrow R-C\overset{O}{\underset{O^- Na^+}{\diagup}} + H_2O \qquad (18.13)$$

$$H-C\overset{O}{\underset{O-H}{\diagup}} + NaHCO_3 \longrightarrow H-C\overset{O}{\underset{O^- Na^+}{\diagup}} + H_2O + CO_2(g) \qquad (18.14)$$
formic acid

The latter reaction illustrates how baking soda ($NaHCO_3$) can be used to treat bee or ant stings. The bicarbonate ion neutralizes the formic acid in the venom and eases the pain and swelling.

Soap is made from carboxylic acids with carbon chains 14 to 16 carbons long by neutralization of the carboxylic acid to give the *carboxylic acid salt* (see end of Chapter 14):

$$CH_3-CH_2-\cdots-CH_2-C\overset{O}{\underset{O^- Na^+}{\diagup}}$$
a typical soap

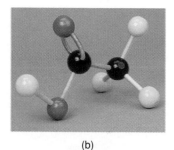

FIGURE 18.16
(a) Structural formula and (b) ball-and-stick model of acetic acid, an example of a carboxylic acid.

Esters

Derived from carboxylic acids, **esters** are compounds in which an —**OR** or —**OAr** group replaces the **OH** group of a carboxylic acid. The general formulas for esters are

$$R-C{\overset{O}{\underset{OR'}{}}} \ , R-C{\overset{O}{\underset{OAr}{}}} \ , Ar-C{\overset{O}{\underset{OR}{}}} \ , \text{ or } Ar-C{\overset{O}{\underset{OAr'}{}}}$$

R = R′ or R ≠ R′   **Ar = Ar′ or Ar ≠ Ar′**

Esters Class of hydrocarbon derivatives in which an —OR or OAr group replaces the —OH group of a carboxylic acid.

The functional group in esters is $-C{\overset{O}{\underset{OR'}{}}}$ or $-C{\overset{O}{\underset{OAr}{}}}$. Ethyl acetate, shown in Figure 18.17, is an ester that is a good solvent and is found in many varnishes and lacquers. The relationship between a carboxylic acid and its derived ester is illustrated by the following reaction. A carboxylic acid and an alcohol react to give an ester and water:

$$CH_3-C{\overset{O}{\underset{OH}{}}} + CH_3-CH_2-OH \longrightarrow CH_3-C{\overset{O}{\underset{O-CH_2-CH_3}{}}} + H_2O \qquad (18.15)$$

Polyesters are fibers made from polymers in which the individual units are held together by ester groups. An example is Dacron, where ester groups hold together long chains:

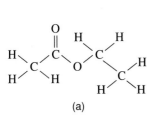

Dacron or Mylar

(*n* denotes that a very large number of these units are linked together)

Esters have very pleasant odors and some are used in perfumery. Certain esters produce the aroma and taste of fine wines. The presence of mixtures of esters

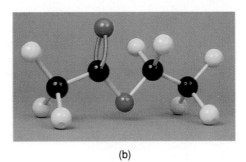

(a)   (b)

FIGURE 18.17
(a) Structural formula and (b) ball-and-stick model of ethyl acetate, an example of an ester.

accounts for the fragrance and flavor of many fruits and flowers. Artificial flavoring essences are generally composed of mixtures of selected esters chosen to duplicate as closely as possible the flavor and aroma of the natural fruits. The following esters have characteristic aromas (in parentheses): ethyl formate (rum), pentyl acetate (banana), octyl acetate (orange), butyl butyrate (pineapple), methyl salicylate (wintergreen), and pentyl butyrate (apricot).

Esters are an important part of our daily diet. *Fats* and *oils* are esters formed from long-chain aliphatic carboxylic acids (C_{12} to C_{22}, in red) and glycerol, an alcohol with three different OH groups. These esters are commonly called *glycerides*. *Fats* are glycerides that are generally solid or semisolid at room temperature; *oils* are liquids at room temperature.

$$
\begin{array}{ll}
\begin{array}{l}
CH_2\text{—}OH \\
| \\
CH\text{—}OH \\
| \\
CH_2\text{—}OH \\
\text{glycerol}
\end{array}
&
\begin{array}{l}
\quad\quad\quad\quad O \\
\quad\quad\quad\quad \| \\
CH_2\text{—}OC\text{—}R'' \\
\quad\quad\quad O \\
\quad\quad\quad \| \\
CH\text{—}OC\text{—}R' \\
\quad\quad\quad O \\
\quad\quad\quad \| \\
CH_2\text{—}OC\text{—}R
\end{array}
\end{array}
$$

A fat or oil; R, R′, and R″ may be the same or different alkyl groups.

The R, R′, and R″ groups in fats and oils can have double bonds present in the chain or they can be saturated (no carbon–carbon double bonds). *Oils* contain larger amounts of *unsaturation* than fats do. Thus, products like *oleomargarine* and *Crisco* have fewer double bonds in the R groups than do vegetable oils from corn, soybeans, or safflowers.

Considerable research has been done on the relationship of fat consumption in humans and diseases of the arteries (primarily atherosclerosis). The relationship is complex, and many factors, especially heredity, have a strong influence. In general, the American Heart Association recommends that we reduce the total amount of fat and oil in our diet and that the fat and oil that we do eat have a large component of *polyunsaturated* fats (fats and oils with *more* than one double bond) and a minimum of saturated fats. The best oils are safflower, soybean, corn, and cottonseed oils. Not all vegetable oils are good, however. Avoid coconut and palm oils, which contain large amounts of saturated fats.

Amines

Amines Class of hydrocarbon derivatives in which one or more of the three hydrogen atoms in ammonia are replaced by alkyl or aryl groups.

Alkyl or aryl derivatives of ammonia (NH_3) are called **amines** (am′ins). *One*, *two*, or all *three* hydrogen atoms in ammonia may be replaced by alkyl or aryl groups to form amines. The functional group is $-NH_2$, $>N-H$, or $>N-$. Ethylamine, shown in Figure 18.18, is an example of a simple amine. Amines with one hydrogen replaced are known as *primary amines*, amines with two hydrogens replaced are called *secondary amines*, and amines with all three hydrogens replaced are termed *tertiary amines*. Table 18.8 summarizes the general formulas for the three classes of amines.

Compounds containing the amine group can strongly affect our body's activities. Examples with which you may be familiar include adrenaline, dopamine, morphine, strychnine, cocaine, and heroin. Quinine (antimalarial), sulfa drugs,

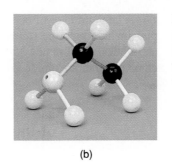

FIGURE 18.18
(a) Structural formula and
(b) ball-and-stick model of
ethylamine, an example of
an amine.

(a)   (b)

ephedrine (the decongestant in Sudafed), Demerol, and countless other medicinal products also contain amines.

Amides

Finally, we come to the **amides** (am′ids), derivatives of carboxylic acids in which an $-NH_2$* replaces the $-OH$ of the carboxyl group. The general formulas for amides are

$$R-C\underset{NH_2}{\overset{O}{\lessgtr}} \quad \text{and} \quad Ar-C\underset{NH_2}{\overset{O}{\lessgtr}}$$

The functional group is $-C\underset{NH_2}{\overset{O}{\lessgtr}}$. Acetamide, shown in Figure 18.19, is an example of an amide.

Amides Class of hydrocarbon derivatives in which an $-NH_2$ replaces the $-OH$ of the carboxyl group in a carboxylic acid.

---

*Alkyl or aryl groups may replace the H in the $-NH_2$, such as $-NHR$, $-NR_2$, NHAr, $-NAr_2$. The compounds are still referred to as amides, as we will see in nylon 66 and proteins.

TABLE 18.8	**General Formulas for Amines**[a]		
PRIMARY AMINES	SECONDARY AMINES	TERTIARY AMINES	
$R-NH_2$	$\underset{}{\overset{R}{\mid}}$ $R-N-H$	$\underset{}{\overset{R}{\mid}}$ $R-N-R$,	$\underset{}{\overset{R}{\mid}}$ $Ar-N-R$
$Ar-NH_2$	$\underset{}{\overset{R}{\mid}}$ $Ar-N-H$	$\underset{}{\overset{R}{\mid}}$ $Ar-N-Ar$,	$\underset{}{\overset{Ar}{\mid}}$ $Ar-N-Ar$
	$\underset{}{\overset{Ar}{\mid}}$ $Ar-N-H$		

[a] The R and Ar groups may be the same or different.

A large number of compounds contain the amide group. A few relevant examples include piperine (the active ingredient in pepper), penicillins, cephalosporin antibiotics, and Tylenol. Nylon is a polymer in which the linkages that hold the long chain together are amide groups:

$$\left[-NH-(CH_2)_6 \!\!\overbrace{\left. -NH-\overset{\overset{\displaystyle O}{\|}}{C} \right|}^{\text{amide}}\!\!(CH_2)_6-\overset{\overset{\displaystyle O}{\|}}{C}-\right]_n$$

nylon 66 (*n* denotes a large number of units)

The amide group is also heavily involved in the process of life. *Proteins*, a major food in our diet, are long polymers of alpha(α)-amino acids that are bonded together by amide groups. An α-amino acid contains both a carboxylic acid group ($-$COOH) and an amine group ($-$NH$_2$). The alpha (α) means that the amino group ($-$NH$_2$) is located on the carbon next to the carboxylic acid group. The carboxylic acid group ($-$COOH) of one α-amino acid may link together with the amino ($-$NH$_2$) of another α-amino acid to form a long chain, similar to nylon. The general formulas for an α-amino acid and a protein are

$$R-\underset{\underset{\displaystyle NH_2}{|}}{CH}-\overset{\displaystyle O}{\underset{\displaystyle O-H}{C}}$$

$$\left[-\overset{\overset{\displaystyle O}{\|}}{O}-\underset{\underset{}{}}{CH}\!\!\overbrace{\left|-NH-\overset{\overset{\displaystyle O}{\|}}{C}\right.}^{\text{amide}}\!\!\overset{\overset{\displaystyle R}{|}}{CH}-NH-\right]_n$$

an alpha amino acid (α-amino acid)   a protein (*n* denotes a large number of units)

Proteins are essential to life because they are involved in structure (connective tissue), movement (muscle), regulation (the immune system) and chemical processes (enzymes). A simple relatively *inexpensive* amide has recently been used in the treatment of pain associated with sickle-cell anemia, a blood disease that afflicts some African-Americans in the United States. This amide is hydroxyurea and has the following structural formula:

$$HO-NH-\overset{\overset{\displaystyle O}{\|}}{C}-NH_2$$

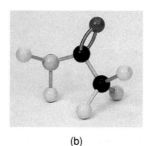

FIGURE 18.19
(a) Structural formula and (b) ball-and-stick model of acetamide, an example of an amide.

(a)   (b)

Study Exercise 18.15
Identify the hydrocarbon derivatives in each of the following compounds as
(1) organic halide, (2) alcohol, (3) ether, (4) aldehyde, (5) ketone, (6) carboxylic
acid, (7) ester, (8) amine, or (9) amide by circling each group and labeling it:

a.

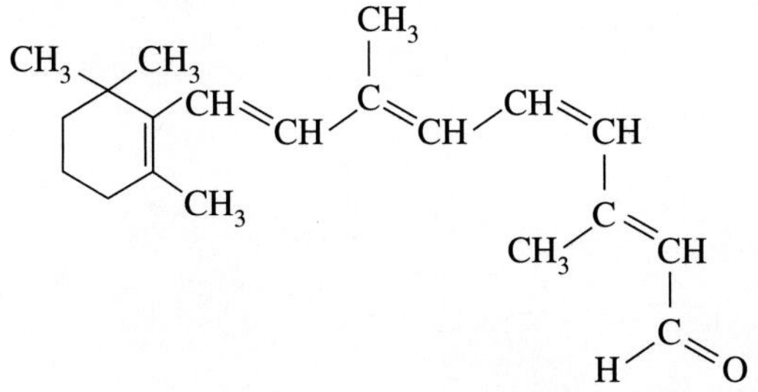

cis-retinal   (aldehyde)

b.

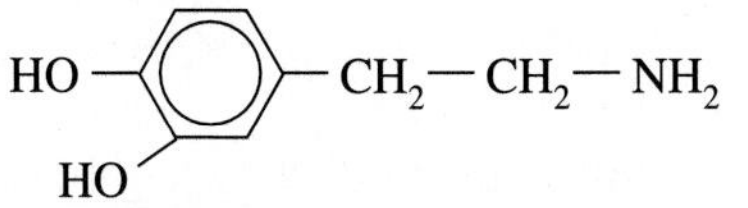

dopamine

(*Hint:* A compound with an — OH group attached to a benzene ring is *not* an alcohol;
it belongs to a class not discussed here, called phenols.)   (amine)

c.

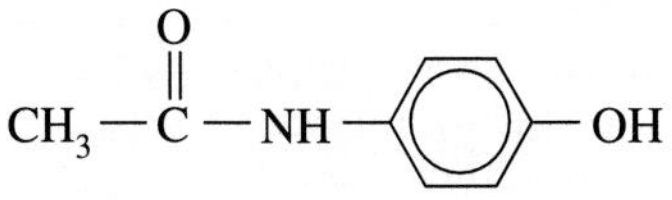

Tylenol

(*Hint:* A compound with an — OH group attached to a benzene ring is *not* an alcohol;
it belongs to a class not discussed here, called phenols.)   (amide)

d.

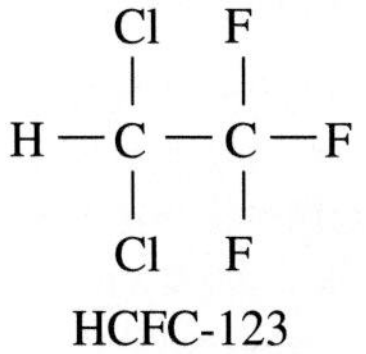

HCFC-123
(hydrochlorofluorocarbon—substitute for chlorofluorocarbons)

(5 organic halides)

e.   HO — CH — CH$_2$— NH — CH$_3$

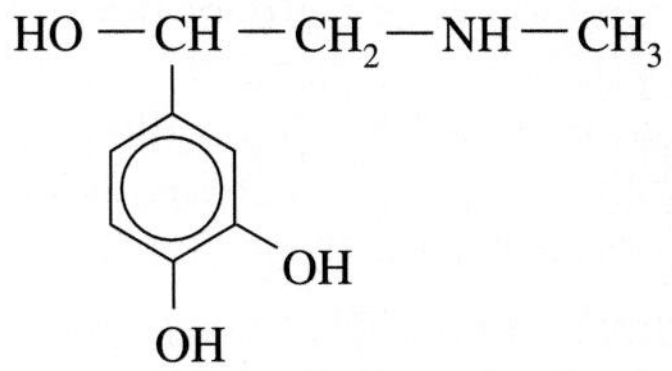

adrenaline (epinephrine)
(*Hint:* A compound with an — OH group attached to a benzene ring is *not* an alcohol;
it belongs to a class not discussed here, called phenols.)   (alcohol and amine)   Work Problem 19.

CHEMISTRY OF THE ATMOSPHERE

Photochemical Smog

Smog has become a fact of life in many American cities. Actually, the term *smog*, derived from "smoke" and "fog," was first applied to fogs in nineteenth-century London that resulted from abnormally high concentrations of smoke particles, sulfur dioxide (SO_2), and sulfuric acid (H_2SO_4). These pollutants, which are very irritating to the lungs, arose from the burning of sulfur-containing coal for home heating and industrial purposes. In nineteenth-century London, smog occurred in the winter when weather conditions trapped the pollutants at ground level in the city. Today, such smog has been minimized in most communities by the use of natural gas for heating and more efficient ways of burning carbon-based fuels.

The smog found in modern cities is more properly termed *photochemical smog*, and it differs in a number of ways from the London-type smogs of the 1800s. London smog was dense and choking and occurred in the winter. In contrast, smog in Los Angeles is typically more of a haze and occurs in the hot months. Photochemical smog is more subtle in its effects on people, but it can be very irritating to lung tissues. Three key elements combine to produce photochemical smog: (1) increased concentrations of pollutants, (2) weather conditions that trap the pollutants at ground level, and (3) sunlight to promote the necessary chemical reactions, as the figure shows. Indeed, photochemical smog forms less often in the winter because the sun is

The Los Angeles basin on a smoggy summer day can be a depressing sight.

lower on the horizon and less light and heat penetrate the atmosphere.

Pollutants are the raw materials from which photochemical smog is produced. The most important pollutants in smog are nitrogen monoxide, nitrogen dioxide, and unburned hydrocarbons and hydrocarbon derivatives. As noted in earlier essays on the atmosphere, these pollutants are released into the atmosphere when humans burn coal or hydrocarbon fuels to produce energy. One of the most important single sources of such pollution is the automobile engine. In addition, hydrocarbons are released into the atmosphere during the refining and transportation of petroleum products and even during the filling of fuel tanks at gas stations. In large cities with heavy commuter traffic populations, these emissions can build up quickly each morning and provide the fuel for smog production.

Weather and geographical considerations can conspire to trap these pollutants over the city. Large cities that lie in a basin (such as Los Angeles) or are surrounded by mountains (such as Las Vegas,

Nevada) provide geographical traps for the polluted air. In addition, when air near the ground is cooler than air at higher altitudes (the reverse of normal), the less dense warm air does not mix well with the denser, cooler air below, where the pollutants are generated. This condition is called a *temperature inversion*. Los Angeles suffers temperature inversions regularly because the cooler sea air moves in during the day and pushes warmer air up to higher altitudes.

The last key element is sunlight. Sunlight promotes the reactions that transform nitrogen oxides (NO and NO_2), oxygen, and hydrocarbons into ozone and the hydrocarbon derivatives that constitute photochemical smog. Scientists do not yet completely understand these processes, but they have identified the three basic features involved. First, nitrogen monoxide is converted into nitrogen dioxide by the action of hydrocarbons, OH radicals ($\cdot\ddot{O}H$), and oxygen gas:

$$NO(g) + O_2(g) + \text{hydrocarbons} \longrightarrow$$
$$NO_2(g) + \text{hydrocarbon derivatives}$$

Second, sunlight acts on the resulting nitrogen dioxide to give oxygen atoms:

$$NO_2(g) + \text{light} \longrightarrow NO(g) + O(g)$$

Third, the oxygen atoms react with oxygen gas to give ozone:

$$O(g) + O_2(g) \longrightarrow O_3(g)$$

A variety of hydrocarbon derivatives is produced in the course of a day. The most offensive are aldehydes, of which formaldehyde (H — CHO) and acetaldehyde (CH_3 — CHO) are the principal components. The ozone and aldehydes cause most of the irritation to your lungs on smoggy days.

Monitoring the concentration of nitrogen monoxide, nitrogen dioxide, ozone, and aldehydes reveals an interesting pattern. The concentration of nitrogen monoxide peaks early (7 A.M.) during the morning commute. The nitrogen dioxide concentration grows more slowly and peaks at about 10 A.M. Meanwhile, the levels of ozone and aldehydes rise later in the day (12 noon to 2 P.M.). This pattern is just what we would expect, given the smog-formation processes described above.

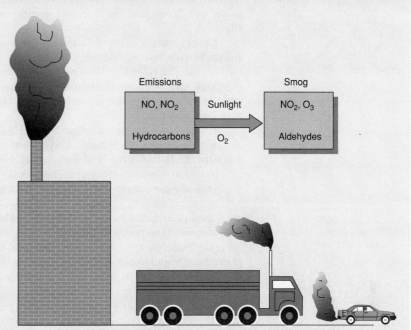

Industrial and automobile emissions generate air pollutants that are trapped in the lower atmosphere. Light from the sun then promotes photochemical reactions that convert nitrogen monoxide, nitrogen dioxide, and hydrocarbons into a mixture of nitrogen dioxide, ozone, and aldehydes (smog).

Photochemical smog has almost become an accepted fact of life for people living in large cities. More efficient automobiles, catalytic converters, and emission controls on industrial sources can have some effect on smog levels, but automobile exhaust remains a very real problem. Until societies are willing to reduce the number of cars on the road, significant reductions in air pollution will not occur.

 # Summary

Organic chemistry is the chemistry of compounds containing the element carbon. Organic compounds are classed as *hydrocarbons* (compounds containing only carbon and hydrogen) or *derivatives of hydrocarbons*. The hydrocarbons are further categorized as *aliphatic hydrocarbons* (*alkanes, alkenes,* and *alkynes*) and *aromatic hydrocarbons* (Section 18.1).

Carbon can bond in three basic ways: (1) *tetrahedral*, where carbon has bond angles of 109.5° and forms four single bonds; (2) *planar*, where carbon has bond angles of 120° and forms a double bond and two single bonds; and (3) *linear*, where carbon has 180° bond angles and forms either a triple bond and a single bond or two double bonds (Section 18.2).

Alkanes have a general formula of C_nH_{2n+2} (open-chain compounds) where all carbon atoms have four single bonds. Organic compounds with the same molecular formula but different structures are called *isomers*. Alkanes are named according to their structural formulas in the IUPAC system of nomenclature. Alkanes are used daily in our world, especially as fuels and in combustion and substitution reactions (Section 18.3).

Alkenes have a general formula of C_nH_{2n} (open-chain compounds) and contain one carbon–carbon double bond. Alkenes are also named according to their structural formulas in the IUPAC system of nomenclature. Alkenes generally undergo addition reactions, one of the most important of which is *polymerization*. Such addition polymers as polyethylene, Teflon, and polyvinyl chloride have a huge impact on our daily lives (Section 18.4).

Alkynes have a general formula of C_nH_{2n-2} (open-chain compounds) and contain one carbon–carbon triple bond. Alkynes are named according to their structural formulas in the IUPAC system of nomenclature. Alkynes generally undergo addition reactions and are important intermediates in the chemical industry (Section 18.5).

Aromatic compounds are related to a benzene (C_6H_6). All aromatic hydrocarbons have a six-carbon ring with three double bonds alternating around the ring. Some have other groups attached to the benzene ring. These compounds are named according to their structural formulas in the IUPAC system of nomenclature. Aromatic compounds generally undergo substitution reactions, and benzene rings can be found in thousands of products found in our daily lives (Section 18.6).

Hydrocarbon derivatives are derived from hydrocarbons by replacing one hydrogen with a *functional group*. A functional group is a characteristic set of atoms bonded together in a particular way. The properties of the hydrocarbon derivatives depend on the nature of the functional group. Such derivatives include alkenes, alkynes, organic halides, alcohols, ethers, aldehydes, ketones, carboxylic acids, esters, amines, and amides (Section 18.7).

Exercises

1. Define or explain the following terms (the number in parentheses refers to the section in the text where the term is mentioned):

   a. organic chemistry (18.1)

   b. hydrocarbons (18.1)

   c. aliphatic hydrocarbons (18.1)

   d. alkanes (saturated hydrocarbons) (18.3)

   e. homologous series (18.3)

   f. isomer (18.3)

   g. alkyl groups (18.3)

   h. halogenation (18.3)

   i. alkene (18.4)

   j. unsaturated hydrocarbons (18.4)

   k. polymers (18.4)

   l. addition polymers (18.4)

   m. addition reaction (18.5)

   n. alkynes (18.5)

   o. aromatic hydrocarbons (18.6)

   p. benzene ring (18.6)

   q. functional group (18.7)

   r. organic halides (18.7)

   s. alcohols (18.7)

   t. ethers (18.7)

   u. aldehydes (18.7)

   v. ketones (18.7)

   w. carboxylic acids (18.7)

   x. carboxyl group (18.7)

   y. esters (18.7)

   z. amines (18.7)

   aa. amides (18.7)

2. Distinguish between the following:
   a. double and triple bond
   b. tetrahedral, planar, and linear arrangements of atoms
   c. *ortho* and *meta* positions
   d. *ortho* and *para* positions
   e. a carbonyl group and a carboxyl group
   f. alcohol and hydroxide ion
   g. alcohol and ether
   h. aldehyde and ketone
   i. carboxylic acid and aldehyde
   j. amine and amide

3. Give the functional group for the following:
   a. alkene (18.4)   b. alkyne (18.5)
   c. organic halide (18.7)   d. alcohol (18.7)
   e. ether (18.7)   f. aldehyde (18.7)
   g. ketone (18.7)   h. carboxylic acid (18.7)
   i. ester (18.7)   j. amine (18.7)
   k. amide (18.7)

4. Write the general formulas for the following:
   a. alkane (18.3)   b. alkene (18.4)
   c. alkyne (18.5)   d. aromatic hydrocarbon (18.6)
   e. alkyl halide (18.7)   f. aryl halide (18.7)
   g. alcohol (18.7)   h. ether (18.7)
   i. aldehyde (18.7)   j. ketone (18.7)
   k. carboxylic acid (18.7)   l. ester (18.7)
   m. amine (18.7)   n. amide (18.7)

✓ Problems

Shapes and Structures of Organic Molecules (See Section 18.2)

5. Identify the geometry (tetrahedral, 109.5°; planar, 120°; linear, 180°) at all carbon atoms in each of the following molecules.

a. 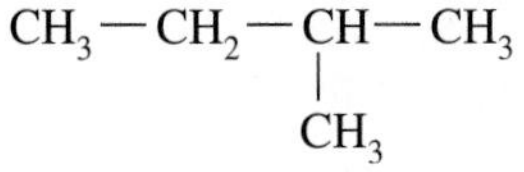 $CH_3-CH_2-CH-CH_3$
                      |
                     CH_3

b. 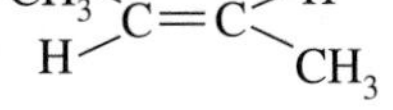
$$CH_3 \atop H{\Large\diagdown}C{=}C{\diagup H \atop \diagdown CH_3}$$

c.
$$H \atop H{\Large\diagdown}C{=}C \quad C{=}C \atop H \quad H$$

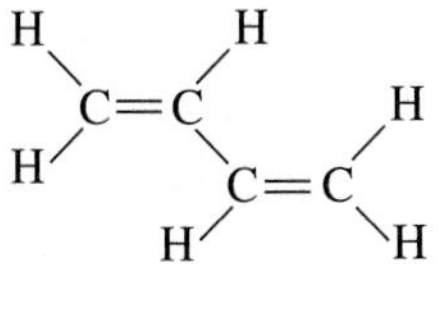

d. 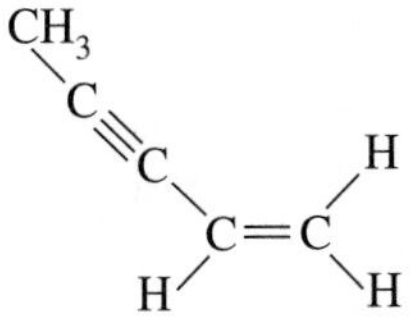
$$CH_3 \atop C{\equiv}C \quad C{=}C$$

e.

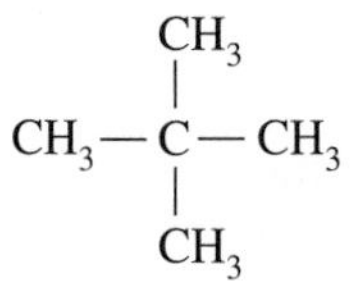

$$CH_3-\underset{\underset{CH_3}{|}}{\overset{\overset{CH_3}{|}}{C}}-CH_3$$

f.

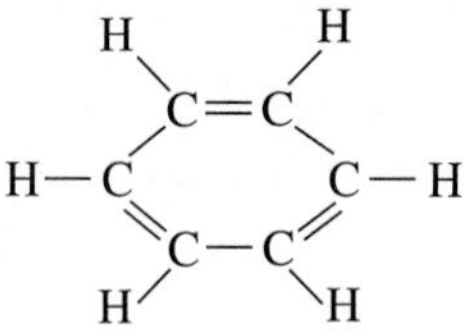

Isomers (See Section 18.3)

6. Write *condensed* structural formulas for the isomers of the following. (The number in parentheses is the number of isomers for the compound.)

   a. C_6H_{14} (5)   **b.** C_7H_{16} (9)

Alkanes (See Section 18.3)

7. Write the IUPAC name for each of the following compounds:

a. 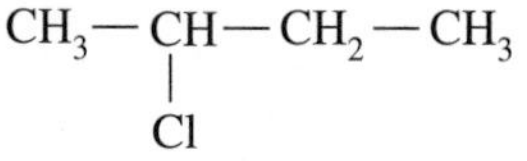 $CH_3-\underset{\underset{Cl}{|}}{CH}-CH_2-CH_3$ **b.** 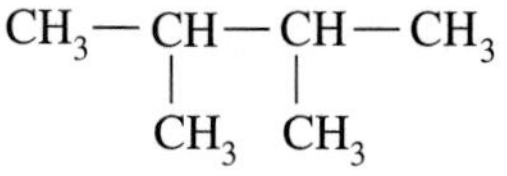 $CH_3-\underset{\underset{CH_3}{|}}{CH}-\underset{\underset{CH_3}{|}}{CH}-CH_3$

c. 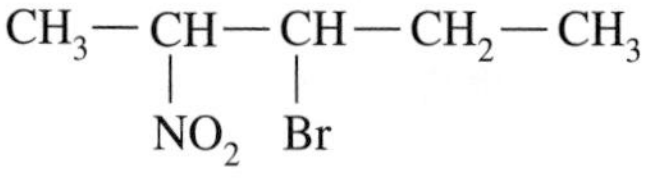 $CH_3-\underset{\underset{NO_2}{|}}{CH}-\underset{\underset{Br}{|}}{CH}-CH_2-CH_3$ **d.** 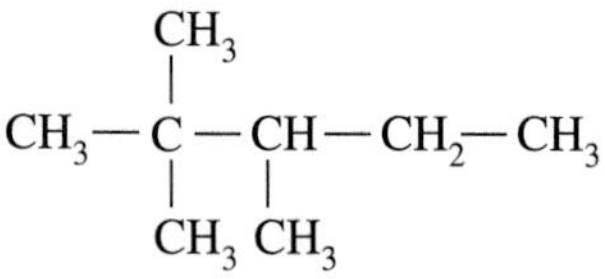 $CH_3-\underset{\underset{CH_3}{|}}{\overset{\overset{CH_3}{|}}{C}}-\underset{\underset{CH_3}{|}}{CH}-CH_2-CH_3$

e. 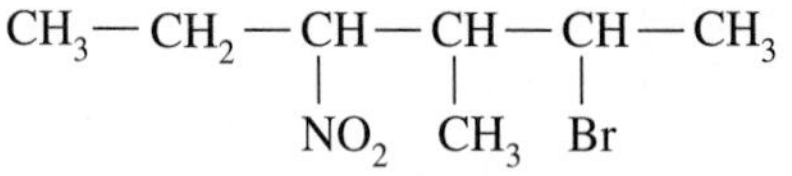 $CH_3-CH_2-\underset{\underset{NO_2}{|}}{CH}-\underset{\underset{CH_3}{|}}{CH}-\underset{\underset{Br}{|}}{CH}-CH_3$

f. 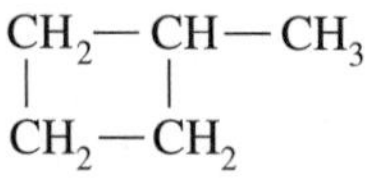 $\underset{\underset{CH_2-CH_2}{|}}{CH_2}-\overset{|}{CH}-CH_3$

8. Write the structural formula for each of the following compounds:

  a. 2-bromobutane   **b.** 3-iodoheptane

  c. 2-nitropropane   **d.** 3-ethylpentane

  e. 2-bromo-3,3-dimethylhexane   **f.** isopropylcyclopentane

9. Complete and balance the reaction equation for *mono*halogenation in each of the following chemical reactions:

  a. $CH_4 + Cl_2 \xrightarrow{\Delta \text{ or light}}$

  b. $CH_3-CH_3 + Br_2 \xrightarrow{\Delta \text{ or light}}$

Alkenes (See Section 18.4)

10. Write the IUPAC name for each of the following compounds:

a. $CH_2{=}CH-CH_2-CH_3$ **b.** 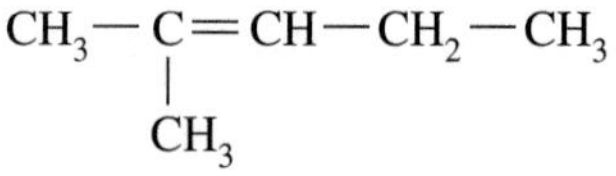 $CH_3-\underset{\underset{CH_3}{|}}{C}{=}CH-CH_2-CH_3$

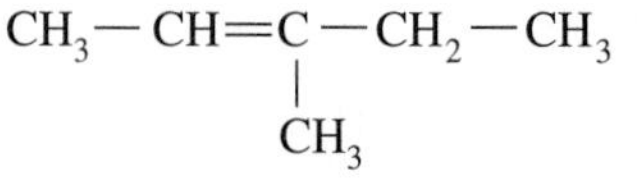

c. $CH_3-CH{=}\underset{\underset{CH_3}{|}}{C}-CH_2-CH_3$ **d.** 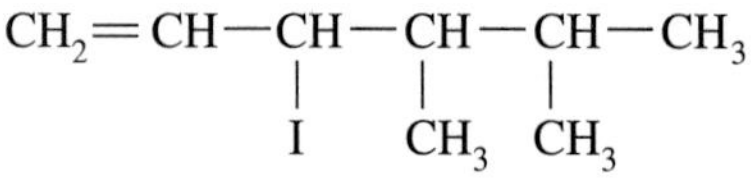 $CH_2{=}CH-\underset{\underset{I}{|}}{CH}-\underset{\underset{CH_3}{|}}{CH}-\underset{\underset{CH_3}{|}}{CH}-CH_3$

e. 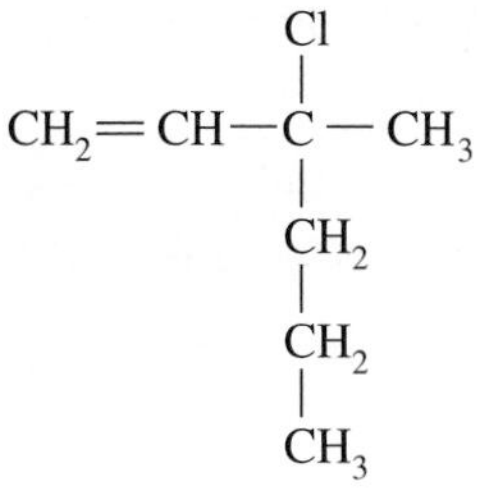 CH_2=CH—$\overset{\overset{\displaystyle Cl}{|}}{\underset{\underset{\displaystyle CH_3}{|}}{\underset{\underset{\displaystyle CH_2}{|}}{\underset{CH_2}{|}}}}$C—$CH_3$

f. 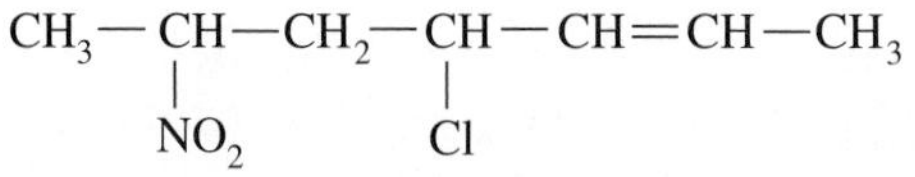 CH_3—$\underset{\underset{\displaystyle NO_2}{|}}{CH}$—$CH_2$—$\underset{\underset{\displaystyle Cl}{|}}{CH}$—CH=CH—$CH_3$

11. Write the structural formula for each of the following compounds:

   a. 1-pentene
   b. 5-methyl-2-hexene
   c. 2,3-dimethyl-1-hexene
   d. cyclobutene
   e. 1-bromo-3-ethyl-1-heptene
   f. 2,4,5-trimethyl-5-nitro-2-heptene

12. Complete and balance the reaction equation for each of the following chemical reactions:

   a. CH_3—CH=CH_2 + Br_2 $\xrightarrow{\hspace{2cm}}$
   (in CCl_4 solvent)

   b. 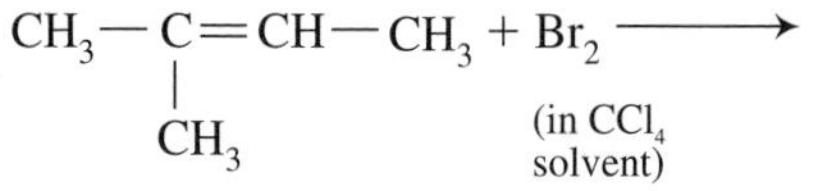 CH_3—$\underset{\underset{\displaystyle CH_3}{|}}{C}$=CH—$CH_3$ + Br_2 $\xrightarrow{\hspace{2cm}}$
   (in CCl_4 solvent)

Alkynes (See Section 18.5)

13. Write the IUPAC name for each of the following compounds:

   a. HC≡C—CH_2—CH_2—CH_3
   b. 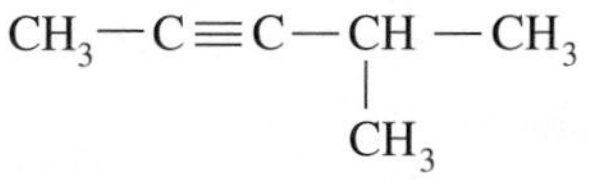 CH_3—C≡C—$\underset{\underset{\displaystyle CH_3}{|}}{CH}$—$CH_3$

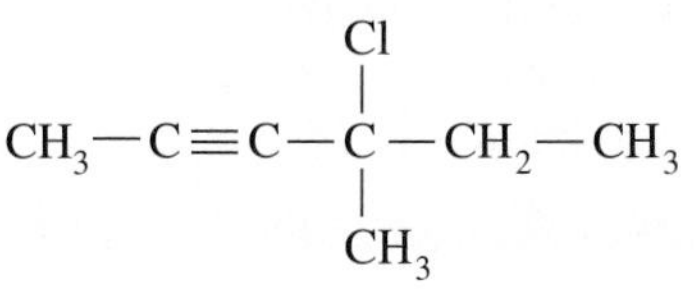

   c. CH_3—C≡C—$\overset{\overset{\displaystyle Cl}{|}}{\underset{\underset{\displaystyle CH_3}{|}}{C}}$—$CH_2$—$CH_3$
   d. 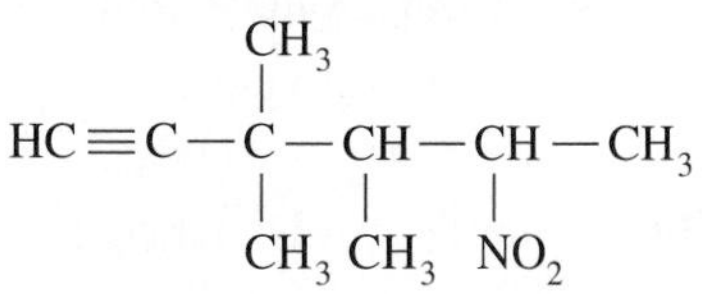 HC≡C—$\overset{\overset{\displaystyle CH_3}{|}}{\underset{\underset{\displaystyle CH_3}{|}}{C}}$—$\underset{\underset{\displaystyle CH_3}{|}}{CH}$—$\underset{\underset{\displaystyle NO_2}{|}}{CH}$—$CH_3$

   e. 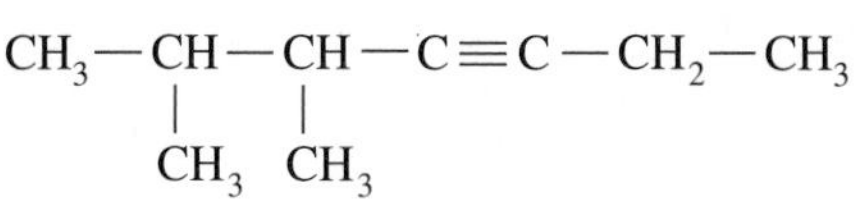 CH_3—$\underset{\underset{\displaystyle CH_3}{|}}{CH}$—$\underset{\underset{\displaystyle CH_3}{|}}{CH}$—C≡C—$CH_2$—$CH_3$

   f. 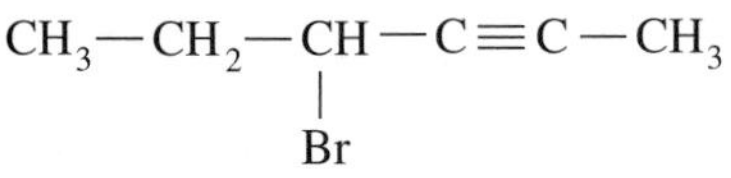 CH_3—CH_2—$\underset{\underset{\displaystyle Br}{|}}{CH}$—C≡C—$CH_3$

14. Write the structural formula for each of the following compounds:

   a. 1-heptyne
   b. 2-heptyne
   c. 4-nitro-1-heptyne
   d. 1-bromo-4,5-dimethyl-2-hexyne
   e. 1-chloro-3-methyl-1-hexyne
   f. 1,2-dichloro-5,5,6-trimethyl-3-octyne

15. Complete and balance the reaction equation for each of the following chemical reactions:

   a. $CH_3-C\equiv C-CH_3 + Br_2 \longrightarrow$
   (excess in CCl_4 solvent)

   b.
   $$CH_3-C\equiv C-\overset{\overset{\displaystyle CH_3}{|}}{\underset{\underset{\displaystyle CH_3}{|}}{C}}-CH_3 + Br_2 \longrightarrow$$
   (excess in CCl_4 solvent)

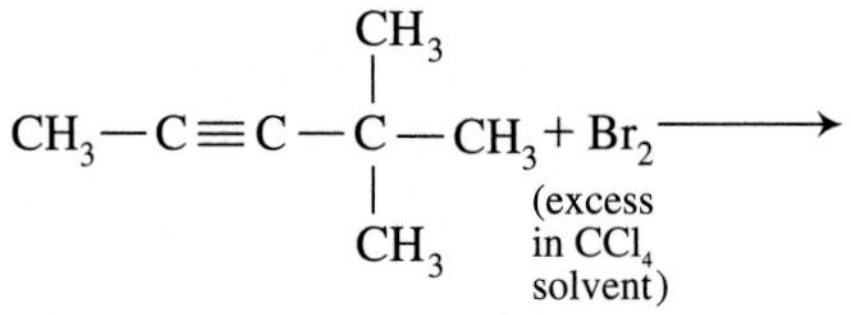

Aromatic Hydrocarbons (See Section 18.6)

16. Write the IUPAC name for each of the following compounds:

a. 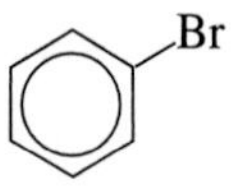   b. 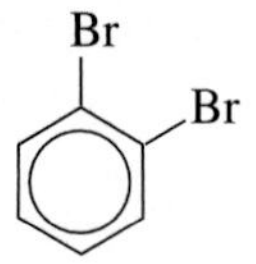   c.

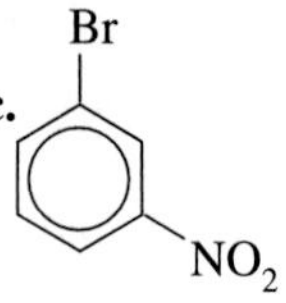

d.    e. 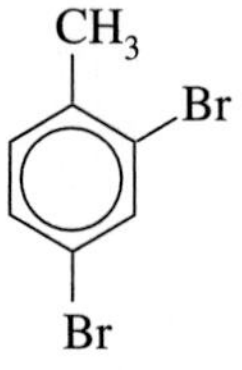   f.

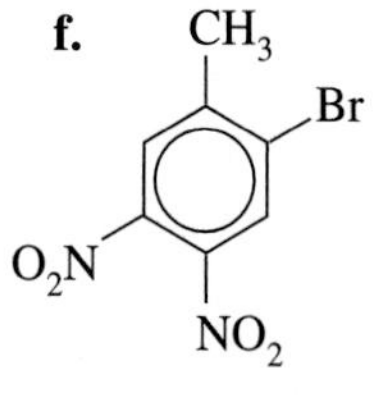

17. Write the structural formula for each of the following compounds:

   a. toluene
   b. iodobenzene
   c. m-nitroiodobenzene
   d. p-iodotoluene
   e. o-chlorotoluene
   f. 2,4,6-trinitrotoluene (the explosive known as TNT)

18. Complete and balance the reaction equation for each of the following chemical reactions:

   a. 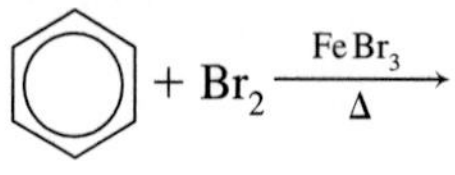 $\bigcirc + Br_2 \xrightarrow[\Delta]{FeBr_3}$

   b. 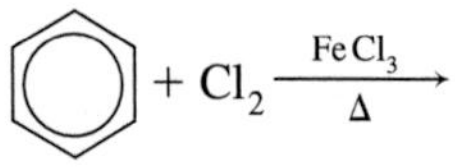 $\bigcirc + Cl_2 \xrightarrow[\Delta]{FeCl_3}$

Hydrocarbon Derivatives (See Section 18.7)

19. Identify the hydrocarbon derivatives in each of the following compounds as an (1) organic halide, (2) alcohol, (3) ether, (4) aldehyde, (5) ketone, (6) carboxylic acid, (7) ester, (8) amine, or (9) amide by circling each group and labeling it:

a. 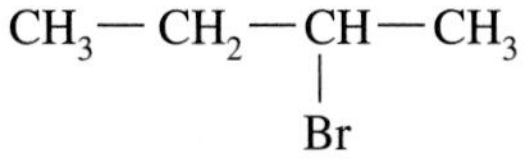 $CH_3-CH_2-CH-CH_3$
        |
        Br

b. 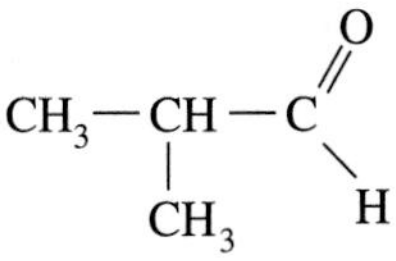 $CH_3-CH-C\overset{\displaystyle O}{\underset{\displaystyle H}{\diagdown}}$
        |
        CH_3

c. $CH_3-CH_2-CH-O-\overset{\displaystyle O}{\overset{\|}{C}}-CH_3$
        |
        CH_3

d. $CH_3-C\equiv C-CH-CH_3$
        |
        CH_3

e. $CH_3-CH_2-CH-OH$
        |
        CH_3

f. $CH_3-CH-C\overset{\displaystyle O}{\underset{\displaystyle OH}{\diagdown}}$
        |
        CH_3

g. CH_3-O-CH_3

h. $CH_3-CH_2-NH-CH_3$

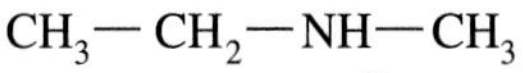

i. 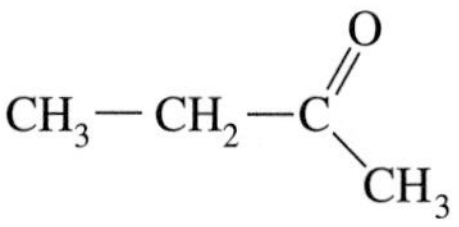 $CH_3-CH_2-C\overset{\displaystyle O}{\underset{\displaystyle CH_3}{\diagdown}}$

j. 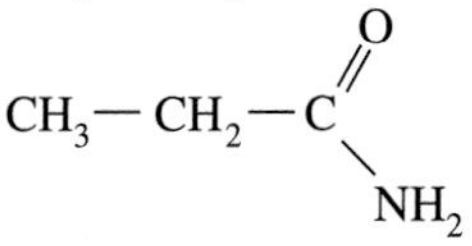 $CH_3-CH_2-C\overset{\displaystyle O}{\underset{\displaystyle NH_2}{\diagdown}}$

General Problems

20. Write the IUPAC name for each of the following compounds:

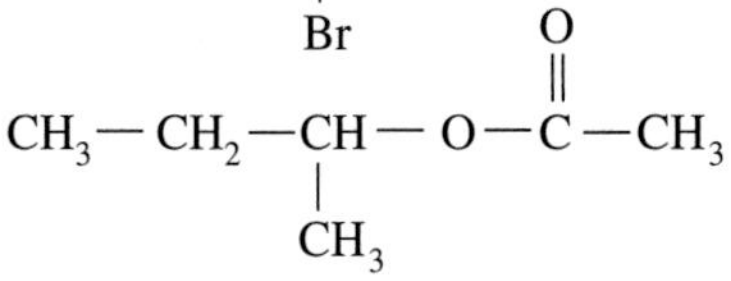

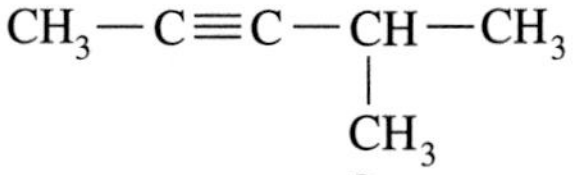

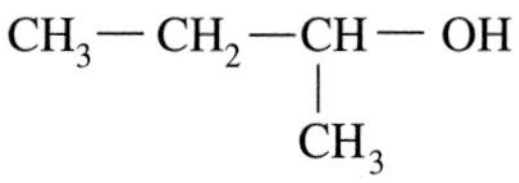

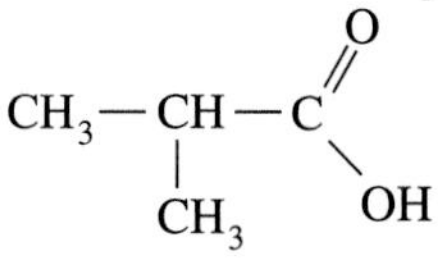

a. 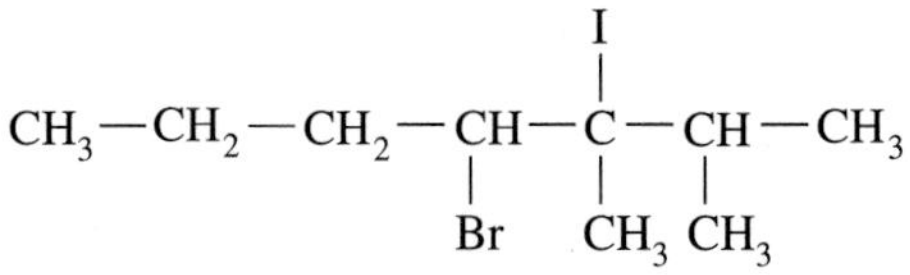 $CH_3-CH_2-CH_2-CH-\overset{\displaystyle I}{\underset{\displaystyle CH_3}{C}}-CH-CH_3$
              | |
              Br CH_3

b. $CH_3-CH_2-CH-CH-\overset{\displaystyle Br\ \ \ Br}{C}-CH-CH_3$
              | |
              CH_3 Br CH_3

c. $CH_2=CH-CH-\overset{\displaystyle NO_2}{C}-CH_3$
                        |
                  CH_3 CH_3

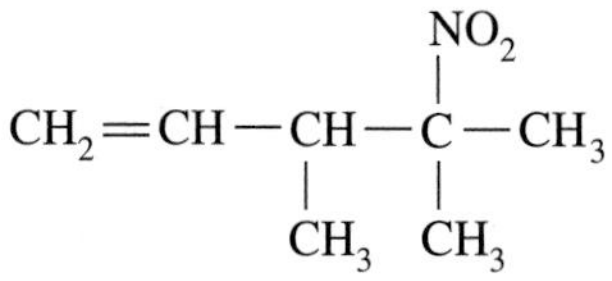

d. 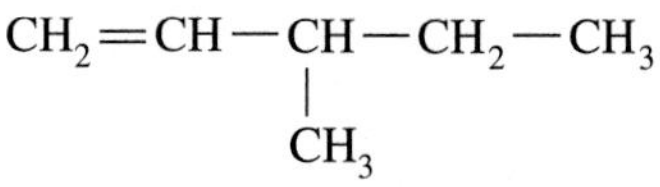 $CH_2=CH-CH-CH_2-CH_3$
              |
              CH_3

e. 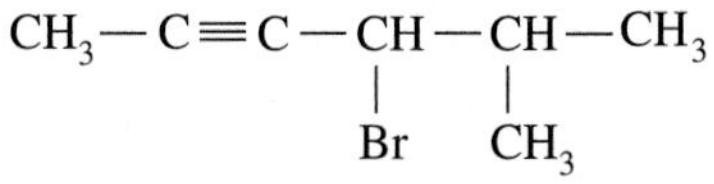 $CH_3-C\equiv C-CH-CH-CH_3$
                    | |
                  Br CH_3

f. 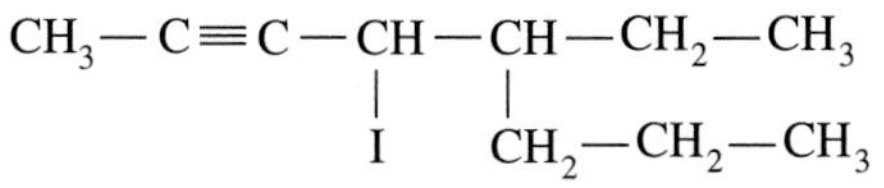 $CH_3-C\equiv C-CH-CH-CH_2-CH_3$
                    | |
                  I $CH_2-CH_2-CH_3$

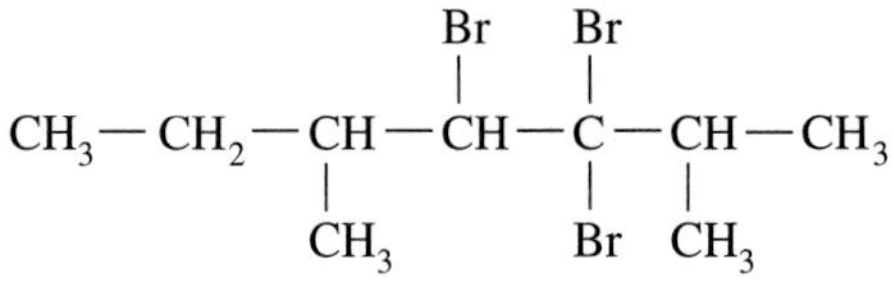

g.

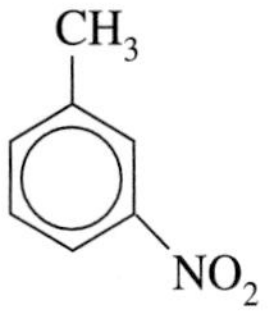

h.

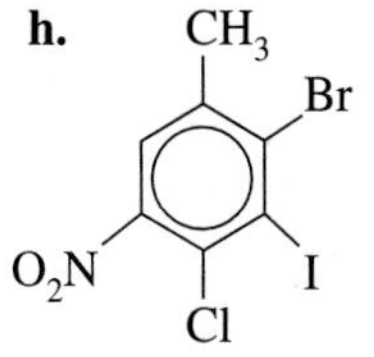

21. Identify the (1) alkene, (2) alkyne, and (3) aromatic groups in each of the following compounds by circling each group and labeling it:

a.

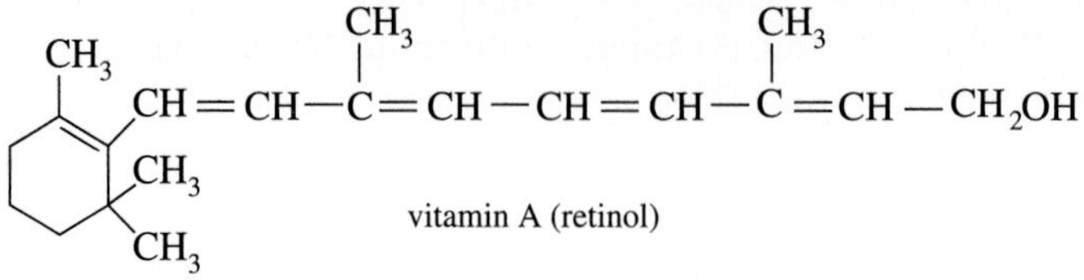

CH3
CH3
CH3
CH=CH—C=CH—CH=CH—C=CH—CH2OH
CH3
CH3

vitamin A (retinol)

b.

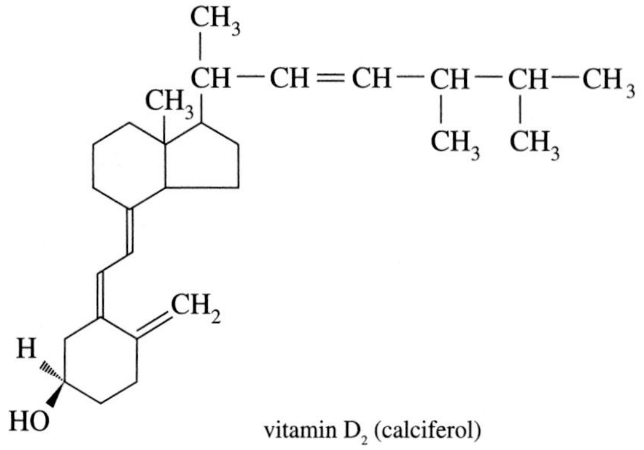

CH3
CH—CH=CH—CH—CH—CH3
CH3
CH3 CH3
CH2
H
HO

vitamin D_2 (calciferol)

(► indicates a bond coming out toward you; ⦙⦙⦙⦙ indicates a bond going away from you)

c.

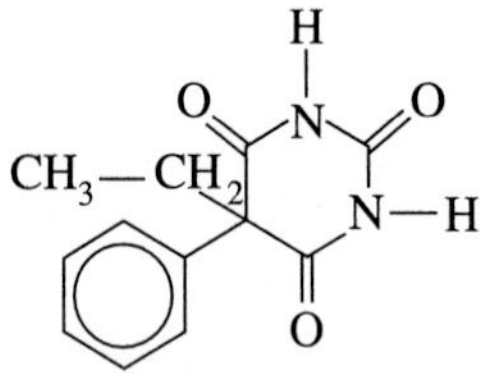

H
O
N
O
CH3—CH2
N—H
O

phenobarbital

d.

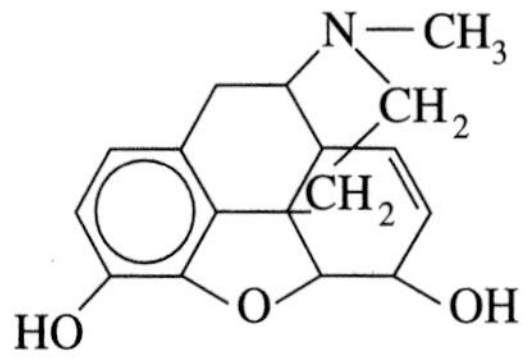

N—CH3
CH2
CH2
HO
O
OH

morphine

e.

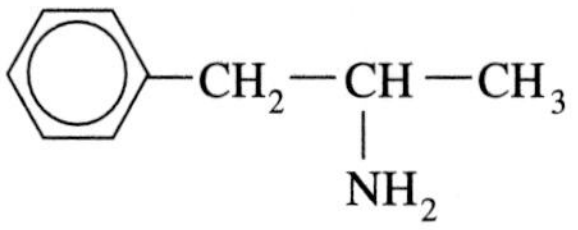

CH2—CH—CH3
NH2

amphetamine (Benzedrine)

f.

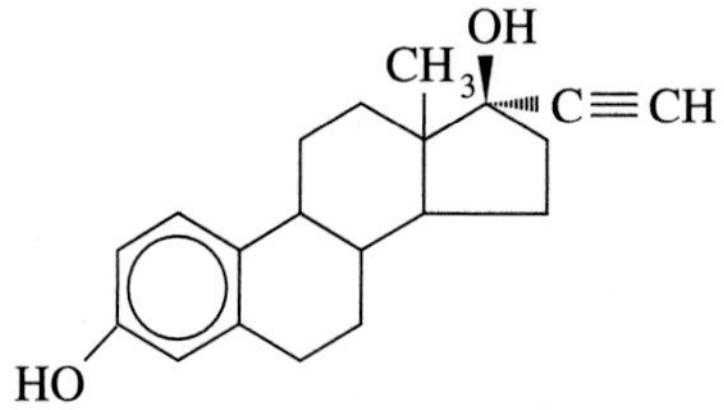

OH
CH3
C≡CH
HO

ethynylestradiol (Estinyl, a female sex hormone)

g.

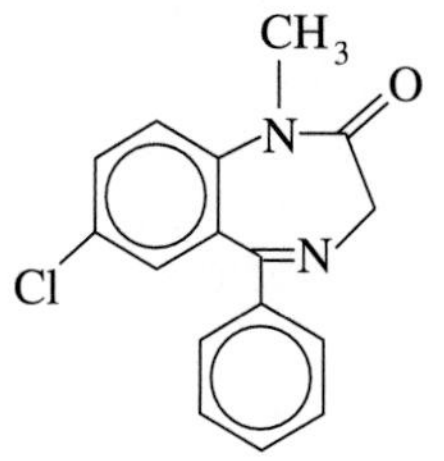

diazepam (valium)

h.

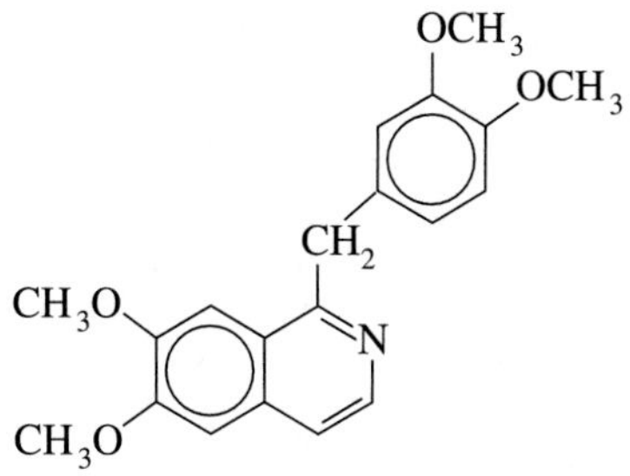

papaverine

22. Identify the hydrocarbon derivatives in each of the following compounds as an (1) organic halide, (2) alcohol, (3) ether, (4) aldehyde, (5) ketone, (6) carboxylic acid, (7) ester, (8) amine, or (9) amide by circling each group and labeling it:

a.

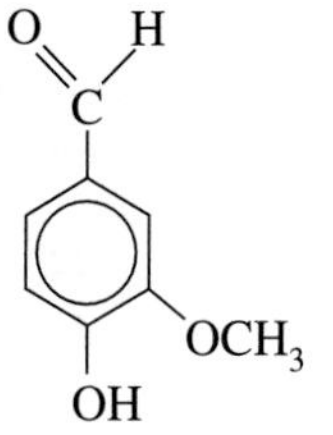

vanillin (a flavoring agent in vanilla)

b.

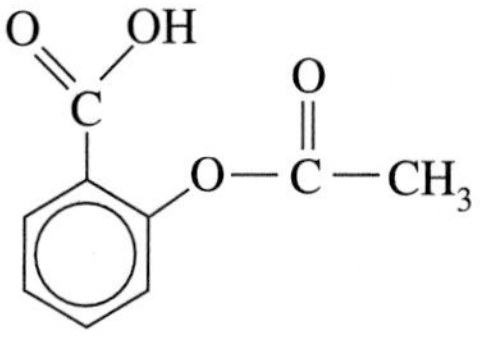

acetylasalicylic acid (aspirin)

(*Hint:* A compound with an —OH group to a benzene ring is *not* an alcohol; it belongs to a class not discussed here, called phenols.)

c.

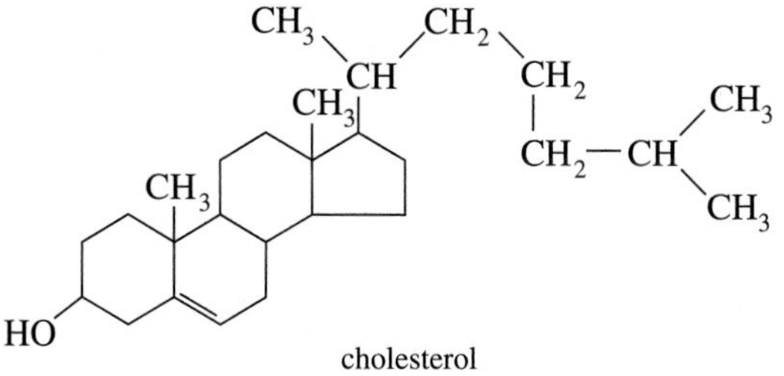

cholesterol

d.

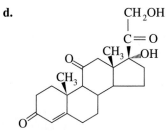

cortisone (an antiinflammatory drug)

(► indicates a bond coming out toward you; ⅲⅲⅲ indicates a bond going away from you)

e.

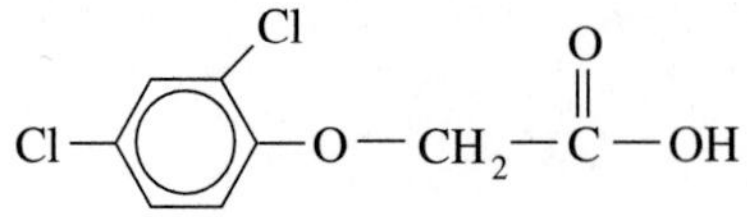

2,4-D (a herbicide which
kills broadleaf plants)

f.

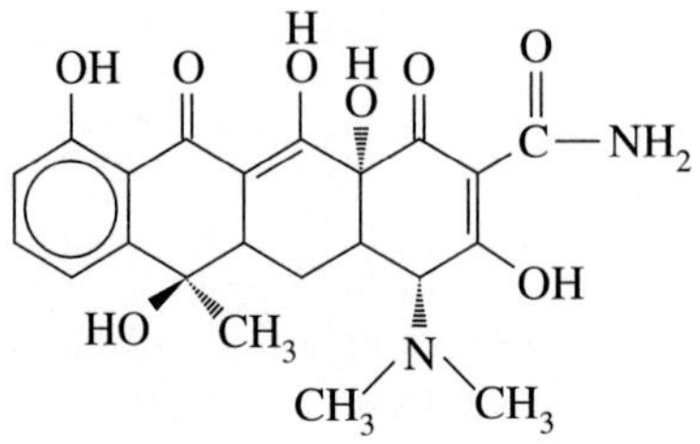

tetracycline (an antibiotic)

(▶ indicates a bond coming out
toward you; ▬ indicates a bond
going away from you)

(*Hint:* A compound with an —OH group to a benzene ring is *not* an alcohol;
it belongs to a class not discussed here, called phenols.)

g.

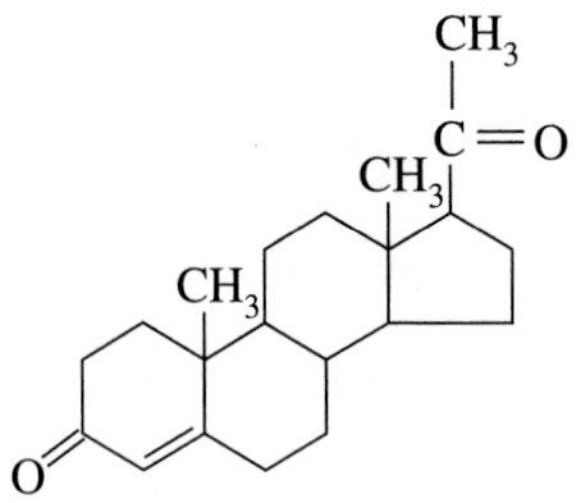

progesterone (a hormone)

h.

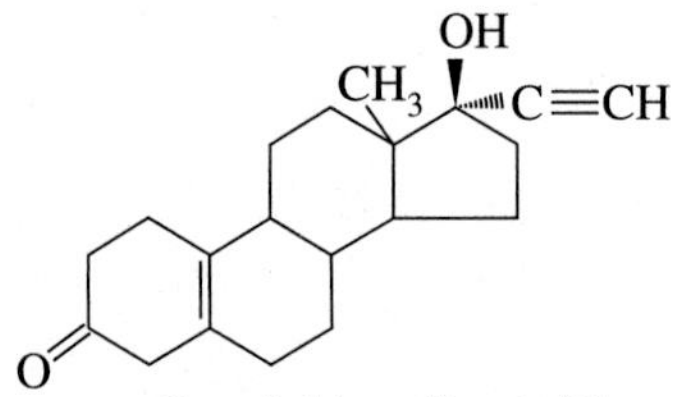

norethynodrel (constituent of the
oral contraceptive Enovid)

(▶ indicates a bond coming out
toward you; ▬ indicates a bond
going away from you)

23. A certain aliphatic hydrocarbon gave on analysis 80.0% carbon and 20.0%
hydrogen. At $3\overline{0}°C$ and $64\overline{0}$ torr pressure, 0.285 g of the gas occupied a
volume of 281 mL. Calculate the molecular formula, write the structural
formula, and give the IUPAC name for the aliphatic hydrocarbon. (*Hint:* See
Sections 8.5 and 11.9.)

24. An important constituent of orange or lemon peel is limonene. Limonene
($C_{10}H_{16}$, molar mass = 136.0 g) is responsible for the fragrance of such peels.
The reaction between bromine and compounds containing double bonds can
be performed just like a titration (see Section 15.4). Bromine (Br_2, dark red) is
added to a solution of the organic compound until the dark red color persists,
at which point we have added just enough bromine to react with all the

double bonds. A 0.476-g sample of limonene reacts with exactly 1.12 g of bromine before the red color of bromine persists. How many double bonds are present in limonene?

25. Treatment of acute methanol poisoning uses isotonic sodium hydrogen carbonate given intravenously to neutralize the action of formic acid (HCO_2H) formed in the metabolism of methanol. Write a balanced chemical equation for this neutralization reaction.

26. Write condensed structural formulas for the isomers of the following (the number in parentheses is the number of isomers of the compound).

    a. C_3H_7Cl (2)

    b. $C_6H_4Cl_2$, a substituted benzene (3)

    c. C_2H_6O; consider different functional groups (2)

    d. C_3H_8O; consider different functional groups (3)

27. A certain alkene gave on analysis 85.7% carbon and 14.3% hydrogen. At 27°C and $63\overline{0}$ torr, the gas had a density of 1.42 g/L. Calculate the molecular formula, write the structural formula, and give the IUPAC name for the alkene.

28. Cholesterol ($C_{27}H_{46}O$) is an important constituent of cells in human beings. However, high dietary levels of cholesterol have been implicated in atherosclerosis (hardening of the arteries) and heart disease. In the isolation of cholesterol from gallstones, part of the procedure includes the reaction between cholesterol, which has a carbon–carbon double bond, and bromine to give cholesterol dibromide ($C_{27}H_{46}Br_2O$).

$$C_{27}H_{46}O + Br_2 \longrightarrow C_{27}H_{46}Br_2O$$

Molar masses: 386.0 g   159.8 g   545.8 g

Calculate (a) the theoretical yield and (b) the percent yield if 1.27 g of cholesterol reacts with 0.789 g of bromine to give 1.35 g of cholesterol dibromide. (*Hint:* See the limiting reagent examples in Section 10.4.)

✓ Chapter Quiz 18A (Sections 18.1 to 18.3)

1. Identify the geometry (tetrahedral, 109.5°; planar, 120°; linear, 180°) at all carbon atoms in each of the following molecules.

    a. $CH_3-C\equiv C-CH-CH_3$
                      |
                     CH_3

    b. CH_3 and CH_3 on one carbon, H and CH_3 on the other, with a $C=C$ double bond between them

2. Write condensed structural formulas for the five isomers of C_6H_{14}.

3. Give the IUPAC name for each of the following compounds:

            Cl
            |
    a. $CH_3-C-CH_2-CH_2-CH_3$
            |
            Cl

b. 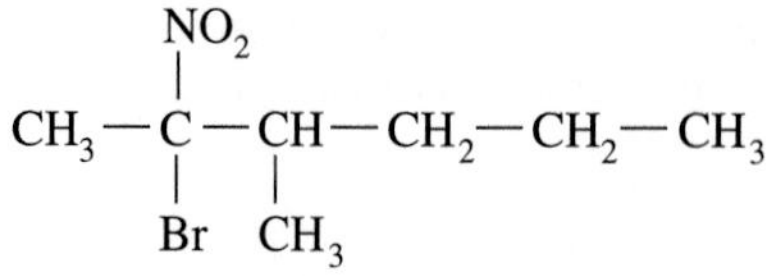
$$CH_3-\overset{\overset{\displaystyle NO_2}{|}}{\underset{\underset{\displaystyle Br}{|}}{C}}-\overset{\overset{}{}}{\underset{\underset{\displaystyle CH_3}{|}}{CH}}-CH_2-CH_2-CH_3$$

c.
$$CH_3-CH_2-\overset{}{\underset{\underset{\displaystyle CH_3}{|}}{CH}}-CH_2-\overset{}{\underset{\underset{\displaystyle CH_2-CH_2-CH_3}{|}}{CH}}-CH_2-CH_3$$

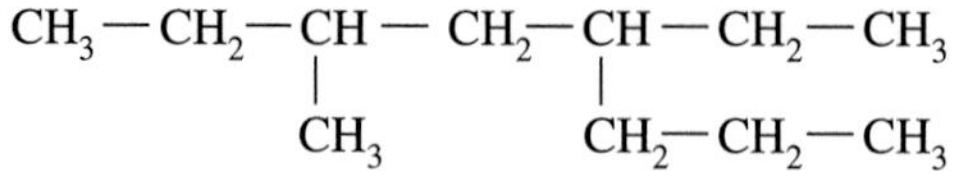

d.

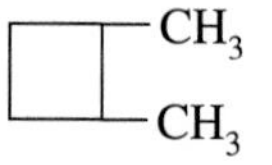

4. Write the structural formula for each of the following compounds:

   a.   1,3-dichlorohexane

   b.   chlorocyclohexane

5. Complete and balance the reaction equation for *mono*halogenation of the following chemical reaction:

$$CH_3-CH_3 \; + \; Cl_2 \; \xrightarrow[\text{light}]{\Delta \; \text{or}}$$

✓ Chapter Quiz 18B (Sections 18.4 to 18.6)

1. Write the IUPAC name for each of the following compounds:

   a. 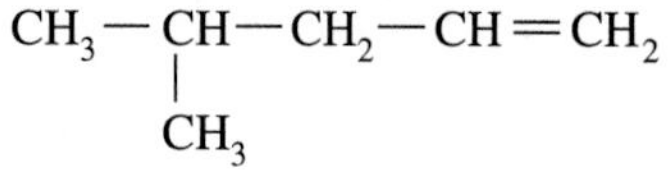
$$CH_3-\overset{}{\underset{\underset{\displaystyle CH_3}{|}}{CH}}-CH_2-CH=CH_2$$

   b. 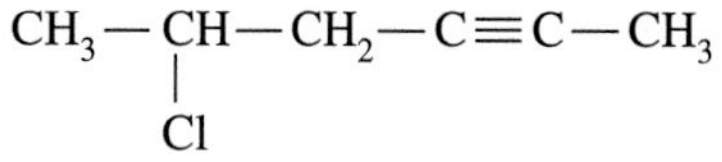
$$CH_3-\overset{}{\underset{\underset{\displaystyle Cl}{|}}{CH}}-CH_2-C\equiv C-CH_3$$

   c.

   CH_3

   (benzene ring with CH_3 and Cl)

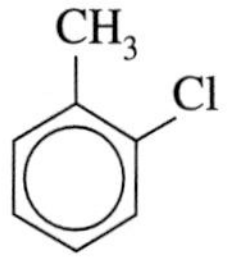

   d.

   CH_3

   (benzene ring with Br and Br)

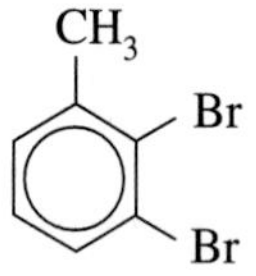

2. Write the structural formula for each compound:

   a.   cyclopentene

   b.   1-bromo-4,5-dimethyl-1-hexyne

   c.   2,4-dichlorotoluene

   d.   2,3-dichloro-4-nitrotoluene

3. Complete and balance the reaction equation for each of the following chemical reactions:

a. 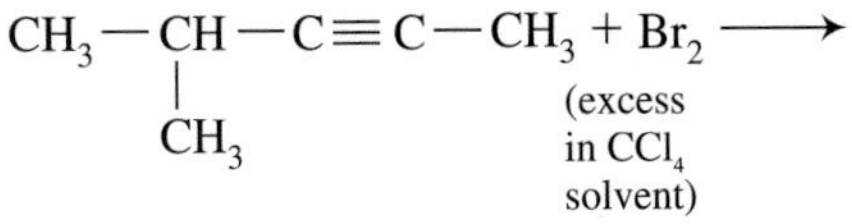 $CH_3-CH-C\equiv C-CH_3 + Br_2 \longrightarrow$
       |
       CH_3

   (excess in CCl_4 solvent)

b. 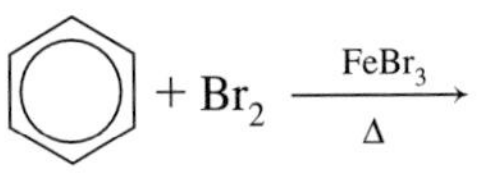  $+ Br_2 \xrightarrow[\Delta]{FeBr_3}$

✓ Chapter Quiz (18C)
(Section 18.7)

1. Circle the functional group in each of the following molecules and label it.

a. 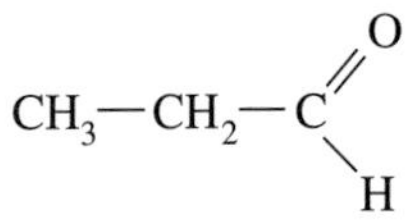 $CH_3-CH_2-C\overset{\displaystyle O}{\underset{\displaystyle H}{\diagdown}}$

b. 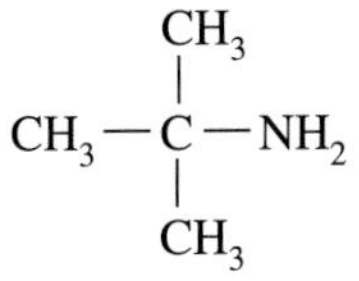 $CH_3-\overset{\displaystyle CH_3}{\underset{\displaystyle CH_3}{\overset{|}{\underset{|}{C}}}}-NH_2$

2. Draw the functional group associated with each of the following classes of hydrocarbon derivatives.

a. ketone
b. alcohol
c. amide
d. ether

3. Identify the hydrocarbon derivatives in each of the following compounds as (1) organic halide, (2) alcohol, (3) ether, (4) aldehyde, (5) ketone, (6) carboxylic acid, (7) ester, (8) amine, or (9) amide by circling each group and labeling it:

a. 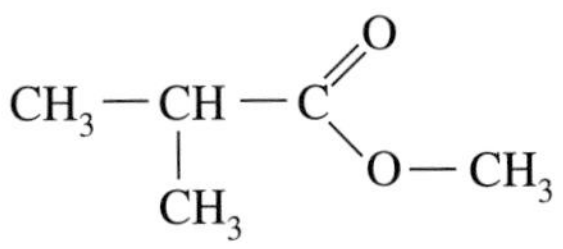 $CH_3-\overset{\displaystyle }{\underset{\displaystyle CH_3}{\overset{|}{CH}}}-C\overset{\displaystyle O}{\underset{\displaystyle O-CH_3}{\diagdown}}$

b. $CH_3-CH_2-C\overset{\displaystyle O}{\underset{\displaystyle OH}{\diagdown}}$

c. $HO-CH-CH-CH_3$ with phenyl below CH and $NH-CH_3$

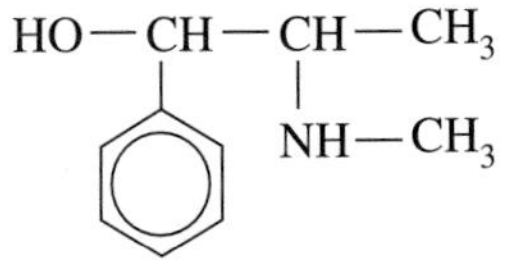

ephedrine

d. 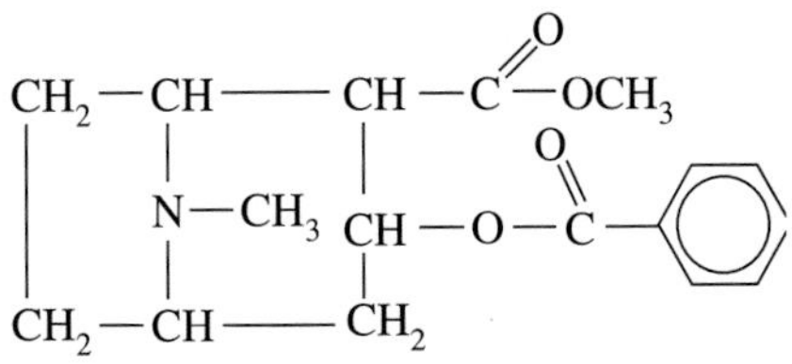 cocaine structure

cocaine

e. 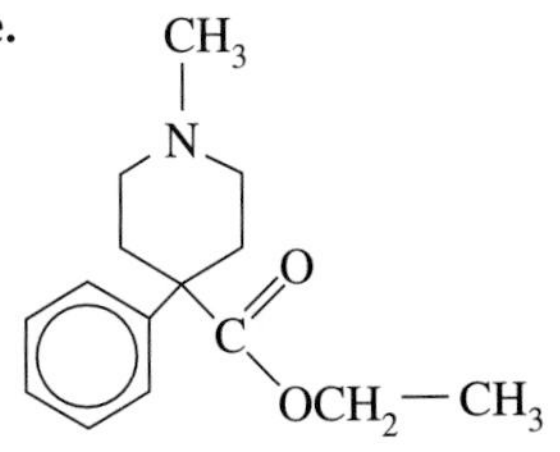 Demerol structure

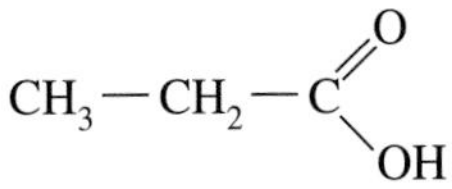

Demerol

f. 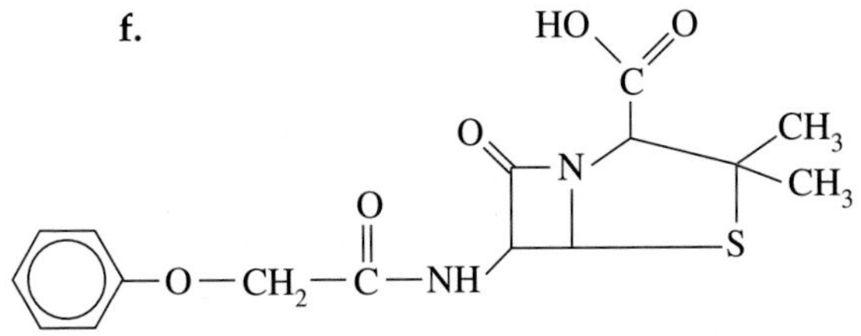 pencillin V structure

pencillin V

Carbon (Symbol: C)

The Element CARBON: From Jewelry to Golf Clubs

FROM

The two stable forms of pure carbon are very different in appearance and are probably familiar to you: graphite (right) and diamond (left).

TO

A golfer with a carbon fiber golf club shaft. No more double bogeys, all birdies!

Name: The name carbon derives from the Latin *carbo*, meaning "coal" or "charcoal." The name for graphite comes from the Greek *grāphein*, meaning "to write."

Forms: Pure carbon occurs in two forms, graphite and diamond. Impure carbon takes many other forms (see Occurrence). Recently, compounds with the formulas C_{60} and C_{70} have been prepared and studied. These materials are examples of pure carbon that were previously unknown. The C_{60} molecule has the shape of a soccer ball (12 pentagons and 20 hexagons) and has unusual electrical and magnetic properties.

Appearance: Graphite is a soft, black substance in which the carbons link to form large sheets. Diamonds, which can be colorless crystals, are the hardest natural substances known.

Occurrence: Deposits of coal, graphite, and diamond are mostly carbon, while oil and natural gas contain carbon in combination with hydrogen. Carbon occurs as $CaCO_3$ (limestone, marble, chalk) and $CaMg(CO_3)_2$ (dolomite). Living organisms contain significant amounts of carbon. Carbon is found in air as CO_2 to the extent of approximately 0.035% by volume.

Source: Some forms of carbon (natural graphite and diamonds) are obtained from natural sources. Other forms (activated carbon, artificial graphite, artificial diamonds, carbon black) are prepared from coal, oil, wood, or natural gas by charring and heating with various fillers and binders. The additives in artificial graphite give it a variety of important properties (heat-resistant, inert to acids and bases).

Common Uses: Diamonds are used as gems, but they also find use as drill tips, abrasives, and cutting tools because of their hardness. The biggest use of carbon is in the steel industry. Coke (a form of carbon), oxygen, and iron oxides are heated in a blast furnace where coke and oxygen form carbon monoxide, which reduces iron(II) and iron(III) to iron metal:

$$2\,C(s) \;+\; O_2(g) \;\longrightarrow\; 2\,CO(g)$$

$$Fe_2O_3(l) \;+\; 3\,CO(g) \;\longrightarrow\; 2\,Fe(l) \;+\; 3\,CO_2(g)$$

$$FeO(l) \;+\; CO(g) \;\longrightarrow\; Fe(l) \;+\; CO_2(g)$$

Graphite is relatively inert and has a high electrical conductivity, and so it is suitable for making electrodes. Graphite electrodes are used to heat furnaces to high temperatures in the production of stainless steels and aluminum. Graphite electrodes are also used in the electrolytic production of fluorine (F_2), sodium (Na), and lithium (Li).

Artificial graphite, a good heat insulator that retains strength at high temperatures, is used to line blast furnaces and rocket nozzles.

Carbon fibers made from rayon or polyacrylonitrile are strong but relatively light. They give strength and flexibility to golf club shafts, tennis rackets, skis, bows, and fishing rods.

Carbon black is used to add strength to tires and as a pigment in inks and paints. Other uses of carbon include water and gas purification (activated charcoal), dry cell electrodes (graphite), and pencil lead (graphite).

Unusual Facts: Passing large amounts of current between two graphite electrodes creates a brilliant, high-intensity light. Such carbon arc lamps are used in searchlights and spotlights and as a light source in movie theater projectors.

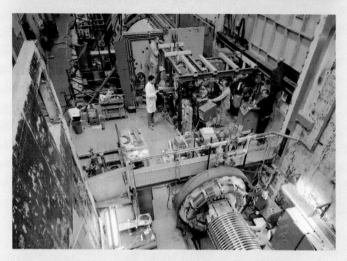

CHAPTER 19   Nuclear Chemistry

The Super-HILAC (Heavy Ion Linear Accelerator) at the Lawrence Berkeley Laboratory of the University of California. Elements with atomic numbers 102 (nobelium, No), 103 (lawrencium, Lr), 104 (rutherfordium, Rf), and 105 (hahnium, Ha) were prepared with this instrument. Element 106 (seaborgium, Sg) was prepared using an upgraded version of this instrument. (Courtesy of Lawrence Berkeley Laboratory, University of California, Glenn T. Seaborg)

GOALS FOR CHAPTER 19

1. To define nuclear reaction and natural radioactivity and to describe the three basic types of natural radioactivity and their behavior when they pass through an electric field (Section 19.1).
2. To define alpha and beta particles and gamma rays and to write balanced nuclear equations involving alpha and beta emissions (Section 19.2).
3. To explain how scientists create radioactive elements not found in nature and to define artificial radioactivity, the positron, and electron capture and to write balanced nuclear equations involving decay by positron emission or electron capture (Section 19.3).
4. To define the half-life of a radioactive isotope and curie and becquerel units in the measurement of radioactivity. To use half-life to solve problems involving the decay of an isotope over a period of time (Section 19.4).
5. To list four uses of radioactive isotopes and explain their benefits (Section 19.5).
6. To define nuclear fission and to explain how it is applied in a nuclear fission reaction—an atomic bomb (Section 19.6).
7. To define nuclear fusion and to explain how it is applied in nuclear fusion reactions—the sun and the hydrogen bomb (Section 19.7).
8. To explain how nuclear fission power plants generate electrical energy, along with the problems associated with these plants. To explain how nuclear fusion power plants could possibly generate electrical energy in the future, along with the problems facing scientists now in developing these plants (Section 19.8).

Countdown

You may use the periodic table.

5. Give the abbreviation and relative charge on the following subatomic particles (Section 4.3):
   a. electron $(e^-$ and $-1)$
   b. proton $(p$ or p^+ and $+1)$
   c. neutron $(n$ or n^0 and $0)$

4. For each of the following *atoms*, calculate the number of protons and neutrons in the nucleus and the number of electrons outside the nucleus (Section 4.4):
   a. $^{238}_{92}U$ $\left(\begin{smallmatrix}92p\\146n\end{smallmatrix}\right.$ $92e^-\left.\right)$

   b. $^{99}_{43}Tc$ $\left(\begin{smallmatrix}43p\\56n\end{smallmatrix}\right.$ $43e^-\left.\right)$

3. For each of the following *ions*, calculate the number of protons and neutrons in the nucleus and the number of electrons outside the nucleus (Section 6.4):
   a. $^{238}_{92}U^{4+}$ $\left(\begin{smallmatrix}92p\\146n\end{smallmatrix}\right.$ $88e^-\left.\right)$

   b. $^{99}_{43}Tc^{7+}$ $\left(\begin{smallmatrix}43p\\56n\end{smallmatrix}\right.$ $36e^-\left.\right)$

2. Carry out the following conversions and express your answer in scientific notation (Sections 2.6 through 2.8):
   a. 0.400 mg to micrograms
      $(4.00 \times 10^2\ \mu g)$
   b. 6.25×10^{-8} g to micrograms
      $(6.25 \times 10^{-2}\ \mu g)$

1. A 15.0-kg sample of uranium ore contains 106 g of uranium-235. What is the percent uranium-235 in the uranium ore sample? (Sections 2.2 and 8.4)
   (0.707%)

Every schoolchild knows that in 1492 an Italian named Christopher Columbus "discovered" the New World. But did you know that is was another Italian, Enrico Fermi, who in 1942 brought chemistry—and the world in general—into a new age, the nuclear age? In that year, Fermi and his colleagues successfully ran a controlled nuclear reaction establishing the importance of nuclear chemistry.

Although we noted in Chapter 4 that atoms have a nucleus containing protons and neutrons, as well as an outer region occupied by electrons, we have largely ignored the nucleus in favor of the electrons. After all, it is the activity of the electrons that determines most of the characteristics of an element and how it reacts with other substances. In this chapter, however, we will consider reactions involving the nuclei of elements and the ways in which nuclear chemistry affects our lives. Only by understanding the forces in the nucleus can we begin to understand the impact of nuclear chemistry on the world in which we live. From the beneficial, such as nuclear medicine, to the disturbing, such as nuclear weapons, nuclear chemistry touches all of us.

19.1 Natural Radioactivity

Although Fermi was the first to carry out an *artificial self-sustaining* reaction in which the nucleus of the atom changed composition—that is, to cause a **nuclear reaction**—his achievement would not have been possible had it not been for the hard work (and luck) of earlier scientists. In particular, the science of nuclear chemistry owes much to German physicist Wilhelm Roentgen (1845–1923) and French physicist Antoine Henri Becquerel (Be krel′) (1852–1908). Roentgen found that X-rays could penetrate the human body and leave an image on a photographic plate. At that time Becquerel was studying substances that emit "light rays" after exposure to the sun. After Roentgen's discovery, Becquerel took some uranium salts used by Roentgen in his experiments and attempted to determine if Roentgen's X-rays were produced by exposure to the sun.

To test his ideas, Becquerel placed uranium salts atop a photographic plate wrapped in black paper and exposed it to sunlight. As expected, the plate showed that rays had penetrated the paper. But when Becquerel tried to repeat his experiment, he ran into a stretch of cloudy weather and placed the bundle in a drawer. Several days later, with no sunlight in sight, he decided to develop the plate anyway. Much to his surprise, the plate had a very strong image on it. Becquerel was forced to conclude that the uranium salts themselves emitted some form of rays similar to X-rays. These spontaneous emissions accompanying changes in the nuclei of atoms are called **radioactivity**. Two years later, in 1898, Polish-French chemist Marie Sklodowska Curie (1867–1934) and her French husband Pierre Curie (1859–1906) discovered two other naturally occurring radioactive substances: polonium and radium.*

Nuclear reaction   Any reaction in which the nucleus of the atom changes composition.

Radioactivity   Spontaneous emissions accompanying changes in the nuclei of atoms.

---

*In isolating radium, Marie Curie isolated 100 mg of almost pure radium chloride from a ton (about 900 kg) of uranium ore by repeated crystallizations—truly an arduous task. Pierre Curie, a physicist, did most of the physical measurements involved in their research, while Marie Curie did the chemical research. In 1903 they received the Nobel prize in physics as a team along with Becquerel. In 1911 after the death of Pierre, Marie Curie received the Nobel prize in chemistry for her isolation of the element polonium and description of the chemical properties of radium and polonium. She was the first woman to receive the Nobel prize, and the first *person* to receive the Nobel prize *twice*.

Scientists do not understand the exact reasons why some isotopes are radioactive and other isotopes are not. One important factor is the ratio of neutrons to protons in a given nucleus. Protons are positively charged and hence repel each other. The presence of neutrons is required to keep the nucleus from flying apart. As the number of protons in the atom increases, a larger number of neutrons is required to maintain the stability of the isotope.

For most elements, there is a range over which the number of neutrons is just right to hold the nucleus together. Too many or too few neutrons can cause an isotope to be radioactive. For example, ^{208}Pb (lead-208) has 82 protons and 126 neutrons and is not radioactive. However, both lead-200 and lead-212 are unstable. Lead-200 has *too few neutrons* (118), and lead-212 has *too many* (130). Most naturally occurring isotopes are not radioactive and do not emit nuclear radiation. Further research confirmed Becquerel's conclusion that different types of radioactive rays exist. In 1899 Ernest Rutherford (1871–1937), an English physicist,* performed an experiment using a device like that depicted in Figure 19.1. As we can see, he found that one type of ray, which he called *alpha rays* (now called *alpha particles*), was positively charged and was attracted to the negative electric field. Another type of ray, which he called *beta rays* (now called *beta particles*), was negatively charged and was attracted to the positive electric field. A third type of ray, called *gamma rays*, was electrically neutral and was unaffected by the electric field.

Alpha and beta particles and gamma rays are often abbreviated by using the first three letters of the Greek alphabet, α, β, and γ. Naturally occurring radioactive substances emit only these three types of nuclear radiation, collectively called

FIGURE 19.1
Nuclear radiation from a small amount of uranium mineral under the influence of an electric field.

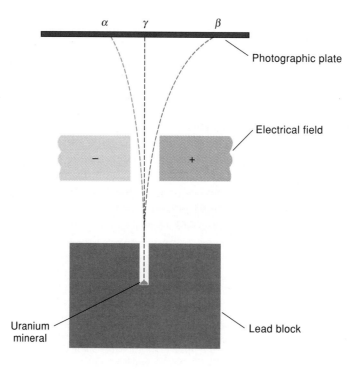

---

*We usually consider Ernest Rutherford to be a British physicist since he did most of his research in England. He was actually born in New Zealand and received most of his education in that country.

natural radioactivity. These substances do not emit alpha and beta particles simultaneously. *Only one* of the two is emitted in a given process. Gamma rays, however, are emitted along with alpha and beta particles.

In addition to differing in electric charge, alpha and beta particles and gamma rays differ in their penetrating power, their effect on human health, and their ability to ionize gases. For example, alpha particles have very low penetrating power. A piece of paper on your skin is enough to stop these rays. Beta particles have slightly greater penetrating power but can be stopped by a sheet of aluminum 4 mm thick. In contrast, both gamma rays and X-rays that *resemble gamma rays* have a high penetration power and can be stopped only by lead and concrete.

Penetrating power has serious consequences for human health since you can develop *radiation sickness* (a potentially fatal ailment) if you are exposed to enough of any form of radioactivity. Fortunately, the most dangerous radiation—alpha particles—has the least penetrating power although it can penetrate your body if you inhale radioactive dust particles, for example. But even small amounts of gamma rays may affect the human genetic code by causing mutations (99% of which are harmful). Thus, physicians use X-rays with discretion, especially if the sex organs are to be exposed to the radiation.

Finally, these types of naturally emitted radioactivity differ in their ability to ionize the gases they pass through. Alpha particles have the greatest ionization effect, and gamma rays the least. Some home smoke detectors take advantage of the ionization effect of alpha particles. They contain a small piece of radioactive americium-241 that ionizes the air in the detector and carries a small electric current. Smoke disrupts the ionization effect of the alpha particles, breaks the current, and triggers the alarm.

Natural radioactivity   Emission of alpha particles, beta particles, and gamma rays by naturally occurring radioactive substances.

19.2 Equations for Natural Radioactivity

Now that we know about the effects of alpha and beta particles and gamma rays, let's consider how they affect the elements in a nuclear reaction.

Among Rutherford's findings was the fact that, as radioactive elements give off these radiations, the radioactive elements themselves change into other elements, a process called **transmutation**. This change is the result of the **radioactive decay** that causes radioactive elements to give off emissions in the first place. Before we can write equations showing this decay, however, we must first know how to write these types of radioactive emissions in chemical form.

Chemists express alpha and beta particles and gamma rays in terms of their atomic number and mass number (number of protons plus number of neutrons; see Section 4.4). An **alpha (α) particle**, which is identical to the nucleus of a helium-4 atom, is written as ^4_2He because it has an atomic number of 2 and a mass number of 4 (2 protons + 2 neutrons). Thus, when an element gives off an alpha particle, its *atomic number decreases by 2*, its *mass number decreases by 4*, and a *new element is formed*. For example, consider the decay of uranium-238 by alpha particle emission:

$$^{238}_{92}\text{U} \longrightarrow {}^4_2\text{He} + {}^{234}_{90}\text{Th} \tag{19.1}$$

The mass number of uranium-238 decreases by 4 (238 − 4 = 234), and the atomic number decreases by 2 (92 − 2 = 90).

Transmutation   Change of a radioactive element into another element.

Radioactive decay   Changes that occur in the composition of radioactive materials as they give off alpha and beta particles and gamma rays.

Alpha particle (α)   Form of radioactive emission that is identical to the nuclei of helium-4 atoms, having an atomic number of 2 and a mass number of 4; thus, emission of an alpha particle causes an element's atomic number to decrease by 2 and its mass number to decrease by 4; written ^4_2He.

Beta particle (β)   Form of radioactive emission that is identical to an electron, having a charge of −1 and a mass number of 0; thus, emission of a beta particle causes an element's atomic number to increase by 1 and its mass number to stay the same; written $_{-1}^{0}$e.

In contrast, a **beta (β) particle** is identical to an electron. A beta particle is written as $_{-1}^{0}$e. It has a charge of −1 and a mass number of 0 (electrons have negligible mass). Beta particles are probably produced when a neutron in the nucleus is transformed into a proton ($_{1}^{1}$H):

$$_{0}^{1}\text{n} \longrightarrow _{1}^{1}\text{H} + _{-1}^{0}\text{e} \tag{19.2}$$

A beta particle is released from the nucleus because electrons do not exist in the nucleus of atoms. Thus, when an element gives off a beta particle, (1) its atomic number *increases* by 1, (2) its mass number remains the same, and (3) again a new element is formed. For example, consider the decay of throrium-234 by beta particle emission:

$$_{90}^{234}\text{Th} \longrightarrow _{-1}^{0}\text{e} + _{91}^{234}\text{Pa} \tag{19.3}$$

Decay series   Series of changes undergone by radioactive substances as they give off alpha and beta particles until they finally become stable (nonradioactive) substances.

The mass number of the new element, protactinium-234 (^{234}Pa), is the same, but it has an atomic number 1 more than that of thorium-234. The resulting protactinium nucleus continues to decay through a series of alpha and beta particle emissions until it becomes a *stable* substance, lead-206 ($_{82}^{260}$Pb). A stable substance is not radioactive and does not emit radiation. Figure 19.2 shows this *decay series*. A **decay series** is a series of changes undergone by radioactive substances as they give off alpha and beta particles until they finally become stable (nonradioactive) substances. Similar series exist for other radioactive isotopes.

Gamma rays   Form of radioactive emission that has no charge or mass and thus has no effect on the atomic number or mass number of an element, despite their high energy; written as γ.

Finally, a **gamma (γ) ray** has no charge or mass and thus has no effect on the atomic number or mass number of a substance. Gamma rays are short-wavelength,

FIGURE 19.2
Decay series for uranium-238. Red arrows (⟶) indicate α decays, and blue arrows (⟶) indicate β decays.

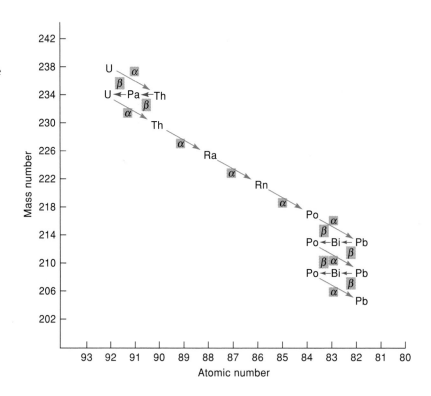

high-energy light rays that are similar to X-rays. They are emitted along with alpha and beta particles.

Rule for Balancing Nuclear Reactions

The sum of the atomic numbers of the reactants must equal the sum of the atomic numbers of the products, and the sum of the mass numbers of the reactants must equal the sum of the mass numbers of the products.

EXAMPLE 19.1 Radium-226 (Ra) decays by alpha emission. Complete and balance a nuclear chemical equation for this nuclear reaction.

SOLUTION   From the periodic table (on the inside front cover of this text), the atomic number for radium is 88. Write an alpha particle as $^4_2$He.

$$^{226}_{88}\text{Ra} \longrightarrow \ ^4_2\text{He} \ + \ ?$$

For the atomic number in the reactant to equal the sum of the atomic numbers in the products, the atomic number of the new element (?) must be 86.

$$88 = 2 + x$$
$$x = 86$$

Applying the same method to the mass number, we will find that the mass number for the new element (?) is 222.

$$226 = 4 + y$$
$$y = 222$$

Refer again to the periodic table and look for the symbol of the element with an *atomic number* of 86. The symbol is Rn, radon (see the list of elements and their symbols on the inside front cover). Therefore, the nuclear chemical equation for this nuclear reaction is

$$^{226}_{88}\text{Ra} \longrightarrow \ ^4_2\text{He} \ + \ ^{222}_{86}\text{Rn} \qquad Answer$$

> **Study Hint:** Be sure that the *sum of the mass numbers* and the *sum of the atomic numbers* on each side of the equation are *equal*.

EXAMPLE 19.2 Bromine-82 decays by beta emission. Complete and balance a nuclear chemical equation for this nuclear reaction.

SOLUTION   Use the periodic table to determine the atomic number of bromine. Write the symbol for a beta particle as $^0_{-1}$e.

$$^{82}_{35}\text{Br} \longrightarrow \ ^0_{-1}\text{e} \ + \ ?$$

The atomic number for the new element is 36, and the mass number is 82:

$$35 = -1 + x \qquad 82 = 0 + y,$$
$$x = 36 \qquad\qquad y = 82$$

Refer again to the periodic table for the symbol of the element with an *atomic number* of 36. The symbol is Kr, krypton. Therefore, the nuclear chemical equation for this nuclear reaction is

$$^{82}_{35}\text{Br} \longrightarrow \ ^0_{-1}\text{e} \ + \ ^{82}_{36}\text{Kr} \qquad Answer$$

> **Study Hint:** In this reaction a neutron changes into a proton and an electron. The proton stays in the nucleus and increases the atomic number, and the electron is emitted as a β particle.

Study Exercise 19.1
Complete and balance a nuclear chemical equation for the following nuclear reactions:

a. Bismuth-211 decays by alpha emission $\left(^{211}_{83}\text{Bi} \longrightarrow {}^{4}_{2}\text{He} + {}^{207}_{81}\text{Tl}\right)$

b. Platinum-197 (Pt) decays by beta emission $\left(^{197}_{78}\text{Pt} \longrightarrow {}^{0}_{-1}\text{e} + {}^{197}_{79}\text{Au}\right)$

Work Problems 9 and 10.

19.3 Artificial Radioactivity

Not all radioactivity is emitted by naturally occurring isotopes. In some cases, scientists create transmutations (and the radioactivity this process gives off) by bombarding various elements with fast-moving particles. These particles are called high-energy particles because they move very rapidly and have high kinetic energies. The first such experiment was performed by Ernest Rutherford, who bombarded a nitrogen nucleus with high-energy alpha particles to produce a proton and an oxygen atom with mass number 17:

$$ {}^{4}_{2}\text{He} + {}^{14}_{7}\text{N} \longrightarrow {}^{1}_{1}\text{H} + {}^{17}_{8}\text{O} \tag{19.4}$$

Since then, many transmutations have been made by bombarding various elements with high-energy alpha particles, neutrons, protons, and deuterons ($^{2}_{1}\text{H}$ nuclei).

Achieving an artificial transmutation is not easy. In Rutherford's case, the particle had to hit the small nucleus of the nitrogen atom precisely for the reaction to occur. Many alpha particles missed their target before one hit right and the reaction took place. In fact, in receiving the 1908 Nobel prize in *chemistry*, Rutherford, a physicist by training, remarked that the fastest transmutation he'd ever achieved was his own change from physicist to chemist!

Cyclotron

Cyclotron Device used to accelerate charged particles.

Many artificial transmutations require particles with *very high* kinetic energies. When very high-energy particles are needed, an instrument must be used to accelerate the bombarding particles. One such particle accelerator for charged particles is the **cyclotron**. As Figure 19.3 shows, a cyclotron consists of two D-shaped electrodes covered by two oppositely charged electromagnets that move the particles in a spiral path. The entire apparatus is contained within a vacuum.

To see how a cyclotron works, suppose that positively charged alpha particles are introduced into the center between the D-shaped electrodes. If one of the electrodes is given a negative charge and the other a positive charge, the alpha particle will travel toward the negative electrode because negative attracts positive. The alpha particle will be repelled by the positive electrode because like charges repel each other. After the alpha particle has completed the half-circle to its oppositely charged electrode, the charge on the electrode changes and the negative electrode becomes positive, and the positive, negative. This process continues until the alpha particle reaches the edge of the electrodes and escapes with a high enough energy to produce a reaction with the target nucleus. The passage of the particle across the gap between the D-shaped electrodes is analogous to a person jumping across a creek with two people assisting. One helper pushes the person and the other helper grabs the person's hand and pulls him, as Figure 19.4 shows.

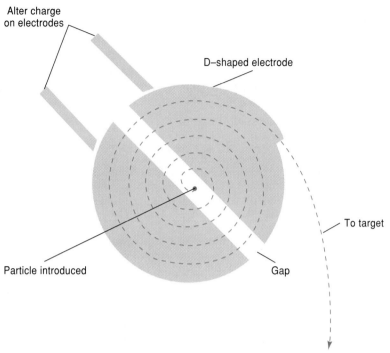

Alter charge on electrodes

D–shaped electrode

Particle introduced

Gap

To target

FIGURE 19.3
Path (– – –) of a particle in a cyclotron. Within a vacuum, two oppositely charged electromagnets above and below the D-shaped electrodes attract and repel particles.

Rutherford's bombardment of nitrogen produced a stable (nonradioactive) substance, oxygen-17. But some artificial transmutations result in radioactive isotopes of an element not found in nature. The first example of such *artificial radioactivity* was discovered in 1934 by two French scientists, Irene Joliot-Curie (1897–1956), daughter of Pierre and Marie Curie, and her husband Frederic Joliot-Curie (1900–

FIGURE 19.4
Crossing a creek or the "gap" in a cyclotron. Like charges repel each other, and opposite charges attract.

Artificial radioactivity
Spontaneous radioactive emission from a substance not found in nature; this form of radioactivity comes from laboratory bombardment of natural or synthetic substances with particles emitted by radioactive isotopes.

1958). **Artificial radioactivity** is the spontaneous radioactive emission from a substance not found in nature; this form of radioactivity comes from laboratory bombardment of natural or synthetic substances with particles emitted by radioactive isotopes. The Joliot-Curies discovered that if boron, magnesium, or aluminum (all nonradioactive) were bombarded with alpha particles, a neutron ($_0^1$n) would be emitted and other elements would form. They both received the Nobel prize in 1935 for this work.

The nuclear equation for the reaction of boron and an alpha particle is

$$_5^{10}\text{B} + {}_2^4\text{He} \longrightarrow {}_7^{13}\text{N} + {}_0^1\text{n} \tag{19.5}$$

Positrons

In creating nitrogen-13, the Joliot-Curies also made another interesting discovery. Artificial radioactive isotopes sometimes emit particles not given off by naturally occurring radioactive substances. For example, nitrogen-13 decays by giving off a *positron* (also called a positron particle). A **positron** is identical to a positive electron. It is written as $_{+1}^0$e because it has a charge of plus one and a mass number of zero (like the electron, the positron has a negligible mass). A positron is probably formed in the nucleus when a proton is transformed into a neutron:

Positron Form of radioactive emission that is identical to a positive electron, having a charge of +1 and a mass number of 0; thus, emission of a positron causes an element's atomic number to decrease by 1 and its mass number to stay the same: written as $_{+1}^0$e.

$$_1^1\text{H} \longrightarrow {}_0^1\text{n} + {}_{+1}^0\text{e} \tag{19.6}$$

Like a beta particle, a positron particle has a small ionizing effect and a low penetrating power. When an isotope gives off a positron particle, its atomic number decreases by 1 and its mass number remains the same. A new element is produced, as in the decay of nitrogen-13 made by Irene and Frederic Joliot-Curie [see Equation (19.5)]:

$$_7^{13}\text{N} \longrightarrow {}_{+1}^0\text{e} + {}_6^{13}\text{C} \tag{19.7}$$

Electron Capture

Electron capture Process in which a nucleus "grabs" an electron outside the nucleus and converts a proton to a neutron.

Another process that generates artificial radioactivity is called *electron capture*. **Electron capture** is a process in which a nucleus "grabs" an electron outside the nucleus and converts a proton to a neutron:

$$_1^1\text{H} + {}_{-1}^0\text{e} \longrightarrow {}_0^1\text{n} \tag{19.8}$$

This process is just the reverse of the one described earlier (see Equation 19.2) for the production of a beta particle. When an element undergoes the electron capture process, its atomic number decreases by 1, while its mass number remains unchanged. A new element is produced, as in the following example:

$$_{92}^{231}\text{U} + {}_{-1}^0\text{e} \longrightarrow {}_{91}^{231}\text{Pa} \tag{19.9}$$

Table 19.1 summarizes various nuclear radiations and the characteristics of each type.

TABLE 19.1		Symbols and Properties of Nuclear Radiation		
RADIATION	SYMBOL	MASS NUMBER	CHARGE	NATURE
Natural artificial radioactivity				
Alpha	^4_2He	4	+2	Helium-4 nucleus
Beta	$^0_{-1}\text{e}$	0	−1	Electron emitted by an unstable nucleus
Gamma	γ	0	0	High-energy radiation
Artificial radioactivity				
Positron	$^0_{+1}\text{e}$	0	+1	Positive electron
Neutron	^1_0n	1	0	Neutron
Other important particles				
Proton	^1_1H	1	+1	Proton
Electron	$^0_{-1}\text{e}$	0	−1	Electron outside the nucleus
Deuteron	^2_1H	2	+1	Hydrogen-2 nucleus
Triton	^3_1H	3	+1	Hydrogen-3 nucleus

Consider the following examples involving artificial radioactivity.

EXAMPLE 19.3   Oxygen-15 decays by positron emission. Complete and balance a nuclear chemical equation for this nuclear reaction.

SOLUTION   Determine the atomic number of oxygen from the periodic table. The symbol for a positron particle is $^0_{+1}\text{e}$, and so

$$^{15}_8\text{O} \longrightarrow \ ^0_{+1}\text{e} + \ ?$$

The atomic number for the new element is 7, and the mass number is 15:

$$8 = 1 + x$$

$$x = 7$$

From the periodic table, the new element is nitrogen (N), and the nuclear chemical equation for this nuclear reaction is

$$^{15}_8\text{O} \longrightarrow \ ^0_{+1}\text{e} + \ ^{15}_7\text{N} \quad \textit{Answer}$$

> **Study Hint:** In this reaction a proton changes into a neutron and a positron. Thus, the atomic number of the nucleus decreases and the positron is emitted from the nucleus.

EXAMPLE 19.4   Strontium-82 undergoes electron capture to give a new element. Complete and balance a nuclear chemical equation for this nuclear reaction.

SOLUTION   Determine the atomic number of strontium from the periodic table. The symbol for an electron is $^0_{-1}\text{e}$, and so

$$^{82}_{38}\text{Sr} + \ ^0_{-1}\text{e} \longrightarrow \ ?$$

> **Study Hint:** In electron capture, the electron ($^0_{-1}\text{e}$) is on the *reactants* side, not the products side. The element in question *captures* the electron.

The atomic number for the new element is 37, and the mass number is 82:

$$38 + (-1) = x \qquad 82 + 0 = y$$
$$x = 37 \qquad y = 82$$

From the periodic table, the new element is rubidium, and the complete nuclear chemical equation for this nuclear reaction is

$$^{82}_{38}\text{Sr} + {}^{0}_{-1}\text{e} \longrightarrow {}^{82}_{37}\text{Rb} \qquad Answer$$

Study Exercise 19.2

Complete and balance a nuclear chemical equation for each of the following nuclear reactions:

a. Krypton-81 undergoes electron capture to give a new element

$$({}^{81}_{36}\text{Kr} + {}^{0}_{-1}\text{e} \longrightarrow {}^{81}_{35}\text{Br})$$

b. Ruthenium-93 (Ru) decays by positron emission

$$({}^{93}_{44}\text{Ru} \longrightarrow {}^{0}_{+1}\text{e} + {}^{93}_{43}\text{Tc})$$

Work Problems 11 through 14.

Synthetic elements Human-made elements that do not exist in nature; several of these elements fill gaps that formerly existed in the periodic table. They have extended the periodic table.

Transuranium elements Synthetic elements whose atomic numbers exceed 92 (the atomic number of uranium). They have extended the periodic table.

In addition to producing artificial radioactive isotopes, bombardment has also produced several **synthetic elements**, human-made elements not found in nature. It is interesting to note that several of these elements fill gaps that formerly existed in the periodic table: technetium (Tc, element 43), promethium (Pm, element 61), astatine (At, element 85), and francium (Fr, element 87). The remaining synthetic elements are called the **transuranium elements** because they all have atomic numbers above 92, the atomic number of uranium, the highest-numbered naturally occurring element. All the transuranium elements, including the best known, plutonium, are radioactive. So far, scientists have produced (see opening photograph in this chapter and Figure 19.5) elements 93 (neptunium, Np) through recently prepared element 111*. For example, to prepare meitnerium (Mt, element 109), scientists bombarded bismuth-209 with iron-58 according to the following chemical nuclear equation:

$$^{209}_{83}\text{Bi} + {}^{58}_{26}\text{Fe} \longrightarrow {}^{266}_{109}\text{Mt} + {}^{1}_{0}\text{n} \qquad (19.10)$$

FIGURE 19.5
The Fermi National Accelerator Laboratory (Fermilab) at Batavia, Illinois (near Chicago) has a gigantic particle accelerator that is over 6 km in diameter.

---

*Element 111, yet unnamed, was prepared in the fall of 1994 by a group of German scientists.

Although the synthesis of each new element is usually more difficult than that of the last, the list of transuranium elements will probably continue to expand.

19.4 Half-Life and Measurement of Radioactivity

In addition to being able to predict how a radioactive isotope will decay, we need to be able to calculate the rate at which this decay occurs. All radioactive isotopes (whether they occur naturally or are created artificially) decay at predictable rates. Scientists express the rate of decay of a radioactive isotope as the **half-life** of the isotope, the amount of time required for one-half of any *given mass* of a radioactive isotope to decay. The half-life of any given radioactive isotope does not appear to vary with temperature, pressure, or the compound in which a radioactive isotope is incorporated. Nothing can speed up or slow down the rate at which a given isotope decays.

The specific rate of decay *does* vary widely from isotope to isotope, however. The half-life of some radioactive isotopes is a fraction of a second. Other such isotopes have half-lives of billions of years. For example, uranium-238 has a half-life of 4.5×10^9 years, while oxygen-19 has a half-life of 29.4 seconds. Thus, *half* of a sample of oxygen-19 will decay before you can read this paragraph, while it will take billions of years for half of a sample of uranium-238 to decay.

To understand what chemists mean by half-life, consider what would happen to a $1\overline{0}0$-g sample of lead-210, which has a half-life of 22 years. At the end of the first 22 years, only half the sample ($5\overline{0}$ g) would be lead-210. At the end of another 22 years, half of the remaining half (25 g) would remain. This process would continue to infinity, with some infinitesimal amount of the lead-210 remaining, although eventually the amount would be too small to see. (For example after 198 years, less than two-tenths of a gram would remain.) Figure 19.6 shows the half-life decay curve for lead-210.

Half-life Amount of time required for one-half of any given mass of a radioactive isotope to decay.

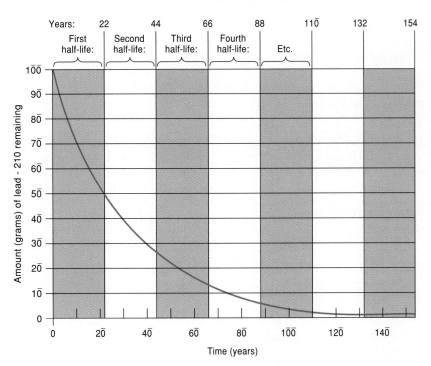

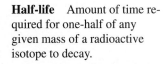

FIGURE 19.6
Decay of $1\overline{0}0$-g sample of lead-210 by beta emission. The half-life is 22 years. Note that after 22 years (one half-life) one-half of the original $1\overline{0}0$ g has decayed to bismuth-210, leaving only $5\overline{0}$ g of lead-210. After two half-lives (44 years), only 25 g of lead-210 remain.

A word of caution: The decrease in lead-210 does *not* mean that the lead has *disappeared*. It simply means that only $5\overline{0}$ or 25 g or a very small number of grams of the material is still lead-210. The other atoms are bismuth-210 (a decay product of ^{210}Pb), polonium-210 (a decay product of ^{210}Bi), and lead-206 (a decay product of ^{210}Po). (See Figure 19.2, the final decay series on the graph, Pb $\longrightarrow$ Bi $\longrightarrow$ Po $\longrightarrow$ Pb.)

Now let us consider some problems involving half-life calculations:

EXAMPLE 19.5	A radioactive isotope of xenon, xenon-125, has a half-life of 17 hours. If we start with 0.2500 g of xenon-125, how many milligrams will remain after 2 days and 3 hours?

SOLUTION   The time of 2 days and 3 hours is equivalent to 51 hours (h).

$$2\ \text{days} \times \frac{24\ \text{h}}{1\ \text{day}} + 3\ \text{h} = 51\ \text{h}$$

At the start (0 h), we have 0.2500 g of xenon-125. After 17 h we will have one-half of the original amount (0.1250 g). After 34 h we will have one-half of 0.1250 g (0.0625 g). Finally, after 51 h we will have one-half of 0.0625 g (0.03125 g), which is 31.2 mg when expressed to three significant digits.

	Time (h)	Amount (g)	
	0	0.2500	
Add 17 h	17	0.1250	Divide the amount that
(half-life)	34	0.0625	remains by 2 each time.
each time.	51	0.03125 = 31.2 mg	Answer

EXAMPLE 19.6	Strontium-90 (2.500 g) was formed in a 1960 atomic explosion at Johnson Island at the Pacific test site. The half-life of strontium-90 is 28 years. In what year will only 0.312 g of this strontium-90 remain?

SOLUTION   At 0 years (1960), there were 2.500 g. After 28 years (1960 + 28 = 1988), there were 1.250 g ($\frac{1}{2} \times$ 2.500 g). After 56 years (1988 + 28 = 2016), there will be 0.625 g ($\frac{1}{2} \times$ 1.250 g). And after 28 more years (2016 + 28 = 2044), there will be 0.3125 g ($\frac{1}{2} \times$ 0.625 g) remaining, or 0.312 g to three significant digits.

	Amount (g)	Time and year	
	2.500	0 (1960)	
Divide the amount	1.250	28 (1988)	Add 28 years (half-life)
that remains by	0.625	56 (2016)	each time.
2 each time.	0.312	84 (2044)	Answer

Study Exercise 19.3
Seaborgium-263 (Sg) has a half-life of 0.9 s. If 1.5000 mg decays over a period of 3.6 s, how many micrograms of seaborgium-263 will remain?

Work Problems 15 through 20.

$(93.8\ \mu\text{g})$

Measuring Radioactivity

Because the half-lives of some highly radioactive substances are so long, and because the danger to human life from exposure to radioactivity is so great, there is

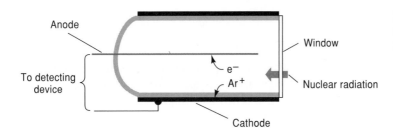

FIGURE 19.7
Geiger–Müller counter
(Geiger counter).

great need to measure the amount of such radiation in various environments. The most common detector is the Geiger–Müller counter (the Geiger counter), which can detect radiations that ionize gases, that is, alpha, beta, and positron particles.

As Figure 19.7 shows, a Geiger counter consists of a tube containing two electrodes, a positive anode and a negative cathode, with a high potential of about 1000 volts between them. The tube contains air or argon gas at a pressure of about 5 to 12 torr. Radiation enters the tube through the window, which is usually covered with thin glass, and ionizes the gas molecules in the tube, forming ions such as Ar^+ ions (if the gas is argon). The electrons and the ions produce a short-lived flow of electric current in the circuit. This short flow of current can be amplified to produce a sound or to record counts on an automatic counting device.

One unit used to measure the quantity of radioactivity is the *curie*, named for Marie and Pierre Curie, the discovers of polonium (Po) and radium (Ra) (see Section 19.1). A **curie (Ci)** is equal to 3.7×10^{10} nuclear disintegrations per second. Because this value is very large, a microcurie (μCi), $1/1,000,000$ ($3.7 \times 10^{10}/10^6$) Ci, is often used. A radioactive sample with one or more curies is considered to have high activity. For example, 1.00 g of radium gives 1.00 Ci. The newer SI unit to measure radioactivity is the *becquerel* (Bq), named for Antoine Henri Becquerel (see Section 19.1). A **becquerel (Bg)** is equal to one nuclear disintegration per second. Therefore, 1 Ci equals 3.7×10^{10} Bq.

In the eruption of Mount St. Helens, 3 million curies of radon gas was given off. In the Three Mile Island incident in Pennsylvania in 1979, 2.5 million curies of radioactive xenon gas was released into the atmosphere. Radon gas is considered 1000 times more hazardous to human health than radioactive xenon.

Curie (Ci)   Measure of radioactivity equal to 3.7×10^{10} nuclear disintegrations per second.

Becquerel (Bq)   Measure of radioactivity and equal to one nuclear disintegration per second.

19.5 Uses of Radioactive Isotopes

Naturally occurring and artificially created radioactive isotopes have a wide range of practical uses in fields as diverse as archeology, medicine, and agriculture.

Radioactive Dating

By using the half-life of the radioactive isotope ^{14}C, we can determine the age of various objects that were once living, such as bones. In *living* material, the ratio of ^{14}C to ^{12}C (nonradioactive) remains relatively *constant*. The ^{14}C in our atmosphere arises from the bombardment of ^{14}N atoms with a neutron from the upper atmosphere:

$$^{14}_{7}N + ^{1}_{0}n \longrightarrow ^{14}_{6}C + ^{1}_{1}H \qquad (19.11)$$

$^{14}_{6}C$ decays to form nitrogen-14 and a beta particle:

$$^{14}_{6}C \longrightarrow \ ^{0}_{-1}e \ + \ ^{14}_{7}N \qquad (\text{half-life } = \ 5730 \text{ years}) \qquad (19.12)$$

The ratio of ^{14}C to ^{12}C in living tissue is believed to have been constant over the millennia. When the tissue in an animal or plant dies, the amount of ^{14}C *decreases* because ^{14}C decays and is no longer replaced by the intake of ^{14}C from the environment. Therefore, the ratio of ^{14}C to ^{12}C in dead tissue *decreases with time*, and the ratio is a measure of the age of the sample. By measuring this ratio in a sample, scientists can estimate how long ago the plant or animal lived.

Similarly, scientists use the half-life of uranium-238 to determine the age of various *nonliving* objects, such as rocks. In such calculations they use the ratio of radioactive uranium-238 to nonradioactive lead-206 and a half-life of 4.5×10^{9} years. Using this method, the age of the oldest rocks on the earth appears to be nearly 4×10^{9} (billion) years.

Medical Imaging

Radiologists can learn a lot about what is going on in a human by injecting very small amounts of radioactive isotopes into the person. These isotopes move through the body, often concentrating in specific organs or certain parts of the body. There the isotopes emit characteristic radiations that allow radiologists to obtain a "picture" of that portion of the body by using X-ray films. By comparing the films obtained from a patient with films from people who have normal organs or tissues physicians can often tell a great deal about what is wrong with the patient. The small amounts of radioactive isotopes used in modern procedures leave the patient's system in the normal excretory process (see Figure 19.8).

A number of radioactive isotopes are currently in use. By far the most important is technetium-99 (^{99}Tc), which accounts for about 85% of all radioactive isotope imaging processes (see The Element TECHNETIUM: Noninvasive Diagnosis). Iodine-123 and -131, gallium-67, chromium-51, selenium-75, thalium-201, indium-111, and xenon-133 are also currently in use (see Table 19.2).

Radiolabeling in Genetic Tracing

Radioactivity has also added another weapon to the medical arsenal: *radiolabeling*. This powerful technique involves the use of radiation detectors that are *very* sensitive to even the *smallest* amounts of radioactive material. In radiolabeling, scientists attach a radioactive atom to a material of interest so that they can always know where the substance is by tracing where the radioactive signal goes. It is analogous to attaching a radio transmitter to a grizzly bear so that rangers can follow the bear's movements in the wilderness.

An example of this technique can be found in the molecular biology lab. DNA, or deoxyribonucleic acid, is the substance that carries the genetic code of a living organism from cell to cell. If a scientist wants to know where a specific piece of DNA is at all times during a laboratory procedure, an atom of radioactive phospho-

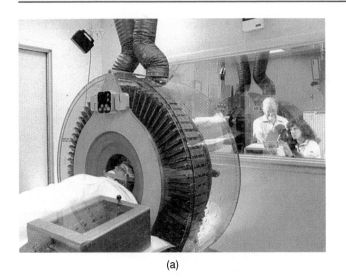
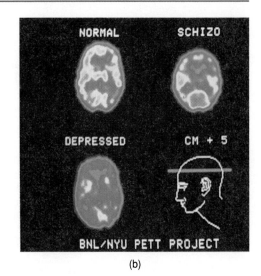

(a)

(b)

Figure 19.8
Imaging with radioactive isotopes has revolutionized modern medicine. (a) A radioactive compound is injected into a person, and radioactive nuclei localize in a particular organ where they emit distinctive radiations that produce images on X-ray film. (b) The resulting images give physicians a dramatically different way to view our bodies. The amount of radiation involved is very small and offers little risk to the patient.

rus (phosphorus-32) is attached to the piece of DNA. Then, the scientist can follow the piece of DNA by detecting the radioactivity emitted by the radioactive phosphorus. Phosphorus-32 decays by beta particle emission:

$$^{32}_{15}\text{P} \longrightarrow {}^{32}_{16}\text{S} + {}^{0}_{-1}\text{e} (\text{half-life} = 14.4 \text{ days}) (19.13)$$

TABLE 19.2		Radioactive Isotopes Used in Medical Imaging and Diagnosis	
Radioactive Isotope	**Decay Mode**	**Form**	**Use**
^{99}Tc	Gamma	NaTcO$_4$ or Tc(IV) bound to another molecule	Imaging of brain, heart, thyroid, stomach, liver, lung, bone marrow, spleen, and kidney
^{131}I	Beta, gamma	NaI	Diagnosis of diseases of thyroid gland
^{123}I	Gamma	NaI	Imaging of thyroid gland
^{67}Ga	Gamma	gallium(III) citrate	Locating soft tumors and internal infections
^{51}Cr	Gamma	Na$_2$CrO$_4$ or Cr(III) attached to blood cell	Blood studies and spleen scanning
^{75}Se	Gamma	Selenium atom in an amino acid	Imaging of pancreas
^{201}Tl	Gamma	TlCl	Imaging of heart
^{111}In	Gamma	In(III) bound to molecule	Blood studies
^{133}Xe	Beta, gamma	Xe gas	Imaging of lung

Irradiation of Produce

A far more common use of radiation is in the irradiation of fresh fruits and vegetables with radiation from cobalt-60. Such irradiation, approved by the Food and Drug Administration (FDA), inhibits the growth and maturation of bacteria and fungi and kills insect eggs and larvae. Thus treated, the produce keeps for long periods of time without spoiling.

Irradiated food does *not* become radioactive. It is as safe as nonirradiated food. The alternative is to use chemicals such as ethylene dibromide (banned by the FDA in 1984) and methyl bromide as insecticides and fumigants. Papayas were the first fruit to be irradiated.

19.6 Nuclear Fission

Although they have enriched human life, the development and use of new radioactive isotopes and elements are only one part of nuclear chemistry. Equally significant, if not always equally beneficial, has been the work of chemists in the areas of nuclear fission and nuclear fusion.

Nuclear fission   Splitting of an atomic nucleus to give two smaller nuclei, neutrons, and energy.

The first of these areas to receive attention was **nuclear fission**, the splitting of an atomic nucleus into two smaller nuclei plus neutrons and energy. The explosion of an atomic bomb, the best known example of nuclear fission, results from bombarding uranium-235 with *neutrons*. The uranium-235 nucleus splits into two smaller nuclei, two or three neutrons, and *energy:*

$$^{235}_{92}\text{U} + ^{1}_{0}\text{n} \longrightarrow ^{94}_{38}\text{Sr} + ^{139}_{54}\text{Xe} + 3\,^{1}_{0}\text{n} + \text{energy} \qquad (19.14)$$

During this reaction, neutrons are released (approximately 2.5 for each uranium-235 atom). These neutrons continue to react with other uranium-235 nuclei. Since more neutrons are released in each reaction than are consumed in the process, the number of reactions can quickly increase to the point where an explosion occurs (see Figure 19.9a). This type of *chain reaction* is similar to the stacking of dominoes shown in Figure 19.9b. When the first domino is knocked over, the rest quickly follow. A **chain reaction** is, therefore, a self-sustaining reaction in which a product is one of the reactants.

Chain reaction   Self-sustaining reaction in which a product is one of the reactants; in a nuclear fission reaction, the neutrons released by the original splitting of the atom continue to react with other uranium-235 nuclei, releasing more neutrons.

In an actual atomic bomb, two masses of uranium-235 and the neutrons take the place of the dominoes, while an ordinary explosive charge takes the place of the finger, forcing the uranium masses together. The result is a *critical mass* of uranium-235, just enough of the substance to sustain a chain reaction, and an atomic explosion. Therefore, a **critical mass** is the minimum amount of a radioactive substance necessary to sustain a chain reaction. For an atomic bomb, the critical mass of uranium-235 is a sphere with a radius of about 8 cm and a mass of about 40 kg.

Critical mass   Minimum amount of a radioactive substance necessary to sustain a chain reaction.

When the conventional explosive charge forces the uranium-235 into a critical mass, the uranium releases a neutron that sets off the chain reaction. The atoms of the uranium-235 shatter into fragments with smaller atomic numbers. These fragments are radioactive and emit nuclear radiation, accounting for the radiation burns and sickness following a nuclear fission explosion (see Figure 19.10).

Uranium-235 is not the only substance capable of sustaining a nuclear fission chain reaction. Plutonium-239 is also capable of such reactions. Plutonium-239 can be formed from uranium-238 (99.3% of naturally occurring uranium):

$$^{238}_{92}\text{U} + ^{1}_{0}\text{n} \longrightarrow ^{239}_{92}\text{U} \qquad (19.15)$$

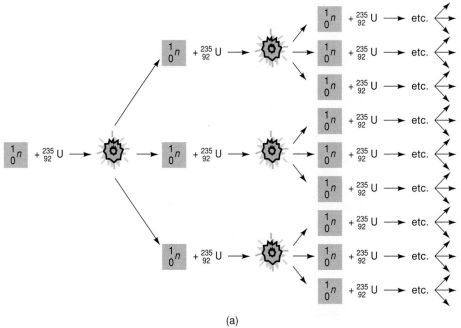

(a)

FIGURE 19.9
Nuclear fission chain reaction. (a) Each reaction between a neutron and a $^{235}_{92}U$ atom gives two new smaller nuclei (not shown), three neutrons, and energy. Thus, 1 neutron produces 3 neutrons, which produce 9 neutrons, which produce 27 neutrons, and so on. (b) A set of dominoes stacked to imitate a chain reaction. When the first one is knocked over, the rest soon follow. Note that in a real fission explosion, not every neutron hits a uranium-235 atom, but enough do to produce an explosion.

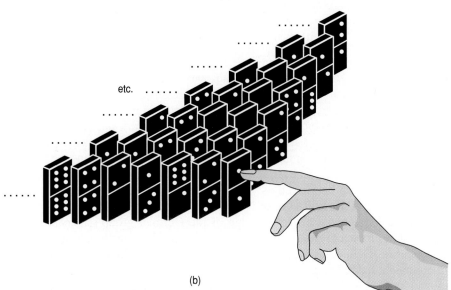

(b)

$$^{239}_{92}U \longrightarrow \; ^{239}_{93}Np \; + \; _{-1}^{0}e, \quad \text{(half-life} = 23.5 \text{ min)} \qquad (19.16)$$

$$^{239}_{93}Np \longrightarrow \; ^{239}_{94}Pu \; + \; _{-1}^{0}e, \quad \text{(half-life} = 2.35 \text{ days)} \qquad (19.17)$$

During a nuclear explosion (see Equation 19.14), a small but definite loss of mass occurs. This mass loss is a fraction of the total mass and hence is not shown in Equation 19.14. The lost mass is converted to the energy given off in the fission process, approximately 200 million electron volts (meV) for every uranium atom split. This energy, expressed in more familiar terms, is 4.6×10^9 kcal/mol of uranium atoms split, or approximately 2.5 million times the energy released by burning a comparable mass of coal. Nuclear fission energy is given off primarily as *heat* energy.

The history of the development of nuclear fission is an international one. It began in Europe in the 1930s in laboratories in both Italy and Germany. The Italian research group was headed by Enrico Fermi (1901–1954), who later moved to the United States. The German research team was composed of Otto Hahn (1879–1968), Fritz Strassman (1902–1980), and Lise Meitner (1878–1968).

Nuclear fission may be the first instance in modern times in which the scientific world was awakened to the consequences of its work on the sociopolitical level. Lise Meitner was of Jewish ancestry and so was forced to leave Nazi Germany in 1938. Hitler's attitude toward Meitner may have cost him the war. Although he allowed Hahn and Strassman to continue their work, he gave them no support for research he viewed as "Jewish-tainted." Meanwhile, Meitner reached Holland and, with the assistance of Niels Bohr, eventually got to Stockholm, Sweden. Meitner conveyed her knowledge of the German group's research to her nephew Otto Frisch (1904–1979), who was then working in Bohr's laboratory in Copenhagen, Denmark. They realized the tremendous military potential nuclear fission might have. During a trip to the United States, Bohr discussed the possibility with many European scientists then working in the United States. One of these scientists, Leo Szilard (1898–1964), a Hungarian, spoke to Albert Einstein about the military potential of nuclear fission and about the possibility that the Germans might at that time be developing it. In August 1939, Szilard wrote and Einstein signed a letter to President Franklin D. Roosevelt expressing his concern (see Figure 19.11). Intrigued, Roosevelt authorized the construction of four facilities in isolated areas in the United States. One was in the mountains of New Mexico near Santa Fe at Los Alamos, and another in the mountains of Tennessee near Knoxville at Oak Ridge. Other research for the project, code-named the Manhattan Project, was done at Hanford, Washington, and at the Metallurgical Laboratory of the University of Chicago. The project, fully funded and organized by the U.S. Army, included many distinguished scientists, among them Vannevar Bush, James B. Conant, Leslie R. Groves, J. Robert Oppenheimer, and Glenn T. Seaborg, all of whom have since died except Seaborg. The first nuclear fission explosion—an atomic bomb—was tested in the desert at Trinity Flats near Alamogordo, New Mexico, on July 16, 1945.

Szilard campaigned not to have the atomic bomb used as a military weapon. He proposed that it be tested openly before the Japanese and an international audience in an attempt to persuade the Japanese to surrender before the bomb was used to kill people. President Harry S. Truman, in accord with his advisors (the Interim Committee on S-1), made the final decision to drop the atomic bomb. On August 6,

FIGURE 19.11
Einstein's letter to President Franklin D. Roosevelt, which was written by Szilard.

```
                                    Albert Einstein
                                    Old Grove Rd.
                                    Nassau Point
                                    Peconic, Long Island

                                    August 2nd, 1939

F.D. Roosevelt,
President of the United States,
White House
Washington, D.C.


Sir:

     Some recent work by E.Fermi and L. Szilard, which has been com-
municated to me in manuscript, leads me to expect that the element uran-
ium may be turned into a new and important source of energy in the im-
mediate future. Certain aspects of the situation which has arisen seem
to call for watchfulness and, if necessary, quick action on the part
of the Administration. I believe therefore that it is my duty to bring
to your attention the following facts and recommendations:

     In the course of the last four months it has been made probable -
through the work of Joliot in France as well as Fermi and Szilard in
America - that it may become possible to set up a nuclear chain reaction
in a large mass of uranium,by which vast amounts of power and large quant-
ities of new radium-like elements would be generated. Now it appears
almost certain that this could be achieved in the immediate future.

     This new phenomenon would also lead to the construction of bombs,
and it is conceivable - though much less certain - that extremely power-
ful bombs of a new type may thus be constructed. A single bomb of this
type, carried by boat and exploded in a port, might very well destroy
the whole port together with some of the surrounding territory. However,
such bombs might very well prove to be too heavy for transportation by
air.
```

Figure 19.11
(*Continued*)

-2-

The United States has only very poor ores of uranium in moderate quantities. There is some good ore in Canada and the former Czechoslovakia, while the most important source of uranium is Belgian Congo.

In view of this situation you may think it desirable to have some permanent contact maintained between the Administration and the group of physicists working on chain reactions in America. One possible way of achieving this might be for you to entrust with this task a person who has your confidence and who could perhaps serve in an inofficial capacity. His task might comprise the following:

a) to approach Government Departments, keep them informed of the further development, and put forward recommendations for Government action, giving particular attention to the problem of securing a supply of uranium ore for the United States;

b) to speed up the experimental work,which is at present being carried on within the limits of the budgets of University laboratories, by providing funds, if such funds be required, through his contacts with private persons who are willing to make contributions for this cause, and perhaps also by obtaining the co-operation of industrial laboratories which have the necessary equipment.

I understand that Germany has actually stopped the sale of uranium from the Czechoslovakian mines which she has taken over. That she should have taken such early action might perhaps be understood on the ground that the son of the German Under-Secretary of State, von Weizsäcker, is attached to the Kaiser-Wilhelm-Institut in Berlin where some of the American work on uranium is now being repeated.

Yours very truly,

A. Einstein

(Albert Einstein)

1945, the United States dropped an atomic bomb on Hiroshima, Japan. Three days later, it dropped another on Nagasaki. Together, the two bombs caused an estimated 200,000 Japanese deaths. On August 14, the Japanese government surrendered unconditionally, and the war was finally over.

The intermingling of sociopolitical climate with scientific discovery is an interesting historical footnote. If Hitler had not used a person's ancestry as the basis for decision making, Nazi Germany might have been the first to develop nuclear fission explosives, with disastrous consequences to the rest of the world.

Nuclear Fusion

While nuclear fission produces energy by *splitting* the nucleus of an atom, **nuclear fusion** produces energy by *combining* two or more atomic nuclei to form a heavier nucleus. Two examples that involve nuclear fusion reactions are the sun and the hydrogen or thermonuclear bomb.

Nuclear fusion   Combination of two atomic nuclei to form a heavier nucleus and produce energy.

The Sun

We owe our continued existence on Earth to the nuclear fusion reactions occurring on the sun, which heats our planet. The primary nuclear reaction on the sun probably involves the fusion of two isotopes of hydrogen, deuterium ($_1^2$H) and tritium ($_1^3$H), to form helium ($_2^4$He):

$$_1^2\text{H} + _1^3\text{H} \longrightarrow _2^4\text{He} + _0^1\text{n} \qquad (19.18)$$

The deuterium and tritium isotopes are formed indirectly through bombardment of hydrogen nuclei ($_1^1$H). During the fusion process, there is considerable loss in mass, which converts to energy primarily, in the form of heat. A large amount of heat energy must be supplied for a nuclear fusion reaction to occur. This type of heat energy is available in the sun because the sun's interior temperature is estimated to be 2×10^7°C.

Hydrogen or Thermonuclear Bomb

The hydrogen or thermonuclear bomb is based on a reaction similar to the one that fuels the sun. In a hydrogen bomb, however, a nuclear *fission* explosion—an atomic bomb—provides the necessary heat energy to start the *fusion* reaction. The heat energy released from a nuclear fusion explosion (a hydrogen bomb) is estimated to be about 15 times the energy released from a large nuclear fission explosion (an atomic bomb).

The past 50 years have seen dramatic "improvements" in the technology of mass destruction. Approximately *50,000* thermonuclear warheads exist. These bombs can be delivered to their targets from the air, from the sea, or from land sites 8000 miles away with ever-increasing accuracy. An especially dangerous trend is the development of systems that will deliver warheads with little or no warning. Only now are people beginning to awaken to the fact that these preparations for global war are wrecking their economies and polluting their countrysides. Recently, the trend in the United States and the former Soviet Union is to destroy some of these weapons of mass destruction.

Peaceful Uses of Nuclear Reactions

Nuclear fission and fusion reactions can be used for more constructive purposes than making bombs. Indeed, fission and (in theory) fusion can be used to produce power in nuclear reactors. Nuclear fission technology is well developed, and nuclear reactors producing electric power have been in continuous operation in the

United States since 1957, although no new reactors have been ordered recently. Modern nuclear fission power plants consist of three essential parts: a nuclear fission reactor, a turbine and generator, and a condenser, as Figure 19.12 shows. In the reactor, nuclear fission produces heat which converts water to steam. The steam drives the turbine, which runs the electric generator.

Nuclear Fission Reactor

The uranium used in a nuclear fission reactor is uranium dioxide (UO_2). The uranium has been processed so that 2 or 3% of it is ^{235}U, the rest being ^{238}U. Because the uranium-235 concentration is much lower than the 90% uranium-235 required for an atomic bomb, a nuclear explosion cannot occur in a nuclear reactor.

Both uranium-235 and plutonium-239 can be used in nuclear reactors, though uranium-235 produces more heat. To prevent overheating and to slow the production of heat, modern reactors contain cadmium rods that control the production of neutrons and absorb some of those produced. Water in the reactor acts to slow down the neutrons. (If neutrons move too quickly, fission will not occur.) The controlled fission converts water into steam.

Turbine and Generator

The steam from the secondary coil (see Figure 19.12) then passes to the turbine, as in a nonnuclear power plant. The turbine drives the generator, which produces elec-

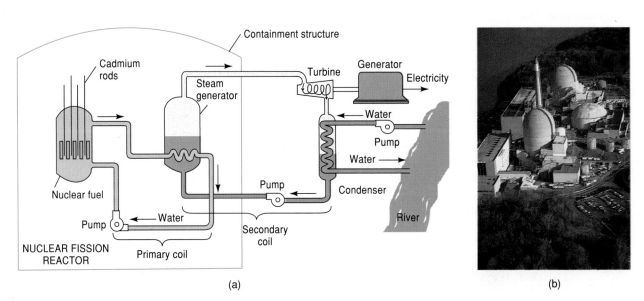

(a)   (b)

FIGURE 19.12
Nuclear fission power plant (pressurized water reactor). (a) Diagram of a nuclear fission power plant. (b) Nuclear power plant at Indian Point, Buchanan, New York.

tric power. Some of the heat energy from a nuclear reaction is thus converted to mechanical energy in the turbine and to electrical energy in the generator.

The difference between a nuclear power plant and a nonnuclear power plant is the source of the steam. In a nuclear power plant, the steam is produced by heat from a nuclear reaction; in a nonnuclear power plant, it is produced by heat from burning fossil fuels, such as coal, oil, or gas.

Condenser

The steam from the secondary coil, after turning the turbine, is liquefied in the condenser and returned to the primary coil again to continue the process. The water used to cool the condenser is obtained from nearby rivers and allowed to cool somewhat before being returned to the river. Thus, nuclear power plants do produce some thermal pollution. Scientists are studying constructive uses of this excess heat. One beneficial solution may be to use the heated river water (about 10°C, from 28°C to 38°C) to irrigate farm crops and lengthen the growing season.

Problems with Nuclear Fission

Nuclear fission power plants have been used to supply electricity to many cities throughout the world. In the United States, nuclear fission produces about 16% of the electric power, a figure that may increase in the future if the problems associated with nuclear fission power plants can be solved.

These problems are (1) disposal of the radioactive waste, (2) the possibility of theft of fissionable material during transportation, and (3) the possibility of a nuclear accident. At present there is no permanent site for the storage of the radioactive waste that accumulates through the nuclear fission process. The Department of Energy (DOE) is building a facility near Carlsbad, New Mexico, to store certain defense nuclear waste by solidifying the material and storing it as fused glass in salt domes. DOE plans call for the site to be operational by 1998. Congress has also approved building a permanent underground storage facility. This first facility may be located in one of a number of places: a salt formation in Texas, Mississippi, or Utah; a volcanic rock structure in Nevada; or a basalt rock formation in Washington (Hanford). The second facility may be built in granite rock formations in Wisconsin or New England. Plans call for the first permanent site to be operational by 1997 to 2006.

There is another problem in addition to the normal hazards associated with the handling and processing of radioactive materials. We need to be careful that no uranium-235 or plutonium-239 is diverted to irresponsible individuals such as terrorists or leaders of terrorist countries. Uranium-235 and plutonium-239 can be used to construct atomic bombs, and such organizations could conceivably develop a bomb and use it to bring about the downfall of governments through blackmail or to cause widespread physical destruction. The possibility of this occurring was evident in 1991 and 1994 with Iraq and North Korea, respectively. Many observers have called for stricter security measures during the processing and transportation of all nuclear materials.

Finally, there is always the possibility of a nuclear accident in a nuclear fission power plant. Temperatures of over 5080°C are required for the fuel material (UO_2) to melt. However, if the reactor fuel (see Figure 19.12) becomes too hot and the method for cooling it does not work or is not used, a *meltdown* can occur. The molten fuel can damage the reactor vessel and even rupture it and release radioactive materials into the containment structure. A catastrophic accident can result in the release of radioactive materials from the containment structure into the environment.

Two major incidents, one at Three Mile Island in Pennsylvania in 1979 and one at Chernobyl in the former Soviet Union in 1986, have brought this problem to life all too vividly. At Three Mile Island, the reactor temperature in some places must have exceeded 5100° because part of the fuel core did melt. During the accident, xenon gas was released into the atmosphere to reduce pressures in the reactor and minimize the severity of the accident. Otherwise, very little radioactive material was released. The cleanup, however, remains a problem. It will eventually cost over a billion dollars to do even a minimal job.

The accident at Chernobyl was caused by design flaws and human error (see Figure 19.13). The accident occurred during a series of dangerous reactor tests with many safety controls turned off. A power surge in the reactor core overheated the fuel and ruptured the core so that cooling water was instantly turned into steam, causing an explosion that broke open the reactor. Subsequent additional damage may have been caused by an explosion, as hydrogen gas, produced by the reaction between zirconium-clad fuel and steam, ignited in the air. The fire from the explosion then ignited the graphite moderator, which burned just like the charcoal in your barbecue. The burning of the graphite destroyed the plant and dispersed radioactive material (^{131}I, ^{137}Cs, and others) over Europe and Scandinavia. The plant at Cher-

FIGURE 19.13
The most serious nuclear accident on record is the 1986 accident and explosion at the Chernobyl nuclear power plant in the Soviet Union.

nobyl did not have a containment structure (see Figure 19.12) like those found in most U.S. nuclear power plants. Such a structure might have kept a large amount of the radioactive material inside the building. The moderator used in U.S. nuclear power plants is water and not graphite as was used at the Chernobyl plant. Therefore, an accident like the one that occurred at the Chernobyl plant is not possible in U.S. nuclear power plants. At present the Chernobyl plant is encased in concrete and referred to as a "sarcophagus." The sarcophagus is supposed to last at least 30 years, but recently it appears that the concrete is deteriorating and the sarcophagus may collapse. The problem would then have to be dealt with once again.

Clearly, we must develop adequate safeguards against serious accidents like these if nuclear fission power plants are to meet the world's energy needs. A new type of reactor now in operation in West Germany (Schmehausen), called a *pebble-bed* reactor, prevents the rapid buildup of heat and hence makes a meltdown less likely. Other new reactor designs and cooling technologies are being examined to try to build in features that reduce the possibility of such accidents. If these problems can be solved, nuclear energy could be the answer to our energy needs for the future.

Nuclear Fusion

Nuclear fusion might also be used to produce electric power in the future. Scientists believe that nuclear fusion power plants would produce less thermal pollution and nuclear radiation and pose a lower risk of nuclear accident than nuclear fission power plants.

Two approaches are being used to develop nuclear fusion to produce electricity: (1) magnetic confinement of the hydrogen and (2) the use of lasers or electron beams to unite the hydrogen atoms. The two isotopes of hydrogen used are deuterium ($^2_1$H) and tritium ($^3_1$H). Deuterium is readily available from seawater. Tritium can be produced by bombarding lithium with neutrons. Both rocks and seawater contain lithium salts.

To obtain a fusion of the hydrogen isotopes, three conditions must be met: (1) High temperatures of about 100 million degrees Celsius are necessary. (2) A high density of about 10^{14} to 10^{16} particles per cubic centimeter is also required. (3) The hydrogen isotopes must be confined long enough (about 1 s) at high temperature and density to enable the fusion reaction to occur and sustain itself. So far, scientists have not been able to meet all three of these conditions at the same time.

Periodically, scientists propose ways of achieving fusion under less vigorous conditions. In 1989, "cold fusion" was reported at temperatures of about 100°C during electrolysis reactions. However, none of these reports has held up under vigorous scrutiny and none has been reproduced in other laboratories.

Meanwhile an international effort among Europe, Japan, the former Soviet Union, and the United States called the International Thermonuclear Experimental Reactor (ITER) may bring needed cooperation to the quest for fusion energy. Experts believe it will be 2020 before we will see a commercial fusion reactor that converts fusion energy to electric power. Strangely enough, the limiting factor on this technology may be economics, not science. Such a program costs an enormous amount of money to develop and there is simply not enough money to go around.

CHEMISTRY OF THE ATMOSPHERE

Radon in the Atmosphere and in Your Home

The element radon (Rn) is a member of the noble gases (VIIIA, 18) that results from normal radioactive processes within the earth's crust. Because radon results from naturally occurring uranium-238, it is in the atmosphere as a result of *natural processes and not human activities.*

   Radon is one of the elements in the ^{238}U decay series described in Figure 19.2. Part of the decay series for radon is reproduced below, along with the associated half-lives in parentheses.

$^{226}_{88}Ra \longrightarrow\ ^{222}_{86}Rn +\ ^{4}_{2}He$ (1620 yr)

$^{222}_{86}Rn \longrightarrow\ ^{218}_{84}Po +\ ^{4}_{2}He$ (3.8 days)

$^{218}_{84}Po \longrightarrow\ ^{214}_{82}Pb +\ ^{4}_{2}He$ (3 min)

$^{214}_{82}Pb \longrightarrow\ ^{214}_{83}Bi +\ ^{0}_{-1}e$ (27 min)

$^{214}_{83}Bi \longrightarrow\ ^{214}_{84}Po +\ ^{0}_{-1}e$ (20 min)

$^{214}_{84}Po \longrightarrow\ ^{210}_{82}Pb +\ ^{4}_{2}He$ (2×10^{-4} s)

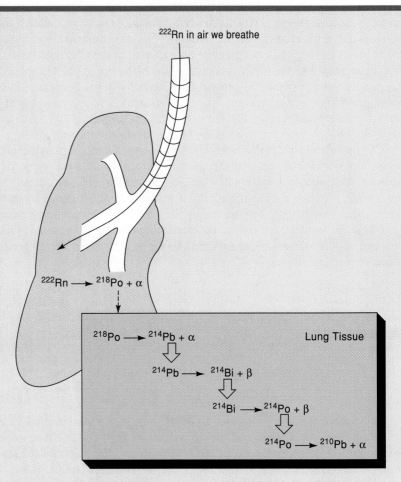

Radon gas in the air is taken into the lungs, where a small amount of the isotope decays to produce $^{218}_{84}Po$, which is deposited in the lung tissue and continues to decay and emit radioactivity.

If we look carefully at this decay series, we will note that most of the products [polonium (Po), lead (Pb), and bismuth (Bi)] are either metals or metalloids. As such, they are nongaseous atoms of limited mobility. The radiation from the decay of these materials is dangerous, but the materials remain in the rock formations and away from humans. In contrast, radon is a noble gas of very low reactivity. Thus, it can diffuse out of mineral and soil formations and into the atmosphere. A careful analysis of the atmosphere reveals low but detectable amounts of radon in the air we breathe. Radon has a short half-life (about 4 days), and some radon atoms decay to $^{218}_{84}Po$ while in the lungs. Polonium is not a gas and is therefore trapped in the lung tissue. Subsequent decays to give $^{214}_{82}Pb$, $^{214}_{83}Bi$, $^{214}_{84}Po$, and $^{210}_{82}Pb$ produce local irradiation of tissues that can promote tumor formation, as the figure shows.

   Radon and the products of its decay appear to be the largest single source of routine human exposure to radioactivity, much larger than doses due to nuclear power plants, X-ray diagnoses, or fallout from nuclear testing. Typically, only uranium miners, people receiving radiation therapy for cancer, or people associated with a nuclear mishap (like Three Mile Island or Chernobyl) receive doses of radioactivity larger than those received from radon exposure.

   For many years, scientists assumed that radon concentrations were naturally so low as to pose no threat to human life. Then, in 1984, came the discovery of homes in eastern Pennsylvania in which the concentration of radioactive radon gas was 1000 times the normal level. In some cases the radiation exposure suffered by the homes' inhabitants was comparable to the exposure received by underground uranium miners. The biggest problems occur in areas where the natural abundance of uranium is higher than normal. Homes in Scandinavian countries have radon concentrations that are typically twice those found in U.S. homes.

Since 1984, studies have explored the causes of severe radon exposure. It appears that radon diffuses from the earth into structures through minute cracks and gaps in the foundations and walls. Whether a large concentration of radon builds up in a home depends on the ventilation of the home, the season, the amounts of dust particles in the air, and the materials from which the home is constructed.

A truly systematic study of American homes has not yet been done. Such a study would provide a better understanding of the geographical areas that have higher risks and help to establish strategies for identifying and fixing homes that have unacceptably high radon levels.

Exposure to radiation from radon gas in some homes can be comparable to the exposure suffered by uranium miners.

✓ Summary

Nuclear reactions are chemical reactions that involve the nuclei of atoms. Changes in the nucleus of an atom usually result in the formation of a new element and are always accompanied by *radioactivity*. Radioactivity is the emission of certain particles (alpha or beta) and gamma rays from a nucleus as it undergoes a nuclear reaction. Only specific isotopes of certain elements are radioactive. *Natural radioactivity* is emitted by naturally occurring radioactive substances (Section 19.1).

When a radioactive isotope *decays*, it produces a different isotope of an element. Often, this product is another radioactive isotope. In this way, a *decay series* can occur in which a whole sequence of radioactive decay processes finally end with a stable (nonradioactive) isotope. From a knowledge of the symbols for nuclei and the various radiations (alpha, beta, and gamma) involved, we can write balanced equations for these nuclear reactions (Section 19.2).

By bombarding nuclei with fast-moving particles (other nuclei or neutrons) in a cyclotron, scientists can induce nuclear reactions and may often produce isotopes of elements that do not occur in nature. These new isotopes of elements decay through *positron emission* and *electron capture* as well as alpha and beta decay (Section 19.3).

The rate of decay of any radioactive isotope is independent of external conditions and depends only on the identity of the isotope itself. A quantitative expression of this rate is the *half-life*, the time required for one-half of any given mass of the radioactive isotope to decay (Section 19.4).

There are many practical uses of radioactivity, including the dating of objects (both living and nonliving), medical imaging, radiolabeling, and irradiating foods (Section 19.5).

Nuclear fission is the splitting of an atomic nucleus into two lighter nuclei and several neutrons. A small amount of mass is converted to heat energy in the process. The atomic bomb is based on nuclear fission (Section 19.6). Nuclear fission has also been used to generate power in nuclear power plants (Section 19.8).

Nuclear fusion is the combination of two atomic nuclei to form a heavier nucleus. Again, a small amount of mass is converted to heat energy during the fusion process. The hydrogen bomb is an example of nuclear fusion (Section 19.7).

Exercises

1. Define or explain the following terms (the number in parentheses refers to the section number in the text where the term is mentioned).

    a. nuclear reaction (19.1)
    b. radioactivity (19.1)
    c. natural ratioactivity (19.1)
    d. transmutation (19.2)
    e. radioactive decay (19.2)
    f. alpha particle (19.2)
    g. beta particle (19.2)
    h. decay series (19.2)
    i. gamma ray (19.2)
    j. cyclotron (19.3)
    k. artificial radioactivity (19.3)
    l. positron (19.3)
    m. electron capture (19.3)
    n. synthetic elements (19.3)
    o. transuranium elements (19.3)
    p. half-life (19.4)
    q. curie (Ci) (19.4)
    r. becquerel (Bq) (19.4)
    s. nuclear fission (19.6)
    t. chain reaction (19.6)
    u. critical mass (19.6)
    v. nuclear fusion (19.7)

2. Distinguish between:

    a. an alpha particle and a beta particle
    b. a beta particle and a positron
    c. natural and artificial radioactivity
    d. nuclear fission and nuclear fusion

Equations for Natural Radioactivity (See Section 19.2)

3. Give the symbols of the following particles or rays:

    a. alpha
    b. beta
    c. gamma

4. List a naturally occurring radioactive decay series and give its final nonradioactive isotope.

Artificial Radioactivity (See Section 19.3)

5. Give the symbols for the following particles:

   a. proton   b. neutron

   c. positron   d. deuteron

6. In your own words, explain the operation of a cyclotron.

Half-Life and Measurement of Radioactivity (See Section 19.4)

7. In your own words, explain the operation of a Geiger–Müller counter.

Peaceful Uses of Nuclear Reactions (See Section 19.8)

8. In your own words, list and explain the operation of the three essential parts of a nuclear fission power plant.

 Problems

Equations for Natural Radioactivity (See Section 19.2)

Refer to the periodic table on the inside front cover of this text to answer these questions.

9. Complete and balance a nuclear chemical equation for each of the following nuclear reactions:

   a. Krypton-87 (Kr) decays by beta emission

   b. Curium-240 (Cm) decays by alpha emission

   c. Uranium-232 (U) decays by alpha emission

   d. Zinc-71 decays by beta emission

   e. Silicon-32 decays by beta emission

   f. Americium-243 (Am) decays by alpha emission

   g. Cobalt-60 decays by beta emission

   h. Chlorine-36 is formed in sodium chloride after a nuclear fission explosion. Chlorine-36 decays by beta emission

10. Phosphorus-32 is formed in milk, butter, seafood, baby food, and pork and beans if the food is exposed to a nuclear fission explosion one-quarter of a mile from ground zero. Phosphorus-32 decays by beta emission. Complete and balance a nuclear chemical equation for this nuclear reaction.

Artificial Radioactivity (See Section 19.3)

Refer to the periodic table on the inside front cover of this text to answer these questions.

11. Complete and balance a nuclear chemical equation for each of the following nuclear reactions:

   a. Praseodymium-140 (Pr) decays by positron emission

   b. Oxygen-16 plus a neutron results in the formation of another element and the release of an alpha particle

   c. Boron-10 plus a neutron results in the formation of another element and the release of an alpha particle

   d. Zinc-65 undergoes electron capture to give a new element

   e. Beryllium-9 plus a proton results in the formation of another element and the release of an alpha particle

    f. Iron-65 undergoes electron capture to give a new element

    g. Neon-18 decays by positron emission

    h. Copper-59 decays by positron emission

    i. Einsteinium-253 (Es) plus an alpha particle results in the formation of another element and the release of a neutron

    j. Cadmium-113 in a nuclear reactor absorbs a neutron to form an isotope of cadmium and gamma rays

    k. Lithium-7 plus a proton results in the formation of another element and the release of a neutron

    l. Selenium-75 undergoes electron capture to give a new element

12. Palladium-108 (Pd) is bombarded with a high-speed alpha particle and a proton is emitted in the transmutation process. Complete and balance a nuclear chemical equation for this nuclear reaction.

13. Tritium ($^3_1$H), a radioactive isotope of hydrogen, is prepared by bombarding lithium-6 with a neutron. Complete and balance a nuclear chemical equation for this nuclear reaction.

14. Nickel-58 is bombarded with a proton, and an alpha particle is emitted in the transmutation process. Complete and balance a nuclear chemical equation for this nuclear reaction.

Half-life and Measurement of Radioactivity (See Section 19.4)

15. Actinium-226 has a half-life of 29 h. If 100.0 mg of actinium-226 decays over a period of 58 h, how many milligrams of actinium-226 will remain?

16. Thallium-201 has a half-life of 73 h. If 4.00 mg of thallium-201 decays over a single period of 6.00 days and 2 h, how many milligrams of thallium-201 will remain?

17. Sodium-25 was to be used in an experiment, but it took 3.00 min to get the sodium from the reactor to the laboratory vessel. If 5.000 mg of sodium-25 was removed from the reactor, how many milligrams of sodium-25 will be placed in the reaction vessel if the half-life of sodium-25 is $6\overline{0}$ s?

18. The half-life of a radioactive isotope, X, is 2.00 years. How many years will it take a 4.000-g sample of X to decay and have only 0.500 g of X remain?

19. Selenium-83 has a half-life of 23 min. How many hours will it take a 10.00-mg sample of selenium-83 to decay and have only 1.25 mg of it remain?

20. Tin-113 is believed to be formed on tin cans (used to store food) in a nuclear fission explosion. If 8.00×10^{-6} g of tin-113 is formed on a tin can following a nuclear fission explosion, how many days will it take to have just 1.00×10^{-6} g of tin-113 if the half-life of tin-113 is 115 days?

General Problems

21. Following a nuclear fission explosion, the sodium in glass containers formed sodium-24. Sodium-24 decays with emission of a beta particle and has a half-life of 15.0 h.

    a. Complete and balance a nuclear chemical equation for this nuclear reaction.

    b. If 5.000×10^{-6} g of sodium-24 is formed in a glass container, how many grams will remain in 2.5 days?

    c. What percent of the original sodium-24 will remain after 2.5 days?

22. Zinc-65 accounts for 50% or more of the total radioactivity in fish exposed to a nuclear fission explosion. Zinc-65 decays by positron emission and has a half-life of 244 days.

    a. Complete and balance a nuclear chemical equation for this nuclear reaction.

    b. If 2.00 μg of zinc-65 forms in a fish, how many days will it take to have just 0.50 μg of zinc-65? Note that the 0.50 μg of zinc-65 remains whether the fish lives or dies, but if it is caught and eaten by people, the zinc-65 will reside in the people who eat the fish.

23. A 4.00×10^{-6} percent solution of sodium-24 chloride is prepared at 9:00 A.M. on Tuesday. What is the percent of sodium-24 chloride at 3:00 P.M. on Wednesday if sodium-24 has a half-life of 15.0 h?

24. A 1.500×10^{-6} M solution of cesium-137 chloride was prepared by a research chemist in 1966. What is the molar concentration of cesium-137 chloride in 1996? (The half-life of cesium-137 is 30.0 years.)

25. Thermonuclear weapons produce new radioactive isotopes as fallout when they detonate. Strontium-90, one of the most biologically hazardous isotopes, is synthesized in the early moments of the detonation from bromine-90 by a succession of three beta emissions. Complete and balance nuclear chemical equations for the three steps in this process during which bromine-90 is converted to strontium-90. Strontium-90 has a long half-life of 28.1 years, and it tends to localize in the bones of living organisms. It then remains in the bones and damages local tissue as it slowly undergoes beta emission.

26. The isotope of technetium that is used in imaging studies, ^{99}Tc, is prepared in the following way. Molybdenum-98 is bombarded with neutrons to give molybdenum-99, which decays by beta emission to give ^{99}Tc with a half-life of 67 h. Complete and balance nuclear equations for each of these processes.

27. Gallium-67 is an excellent radioactive isotope for locating the exact position of soft tumors by imaging techniques. It is prepared by the nuclear reaction between a zinc-68 atom and a proton. The products are gallium-67 and neutrons.

    a. How many neutrons are liberated?

    b. What is the balanced nuclear equation for this process?

✓ Chapter Quiz 19

You may use the periodic table.

1. Complete and balance the following nuclear equations:

    a. Uranium-232 decays by alpha emission

    b. Lead-201 decays by positron emission

    c. Copper-63 plus a bombarding particle results in the formation of a neutron, sodium-24, and potassium-39

2. An isotope of zinc, zinc-72, has a half-life of 46.5 h. If 2.000 mg of zinc-72 decays over a period of 7 days and 18 h, how many milligrams of zinc-72 will remain?

3. An isotope of iridium (Ir), iridium-192, is used in treating brain tumors by implantation techniques. The isotope has a half-life of 74 days. How many days will it take for a 5.000-mg sample of iridium-192 to decay to 1.250 mg?

This chapter concludes your study of basic chemistry. Throughout this book, our goal has been to introduce you to what we consider the fascinating science of chemistry and to point out that *many people* from diverse backgrounds, nationalities, and races have been able to *work together* to solve some of the chemical problems related to our existence on the planet Earth.

Technetium (Symbol: Tc)

The Element TECHNETIUM: Noninvasive Diagnosis

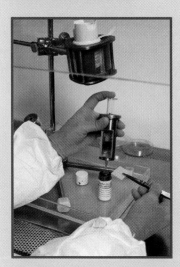

Solution of sodium pertechnetate (NaTcO₄) are used extensively in medical diagnosis. The picture shows the preparation of an injectable technetium solution.

Name: The name derives from the Greek *technetos*, which means "artificial."

Appearance: Technetium metal is not commonly available, but it is a silver-gray powder. The most useful form of technetium is the anion technetate, TcO_4^-, which is colorless in solution.

Occurrence: Technetium is found in the environment only in trace amounts as a by-product of natural ^{235}U fission. It is most commonly found as a component of spent nuclear reactor fuel.

Source: Most technetium is used as a radioactive tracer in medicine (see picture). The ^{99}Tc is prepared by the beta decay of ^{99}Mo, which is obtained from nuclear reactors or by bombarding ^{98}Mo with neutrons. In all cases, the technetium is present as $NaTcO_4$, and the molybdenum is present as Na_2MoO_4.

$$^{98}_{42}Mo + {}^{1}_{0}n \longrightarrow {}^{99}_{42}Mo$$

$$^{99}_{42}Mo \longrightarrow {}^{99}_{43}Tc + {}^{0}_{-1}e$$

Common Uses: Technetium-99 is the single most important radioactive isotope in medical diagnosis. It can be used to image the brain, heart, thyroid, stomach, liver, lung, bone marrow, spleen, and kidney. The development of radioactive tracer technology may be as important to medicine as the invention of the microscope. It is a relatively noninvasive way of looking at an organ and often gives information that can be obtained in no other practical way.

The nuclear chemistry involved is actually a bit more complex than the chapter suggests. The beta decay of ^{99}Mo to ^{99}Tc actually gives a species called ^{99m}Tc, where the m stands for *metastable*. This means that the ^{99}Tc nucleus is formed in a high-energy or metastable state. This metastable state decays to normal ^{99}Tc with a half-life of 6 h by gamma decay. It is these gamma rays that are detected in the imaging process. The normal ^{99}Tc that results decays by beta emission to ruthenium-99 (^{99}Ru) with a half-life of 212,000 years.

$$^{99m}_{43}\text{Tc} \longrightarrow {}^{99}_{43}\text{Tc} + \gamma$$

$$^{99}_{43}\text{Tc} \longrightarrow {}^{99}_{44}\text{Ru} + {}^{0}_{-1}\text{e}$$

Unusual Facts: Technetium had been predicted on the basis of the periodic table, but was unknown until 1937, when C. Perrier and E. G. Segrè prepared small amounts by bombarding molybdenum with deuterons. Technetium was thus the first unknown element prepared by artificial means.

APPENDIXES

The Interconversion of Units

There are two major systems of units in the world, the metric system and the English-based system. The English-based system is primarily used in the United States, while the metric system is used throughout almost all the rest of the world. The metric system is gradually replacing English-based units in the United States. The International System of Units (SI) is a modification of the metric system and has been adopted as the official system of units in the scientific community.

Interconversion in the International System of Units

We have introduced the International System of Units (SI) in this text; the scientific community uses this system to a limited extent now, and in a few years it probably will be the only system in use. The SI is derived from seven *base units*, as Table I.1 shows.

In addition to these base units, there are two *supplementary units*, the radian (rad) and steradian (sr), used to define angular measurement.

Multiple and submultiple *prefixes* indicate orders of magnitude. These prefixes define either a fractional or a multiple value of the base unit; thus, a kilometer is 1000 meters, and a millimeter is 0.001 (or 10^{-3}) meter. Table I.2 shows these prefixes and their orders of magnitude.

A series of *derived units* that defines various physical quantities used in scientific measurements are derived from the seven base and two supplementary units. Table I.3 shows some of these.

Units currently being used in chemical measurements (and used in this text) that are not *exactly* defined in terms of SI units are the *atmosphere*, *torr*, *mm Hg*, and

TABLE I.1	Base Units	
QUANTITY MEASURED	UNIT NAME	SI SYMBOL FOR UNIT
Length	meter	m
Mass	kilogram	kg
Time	second	s
Electric current	ampere	A
Thermodynamic temperature	kelvin	K
Amount of substance	mole	mol
Luminous intensity	candela	cd

TABLE I.2		
SI Prefixes		

		SI
FACTOR	PREFIX	SYMBOL
10^{12}	tera	T
10^{9}	giga	G
10^{6}	mega	M
10^{3}	kilo	k
10^{2}	hecto[a]	h
10^{1}	deka[a]	da
10^{-1}	deci[a]	d
10^{-2}	centi[a]	c
10^{-3}	milli	m
10^{-6}	micro	μ
10^{-9}	nano	n
10^{-12}	pico	p
10^{-15}	femto	f
10^{-18}	atto	a

[a] Avoid these prefixes where possible.

TABLE I.3	**Derived Units**		
PHYSICAL QUANTITY	UNIT	SI SYMBOL	DEFINITION
Acceleration	meter/second^2	m/s^2	m/s^2
Area	meter^2	m^2	m^2
Density	kilogram/meter^3	kg/m^3	kg/m^3
Electric capacitance	farad	F	A·s/V
Electric potential difference	volt[a]	V	J/(A·s) = W/A
Electric resistance	ohm	Ω	V/A = kg·m^2/(s^3·A^2)
Energy	joule	J	N·m = kg·m^2/s^2
Force	newton	N	kg·m/s^2
Power	watt	W	J/s = kg·m^2/s^3
Quantity of electricity	coulomb	C	A·s
Pressure	pascal	Pa	N/m^2 = kg/(m·s^2)
Quantity of heat	joule	J	N·m = kg·m^2/s^2
Specific heat	joule/(kilogram·kelvin)	J/(kg·K)	J/(kg·K)
Velocity	meter/second	m/s	m/s
Volume[b]	cubic meter	m^3	m^3

[a] Also the unit for expressing electromotive force.
[b] In 1964, the liter was adopted as a special name for the cubic decimeter (dm^3), but its use for measuring extremely precise volumes is discouraged.

calorie. The SI committee recommends that these units be abandoned; however, many chemists will probably use them for some time.

Table I.4 lists conversion factors that may be used to convert from the non-SI unit to the recognized SI unit.

Interconversion and English-Based Units

English-based units consist of the ounce, pound, and ton (mass); the fluid ounce, pint, quart, and gallon (volume); the inch, foot, yard, and mile (length); and the second (time). Table I.5 summarizes English-based units of measurements.

TABLE I.4	**Selected Conversion Factors**	
TO CONVERT FROM:	TO:	MULTIPLY BY:
calorie (cal)	joule (J)	4.184
atmosphere (atm)	pascal (Pa)	1.013×10^5
torr (torr)	pascal (Pa)	1.333×10^2
inch (in.)	meter (m)	2.54×10^{-2}
pound-mass (lbm)	kilogram (kg)	4.536×10^{-1}

TABLE I.5	Summary of English-Based Units of Measurements		

MASS	LENGTH	VOLUME
16 ounces (oz) = 1 pound (lb)	12 inches (in.) = 1 foot (ft)	16 fluid ounces (fl oz) = 1 pint (pt)
2000 pounds = 1 ton	3 feet = 1 yard (yd)	2 pints = 1 quart (qt)
	5280 feet = 1 mile (mi)	4 quarts = 1 gallon (gal)

To convert from the metric system to English-based units, and vice versa, we need to know certain conversion factors (see Table I.6). You must either memorize these factors or have them given to you to solve problems involving these units.

EXAMPLE I.1   Convert 385 g to pounds.

SOLUTION   Using the conversion factor 454 g = 1 lb, solve the problem as follows:

$$385 \, g \times \frac{1 \, \text{lb}}{454 \, g} = 0.848 \, \text{lb} \qquad Answer$$

EXAMPLE I.2   Convert 1.00 gal to liters.

SOLUTION   To use the conversion factor 1 L = 1.06 qt, convert the gallon to quarts using 4 qt = 1 gal. The solution is

$$1.00 \, \text{gal} \times \frac{4 \, \text{qt}}{1 \, \text{gal}} \times \frac{1 \, \text{L}}{1.06 \, \text{qt}} = 3.77 \, \text{L} \qquad Answer$$

EXAMPLE I.3   Convert 3.00 ft to meters.

SOLUTION   To use the factor 2.54 cm = 1 in., convert the feet to inches using 12 in. = 1 ft. Then convert the centimeters to meters using the metric conversion 100 cm = 1 m (see Section 2.1). The solution is

$$3.00 \, \text{ft} \times \frac{12 \, \text{in.}}{1 \, \text{ft}} \times \frac{2.54 \, \text{cm}}{1 \, \text{in.}} \times \frac{1 \, \text{m}}{100 \, \text{cm}} = 0.914 \, \text{m} \qquad Answer$$

TABLE I.6	Metric–English Unit Equivalents	

DIMENSION	METRIC UNIT	ENGLISH EQUIVALENT
Mass[a]	454 grams (g)	= 1 pound (lb)
Volume	1 liter (L)	= 1.06 quarts (qt)
Length	2.54 centimeters (cm)	= 1 inch (in.)
Time	1 second (s)	= 1 second (s)

[a] Although gram is a unit of mass and pound is a unit of weight, the two units are used interchangeably in most calculations in chemistry.

EXAMPLE I.4 A box has the following dimensions: 15.0 in., 20.0 cm, and 2.00 ft. Calculate its volume in (a) cubic centimeters and (b) cubic inches.

SOLUTION

a. Convert the 15.0 in. and 2.00 ft to centimeters and then calculate the volume in cubic centimeters as side $\times$ side $\times$ side:

$$15.0 \text{ in.} \times \frac{2.54 \text{ cm}}{1 \text{ in.}} \times 20.0 \text{ cm} \times 2.00 \text{ ft} \times \frac{12 \text{ in.}}{1 \text{ ft}} \times \frac{2.54 \text{ cm}}{1 \text{ in.}}$$

$$= 46{,}500 \text{ cm}^3 \text{ (to three significant digits)} \qquad Answer$$

Note that $cm^1 \times cm^1 \times cm^1 = cm^3$, with the exponents added algebraically as in the multiplication of exponential numbers (see Section 2.7).

b. Convert the 20.0 cm and 2.00 ft to inches and then calculate the volume in cubic inches:

$$15.0 \text{ in.} \times 20.0 \text{ cm} \times \frac{1 \text{ in.}}{2.54 \text{ cm}} \times 2.00 \text{ ft} \times \frac{12 \text{ in.}}{1 \text{ ft}}$$

$$= 2830 \text{ in.}^3 \text{ (to three significant digits)} \qquad Answer$$

EXAMPLE I.5 Carry out the following conversions:

a. a density of 2.60 g/mL to lb/gal
b. a density of 5.00 g/cm^3 to lb/ft^3

SOLUTION

a. Convert the mass in grams to pounds using the factor 1 lb = 454 g and then convert the volume in milliliters to gallons using the factors 1000 mL = 1 L, 1 L = 1.06 qt, and 4 qt = 1 gal. The solution is

$$\frac{2.60 \text{ g}}{mL} \times \frac{1 \text{ lb}}{454 \text{ g}} \times \frac{1000 \text{ mL}}{1 \text{ L}} \times \frac{1 \text{ L}}{1.06 \text{ qt}} \times \frac{4 \text{ qt}}{1 \text{ gal}} = 21.6 \text{ lb/gal} \qquad Answer$$

b. Convert the mass in grams to pounds using the factor 1 lb = 454 g and then convert the volume in cubic centimeters to cubic feet; this operation involves $(1 \text{ in.})^3 = (2.54 \text{ cm})^3$ and then $(12 \text{ in.})^3 = (1 \text{ ft})^3$. To perform this operation, multiply the 2.54 by itself three times, that is, 2.54 $\times$ 2.54 $\times$ 2.54, and do the same for the 12, that is, 12 $\times$ 12 $\times$ 12 (see Section 2.7, Positive Powers of Exponential Numbers). The solution is

$$\frac{5.00 \text{ g}}{cm^3} \times \frac{1 \text{ lb}}{454 \text{ g}} \times \frac{(2.54)^3 \text{ cm}^3}{1 \text{ in.}^3} \times \frac{(12)^3 \text{ in.}^3}{1 \text{ ft}^3} = 312 \text{ lb/ft}^3 \qquad Answer$$

✓ Problems

1. Carry out each of the following conversions:

   a. 2.00 kg to pounds
   b. 3.25 lb to kilograms
   c. 25$\overline{0}$ mL to pints
   d. 6.00 qt to liters
   e. 16.8 in. to meters
   f. 6.00 m to feet

2. Mt. Everest on the Nepal–Tibet border is 29,028 ft above sea level. What is this height in meters to three significant digits?

3. A box has the following dimensions: 25.0 cm, 11.0 in., and 2.00 ft. Calculate its volume in (a) cubic centimeters and (b) cubic inches.

4. Carry out the following conversions:

   a. a density of 3.65 g/mL to lb/gal

   b. a density of 295 lb/ft^3 to g/mL

5. Calculate the density in lb/ft^3 of a piece of metal having a mass of 246 kg and the following dimensions: 15.0 cm, 30.0 in., and 0.500 ft.

APPENDIX II   Your Calculator

Modern technology has made it possible for you to purchase highly reliable and convenient calculators at a modest cost. As a result, longhand calculations, slide rules, and logarithm tables have become a thing of the past. Calculators equipped with the functions $+$, $-$, $\times$, $\div$, $\sqrt{}$, log x, and 10^x or y^x are sufficient for calculations in this text.

Although a number of calculator designs are available, we will describe the use of the two most common types. The first is most easily recognized by the presence of an $=$ key. The second most common type has an ENTER key.

Two notes of caution: First, calculators are fast and convenient, and they do not make mistakes. You, however, can make a mistake in using a calculator. Always check to see if your answer is reasonable. For example, let's say you multiply 8.2×2.3 and get 188.6. Is this reasonable? Well, you know that $8 \times 2 = 16$, and so 8.2×2.3 should be a bit more than 16. Your answer, 188.6, *makes no sense*—you must have made an error in using the calculator. Repeat your calculation if you have any doubts. The correct answer is 18.86. In the original calculation, you almost certainly misplaced a decimal point and entered 82×2.3 or 8.2×23!

Second, calculators give you far more significant digits than you can justify. Always round off your answers to the correct number of significant digits when you are done. *Do not record the answer shown on the calculator without considering what the correct number of significant digits should be.* In the following examples, the tables show the numbers as displayed on the calculator. Round off any displayed number before presenting it as an answer.

Entering Numbers, Exponential Notation

Entering numbers into a calculator is very simple. You just key them in as they appear on the paper, being sure to enter any decimals points in the correct position. Practice a few times by entering 12.41, 12000.1, 273, and 0.000572.

You might wish to enter the second and fourth numbers in the last paragraph in exponential or scientific notation. To do this, you will need to identify two keys on your particular brand of calculator:

▶ Exponent key: On the = type of calculator, two common labels for this key are EE (**E**nter **E**xponent) and EXP (**EXP**onent). On the ENTER type of calculator, the key is often labeled EEX (**E**nter **EX**ponent). Consult the manual for your calculator if you have any doubts.

▶ Change sign key: On the = type of calculator, the most common label for this key is +/−. On the ENTER type of calculator, the key is often labeled CHS (**CH**ange **S**ign).

In either case, you key in the number and then press the exponent key. Two zeros should appear to the right of the entered number. To complete the operation, just key in the desired exponent (or power) to which 10 is to be taken, and, if the exponent to which 10 is taken is a *negative* number, press the change sign key after the exponent has been keyed in. Examples are shown in the following table:

= or ENTER Type			
DESIRED NUMBER	PRESS	ON DISPLAY	
1.765×10^4	1.765	*1.765*	
	EXP, EE, or EEX	*1.765*	*00*
	4	*1.765*	*04*
1.241×10^{-7}	1.241	*1.241*	
	EXP, EE, or EEX	*1.241*	*00*
	7	*1.241*	*07*
	+/− or CHS	*1.241*	*−07*

Addition, Subtraction, Multiplication, and Division

To perform an arithmetic operation, you must press a series of keys. The two basic calculator types differ slightly as shown below:

= TYPE	ENTER TYPE
On this type of calculator, you key in the operation as it is written. To determine 2 + 3 you press 2, then +, then 3, and finally =. Examples of addition, subtraction, multiplication, and division are given below.	On this type of calculator, you enter the first number, key in the second number, and *then* press the appropriate operation key. To determine 2 + 3 you press the following keys in order: 2, enter, 3, and +. Examples of addition, subtraction, multiplication, and division are given below.

= Type			ENTER Type		
DESIRED OPERATION	PRESS	ON DISPLAY	DESIRED OPERATION	PRESS	ON DISPLAY
176.5 + 12.41	176.5	*176.5*	176.5 + 12.41	176.5	*176.5*
	+	*176.5*		ENTER	*176.5*
	12.41	*12.41*		12.41	*12.41*
	=	*188.91*		+	*188.91*
176.5 − 12.41	176.5	*176.5*	176.5 − 12.41	176.5	*176.5*
	−	*176.5*		ENTER	*176.5*
	12.41	*12.41*		12.41	*12.41*
	=	*164.09*		−	*164.09*
176.5 × 12.41	176.5	*176.5*	176.5 × 12.41	176.5	*176.5*
	×	*176.5*		ENTER	*176.5*
	12.41	*12.41*		12.41	*12.41*
	=	*2190.365*		×	*2190.365*
176.5 ÷ 12.41	176.5	*176.5*	176.5 ÷ 12.41	176.5	*176.5*
	÷	*176.5*		ENTER	*176.5*
	12.41	*12.41*		12.41	*12.41*
	=	*14.22240129*		÷	*14.22240129*

Now try the following examples to check your technique. The answers are given as they should be displayed on your calculator and without regard to significant digits.

$$39.21 + 9.57 = 48.78 \qquad 39.21 - 9.57 = 29.64$$

$$1.021 + 0.24 = 1.261 \qquad 1.021 - 0.24 = 0.781$$

$$39.21 \times 9.57 = 375.2397 \qquad 39.21 \div 9.57 = 4.09717868$$

$$1.021 \times 0.24 = 0.24504 \qquad 1.021 \div 0.24 = 4.25416667$$

Series or Chain Calculations

Many times, problems in this text require that you perform a series or chain of arithmetic operations. An example of this type of calculation is summing up a list of numbers or performing a series of multiplications and divisions. You should generally perform these operations in the calculator *without* removing any intermediate answers. This approach minimizes round-off errors and copying errors.

= Type				ENTER Type			
DESIRED OPERATION	PRESS	ON DISPLAY		DESIRED OPERATION	PRESS	ON DISPLAY	
12.3 + 35.6 − 1.55	12.3	*12.3*		12.3 + 35.6 − 1.55	12.3	*12.3*	
	+	*12.3*			ENTER	*12.3*	
	35.6	*35.6*			35.6	*35.6*	
	=	*47.9*			+	*47.9*	
	−	*47.9*			ENTER	*47.9*	
	1.55	*1.55*			1.55	*1.55*	
	=	*46.35*			−	*46.35*	
$\frac{12.3 \times 35.6}{1.55 \times 2.68}$	12.3	*12.3*		$\frac{12.3 \times 35.6}{1.55 \times 2.68}$	12.3	*12.3*	
	×	*12.3*			ENTER	*12.3*	
	35.6	*35.6*			35.6	*35.6*	
	=	*437.88*			×	*437.88*	
	÷	*437.88*			ENTER	*437.88*	
	1.55	*1.55*			1.55	*1.55*	
	=	*282.5032258*			÷	*282.5032258*	
	÷	*282.5032258*			ENTER	*282.5032258*	
	2.68	*2.68*			2.68	*2.68*	
	=	*105.4116514*			÷	*105.4116514*	

Now try the following examples to check your technique. The answers are given as they should be displayed on your calculator and without regard to significant digits.

$$39.21 - 9.57 + 47.09 - 2.675 = 74.055$$

$$-39.21 + 9.57 - 47.09 - 2.675 = -79.405 \quad \text{(\textit{Hint:} Key in 39.21 and use the}$$
$$\text{CHS or } +/- \text{ key to get } -39.21.\text{)}$$

$$\frac{39.21 \times 9.57}{47.09 \times 2.675} = 2.97890260 \qquad \frac{39.21 \times 2.675}{47.09 \times 9.57} = 0.23274481$$

Square Roots, Squares, Logarithms, and Antilogarithms

Four other functions can be useful in working the problems in this book: taking a square root, squaring a number, taking a logarithm of a number, and taking an antilogarithm of a number. There is usually no difference in these operations from one calculator to the next, as long as you are able to identify the correct key for each operation. Some calculators designate two or more operations to a given key and require two keystrokes to accomplish these operations. In such cases, a shift or function key (often color-coded with the label) should precede the desired operation.

Check your manual if you are unsure. The typical key labels for these operations are

Square root key: $\sqrt{}$
Squaring a number: x^2
Taking a logarithm: log
Taking an antilogarithm: 10^x or y^x

The following table illustrates examples of using these keys both with regular numbers and exponential numbers.

= or ENTER TYPE

DESIRED OPERATION	PRESS	ON DISPLAY
$\sqrt{176.5}$	176.5	176.5
	$\sqrt{}$	13.28533026
$(12.41)^2$	12.41	12.41
	x^2	154.0081
log (176.5)	176.5	176.5
	log	2.24674471
antilog (12.41)	12.41	12.41
	10^x	2.570395 12
	or: 10	10
	y^x	10
	12.41	12.41
	=	2.570395 12
$\sqrt{1.241 \times 10^{-7}}$	1.241	1.241
	EXP, EE, or EEX	1.241 00
	7	1.241 07
	+/− or CHS	1.241 −07
	$\sqrt{}$	0.000352278
log (1.765×10^4)	1.765	1.765
	EXP, EE, or EEX	1.765 00
	4	1.765 04
	log	4.24674471

Now try the following examples to check your technique. The answers are given as they should be displayed on your calculator and without regard to significant digits.

$$\sqrt{72.471} = 8.51299007 \qquad (72.471)^2 = 5{,}252.045841$$
$$\log 72.471 = 1.86016425 \qquad \text{antilog } 72.471 = 2.958012 \times 10^{72}$$

Atomic Number	Isotope	Percent Abundance	Atomic Number	Isotope	Percent Abundance
1	$^{1}_{1}H$	99.98	18	$^{36}_{18}Ar$	0.34
	$^{2}_{1}H$	0.02		$^{38}_{18}Ar$	0.06
2	$^{3}_{2}He$	Trace		$^{40}_{18}Ar$	99.60
	$^{4}_{2}He$	100.00	19	$^{39}_{19}K$	93.10
3	$^{6}_{3}Li$	7.42		$^{40}_{19}K$	0.01
	$^{7}_{3}Li$	92.58		$^{41}_{19}K$	6.88
4	$^{9}_{4}Be$	100.00	20	$^{40}_{20}Ca$	96.97
5	$^{10}_{5}B$	19.6		$^{42}_{20}Ca$	0.64
	$^{11}_{5}B$	80.4		$^{43}_{20}Ca$	0.14
6	$^{12}_{6}C$	98.89		$^{44}_{20}Ca$	2.06
	$^{13}_{6}C$	1.11		$^{46}_{20}Ca$	Trace
7	$^{14}_{7}N$	99.63		$^{48}_{20}Ca$	0.18
	$^{15}_{7}N$	0.37	21	$^{45}_{21}Sc$	100.00
8	$^{16}_{8}O$	99.76	22	$^{46}_{22}Ti$	7.93
	$^{17}_{8}O$	0.04		$^{47}_{22}Ti$	7.28
	$^{18}_{8}O$	0.20		$^{48}_{22}Ti$	73.94
9	$^{19}_{9}F$	100.00		$^{49}_{22}Ti$	5.51
10	$^{20}_{10}Ne$	90.92		$^{50}_{22}Ti$	5.34
	$^{21}_{10}Ne$	0.26	23	$^{50}_{23}V$	0.24
	$^{22}_{10}Ne$	8.82		$^{51}_{23}V$	99.76
11	$^{23}_{11}Na$	100.00	24	$^{50}_{24}Cr$	4.31
12	$^{24}_{12}Mg$	78.70		$^{52}_{24}Cr$	83.76
	$^{25}_{12}Mg$	10.13		$^{53}_{24}Cr$	9.55
	$^{26}_{12}Mg$	11.17		$^{54}_{24}Cr$	2.38
13	$^{27}_{13}Al$	100.00	25	$^{55}_{25}Mn$	100.00
14	$^{28}_{14}Si$	92.21	26	$^{54}_{26}Fe$	5.82
	$^{29}_{14}Si$	4.70		$^{56}_{26}Fe$	91.66
	$^{30}_{14}Si$	3.09		$^{57}_{26}Fe$	2.19
15	$^{31}_{15}P$	100.00		$^{58}_{26}Fe$	0.33
16	$^{32}_{16}S$	95.00	27	$^{59}_{27}Co$	100.00
	$^{33}_{16}S$	0.76	28	$^{58}_{28}Ni$	67.88
	$^{34}_{16}S$	4.22		$^{60}_{28}Ni$	26.23
	$^{36}_{16}S$	0.01		$^{61}_{28}Ni$	1.19
17	$^{35}_{17}Cl$	75.53		$^{62}_{28}Ni$	3.66
	$^{37}_{17}Cl$	24.47		$^{64}_{28}Ni$	1.08

ATOMIC NUMBER	ISOTOPE	PERCENT ABUNDANCE	ATOMIC NUMBER	ISOTOPE	PERCENT ABUNDANCE
29	$^{63}_{29}\text{Cu}$	69.09		$^{94}_{42}\text{Mo}$	9.04
	$^{65}_{29}\text{Cu}$	30.91		$^{95}_{42}\text{Mo}$	15.72
30	$^{64}_{30}\text{Zn}$	48.89		$^{96}_{42}\text{Mo}$	16.53
	$^{66}_{30}\text{Zn}$	27.81		$^{97}_{42}\text{Mo}$	9.46
	$^{67}_{30}\text{Zn}$	4.11		$^{98}_{42}\text{Mo}$	23.78
	$^{68}_{30}\text{Zn}$	18.57		$^{100}_{42}\text{Mo}$	9.13
	$^{70}_{30}\text{Zn}$	0.62	44	$^{96}_{44}\text{Ru}$	5.51
31	$^{69}_{31}\text{Ga}$	60.4		$^{98}_{44}\text{Ru}$	1.87
	$^{71}_{31}\text{Ga}$	39.6		$^{99}_{44}\text{Ru}$	12.72
32	$^{70}_{32}\text{Ge}$	20.52		$^{100}_{44}\text{Ru}$	12.62
	$^{72}_{32}\text{Ge}$	27.43		$^{101}_{44}\text{Ru}$	17.07
	$^{73}_{32}\text{Ge}$	7.76		$^{102}_{44}\text{Ru}$	31.61
	$^{74}_{32}\text{Ge}$	36.54		$^{104}_{44}\text{Ru}$	18.58
	$^{76}_{32}\text{Ge}$	7.76	45	$^{103}_{45}\text{Rh}$	100.00
33	$^{75}_{33}\text{As}$	100.00	46	$^{102}_{46}\text{Pd}$	0.96
34	$^{74}_{34}\text{Se}$	0.87		$^{104}_{46}\text{Pd}$	10.97
	$^{76}_{34}\text{Se}$	9.02		$^{105}_{46}\text{Pd}$	22.23
	$^{77}_{34}\text{Se}$	7.58		$^{106}_{46}\text{Pd}$	27.33
	$^{78}_{34}\text{Se}$	23.52		$^{108}_{46}\text{Pd}$	26.71
	$^{80}_{34}\text{Se}$	49.82		$^{110}_{46}\text{Pd}$	11.81
	$^{82}_{34}\text{Se}$	9.19	47	$^{107}_{47}\text{Ag}$	51.82
35	$^{79}_{35}\text{Br}$	50.54		$^{109}_{47}\text{Ag}$	48.18
	$^{81}_{35}\text{Br}$	49.46	48	$^{106}_{48}\text{Cd}$	1.22
36	$^{78}_{36}\text{Kr}$	0.35		$^{108}_{48}\text{Cd}$	0.88
	$^{80}_{36}\text{Kr}$	2.27		$^{110}_{48}\text{Cd}$	12.39
	$^{82}_{36}\text{Kr}$	11.56		$^{111}_{48}\text{Cd}$	12.75
	$^{83}_{36}\text{Kr}$	11.55		$^{112}_{48}\text{Cd}$	24.07
	$^{84}_{36}\text{Kr}$	56.90		$^{113}_{48}\text{Cd}$	12.26
	$^{86}_{36}\text{Kr}$	17.37		$^{114}_{48}\text{Cd}$	28.86
37	$^{85}_{37}\text{Rb}$	72.15		$^{116}_{48}\text{Cd}$	7.58
	$^{87}_{37}\text{Rb}$	27.85	49	$^{113}_{49}\text{In}$	4.28
38	$^{84}_{38}\text{Sr}$	0.56		$^{115}_{49}\text{In}$	95.72
	$^{86}_{38}\text{Sr}$	9.86	50	$^{112}_{50}\text{Sn}$	0.96
	$^{87}_{38}\text{Sr}$	7.02		$^{114}_{50}\text{Sn}$	0.66
	$^{88}_{38}\text{Sr}$	82.56		$^{115}_{50}\text{Sn}$	0.35
39	$^{89}_{39}\text{Y}$	100.00		$^{116}_{50}\text{Sn}$	14.30
40	$^{90}_{40}\text{Zr}$	51.46		$^{117}_{50}\text{Sn}$	7.61
	$^{91}_{40}\text{Zr}$	11.23		$^{118}_{50}\text{Sn}$	24.03
	$^{92}_{40}\text{Zr}$	17.11		$^{119}_{50}\text{Sn}$	8.58
	$^{94}_{40}\text{Zr}$	17.40		$^{120}_{50}\text{Sn}$	32.85
	$^{96}_{40}\text{Zr}$	2.80		$^{122}_{50}\text{Sn}$	4.92
41	$^{93}_{41}\text{Nb}$	100.00		$^{124}_{50}\text{Sn}$	5.94
42	$^{92}_{42}\text{Mo}$	15.84	51	$^{121}_{51}\text{Sb}$	57.25

ATOMIC NUMBER	ISOTOPE	PERCENT ABUNDANCE	ATOMIC NUMBER	ISOTOPE	PERCENT ABUNDANCE
	$^{123}_{51}\text{Sb}$	42.75		$^{148}_{62}\text{Sm}$	11.24
52	$^{120}_{52}\text{Te}$	0.09		$^{149}_{62}\text{Sm}$	13.83
	$^{122}_{52}\text{Te}$	2.46		$^{150}_{62}\text{Sm}$	7.44
	$^{123}_{52}\text{Te}$	0.87		$^{152}_{62}\text{Sm}$	26.72
	$^{124}_{52}\text{Te}$	4.61		$^{154}_{62}\text{Sm}$	22.71
	$^{125}_{52}\text{Te}$	6.99	63	$^{151}_{63}\text{Eu}$	47.82
	$^{126}_{52}\text{Te}$	18.71		$^{153}_{63}\text{Eu}$	52.18
	$^{128}_{52}\text{Te}$	31.79	64	$^{152}_{64}\text{Gd}$	0.20
	$^{130}_{52}\text{Te}$	34.48		$^{154}_{64}\text{Gd}$	2.15
53	$^{127}_{53}\text{I}$	100.00		$^{155}_{64}\text{Gd}$	14.73
54	$^{124}_{54}\text{Xe}$	0.10		$^{156}_{64}\text{Gd}$	20.47
	$^{126}_{54}\text{Xe}$	0.09		$^{157}_{64}\text{Gd}$	15.68
	$^{128}_{54}\text{Xe}$	1.92		$^{158}_{64}\text{Gd}$	24.87
	$^{129}_{54}\text{Xe}$	26.44		$^{160}_{64}\text{Gd}$	21.90
	$^{130}_{54}\text{Xe}$	4.08	65	$^{159}_{65}\text{Tb}$	100.00
	$^{131}_{54}\text{Xe}$	21.18	66	$^{156}_{66}\text{Dy}$	0.05
	$^{132}_{54}\text{Xe}$	26.89		$^{158}_{66}\text{Dy}$	0.09
	$^{134}_{54}\text{Xe}$	10.44		$^{160}_{66}\text{Dy}$	2.29
	$^{136}_{54}\text{Xe}$	8.87		$^{161}_{66}\text{Dy}$	18.88
55	$^{133}_{55}\text{Cs}$	100.00		$^{162}_{66}\text{Dy}$	25.53
56	$^{130}_{56}\text{Ba}$	0.10		$^{163}_{66}\text{Dy}$	24.97
	$^{132}_{56}\text{Ba}$	0.10		$^{164}_{66}\text{Dy}$	28.18
	$^{134}_{56}\text{Ba}$	2.42	67	$^{165}_{67}\text{Ho}$	100.00
	$^{135}_{56}\text{Ba}$	6.59	68	$^{162}_{68}\text{Er}$	0.14
	$^{136}_{56}\text{Ba}$	7.81		$^{164}_{68}\text{Er}$	1.56
	$^{137}_{56}\text{Ba}$	11.32		$^{166}_{68}\text{Er}$	33.41
	$^{138}_{56}\text{Ba}$	71.66		$^{167}_{68}\text{Er}$	22.94
57	$^{138}_{57}\text{La}$	0.09		$^{168}_{68}\text{Er}$	27.07
	$^{139}_{57}\text{La}$	99.91		$^{170}_{68}\text{Er}$	14.88
58	$^{136}_{58}\text{Ce}$	0.19	69	$^{169}_{69}\text{Tm}$	100.00
	$^{138}_{58}\text{Ce}$	0.25	70	$^{168}_{70}\text{Yb}$	0.14
	$^{140}_{58}\text{Ce}$	88.48		$^{170}_{70}\text{Yb}$	3.03
	$^{142}_{58}\text{Ce}$	11.07		$^{171}_{70}\text{Yb}$	14.31
59	$^{141}_{59}\text{Pr}$	100.00		$^{172}_{70}\text{Yb}$	21.82
60	$^{142}_{60}\text{Nd}$	27.11		$^{173}_{70}\text{Yb}$	16.13
	$^{143}_{60}\text{Nd}$	12.17		$^{174}_{70}\text{Yb}$	31.84
	$^{144}_{60}\text{Nd}$	23.85		$^{176}_{70}\text{Yb}$	12.73
	$^{145}_{60}\text{Nd}$	8.30	71	$^{175}_{71}\text{Lu}$	97.41
	$^{146}_{60}\text{Nd}$	17.22		$^{176}_{71}\text{Lu}$	2.59
	$^{148}_{60}\text{Nd}$	5.73	72	$^{174}_{72}\text{Hf}$	0.18
	$^{150}_{60}\text{Nd}$	5.62		$^{176}_{72}\text{Hf}$	5.20
62	$^{144}_{62}\text{Sm}$	3.09		$^{177}_{72}\text{Hf}$	18.50
	$^{147}_{62}\text{Sm}$	14.97		$^{178}_{72}\text{Hf}$	27.14

Atomic Number	Isotope	Percent Abundance	Atomic Number	Isotope	Percent Abundance
	$^{179}_{72}\text{Hf}$	13.75		$^{194}_{78}\text{Pt}$	32.90
	$^{180}_{72}\text{Hf}$	35.24		$^{195}_{78}\text{Pt}$	33.80
73	$^{180}_{73}\text{Ta}$	0.01		$^{196}_{78}\text{Pt}$	25.30
	$^{181}_{73}\text{Ta}$	99.99		$^{198}_{78}\text{Pt}$	7.21
74	$^{180}_{74}\text{W}$	0.14	79	$^{197}_{79}\text{Au}$	100.00
	$^{182}_{74}\text{W}$	26.41	80	$^{196}_{80}\text{Hg}$	0.15
	$^{183}_{74}\text{W}$	14.40		$^{198}_{80}\text{Hg}$	10.02
	$^{184}_{74}\text{W}$	30.64		$^{199}_{80}\text{Hg}$	16.84
	$^{186}_{74}\text{W}$	28.41		$^{200}_{80}\text{Hg}$	23.13
75	$^{185}_{75}\text{Re}$	37.07		$^{201}_{80}\text{Hg}$	13.22
	$^{187}_{75}\text{Re}$	62.93		$^{202}_{80}\text{Hg}$	29.80
76	$^{184}_{76}\text{Os}$	0.02		$^{204}_{80}\text{Hg}$	6.85
	$^{186}_{76}\text{Os}$	1.59	81	$^{203}_{81}\text{Tl}$	29.50
	$^{187}_{76}\text{Os}$	1.64		$^{205}_{81}\text{Tl}$	70.50
	$^{188}_{76}\text{Os}$	13.30	82	$^{204}_{82}\text{Pb}$	1.48
	$^{189}_{76}\text{Os}$	16.10		$^{206}_{82}\text{Pb}$	23.60
	$^{190}_{76}\text{Os}$	26.40		$^{207}_{82}\text{Pb}$	22.60
	$^{192}_{76}\text{Os}$	41.00		$^{208}_{82}\text{Pb}$	52.30
77	$^{191}_{77}\text{Ir}$	37.3	83	$^{209}_{83}\text{Bi}$	100.00
	$^{193}_{77}\text{Ir}$	62.7	92	$^{234}_{92}\text{U}$	0.01
78	$^{190}_{78}\text{Pt}$	0.01		$^{235}_{92}\text{U}$	0.72
	$^{192}_{78}\text{Pt}$	0.78		$^{238}_{92}\text{U}$	99.27

APPENDIX IV

Electronic Configurations of the Elements Showing Sublevels

ATOMIC NUMBER	ELEMENT SYMBOL	1 s	2 s p	3 s p d	4 s p d f	5 s p d f	6 s p d f	7 s
1	H	1						
2	He	2						
3	Li	2	1					
4	Be	2	2					
5	B	2	2 1					
6	C	2	2 2					
7	N	2	2 3					
8	O	2	2 4					
9	F	2	2 5					
10	Ne	2	2 6					
11	Na	2	2 6	1				
12	Mg	2	2 6	2				
13	Al	2	2 6	2 1				
14	Si	2	2 6	2 2				
15	P	2	2 6	2 3				
16	S	2	2 6	2 4				
17	Cl	2	2 6	2 5				
18	Ar	2	2 6	2 6				
19	K	2	2 6	2 6	1			
20	Ca	2	2 6	2 6	2			
21	Sc	2	2 6	2 6 1	2			
22	Ti	2	2 6	2 6 2	2			
23	V	2	2 6	2 6 3	2			
24	Cr	2	2 6	2 6 5	1			
25	Mn	2	2 6	2 6 5	2			
26	Fe	2	2 6	2 6 6	2			
27	Co	2	2 6	2 6 7	2			
28	Ni	2	2 6	2 6 8	2			
29	Cu	2	2 6	2 6 10	1			
30	Zn	2	2 6	2 6 10	2			

ATOMIC NUMBER	ELEMENT SYMBOL	1 s	2 s p	3 s p d	4 s p d f	5 s p d f	6 s p d f	7 s
31	Ga	2	2 6	2 6 10	2 1			
32	Ge	2	2 6	2 6 10	2 2			
33	As	2	2 6	2 6 10	2 3			
34	Se	2	2 6	2 6 10	2 4			
35	Br	2	2 6	2 6 10	2 5			
36	Kr	2	2 6	2 6 10	2 6			
37	Rb	2	2 6	2 6 10	2 6	1		
38	Sr	2	2 6	2 6 10	2 6	2		
39	Y	2	2 6	2 6 10	2 6 1	2		
40	Zr	2	2 6	2 6 10	2 6 2	2		
41	Nb	2	2 6	2 6 10	2 6 4	1		
42	Mo	2	2 6	2 6 10	2 6 5	1		
43	Tc	2	2 6	2 6 10	2 6 6	1		
44	Ru	2	2 6	2 6 10	2 6 7	1		
45	Rh	2	2 6	2 6 10	2 6 8	1		
46	Pd	2	2 6	2 6 10	2 6 10			
47	Ag	2	2 6	2 6 10	2 6 10	1		
48	Cd	2	2 6	2 6 10	2 6 10	2		
49	In	2	2 6	2 6 10	2 6 10	2 1		
50	Sn	2	2 6	2 6 10	2 6 10	2 2		
51	Sb	2	2 6	2 6 10	2 6 10	2 3		
52	Te	2	2 6	2 6 10	2 6 10	2 4		
53	I	2	2 6	2 6 10	2 6 10	2 5		
54	Xe	2	2 6	2 6 10	2 6 10	2 6		
55	Cs	2	2 6	2 6 10	2 6 10	2 6	1	
56	Ba	2	2 6	2 6 10	2 6 10	2 6	2	
57	La	2	2 6	2 6 10	2 6 10	2 6 1	2	
58	Ce	2	2 6	2 6 10	2 6 10 1	2 6 1	2	
59	Pr	2	2 6	2 6 10	2 6 10 3	2 6	2	
60	Nd	2	2 6	2 6 10	2 6 10 4	2 6	2	
61	Pm	2	2 6	2 6 10	2 6 10 5	2 6	2	
62	Sm	2	2 6	2 6 10	2 6 10 6	2 6	2	
63	Eu	2	2 6	2 6 10	2 6 10 7	2 6	2	
64	Gd	2	2 6	2 6 10	2 6 10 7	2 6 1	2	
65	Tb	2	2 6	2 6 10	2 6 10 9	2 6	2	
66	Dy	2	2 6	2 6 10	2 6 10 10	2 6	2	
67	Ho	2	2 6	2 6 10	2 6 10 11	2 6	2	
68	Er	2	2 6	2 6 10	2 6 10 12	2 6	2	
69	Tm	2	2 6	2 6 10	2 6 10 13	2 6	2	
70	Yb	2	2 6	2 6 10	2 6 10 14	2 6	2	
71	Lu	2	2 6	2 6 10	2 6 10 14	2 6 1	2	
72	Hf	2	2 6	2 6 10	2 6 10 14	2 6 2	2	
73	Ta	2	2 6	2 6 10	2 6 10 14	2 6 3	2	

ATOMIC NUMBER	ELEMENT SYMBOL	1 — s	2 — s p	3 — s p d	4 — s p d f	5 — s p d f	6 — s p d f	7 — s
74	W	2	2 6	2 6 10	2 6 10 14	2 6 4	2	
75	Re	2	2 6	2 6 10	2 6 10 14	2 6 5	2	
76	Os	2	2 6	2 6 10	2 6 10 14	2 6 6	2	
77	Ir	2	2 6	2 6 10	2 6 10 14	2 6 7	2	
78	Pt	2	2 6	2 6 10	2 6 10 14	2 6 9	1	
79	Au	2	2 6	2 6 10	2 6 10 14	2 6 10	1	
80	Hg	2	2 6	2 6 10	2 6 10 14	2 6 10	2	
81	Tl	2	2 6	2 6 10	2 6 10 14	2 6 10	2 1	
82	Pb	2	2 6	2 6 10	2 6 10 14	2 6 10	2 2	
83	Bi	2	2 6	2 6 10	2 6 10 14	2 6 10	2 3	
84	Po	2	2 6	2 6 10	2 6 10 14	2 6 10	2 4	
85	At	2	2 6	2 6 10	2 6 10 14	2 6 10	2 5	
86	Rn	2	2 6	2 6 10	2 6 10 14	2 6 10	2 6	
87	Fr	2	2 6	2 6 10	2 6 10 14	2 6 10	2 6	1
88	Ra	2	2 6	2 6 10	2 6 10 14	2 6 10	2 6	2
89	Ac	2	2 6	2 6 10	2 6 10 14	2 6 10	2 6 1	2
90	Th	2	2 6	2 6 10	2 6 10 14	2 6 10	2 6 2	2
91	Pa	2	2 6	2 6 10	2 6 10 14	2 6 10 2	2 6 1	2
92	U	2	2 6	2 6 10	2 6 10 14	2 6 10 3	2 6 1	2
93	Np	2	2 6	2 6 10	2 6 10 14	2 6 10 4	2 6 1	2
94	Pu	2	2 6	2 6 10	2 6 10 14	2 6 10 6	2 6	2
95	Am	2	2 6	2 6 10	2 6 10 14	2 6 10 7	2 6	2
96	Cm	2	2 6	2 6 10	2 6 10 14	2 6 10 7	2 6 1	2
97	Bk	2	2 6	2 6 10	2 6 10 14	2 6 10 9	2 6	2
98	Cf	2	2 6	2 6 10	2 6 10 14	2 6 10 10	2 6	2
99	Es	2	2 6	2 6 10	2 6 10 14	2 6 10 11	2 6	2
100	Fm	2	2 6	2 6 10	2 6 10 14	2 6 10 12	2 6	2
101	Md	2	2 6	2 6 10	2 6 10 14	2 6 10 13	2 6	2
102	No	2	2 6	2 6 10	2 6 10 14	2 6 10 14	2 6	2
103	Lr	2	2 6	2 6 10	2 6 10 14	2 6 10 14	2 6 1	2
104	Rf	2	2 6	2 6 10	2 6 10 14	2 6 10 14	2 6 2	2
105	Ha	2	2 6	2 6 10	2 6 10 14	2 6 10 14	2 6 3	2
106	Sg	2	2 6	2 6 10	2 6 10 14	2 6 10 14	2 6 4	2
107	Ns	2	2 6	2 6 10	2 6 10 14	2 6 10 14	2 6 5	2
108	Hs	2	2 6	2 6 10	2 6 10 14	2 6 10 14	2 6 6	2
109	Mt	2	2 6	2 6 10	2 6 10 14	2 6 10 14	2 6 7	2
110	—	2	2 6	2 6 10	2 6 10 14	2 6 10 14	2 6 8	2
111	—	2	2 6	2 6 10	2 6 10 14	2 6 10 14	2 6 9	2

APPENDIX V

Vapor Pressure of Water at Various Temperatures

TEMPERATURE (°C)	PRESSURE[a] mm Hg (torr)	Atm	Pa	TEMPERATURE (°C)	PRESSURE[a] mm Hg (torr)	Atm	Pa
0	4.6	0.0061	610	33	37.7	0.0496	5030
5	6.5	0.0086	872	34	39.9	0.0525	5319
10	9.2	0.0121	1227	35	42.2	0.0555	5622
15	12.8	0.0168	1705	36	44.6	0.0586	5941
16	13.6	0.0179	1818	37	47.1	0.0619	6275
17	14.5	0.0191	1937	38	49.7	0.0654	6625
18	15.5	0.0204	2063	39	52.4	0.0690	6992
19	16.5	0.0217	2197	40	55.3	0.0728	7376
20	17.5	0.0231	2338	45	71.9	0.0946	9583
21	18.6	0.0245	2486	50	92.5	0.1217	12,333
22	19.8	0.0261	2643	55	118.0	0.1553	15,737
23	21.1	0.0277	2809	60	149.4	0.1965	19,915
24	22.4	0.0294	2983	65	187.5	0.2468	25,002
25	23.8	0.0313	3167	70	233.7	0.3075	31,157
26	25.2	0.0332	3360	75	289.1	0.3804	38,543
27	26.7	0.0352	3564	80	355.1	0.4672	47,342
28	28.3	0.0373	3779	85	433.6	0.5705	57,808
29	30.0	0.0395	4005	90	525.8	0.6918	70,094
30	31.8	0.0419	4242	95	633.9	0.8341	84,512
31	33.7	0.0443	4492	100	760.0	1.0000	101,325
32	35.7	0.0469	4755				

[a] Units in mm Hg (torr) to nearest tenth; atm to nearest ten-thousandth; Pa to nearest unit.

APPENDIX VI   Algebra Review

The most common type of equation you will have to solve in this book is an equation that has one unknown whose highest power is equal to 1. Such equations are called *linear equations* and are easily solved by simple algebraic operations. A general example of a linear equation is

$$ax = b \quad \text{or} \quad ax^1 = b$$

The unknown or variable is x raised to the first power (there are no x^2 or x^3 terms), a is a coefficient ($a \neq 0$), and b is a number. We can use the solutions of linear equations in chemistry for substituting into the ideal gas equation (see Section 11.8) for the variables P (pressure), V (volume), n (moles), and T (temperature) and for calculating oxidation numbers of elements in compounds and ions (see Sections 6.3 and 16.2).

Solving a Linear Equation

To solve for the unknown quantity in a linear equation, we need to carry out algebraic transformations or changes.

1. Clear any parentheses.
2. Collect similar terms. Place all unknowns on one side of the equation (usually the left) and all numbers on the other side (usually the right).
3. Rearrange. To place unknowns on one side of the equation and numbers on the other side, we must rearrange the equation according to the following rules:
   a. Adding or subtracting the same number on *both* sides of the equation does not change the equation.
   b. Multiplying or dividing *both* sides of the equation by the same number does not change the equation.
4. Solve for *one* unit of the unknown.

EXAMPLE VI.1 | Solve the following linear equations for the unknown (x).

a. $2x = 4$
b. $2x = 5 - 1$
c. $2x + 1 = 5$
d. $2(x + 1) = 8$
e. $3x = -18$

617

SOLUTION

a. $2x = 4$ Divide both sides of the equation by 2.

$$\frac{2x}{2} = \frac{4}{2}$$

$x = 2$ *Answer*

b. $2x = 5 - 1$ Collect similar terms.

$2x = 5 - 1 = 4$

$$\frac{2x}{2} = \frac{4}{2}$$

$x = 2$ *Answer*

c. $2x + 1 = 5$ Subtract 1 from both sides of the equation.

$2x + 1 - 1 = 5 - 1$

$2x = 4$

$$\frac{2x}{2} = \frac{4}{2}$$

$x = 2$ *Answer*

d. $2(x + 1) = 8$ Clear the parentheses.

$2x + 2 = 8$ Subtract 2 from both sides of the equation.

$2x + 2 - 2 = 8 - 2$

$2x = 6$ Divide both sides of the equation by 2.

$$\frac{2x}{2} = \frac{6}{2}$$

$x = 3$ *Answer*

e. $3x = -18$ Divide both sides of the equation by 3.

$$\frac{3x}{3} = \frac{-18}{3}$$

$x = -6$ *Answer*

Substituting into a Linear Equation

We can check a linear equation by substituting the value obtained for the unknown into the original equation. Checking the linear equations of Example VI.1 gives the following solutions:

a. $2x = 4; \quad x = 2$
$2(2) = 4$
$4 = 4$

b. $2x = 5 - 1; \quad x = 2$
$2(2) = 5 - 1$
$4 = 4$

c. $2x + 1 = 5; \quad x = 2$
$2(2) + 1 = 5$
$4 + 1 = 5$
$5 = 5$

d. $2(x + 1) = 8; \quad x = 3$
$2(3 + 1) = 8$
$2(4) = 8$
$8 = 8$

e. $3x = -18; \quad x = -6$
$3(-6) = -18$
$-18 = -18$

EXAMPLE VI.2   Given the following ideal-gas equation:

$$PV = nRT$$

a. Solve for pressure (P) in atmospheres if the volume (V) is 6.00 L, the temperature (T) is $3\overline{0}0$ K, the moles of gas (n) is 0.900 mol, and the universal gas constant (R) is 0.0821 atm·L/(mol·K).

b. Solve for the moles of a gas (n) if the pressure (P) is 1.25 atm, the volume (V) is 2.00 L, the temperature (T) is 273 K, and the universal gas constant (R) is 0.0821 atm·L/(mol·K).

c. Solve for the temperature in degrees Celsius if the pressure (P) is 1.10 atm, the volume (V) is 30.0 L, the moles of gas (n) is 1.25 mol, and the universal gas constant (R) is 0.0821 atm·L/(mol·K).

SOLUTION:

a. Use the ideal-gas equation and solve for P as follows:

$$PV = nRT \qquad \text{Divide both sides of the equation by } V.$$

$$\frac{P\cancel{V}}{\cancel{V}} = \frac{nRT}{V}$$

$$P = \frac{nRT}{V}$$

Substitution into the linear equation above for $V = 6.00$ L, $T = 3\overline{00}$ K, $n = 0.900$ mol, and $R = 0.0821$ atm·L/(mol·K) gives

$$P = \frac{0.900 \cancel{\text{mol}} \times 0.0821 \, \dfrac{\text{atm} \cdot \cancel{L}}{\cancel{\text{mol}} \cdot \cancel{K}} \times 3\overline{00} \, \cancel{K}}{6.00 \, \cancel{L}}$$

$$= 3.69 \text{ atm} \qquad \textit{Answer}$$

Note that the units of R cancel, leaving only atmospheres, the units of pressure.

b. Use the ideal-gas equation and solve for n as follows:

$$PV = nRT \qquad \text{Place the unknown } n \text{ on the left.}$$

$$nRT = PV \qquad \text{Divide both sides of the equation by } RT.$$

$$\frac{n\cancel{RT}}{\cancel{RT}} = \frac{PV}{RT}$$

$$n = \frac{PV}{RT}$$

Substitution into the linear equation above for $P = 1.25$ atm, $V = 2.00$ L, $T = 273$ K, and $R = 0.0821$ atm·L/(mol·K) gives

$$n = \frac{1.25 \, \cancel{\text{atm}} \times 2.00 \, \cancel{L}}{0.0821 \, \dfrac{\cancel{\text{atm}} \cdot \cancel{L}}{\text{mol} \cdot \cancel{K}} \times 273 \, \cancel{K}}$$

$$= 0.112 \text{ mol} \qquad \textit{Answer}$$

Note that the units of R cancel again, leaving only moles. In the division of fractions, you invert and multiply; therefore,

$$\frac{1}{\dfrac{1}{\text{mol}}} = 1 \times \frac{\text{mol}}{1} = \text{mol}$$

c. Use the ideal-gas equation and solve for *T* as follows:

$$PV = nRT \qquad \text{Place the unknown } T \text{ on the left.}$$

$$nRT = PV \qquad \text{Divide both sides of the equation by } nR.$$

$$\frac{\cancel{n}R T}{\cancel{n}R} = \frac{PV}{nR}$$

$$T = \frac{PV}{nR}$$

Substitution into the linear equation above for $P = 1.10$ atm, $V = 30.0$ L, $n = 1.25$ mol, and $R = 0.0821$ atm·L/(mol·K) gives

$$T = \frac{1.10\,\cancel{\text{atm}} \times 30.0\,\cancel{L}}{1.25\,\cancel{\text{mol}} \times 0.0821\,\dfrac{\cancel{\text{atm}} \cdot \cancel{L}}{\cancel{\text{mol}} \cdot K}}$$

$$= 322 \text{ K}$$

Note that the units of *R* cancel again, leaving only kelvins. We are to express the temperature requested in degrees Celsius. Therefore, °C = K − 273 from Equation (2.4) in Section 2.8; the temperature in degrees Celsius is

$$322 \text{ K} = (322 - 273)°\text{C} = 49°\text{C} \qquad \textit{Answer}$$

✓ Problems

1. Solve the following linear equations for the unknown (*x*):

   a. $2x = 5 + 3$ **b.** $5x + 6 = 4x + 2$

   c. $2(x + 2) = 10$ **d.** $3(x + 1) = 2(x + 4)$

2. Given the following ideal-gas equation:

$$PV = nRT$$

   a. Solve for the volume (*V*) in liters if the pressure (*P*) is 2.00 atm, the temperature (*T*) is $30\overline{0}$ K, the moles of gas (*n*) is 0.750 mol, and the universal gas constant (*R*) is 0.0821 atm·L/(mol·K).

   b. Solve for the temperature in degrees Celsius (see Section 2.8) if the pressure (*P*) is 1.50 atm, the volume (*V*) is 38.0 L, the moles of gas (*n*) is 1.10 mol, and the universal gas constant (*R*) is 0.0821 atm·L/(mol·K).

   c. Solve for the moles of gas (*n*) is the pressure (*P*) is 0.970 atm, the volume (*V*) is 3.00 L, the temperature (*T*) is 303 K, and the universal gas constant (*R*) is 0.0821 atm·L/(mol·K).

Answers to Selected Exercises and Problems

Chapter 1

3. *Experimentation:* driving routes as described; *hypothesis formation:* comparing results and deciding which one faster; *further experimentation:* repeated measurements using initially judged faster route. To make a law, verify selected route fastest regardless of conditions.

4. (a) physical; **(b)** analytical; **(c)** biochemical; **(d)** organic

5. (a) inorganic; **(b)** organic; **(c)** inorganic; **(d)** biochemical; **(e)** inorganic; **(f)** inorganic; **(g)** physical

Chapter Quiz

1. chemistry

2. (a) physical; **(b)** physical; **(c)** organic

3. *Experimentation:* three strikeouts with heavy bat and then a hit with lighter bat; *hypothesis formation:* continued to use lighter bat; *further experimentation:* continued success with lighter bat.

Chapter 2

5. (a) 3; **(b)** 3; **(c)** 5; **(d)** 4; **(e)** 3; **(f)** 1

6. (a) 3; **(b)** 3; **(c)** 4; **(d)** 5; **(e)** 4; **(f)** 3

7. (a) 2.44; **(b)** 2.48; **(c)** 8.68; **(d)** 10.5; **(e)** 13.4; **(f)** 96,800

8. (a) 10.6; **(b)** 3.88; **(c)** 0.00454; **(d)** 0.785; **(e)** 6.99; **(f)** 3.46

9. 0.454

10. (a) 12.6; **(b)** 81.792; **(c)** 12.14; **(d)** 0.334; **(e)** 18; **(f)** 366; **(g)** 0.071; **(h)** 8.1; **(i)** 35.7

11. 3.75×10^4

12. 3.25×10^{-3}

13. (a) 9.75×10^2; **(b)** 9.84×10^6; **(c)** 6.32×10^{-4}; **(d)** 7.28×10^{-3}

14. (a) 3.25×10^{-4}; **(b)** 7.29×10^6; **(c)** 4.78×10^3; **(d)** 5.26×10^{-4}

15. 9.3×10^7 mi

16. (a) 5.74×10^3; **(b)** 5.49×10^2; **(c)** 1.12×10^3; **(d)** 6.51×10^5; **(e)** 8.51×10^5; **(f)** 4.66×10^4; **(g)** 2.35×10^4; **(h)** 4.03×10^{-3}

17. (a) 5.00×10^4; **(b)** 3.54×10^2; **(c)** 8.22×10^4; **(d)** 6.12×10^{-4}

18. (a) 1.59×10^6; **(b)** 3.35×10^{16}; **(c)** 3.18×10^6; **(d)** 4.86×10^{10}

19. (a) 3.1×10^3 g; **(b)** 13 km; **(c)** 2.5×10^{-5} kg; **(d)** 0.6 L; **(e)** 8.75×10^{-3} kL; **(f)** 4.2×10^6 nm

20. (a) 7.4×10^3 m; **(b)** 18 kg; **(c)** 7.5×10^{-5} kg; **(d)** 0.875 L; **(e)** 4.00×10^{-2} nm; **(f)** 4.2 cm^2

21. 5.975 g

22. 6010.7004 m

23. 91.4 cm/yd

24. (a) $10\overline{0}$°F, 311 K; **(b)** 248°F, 393 K; **(c)** −26°F, 241 K; **(d)** −166°F, 163 K;

25. (a) $2\overline{0}$°C, 293 K; **(b)** −24°C, 249 K; **(c)** −43°C, $23\overline{0}$ K; **(d)** $16\overline{0}$°C, 433 K

26. −196°C; −321°F

27. −40.0°

28. −62.1°C

29. −128.6°F

30. 58.0°C

31. (a) 5.0 g/mL; **(b)** 2.4 g/mL; **(c)** 13 g/mL; **(d)** 6.0 g/mL

32. (a) 40.6 mL; **(b)** 305 mL; **(c)** 25.2 mL; **(d)** 2.8×10^3 mL

33. (a) 24.8 g; **(b)** 233 g; **(c)** 103 g; **(d)** 468 g

34. (a) 0.156 L; **(b)** 0.690 L; **(c)** 0.376 L; **(d)** 1.38 L

35. (a) 17.6 g; **(b)** 158 g; **(c)** 2.46×10^3 g; **(d)** 5.36×10^3 g

36. 91.5 m

37. 5.94 g/mL

38. 525 ft to 425 ft, 88 ft to 71 ft, 52 ft to 42 ft

39. (a) 1.6×10^{10} swimming pools; **(b)** 1.3×10^{15} kg

Chapter Quiz A

1. (a) 3; **(b)** 4

2. (a) 12.4; **(b)** 3.75

3. (a) 8.75×10^5; **(b)** 2.95×10^{-3}

4. (a) 4.06×10^3; **(b)** 7.35×10^6; **(c)** 7.90×10^4; **(d)** 1.86×10^6

Chapter Quiz B

1. 9.25×10^{-4} kg

2. 7.25×10^{-4} g

3. −148°F, 173 K

4. $63\overline{0}$ g

5. 268 mL

Chapter 3

3. (a) solid; **(c)** liquid; **(e)** gas

4. O, Si, Al, Fe, Ca, Na, K, Mg, H, Ti

5. (a) element; **(c)** element; **(e)** compound; **(g)** mixture

6. (a) 1 atom carbon, 4 atoms hydrogen, 5;
   (c) 1 atom carbon, 2 atoms chlorine, 2 atoms fluorine, 5;
   (e) 34 atoms carbon, 32 atoms hydrogen, 1 atom iron, 4 atoms nitrogen, 4 atoms oxygen, 75

7. (a) SO_2; (c) Ag_2S;
   (e) $C_{10}H_{16}N_5O_{13}P_3$

9. (a) physical; (c) physical;
   (e) chemical; (g) chemical

10. (a) physical; (c) chemical;
    (e) physical

11. (a) chemical; (c) chemical;
    (e) chemical

12. 7.32×10^2 J/(kg·K)

13. 10.5 kcal

14. 6.36×10^3 J

15. 7.84 g

16. 40.3 g

17. 1.61×10^3 J

26. 0.196 L

27. 9.64 lb

28. (a) 406 cal; (b) 1.70×10^3 J

29. (a) 3.33×10^{-1} cal/(g·°C);
    (b) 1.39×10^3 J/(kg·K)

30. (a) 4.97×10^4 J; (b) 11.9 kcal

31. 0.842 kg

32. (a) 5.17×10^5 J;
    (b) 2.89×10^4 J; (c) water;
    (d) 5.46×10^5 J

33. 6.16 g/mL

34. (a) 2; (b) -333°F

Chapter Quiz

1. (a) Co; (b) Mg
2. (a) 1 atom copper, 2 atoms iodine, 3;
   (b) 2 atoms aluminum, 3 atoms sulfur. 5
3. (a) N_2O_5; (b) $C_8H_9NO_2$
4. (a) physical; (b) chemical;
   (c) chemical; (d) physical
5. (a) physical; (b) chemical;
   (c) physical; (d) chemical
6. 12.6 kcal

Chapter 4

5. (a) 4p 5n, 4e⁻ (c) 22p 24n, 22e⁻
   (e) 27p 32n, 27e⁻

6. (a) 11p 12n, 11e⁻
   (c) 46p 59n, 46e⁻
   (e) 58p 84n, 58e⁻

7. (a) different; (b) same

8. 10.811 amu is nearer to 11.009 amu than 10.013 amu; hence ^{11}B must predominate in nature.

9. 69.72 amu

10. 121.8 amu

11. (a) 2; (c) 18; (e) 72

12. (a) 3p 4n, 2e⁻ 1e⁻
    (c) 8p 8n, 2e⁻ 6e⁻
    (e) 15p 16n, 2e⁻ 8e⁻ 5e⁻

13. (a) 4p 7n, 2e⁻ 2e⁻
    (c) 9p 10n, 2e⁻ 7e⁻
    (e) 12p 12n, 2e⁻ 8e⁻ 2e⁻

14. (a) He:; (c) Be·; (e) :F·

15. (a) ·C·; (c) Na·; (e) ·P·

16. (a) $1s^2, 2s^1$ (1);
    (c) $1s^2, 2s^2 2p^2$ (4);
    (e) $1s^2, 2s^2 2p^6, 3s^2 3p^3$ (5)

17. (a) $1s^2, 2s^2 2p^1$ (3);
    (c) $1s^2, 2s^2 2p^6, 3s^2 3p^4$ (6);
    (e) $1s^2, 2s^2 2p^6, 3s^2 3p^6 3d^{10}, 4s^2 4p^3$ (5) or $1s^2, 2s^2 2p^6, 3s^2 3p^6, 4s^2, 3d^{10}, 4p^3$ (5)

18. (a) 5513°F;
    (b) 2.257×10^4 kg/m³;
    (c) 1.408×10^3 lb/ft³

19. $1s^2, 2s^2 2p^6, 3s^2 3p^6 3d^{10}, 4s^2 4p^6 4d^{10} 4f^{14}, 5s^2 5p^6 5d^6, 6s^2$ or $1s^2, 2s^2 2p^6, 3s^2 3p^6, 4s^2, 3d^{10}, 4p^6, 5s^2, 4d^{10}, 5p^6, 6s^2, 4f^{14}, 5d^6$

20. ^{123}Sb has two more neutrons in its nucleus than ^{121}Sb. Their electronic structures are the same.

21. 2.72×10^{26} protons

Chapter Quiz

1. (a) 6p 7n, 2e⁻ 4e⁻
   (b) 14p 14n, 2e⁻ 8e⁻ 4e⁻
2. (a) 32; (b) 72
3. (a) $1s^2, 2s^2 2p^6, 3s^2 3p^5$;
   (b) $1s^2, 2s^2 2p^6, 3s^2 3p^6 3d^5, 4s^2$ or $1s^2, 2s^2 2p^6, 3s^2 3p^6, 4s^2, 3d^5$
4. (a) ·P·; (b) :Ar:
5. 13.84 amu

Chapter 5

3. (a) metal; (c) metalloid;
   (d) nonmetal
4. (a) nonmetal; (c) metalloid;
   (d) metal
5. (a) 1; (b) 4; (c) 6; (d) 8
6. (a) 8; (b) 6; (c) 3; (d) 5
7. (a) and (d); (b) and (c)
8. (a) I; (c) Mg
9. (a) and (d); (b) and (c)
10. (a) Si; (c) K
11. (a) selenium; (c) aluminum
12. (a) barium; (c) lead
13. (a) bromine; (c) barium
14. (a) silicon; (c) barium
15. (a) metalloid; (b) 4;
    (c) Ge: $1s^2, 2s^2 2p^6, 3s^2 3p^6 3d^{10}, 4s^2 4p^2$ or $1s^2, 2s^2 2p^6, 3s^2 3p^6, 4s^2, 3d^{10}, 4p^2$; Si: $1s^2, 2s^2 2p^6, 3s^2 3p^2$;
    (d) more metallic;
    (e) 5.5×10^3 kg/m³, 5.3×10^3 kg/m³
16. (a) 1.33×10^{-8} cm;
    (b) 5.24×10^{-9} in.

Chapter Quiz

1. (a) metal; (b) metalloid;
   (c) nonmetal; (d) metal
2. (a) 3; (b) 6; (c) 2; (d) 7
3. (a) and (c); (b) and (d)
4. (a) cesium; (b) bismuth
5. (a) polonium; (b) cadmium

Chapter 6

4. (a) +1; **(b)** +5; **(c)** +3;
 (d) −2; **(e)** +4; **(f)** +5;
 (g) +6; **(h)** +5; **(i)** +3;
 (j) +5

5. (a) +7; **(b)** +1; **(c)** +5;
 (d) +3; **(e)** +7; **(f)** +3;
 (g) +6; **(h)** +6

6. (a) $\boxed{\begin{array}{c}1p\\0n\end{array}}$ +1

 (c) $\boxed{\begin{array}{c}12p\\12n\end{array}}$ $2e^-$ $8e^-$ +2

 (e) $\boxed{\begin{array}{c}13p\\14n\end{array}}$ $2e^-$ $8e^-$ +3

 (g) $\boxed{\begin{array}{c}8p\\8n\end{array}}$ $2e^-$ $8e^-$ −2

 (i) $\boxed{\begin{array}{c}7p\\7n\end{array}}$ $2e^-$ $8e^-$ −3

7. (a) $1s^2$; **(c)** $1s^2, 2s^2\, 2p^6$;
 (e) $1s^2, 2s^2\, 2p^6, 3s^2\, 3p^6$;
 (g) $1s^2, 2s^2\, 2p^6, 3s^2\, 3p^6$;
 (i) $1s^2, 2s^2\, 2p^6, 3s^2\, 3p^6$;
 (j) $1s^2, 2s^2\, 2p^6, 3s^2\, 3p^6\, 3d^{10}$,
 $4s^2\, 4p^6$ or $1s^2, 2s^2\, 2p^6$,
 $3s^2\, 3p^6, 4s^2, 3d^{10}, 4p^6$

8. (1) Loss of electrons in principal
 energy level and (2) greater
 nuclear attraction

9. (1) Smaller nuclear attraction and
 (2) repulsion of electrons

10. 3.42×10^3 J

11. (a) $H^{\delta+}F^{\delta-}$; **(c)** $H_2^{\delta+}O^{\delta-}$;
 (e) $B^{\delta+}Cl_3^{\delta-}$; **(g)** $P^{\delta+}Cl_5^{\delta-}$;
 (i) $O^{\delta+}F_2^{\delta-}$

12. Structural formulas only

 (a)   H — Cl

 (c)
```
        Cl
        |
  Cl — C — Cl
        |
        Cl
```

 (e)   N ≡ N

 (g)   H — C ≡ C — H

 (i)   [C ≡ N]⁻

 (j)
$$\left[\begin{array}{c}O\\|\\O-S-O\\\end{array}\right]^{2-}$$

13. Structural formulas only

(a)   F — F

(c)
```
        H
        |
  Cl — C — Cl
        |
        Cl
```

(e)   Cl — Cl

(g)
```
        O
        ‖
  H — O — C — O — H
```

(i)
$$\left[\begin{array}{c}O\\|\\O-P-O\\|\\O\end{array}\right]^{3-}$$

(j)
$$\left[\begin{array}{cc}O & O\\| & |\\O-P-O-P-O\\| & |\\O & O\end{array}\right]^{4-}$$

14. (a) linear, 180°; **(b)** tetrahedral,
 109.5°; **(c)** trigonal planar, 120°;
 (d) pyrimidal, 109.5°

15. (a) NaCl; **(c)** Mg_3N_2; **(e)** CdO;
 (g) LiH; **(i)** $Al(ClO_4)_3$;
 (j) $Ba_3(PO_4)_2$

16. (a) AgCl; **(c)** $CuBr_2$;
 (e) $Zn(HCO_3)_2$; **(g)** $Fe_3(PO_4)_2$;
 (i) $Hg_2(CN)_2$; **(j)** $(NH_4)_2Cr_2O_7$

17. (a) +2; **(c)** +6, −2; **(e)** +3;
 (g) +7, −1; **(i)** +3; **(j)** +8

18. (a) BaO; **(c)** Na_3N; **(e)** In_2O_3;
 (g) Al_2S_3; **(i)** Tl_2S_3; **(j)** Na_2Te

19. (a) $23\overline{0}$ pm; **(b)** 3.66 g/mL;
 (c) 778°C; **(d)** 4.63 g/mL;
 (e) 212 kcal/mol

20. (a) K_2SO_4; **(c)** $Mg_3(AsO_4)_2$;
 (e) Cs_2SeO_4; **(g)** Tl_2S_3;
 (i) Rb_2SeO_4; **(j)** In_2S_3

21. (a) ionic; **(c)** covalent;
 (e) covalent; **(g)** ionic;
 (i) covalent

22. (a) IVA (14); **(b)** 4,
 (c) more metallic; **(d)** Pb;
 (e)
$$\left[\begin{array}{c}O\\‖\\O-X-O\end{array}\right]^{2-}$$
 structural formula only

23. (a) VIIA (17); **(b)** 7; **(c)** At;
 (d) metalloid; **(e)** 1⁻; **(f)** NaY;
 (g) larger

24. (a) IA (1); **(b)** 1; **(c)** 1⁺; **(d)** Fr;
 (e) ZBr, ionic

Chapter Quiz

1. (a) +4; **(b)** +6
2. (a) $1s^2, 2s^2\, 2p^6$; **(b)** $1s^2, 2s^2\, 2p^6$
3. (a) $CaCl_2$; **(b)** K_2SO_4;
 (c) $Fe_2(CO_3)_3$; **(d)** $Sn(SO_3)_2$
4. Structural formulas only

 (a)
```
        H
        |
  H — C — H
        |
        H
```

 (b)
```
   H         H
    \       /
     C  =  C
    /       \
   H         H
```

5. (a) pyramidal, 109.5°
 (b) tetrahedral, 109.5°
6. (a) ionic; **(b)** covalent;
 (c) ionic; **(d)** ionic

Chapter 7

6. Question 7 in the order j–a
7. Question 6 in the order j–a
8. Question 9 in the order j–a
9. Question 8 in the order j–a
10. Question 11 in the order j–a
11. Question 10 in the order j–a
12. Question 13 in the order j–a
13. Question 12 in the order j–a
14. Question 15 in the order h–a
15. Question 14 in the order h–a
16. (a) (1); **(b)** (3); **(c)** (3); **(d)** (3);
 (e) (1); **(f)** (2); **(g)** (1); **(h)** (3)
17. (a) (3); **(b)** (3); **(c)** (3); **(d)** (2);
 (e) (3); **(f)** (1); **(g)** (3); **(h)** (2)
18. (a) $HC_2H_3O_2$; **(b)** $CaCO_3$;
 (c) NaCl; **(d)** $NaHCO_3$;
 (e) $Mg(OH)_2$; **(f)** NH_3;
 (g) N_2O; **(h)** HCl
19. KCl, K_2CO_3, K_2SO_4, K_3PO_4,
 $BaCl_2$, $BaCO_3$, $BaSO_4$,
 $Ba_3(PO_4)_2$, $FeCl_3$, $Fe_2(CO_3)_3$,
 $Fe_2(SO_4)_3$, $FePO_4$, $AlCl_3$,
 $Al_2(CO_3)_3$, $Al_2(SO_4)_3$, $AlPO_4$
20. (a) $Sn_3(PO_4)_2$; **(b)** $AgMnO_4$;
 (c) $Ca(IO)_2$; **(d)** $MgSO_4$;
 (e) $Na_2C_2O_4$;
 (f) $HClO_4$ in water; **(g)** PF_3;
 (h) $Cd(NO_3)_2$; **(i)** $Pb_3(PO_4)_2$;

(**j**) $Sr(HCO_3)_2$; (**k**) Ca_3N_2;
(**l**) $HgCl_2$; (**m**) PCl_5; (**n**) PbO_2

21. (**a**) calcium phosphate;
(**b**) magnesium chlorate;
(**c**) lead(II) sulfate or plumbous sulfate;
(**d**) calcium dichromate;
(**e**) barium hydroxide;
(**f**) tin(II) or stannous hydrogen carbonate or bicarbonate;
(**g**) potassium oxalate;
(**h**) lithium dichromate;
(**i**) acetic acid or vinegar;
(**j**) zinc chloride;
(**k**) cadmium phosphide;
(**l**) lithium hydroxide;
(**m**) potassium hydride;
(**n**) tin(IV) sulfide or stannic sulfide

22. $CaCO_3$

23. K^+: $1s^2, 2s^2 2p^6, 3s^2 3p^6$;
I^-: $1s^2, 2s^2 2p^6, 3s^2 3p^6 3d^{10}$,
$4s^2 4p^6 4d^{10}, 5s^2 5p^6$ or $1s^2$,
$2s^2 2p^6, 3s^2 3p^6, 4s^2, 3d^{10}, 4p^6$,
$5s^2, 4d^{10}, 5p^6$

Chapter Quiz

1. (**a**) potassium iodide;
(**b**) copper(I) iodide or cuprous iodide;
(**c**) iron(II) sulfite or ferrous sulfite;
(**d**) diphosphorus pentasulfide;
(**e**) potassium chlorate;
(**f**) tin(IV) oxide or stannic oxide;
(**g**) ammonium sulfate;
(**h**) sodium phosphate

2. (**a**) Ca_3N_2; (**b**) Cu_2O; (**c**) NH_3;
(**d**) Li_2CrO_4; (**e**) HNO_3;
(**f**) NH_4ClO_4; (**g**) $Sn(SO_4)_2$;
(**h**) $MgCr_2O_7$

3. (**a**) (3); (**b**) (1); (**c**) (2); (**d**) (3)

Chapter 8

3. (**a**) 44.0 amu; (**b**) $18\overline{0}$ amu;
(**c**) 17.0 amu; (**d**) 16.0 amu;
(**e**) 80.1 amu; (**f**) 76.0 amu

4. (**a**) 102.0 amu; (**b**) 103.4 amu;
(**c**) 74.1 amu; (**d**) 324.6 amu;
(**e**) $31\overline{0}$ amu; (**f**) 342 amu

5. (**a**) 28.0 g; (**c**) 111.1 g;
(**d**) 164.1 g

6. (**a**) 6 mol C atoms, 6 mol H atoms;
(**b**) 2 mol N atoms, 4 mol O atoms;
(**c**) 2 mol Al atoms, 3 mol S atoms, 12 mol O atoms;
(**d**) 1 mol Ca atoms, 2 mol O atoms, 2 mol H atoms;
(**e**) 2 mol K atoms, 1 mol C atoms, 3 mol O atoms;
(**f**) 1 mol Ba atoms, 4 mol C atoms, 6 mol H atoms, 4 mol O atoms

7. (**a**) 0.133 mol; (**b**) 3.12 mol;
(**c**) 1.56 mol; (**d**) 0.384 mol;
(**e**) 0.0350 mol; (**f**) 1.27 mol;
(**g**) 1.50 mol Fe atoms, 4.50 mol Cl atoms;
(**h**) 8.10 mol Mg^{2+} ions, 5.40 mol P atoms, 21.6 mol O atoms;
(**i**) 0.266 mol;
(**j**) 12.5 mol

8. (**a**) 0.957 mol; (**b**) 2.65 mol;
(**c**) 6.75 mol; (**d**) 0.0726 mol;
(**e**) 42.0 mol; (**f**) 0.467 mol;
(**g**) 1.05 mol; (**h**) 9.39 mol;
(**i**) 2.49 mol; (**j**) 114 mol

9. (**a**) 0.235 mol; (**b**) 0.159 mol;
(**c**) 6.40 mol; (**d**) 8.72 mol;
(**e**) 69.9 mol

10. (**a**) 55.0 g; (**b**) 246 g; (**c**) 29.9 g;
(**d**) 8.78 g; (**e**) 436 mg;
(**f**) 445 g; (**g**) $34\overline{0}$ mg;
(**h**) 43.4 g; (**i**) 80.0 g; (**j**) 13.9 g

11. (**a**) 112 g; (**b**) 159 g; (**c**) 78.9 g;
(**d**) 96.0 mg; (**e**) 9.46 g;
(**f**) 173 mg; (**g**) 196 g;
(**h**) 262 g; (**i**) 0.0338 g;
(**j**) 0.168 g

12. (**a**) 6.32 g; (**b**) 14.6 g;
(**c**) 79.5 g; (**d**) 259 g;
(**e**) $91\overline{0}$ g

13. (**a**) 3.61×10^{23} atoms;
(**b**) 2.41×10^{22} atoms;
(**c**) 4.70×10^{24} molecules;
(**d**) 2.05×10^{23} molecules

14. (**a**) 1.87×10^{24} molecules;
(**b**) 6.02×10^{24} molecules;
(**c**) 3.01×10^{24} atoms;
(**d**) 2.13×10^{23} atoms

15. (**a**) 6.64×10^{-24} g;
(**b**) 1.03×10^{-22} g;

(**c**) 1.41×10^{-22} g;
(**d**) 3.39×10^{-22} g;

16. (**a**) 0.670 mol; (**b**) 0.0391 mol;
(**c**) 2.14 mol; (**d**) 27.5 g;
(**e**) 4.64 g; (**f**) 7.06 g

17. (**a**) 3.94 amu, 3.94 g;
(**b**) 27.9 amu, 27.9 g;
(**c**) 67.7 amu, 67.7 g;
(**d**) 16.2 amu, 16.2 g;
(**e**) 37.0 amu, 37.0 g;
(**f**) 40.3 amu, 40.3 g

18. (**a**) 0.759 g/L; (**b**) 1.34 g/L;
(**c**) 1.16 g/L; (**d**) 1.96 g/L;
(**e**) 5.71 g/L; (**f**) 0.500 g/L

19. (**a**) 7.20 L; (**b**) 5.25 L;
(**c**) 2.00 L; (**d**) 1.45 L;
(**e**) 5.16 L; (**f**) 3.21 L

20. (**a**) 39.3% Na, 60.7% Cl;
(**b**) 5.9% H, 94.1% S;
(**c**) 69.6% Ba, 6.08% C, 24.3% O;
(**d**) 38.7% Ca, 20.0% P, 41.3% O;
(**e**) 52.2% C, 13% H, 34.8% O;
(**f**) 24.0% Fe, 30.9% C, 3.9% H, 41.2% O

21. (**a**) 57.9%; (**b**) 51.3%;
(**c**) 58.7%; (**d**) 67.1%

22. 30.3%

23. (**a**) 21.8g; (**b**) 49.1 g;
(**c**) 12.1 g; (**d**) 6.91 g

24. (**a**) $ZnCl_2$; (**b**) SnI_4; (**c**) $FeBr_2$;
(**d**) $Pb(NO_3)_2$; (**e**) $MgCO_3$;
(**f**) $Ca_3(PO_4)_2$; (**g**) Na_2SO_3;
(**h**) K_2SO_4; (**i**) PBr_5; (**j**) Ga_2O_3

25. (**a**) Al_2O_3; (**b**) CO; (**c**) Na_2S;
(**d**) SO_2

26. (**a**) C_2H_6; (**b**) C_6H_{14};
(**c**) C_2H_2; (**d**) $C_4H_4O_4$;
(**e**) $C_4H_8Cl_2$

27. $C_{10}H_{10}N_4SO_2$

28. $C_{18}H_{22}O_2$

29. $C_{10}H_{14}N_2$

30. C_2N_2

31. 3.40 mol

32. 246 mL

33. 1.31×10^{19} molecules/mL

34. Ammonia

35. (**a**) 4.09 mmol; (**b**) 3.83 g;
(**c**) 4.97×10^{21} molecules

36. (**a**) 0.0050 mmol/mL,
0.00778 mmol/mL,
3.0×10^{18} molecules/mL,
4.68×10^{18} molecules/mL;
(**b**) 28 mmol, 42.8 mmol,
5.0 g, 7.70 g

37. (a) 0.00750 mmol/mL,
0.0128 mmol/mL,
4.52×10^{18} molecules/mL,
7.71×10^{18} molecules/mL;
   (b) 41.2 mmol, 70.4 mmol,
7.42 g, 12.7 g
38. C_2H_4
39. C_3H_6
40. 5.84×10^{20} molecules
41. CO
42. Mostly CH_4

Chapter Quiz A

1. 113 g
2. 9.34×10^{22} molecules
3. 7.34 L
4. 32.5 amu, 32.5 g
5. 1.25 g/L

Chapter Quiz B

1. 40.0% C, 6.7% H, 53.3% O
2. 16.0 g O
3. PBr_5
4. $C_3H_6O_3$
5. C_2H_6

Chapter 9

(The numbers represent the
coefficients in front of the formulas
in the balanced equation.)

4. (a) $1 + 1 \longrightarrow 1 + 2$
   (b) $2 \longrightarrow 2 + 3$
   (c) $1 + 1 \longrightarrow 1 + 2$
   (d) $2 + 3 \longrightarrow 1 + 6$
   (e) $2 + 2 \longrightarrow 2 + 1$
   (f) $3 + 1 \longrightarrow 1$
   (g) $2 + 1 \longrightarrow 2$
   (h) $2 + 1 \longrightarrow 2 + 1$
   (i) $1 + 3 \longrightarrow 2 + 3$
   (j) $3 + 2 \longrightarrow 1 + 6$
5. (a) $1 + 2 \longrightarrow 1 + 1$
   (b) $3 + 4 \longrightarrow 2 + 3$
   (c) $3 + 2 \longrightarrow 1 + 3 + 3$
   (d) $2 + 3 \longrightarrow 1 + 3$
   (e) $1 + 6 \longrightarrow 4$
   (f) $1 + 5 \longrightarrow 3 + 4$
   (g) $6 + 1 \longrightarrow 4$
   (h) $1 + 4 \longrightarrow 1 + 5$
   (i) $1 + 2 \longrightarrow 2 + 1$
   (j) $1 + 2 \longrightarrow 1 + 4$
6. (a) $2\,NaCl + Pb(NO_3)_2 \longrightarrow$
$PbCl_2 + 2\,NaNO_3$

(c) $3\,NaHCO_3 + H_3PO_4 \longrightarrow$
$Na_3PO_4 + 3\,CO_2 + 3\,H_2O$
(e) $CaI_2 + H_2SO_4 \longrightarrow$
$2\,HI + CaSO_4$
(g) $Mg(CN)_2 + 2\,HCl \longrightarrow$
$2\,HCN + MgCl_2$
(i) $2\,NaHSO_3 + H_2SO_4 \longrightarrow$
$Na_2SO_4 + 2\,SO_2 + 2\,H_2O$
(j) $Al_2(SO_4)_3 + 6\,NaOH \longrightarrow$
$2\,Al(OH)_3 + 3\,Na_2SO_4$
7. (a) $2\,Fe + 3\,Cl_2 \longrightarrow 2\,FeCl_3$
   (c) $Ba + 2\,H_2O \longrightarrow$
$Ba(OH)_2 + H_2$
   (e) $(NH_4)_2S + HgBr_2 \longrightarrow$
$2\,NH_4Br + HgS$
   (g) $SnO + 2\,HCl \longrightarrow$
$SnCl_2 + H_2O$
   (i) $2\,HBr + Ca(OH)_2 \longrightarrow$
$CaBr_2 + 2\,H_2O$
8. (a) $4 + 3 \longrightarrow 2$; **(c)** $1 + 1 \longrightarrow 1$;
   (e) $2 + 1 \longrightarrow 2$; **(g)** $1 + 3 \longrightarrow 2$;
   (i) $1 + 1 \longrightarrow 1$
9. (a) $2 \longrightarrow 2 + 1$; **(c)** $1 \longrightarrow 1 + 1$;
   (e) $2 \longrightarrow 1 + 1 + 1$;
   (g) $1 \longrightarrow 1 + 1 + 1$;
   (i) $1 \longrightarrow 1 + 7$
10. (a) $2 + 6 \longrightarrow 2\,AlCl_3 + 3\,H_2$;
   (c) $2 + 6 \longrightarrow$
$2\,Al(C_2H_3O_2)_3 + 3\,H_2$;
   (e) $2 + 3 \longrightarrow 3\,Sn + 2\,AlCl_3$;
   (g) $1 + 1 \longrightarrow Hg + PbBr_2$;
   (i) $1 + 2 \longrightarrow BaCl_2 + H_2$
11. (a) $2 + 1 \longrightarrow Ag_2S(s) + 2\,HNO_3$;
   (c) $1 + 2 \longrightarrow 2\,NaC_2H_3O_2 +$
$CO_2(g) + H_2O$;
   (e) $1 + 1 \longrightarrow SnS(s) + 2\,HCl$;
   (g) $1 + 1 \longrightarrow$
$FeSO_4 + CO_2(g) + H_2O$;
   (i) $1 + 1 \longrightarrow PbS(s) + 2\,HNO_3$;
12. (a) $1 + 2 \longrightarrow Zn(NO_3)_2 + 2\,H_2O$;
   (c) $3 + 2 \longrightarrow Fe_2(SO_4)_3 + 3\,H_2O$;
   (e) $1 + 2 \longrightarrow Na_2SO_4 + 2\,H_2O$;
   (g) $1 + 2 \longrightarrow Na_2CO_3 + H_2O$;
   (i) $2 + 1 \longrightarrow Sr(NO_3)_2 + 2\,H_2O$
13. (a) $2 + 1 \longrightarrow 2$, combination;
   (c) $1 + 1 \longrightarrow 1 + 1$, single
replacement;
   (e) $1 + 2 \longrightarrow 1 + 2$, double
replacement;
   (g) $1 + 2 \longrightarrow 1 + 1$, double
replacement;
   (i) $1 + 1 \longrightarrow 1 + 1$,
neutralization

14. (a) $1 + 2 \longrightarrow 1 + 1$, single
replacement;
   (c) $1 + 1 \longrightarrow 1 + 3$,
neutralization;
   (e) $1 + 3 \longrightarrow 1 + 3$, double
replacement;
   (g) $1 + 1 \longrightarrow 1 + 1$,
neutralization
   (i) $1 \longrightarrow 1 + 1$, decomposition
15. (a) $1 \longrightarrow 1 + 1$, decomposition;
   (c) $1 + 1 \longrightarrow 1$, combination;
   (e) $1 + 1 \longrightarrow 1 + 2$, double
replacement;
   (g) $1 + 2 \longrightarrow 1 + 1$,
neutralization;
   (i) $1 + 2 \longrightarrow 1 + 1$, single
replacement;
16. (e) $1 + 2 \longrightarrow 1 + 1 + 1$, double
replacement;
   (c) $1 + 1 \longrightarrow 1 + 1$, single
replacement;
   (e) $1 + 1 \longrightarrow 1 + 2$, double
replacement;
   (g) $1 + 1 \longrightarrow 1$, combination;
   (i) $1 + 6 \longrightarrow 4$, combination;
17. (a) $1 + 2 \longrightarrow 1 + 1$,
neutralization;
   (c) $1 + 2 \longrightarrow 1 + 1 + 1$, double
replacement;
   (e) $2 \longrightarrow 2 + 1$, decomposition;
   (g) $4 + 1 \longrightarrow 2$, combination;
   (i) $1 + 2 \longrightarrow 1 + 2$,
neutralization
18. (i) (a) $1 + 2 \longrightarrow CdCl_2 + H_2(g)$
   (c) $1 + 2 \longrightarrow Ca(NO_3)_2 +$
$H_2O + CO_2(g)$
   (ii) (a) single-replacement
   (c) double replacement
19. (i) (a) $2\,Al + 3\,PbCl_2 \longrightarrow$
$2\,AlCl_3 + 3\,Pb(s)$;
   (c) $BaCl_2 + Na_2CO_3 \longrightarrow$
$2\,NaCl + BaCO_3(s)$
   (ii) (a) single-replacement
   (c) double replacement
20. $H_3C_6H_5O_7 + 3\,NaHCO_3 \longrightarrow$
$3\,CO_2(g) + Na_3C_6H_5O_7 + 3\,H_2O$
21. (1) $CaCO_3 + H_2SO_4 \longrightarrow$
$CaSO_4 + H_2O + CO_2$;
   (2) $SO_3 + H_2O \longrightarrow H_2SO_4$
22. $PCl_3 + 3\,H_2O \longrightarrow$
$H_3PO_3 + 3\,HCl$
23. $TiCl_4 + 2\,BCl_3 + 5\,H_2 \longrightarrow$
$TiB_2 + 10\,HCl$

24. $Na_2CO_3 + 2\,HNO_3 \longrightarrow$
    $\qquad 2\,NaNO_3 + H_2O + CO_2$
25. $Mg_3N_2 + 6\,H_2O \longrightarrow$
    $\qquad 3\,Mg\,(OH)_2 + 2\,NH_3(g)$
26. $Fe_2O_3 + 6\,HC_2H_3O_2 \longrightarrow$
    $\qquad 2\,Fe(C_2H_3O_3)_3 + 3\,H_2O$

Chapter Quiz

1. (a) $3 + 1 \longrightarrow 1$;
   (b) $2 + 7 \longrightarrow 4 + 6$
2. $Al_2(SO_4)_3 + 6\,NaOH \longrightarrow$
   $\qquad 2\,Al(OH)_3 + 3\,Na_2SO_4$
3. (a) combination;
   (b) single replacement;
   (c) decomposition;
   (d) double replacement
4. (a) $1 + 2 \longrightarrow$
   $\qquad Ba(C_2H_3O_2)_2 + 2\,H_2O$;
   (b) $1 + 2 \longrightarrow MgCl_2 + H_2(g)$;
   (c) $3 + 2 \longrightarrow$
   $\qquad Ca_3(PO_4)_2(s) + 6\,HCl$;
   (d) $1 + 2 \longrightarrow PbCl_2(s) + 2\,HNO_3$

Chapter 10

3. 55.3 g
4. 0.600 g
5. 5.23 g
6. 0.785 kg
7. 8.88 g
8. 0.242 kg
9. 13.3 kg
10. 1.37 g
11. 0.202 mol
12. 0.648 mol
13. 165 g
14. 35.1 g
15. 0.325 mol
16. 0.117 mol
17. 1.05 mol
18. 0.500 mol
19. (a) 50.2 g; (b) 90.0%
20. (a) 0.525 mol; (b) 95.2%;
    (c) 0.075 mol
21. (a) 0.12 g; (b) $5\overline{0}$%;
    (c) 0.025 mol
22. (a) 38.2 g; (b) 89.0%;
    (c) 0.037 mol
23. 1.02 L
24. 4.98 L
25. 0.192 L
26. 5.52 g, 0.164 mol
27. 9.24 L
28. 0.0246 mol

29. (a) 15.1 L; (b) 0.151 mol
30. (a) 44.2 g; (b) 0.099 mol
31. 1.75 L
32. 4.00 L
33. 4.25 L
34. 2.62 L
35. (a) 3.10 L; (b) 2.70 L
36. (a) 8.25 L; (b) 0.13 L
37. (a) exothermic; (b) 12.5 kcal
38. (a) endothermic; (b) 12.4 g
39. (a) exothermic; (b) 433 kJ
40. (a) $CH_4(g) + 2\,O_2(g) \overset{\Delta}{\longrightarrow}$
    $\qquad CO_2(g) + 2\,H_2O(g)$;
    (b) 2.25 mol; (c) 12.5 mol;
    (d) 32.0 g; (e) 16.8 L;
    (f) 11.2 L; (g) 36.3 g;
    (h) 86.2%
41. 95.0 mL
42. 194 mL
43. (a) 0.026 g; (b) 77%
44. (a) 0.20 g; (b) $9\overline{0}$%
45. 7.32×10^5 L

Chapter Quiz

1. 0.282 mol
2. 5.71 L
3. 186 kcal
4. (a) 39.6 g; (b) 81.1%;
   (c) 0.89 mol

Chapter 11

8. 1110 mL
9. 81.2 mL
10. 238 torr
11. 202 atm
12. 192 mL
13. 80.6 mL
14. $-88°C$
15. $25\overline{0}°F$
16. 145 torr
17. 663 torr
18. 97°C
19. $-7°C$
20. 439 mL
21. 295 mL
22. 192 mL
23. 1920 mm Hg
24. 0.840 atm
25. 0.801 atm
26. 455°C
27. 518°C
28. $-75°C$
29. (a) 746 torr; (b) 1.23 L

30. (a) 385 torr; (b) 163 mL
31. 115 mL
32. 208 mL
33. 611 mL
34. 1.52 atm
35. $1\overline{0}0°C$
36. 0.154 mol
37. 8.19 g
38. 24.9 amu, 24.9 g
39. 15.5 amu, 15.5 g
40. 32.1 amu, 32.1 g
41. 5.21 g/L
42. 0.337 g/L
43. 1.09 g/L
44. 625 mL
45. 0.320 mol
46. 1.09 g
47. 89.4 amu, 89.4 g
48. 1.16×10^{23} molecules
49. 86.0 amu, C_6H_{14}
50. 30.0 amu, C_2H_6
51. 197 amu, $C_2HBrClF_3$
52. 8.31 Pa·m³/(mol·K)
53. 170 lb
54. 2.21×10^4 mL

Chapter Quiz

1. 34.2 mL
2. 377 mL
3. 58.3 amu, 58.3 g
4. 1.42 g/L
5. 4.97 L

Chapter 12

4. Higher-kinetic-energy molecules escape from surface of the liquid and evaporate. This cools skin.
5. At higher elevations (mountains), atmospheric pressure is less, so boiling temperature is lower.
6. Lower surface tension.
7. Temperature increases; average KE increases; breaks attractive forces between molecules; viscosity decreases.
8. Strongest, solid; weakest, gas.
9. (1) Increased pressure and (2) friction.
10. Alcohol ($-117.3°C$) has lower melting point than mercury ($-38.9°C$).
11. Sublimation occurs at slower rate in covered box.

12. Solid sublimed.

13. (a) 90°C; (b) 85°C; (c) 70°C;
(d) 65°C

14. 8.64 kcal

15. 305 kJ

16. 5.21 kJ

17. 5010 cal for CCl_4; 25,500 cal for
$NaCl$; $NaCl$ greater attractive
forces than CCl_4

18. −0.13°C

19. −0.22°C

20. 4.4 kcal

21. 1.1 kcal

22. 0.601 kJ

23. 1.02 kcal

24. 18.0 kcal

25. 105.4 kJ

26. 14.3 kcal

27. 291 kJ

28. 10,896 cal

29. $5\overline{0}00$ g

30. (a) 1.07×10^{12} m³;
(b) 1.7×10^{11} kcal

Chapter Quiz

1. 35.1 kcal

2. 8.95 kJ

3. 11 kcal

4. 226 kJ

Chapter 13

3. (a) and (b): no; (c): yes

8. (a) H—O; (b) 0; (c) $\overset{\longrightarrow}{\text{H}-\text{Cl}}$;
    ↗ |
   ✗ H
(d) $\overset{\longrightarrow}{\text{Br}-\text{Cl}}$

9. Bp of HF is higher than that of
HCl and HBr. Because of the high
electronegativity of F, HF
hydrogen bonds. Cl and Br have
lower electronegativities and do not
hydrogen bond. Energy required
to separate molecules of HF
includes energy to break hydrogen
bonds. HCl and HBr do not have
hydrogen bonds to break and thus
separate with less energy needed.

10. (a) $1 + 6 \longrightarrow$
    $6 CO_2(g) + 6 H_2O(g)$
(c) $1 + 1 \longrightarrow 2 H_2O + CaSO_4(s)$
(e) $1 + 2 \longrightarrow K_2CO_3 + H_2O$
(g) $2 + 2 \longrightarrow 2 KOH + H_2(g)$

(i) $1 + 1 \longrightarrow MgO(s) + H_2(g)$
(j) $2 + 3 \longrightarrow Al_2O_3(s) + 3 H_2(g)$

11. (a) $2 + 7 \longrightarrow$
    $4 CO_2(g) + 6 H_2O(g)$
(c) $1 + 1 \longrightarrow CaCO_3(s) + H_2O$
(e) $2 + 1 \longrightarrow K_2SO_4 + H_2O$
(g) $1 + 2 \longrightarrow ZnCl_2 + H_2O$
(h) $1 + 2 \longrightarrow Ca(OH)_2 + H_2(g)$
(i) $2 + 2 \longrightarrow 2 LiOH + H_2(g)$
(j) $1 + 1 \longrightarrow ZnO(s) + H_2(g)$

12. (a) 51.1%; (b) 20.9%;
(c) 43.8%; (d) 44.8%;
(e) 24.5%

13. (a) $CuSO_4 \cdot H_2O$;
(b) $CuSO_4 \cdot 3 H_2O$;
(c) $CuSO_4 \cdot 5 H_2O$;
(d) $Na_2CO_3 \cdot H_2O$;
(e) $Na_2CO_3 \cdot 10H_2O$

14. $CaCl_2 \cdot 6 H_2O$

15. Structural formulas only
(a) H—O
         \
          H
(b) H
     \
      O—O
          \
           H

16. (a) $2 \longrightarrow 2 H_2O + O_2(g)$

17. 5.34 g

18. 4.36 L

19. 6.93 L

20. 59.3 mL

21. 0.607 mL

22. $(CaSO_4)_2 \cdot H_2O(s) +$
$3 H_2O(l) \longrightarrow 2 CaSO_4 \cdot 2 H_2O(s)$;
combination reaction

Chapter Quiz

1. Bp of NH_3 is higher than that of
PH_3 and AsH_3. Because of the
high electronegativity of N, NH_3
hydrogen bonds. P and As have
low electronegativities and do not
hydrogen bond. Energy required to
separate molecules of NH_3
includes energy to break hydrogen
bonds. PH_3 and AsH_3 do not have
hydrogen bonds to break and thus
separate with less energy needed.

2. (a) $1 + 1 \longrightarrow BaSO_4(s) + H_2O_2$;
(b) $1 + 2 \longrightarrow Na_2CO_3 + H_2O$;
(c) $3 + 2 \longrightarrow$
    $Ca_3(PO_4)_2(s) + 6 H_2O$
(d) $2 + 3 \overset{\Delta}{\longrightarrow}$
    $Al_2O_3 + 3 H_2(g)$

3. 42.4%

4. $LiBr \cdot 2 H_2O$

Chapter 14

3. (a), (c), (e), (j): solid in liquid;
(b), (g), (h), (i): gas in liquid;
(d), (f): liquid in liquid

4. (a), (b), (d) water

6. Add crystal of solute

7. (a) solid in liquid—sol;
(c) liquid in gas—liquid aerosol

8. A forms a solution

9. The colloid comes in contact with
ocean water containing ionic
compounds at the river delta and
silt deposits.

10. (a) 60 g/100 g; (b) 5 g/100 g;
(c) 36 g/100 g; (d) 34 g/100 g;
(e) 49 g/100 g; (f) 50 g/100 g

11. 0.184 g/L

12. 0.294 g/L

13. (a) 7.63%; (b) 20.1%; (c) 4.13%

14. (a) 11.8%; (b) 13.3%; (c) 1.02%

15. (a) 61.8 g; (b) 2.05 g; (c) 30.6 g

16. (a) 64.0 g; (b) 207 g; (c) 74.0 g

17. (a) 455 g; (b) 20.0 g; (c) 50.0 g

18. (a) 233 ppm; (b) 202 ppm;
(c) 1.7 ppm

19. (a) $75\overline{0}$ ppm; (b) 130 ppm;
(c) 6.0×10^{-3} ppm

20. (a) 82 mg; (b) 640 mg;
(c) 4.5×10^{-3} mg

21. (a) 3.99 M;
(b) 1.13 M, 1.13 M Cl⁻;
(c) 0.221 M

22. (a) 0.309 M, 0.309 M Br⁻;
(b) 0.169 M, 0.338 M Cl⁻;
(c) 0.0439 M, 0.0878 M Br⁻

23. (a) 1.98 g NaOH; the NaOH (1.98
g) is dissolved in *sufficient*
water to make the total
volume of the solution equal
to $45\overline{0}$ mL;
(b) 6.11 g; (c) 2.20 g

24. (a) 534 mL; (b) 49.2 mL;
(c) $2\overline{0}0$ mL

25. (a) $39\overline{0}$ mL; $39\overline{0}$ mL is added to
1-L volumetric flask contain-
ing some water (about 400 mL)
and swirled. Then more water
added to 1-L mark. The flask
is capped and turned upside-
down a number of times.

(b) 519 mL, similar to (a)

(c) 97.4 mL, similar to (a), except 97.4 mL is added to water (about 200 mL) and more water added to 500-mL mark.

26. (a) 0.356 M; (b) 0.890 M; (c) 1.60 M

27. (a) 0.528 N; (b) 0.0491 N; (c) 8.01 N

28. (a) 0.495 N; (b) 0.402 N; (c) 0.0293 N

29. (a) 0.172 g; (b) 3.79 g; (c) 0.185 g

30. (a) 383 ml; (b) 54$\overline{0}$ mL; (c) 386 mL

31. (a) 3.00 N; (b) 7.50 N; (c) 0.0300 N

32. (a) 1.50 M; (b) 2.50 M; (c) 0.0500 M

33. (a) 4.62 m; (b) 2.97 m; (c) 0.131 m

34. (a) 0.499 m; (b) 0.406 m; (c) 2.79 m

35. (a) 8.13 g; (b) 32.7 g; (c) 70.2 g

36. (a) 236 g; (b) 34.7 g; (c) 342 g

37. (a) 100.96°C, −3.44°C; (b) 100.21°C, −0.74°C (c) 100.35°C, −1.24°C

38. (a) 100.42°C, −1.49°C; (b) 100.34°C, −1.23°C (c) 100.37°C, −1.32°C

39. (a) 93.7 amu, 93.7 g (b) 41.1 amu, 41.1 g (c) 197 amu, 197 g

40. $C_6H_{12}O_6$

41. 0.398 M

42. 103.0°C, −10.8°C

Chapter Quiz A

1. 68.0 g
2. 6$\overline{0}$0 mg
3. 1.51 M
4. 22.4 g
5. 86.2 mL

Chapter Quiz B

1. 0.780 N
2. 12.0 N
3. 11$\overline{0}$ g
4. −0.44°C
5. 1 (F); 2 (C); 3(D); 4(I)

Chapter 15

6. (a), (d), (e): acid

7. (a), (d), (e): acid; (b), (c), (f): amphoteric

8. (a) conjugate acid, H_3O^+; conjugate base, HSO_4^-;
   (b) conjugate acid, H_3O^+; conjugate base, SO_4^{2-}

9. (a) strong acid; (b) weak base; (c) weak acid; (d) strong base

10. (a) $HClO_4(aq) + H_2O(l) \longrightarrow$ $H_3O^+(aq) + ClO_4^-(aq)$
    (b) $HClO(aq) + H_2O(l) \rightleftharpoons$ $H_3O^+(aq) + ClO^-(aq)$

11. (a) $2\,HCl(aq) + Zn(s) \longrightarrow$ $ZnCl_2(aq) + H_2(g)$
    (b) $2\,HBr(aq) + Cd(s) \longrightarrow$ $CdBr_2(aq) + H_2(g)$
    (c) No reaction

12. (a) $3\,NaOH(aq) + Al(NO_3)_3(aq)$ $\longrightarrow Al(OH)_3(s) + 3\,NaNO_3(aq)$
    (b) $3\,Ba(OH)_2(aq) + 2\,FeCl_3(aq)$ $\longrightarrow 3\,BaCl_2(aq) + 2\,Fe(OH)_3(s)$
    (c) $2\,KOH(aq) + NiBr_2(aq) \longrightarrow$ $2\,KBr(aq) + Ni(OH)_2(s)$

13. (a) $HCl(aq) + KOH(aq) \longrightarrow$ $KCl(aq) + H_2O(l)$
    (b) $2\,HNO_3(aq) + Ba(OH)_2(aq)$ $\longrightarrow Ba(NO_3)_2(aq) + 2\,H_2O(l)$
    (c) $H_2SO_4(aq) + 2\,NaOH(aq)$ $\longrightarrow Na_2SO_4(aq) + 2\,H_2O(l)$

14. 0.0455 M

15. 0.0194 M

16. 0.112 M

17. (a) 0.752 M; (b) 2.71%

18. (a) 0.792 M; (b) 4.72%

19. 0.399 M

20. (a) 0.877 M; (b) 3.15%

21. 35.3 mL

22. (a) 0.160 N; (b) 0.160 M

23. (a) 9.00, 5.00; (b) 11.70, 2.30; (c) 6.70, 7.30; (d) 6.20, 7.80; (e) 3.60, 10.40

24. (a) 5.62, 8.38; (b) 3.10, 10.90; (c) 2.85, 11.15; (d) 8.28, 5.72; (e) 5.07, 8.93

25. (a) 7.49 to 8.00;
    (b) bitter, soapy taste due to properties of bases

26. (a) 4.49, 9.51; (b) 10.00, 4.00

27. (a) 6.3×10^{-7} mol/L;
    (b) 2.5×10^{-10} mol/L;
    (c) 3.2×10^{-12} mol/L;

(d) 1.0×10^{-2} mol/L;
(e) 5.9×10^{-10} mol/L

28. (a) 1.0×10^{-5} mol/L;
    (b) 2.5×10^{-10} mol/L;
    (c) 5.0×10^{-9} mol/L;
    (d) 1.6×10^{-6} mol/L;
    (e) 2.5×10^{-9} mol/L

29. 1.0×10^{-4} mol/L. Nonmetal oxides dissolved in water form acids.

30. 1.3×10^{-2} mol/L

31. Net ionic equations only
    (a) $Ba^{2+} + CO_3^{2-} \longrightarrow BaCO_3(s)$
    (b) $Fe^{3+} + 3\,NH_3 + 3\,H_2O \longrightarrow$ $Fe(OH)_3(s) + 3\,NH_4^+(aq)$
    (c) $Sr^{2+} + CO_3^{2-} \longrightarrow SrCO_3(s)$
    (d) $CO_3^{2-} + 2\,H^+ \longrightarrow$ $H_2O + CO_2(g)$
    (e) $Cl^- + Ag^+ \longrightarrow AgCl(s)$
    (f) $2\,Al(s) + 6\,H^+ \longrightarrow$ $2\,Al^{3+} + 3\,H_2(g)$
    (g) $CO_2(g) + Ca^{2+} + 2\,OH^- \longrightarrow$ $CaCO_3(s) + H_2O$
    (h) $SrCO_3(s) + 2\,HC_2H_3O_2 \longrightarrow$ $Sr^{2+} + 2\,C_2H_3O_2^-$ $+ H_2O + CO_2(g)$
    (i) $Fe(s) + Cu^{2+} \longrightarrow$ $Fe^{2+} + Cu(s)$
    (j) $Cd^{2+} + H_2S \longrightarrow$ $CdS(s) + 2\,H^+$

32. Net ionic equations only
    (a) $Cu^{2+} + H_2S \longrightarrow$ $CuS(s) + 2\,H^+$
    (b) $Mg^{2+} + 2\,OH^- \longrightarrow$ $Mg(OH)_2(s)$
    (c) $CaO(s) + 2\,H^+ \longrightarrow$ $Ca^{2+} + H_2O$
    (d) $Fe^{3+} + 3\,NH_3 + 3\,H_2O \longrightarrow$ $Fe(OH)_3(s) + 3\,NH_4^+$
    (e) $Fe^{2+} + S^{2-} \longrightarrow FeS(s)$
    (f) $Al(OH)_3(s) + 3\,H^+ \longrightarrow$ $Al^{3+} + 3\,H_2O$
    (g) $H_3PO_4 + 3\,OH^- \longrightarrow$ $PO_4^{3-} + 3\,H_2O$
    (h) $Mg^{2+} + CO_3^{2-} \longrightarrow$ $MgCO_3(s)$
    (i) $Cl_2(g) + 2\,Br^- \longrightarrow$ $2\,Cl^- + Br_2$
    (j) $2\,Bi^{3+} + 3\,H_2S \longrightarrow$ $Bi_2S_3(s) + 6\,H^+$

33. (a) 2.14 M; (b) 7.53%;
    (c) 2.14 N

34. Net: $AlZ_3(s) + 3 H^+ \longrightarrow$
$$Al^{3+} + 3 HZ$$

35. (a) Three;
(b) Net: $C_6H_8O_7(aq) +$
$3 OH^-(aq) \longrightarrow C_6H_5O_7^{3-}(aq)$
$+ 3 H_2O(l)$

36. $23\overline{0}$ amu, $23\overline{0}$ g

Chapter Quiz A

1. (a) acid;
(b) and (d) amphoteric
(c) base
2. 0.270 M
3. 0.352 M
4. 0.0801 N

Chapter Quiz B

1. (a) 7.22, 6.78; (b) 6.89, 7.11
2. (a) 3.2×10^{-5} mol/L;
(b) 2.5×10^{-10} mol/L
3. (a) Net: $Cd^{2+} + 2 OH^- \longrightarrow$
$$Cd(OH)_2(s);$$
(b) Net: $Zn + 2 H^+ \longrightarrow$
$$Zn^{2+} + H_2(g);$$
(c) Ionic and net: $Zn^{2+} + H_2S$
$$\longrightarrow ZnS(s) + 2 H^+$$

Chapter 16

5. (a) +4; (b) +5; (c) +3;
(d) +5; (e) +6; (f) +6
The numbers represent the coefficients in front of the formulas in the balanced equation.
6. (a) $2 + 6 \longrightarrow 2 + 3 + 4$;
(b) $8 + 5 \longrightarrow 1 + 4 + 4 + 4$;
(c) $3 + 8 \longrightarrow 3 + 2 + 4$;
(d) $1 + 7 + 8 \longrightarrow 8 + 4 + 4$;
(e) $10 + 1 \longrightarrow 10 + 4 + 2$;
(f) $2 + 2 \longrightarrow 1 + 1 + 2$;
(g) $1 + 6 + 14 \longrightarrow 2 + 6 + 7$;
(h) $1 + 4 + 1 \longrightarrow 1 + 1 + 2$;
(i) $1 + 1 + 2 \longrightarrow 2 + 1 + 1$;
(j) $3 + 1 + 14 \longrightarrow 3 + 2 + 7$
7. Oxidized, reducing agent:
(a) HI; (b) KI; (c) Cu; (d) KI;
(e) I_2; (f) Ag; (g) Fe^{2+};
(h) Cu; (i) CdS; (j) Zn
8. (a) $2 + 3 \longrightarrow 2 + 3 + 3$;
(b) $6 + 10 \longrightarrow 3 + 10 + 5$;
(c) $1 + 4 + 6 \longrightarrow 6 + 1 + 3 + 2$;
(d) $2 + 5 + 16 \longrightarrow 2 + 5 + 5 + 7$;
(e) $3 + 5 + 1 + 3 \longrightarrow 3 + 3 + 3$;
(f) $1 + 5 \longrightarrow 1 + 5$;

(g) $1 + 14 + 6 \longrightarrow 2 + 3 + 7$;
(h) $10 + 2 + 16 \longrightarrow 2 + 5 + 8$;
(i) $2 + 5 + 2 \longrightarrow 2 + 5 + 4$;
(j) $3 + 1 + 8 \longrightarrow 3 + 2 + 7$;
9. Oxidized, reducing agent:
(a) K_2SnO_2; (b) Sb; (c) NaI;
(d) $Mn(NO_3)_2$; (e) KI; (f) CO;
(g) Cl^-; (h) I^-; (i) SO_2;
(j) H_2S
10. (a) $3 + 1 + 6 H^+ \longrightarrow$
$$3 + 1 + 3 H_2O;$$
(b) $5 + 2 + 6 H^+ \longrightarrow$
$$5 + 2 + 3 H_2O;$$
(c) $5 + 2 + 16 H^+ \longrightarrow$
$$10 + 2 + 8 H_2O;$$
(d) $2 + 5 + 14 H^+ \longrightarrow$
$$2 + 5 + 7 H_2O;$$
(e) $2 + 5 + 6 H^+ \longrightarrow$
$$2 + 5 + 8 H_2O;$$
(f) $1 + 1 + 4 H^+ \longrightarrow$
$$1 + 1 + 2 H_2O;$$
(g) $3 + 6 OH^- \longrightarrow$
$$1 + 5 + 3 H_2O;$$
(h) $1 + 2 OH^- \longrightarrow 1 + 1 + H_2O$;
(i) $2 + 1 + 4 OH^- \longrightarrow 2 + 2$;
(j) $1 + 4 \longrightarrow 1 + 4$
11. Oxidized, reducing agent:
(a) Sn^{2+}; (b) AsO_2^-;
(c) $C_2O_4^{2-}$; (d) Mn^{2+};
(e) H_2O_2; (f) Fe; (g) Cl_2;
(h) Cl_2; (i) MnO_2; (j) PbS
12. (a) $1 + 3 + 14 H^+ \longrightarrow$
$$2 + 6 + 7 H_2O;$$
(b) $3 + 2 + 8 H^+ \longrightarrow$
$$3 + 2 + 4 H_2O;$$
(c) $5 + 2 + 6 H^+ \longrightarrow$
$$5 + 2 + 3 H_2O;$$
(d) $1 + 1 + H_2O \longrightarrow$
$$1 + 2 + 2 H^+;$$
(e) $1 + 2 + 2 H^+ \longrightarrow$
$$1 + 1 + H_2O;$$
(f) $2 + 3 \longrightarrow 3 + 2 + 3 H_2O$;
(g) $1 + 1 + 2 OH^- \longrightarrow 1 + 2$;
(h) $4 + 3 + 2 H_2O \longrightarrow$
$$4 + 3 + 4 OH^-;$$
(i) $3 + 8 H^+ + 2 NO_3^- \longrightarrow$
$$3 Ni^{2+} + 2 + 3 + 4 H_2O;$$
(j) $SbCl_6^- + 1 \longrightarrow$
$$SbCl_6^{3-} + 1 + 2 H^+$$
13. Oxidized, reducing agent:
(a) $C_2O_4^{2-}$; (b) S^{2-}; (c) SO_3^{2-};
(d) AsO_3^{3-}; (e) I^-; (f) SnO_2^{2-};
(g) Mn^{2+}; (h) ClO_2^-; (i) NiS;
(j) H_2S

14. (1) oxidation half-reaction:
$Zn + 2 OH^- \longrightarrow Zn(OH)_2 + 2e^-$
Reduction half-reaction:
$O_2(g) + 4e^- + 2 H_2O \longrightarrow 4 OH^-$
(2) Reduction at cathode; oxidation at anode
15. (a) 31.4 g; (b) 23.5 L;
(c) 28.0 L; (d) 21.9 g
16. 5.10 g
17. (a) $2 + 10 + 8 \longrightarrow 2 + 5 + 1 + 8$;
(b) 0.00303 mol; (c) 30.3 mL
18. (a) $3 + 2 + 8 \longrightarrow 3 + 2 + 2 + 11$;
(b) $K_2Cr_2O_7$
19. (a) $2 + 5 \longrightarrow$
$$1 + 5 + H_2O + 3 H^+;$$
(b) 513 g
20. (a) $4 H^+ + 2 ClO^- + 2e^- \longrightarrow$
$$Cl_2 + 2 H_2O;$$
(b) reduction

Chapter Quiz

1. (a) +4; (b) +2
2. $3 + 2 \longrightarrow 3 + 2 + 4$
3. HNO_3
4. (a) $5 + 2 + 12 H^+ \longrightarrow$
$$5 + 1 + 6 H_2O$$
(a) $4 + 1 + 6 H_2O \longrightarrow$
$$4 + 1 + 9 OH^-$$
5. (a) Zn; (b) Zn
6. (1) Oxidation: $Fe(s) + 2 OH^-$
$$\longrightarrow Fe(OH)_2(s) + 2e^-$$
Reduction: $NiO_2(s) + 2 H_2O +$
$2e^- \longrightarrow Ni(OH)_2(s) + 2 OH^-$
(2) Oxidation at anode;
reduction at cathode

Chapter 17

4. A catalyst does not alter the point of equilibrium.
7. (a) 4 times; (b) 9 times;
(c) 2 times, the original rates.
8. (a) $K = \dfrac{[CH_3Cl][HCl]}{[CH_4][Cl_2]}$

(c) $K = \dfrac{[NO_2]^2}{[NO]^2[O_2]}$

(e) $K = \dfrac{[H_2]^4}{[H_2O]^4}$
(Both Fe_3O_4 and Fe are solids.)

(g) $K = \dfrac{[SO_2][Cl_2]}{[SO_2Cl_2]}$

(i) $K = \dfrac{[NO_2]^4[O_2]}{}$
[Both PbO and Pb(NO$_3$)$_2$ are solids.]

(j) $K = \dfrac{[H_2O]^2}{[H_2]^2[O_2]}$

9. (a) $K = \dfrac{[NO]^2}{[N_2][O_2]}$

(c) $K = [SO_2]$
(BaSO$_3$ and BaO are solids)

(e) $K = \dfrac{[NO]^2[Cl_2]}{[NOCl]^2}$

(g) $K = [Hg]^2[O_2]$
(HgO is a solid)

(i) $K = \dfrac{[CO_2][H_2O]^2}{[CH_4][O_2]^2}$

(j) $K = \dfrac{1}{[Cl_2]}$

(Both Ag and AgCl are solids.)

10. Sulfurous acid > chloroacetic acid > formic acid > lactic acid > barbituric acid > acetic acid

11. Silver hydroxide > ammonia > novocain > codeine > nicotine > urea

12. (a), (c), (f), (h), (i): right

13. (a), (c), (f), (h): right:
(i) and (j): no effect

14. (a) (1) left, (2) right, (3) right, (4) left

(b) (1) left, (2) right, (3) no effect, (4) no effect

(c) (1) right, (2) left, (3) no effect, (4) no effect

(d) (1) right, (2) left, (3) right, (4) left

(e) (1) left, (2) right, (3) right, (4) left

15. (a) (1) left, (2) right, (3) right, (4) left

(b) (1) left, (2) right, (3) no effect, (4) no effect

(c) (1) left, (2) right, (3) right, (4) left

(d) (1) right, (2) left, (3) right, (4) left

(e) (1) left, (2) right, (3) no effect, (4) no effect

16. (a) 1.81×10^{-5} (mol/L);
(b) 7.38×10^{-4} (mol/L);

(c) 7.20×10^{-10} (mol/L);
(d) 7.99×10^{-5} (mol/L)

17. (a) 1.81×10^{-5} (mol/L);
(b) 1.97×10^{-4} (mol/L);
(c) 7.89×10^{-5} (mol/L);
(b) 1.78×10^{-4} (mol/L)

18. (a) 1.84×10^{-3} mol/L, 0.920%;
(b) 2.92×10^{-3} mol/L, 0.584%;
(c) 1.82×10^{-2} mol/L, 0.910%;
(d) 4.00×10^{-4} mol/L, 0.160%;

19. (a) 1.87×10^{-5} mol/L, 0.0187%;
(b) 8.00×10^{-7} mol/L, 0.00288%;
(c) 2.35×10^{-4} mol/L, 0.470%;
(d) 9.49×10^{-5} mol/L, 0.0316%

20. (a) Weak acid: HC$_2$H$_3$O$_2$;
conjugate base: C$_2$H$_3$O$_2{}^-$
HC$_2$H$_3$O$_2$(aq) $\rightleftharpoons$ H$^+$(aq) + C$_2$H$_3$O$_2{}^-$(aq)
H$^+$(aq) + C$_2$H$_3$O$_2{}^-$(aq) $\rightarrow$ HC$_2$H$_3$O$_2$(aq)
OH$^-$(aq) + HC$_2$H$_3$O$_2$(aq) $\rightarrow$ H$_2$O(l) + C$_2$H$_3$O$_2{}^-$(aq)

(b) Weak acid: NH$_4{}^+$; conjugate base: NH$_3$
NH$_4{}^+$(aq) $\rightleftharpoons$ NH$_3$(aq) + H$^+$(aq)
H$^+$(aq) + NH$_3$(aq) $\rightarrow$ NH$_4{}^+$(aq)
OH$^-$(aq) + NH$_4{}^+$(aq) $\rightarrow$ H$_2$O(l) + NH$_3$(aq)

(c) Weak acid: HCN; conjugate base: CN$^-$
HCN(aq) $\rightleftharpoons$ H$^+$(aq) + CN$^-$(aq)
H$^+$(aq) + CN$^-$(aq) $\rightarrow$ HCN
OH$^-$(aq) + HCN(aq) $\rightarrow$ H$_2$O(l) + CN$^-$(aq)

(d) Weak acid: HCHO$_2$; conjugate base: CHO$_2{}^-$
HCHO$_2$(aq) $\rightleftharpoons$ H$^+$(aq) + CHO$_2{}^-$(aq)
H$^+$(aq) + CHO$_2{}^-$(aq) $\rightarrow$ HCHO$_2$(aq)
OH$^-$(aq) + HCHO$_2$(aq) $\rightarrow$ H$_2$O(l) + CHO$_2{}^-$(aq)

21. (a) Weak acid: H$_2$CO$_3$; conjugate base: HCO$_3{}^-$
H$_2$CO$_3$(aq) $\rightleftharpoons$ H$^+$(aq) + HCO$_3{}^-$(aq)
H$^+$(aq) + HCO$_3{}^-$(aq) $\rightarrow$ H$_2$CO$_3$(aq)
OH$^-$(aq) + H$_2$CO$_3$(aq) $\rightarrow$ H$_2$O(l) + HCO$_3{}^-$(aq)

(b) Weak acid: H$_2$PO$_4{}^-$; conjugate base: HPO$_4{}^{2-}$
H$_2$PO$_4{}^-$(aq) $\rightleftharpoons$ H$^+$(aq) + HPO$_4{}^{2-}$(aq)

H$^+$(aq) + HPO$_4{}^{2-}$(aq) $\rightarrow$ H$_2$PO$_4{}^-$(aq)
OH$^-$(aq) + H$_2$PO$_4{}^-$(aq) $\rightarrow$ H$_2$O(l) + HPO$_4{}^{2-}$(aq)

22. Silver acetate > silver bromate > silver chloride > silver bromide > silver iodide

23. (a) $K_{sp} = [Ag^+][Cl^-]$
(b) $K_{sp} = [Al^{3+}][OH^-]^3$
(e) $K_{sp} = [Ag^+]^2[CO_3{}^{2-}]$

24. (a) $K_{sp} = [Ag^+][OH^-]$
(c) $K_{sp} = [Fe^{3+}][OH^-]^3$
(e) $K_{sp} = [Hg^{2+}][I^-]^2$

25. (a) 7.2×10^{-13} (mol^2/L^2);
(b) 8.1×10^{-9} (mol^2/L^2);
(c) 1.0×10^{-8} (mol^2/L^2);
(d) 2.2×10^{-10} (mol^2/L^2)

26. (a) 1.6×10^{-9} (mol^2/L^2);
(b) 4.8×10^{-12} (mol^2/L^2);
(c) 2.0×10^{-10} (mol^2/L^2);
(d) 6.6×10^{-7} (mol^2/L^2)

27. (a) 9.3×10^{-3} M, 1.0 g/L
(b) 8.4×10^{-5} M, 1.7×10^{-2} g/L
(c) 2.0×10^{-3} M, 0.57 g/L
(d) 4.0×10^{-5} M, 5.9×10^{-3} g/L, yes

28. (a) 1.2×10^{-8} M, 2.8×10^{-6} g/L
(b) 1.8×10^{-7} M, 4.8×10^{-5} g/L
(c) 0.15 M, 23 g/L
(d) 1.5×10^{-5} M, 3.8×10^{-3} g/L, no

29. (a) 2.52, 11.48; **(b)** 11.11, 2.89

30. (a) 3.38, 10.62; **(b)** 11.48, 2.52

31. Lower pH means increase in [H$^+$] which shifts equilibrium to left, increasing concentration of Cl$_2$.

Chapter Quiz A

1. Trichloracetic acid > dichloroacetic acid > chloroacetic acid > acetic acid

2. (a) $K = \dfrac{[NO_2]^2}{[NO]^2[O_2]}$

(b) $K = \dfrac{[NO_2]^2}{[N_2O_4]}$

(c) $K = \dfrac{[CO]^2}{[CO_2]}$

(d) $K = [SO_2]$

3. (a) right; **(b)** left

4. (a) right; **(b)** right

Chapter Quiz B

1. (a) $K_{sp} = [Pb^{2+}][I^-]^2$;
   (b) $K_{sp} = [Ag^+]^2[Cr_2O_7^{2-}]$
2. 8.55×10^{-5} (mol/L)
3. 2.96×10^{-4} mol/L
4. 7.2×10^{-13} (mol^2/L^2)
5. 3.8 g/L

Chapter 18

5. (a) all five carbons tetrahedral (109.5°);
   (b) —CH_3 group tetrahedral (109.5°), $\rangle C =$ carbons planar (120°);
   (c) all carbons planar (120°);
   (d) —CH_3 group tetrahedral (109.5°), $\rangle C =$ carbons planar (120°), —$C \equiv$ carbons linear (180°);
   (e) all five carbons tetrahedral (109.5°);
   (f) all six carbons planar (120°)

6. (a) $CH_3CH_2CH_2CH_2CH_2CH_3$

   $CH_3CH_2CHCH_2CH_3$
   |
   CH_3

   CH_3
   |
   $CH_3C-CH_2CH_3$
   |
   CH_3

   $CH_3CHCH_2CH_2CH_3$
   |
   CH_3

   $CH_3CH-CHCH_3$
   | |
   CH_3 CH_3

   (b) $CH_3CH_2CH_2CH_2CH_2CH_2CH_3$

   $CH_3CH_2CHCH_2CH_2CH_3$
   |
   CH_3

   $CH_3CHCH_2CHCH_3$
   | |
   CH_3 CH_3

   CH_3
   |
   $CH_3CH_2CCH_2CH_3$
   |
   CH_3

   $CH_3CH_2CHCH_2CH_3$
   |
   CH_2-CH_3

   $CH_3CHCH_2CH_2CH_2CH_3$
   |
   CH_3

   $CH_3CH-CHCH_2-CH_3$
   | |
   CH_3 CH_3

   CH_3
   |
   $CH_3CCH_2CH_2CH_3$
   |
   CH_3

   CH_3
   |
   $CH_3CH-C-CH_3$
   | |
   CH_3 CH_3

7. (a) 2-chlorobutane
   (c) 3-bromo-2-nitropentane
   (e) 2-bromo-3-methyl-4-nitrohexane
   (f) methylcyclobutane

8. (a) CH_3CH_2CHBr
      |
      CH_3

   (c) CH_3CHCH_3
      |
      NO_2

   (e) CH_3
      |
   $CH_3CH-CCH_2CH_2CH_3$
      | |
      Br CH_3

   (f)

9. (a) $1 + 1 \xrightarrow[\text{or light}]{\Delta} CH_3-Cl + HCl(g)$
   (b) $1 + 1 \xrightarrow[\text{or light}]{\Delta} CH_3-CH_2-Br + HBr(g)$

10. (a) 1-butene
    (c) 3-methyl-2-pentene
    (e) 3-chloro-3-methyl-1-hexene
    (f) 4-chloro-6-nitro-2-heptene

11. (a) $CH_2 = CHCH_2CH_2CH_3$
    (c) $CH_2 = C-CHCH_2CH_2CH_3$
                | |
               CH_3 CH_3
    (e) $CH = CHCHCH_2CH_2CH_2CH_3$
         | |
         Br CH_2-CH_2
    (f) NO_2
                       |
    $CH_3C = CH-CH-CCH_2CH_3$
         | | |
        CH_3 CH_3 CH_3

12. (a) $1 + 1 \rightarrow CH_3-CH-CH_2$
                          | |
                         Br Br
    (b) $1 + 1 \rightarrow$
       Br Br
       | |
    $CH_3-C-CH-CH_3$
       |
      CH_3

13. (a) 1-pentyne
    (c) 4-chloro-4-methyl-2-hexyne
    (e) 5,6-dimethyl-3-heptyne
    (f) 4-bromo-2-hexyne

14. (a) $HC \equiv C(CH_2)_4CH_3$
    (c) $HC \equiv C-CH_2CHCH_2CH_2CH_3$
                          |
                         NO_2
    (e) $ClC \equiv CCHCH_2CH_2CH_3$
                    |
                   CH_3
    (f) CH_3
                          |
    $ClCH_2CHC \equiv CC-CHCH_2CH_3$
         | | |
        Cl CH_3 CH_3

15. (a) Br Br
        | |
    $1 + 2 \rightarrow CH_3C-CCH_3$
        | |
        Br Br
    (b) Br Br CH_3
        | | |
    $1 + 2 \rightarrow CH_3C-C-CCH_3$
        | | |
        Br Br CH_3

16. (a) bromobenzene
    (c) *m*-bromonitrobenzene
    (e) 2,4-dibromotoluene
    (f) 2-bromo-4,5-dinitrotoluene

17. (a)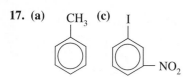

(c) I / NO$_2$ (structure)

(e) CH$_3$ / Cl (structure)

(f) O$_2$N — CH$_3$ / —NO$_2$ / NO$_2$ (structure)

18. (a) 1 + 1 $\xrightarrow[\Delta]{FeBr_3}$ Br (structure) + HBr(g)

(b) 1 + 1 $\xrightarrow[\Delta]{FeCl_3}$ Cl (structure) + HCl(g)

19. (a) alkyl halide; **(b)** aldehyde;
(c) ester; **(d)** alkyne;
(e) alcohol; **(f)** carboxylic acid;
(g) ether; **(h)** amine; **(i)** ketone;
(j) amide

20. (a) 4-bromo-3-iodo-2,3-dimethylheptane
(c) 3,4-dimethyl-4-nitro-1-pentene
(e) 4-bromo-5-methyl-2-hexyne
(g) *m*-nitrotoluene
(h) 2-bromo-4-chloro-3-iodo-5-nitrotoluene

21. (a) five alkenes;
(b) four alkenes;
(c) one aromatic;
(d) one alkene, one aromatic;
(e) one aromatic;
(f) one alkyne, one aromatic;
(g) two aromatics;
(h) two aromatics, one alkene

22. (a) one aldehyde, one ether
(b) one carboxylic acid, one ester;
(c) one alcohol;
(d) three ketones, two alcohols;
(e) one ether, one carboxylic acid, two organic halides;
(f) four alcohols, two ketones, one amide, one amine;

(g) two ketones;
(h) one alcohol, one ketone
23. C$_2$H$_6$, CH$_3$CH$_3$, ethane
24. Two
25.

H — C(=O)OH + NaHCO$_3$ $\longrightarrow$

H — C(=O)O$^-$Na$^+$ + H$_2$O + CO$_2$

26. (a) CH$_3$CH$_2$CH$_2$Cl,

CH$_3$CHCH$_3$ / Cl

(b) Cl / Cl (structure)   Cl / Cl (structure)

Cl / Cl (structure)

(c) CH$_3$CH$_2$—OH,

CH$_3$—O—CH$_3$

(d) CH$_3$CH$_2$CH$_2$—OH,

CH$_3$CH—OH / CH$_3$

CH$_3$CH$_2$—O—CH$_3$

27. C$_3$H$_6$, CH$_3$CH=CH$_2$, propene
28. (a) 1.80 g; **(b)** 75.0%

Chapter Quiz A

1. (a) CH$_3$ group carbons and

>CH— carbon tetrahedral

(109.5°), —C≡ carbons linear (180°);

(b) CH$_3$ group carbons tetrahedral

(109.5°), >C= carbons planar (120°)

2. CH$_3$(CH$_2$)$_4$CH$_3$,

CH$_3$CH(CH$_2$)$_2$CH$_3$, / CH$_3$

CH$_3$CH$_2$CHCH$_2$CH$_3$, / CH$_3$

CH$_3$CH — CHCH$_3$, / CH$_3$ CH$_3$

CH$_3$ / CH$_3$CCH$_2$CH$_3$ / CH$_3$

3. (a) 2,2-dichloropentane;
(b) 2-bromo-3-methyl-2-nitrohexane;
(c) 5-ethyl-3-methyloctane;
(d) 1,2-dimethylcyclobutane

4. (a) CH$_2$CH$_2$CHCH$_2$CH$_2$CH$_3$ / Cl   Cl

(b) Cl (cyclohexane structure)

5. 1 + 1 $\longrightarrow$ CH$_3$CH$_2$Cl + HCl(g)

Chapter Quiz B

1. (a) 4-methyl-1-pentene;
(b) 5-chloro-2-hexyne;
(c) *o*-chlorotoluene
(d) 2,3-dibromotoluene

2. (a) (cyclopentene structure)

(b) BrC≡CCH$_2$CH — CHCH$_3$ / CH$_3$ CH$_3$

(c) CH$_3$ / Cl / Cl (structure)

(d) CH$_3$ / Br / Br / NO$_2$ (structure)

3. (a) $1 + 2 \longrightarrow$ CH$_3$CH$-$C$-$CCH$_3$ (with Br Br on top, CH$_3$ Br Br on bottom)

(b) $1 + 1 \xrightarrow[\Delta]{\text{FeBr}_3}$ [benzene ring with Br] $+ \text{HBr}(g)$

Chapter Quiz C

1. (a) $-\text{C}\overset{\text{O}}{\underset{\text{H}}{\diagdown}}$; **(b)** $-\text{NH}_2$

   aldehyde   amine

2. (a) $\overset{\text{O}}{\underset{\|}{-\text{C}-}}$ **(b)** $-\text{OH}$

(c) $-\text{C}\overset{\text{O}}{\underset{\text{NH}_2}{\diagup}}$

(d) $-\text{O}-$

3. (a) ester; **(b)** carboxylic acid;
(c) alcohol and amine;
(d) amine and 2 esters;
(e) amine and ester;
(f) carboxylic acid, 2 amides
   (1 in ring), and ether

Chapter 19

9. (a) $^{87}_{36}\text{Kr} \longrightarrow \, ^{0}_{-1}\text{e} + \, ^{87}_{37}\text{Rb}$
(c) $^{232}_{92}\text{U} \longrightarrow \, ^{4}_{2}\text{He} + \, ^{228}_{90}\text{Th}$
(e) $^{32}_{14}\text{Si} \longrightarrow \, ^{32}_{15}\text{P} + \, ^{0}_{-1}\text{e}$
(g) $^{60}_{27}\text{Co} \longrightarrow \, ^{60}_{28}\text{Ni} + \, ^{0}_{-1}\text{e}$
10. $^{32}_{15}\text{P} \longrightarrow \, ^{0}_{-1}\text{e} + \, ^{32}_{16}\text{S}$
11. (a) $^{140}_{59}\text{Pr} \longrightarrow \, ^{140}_{58}\text{Ce} + \, ^{0}_{+1}\text{e}$
(c) $^{10}_{5}\text{B} + \, ^{1}_{0}\text{n} \longrightarrow \, ^{7}_{3}\text{Li} + \, ^{4}_{2}\text{He}$
(e) $^{9}_{4}\text{Be} + \, ^{1}_{1}\text{H} \longrightarrow \, ^{6}_{3}\text{Li} + \, ^{4}_{2}\text{He}$
(g) $^{18}_{10}\text{Ne} \longrightarrow \, ^{18}_{9}\text{F} + \, ^{0}_{+1}\text{e}$
(i) $^{253}_{99}\text{Es} + \, ^{4}_{2}\text{He} \longrightarrow \, ^{256}_{101}\text{Md} + \, ^{1}_{0}\text{n}$
(k) $^{7}_{3}\text{Li} + \, ^{1}_{1}\text{H} \longrightarrow \, ^{7}_{4}\text{Be} + \, ^{1}_{0}\text{n}$
12. $^{108}_{46}\text{Pd} + \, ^{4}_{2}\text{He} \longrightarrow \, ^{1}_{1}\text{H} + \, ^{111}_{47}\text{Ag}$
13. $^{6}_{3}\text{Li} + \, ^{1}_{0}\text{n} \longrightarrow \, ^{3}_{1}\text{H} + \, ^{4}_{2}\text{He}$
14. $^{58}_{28}\text{Ni} + \, ^{1}_{1}\text{H} \longrightarrow \, ^{4}_{2}\text{He} + \, ^{55}_{27}\text{Co}$
15. 25.0 mg
16. 1.00 mg
17. 0.625 mg
18. 6.00 years
19. 1.2 h
20. 345 days
21. (a) $^{24}_{11}\text{Na} \longrightarrow \, ^{0}_{-1}\text{e} + \, ^{24}_{12}\text{Mg}$;
(b) 3.12×10^{-7} g; **(c)** 6.24%
22. (a) $^{65}_{30}\text{Zn} \longrightarrow \, ^{0}_{+1}\text{e} + \, ^{65}_{29}\text{Cu}$;
(b) 488 days
23. $1.00 \times 10^{-6}\%$
24. $7.50 \times 10^{-7}\,M$

25. $^{90}_{35}\text{Br} \longrightarrow \, ^{90}_{36}\text{Kr} + \, ^{0}_{-1}\text{e}$;
$^{90}_{36}\text{Kr} \longrightarrow \, ^{90}_{37}\text{Rb} + \, ^{0}_{-1}\text{e}$;
$^{90}_{37}\text{Rb} \longrightarrow \, ^{90}_{38}\text{Sr} + \, ^{0}_{-1}\text{e}$;
26. $^{98}_{42}\text{Mo} + \, ^{1}_{0}\text{n} \longrightarrow \, ^{99}_{42}\text{Mo}$;
$^{99}_{42}\text{Mo} + \, ^{0}_{-1}\text{e} \longrightarrow \, ^{99}_{43}\text{Tc}$
27. (a) 2 neutrons;
(b) $^{68}_{30}\text{Zn} + \, ^{1}_{1}\text{H} \longrightarrow \, ^{67}_{31}\text{Ga} + 2\, ^{1}_{0}\text{n}$

Chapter Quiz

1. (a) $^{232}_{92}\text{U} \longrightarrow \, ^{4}_{2}\text{He} + \, ^{228}_{90}\text{Th}$
(b) $^{201}_{82}\text{Pb} \longrightarrow \, ^{0}_{+1}\text{e} + \, ^{201}_{81}\text{Tl}$
(c) $^{63}_{29}\text{Cu} + (^{1}_{1}\text{H}) \longrightarrow$
$^{1}_{0}\text{n} + \, ^{24}_{11}\text{Na} + \, ^{39}_{19}\text{K}$
2. 0.125 mg
3. 148 days

Appendix I

1. (a) 4.41 lb; **(b)** 1.48 kg;
(c) 0.530 pt; **(d)** 5.66 L;
(e) 0.427 m; **(d)** 19.7 ft
2. 8850 m
3. (a) 42,600 cm^3; **(b)** $26\overline{0}0$ in.^{3}
4. (a) 30.3 lb/gal; **(b)** 4.73 g/mL
5. 881 lb/ft^3

Appendix VI

1. (a) 4; **(b)** -4; **(c)** 3; **(d)** 5
2. (a) 9.24 L; **(b)** 358°C;
(c) 0.117 mol

GLOSSARY

Number in parentheses refers to the section(s) in the text where the term is mentioned.

Acid (Arrhenius definition) Any substance that releases hydrogen ions (H^+) when dissolved in water (15.1).

Acid (Brønsted-Lowry definition) Any substance that can donate a proton (H^+) to some other substance (15.1).

Acid (simplified definition) Hydrogen compound that yields hydrogen ion (H^+) in aqueous solution (7.6).

Acid oxide (acid anhydride) Nonmetal oxide that reacts with water in a combination reaction to form an acid (9.6).

Actinide series Elements 90 to 103 in the periodic table, so named because they follow the element actinium (5.2).

Activation energy Energy barrier that must be overcome before molecules can react (17.1).

Actual yield Amount of product obtained in an actual chemical reaction; it is always less than the theoretical yield because of losses in the isolation and purification stages and the production of minor by-products (10.4).

Addition polymers Polymers formed from alkene monomers that bond together by breaking one part of the double bond and forming two new single bonds (18.4).

Addition reaction Form of combination reaction in which an alkene and either bromine or chlorine reacts to give dibromo- or dichloro- compounds, respectively (18.4).

Alcohols Class of hydrocarbon derivatives in which a hydroxy (—OH) group replaces a hydrogen atom on a carbon atom of an aliphatic hydrocarbon (18.7).

Aldehydes Class of hydrocarbon derivatives in which *one* hydrocarbon (alkyl or aryl) group is attached to a carbonyl group $\left(\begin{array}{c} \diagdown \\ \diagup \end{array} C = O \right)$ on one side and a *hydrogen atom* on the other (18.7).

Aliphatic hydrocarbons Hydrocarbons composed of alkanes, alkenes, and alkynes (18.1)

Alkali metals All the elements in group IA (1) of the periodic table *except* for hydrogen: lithium, sodium, potassium, rubidium, cesium, and francium (5.2).

Alkaline earth metals All the elements in group IIA (2) of the periodic table: beryllium, magnesium, calcium, strontium, barium, and radium (5.2).

Alkanes (saturated hydrocarbons) Form of aliphatic hydrocarbons; they have the general molecular formula C_nH_{2n+2} for open-chain compounds. Each carbon atom has a tetrahedral structure and four single bonds (18.3).

Alkenes Form of unsaturated aliphatic hydrocarbon; they have the general molecular formula C_nH_{2n} for open-chain compounds, a carbon–carbon double bond, and a planar structure (18.4).

Alkyl group Hydrocarbon group obtained by removing a hydrogen atom from an alkane. They are usually represented by the symbol R (18.3).

Alkynes Form of unsaturated aliphatic hydrocarbon; they have the general molecular formula C_nH_{2n-2} for open-chain compounds, a carbon–carbon triple bond, and a linear structure (18.5).

Alpha particle (α) Form of radioactive emission that is identical to the nuclei of helium-4 atoms, having an atomic number of 2 and a mass number of 4; thus, emission of an alpha particle causes an element's atomic number to decrease by 2 and its mass number to decrease by 4; written ^4_2He (19.2).

Amides Class of hydrocarbon derivatives in which an —NH_2 replaces the —OH of the carboxyl group in a carboxylic acid (18.7).

Amines Class of hydrocarbon derivatives in which one or more of the three hydrogen atoms in ammonia are replaced by alkyl or aryl groups (18.7).

Amorphous solid Any solid that consists of particles arranged in an irregular manner and thus lacks the regular structure of a crystalline solid (12.8).

Amphoteric substance Any substance that can act as an acid or base, depending on the nature of the solution; water is the best known amphoteric substance (15.1).

Analytical chemistry Study of the quantitative and qualitative analysis (examination) of elements and compound substances (1.2).

Anion Any ion carrying a negative charge (6.2).

Anode Positively charged electrode in an electrolytic cell; it is the site of oxidation reactions since it attracts the negatively charged ions in a solution (16.5).

Aqueous solution Solution in which a gas, solid, or liquid is dissolved in water (7.6).

Aromatic hydrocarbons Hydrocarbons that have a ring of six carbon atoms (*benzene ring*) and alternating carbon–carbon double bonds within that ring (18.6).

Artificial radioactivity Spontaneous radioactive emission from a substance not found in nature; this form of radioactivity comes from laboratory bombardment of natural or synthetic substances with particles emitted by radioactive isotopes (19.3).

Atom Smallest "piece" of an element that can exist and still exhibit the properties of that element, including the ability to react with other atoms (3.2).

Atomic mass (atomic weight) scale Scale of the relative masses of atoms; it is based on an arbitrarily assigned value of exactly 12 atomic mass units for a single atom of carbon-12 (4.1).

Atomic mass units (amu) Units used to express the relative masses of atoms on the atomic mass scale; 1 amu is equal to exactly one-twelfth the mass of a carbon-12 atom (4.1).

Atomic number Number of protons in the nucleus of an element's atom (4.4).

Avogadro's number Number of atoms in exactly 12 g of carbon-12 (approximately 6.02×10^{23}); it is equivalent to 1 mol of a substance (8.2).

Base (Arrhenius definition) Any substance that releases hydroxide ions (OH^-) when dissolved in water (15.1).

Base (Brønsted–Lowry definition) Any substance that can accept a proton (H^+) from another substance (15.1).

Base (simplified definition) A compound that contains a metal ion and one or more hydroxide (OH^-) ions (7.6).

Basic oxide (basic anhydride) Metal oxide that reacts with water in a combination reaction to form a base (9.6).

Becquerel (Bq) Measure of radioactivity and equal to one nuclear disintegration per second (19.4).

Benzene ring Structure characteristic of aromatic hydrocarbons; it is a regular (equal-sided) hexagon with alternating double bonds around the ring (18.6).

Beta particle (β) Form of radioactive emission that is identical to an electron, having a charge of -1 and a mass number of 0; thus, emission of a beta particle causes an element's atomic number to increase by 1 and its mass number to stay the same; written $_{-1}^{0}e$ (19.2).

Binary compounds Compounds containing two different elements (6.9 and 7.1).

Biochemistry Study of the chemical reactions that occur in living organisms (1.2).

Boiling point Temperature at which the vapor pressure of a liquid is equal to the external pressure above the surface of the liquid (12.4).

Bond angle Angle defined by three atoms and the two covalent bonds that connect them (6.7).

Bond length Distance between the nuclei of two atoms that are bonded together (6.5).

Boyle's law Principle that, at constant temperature, the volume of a fixed mass of a given gas is *inversely* proportional to the pressure; thus, if the pressure is doubled, the volume will be halved (11.3).

Buffer solution Solution that contains substantial amounts of both a weak acid and its conjugate base and so resists changes in its pH (17.5).

Calorie (cal) Standard unit for the measurement of heat energy; 1 cal is equal to the quantity of heat required to raise the temperature of 1 g of water from 14.5° to 15.5°C (3.5).

Carboxylic acids Class of hydrocarbon derivations in which a carboxyl group ($-COOH$) is attached to a hydrocarbon (alkyl or aryl) group (18.7).

Carboxyl group Functional group that consists of a carbon atom that has a double bond with oxygen on one side and a single bond to a $-OH$ group on the other side (18.7).

Catalyst Any substance that increases the rate of (speeds up) a chemical reaction but is recovered unchanged at the end of the reaction (9.2).

Cathode Negatively charged electrode in an electrolytic cell; it is the site of reduction reactions since it attracts the positively charged ions in a solution (16.5).

Cation Any ion carrying a positive charge (6.2).

Chain reaction Self-sustaining reaction in which the product is one of the reactants; in a nuclear fission reaction, the neutrons released by the original splitting of the atom continue to react with other uranium-235 nuclei, releasing more neutrons (19.6).

Chalcogens All the elements in group VIA (16) of the periodic table: oxygen, sulfur, selenium, tellurium, and polonium (5.2).

Charles's law Principle that, at constant pressure, the volume of a fixed mass of a given gas is *directly* proportional to the temperature in kelvins; thus, if the Kelvin temperature doubles, so does the volume (11.4).

Chemical bonds Attractive forces that hold atoms together (6.1).

Chemical changes Changes that result in changes in the composition of the substance. New substances are formed (3.4).

Chemical equation A shorthand method for expressing a chemical reaction in symbols and formulas (9.1).

Chemical properties Properties of a substance that can be observed only when a substance undergoes a change in composition (3.4).

Chemical reaction Any interaction among chemical substances that brings about some change (9.1).

Chemistry Study of the composition of substances and the changes they undergo (1.2).

Colligative properties Properties of a solution that depend only on the number of particles of solute present in a solution and not on the actual identity of these solute particles (14.11).

Colloid A mixture in which the particles are dispersed without appreciable bonding to the solvent molecules and do not settle out on standing (14.12).

Combination reaction Type of chemical reaction in which two or more substances (either elements or compounds) react to produce one substance (always a compound) (9.6).

Combustion reaction Type of combination reaction in which oxygen reacts with a substance; also called burning (9.6).

Compound Any pure substance that can be broken down by chemical means into two or more different simpler substances (3.2).

Concentration Measure of how much solute is contained in a given amount of solvent or solution (14.4).

Condensation Conversion of vapor (gas) molecules to a liquid; this reverse of the evaporation process is an exothermic change of state (12.2).

Conjugate acid Species formed when a proton (H^+) is added to a base (15.1).

Conjugate base Species formed when a proton (H^+) is removed from an acid (15.1).

Coordinate covalent bond Type of chemical bond, also known as a coordinate bond, formed when one atom supplies both electrons of the electron-pair bond, while the other atoms offers only an empty orbital (6.6).

Corrosion Process in which a metal is eaten away (oxidation) by a chemical reaction with oxygen to give a product that lacks the structural properties of the original metal (16.8).

Coulomb's law Principle that the attractive force between a proton and

an electron increases as the distance between these two particles decreases (6.5).

Covalent bond Type of chemical bond formed by the sharing of electrons between two atoms (6.5).

Critical mass Minimum amount of a radioactive substance necessary to sustain a chain reaction (19.6).

Crystalline solid Any solid that consists of particles arranged in a definite geometric shape or form that is distinctive for that solid (12.8).

Crystallization Method of purifying solids in which a sample is dissolved in a hot solvent, filtered, and cooled to induce crystallization of the purified solid (14.4).

Curie (Ci) Measure of radioactivity equal to 3.7×10^{10} nuclear disintegrations per second (19.4).

Cyclotron Device used to accelerate charged particles (19.3).

Dalton's law of partial pressures Principle that each gas in a mixture of gases exerts a partial pressure equal to the pressure it would exert if it were the only gas present in the same volume; hence, the total pressure of the mixture is the *sum* of the partial pressures of all the gases present (11.7).

Decay series Series of changes undergone by radioactive substances as they give off alpha and beta particles until they finally become stable (nonradioactive) substances (19.2).

Decomposition reaction Type of chemical reaction in which one substance (always a compound) breaks down to form two or more substances (either elements or compounds) (9.7).

Deliquescent substance Any hygroscopic substance that continues to absorb water from the air to the point where it becomes a solution (13.7).

Density (d) Mass of a substance occupying a unit volume, expressed as the mass divided by the volume (2.1).

Deposition Direct conversion of a vapor to a solid without passing

through the liquid state; it is an exothermic change of state (12.10).

Diatomic molecule Molecule composed of two atoms (6.5).

Dimensional analysis (factor-unit method) Method of converting among measures expressed in different units by developing a relationship between these units and expressing this relationship as a factor of both units (2.8).

Dipole moment Property of a molecule where the presence of polar bonds generates a center of positive charge and a center of negative charge that do not coincide; such molecules have regions that are always positively or negatively charged (13.3).

Dispersed particles (dispersed phase) The colloidal particles in a colloid, comparable to the solute in a solution with a diameter range of 1000 to 200,000 pm (14.12).

Dispersing medium (dispersing phase) The substance in a colloid in which the colloidal particles are distributed, comparable to the solvent in a solution (14.12).

Dissociation Process in which the ions in ionic substances separate when acted on by a solvent (15.3).

Distillation Purifying of a liquid by heating it to the boiling point and cooling the vapors in a condenser (12.5).

Double bond Chemical bond in which two atoms share two pairs of electrons (four electrons) (6.7).

Double-replacement reaction Type of chemical reaction in which two compounds react and the cation of one compound exchanges places with the cation of the other compound (9.9).

Ductile Capable of being drawn out into a thin wire (3.7).

Dynamic equilibrium A situation in which the rate of a forward process is equal to the rate of the reverse process occurring simultaneously, as when the rate of evaporation is equal to the rate of condensation for a liquid in a closed container (11.1).

Efflorescent substance   Any hydrate that releases water when simply exposed to the atmosphere (13.7).

Electrochemistry   Study of the relationship between electrical and chemical energy (16.5).

Electrolysis   Application of a direct electric current to a substance to produce a chemical change (13.6).

Electrolytes   Substances that produce *ions* in aqueous solution and solutions that *conduct* an electric current (15.6).

Electrolytic cell   Device that uses electrolytes and electrical energy to produce chemical reactions that would otherwise not occur (16.5).

Electromotive (activity) series   Arrangement of metals in order of descending reactivity; thus, each element in the series displaces any element below it from a salt or acid of the second element (9.8).

Electron   Subatomic particle with a relative charge of -1 and a negligible mass (9.109×10^{-28} g); electrons exist outside the nucleus of the atom in one of many energy levels (4.3).

Electron affinity   Amount of energy given off when an atom or ion gains an extra electron (6.2).

Electron capture   Process in which a nucleus "grabs" an electron outside the nucleus and converts a proton to a neutron (19.3).

Electronegativity   Degree to which an atom attracts a pair of covalently bonded electrons to itself (6.5).

Element   Any pure substance that cannot be broken down by chemical means into two or more different simpler substances; its atoms all have the same atomic number (3.2 and 4.4).

Empirical formula   Simplest formula for a compound; it is the smallest whole-number ratio of the atoms present (8.5).

Endothermic change of state   Any change in the state of matter in which heat is absorbed (12.4).

Endothermic reaction   Any chemical reaction in which heat energy is absorbed (10.7).

End point   Point at which the indicator in an acid–base titration changes color. The end point is often the same as the equivalence point, but not always (15.4).

Energy   Capacity to do work or transfer heat (3.5).

Energy levels   Series of areas outside the nucleus of an atom in which the electrons are located (4.4).

Equilibrium constant expression   For the general reaction $a\text{A} + b\text{B} \rightleftharpoons c\text{C} + d\text{D}$, the equilibrium constant expression is equal to $[\text{C}]^c[\text{D}]^d$ divided by $[\text{A}]^a[\text{B}]^b$; the equilibrium constant is designated as K and is an experimentally determined constant that varies with the temperature (17.2).

Equilibrium position   Equilibrium composition of a reaction mixture; it refers to a new composition formed in response to changes in concentration, temperature, or pressure (17.3).

Equivalence point   Point at which an acid or base is exactly neutralized in the titration process (15.4).

Equivalent of a base   Quantity of a substance that reacts with 1 mol (6.02×10^{23}) of hydrogen ions (H^+) or supplies 1 mol (6.02×10^{23}) of hydroxide ions (OH^-) (14.9).

Equivalent of an acid   Quantity of a substance that reacts to yield 1 mol (6.02×10^{23}) of hydrogen ions (H^+) (14.9).

Esters   Class of hydrocarbon derivatives in which an —OR or —OAr group replaces the OH group of a carboxylic acid (18.7).

Ethers   Class of hydrocarbon derivatives in which *two hydrocarbon groups* [alkyl (R) and/or aryl (Ar), the same or different] are attached to an oxygen atom (18.7).

Evaporation   Escape of molecules from the surface of a liquid to form a vapor in the surrounding space above the liquid; this reverse of the condensation process is an endothermic change of state (12.2).

Exact numbers   Numbers for measurements that are precisely known and can have as many significant digits as a calculation requires; thus they are not used to determine the number of significant digits for an answer (2.6).

Excess reagent   Reactant in a chemical reaction that is *not* completely used up in the reaction, so called because when the last trace of the new compound is formed, some of this reactant is left over (10.4).

Exothermic change of state   Any change in the state of matter in which heat is released (12.4).

Exothermic reaction   Any chemical reaction in which heat energy is given off (evolved) (10.7).

Experimentation   Collection of facts and data by observing natural events under carefully controlled conditions (1.1).

Exponent   Whole number or symbol written as a superscript to a base and denoting the number of times the base is to be multiplied by itself (2.7).

Exponential notation   A form of mathematical expression in which a number is expressed as the product of two numbers, one a decimal and the other a power of 10 (2.7).

Formula unit   Smallest unit of an ionically bonded substance that can exist and undergo chemical changes (6.4).

Free state   Term describing an element existing on its own—not combined with any other element (3.7).

Freezing point   Temperature at which the particles of a liquid begin to form crystals or irregular particles of a solid (12.9).

Functional group   Atom or group of atoms (other than hydrogen) that is attached to a hydrocarbon chain and that confers some distinctive chemical and physical properties on the organic compound (18.7).

Gamma rays   Form of radioactive emission that has no charge or mass and thus has no effect on the atomic number or mass number of an element, despite their high energy; written as γ (19.2).

Gas One of three states of matter; it is characterized by (1) infinite and uniform expansion, (2) indefinite shape or volume, (3) compressibility, (4) low density, and (5) complete and rapid mixing in other gases (11.1).

Gay-Lussac's law Principle that, at constant volume, the pressure of a fixed mass of a given gas is *directly* proportional to the temperature in kelvins. Thus, if the Kelvin temperature doubles, so will the pressure (11.5).

Gay-Lussac's law of combining volumes Principle stating that whenever gases react or are formed, their volumes are in small whole-number ratios, provided these are measured at the same temperature and pressure. The ratio of volumes for such a reaction is directly proportional to the values of the coefficients in the balanced equation (10.6).

Group (family) One of the 18 vertical columns in the periodic table (5.2).

Half-life Amount of time required for one-half of any given mass of a radioactive isotope to decay (19.4).

Halogenation Substitution reaction in which a hydrogen atom on an alkane is replaced by a chlorine or bromine atom (18.3).

Halogens All the elements in group VIIA (17) of the periodic table: fluorine, chlorine, bromine, iodine, and astatine (5.2).

Heat energy Energy transferred from one substance to another when there is a temperature difference between the substances; it is associated with the random motion of molecules (3.5).

Heat of condensation Quantity of heat released in order to condense 1 g of gas to a liquid at its boiling point and at constant pressure (12.4).

Heat of fusion Quantity of heat required to convert 1 g of a solid to liquid at the melting point of the substance; thus, fusion is an endothermic change of state (12.9).

Heat of reaction Number of calories or joules of heat energy given off (evolved) or absorbed in a particular chemical reaction for a given amount of reactant or product (10.7).

Heat of solidification (crystallization) Quantity of heat released by 1 g of a liquid as it becomes a solid at the melting point of the substance; it is an exothermic change of state (12.9).

Heat of vaporization Quantity of heat required to evaporate 1 g of a given liquid at its boiling point and at constant pressure (12.4).

Henry's law Principle that the solubility of a gas in a liquid is directly proportional to the partial pressure of the gas above the liquid (14.3).

Heterogeneous matter Matter that is not uniform in composition and/or properties throughout the sample, but rather consists of two or more distinct substances unequally distributed (3.2).

Homogeneous matter Matter that is uniform in composition and properties throughout the sample (3.2).

Homogeneous mixture Homogeneous throughout and composed of two or more pure substances whose proportions can be varied in some cases without limit (3.2).

Homologous series Series of compounds in which each compound in the series differs from the next compound by a multiple, such as $-CH_2-$ (18.3).

Hydrates Crystalline compounds that contain chemically bound water in definite proportions (13.7).

Hydrocarbons Organic compounds that contain only the elements carbon and hydrogen (18.1).

Hydrogen bond Type of weak chemical bond formed when a hydrogen atom bonded to a highly electronegative atom (F, O, or N) is also bonded partially to another electronegative atom (13.4).

Hygroscopic substance Any substance that readily absorbs moisture from the air (13.7).

Hypothesis Tentative explanation of the results of experimentation; it is subject to confirmation or rejection in further experiments (1.1).

Ideal-gas equation Formula that allows scientists to vary not only the temperature, pressure, and volume of a gas but also its mass; it is expressed mathematically as $PV = nRT$, where P is pressure, V is volume, n is the amount of gas in moles, T is temperature, and R is the universal gas constant (11.8).

Ideal gases Gases that conform to the basic assumptions of the kinetic theory; they are composed of molecules that have no attractive forces for one another and are in constant rapid motion, colliding with one another in a perfectly elastic fashion, and have an average kinetic energy per molecule that is proportional to the temperature in kelvins (11.1).

Immiscible liquids Two liquids that do not mix, but form separate layers; neither liquid dissolves appreciably in the other (14.2).

Indicators Compounds that change color in the presence of certain chemicals, such as acid and bases. Litmus paper is the best known indicator (15.1).

Inorganic chemistry Study of all substances other than those containing carbon (1.2).

Inorganic compounds Compounds not containing carbon (7.1).

Ion Charged entity resulting from loss or gain of electrons by an atom or group of bonded atoms. The number of electrons does not equal the number of protons, therefore it carries a positive or negative charge (4.4).

Ion electron method Method of balancing oxidation–reduction reaction equations which separates the process into two partial reactions representing half-reactions: one describes the oxidation reaction, and the other describes the reduction reaction (16.4).

Ionic bond The force of attraction between ions of opposite charge which holds them together in an ionic compound. These ions of opposite charge are formed by the transfer of electrons from one atom to another (6.4).

Ionic charge Charge on an ion. The ion may consist of a single atom or a group of atoms bonded together (6.3).

Ionic equations Express chemical reactions involving compounds that exist mostly in ionic form in aqueous solution (15.7).

Ionization Process in which covalent compounds dissolve to produce ions in aqueous solution (15.3).

Ionization energy Amount of energy required to remove only the most loosely bound electron from an atom (6.2).

Isomer Any compound that has the same molecular formula but a different structural formula. Each isomer has physical and chemical properties different from those of any other isomer of the same molecular formula (18.3).

Isotope Any atom having the same atomic number but a different mass number. They have the same number of protons and electrons but a different number of neutrons (4.5).

Joule (J) Standard unit for the measurement of heat energy in the Système Internationale (SI); $4.184 \text{ J} = 1 \text{ cal}$ (3.5).

Ketones Class of hydrocarbon derivatives in which *two* hydrocarbon (alkyl or aryl) groups are attached to a carbonyl group $\left(\ \diagdown \!\!\!\! C = O\ \right)$ (18.7).

Kinetic energy Energy possessed by a substance by virtue of its motion (3.5).

Kinetic theory Hypotheis that heat and motion are related, that particles of all matter are in motion to some degree, and that heat is an indication of this motion; this theory helps to explain the different behaviors and properties of various states of matter (11.1).

Lanthanide series Elements 58 to 71 in the periodic table, so named because they follow the element lanthanum. Also called the *rare-earth* elements (5.2).

Law of conservation of energy Principle that energy can neither be created nor destroyed, although it may be transformed from one form to another (3.6).

Law of conservation of mass Principle that mass is neither created nor destroyed and that the total mass of the substances involved in a physical or chemical change remains constant (3.6 and 9.1).

Law of definite proportions (constant composition) Principle that a given pure compound always contains the same elements in exactly the same proportions by mass (3.3).

Law of mass action Principle that for a general reaction $a\text{A} + b\text{B} \longrightarrow c\text{C} + d\text{D}$, the rate of reaction can be defined as $k[\text{A}]^x[\text{B}]^y$, where [A] and [B] are concentrations of A and B in moles per liter, respectively, k is the specific rate constant, and x and y are determined by experimentation (17.1).

Le Châtelier's principle Principle that if an equilibrium system is subjected to a change in conditions, the composition of the reaction mixture will change in such a way as to try to restore the original condition (17.3).

Lewis structure Method of expressing electrons among atoms in a molecule using the rule of eight (octet rule) and dots (:) to represent electrons (6.7).

Limiting reagent Reactant in a chemical reaction that is completely used up in the reaction, so called because the amount of this reactant limits the amount of new compounds that can be formed (10.4).

Liquid One of the three states of matter; it is characterized by (1) limited expansion, (2) lack of characteristic shape, (3) maintenance of volume, (4) slight compressibility, (5) high density, and (6) diffusion in other liquids (usually slow) (12.1).

Malleable Capable of being shaped by beating with a hammer (3.7).

Mass Quantity of matter in particular body (2.1).

Mass number Sum of the number of protons and the number of neutrons in the nucleus of an atom (4.4).

Matter Any substance that has mass and occupies space (2.1).

Melting point Temperature at which the kinetic energy of some of the particles in a solid matches the attractive forces in the solid and the solid begins to liquefy (12.9).

Metalloids (semimetals) Elements (except aluminum) that lie on either side of the *colored* stair step line in the periodic table; they have both metallic and nonmetallic properties: boron, silicon, germanium, arsenic, antimony, tellurium, polonium, and astatine (3.7 and 5.3).

Metals Elements that have a higher luster, conduct electricity and heat well, are malleable and ductile, have high densities and melting points, are hard, and do not readily combine with one another (3.7).

Metric system System of weights and measures in which each unit is a tenth, hundredth, thousandth, and so on, of another unit; it is the standard system in use in every nation except the United States and is used extensively by scientists (2.2).

Miscible liquids Two liquids that mix completely in all proportions to form a solution (14.2).

Mixture Heterogeneous matter composed of two or more pure substances, each of which retains its identity and specific properties (3.2).

Molality (*m*) Measure of the concentration of a solution expressed as the number of moles of solute per kilogram of solvent; it is calculated as the moles of solute divided by the kilograms of solvent (14.10).

Molarity (*M*) Measure of the concentration of a solution expressed as the number of moles of solute per liter of solution; it is calculated as the moles of solute divided by the liters of solution (14.8).

Molar mass   Mass in grams of 1 mol of any substance, element or compound (8.2).

Molar volume of a gas   The 22.4 L occupied by 1 mol of any gas molecules at standard temperature (0°C) and pressure ($76\overline{0}$ mm Hg, torr) (8.3).

Mole (mol)   Amount of a substance containing the same number of atoms, formula units, molecules, or ions as there are atoms in exactly 12 g of carbon-12 (approximately 6.02×10^{23} atoms) (8.2).

Molecular formula   Way of expressing the composition of one molecule of a compound or element by using elemental symbols for each element involved and subscripts reflecting the number of atoms (above 1) of that element in the molecule; it shows the actual number of atoms of each element present in one molecule of the compound (3.3 and 8.5).

Molecule   Smallest particle of a pure substance that can exist and undergo chemical changes (3.3).

Natural radioactivity   Emission of alpha particles, beta particles, and gamma rays by naturally occurring radioactive substances (19.1).

Neutralization reaction   Type of chemical reaction in which an acid or acid oxide reacts with a base or basic oxide, usually producing water as one of the products (9.10).

Neutron   Subatomic particle with no charge but a mass of approximately 1 amu (1.6748×10^{-24} g); it is located in the nucleus of the atom. Also, a form of radioactive emission having an atomic number of 0 and an atomic mass of 1 (written as $_0^1$n); thus, emission of a neutron causes an element's atomic number to stay the same and its mass number to decrease by 1 (4.3 and 19.3).

Noble gases   All the elements in group VIIIA (18) of the periodic table; these relatively unreactive (inert) nonmetal elements include helium, neon, argon, krypton, xenon, and radon (3.7 and 5.2).

Nonelectrolytes   Substances that *do not* produce *ions* in aqueous solution and solutions that *do not* conduct an electric current (15.6).

Nonmetals   Elements whose characteristics are the opposite of metals and combine readily both with metals and with other nonmetals (3.7).

Normality (N)   Measure of the concentration of a solution expressed as the number of equivalents of solute per liter of solution; it is calculated as the equivalents of solute divided by the liters of solution (14.9).

Nuclear fission   Splitting of an atomic nucleus to give two smaller nuclei, neutrons, and energy (19.6).

Nuclear fusion   Combination of two atomic nuclei to form a heavier nucleus and produce energy (19.7).

Nuclear reaction   Any reaction in which the nucleus of the atom changes composition (19.1).

Nucleus   Small, dense region at the center of an atom containing nearly all the mass of the atom—the protons and neutrons; it has a positive electrical charge (4.4).

Octet rule (rule of eight)   Principle that, in forming molecules, most atoms attempt to obtain a stable configuration of eight valence electrons around the atom (4.7 and 6.1).

Orbital   Region of space surrounding the nucleus of an atom in which there is a high probability of finding up to two electrons (4.9).

Organic chemistry   Study of the substances containing carbon (1.2 and 18.1).

Organic halides   Class of hydrocarbon derivatives consisting of the alkyl halides and aryl halides (18.7).

Oxidation   Any chemical reaction in which a substance loses electrons (16.1).

Oxidation number or oxidation state   A positive or negative whole number assigned to an element in a compound or ion. It is based on certain rules (6.3 and 16.2).

Oxidation number method   Method of balancing oxidation–reduction reaction equations which uses the oxidation numbers of the elements involved in the reaction (16.3).

Oxidation–reduction (redox) reaction   Type of chemical reaction in which one substance transfers electrons to another substance (Introduction to Chapter 16).

Oxidizing agent (oxidant)   Substance in an oxidation–reduction chemical reaction that is reduced (gains electrons) (16.1).

Parts per million (ppm)   Measure of the concentration of a solution expressed as parts by mass of solute per 1 million parts by mass of solution; it is calculated as the mass of solute divided by the mass of solution, with the result multiplied by 1,000,000 (14.7).

Percent by mass   Measure of the concentration of a solution expressed as the parts by mass of solute per 100 parts by mass of solution; it is calculated as the mass of solute divided by the mass of solution, with the result multiplied by 100 (14.6).

Percent yield   Percent of the theoretical yield actually obtained; it is expressed as the actual yield divided by the theoretical yield, with the result multiplied by 100 (10.4).

Period   One of the seven horizontal rows in the periodic table (5.2).

Periodic law   Appearance of elements with similar chemical properties at regular, periodic intervals when they are listed in order of increasing atomic number (5.1).

Periodic table   Special table listing all the known elements according to their atomic numbers and arranged so that elements in a given column have similar chemical properties (3.7).

pH   Quantitative way of expressing the acidic or basic nature of solutions using the negative logarithmic values of their hydrogen ion (H^+) concentrations; a substance with $pH < 7$ is acidic, one with $pH > 7$ is basic, and one with $pH = 7$ is neutral (15.5).

Physical changes   All changes in a substance other than changes in its chemical composition (3.4).

Physical chemistry Study of the structure of substances, how fast they change, and the role of heat in chemical changes (1.2).

Physical properties Properties of a substance that can be observed without the composition of the substance changing (3.4).

Physical state Any of three forms in which matter may exist, as a gas, as a liquid, or as a solid; the state depends on the surrounding temperature and the atmospheric pressure, as well as on the specific characteristics of the particular type of matter (3.1).

pOH Quantitative way of expressing the acidic or basic nature of solutions using the negative logarithmic values of their hydroxide ion (OH$^-$) concentrations; a substance with pOH < 7 is basic, one with pOH > 7 is acidic, and one with pOH = 7 is neutral (15.5).

Polar bond Type of chemical bond, also known as a polar covalent bond, formed by the *unequal* sharing of electrons between two atoms whose electronegativities differ (6.5 and 13.3).

Polyatomic ion Any ion made up of more than one atom (6.7).

Polymer Substance made up of thousands of smaller molecules (*monomers*) that have bonded together to form a giant molecule (18.4).

Positron Form of radioactive emission that is identical to a positive electron, having a charge of +1 and a mass number of 0; thus, emission of a positron causes an element's atomic number to decrease by 1 and its mass number to stay the same; written as $_{+1}^{0}e$ (19.3).

Potential energy Energy possessed by a substance by virtue of its position in space or its chemical composition (3.5).

Precipitate Any solid that appears in a solution in the course of a chemical reaction (9.9).

Pressure Force per unit area, whether expressed as pounds per square inch (psi), centimeters of mercury (cm Hg), millimeters of mercury (mm Hg), torr, inches of mercury (in. Hg), atmospheres (atm), pascals (Pa), or millibars (mbar) (11.1).

Primary cell Form of voltaic cell (battery) that may not be reused after it discharges and is typically thrown away (16.7).

Products Substances formed by a chemical reaction; they are written on the right side of a chemical equation (9.2).

Proton Subatomic particle with a relative charge of +1 and a mass of approximately 1 amu (1.6726 $\times$ 10^{-24} g); it is located in the nucleus of the atom (4.3).

Pure substance Substance characterized by definite and constant composition and having definite and constant properties under a given set of conditions (3.2).

Radioactive decay Changes that occur in the composition of radioactive materials as they give off alpha and beta particles and gamma rays (19.2).

Radioactivity Spontaneous emissions accompanying changes in the nuclei of atoms (19.1).

Reactants Substances that interact with one another in a chemical reaction; they are written on the left side of a chemical equation (9.2).

Reaction rate The rate of speed at which products are formed or reactants are consumed (17.1).

Real gases Gases such as hydrogen, oxygen, and nitrogen that behave as ideal gases under moderate conditions of temperature and pressure but deviate from these properties if the temperature is very low or the pressure is very high (11.1).

Reducing agent (reductant) Substance in an oxidation–reduction chemical reaction that is oxidized (loses electrons) (16.1).

Reduction Any chemical reaction in which a substance gains electrons (16.1).

Representative elements All the A group elements in the periodic table (5.3).

Reversible reaction Chemical reaction that is never complete because the products of the reaction also react to reform the original reactants. When the reaction is finished, there are *both* products and reactants present (10.7 and 17.2).

Rule of eight (octet rule) Principle that, in forming molecules, most atoms attempt to obtain a stable configuration of eight valence electrons around the atom (4.7 and 6.1).

Rule of two An exception to the rule of eight (6.1).

Salt Ionic compound made up of a positively charged ion (cation) and a negatively charged ion (anion); it is often produced in neutralization reactions (7.6).

Saturated solution Solution that contains just as much solute as can be dissolved in the solvent by ordinary means. Dissolved solute is in dynamic equilibrium with any undissolved solute. That is, the rate of dissolution (dissolving) of any undissolved solute is equal to the rate of crystallization of dissolved solute (14.4).

Science Organized or systematized knowledge gathered by the scientific method (1.1).

Scientific method Procedure for studying the world in three organized steps: experimentation, hypothesis proposal, and further experimentation (1.1).

Scientific notation Form of exponential notation in which the decimal part must have exactly one nonzero digit to the left of the decimal point; it is widely used by scientists (2.7).

Secondary cell Form of voltaic cell (battery) that may be recharged and used over and over again (16.7).

Semimetals (metalloids) Elements (except aluminum) that lie on the *colored* stair step line in the periodic table; they have both metallic and nonmetallic properties: boron, silicon, germanium, arsenic, antimony, tellurium, polonium, and astatine (3.7 and 5.3).

Significant digits (figures) Digits in a measurement that are known to be precise, along with a final digit about which there is some uncertainty (2.5).

Single-replacement reaction Type of chemical reaction in which an element and a compound react and the element replaces another element in the compound (9.8).

Solid One of the three states of matter; it is characterized by (1) lack of expansion, (2) definite shape, (3) constant volume, (4) lack of compressibility, (5) high density, and (6) severely limited mixability (12.7).

Solubility product constant (K_{sp}) An experimentally determined constant which describes the concentration equilibrium between a slightly soluble salt and its dissolved ions at saturation. In the expression for K_{sp}, the coefficients of the ions become exponents for the concentrations of the ions in the solution (17.6).

Solute Substance dissolved in a solution; it is usually present in lesser quantity than the solvent (14.1).

Solution Homogeneous mixture involving two or more pure substances; its composition can be varied within certain limits (3.2 and 14.1).

Solvent Dissolving substance in a solution; it is usually present in greater quantity than the solute (14.1).

Specific gravity Density of a substance divided by the density of some substance taken as a standard (2.9).

Specific heat Quantity of heat required to raise the temperature of 1.00 kg of a substance by 1.00 K or to raise the temperature of 1.00 g of a substance 1.00°C (3.5).

Specific rate constant Experimentally determined constant factor in the rate of reactions, expressed mathematically as k (17.1).

Standard pressure Pressure that supports a column of mercury at a height of 76.0 cm at 0°C at sea level: 14.7 psi, 76.0 cm of mercury, $76\overline{0}$ mm of mercury, $76\overline{0}$ torr, 29.9 in. of mercury, 1.00 atm, 1.013×10^5 Pa, or 1013 mbar (11.2).

Standard temperature and pressure (STP) Temperature of 0°C and atmospheric pressure of $76\overline{0}$ mm Hg (torr) or 1 atm (8.3).

Stoichiometry Measurement of the relative quantities of chemical reactants and products in a chemical reaction (Introduction to Chapter 10).

Structural formula Method of expressing the chemical bonds among atoms in a molecule using lines to represent the bonds (6.7).

Subatomic particles Particles that make up the atom of an element: the electron, proton, and neutron (4.3).

Sublimation Direct conversion of a solid to a vapor without passing through the liquid state; it is an endothermic change of state (12.10).

Supersaturated solution Solution in which the concentration of the solute is greater than that in a saturated solution under the same conditions. This solution is unstable and reverts to a saturated solution if a seed crystal is added; the excess solute crystallizes out of solution (14.4).

Surface tension Property of a liquid that tends to draw the surface molecules into the body of the liquid and hence to reduce the surface to a minimum (12.6).

Synthetic elements Human-made elements that do not exist in nature; several of these elements fill gaps that formerly existed in the periodic table. They have extended the periodic table (19.3).

Temperature Degree of hotness of matter (2.1).

Ternary compounds Compounds containing three different elements (6.9 and 7.1).

Theoretical yield Amount of product expected if all the limiting reagent forms product with none of it left. This assumes that none of the product is lost in isolation and purification (10.4).

Titration Procedure for determining the concentration of an acid or base by adding a base or acid of *known* concentration (15.4).

Transition elements All the B group elements plus the group VIII (8, 9, 10) elements in the periodic table (5.3).

Transmutation Change of a radioactive element into another element (19.2).

Transuranium elements Synthetic elements whose atomic numbers exceed 92 (the atomic number of uranium). They have extended the periodic table (19.3).

Triple bond Chemical bond in which two atoms share three pairs of electrons (six electrons) (6.7).

Unsaturated hydrocarbons Hydrocarbons that contain less hydrogen than the corresponding alkane. They contain a double or triple bond; for example, alkenes and alkynes (18.4).

Unsaturated solution Solution in which the concentration of solute is less than that in a saturated solution under the same conditions (14.4).

Unshared pair of electrons (nonbonding pair, lone pair) A *pair* of electrons on one atom (6.6).

Valence electrons Electrons occupying the highest principal energy level in an atom (4.7).

Vapor pressure In a closed container, the pressure exerted by a vapor in dynamic equilibrium with its liquid state (12.3).

Viscosity Property of a liquid that describes the resistance of a liquid to flow; highly viscous liquids like honey flow very slowly (12.6).

Voltaic (galvanic) cells Devices that use oxidation–reduction reactions to produce electricity; commonly called batteries (16.6).

Volume Cubic space taken up by matter (2.1).

Weight Measure of the gravitational force of attraction between the body's mass and the mass of the planet or satellite on which it is weighed (2.1).

PHOTO CREDITS

Cover	Superstock
Chapter 1	**Tom Bross/Stock Boston**
1-1	NIH/Science Source/Photo Researchers, Inc.
1-3	Will & Deni McIntyre/Science Source/Photo Researchers, Inc.
1-4	T. Kitchin/Tom Stack & Associates
1-5	United States Postal Service
1-6	Jan Halaska/Photo Researchers, Inc.
1-6	Photo Researchers, Inc.
1-7	Stephen Frisch/Stock Boston
Gold Photo (a)	Paul Silverman/Fundamental Photographs
Gold Photo (b)	Bettmann

Chapter 2	**William Taufic/The Stock Market**
2-3a	Jack Plekan/Fundamental Photographs
2-3b	Laura Dwight/Peter Arnold, Inc.
2-3b	Leonard Lessin/Peter Arnold, Inc.
2-3c	Teri Stratford
2-4	Kip Peticolas/Fundamental Photographs
2-6a	Richard Megna/Fundamental Photographs
2-6	Michael Dalton/Fundamental Photographs
2-6c	Diane Schiumo/Fundamental Photographs
2-6d	Lawrence Migdale/Science Source/Photo Researchers, Inc.
2-8	Laura Dwight/Peter Arnold, Inc.
2-10	Bill Horsman/Stock Boston
2-11b	Manfred Kage/Peter Arnold, Inc.
2-11a	Dr. Guy Worthey, University of Michigan/NASA
2-12	Leonard Lessin/Peter Arnold, Inc.
Chemistry of the Atmosphere Photo	Keith Kent/Peter Arnold, Inc.
Sodium Chloride Photo	Liane Enkelis/Stock Boston

Chapter 3	**David Muench/Tony Stone Images**
3-1	Sven-Olof Lindblade/Photo Researchers, Inc.
3-2a	Leonard Lessin/Peter Arnold, Inc.
3-2b	Leonard Lessin/Peter Arnold, Inc.
3-3a	Tom Bochsler/Tom Bochsler, Photography Ltd.
3-3b	Tom Bochsler, Photography Ltd.
3-3c	Tom Bochsler, Photography Ltd.
3-5a	Richard Megna/Fundamental Photographs
3-5b	Richard Megna/Fundamental Photographs
3-5c	Richard Megna/Fundamental Photographs
3-5d	Richard Megna/Fundamental Photographs
3-5e	Richard Megna/Fundamental Photographs
3-5f	Russ Lappa/Science Source/Photo Researchers, Inc.
3-8	Michael Dalton/Fundamental Photographs
3-9a	Stephen Frisch/Stock Boston
3-9b	Richard Megna/Fundamental Photographs
3-9c	Richard Megna/Fundamental Photographs
3-10	John Maher/Stock Boston
3-11	Ron Glazer/Photo Researchers, Inc.
3-13	J.L. David/The Granger Collection
3-15	Guy Gillette/Photo Researchers, Inc.
3-16	Will McIntyre/Photo Researchers, Inc.

Silicon Photo (a)	Hewlett Packard
Silicon Photo (b)	IBM

Chapter 4	**Oscar Knapp, Ph.D./The Enrico Fermi Institute**
4-2	The Granger Collection
Silver Photo	D. Walker/Gamma-Liaison, Inc.

Chapter 5	**Karen Leeds/The Stock Market**
5-4	Richard Megna/Fundamental Photographs
5-5	Richard Megna/Fundamental Photographs
5-6a	Stephen Frisch/Stock Boston
5-6b	Stephen Frisch/Stock Boston
5-6c	Russ Lappa/Science Source/Photo Researchers, Inc.
Chlorine Photo	Dr. E.R. Degginger

Chapter 6	**Nelson Max/LLNL/Peter Arnold, Inc.**
6-2	Richard Megna/Fundamental Photographs
6-3	Paul Silverman/Fundamental Photographs
6-16a	Laima E. Druskis
6-18b	Laima E. Druskis
6-19a	Laima E. Druskis
6-20	Richard Megna/Fundamental Photographs
Chemistry of the Atmosphere Photo	Joyce Photographics/Photo Researchers, Inc.
Phosphoric Acid Photo (a)	Teri Stratford/Teri Stratford
Phosphoric Acid Photo (b)	Teri Stratford

Chapter 7	**David W. Hamilton/The Image Bank**
7-1	NASA Headquarters
7-2	Richard Megna/Fundamental Photographs
7-4	Richard Megna/Fundamental Photographs
7-5	RichardMegna/Fundamental Photographs
Sucrose Photo	Pedro Coll/The Stock Market

Chapter 8	**Dr. E.R. Degginger**
8-1	Gary Shulfer/C. Marvin Lang
8-3	David Brooks/The Stock Market
8-4	Tom McHugh/Photo Researchers, Inc.
8-5b	Teri Stratford
Copper Photo	Tom McHugh/Photo Researchers, Inc.

Chapter 9	**Lawrence Migdale/Science Source/Photo Researchers, Inc.**
9-1	J.P.Nacivet/Photo Researchers, Inc.
9-2	Tom Bochsler, Photography Ltd.
9-3	General Motors
9-4a	Donald Clegg
9-4b	Donald Clegg
9-4c	Donald Clegg
9-5b	Tom Bochsler, Photography Ltd.
9-5c	Tom Bochsler, Photography Ltd.
9-7a	Tom Bochsler, Photography Ltd.
9-7b	Tom Bochsler, Photography Ltd.
9-7c	Tom Bochsler, Photography Ltd.
9-8	Dr. E.R. Degginger
9-9	Tom Bochsler, Photography Ltd.

9-10   Dr. E.R. Degginger
9-11a   M. Wendler/OKAPIA/Photo Researchers, Inc.
9-11b   Andrew Martinez/Photo Researchers, Inc.
9-11c   Farrell Grehan/Photo Researchers, Inc.
9-11d   Anthony Mercieca Photo/Photo Researchers, Inc.
9-12a   Boutin/Explorer/Photo Researchers, Inc.
9-12b   C.J. Collins/Photo Researchers, Inc.
Chemistry of the Atmosphere Photo   Owen Franken/Stock Boston
Carbon Dioxide Photo   Mathew McVay/Tony Stone Images

Chapter 10   Guido Alberto Rossi/The Image Bank

10-3   Dr. E.R. Degginger
10-5   Richard Megna/Fundamental Photographs
10-6a   Lawrence Migdale/Science Source/Photo Researchers, Inc.
10-6b   Lawrence Migdale/Science Source/Photo Researchers, Inc.
10-8   Richard Megna/Fundamental Photographs
10-8   Richard Megna/Fundamental Photographs
Chromium Photo (a)   Stephen Frisch/Stock Boston/Vaughan Fleming/Science Photo Library/Photo Researchers, Inc.
Chrom um Photo (b)   Rich Buzzelli/Tom Stack & Associates
Chromium Photo (c)   Joseph P. Sinnot/Fundamental Photographs

Chapter 11   Frank Whitney/The Image Bank

11-9   Neal Peters Collection
Chemistry of the Atmosphere Photo   David Austen/Tony Stone Images
Oxygen Photo   Richard Megna/Fundamental Photographs

Chapter 12   Steve Elmore/The Stock Market

12-8a   Teri Stratford
12-8b   Teri Stratford
12-8c   Teri Stratford
12-10a   Runk/Schoenberger/Grant Heilman Photography
12-10b   Runk/Schoenberger/Grant Heilman Photography
12-11a   Fred J. Maroon/Photo Researchers, Inc.
12-11b   Peticolas/Megna/Fundamental Photographs
12-12   Gerard Vandystadt/Photo Researchers, Inc.
12-15   Richard Folwell/Science Photo Library/Photo Researchers, Inc.
Mercury (a)   Richard Megna/Fundamental Photographs
Mercury (b)   Sir John Tenniel/The Granger Collection

Chapter 13   NASA Headquarters

13-1   Tony Stone Images
13-9a   Donald Clegg
13-9b   Donald Clegg
13-11   Tom Bochsler, Photography Ltd.
13-12   Katharine P. Daub and Los Alamos Scientific Laboratory
Chemistry of the Atmosphere Photo   NASA Headquarters
Hydrogen Photo   Bettmann

Chapter 14   Michael Furman/Photographer Ltd./The Stock Market

14-2   Kip Peticolas/Fundamental Photographs
14-3b   Laima Druskis/PH Archives
14-6a   Dr. E.R. Degginger
14-6b   Dr. E.R. Degginger
14-12   Bill Bridge/Uniphoto
14-13b   Kees Van Den Berg/Photo Researchers, Inc.
Soap Photo   Wes Thompson/The Stock Market

Chapter 15   Paul Silverman/Fundamental Photographs

15-2   Tom Bochsler, Photography Ltd.
15-3   D. Clegg and R. Wilson/PH Archives

15-4   Tom Bochsler, Photography Ltd.
15-5a   Dr. E.R. Degginger
15-5b   Dr. E.R. Degginger
15-5c   Dr. E.R. Degginger
15-6a   Beckman Instruments, Inc./PH Archives
15-6b   Richard Megna/Fundamental Photographs
15-9a   Dr. E.R. Degginger
15-9b   Robert Mathena/Fundamental Photographs
Chemistry of the Atmosphere Photo (a)   Regis Bossu/Sygma
Chemistry of the Atmosphere Photo (b)   Regis Bossu/Sygma
Calcium Carbonate Photo (a)   Teri Stratford
Calcium Carbonate Photo (b)   Teri Stratford

Chapter 16   Peticolas/Megna/Fundamental Photographs

16-3   Richard Megna/Fundamental Photographs
16-6   Tom Hollyman/Photo Researchers, Inc.
16-11   Richard Megna/Fundamental Photographs
16-12   Gary Moon/Tony Stone Images
Magnesium Photo   E. Poupinet/Explorer/Photo Researchers, Inc.
Magnesium Photo   N. Yosha/Superstock
Magnesium Photo   Richard Megna/Fundamental Photographs

Chapter 17   Paul Silverman/Fundamental Photographs

17-4   Dr. E.R. Degginger
17-5a   Richard Megna/Fundamental Photographs
17-5b   Richard Megna/Fundamental Photographs
17-5c   Richard Megna/Fundamental Photographs
17-8   D. Clegg and R. Wilson/PH Archives
17-9b   Tom Bochsler, Photography Ltd.
Chemistry of the Atmosphere Photo   Joseph Nettis/Photo Researchers, Inc.
Nitrogen Photo   David Taylor/Science Photo Library/Photo Researchers, Inc.

Chapter 18   Laura Luongo/Gamma-Liaison, Inc.

18-1d   Dr. E.R. Degginger
18-4a   Dr. E.R. Degginger
18-4b   Dr. E.R. Degginger
18-4c   Dr. E.R. Degginger
18-5   Teri Stratford
18-6   Kip and Pat Peticolas/Fundamental Photographs
18-7   Teri Stratford/Teri Stratford
18-8   Laima Druskis/PH Archives
18-9b   Laima Druskis/PH Archives
18-10b   Laima Druskis/PH Archives
18-11   Teri Stratford/Teri Stratford
18-12b   Liama Druskis/PH Archives
18-13b   Laima Druskis/PH Archives
18-14b   Laima Druskis/PH Archives
18-15b   Laima Druskis/PH Archives
18-16b   Laima Druskis/PH Archives
18-17b   Laima Druskis/PH Archives
18-18b   Laima Druskis/PH Archives
T18-1a   Laima Druskis/PH Archives
T18-1b   Laima Druskis/PH Archives
T18-1c   Laima Druskis/PH Archives
18-19b   Laima Druskis/PH Archives
Chemistry of the Atmosphere   Tom McHugh/Photo Researchers, Inc.
Carbon Photo (a)   Paul Silverman/Fundamental Photographs
Carbon Photo (a)   Carbon Fiber Products

Chapter 19   Lawrence Berkeley Laboratory, University of California, Dr. Glenn T. Seaborg

19-10   Former Atomic Energy Commission
19-11   Franklin D. Roosevelt Library, Hyde Park
19-12b   Tom Stock/Photo Researcher
Radan Photo   Lowell Georgia/Photo Researchers
Technetium Photo   Photo Researchers

INDEX

Note: **Boldface** pages locate definitions of entries.

Table of Approximate Atomic Masses

Element	Symbol	Atomic Number	Atomic Mass[a] (amu)	Element	Symbol	Atomic Number	Atomic Mass[a] (amu)
Actinium	Ac	89	(227)	Mendelevium	Md	101	(256)
Aluminum	Al	13	27.0	Mercury	Hg	80	200.6
Americium	Am	95	(243)	Molybdenum	Mo	42	95.9
Antimony	Sb	51	121.8	Neodymium	Nd	60	144.2
Argon	Ar	18	39.9	Neon	Ne	10	20.2
Arsenic	As	33	74.9	Neptunium	Np	93	(237)
Astatine	At	85	(210)	Nickel	Ni	28	58.7
Barium	Ba	56	137.3	Nielsbohrium[b]	Ns	107	(262)
Berkelium	Bk	97	(247)	Niobium	Nb	41	92.9
Beryllium	Be	4	9.0	Nitrogen	N	7	14.0
Bismuth	Bi	83	209.0	Nobelium	No	102	(255)
Boron	B	5	10.8	Osmium	Os	76	190.2
Bromine	Br	35	79.9	Oxygen	O	8	16.0
Cadmium	Cd	48	112.4	Palladium	Pd	46	106.4
Calcium	Ca	20	40.1	Phosphorus	P	15	31.0
Californium	Cf	98	(251)	Platinum	Pt	78	195.1
Carbon	C	6	12.0	Plutonium	Pu	94	(244)
Cerium	Ce	58	140.1	Polonium	Po	84	(209)
Cesium	Cs	55	132.9	Potassium	K	19	39.1
Chlorine	Cl	17	35.5	Praseodymium	Pr	59	140.9
Chromium	Cr	24	52.0	Promethium	Pm	61	(145)
Cobalt	Co	27	58.9	Protactinium	Pa	91	(231)
Copper	Cu	29	63.5	Radium	Ra	88	(226)
Curium	Cm	96	(247)	Radon	Rn	86	(222)
Dysprosium	Dy	66	162.5	Rhenium	Re	75	186.2
Einsteinium	Es	99	(254)	Rhodium	Rh	45	102.9
Element 110	—	110	(269)	Rubidium	Rb	37	85.5
Element 111	—	111	(272)	Ruthenium	Ru	44	101.1
Erbium	Er	68	167.3	Rutherfordium[b]	Rf	104	(261)
Europium	Eu	63	152.0	Samarium	Sm	62	150.4
Fermium	Fm	100	(257)	Scandium	Sc	21	45.0
Fluorine	F	9	19.0	Seaborgium[b]	Sg	106	(263)
Francium	Fr	87	(223)	Selenium	Se	34	79.0
Gadolinium	Gd	64	157.2	Silicon	Si	14	28.1
Gallium	Ga	31	69.7	Silver	Ag	47	107.9
Germanium	Ge	32	72.6	Sodium	Na	11	23.0
Gold	Au	79	197.0	Strontium	Sr	38	87.6
Hafnium	Hf	72	178.5	Sulfur	S	16	32.1
Hahnium[b]	Ha	105	(262)	Tantalum	Ta	73	180.9
Hassium[b]	Hs	108	(265)	Technetium	Tc	43	(97)
Helium	He	2	4.0	Tellurium	Te	52	127.6
Holmium	Ho	67	164.9	Terbium	Tb	65	158.9
Hydrogen	H	1	1.0	Thallium	Tl	81	204.4
Indium	In	49	114.8	Thorium	Th	90	232.0
Iodine	I	53	126.9	Thulium	Tm	69	168.9
Iridium	Ir	77	192.2	Tin	Sn	50	118.7
Iron	Fe	26	55.8	Titanium	Ti	22	47.9
Krypton	Kr	36	83.8	Tungsten	W	74	183.8
Lanthanum	La	57	138.9	Uranium	U	92	238.0
Lawrencium	Lr	103	(257)	Vanadium	V	23	50.9
Lead	Pb	82	207.2	Xenon	Xe	54	131.3
Lithium	Li	3	6.9	Ytterbium	Yb	70	173.0
Lutetium	Lu	71	175.0	Yttrium	Y	39	88.9
Magnesium	Mg	12	24.3	Zinc	Zn	30	65.4
Manganese	Mn	25	54.9	Zirconium	Zr	40	91.2
Meitnerium[b]	Mt	109	(266)				

[a] Based on the assigned relative atomic mass of ^{12}C = exactly 12 amu; parentheses denote the mass number of the isotope with the longest half-life.

[b] American Chemical Society (ACS) recommended names and symbols. International Union of Pure & Applied Chemistry (IUPAC) recommended names and symbols differ as follows: 104 (Dubnium, Db); 105 (Joliotium, Jl); 106 (Rutherfordium, Rf); 107 (Bohrium, Bh); 108 (Hahnium, Hn).